# Lecture Notes in Computer Science      14764

## Advanced Research in Computing and Software Science
Subline of Lecture Notes in Computer Science

Adele Anna Rescigno · Ugo Vaccaro
Editors

# Combinatorial Algorithms

35th International Workshop, IWOCA 2024
Ischia, Italy, July 1–3, 2024
Proceedings

 Springer

*Editors*
Adele Anna Rescigno 🄳
University of Salerno
Fisciano, Italy

Ugo Vaccaro 🄳
University of Salerno
Fisciano, Italy

ISSN 0302-9743          ISSN 1611-3349   (electronic)
Lecture Notes in Computer Science
ISBN 978-3-031-63020-0          ISBN 978-3-031-63021-7   (eBook)
https://doi.org/10.1007/978-3-031-63021-7

This Springer imprint is published by the registered company Springer Nature Switzerland AG
The registered company address is: Gewerbestrasse 11, 6330 Cham, Switzerland

If disposing of this product, please recycle the paper.

# Preface

The 35th International Workshop on Combinatorial Algorithms (IWOCA 2024) was held in Ischia, Italy, from July 1st to July 3rd, 2024.

Since its inception in 1989 as the Australasian Workshop on Combinatorial Algorithms (AWOCA), IWOCA has provided an annual forum for researchers who design algorithms to address the myriad combinatorial problems underlying computer applications in science, engineering, and business. Previous IWOCA and AWOCA meetings have occurred in various countries, including Australia, Canada, the Czech Republic, Finland, France, Germany, India, Indonesia, Italy, Japan, Singapore, South Korea, the UK, Taiwan, and the USA.

The quality of IWOCA 2024 was ensured by a Program Committee comprising 42 researchers of international reputation. IWOCA 2024 received 110 submissions in response to the call for papers. Each submission underwent rigorous single-blind review by at least three Program Committee members and trusted external reviewers, and was evaluated based on its quality, originality, and relevance to the conference. The Program Committee selected 40 high-quality papers for presentation at the conference and inclusion in the proceedings.

The conference program also featured three keynote talks by distinguished speakers: Ralf Klasing (CNRS and University of Bordeaux, France), Lucia Moura (University of Ottawa, Canada), and Nadia Pisanti (University of Pisa, Italy). Abstracts of their talks are included in the conference volume.

Among the highest-scoring contributed papers, the Program Committee selected the paper "On the Structure of Hamiltonian Graphs with Small Independence", by Nikola Jedličková and Jan Kratochvil for the Best Paper Award, and the paper "Lower Bounds for Leaf Rank of Leaf Powers", by Svein Høgemo for the Best Student Paper Award.

We extend our gratitude to the Steering Committee for entrusting us with the roles of Program Chairs for IWOCA 2024. We thank all the authors who submitted papers for allowing us to consider their work, the Program Committee members, and the external reviewers for volunteering their time to review conference papers. We thank Springer for publishing the proceedings in their prestigious ARCoSS/LNCS series and for the financial support for the Best Paper Award and the Best Student Paper Award. We thank the MDPI journals "Algorithms" and "Mathematics" for financial support. We thank the Dipartimento di Informatica: *Dipartimento di Eccellenza*, and the University of Salerno for their financial support.

Last, but certainly not least, we express our sincere thanks to the members of the local Organizing Committee: Roberto Bruno (University of Salerno), Gennaro

Cordasco (University della Campania "L. Vanvitelli"), and Luisa Gargano (University of Salerno), who greatly helped us to manage the conference.

July 2024

Adele Anna Rescigno
Ugo Vaccaro

# Organization

## Program Chairs

Adele Rescigno                 University of Salerno, Italy
Ugo Vaccaro                    University of Salerno, Italy

## Program Committee

Bogdan Alecu                   University of Leeds, UK
Petra Berenbrink               University of Hamburg, Germany
Tiziana Calamoneri             Sapienza University of Rome, Italy
Yi-Jun Chang                   National University of Singapore, Singapore
Gennaro Cordasco               Università della Campania "L. Vanvitelli", Italy
Peter Damaschke                Chalmers University of Technology, Sweden
Adrian Dumitrescu              Algoresearch L.L.C., USA
Klim Efremenko                 Ben-Gurion University of the Negev, Israel
David Eppstein                 University of California, Irvine, USA
Leah Epstein                   University of Haifa, Israel
Thomas Erlebach                Durham University, UK
Amedeo Esposito                Institute of Science and Technology, Austria
Henning Fernau                 University of Trier, Germany
Fedor Fomin                    University of Bergen, Norway
Travis Gagie                   Dalhousie University, Canada
Robert Ganian                  Vienna University of Technology, Austria
Leszek Gasieniec               University of Liverpool, UK
Roberto Grossi                 Università di Pisa, Italy
Ling-Ju Hung                   National Taipei University of Business, Taiwan
Yusuke Kobayashi               Kyoto University, Japan
Evangelos Kranakis             Carleton University, Canada
Danny Krizanc                  Wesleyan University, USA
Manuel Lafond                  Université de Sherbrooke, Canada
Chia-Wei Lee                   University of Taipei, Taiwan
Kitty Meeks                    University of Glasgow, UK
Martin Milanič                 University of Primorska, Slovenia
Neeldhara Misra                Indian Institute of Technology, Gandhinagar, India
Alantha Newman                 Université Grenoble Alpes, France
Aris Pagourtzis                National Technical University of Athens, Greece
Ely Porat                      Bar-Ilan University, Israel
Dieter Rautenbach              University of Ulm, Germany
Saket Saurabh                  Institute of Mathematical Sciences, India
Marinella Sciortino            University of Palermo, Italy
Sagnik Sen                     Indian Institute of Technology Dharwad, India

| | |
|---|---|
| Maria Serna | Universitat Politècnica de Catalunya, Spain |
| Florian Sikora | Université Paris Dauphine, France |
| Ana Silva | Universidade Federal do Ceará, Brazil |
| Tatiana Starikovskaya | École Normale Supérieure, France |
| Teresa Anna Steiner | Technical University of Denmark, Denmark |
| Jukka Suomela | Aalto University, Finland |
| Ugo Vaccaro | University of Salerno, Italy |
| Michal Wlodarczyk | University of Warsaw, Poland |
| Meirav Zehavi | Ben-Gurion University of the Negev, Israel |

## Steering Committee

| | |
|---|---|
| Maria Chudnovsky | Princeton University, USA |
| Henning Fernau | Universität Trier, Germany |
| Costas Iliopoulos | King's College London, UK |
| Ralf Klasing | CNRS and University of Bordeaux, France |
| Bill Smyth | McMaster University, Canada |
| Wing-Kin (Ken) Sung | National University of Singapore, Singapore |

## Additional Reviewers

Amanatidis, Georgios
Antony, Dhanyamol
Arraes, Pedro
Balogh, János
Barequet, Gill
Bastide, Paul
Bathie, Gabriel
Belmonte, Rémy
Bentert, Matthias
Bergougnoux, Benjamin
Bernardini, Giulia
Bhyravarapu, Sriram
Biermeier, Felix
Bliznets, Ivan
Blom, Michelle
Boeckenhauer, H.-Joachim
Boneh, Itai
Bonerath, Annika
Boyar, Joan
Brakensiek, Joshua
Brosse, Caroline
Bruno, Roberto
Bumpus, B. Merlin
Buzzega, Giovanni
Bényi, Beáta

Casselgren, Carl Johan
Chakraborty, Dibyayan
Chen, Ke
Cicerone, Serafino
Cichacz, Sylwia
Cleary, Alan
Cohen, Johanne
Corò, Federico
Crescenzi, Pierluigi
Cunial, Fabio
Danda, Sravan
Das, Angsuman
Das, Arun Kumar
Dondi, Riccardo
Ducoffe, Guillaume
Dupré La Tour, Max
El Ghazi El Houssaini, Taha
Equi, Massimo
Ferraioli, Diodato
Fioravantes, Foivos
Flocchini, Paola
Foucaud, Florent
Fried, Dvir
Fusy, Eric
Gahlawat, Harmender

Gajjala, Rishikesh
Georgiou, Konstantinos
Ghazawi, Samah
Giannopoulou, Archontia
Gibney, Daniel
Goedgebeur, Jan
Golovach, Petr
Guo, Litao
Hahn, Christopher
Harutyunyan, Hovhannes
Higashikawa, Yuya
Hintze, Lukas
Hirvonen, Juho
Hoang, Hung
Hosseinpour, Hamed
Huang, Chien-Chung
Huang, Shenwei
Jaffke, Lars
Jana, Satyabrata
Jansen, Klaus
Jedelský, Jan
Johnson, Matthew
Kakimura, Naonori
Kalachev, Gleb
Kempa, Dominik
Keusch, Ralph
Kiyomi, Masashi
Klasing, Ralf
Klostermeyer, Chip
Knop, Dušan
Korhonen, Tuukka
Kortsarz, Guy
Kozma, Laszlo
Kurita, Kazuhiro
Lampis, Michael
Lecroq, Thierry
Li, Shaohua
Li, Shi
Lima, Carlos Vinicius
Lin, Cheng-Kuan
Lin, Chuang-Chieh
Liotta, Giuseppe
Liu, Jiafei
Lochet, William
Madathil, Jayakrishnan
Maezawa, Shunichi

Malaguti, Enrico
Mann, Kevin
Martin, Barnaby
Mernik, Marjan
Mitsou, Valia
Modanese, Augusto
Morin, Pat
Moura, Phablo
Mukherjee, Joydeep
Mütze, Torsten
Nanoti, Saraswati
O'Rourke, Joseph
Onete, Cristian
Ono, Hirotaka
Paesani, Giacomo
Papaioannou, Ioannis
Pawłowski, Michał
Petsalakis, Stavros
Pettie, Seth
Peyrille, Benjamin
Pierre-Marie, Marcille
Pissis, Solon
Pratt, Kevin
Rahman, Md. Saidur
Rahul, Saladi
Rau, Malin
Rehs, Carolin
Robidou, Lucas
Rodríguez Aldama, René
Rosenbaum, Will
Rucci, Davide
Saghafian, Morteza
Sahu, Abhishek
Saitoh, Toshiki
Salvagnin, Domenico
Salvo, Ivano
Sampaio, Rudini
Sarma, Jayalal
Schirneck, Martin
Schmidt, Daniel R.
Scholz, Guillaume
Seelbach Benkner, Louisa
Sethia, Aditi
Sinaimeri, Blerina
Stamoulis, Giannos
Tan, Jane

Tsai, Meng-Tsung
Tsakalidis, Konstantinos
Tu, Jianhua
Upasana, Anannya
Vasilakis, Manolis
Vötsch, Maximilian
Wang, Hung-Lung
Wasa, Kunihiro
Watrigant, Rémi

Wicke, Kristina
Wolf, Petra
Wu, Kaiyu
Xu, Jinghan
Yang, Hao-Tsung
Yang, Yongjie
Yingchareonthawornchai, Sorrachai
Šparl, Primož

# Abstracts of Invited Talks

# Bamboo Garden Trimming Problem

Ralf Klasing

CNRS and University of Bordeaux, France
ralf.klasing@labri.fr

A garden is populated by n bamboos, each with its own daily growth rate. The Bamboo Garden Trimming Problem (BGT) is to design for a robotic gardener a perpetual schedule of cutting bamboos to keep the elevation of the garden as low as possible. The frequency of cutting is constrained by the time needed to move from one bamboo to the next, which is one day in Discrete BGT and is defined by the distance between the two bamboos in Continuous BGT. The bamboo garden is a metaphor for a collection of machines that have to be serviced, with different frequencies, by a robot that can service only one machine at a time. For Discrete BGT, we show a simple 4-approximation algorithm and, by exploiting the relationship between BGT and the classical Pinwheel Scheduling Problem, we derive a 2-approximation for the general case and a tighter approximation when the growth rates are balanced. For Continuous BGT, we propose approximation algorithms that achieve logarithmic approximation ratios.

# Cover-Free Families and Generalizations Based on Hypergraphs

Lucia Moura

University of Ottawa, Canada
lmoura@uottawa.ca

Cover-free families are families of finite sets where no set is contained in the union of d others. Superimposed codes, disjunct matrices, and strongly selective families are other names for the same combinatorial object. Cover-free families have applications in group testing, information retrieval, cryptography, and communications, among other areas.

Let us consider the incidence matrix representation of a family of sets, where rows correspond to elements and columns correspond to the characteristic vectors of the sets in the family. The cover-free property specifies that the union (logical-and) of any d columns of this binary matrix must not contain (cover) any other column. In our research, the cover-free property is generalized so that the edges of a hypergraph specify the sets of columns whose union must not cover any other column. Using this model, the classical cover-free family corresponds to the special case where the hypergraph is the d-uniform complete hypergraph. This generalization of cover-free families, which we call cover-free families on hypergraphs, is interesting combinatorially and motivated by applications. In this talk, we will begin with classical results on cover-free families, and discuss initial results, constructions, and ongoing research on this new generalization involving hypergraphs. We will also discuss some of their applications in group testing and cryptography.

# ED-Strings Similarities for PanGenomes Comparison

Nadia Pisanti

University of Pisa, Italy
nadia.pisanti@unipi.it

The notion of ED string was introduced as a simple alternative to variation and sequence graphs for representing a pangenome, that is, a collection of genomic sequences to be analyzed jointly or to be used as a reference. It can also be seen as a directed acyclic graph, whose edges are labeled by strings. We define notions of matching statistics of two ED strings, and we show how these can be used to conceive similarity measures for two pangenomes, and also consequently infer distance measures. We then show that both of these measures can be computed efficiently, both in theory and in practice, by employing the notion of an "intersection graph" of two ED strings. We used real SARS-CoV-2 datasets and our matching statistics method to reproduce a well-established clades classification of SARS-CoV-2, thus showing that the classification obtained by our method is in accordance with the existing one.

# Contents

# On Computing Sets of Integers
# with Maximum Number of Pairs
# Summing to Powers of 2

Max A. Alekseyev[✉]

The George Washington University, Washington, DC, USA
maxal@gwu.edu

**Abstract.** We address the problem of finding sets of integers of a given size with the maximum number of pairs summing to powers of 2. By fixing particular pairs, this problem reduces to finding a labeling of the vertices of a given graph with pairwise distinct integers such that the endpoint labels for each edge sum to a power of 2. We propose an efficient algorithm for this problem, which at its core relies on another algorithm that, given two sets of linear homogeneous polynomials with integer coefficients, computes all variable assignments to powers of 2 that nullify all polynomials from the first set but neither one from the second. With the proposed algorithms, we determine the maximum size of graphs of order $n$ that admit such a labeling for all $n \leq 20$. We also identify the minimal forbidden subgraphs of order $\leq 11$, whose presence prevents the graphs from having such a labeling.

## 1   Introduction

In April 2021, Dan Ullman and Stan Wagon published "Problem of the Week 1321" [12] (as a generalization of AMM Problem 12272 [10]), where they introduced a function $f(A)$ of a finite set $A$ of integers as the number of 2-element subsets of $A$ that sum to a power of 2. They gave an example $f(\{-1, 3, 5\}) = 3$ and further defined a function $g(n)$ as the maximum of $f(A)$ over all $n$-element sets $A$. The problem asked for a proof that $g(10) \geq 14$, which was quickly improved to $g(10) \geq 15$ by the readers, and Pratt further showed[1] that, in fact, $g(10) = 15$.

Multiple researchers have noted that computing $g(n)$ has a natural interpretation as finding a maximum graph of order $n$, where the vertices are labeled with pairwise distinct integers, and the sum of the endpoint labels for each edge is a power of 2 (see Fig. 2 for examples). In March 2022, M. S. Smith proved[2] that such a graph cannot contain a cycle $C_4$, limiting the candidate graphs to well-studied squarefree graphs [4]. Smith's result made it easy to establish the values of $g(n)$ for all $n \leq 9$, and led to the creation of the sequence A352178

---

[1] Pratt's proof apparently is unpublished and only mentioned in [12, Solution].
[2] Smith emailed a proof to Neil Sloane, and it is quoted in the OEIS entry A352178 [11].

A. A. Rescigno and U. Vaccaro (Eds.): IWOCA 2024, LNCS 14764, pp. 1–13, 2024.
https://doi.org/10.1007/978-3-031-63021-7_1

in the Online Encyclopedia of Integer Sequences (OEIS) [11]. It also provided a non-trivial upper bound for $g(n)$, namely the maximum size of squarefree graphs of order $n$ (given by sequence A006855 in the OEIS).

The problem received further attention in September 2022 after Neil Sloane presented it on the popular Numberphile Youtube channel [5]. This was followed by a few improvements, including the value[3] $g(10) = 15$ from Matthew Bolan [2], and independently the same value and $g(11) = 17$ from Firas Melaih [7]. These results were obtained via a graph-theoretical treatment of the problem by manually analyzing a few candidate graphs.

At the same time, two methods were proposed to obtain lower bounds for $g(n)$, giving those for the first few hundreds of values of $n$, which are listed in the sequences A347301 and A357574 in the OEIS.

In the present paper, we propose an algorithm for testing the *admissibility* of a given unlabeled graph, i.e., whether its vertices can be labeled with pairwise distinct integers such that the sum of the endpoint labels for each edge is a power of 2. At its core, this algorithm relies on another proposed algorithm that, given two sets of linear homogeneous polynomials with integer coefficients, computes all variable assignments to powers of 2 that nullify all polynomials from the first set but neither one from the second. The polynomials from the first and second sets are referred to as equations and inequations, respectively.

We use our algorithms to bound $g(n)$ from above, and together with the known lower bounds to establish the values of $g(n)$ for $n$ in the interval $[12, 20]$. Also, interpreting Smith's result as saying that the cycle $C_4$ is a *minimal forbidden subgraph* (MFS), i.e., inadmissible graphs in which all proper subgraphs are admissible, we find all MFSs of orders up to 11. Namely, we show that there are no MFSs of order 5, 6, 8, or 9, while there are 2 MFSs of order 7 (Fig. 1), 15 MFSs of order 10, and 77 MFSs of order 11. We also enumerate the *maximum admissible graphs* (MAGs), i.e. admissible graphs of the maximum size, of orders up to 18. We establish that there are 4, 28, 2, 18, and 2 MAGs of order 14, 15, 16, 17, and 18, respectively.

## 2   Algorithm for Testing Graph Admissibility

A given graph $G$ on $n$ vertices with $m$ edges is admissible if and only if the following matrix equation is solvable:

$$M \cdot L = X, \tag{1}$$

where

- $M$ is the $m \times n$ incidence matrix of $G$ with rows and columns indexed by the edges and vertices of $G$, that is, $M$ is a $\{0, 1\}$-matrix with each row containing exactly two 1's;

---

[3] Apparently the exact value of $g(10)$ was not present in the OEIS back then, and most people were unaware that Pratt had already proved $g(10) = 15$.

- $L = (l_1, l_2, \ldots, l_n)^T$ is a column vector of pairwise distinct integer vertex labels;
- $X = (x_1, x_2, \ldots, x_m)^T$ is a column vector formed by powers of 2 representing the sums of edges' endpoint labels;[4]
- both $L$ and $X$ are unknown and have to be determined.

We start with solving (1) for $L$ in terms of $X$, ignoring for a moment the distinctness requirement for the elements of $L$. That is, we compute a (partial) solution of the form $l_i = p_i(x_1, \ldots, x_m)$ for $i \in \{1, 2, \ldots, n\}$, where each $p_i$ is a linear polynomial with rational coefficients. By the Rouche–Frobenius–Capelli theorem, such a solution exists if and only if $K_l \cdot X = 0$, where $K_l$ is an integer matrix with rows forming a basis of the left kernel of $M$. It will be seen that the performance of our algorithm is sensitive to the number of nonzero elements in $K_l$, and thus we assume that $K_l$ is sparse, which may be achieved (to some extent) with LLL reduction [3]. Let $E$ be the set of elements of $K_l \cdot X$, which are homogeneous linear polynomials with integer coefficients representing linear equations in $x_1, \ldots, x_m$.

To fulfill the distinctness requirement for the elements of $L$, we consider an integer matrix $K_r$ whose columns form a basis of the lattice of integer vectors in the right kernel of $M$. For a connected graph $G$, it is known that $K_r$ has size $n \times t$, where $t = 1$ or $t = 0$ depending on whether $G$ is bipartite [13]. We find it convenient to view a $n \times 0$ matrix as being composed of $n$ empty rows (thus all rows are equal). Adding a linear combination of the columns of $K_r$ to a solution $L$ to Eq. (1) turns it into another solution $L'$, and furthermore all solutions with the same $X$ can be obtained this way. The following theorem implies that for any set of pairwise distinct rows of $K_r$, we can find a linear combination of the columns of $K_r$ such that the corresponding elements of $L'$ are also pairwise distinct.

**Theorem 1.** *Let $v$ be an integer column vector of size $k \geq 1$, and $A$ be a $k \times s$ integer matrix with pairwise distinct rows. Then there exists an integer linear combination of the columns of $A$ such that adding it to $v$ results in a vector with pairwise distinct elements.*

*Proof.* If $s = 0$, then with necessity we have $k = 1$, and thus $v$ already has pairwise distinct elements.

Let us prove the statement for $s = 1$. In this case, $A$ represents a column vector with pairwise distinct elements. Let $t$ be the difference between the largest and smallest elements of $v$. It is easy to see that the vector $v + A \cdot (t + 1)$ has pairs of distinct elements.

In the case of $s > 1$, let $d$ be the difference between the largest and smallest elements of $A$. Then the $k \times 1$ matrix $A' := A \cdot (1, (d+1), (d+1)^2, \ldots, (d+1)^{s-1})^T$ has pairwise distinct elements, thus reducing the problem to the case $s = 1$ considered above.                                                                    □

---

[4] The elements of $X$ are not required to be distinct.

Theorem 1 implies that we can restrict our attention only to the pairs of elements of $L$ corresponding to the equal rows in $K_r$. For each pair of equal rows in $K_r$ with indices $i < j$, we compute $q(x_1, \ldots, x_m) := (p_i(x_1, \ldots, x_m) - p_j(x_1, \ldots, x_m))c$, where $c$ is a positive integer factor making all coefficients of $q$ integer. If $q$ is a zero polynomial, then condition $l_i \neq l_j$ is unattainable, and thus graph $G$ is inadmissible. On the other hand, if $q$ consists of just a single term with a nonzero coefficient, then condition $l_i \neq l_j$ always holds, and we ignore such $q$. In the remaining case, when $q$ contains two or more terms with nonzero coefficients, we add $q$ to the set $N$. The resulting set $N$ consists of polynomials in $x_1, \ldots, x_m$ that must take nonzero values on a solution to (1). We refer to such polynomials as *inequations*.[5]

Our next goal is to solve the system of equations $E$ and inequations $N$ in powers of 2, which we address in Sect. 3. For each solution $X = X_0$, we plug it into the system (1) turning it into a standard matrix equation, which we then solve for $L$. The resulting solution can be parametric with parameters representing the free variables in $X_0$ (if any) and the coefficients of a linear combination of the columns of $K_r$ when it is not empty. In either case, such a solution is guaranteed to produce a solution composed of pairwise distinct integers by Theorem 1.

We outline the above description in Algorithm 1. While this algorithm can produce all solutions (i.e., labelings of the graph vertices), when testing graph admissibility we stop it as soon as one solution is obtained.[6]

## 3   Solving a System of (in)equations in Powers of 2

For given finite sets $E$ and $N$ of nonzero linear polynomials in $x_1, x_2, \ldots, x_m$ with integer coefficients, our goal is to find all $n$-tuples of nonnegative integers $(y_1, y_2, \ldots, y_m)$ such that

$$\forall p \in E : \quad p(2^{y_1}, 2^{y_2}, \ldots, 2^{y_m}) = 0,$$
$$\forall p \in N : \quad p(2^{y_1}, 2^{y_2}, \ldots, 2^{y_m}) \neq 0.$$

As simple as it sounds, the following theorem provides a foundation for our algorithm.

**Theorem 2.** *In any nonempty multiset of nonzero integers summing to 0, there exist two elements with equal 2-adic valuations.*[7]

*Proof.* Let $S$ be a nonempty multiset of nonzero integers summing to 0, and let $k$ be an element of $S$ with the smallest 2-adic valuation $q := \nu_2(k)$. If all the

---

[5] We deliberately use the term *inequation* to denote relationship $p \neq 0$ and to avoid confusion with inequalities traditionally denoting relationships $\geq$, $\leq$, $>$, or $<$.

[6] Technically GRAPHSOLVE is implemented as a SageMath/Python's generator, which produces solutions one after another on demand and which we can easily stop after obtaining just one solution (when it exists).

[7] Recall that the 2-adic valuation of a nonzero integer $k$, denoted $\nu_2(k)$, equals the exponent of 2 in the prime factorization of $k$, while $\nu_2(0) = \infty$.

---

**Algorithm 1.** An algorithm for solving system (1) for a given graph $G$.

---

1: **function** GRAPHSOLVE($G$)
2:     Set $E := \emptyset$ and $N := \emptyset$
3:     Let $m$ and $n$ be the size and order of $G$, respectively.
4:     Construct the $m \times n$ incidence matrix $M$ of $G$ with rows and columns indexed by the edges and vertices of $G$
5:     Compute a sparse basis $K_l$ of the left kernel of $M$.         ▷ We have $K_l \cdot M = 0$.
6:     Let $X := (x_1, \ldots, x_m)^T$ be the column-vector formed by indeterminates.
7:     **for** each row $r$ in $K_l$ **do**
8:         Add polynomial $r \cdot X$ to $E$
9:     **end for**
10:    Solve $ML = X$ for $L$ in terms of $X$, let $(p_1, \ldots, p_n)$ be any particular solution.
11:    Compute $K_r$ whose columns form a basis of the right integer kernel of $M$.         ▷ We have $M \cdot K_r = 0$.
12:    **for** each $\{i, j\} \subset \{1, 2, \ldots, n\}$ **do**
13:        **if** $i$th and $j$th rows of $K_r$ are not equal **then**
14:            **continue** to next subset $\{i, j\}$ ▷ Per Theorem 1 we ignore such pair of indices.
15:        **end if**
16:        Set $q$ equal to a multiple of $p_i - p_j$ with integer coefficients
17:        **if** $q = 0$ **then**
18:            **return** $\emptyset$                     ▷ No solutions with $l_i \neq l_j$.
19:        **end if**
20:        **if** $q$ contains two or more terms **then**
21:            Add $q$ to $N$.
22:        **end if**
23:    **end for**
24:    $S := \emptyset$
25:    **for** each $s$ in SOLVEINPOWERS($E$, $N$) **do**         ▷ $s$ is a map from $Y$ to linear polynomials in $Y$
26:        Set $x_i := 2^{s[y_i]}$ for each $i \in \{1, 2, \ldots, m\}$.
27:        Solve $ML = X$ for $L$ composed of pairwise distinct integers and add the solution to $S$.         ▷ Use Theorem 1 if needed.
28:    **end for**
29:    **return** $S$
30: **end function**

---

other elements of $S$ have valuation greater than $q$, then the sum of all elements (which is 0) has valuation $q$, which is impossible. Hence, there exist at least two elements in $S$ that have 2-adic valuation equal to $q$.                     □

Applying Theorem 2 to an equation $c_1 x_1 + \cdots + c_m x_m \in E$, we conclude that if only one of the coefficients $c_1, c_2, \ldots, c_m$ is nonzero, then system $(E, N)$ is insolvable. Otherwise, if there are two or more nonzero coefficients among $c_1, c_2, \ldots, c_m$, then there exists a pair of indices $i < j$ such that $c_i \neq 0$, $c_j \neq 0$, and $\nu_2(c_i x_i) = \nu_2(c_j x_j)$, which implies that we can make a substitution $x_i = 2^{\nu_2(c_j) - \nu_2(c_i)} x_j$ or $x_j = 2^{\nu_2(c_i) - \nu_2(c_j)} x_i$ (we pick one with integer coefficients). Then we proceed with making this substitution in $E$ and $N$, thus reducing the

**Algorithm 2.** An algorithm for solving a given system of linear equations $E$ and inequations $N$ over indeterminates from $X := \{x_1, \ldots, x_m\}$ in powers of 2. It returns a set of maps $s$ from variables $Y := \{y_1, y_2, \ldots, y_m\}$ to linear polynomials in these variables such that $(x_1, \ldots, x_m) = (2^{s[y_1]}, 2^{s[y_2]}, \ldots, 2^{s[y_m]})$ is a solution.

```
 1: function SOLVEINPOWERS(E, N)
 2:     if E = ∅ then
 3:         return {the identity map}              ▷ Every variable in Y is free.
 4:     end if
 5:     Pick c₁x₁ + ⋯ + cₘxₘ ∈ E with the smallest number of nonzero coefficients.
 6:     Let I := {i | 1 ≤ i ≤ m, cᵢ ≠ 0} be the set of indices of nonzero coefficients.
 7:     Set S := ∅                                  ▷ We accumulate solutions in S.
 8:     for each {i, j} ⊆ I do           ▷ We iterate over all 2-element subsets of I.
 9:         Possibly exchanging the values of i and j, ensure that d := ν₂(cᵢ) − ν₂(cⱼ) ≥ 0.
10:         Compute N′ from N by substituting xⱼ ← 2ᵈxᵢ and excluding nonzero constant polynomials.
11:         if 0 ∈ N′ then
12:             continue to the next pair {i, j}.
13:         end if
14:         Add xⱼ − 2ᵈxᵢ to N.        ▷ For future we disallow the equality cⱼxⱼ = cᵢxᵢ.
15:         Compute E′ from E by substituting xⱼ ← 2ᵈxᵢ and excluding zero polynomials.
16:         for each s in SOLVEINPOWERS(E′, N′) do  ▷ s is a map from Y to linear polynomials in Y.
17:             Redefine s[yⱼ] := s[yᵢ] + d.
18:             Add s to set S.
19:         end for
20:     end for
21:     return S
22: end function
```

number of indeterminates. If this substitution does not nullify any polynomial in $N$, we proceed to solve the reduced system recursively. After exploring the pair $(i, j)$, we add a new inequation $2^{\nu_2(c_i)}x_i - 2^{\nu_2(c_j)}x_j$ to $N$ to avoid following the same substitutions in a different order and obtaining the same solutions in future, and proceed to the next pair of indices.

We outline the above description in Algorithm 2. For given sets $E$ and $N$ of linear equations and inequations in $x_1, \ldots, x_m$ with integer coefficients, the function SOLVEINPOWERS($E$, $N$) computes the set of their solutions in powers of 2. Each solution is given in the form of a map $s$ from the set of variables $Y := \{y_1, y_2, \ldots, y_m\}$ to linear polynomials in these variables, representing the exponents in the powers of 2. Namely, $s$ sends every variable from $Y$ either to itself (when it is a free variable), or to a linear polynomial in the free variables. For example, the map $\{y_1 \to y_4+1, \ y_2 \to y_2, \ y_3 \to y_2+3, \ y_4 \to y_4\}$ corresponds to the solution $(x_1, x_2, x_3) = (2^{y_4+1}, 2^{y_2}, 2^{y_2+3}, 2^{y_4})$, where $y_2$ and $y_4$ are free variables taking arbitrary nonnegative integer values.

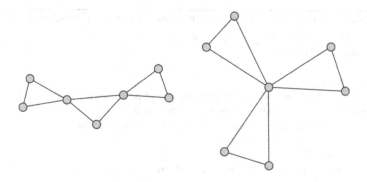

**Fig. 1.** Minimal forbidden subgraphs of order 7.

We note that Algorithm 2 is a recursive branch-and-bound algorithm with a maximum recursion depth equal to $m - 1$. In the worst case without bounding, at depth $d$ it would branch into $\binom{m-d}{2} = \frac{(m-d)(m-d-1)}{2}$ recursive calls, and thus its complexity can be estimated as $O\big((|E| + |N|) \cdot \frac{m!^2}{2^m}\big)$ algebraic operations. However, this naive estimate does not take into account the bounding controlled by the elements of $N$ and the narrowed branching controlled by the choice of sparse elements of $E$. Although it is difficult to give an accurate complexity estimate, our computational results presented in Sects. 4–6 show that Algorithm 2 and thus Algorithm 1 are quite efficient in practice (in particular, see Tables 1, 2, 3).

## 4   Minimal Forbidden Subgraphs

We use the proposed algorithms to find minimal forbidden subgraphs of small order. It is easy to see that each MFS must be connected. Also, if an inadmissible graph contains a vertex of degree 1, then the graph without this vertex is also inadmissible. Therefore, the minimum degree of an MFS must be $\geq 2$.

It is almost trivial to verify that $C_4$ is the smallest MFS and the only one on 4 vertices. Therefore, for $n > 4$ we can restrict our attention to the connected squarefree graphs with the minimum degree $\geq 2$ as candidates, which we generate in SageMath [9] with the function nauty_geng() based on nauty tool [6] supporting both connected (option -c) and squarefree (option -f) graphs as well as a lower bound on the minimum degree (option -d#). This significantly speeds up the algorithm and eliminates the need to test the presence of MFS $C_4$ as a subgraph.

We search for MFSs, other than $C_4$ iteratively increasing their order, accumulating found MFSs in a set $S$ (initially empty). For each order, we iterate over the candidate graphs $G$ in the order of non-decreasing size, and check if $G$ contains any graphs from $S$ as a subgraph using the SageMath function is_subgraph(). If $G$ contains any of the graphs from $S$, we go to the next candidate graph $G$. Otherwise, we test the admissibility of $G$ by calling GraphSolve($G$). If $G$

**Algorithm 3.** An algorithm iteratively computing the minimal forbidden subgraphs, other than $C_4$, of order up to $u$.

---
1: **function** FINDMFS($u$)
2:     $S := \emptyset$
3:     **for** $n = 5, \ldots, u$ **do**
4:         **for** each connected squarefree graph $G$ of order $n$ and minimum degree $\geq 2$, iterated over such that the size of $G$ is nondecreasing **do**
5:             **if** $G$ contains any graph from $S$ as a subgraph **then**
6:                 **continue** to next $G$
7:             **end if**
8:             **if** GRAPHSOLVE($G$) is empty **then**
9:                 Add $G$ to the set $S$.
10:            **end if**
11:        **end for**
12:    **end for**
13:    **return** $S$
14: **end function**

---

Table 1. Performance benchmarks for computing MFSs of orders 5 to 11.

| Graph order | 5 | 6 | 7 | 8 | 9 | 10 | 11 |
|---|---|---|---|---|---|---|---|
| # candidate graphs | 2 | 3 | 10 | 28 | 112 | 533 | 3126 |
| # graphs with an MFS | 0 | 0 | 0 | 1 | 12 | 64 | 528 |
| # graphs tested w. GRAPHSOLVE | 2 | 3 | 10 | 27 | 100 | 469 | 2598 |
| Average test time (sec.) | 1.0 | 1.1 | 1.2 | 1.7 | 2.3 | 3.5 | 2.4 |

is inadmissible, then it represents an MFS and we add it to $S$. The described algorithm is outlined in Algorithm 3.

We determine that the next smallest MFSs after $C_4$ are two graphs of order 7 (Fig. 1). One of these graphs was previously shown to be inadmissible by Bolan while proving that $g(10) = 15$ [2].

There are no MFSs of order 8 or 9, but there are 15 of them of order 10, and there are 77 MFSs of order 11. We evaluate the performance[8] of this computation in Table 1. We use MFSs of order $\leq 10$ for quick filtering of some inadmissible graphs.

## 5   Computing Values of $g(n)$

We use the proposed algorithms for computing values of $g(n)$ (sequence A352178 in the OEIS) for $n$ in [12, 20], relying on the known values $g(n)$ for $n \leq 11$ and the lower bound $\ell(n)$ (sequence A347301 in the OEIS) for $n \geq 12$:

---

[8] Here and below the performance is benchmarked at a desktop computer with Intel Xeon E5-2670v2 2.50GHz CPU.

| $n$ | 1 | 2 | 3 | 4 | 5 | 6 | 7 | 8 | 9 | 10 | 11 |
|---|---|---|---|---|---|---|---|---|---|---|---|
| $g(n)$ | 0 | 1 | 3 | 4 | 6 | 7 | 9 | 11 | 13 | 15 | 17 |

| $n$ | 12 | 13 | 14 | 15 | 16 | 17 | 18 | 19 | 20 |
|---|---|---|---|---|---|---|---|---|---|
| $\ell(n)$ | 19 | 21 | 24 | 26 | 29 | 31 | 34 | 36 | 39 |

**Theorem 3.** *For each integer* $n \in [12, 20]$, *we have* $g(n) = \ell(n)$.

The remainder of this section is devoted to the proof of Theorem 3. Our goal is to show that for each $n \in [12, 20]$ we have $g(n) \leq \ell(n)$, which is equivalent to showing that every graph of order $n$ and size $\ell(n) + 1$ is inadmissible.

**Theorem 4.** *For any integer* $n > 2$:

(I) *if an admissible graph of order* $n$ *and size* $e$ *exists, then its minimum degree is at least* $e - g(n - 1)$;

(II)
$$g(n) \leq \left\lfloor \frac{n \cdot g(n-1)}{n-2} \right\rfloor;$$

(III) *if*
$$\frac{g(n-1)}{n-1} = \max_{k \in \{1, 2, \ldots, n-1\}} \frac{g(k)}{k},$$

*then any admissible graph of order* $n$ *and size* $e > \frac{n}{n-1}g(n-1)$ *is connected.*

*Proof.* Let $G$ be an admissible graph of order $n$ with $e$ edges. If $G$ has a vertex of degree smaller than $e - g(n - 1)$, then removing it from $G$ results in an admissible graph of order $n-1$ with more than $g(n-1)$ edges. The contradiction shows that the degree of any vertex of $G$ is at least $e - g(n - 1)$, proving the statement (I). Furthermore, it implies that $G$ has at least $\frac{n(e-g(n-1))}{2}$ edges, that is, $e \geq \frac{n(e-g(n-1))}{2}$, implying that $e \leq \lfloor \frac{n \cdot g(n-1)}{n-2} \rfloor$. For an admissible graph of order $n$ with $e = g(n)$ edges, it implies $g(n) \leq \lfloor \frac{n \cdot g(n-1)}{n-2} \rfloor$, which proves the statement (II).

To prove the statement (III), suppose that we have an admissible graph $G$ of order $n$ and size $e > \frac{n}{n-1}g(n-1)$. Let $t$ be the number of its connected components, and let $s_1, \ldots, s_t$ be their orders. Let us assume that $t \geq 2$, implying that $\max_{1 \leq i \leq t} s_i \leq n - 1$. Clearly, we have $s_1 + \cdots + s_t = n$ and $e \leq g(s_1) + \cdots + g(s_t)$. It follows that

$$\frac{g(s_1) + \cdots + g(s_t)}{s_1 + \cdots + s_t} \geq \frac{e}{n} > \frac{g(n-1)}{n-1}.$$

On the other hand, since $\frac{g(s_1) + \cdots + g(s_t)}{s_1 + \cdots + s_t}$ is the mediant fraction of $\frac{g(s_1)}{s_1}, \ldots, \frac{g(s_t)}{s_t}$, we have

$$\frac{g(s_1) + \cdots + g(s_t)}{s_1 + \cdots + s_t} \leq \max_{1 \leq i \leq t} \frac{g(s_i)}{s_i} \leq \max_{k \in \{1, 2, \ldots, n-1\}} \frac{g(k)}{k} = \frac{g(n-1)}{n-1}.$$

The contradiction proves that we cannot have $t \geq 2$, i.e., $G$ is connected.  $\square$

**Table 2.** Performance benchmarks for testing (inadmissible) graphs of order $n$, size $\ell(n) + 1$, and minimum degree $\geq 3$, for $n \in \{12, 13, 15, 20\}$.

| Graph order | 12 | 13 | 15 | 20 |
|---|---|---|---|---|
| # candidate graphs | 18 | 173 | 8280 | 15,156 |
| # graphs with an MFS | 16 | 159 | 8252 | 14,591 |
| # graphs tested w. GRAPHSOLVE | 2 | 14 | 28 | 565 |
| Average test time (sec.) | 6.5 | 17.5 | 24.5 | 247.9 |

Iteratively for each $n \in [12, 20]$, Theorem 4(III) from the established values $g(k)$ for all $k < n$ implies that any admissible graph of order $n$ and size $\geq \ell(n)$ is connected. That is, in the venue of proving Theorem 3 we can focus on the connected squarefree graphs only, which we refer to as *candidate graphs*. We test the admissibility of a candidate graph $G$ by first checking the presence of any MFS of order $\leq 10$ as a subgraph in $G$, and then, if no MFS is present, by invoking the GRAPHSOLVE algorithm for $G$.

For order $n = 12$, Theorem 4(II) implies $g(12) \leq 20$. If there is an admissible graph of size 20, its minimum degree should be at least 3. Hence, to prove Theorem 3 for $n = 12$, we generate all candidate graphs of order 12, size $\ell(12) + 1 = 20$, and minimum degree at least $20 - g(11) = 3$, which we then test for admissibility. There are 18 candidate graphs and none are admissible, which proves $g(12) = 19$. Similarly, for $n = 13$, we generate all candidate graphs of order 13, size $\ell(13) + 1 = 22$, and minimum degree at least $22 - g(12) = 3$. There are 173 such candidate graphs, and again none of them are admissible.

From $g(13) = 21$, Theorem 4(II) implies that $g(14) \leq 24$, which matches the lower bound. Therefore, we obtain $g(14) = 24$ without any computation.

For order $n = 15$, we test if there is any admissible graph of size $\ell(15) + 1 = 27$. By Theorem 4(I) such a graph should have minimum degree at least 3. We generate 8,280 such candidate graphs, but our check shows that all of them are inadmissible. Hence, $g(15) = 26$.

For order $n = 16$, Theorem 4(II) implies $g(16) \leq 29 = \ell(16)$ and thus $g(16) = 29$.

For order $n = 17$, Theorem 4(I) implies that any admissible graph of size $\ell(17) + 1 = 32$ should have minimum degree of at least 3. In fact, the minimum degree should be exactly 3, since otherwise the size would be at least $17 \cdot 4/2 = 34$. There are $1,023,100$ such candidate graphs, which, in principle, are possible to inspect directly, although it would be quite time consuming. Instead, we approached this problem from another angle: noticing that the removal of a vertex of degree 3 from an admissible graph of order 17 and size 32 results in a maximum admissible graph (MAG) of order 16. We constructed all MAGs of order 16 as explained in Sect. 6 and established that none of them extends to an admissible graph of order 17 and size 32, thus proving that $g(17) = \ell(17) = 31$.

For order $n = 18$, Theorem 4(II) implies $g(18) \leq 34 = \ell(18)$ and thus $g(18) = 34$.

For order $n = 19$, Theorem 4(I) implies that any admissible graph of size $\ell(19) + 1 = 37$ should have minimum degree of at least 3. With necessity such a graph has a vertex of degree 3 and we proceed similarly to the case $n = 17$ above. We construct MAGs of order 18 (see Sect. 6) and show that none of them can be extended to an admissible graph of order 19 and size 37. It follows that $g(19) = 36$.

**Table 3.** Performance benchmarks for identifying maximum admissible graphs.

| Graph order | 14 | 15 | 16 | 17 | 18 |
|---|---|---|---|---|---|
| # candidate graphs | 2,184 | 33,732 | 243 | 5,847,788 | 287 |
| # graphs with an MFS | 1,976 | 29,251 | 215 | 5,734,238 | 257 |
| # graphs tested w. GRAPHSOLVE | 208 | 4,481 | 28 | 113,550 | 30 |
| Average test time (sec.) | 30.9 | 44.2 | 51.0 | 81.4 | 79.1 |

For order $n = 20$, Theorem 4(I) implies that any admissible graph of size $\ell(20) + 1 = 40$ should have minimum degree $\geq 4$, from where it follows that this graph is regular of degree 4. There are $15,156$ such candidate graphs, all of which are inadmissible. Hence, $g(20) = \ell(20) = 39$. This concludes our proof of Theorem 3. We evaluate the performance of the computations used in this proof in Table 2.

## 6  Maximum Admissible Graphs

From the definition of $g(n)$ it follows that the maximum admissible graphs (MAG) of order $n$ have size $g(n)$ each. All MAGs of each order $n \leq 14$ can be obtained directly from the candidate graphs generated by **nauty**. In particular, for $n = 14$ we can restrict our attention to the $2,184$ candidate graphs of minimum degree 3, among which we identified only 4 admissible graphs.

To construct MAGs of order $n \in \{15, 16, 17, 18\}$, which by Theorem 4(I) have minimum degree at least $g(n) - g(n-1)$, we generate and test for admissibility candidate graphs of two types: *extended* graphs resulted from adding a vertex of degree $g(n) - g(n-1)$ to a MAG of order $n-1$, and *denovo* graphs with minimum degree $\geq g(n) - g(n-1) + 1$ generated by **nauty**. Clearly, denovo candidate graphs may exist only if $n(g(n) - g(n-1) + 1) \leq 2g(n)$, i.e., $g(n) \leq \frac{n}{n-2}(g(n-1) - 1)$.

For $n = 15$, there are 124 extended and $33,608$ denovo candidate graphs, among which we identified 20 and 8 MAGs, respectively.

For $n = 16$, there are 243 extended and no denovo candidate graphs, delivering just two MAGs (Fig. 2). We use these MAGs to establish $g(17) = 31$ as explained in Sect. 5.

For $n = 17$, there are 82 extended and $5,847,706$ denovo candidate graphs, delivering 11 and 7 MAGs, respectively.

For $n = 18$, there are 287 extended and no denovo candidate graphs, delivering just two MAGs. We use these MAGs to establish $g(19) = 36$ as explained in Sect. 5. Benchmarks of the computations used in this section are summarized in Table 3.

## 7   Discussion

In this work, we proposed the SOLVEINPOWERS Algorithm 2 for solving a given system of equations and inequations in powers of 2, and applied it to the problem of determining admissibility of a given graph with respect to labeling its vertices with pairwise distinct integers summing to a power of 2 for each edge (as outlined in GRAPHSOLVE Algorithm 1). Algorithm SOLVEINPOWERS can be easily extended to the powers of primes other than 2, and further to the powers of an arbitrary positive integer $b$ by introducing independent variables for the powers of each prime factor of $b$.

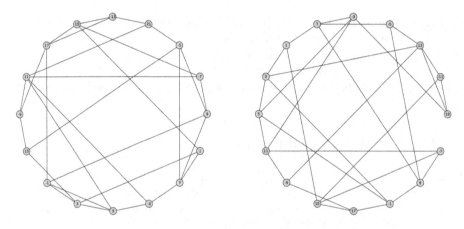

**Fig. 2.** Maximum admissible graphs of order 16 with the corresponding vertex labelings.

The performance of the SOLVEINPOWERS algorithm is sensitive to the number of nonzero coefficients in the equations $E$, which in the underlying the GRAPHSOLVE Algorithm 1 correspond to nonzero elements of $K_l$. While $K_l$ can be any basis of the left kernel of the given graph's incidence matrix, we found that upfront LLL reduction of the basis helps reduce the number of nonzero elements in $K_l$ and noticeably improves the algorithms' performance. We have also explored the idea of performing an LLL reduction of the equations in $E$ after each substitution, but it did not seem to provide much benefit.

Theorem 4 easily generalizes to graphs with a property that is invariant with respect to *induced* subgraphs (i.e., if a graph has such a property, then so does each of its induced subgraphs). In particular, if $h(n)$ denotes the maximum size of a squarefree graph on $n$ vertices (OEIS A006855), then $h(n) \leq \lfloor h(n-1) \cdot n/(n-2) \rfloor$. At the same time, it is known that $h(n)$ grows proportionally to $n^{3/2}$ and satisfies $h(n) \leq \lfloor n(1 + \sqrt{4n-3})/4 \rfloor$ [1,8], which also gives an upper bound for $g(n)$. Therefore, we cannot expect better asymptotic bounds from the inequality in Theorem 4(II) alone.

While the current paper was under review, we have further established that $g(21) = 41$, which we will describe elsewhere.

Supplementary figures and data are available at the GitHub repository https://github.com/maxale/power2_graphs.

**Acknowledgements.** The author thanks Neil Sloane for his nice introduction to the problem [5] and proofreading of the earlier version of this paper. The author also thanks anonymous reviewers for numerous helpful suggestions.

# References

1. Aigner, M., Ziegler, G.M.: Proofs from THE BOOK, 6th edn. Springer, Heidelberg (2018). https://doi.org/10.1007/978-3-662-57265-8
2. Bolan, M.: Stan Wagon 1321 Solution. Memo (2022). https://oeis.org/A352178/a352178.pdf
3. Bremner, M.R.: Lattice Basis Reduction: An Introduction to the LLL Algorithm and Its Applications. CRC Press, Boca Raton (2011)
4. Clapham, C.R.J., Flockhart, A., Sheehan, J.: Graphs without four-cycles. J. Graph Theory **13**(1), 29–47 (1989). https://doi.org/10.1002/jgt.3190130107
5. Haran, B., Sloane, N.J.A.: Problems with Powers of Two. Numberphile Youtube Channel (2022). https://youtu.be/IPoh5C9CcI8
6. McKay, B.D., Piperno, A.: Practical graph isomorphism, II. J. Symb. Comput. **60**, 94–112 (2014). https://doi.org/10.1016/j.jsc.2013.09.003
7. Melaih, F.: On The OEIS Sequence A352178. Memo (2022). https://oeis.org/A352178/a352178_3.pdf
8. Reiman, I.: Über ein Problem von K. Zarankiewicz. Acta Mathematica Academiae Scientiarum Hungarica **9**(3), 269–273 (1958). https://doi.org/10.1007/BF02020254
9. SageMath: version 10.0 (2023). https://www.sagemath.org/
10. ShahAli, H.A., Wagon, S.: Problem 12272. Am. Math. Mon. **128**(8), 755–763 (2021). https://doi.org/10.1080/00029890.2021.1953285
11. The OEIS Foundation: The On-Line Encyclopedia of Integer Sequences (2023). http://oeis.org
12. Ullman, D., Wagon, S.: Problem 1321: Powers of Two. Macalester College Problem of the Week (2021). https://oeis.org/A347301/a347301_1.pdf
13. Van Nuffelen, C.: On the incidence matrix of a graph. IEEE Trans. Circ. Syst. **23**(9), 572–572 (1976). https://doi.org/10.1109/TCS.1976.1084251

# Matchings in Hypercubes Extend to Long Cycles

Jiří Fink[1] and Torsten Mütze[1,2(✉)]

[1] Department of Theoretical Computer Science and Mathematical Logic, Charles University, Prague, Czech Republic
fink@ktiml.mff.cuni.cz
[2] Department of Computer Science, University of Warwick, Coventry, UK
torsten.mutze@warwick.ac.uk

**Abstract.** The $d$-dimensional hypercube graph $Q_d$ has as vertices all subsets of $\{1, \ldots, d\}$, and an edge between any two sets that differ in a single element. The Ruskey-Savage conjecture asserts that every matching of $Q_d$, $d \geq 2$, can be extended to a Hamilton cycle, i.e., to a cycle that visits every vertex exactly once. We prove that every matching of $Q_d$, $d \geq 2$, can be extended to a cycle that visits at least a 2/3-fraction of all vertices.

**Keywords:** Hypercube · matching · cycle · Gray code

## 1 Introduction

Cycles and matchings in graphs are structures of fundamental interest. In this paper, we consider these structures in hypercubes, a family of graphs that has been studied widely in computer science and mathematics. Specifically, the *d-dimensional hypercube* $Q_d$ is the graph whose vertices are all subsets of $[d] := \{1, \ldots, d\}$ and whose edges connect sets that differ in a single element; see Fig. 1(a). It is well-known and easy to show that $Q_d$, $d \geq 2$, admits a *Hamilton cycle*, i.e., a cycle that visits every vertex exactly once. Clearly, any Hamilton cycle in $Q_d$ is the union of two perfect matchings. A *matching* in a graph is a set of edges that are pairwise disjoint, and a matching is *perfect* if it includes every vertex of the graph.

30 years ago, Ruskey and Savage [21] asked whether every matching of $Q_d$ can be extended to a Hamilton cycle. This became known as the *Ruskey-Savage conjecture*.

**Conjecture 1 ([21]).** *Every matching of $Q_d$, $d \geq 2$, can be extended to a Hamilton cycle.*

This work was supported by Czech Science Foundation grant GA 22-15272S. Both authors participated in the workshop 'Combinatorics, Algorithms and Geometry' in March 2024, which was funded by German Science Foundation grant 522790373.

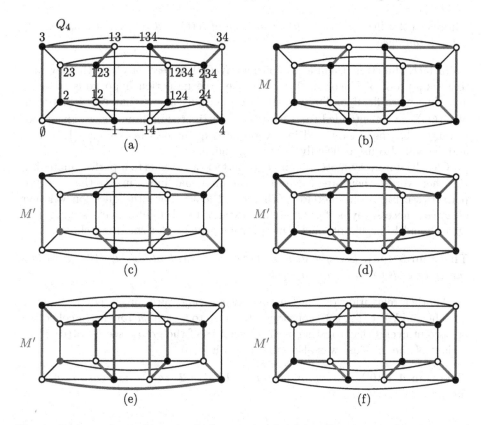

**Fig. 1.** Cycles and matchings in the hypercube $Q_4$: (a) a Hamilton cycle, where the vertex labels omit curly brackets and commas for conciseness; (b) a perfect matching $M$ and an extension of $M$ to a Hamilton cycle; (c) a maximal, but not perfect matching $M'$; (d) an extension of $M'$ to a cycle factor with two cycles; (e) an extension of $M'$ to a (non-Hamilton) cycle; (f) an extension of $M'$ to a Hamilton cycle (obtained by joining the two cycles in (d)).

This problem received considerable attention, and several natural relaxations have been proved. In particular, Fink [9] settled the conjecture affirmatively for the case when the prescribed matching is perfect, thereby answering a problem due to Kreweras [17]; see Fig. 1(b). In fact, Fink established a considerable strengthening, obtained by considering the graph $K(Q_d)$, which is the complete graph on the vertex set of $Q_d$. In this context, we say that a matching $M$ of $K(Q_d)$ *extends to* a Hamilton cycle $C$ (or some other structure) if all edges in $C \setminus M$ belong to $Q_d$. Put differently, while the matching $M$ may use arbitrary edges of $K(Q_d)$, including edges not present in $Q_d$, the edges $C \setminus M$ used for the extension are required to be edges of the hypercube $Q_d$ (otherwise $C$ would trivially be present in the complete graph).

**Theorem 2** ([9]). *Every perfect matching of $K(Q_d)$, $d \geq 2$, can be extended to a Hamilton cycle.*

In fact, the strengthening to consider $K(Q_d)$ instead of $Q_d$ is the key idea to Fink's proof of Theorem 2, as it makes the induction hypothesis stronger and thus more flexible. This construction actually shows that every matching of $K(Q_d)$ can be extended to at least $2^{2^{d-4}}$ distinct Hamilton cycles. This proof technique has been exploited in several subsequent papers (see e.g. [1,14,15]), and we will also use it heavily in our arguments.

Clearly, not every matching in $Q_d$ is perfect or extends to a perfect matching. In other words, there are (inclusion-)maximal matchings in $Q_d$ that are not perfect; see Fig. 1(c). Therefore, Theorem 2 leaves open the question whether every (not necessarily perfect) matching extends to a Hamilton cycle; see Fig. 1(f). Dvořák and Fink [6] obtained some positive evidence for small matchings.

**Theorem 3** ([6]). *Every matching of $Q_d$, $d \geq 2$, with at most $d^2/16 + d/4$ edges can be extended to a Hamilton cycle.*

Another relaxation of the Ruskey-Savage conjecture, proposed by Vandenbussche and West [24], is to consider extensions to a *cycle factor*, i.e., a collection of disjoint cycles that together visit all vertices of the graph; see Fig. 1(d). This variant of the problem was also settled by Fink [10].

**Theorem 4** ([10]). *Every matching of $Q_d$, $d \geq 2$, can be extended to a cycle factor.*

## 1.1  Our Results

In this paper we prove that every matching of $Q_d$ extends to a single cycle $C$. However, $C$ possibly omits some vertices of $Q_d$, i.e., $C$ is not necessarily a Hamilton cycle; see Fig. 1(e). Our theorem also holds in the stronger setting of the complete graph $K(Q_d)$.

**Theorem 5.** *Every matching of $K(Q_d)$, $d \geq 2$, can be extended to a cycle.*

Furthermore, we can give some guarantee about the length of the cycle obtained from our construction, namely that it visits a constant fraction of the vertex set $V(Q_d)$.

**Theorem 6.** *Every matching of $Q_d$, $d \geq 2$, can be extended to a cycle of length at least $\frac{2}{3}|V(Q_d)| = 2^{d+1}/3$.*

**Theorem 7.** *Every matching of $K(Q_d)$, $d \geq 2$, can be extended to a cycle of length at least $\frac{1}{2}|V(Q_d)| = 2^{d-1}$.*

Our results are another step towards the Ruskey-Savage conjecture, and we want to point out an analogy to the *middle levels conjecture*, which asserts that the subgraph of $Q_{2d+1}$ induced by all sets of size $d$ or $d + 1$ admits a

Hamilton cycle. Historically, Felsner and Trotter [8], Savage and Winkler [22], and Johnson [16] first established the existence of long cycles in this graph, and their work subsequently led to a proof of the middle levels conjecture [20].

We also note that not every matching of $K(Q_d)$ can be extended to a Hamilton cycle, nor a cycle factor. In fact, any matching in $K(Q_d)$ between all vertices of the same parity has the property that any cycle extending it includes only half of the vertices from the other parity, so it has only length $\frac{3}{4}|V(Q_d)|$. The *parity* of a vertex $u$ of $Q_n$ is the parity of $|u|$, i.e., of the size of the set $u$. In particular, in $K(Q_2)$, the matching with the single edge $(\emptyset, \{1, 2\})$ can only be extended to a cycle of length 3.

Clearly, if a matching $M$ of $K(Q_d)$ is extendable to a Hamilton cycle, then $M$ has the same number of vertices of each parity. Note that every matching of $B(Q_d)$ satisfies this condition, where $B(Q_d)$ is the complete bipartite graph obtained from $Q_d$ by adding all edges between vertices of opposite parity. However, Dvořák and Fink [6] showed that this condition is not sufficient, by constructing a matching of $B(Q_d)$ for $d \geq 9$ which cannot be extended to a Hamilton cycle, nor a cycle factor.

To prove Theorem 5 inductively, we will need the following theorem, which yields a cycle that extends a given matching but also avoids one 'forbidden' vertex $z$, i.e., avoiding $z$ is an additional constraint imposed on the cycle. Using that $Q_d$ is vertex-transitive, we may assume w.l.o.g. that the forbidden vertex is $z = \emptyset$. The following theorem requires some additional conditions, and to state them we need to introduce some notation; see Fig. 3(a): We split $Q_d$ in a *direction* $i \in [d]$ into subgraphs $Q_0^i$ and $Q_1^i$, where $Q_1^i$ is the subgraph induced by vertices of $Q_d$ containing $i$ and $Q_0^i$ is the subgraph induced by vertices of $Q_d$ not containing $i$. Note that $z \in V(Q_0^i)$ for every $i \in [d]$. The set of edges of $Q_d$ that have one end vertex in $Q_0^i$ and the other in $Q_1^i$ is called a *layer* in direction $i$. This set splits into two equal-sized sets of edges, depending on the parity of their end vertices in $Q_0^i$, and we refer to each of these two sets of edges as a *half-layer*.

**Theorem 8.** *Let $d \geq 5$. A matching $M$ of $K(Q_d)$ that avoids $z = \emptyset$ can be extended to a cycle that avoids $z$ if and only it satisfies the following condition:*

(H) *for every $i \in [d]$, if $M$ contains a half-layer in direction $i$, then there is a vertex $u \in V(Q_0^i) \setminus \{z\}$ that is avoided by $M$.*

We remark that the statement of Theorem 8 is false for $d = 4$, and all counterexamples are listed in [15].

In the following we will repeatedly refer to the condition on $M$ stated in Theorem 8 as *property* (H). Note that this property depends on $Q_d$, $M$ and $z$, but for simplicity we do not make this dependence explicit, as the context should always be clear later when we apply the theorem to subcubes $Q_0^i$ and $Q_1^i$ for some $i \in [d]$, certain matchings in those subcubes, and certain vertices to avoid in them. Note that the vertex $u$ must have the same (even) parity as $z$, as all vertices in $Q_0^i$ of the opposite (odd) parity are covered by the half-layer.

We prove Theorems 5 and 8 by induction on the dimension $d$. Specifically, in the induction step we assume that both statements hold in dimension $d - 1$, and we use this to prove that they also hold in dimension $d$.

## 1.2    Applications to Hamilton-Laceability

The hypercube $Q_d$ is *Hamilton-laceable*, i.e., it admits a Hamilton path between any two prescribed end vertices of opposite parity [23]. Gregor, Novotný, and Škrekovski [15] considered laceability combined with matching extensions. Specifically, they considered the problem of extending a perfect matching of $Q_d$ to a Hamilton path between two prescribed end vertices with opposite parity. Their proof works in the more general setting of the complete bipartite graph $B(Q_d)$. For a matching $M$ of $Q_d$ and one of its vertices $x$, we write $x^M$ for the other end vertex of the edge of $M$ incident with $x$.

**Theorem 9** ([15, Thm. 2]). *Let $d \geq 5$, and let $x, y$ be two vertices of opposite parity in $Q_d$. A perfect matching $M$ of $B(Q_d)$ with $xy \notin M$ can be extended to a Hamilton path from $x$ to $y$ if and only if $(M \setminus \{xx^M, yy^M\}) \cup \{x^M y^M\}$ contains no half-layers.*

The following is an equivalent reformulation of this theorem.

**Theorem 10** ([15, Thm. 3]). *Let $d \geq 5$, and let $x, y$ be two vertices of opposite parity in $Q_d$. A perfect matching $M$ of $B(Q_d \setminus \{x, y\})$ can be extended to a cycle that avoids $x$ and $y$ if and only if $M$ contains no half-layers.*

The authors also conjectured the strengthenings of Theorems 9 and 10 obtained by replacing $B(Q_d)$ and $B(Q_d \setminus \{x, y\})$ by $K(Q_d)$ and $K(Q_d \setminus \{x, y\})$, respectively. In fact, these stronger versions follow easily from our Theorem 8, settling the conjecture raised by Gregor, Novotný, and Škrekovski.

**Theorem 11.** *Let $d \geq 5$, and let $x, y$ be two vertices of opposite parity in $Q_d$. A perfect matching $M$ of $K(Q_d)$ with $xy \notin M$ can be extended to a Hamilton path from $x$ to $y$ if and only if $(M \setminus \{xx^M, yy^M\}) \cup \{x^M y^M\}$ contains no half-layers.*

We prove the following equivalent reformulation of this theorem.

**Theorem 12.** *Let $d \geq 5$, and let $x, y$ be two vertices of opposite parity in $Q_d$. A perfect matching $M$ of $K(Q_d \setminus \{x, y\})$ can be extended to a cycle that avoids $x$ and $y$ if and only if $M$ contains no half-layers.*

## 1.3    Avoiding and Including Other Structures

There has also been substantial work on extending sets of edges $E$ other than matchings to Hamilton cycles, and on avoiding certain sets $F$ of 'forbidden' edges. These questions are motivated by applications in computer networks, where the hypercube is frequently used as a network topology with a number of desirable properties, such as small degree and diameter. Including or avoiding certain edges corresponds to prescribed connections or faulty connections, respectively.

Dvořák [5] showed that for any set $E$ of at most $2d - 3$ edges in $Q_d$, $d \geq 2$, that form disjoint paths, there is a Hamilton cycle that contains all of $E$; see

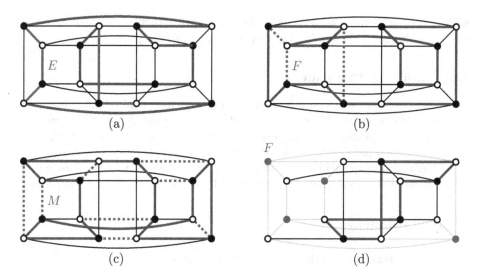

(a)    (b)

(c)    (d)

**Fig. 2.** Illustration of Hamilton cycles in $Q_d$, $d = 4$, subject to various constraints: (a) a set $E$ of $2d - 3 = 5$ edges and a Hamilton cycle extending it; (b) a set $F$ of $2d - 5 = 3$ edges and a Hamilton cycle avoiding it; (c) a perfect matching $M$ and a Hamilton cycle avoiding it; (d) a set $F$ of $\binom{d}{2} - 2 = 4$ vertices and a cycle of length $2^d - 2|F| = 8$ avoiding it.

Fig. 2(a). Dvořák and Gregor [7] proved that for any set $E$ of at most $2d - 4$ edges in $Q_d$, $d \geq 5$, that form disjoint paths and two vertices $x$ and $y$ of opposite parity that are neither internal vertices of the paths nor end vertices of the same path, there is a Hamilton path with end vertices $x$ and $y$ that contains all of $E$.

We now consider the problem of avoiding a set $F$ of 'forbidden' edges. In this direction, Latifi, Zheng and Bagherzadeh [18] showed that for any set $F$ of at most $d - 2$ edges in $Q_d$, there is a Hamilton cycle that avoids $F$. Chan and Lee [3] proved that for any set $F$ of at most $2d - 5$ edges in $Q_d$, $d \geq 3$, such that every vertex of $Q_d$ is incident to at least two edges not in $F$, there is a Hamilton cycle that avoids $F$ (see also [19]); see Fig. 2(b). With regards to perfect matchings, Dimitrov, Dvořák, Gregor and Škrekovski [4] proved that for a given perfect matching $M$ of $Q_d$, there is a Hamilton cycle that avoids $M$ if and only if $Q_d \setminus M$ is a connected graph; see Fig. 2(c).

The bounds $2d - 3$, $2d - 4$, $d - 2$, $2d - 5$ in the aforementioned results on extending or avoiding sets of edges are best possible, i.e., prescribing or forbidding more edges leads to situations where the desired cycle or path does not exist.

Instead of forbidding edges, one may also consider the problem of forbidding certain vertices. To this end, Fink and Gregor [12] proved that for any set $F$ of at most $\binom{d}{2} - 2$ vertices of $Q_d$, $d \geq 3$, there is a cycle of length at least $2^d - 2|F|$ that avoids all vertices in $F$; see Fig. 2(d). Their result answer a conjecture by Castañeda and Gotchev [2], and the bound $\binom{d}{2} - 2$ is best possible. Note that if we forbid some number of vertices of the same parity, then any extending cycle will also skip the same number of vertices of the opposite parity.

## 1.4   Outline of This Paper

In Sect. 2 we present some notations and terminology used throughout this paper, as well as some auxiliary lemmas needed later. In Sects. 3 and 4 we present the inductive arguments for the proofs of Theorems 5 and 8, respectively. We conclude with some open problems in Sect. 5. The proofs of Theorems 6, 7, and 12, of the auxiliary lemmas, as well as some details in the proof of Theorem 8 are omitted in this extended abstract; see [13] for the full version of the paper.

## 2   Preliminaries

We need the following definitions and auxiliary lemmas.

### 2.1   Notation and Definitions

For a graph $G$, the sets of vertices and edges of $G$ are denoted by $V(G)$ and $E(G)$, respectively. Furthermore, we write $K(G)$ for the complete graph on the vertex set of $G$. If $G$ is connected and bipartite, then there is a unique bipartition of its vertices, and we write $B(G)$ for the complete bipartite graph with that bipartition. In other words, $K(G)$ is obtained from $G$ by adding all missing edges to $G$, and $B(G)$ is obtained from $G$ by adding all possible edges to $G$, while preserving the property that the graph remains bipartite.

For two sets $A$ and $B$, the symmetric difference of $A$ and $B$ is denoted by $A \triangle B$. Recall that $Q_d$ has an edge between any two vertices $u$ and $v$ with $|u \triangle v| = 1$. For a vertex $u$ of $Q_d$ and $i \in [d]$, we define $u^i := u \triangle \{i\}$, and we refer to $i$ as the *direction* of the edge $uu^i$. We say that $u$ is *even* or *odd*, if $|u|$ is even or odd, respectively.

A *linear forest* is collection of vertex-disjoint paths. The *terminals* of a linear forest are its vertices of degree 1, i.e., the end vertices of the paths. A *shortcut* of a linear forest in a graph $G$ is the set of edges of $K(G)$ that connect the two terminals of every path to each other.

For a graph $G$ and a subset of edges $M$ of $G$, we say that a vertex $u$ of $G$ is *covered* by $M$, if $u$ is end vertex of one of the edges of $M$. We write $V(M)$ for the set of all vertices covered by $M$. If $u$ is not covered by $M$ we say that it is *avoided* by $M$. Recall that for a vertex $u$ covered by an edge of a matching $M$, we write $u^M$ for the other end vertex of that edge of $M$. Furthermore, for $i \in [d]$ we use the abbreviation $u^{iM} := (u^i)^M$.

For a set of edges $F \subseteq E(K(Q_d))$ and a direction $i \in [d]$ we define $F_0^i := F \cap E(K(Q_0^i))$ and $F_1^i := F \cap E(K(Q_1^i))$, and we write $F_-^i$ for the edges of $F$ joining vertices of $Q_0^i$ and $Q_1^i$. This partitions $F$ into three disjoint sets $F_0^i$, $F_1^i$ and $F_-^i$.

The following definitions are illustrated in Fig. 3. Recall the definition of half-layers given in Sect. 1.1. Given some direction $i \in [d]$, a half-layer in $Q_0^i$ or in $Q_1^i$ is referred to as a *quad-layer* in $Q_d$. Clearly, a half-layer has $2^{d-2}$ edges, and a quad-layer has $2^{d-3}$ edges. A *near half-layer* (*near quad-layer*) is a half-layer (quad-layer) minus one edge. We refer to the two end vertices of the removed edge, which are not part of the near half-layer (near quad-layer), as *extension vertices* of the near half-layer (near quad-layer). Given a matching $M$ that includes a near half-layer (near quad-layer), we say that this near half-layer (near quad-layer) is *covered* if both of its extension vertices are covered by two distinct edges of $M$.

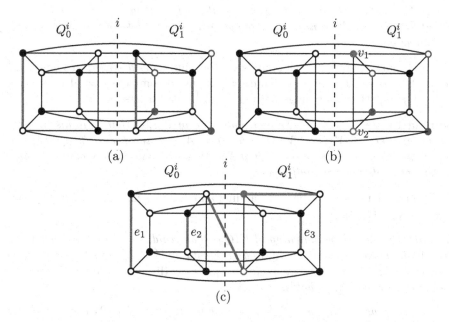

**Fig. 3.** The red edges in $Q_4$ form a (a) half-layer, (b) near half-layer with extension vertices $v_1$ and $v_2$, or (c) matching that includes the covered near half-layer $\{e_1, e_2, e_3\}$. The subset of red edges incident with a vertex in $Q_1^i$ form a (a) quad-layer, (b) near quad-layer, or (c) matching that includes the covered near quad-layer $\{e_3\}$. Every vertex $x$ in (a) and (b) colored blue is one for which the (near) quad-layer contained in $Q_1^i$ is $x$-dangerous. The dashed vertical line separates $Q_0^i$ and $Q_1^i$. (Color figure online)

A half-layer of $Q_d$ is $x$-*dangerous* for every vertex $x$ not incident with any edge from the half-layer. An $x$-*dangerous* near half-layer is an $x$-dangerous half-layer minus one edge. Note that $x$ cannot be one of the extension vertices of the near half-layer. For a vertex $x$ of $Q_d$, a quad-layer is $x$-*dangerous* if $x$ is not incident with any edge from the quad-layer and $x$ belongs to the same $(d-1)$-dimensional subcube as the quad-layer. Furthermore, we define an $x$-*dangerous* near quad-layer as an $x$-dangerous quad-layer minus one edge. Similarly to before, $x$ cannot be one of the extension vertices of the near quad-layer.

## 2.2   Auxiliary Lemmas

In the following we collect a few lemmas we need for proving our theorems. The proofs of those lemmas can be found in [13].

The first lemma asserts that half-layers and quad-layers cannot exist simultaneously in several different directions.

**Lemma 13.** *Let $M$ be a matching of $K(Q_d)$ and let $z := \emptyset$.*

*(i) If $d \geq 4$, then $M$ contains a (near) half-layer in at most one direction.*
*(ii) If $d \geq 6$, then $M$ contains a $z$-dangerous (near) quad-layer in at most one direction.*

We remark that the lower bounds on the dimension $d$ stated in Lemma 13 are all best possible. The next lemma allows us to complete partial matchings of $K(Q_d)$ to perfect matchings without creating half-layers.

**Lemma 14.** *Let $d \geq 4$, let $M$ be a matching of $K(Q_d)$ that contains no half-layers, and let $A \subseteq V(Q_d) \setminus V(M)$ such that $|A| \geq 4$ and even. Then there is a perfect matching $P$ on $K(A)$ such that $M \cup P$ contains no half-layers. The same statement holds for near half-layers.*

The next lemma describes how adding an edge to a matching of $K(Q_d)$ can violate property (H).

**Lemma 15.** *Let $M$ be matching of $K(Q_d)$ that avoids $z = \emptyset$ and satisfies property (H) in Theorem 8. Let $i \in [d]$ and $u, u^i$ with $u \in V(Q_0^i) \setminus \{z\}$ be two vertices avoided by $M$. Then $M \cup \{uu^i\}$ violates property (H) if and only if one of the following two conditions holds:*

*(i) $M$ contains a half-layer in direction $i$ and all vertices of $Q_0^i$ except $z$ and $u$ are covered by $M$;*
*(ii) $M$ contains a near half-layer in direction $i$ with extension vertices $u$ and $u^i$, and all vertices of $Q_0^i$ except $z$ and $u$ are covered by $M$.*

*In both cases $|M_-^i|$ is even. Furthermore, we have $|M_-^i| \geq \frac{1}{4}|V(Q_d)| = 2^{d-2}$ and $|M| \geq \frac{3}{8}|V(Q_d)| - 1 = 3 \cdot 2^{d-3} - 1$.*

We say that a matching $M$ of $K(Q_d)$ that avoids $z = \emptyset$ and satisfies property (H) is (H)-*maximal* if for any two vertices $u, u^i$ for some $i \in [d]$ that are avoided by $M$ and different from $z$ the matching $M \cup \{uu^i\}$ violates property (H).

The next lemma yields a direction such that many edges of an (H)-maximal matching belong to the layer in this direction. In the proof of Theorem 8 we will choose this direction to split the cube into two subcubes and apply induction.

**Lemma 16.** *For every (H)-maximal matching $M$ of $K(Q_d)$ there is a direction $i \in [d]$ such that $|M_-^i| \geq 3$ for $d = 5$ and $|M_-^i| \geq 4$ for $d \geq 6$.*

# 3   Proof of Theorem 5

*Proof (Proof of Theorem 5).* For $2 \leq d \leq 4$, we verified the theorem with computer help. For $d \geq 5$, we prove Theorem 5 assuming that Theorem 8 holds for dimension $d$. If $M$ is perfect, we apply Theorem 2. Otherwise, if $M$ has no half-layers in any direction, then we choose a vertex $u$ avoided by $M$. Otherwise, if $M$ contains a half-layer in a direction $i \in [d]$ such that $M$ avoids at least two vertices of $Q_0^i$ or $Q_1^i$, then we choose a vertex $u$ avoided by $M$ in $Q_0^i$ or $Q_1^i$, respectively. By Lemma 13 (i) and the fact that $M$ is not perfect, $M$ contains no other half-layer. Consequently, in the latter two cases, $M$ and $u$ satisfy property (H). We may thus apply Theorem 8 to obtain a cycle that extends $M$ (and avoids $u$).

In the remaining case, $M$ contains a half-layer in a direction $i \in [d]$ and avoids exactly one vertex $u$ in $Q_0^i$ and one vertex $v$ in $Q_1^i$. If $v = u^i$, we apply Theorem 2 to the perfect matching $M \cup \{uv\}$. Otherwise consider the modified matching $N := (M \setminus \{u^i u^{iM}\}) \cup \{uu^{iM}\}$, which avoids two vertices in $Q_1^i$, namely $u^i$ and $v$. Consequently, by Theorem 8 there is a cycle $C$ that extends $N$ and avoids $u^i$. Then, the cycle $(C \setminus \{uu^{iM}\}) \cup \{uu^i, u^i u^{iM}\}$ extends $M$.   □

# 4   Proof of Theorem 8

## 4.1   Forward Implication

*Proof (Proof of Theorem 8 ($\Rightarrow$)).* For the sake of contradiction suppose that $M$ violates property (H) and that there is a cycle $C$ that extends $M$ and avoids $z$. Then for some $i \in [d]$ there is a half-layer in direction $i$ and all vertices of $Q_0^i$ except $z$ are covered by $M$. Every odd vertex of $Q_0^i$ is end vertex of an edge of the half-layer, and is therefore connected in $C \setminus M$ to an even vertex of $Q_0^i$. As every vertex except $z$ is covered by $M$, no two of these even vertices are the same. It follows that $C$ visits all odd vertices of $Q_0^i$ and at least as many even vertices, but at the same time it avoids the even vertex $z$, a contradiction.   □

## 4.2   Reverse Implication (Induction Step $d - 1 \to d$ for $d \geq 6$)

Our proof for the reverse implication in Theorem 8 uses induction on the dimension $d$, and it follows a similar strategy as Fink's proof of Theorem 2, namely to choose a direction $i \in [d]$ and to split $Q_d$ into subcubes $Q_0^i$ and $Q_1^i$, to which we apply induction. However, the requirement for the extending cycle to avoid a prescribed vertex causes substantial additional technical complications. In particular, we need to deal with the case that an odd number of edges is present in $M_-^i$, a case that never occurs if the matching $M$ is perfect. In that case, we need to add one additional edge in direction $i$ to be included in the extending cycle (as the cycle has to cross between $Q_0^i$ and $Q_1^i$ an even number of times). We thus need to choose the direction $i$ and the added edge so that no obstacles for applying induction in the subcubes, namely half-layers of $Q_0^i$ or $Q_1^i$, are created.

In this extended abstract we present a proof sketch for the induction step $d - 1 \to d$ for $d \geq 6$, with some details being omitted; see [13] for those. The base case $d = 5$ is settled with computer help (see [13] for details).

*Proof (Proof sketch of Theorem 8 ($\Leftarrow$) for $d \geq 6$).* We prove Theorem 8 inductively for $d \geq 6$, assuming that Theorems 5 and 8 hold for dimension $d - 1$.

Let $M$ be a matching of $K(Q_d)$ that avoids $z$ and satisfies property (H). We assume w.l.o.g. that $M$ is (H)-maximal (recall the definition from before Lemma 16). We select a direction $i \in [d]$ according to the following rules applied in this order:

(1) If $M$ contains a $z$-dangerous (covered near) quad-layer, we choose $i$ to be its direction;
(2) otherwise, if $M$ contains a quad-layer, we choose $i$ to be its direction;
(3) otherwise, we choose a direction $i \in [d]$ that maximizes the quantity $|M_-^i|$.

These rules guarantee the following properties:

(i) $M_0^i$ contains no (covered near) half-layers of $Q_0^i$.
(ii) If $M_1^i$ contains a half-layer of $Q_1^i$, then $M_-^i$ contains a (covered near) quad-layer of $Q_d$.
(iii) If rule (1) or (2) applies we have $|M_-^i| \geq 7$, and if rule (3) applies we have $|M_-^i| \geq 4$.

Proof of (i): If $M_0^i$ did contain a (covered near) half-layer of $Q_0^i$, then there would be a $z$-dangerous (covered near) quad-layer of $Q_d$ contained in $M$, and by Lemma 13 (ii) these can occur in at most one direction, which would have been selected by rule (1) with the highest priority.

Proof of (ii): Suppose that $M_1^i$ contains a half-layer $L$ of $Q_1^i$. The direction $i$ was chosen differently from the direction of $L$, which means that rule (1) or (2) were applied, and this proves the claim.

Proof of (iii): The first part follows as a near quad-layer of $Q_d$, $d \geq 6$, has $2^{d-3}-1 \geq 7$ edges. The second part follows from Lemma 16, using the assumption that $M$ is (H)-maximal.

Let $A_0 := V(Q_0^i) \cap V(M_-^i)$. From (iii) we obtain $|A_0| \geq 4$.

**Case 1:** $|M_-^i| \geq 4$ is even. By (i), we can apply Lemma 14 and obtain a perfect matching $P_0$ on $K(A_0)$ such that $M_0^i \cup P_0$ contains no half-layers. Applying Theorem 8 inductively to $K(Q_0^i)$, we obtain that $M_0^i \cup P_0$ can be extended to a cycle $C_0$ that avoids $z$. Let $P_1$ be the shortcut edges of the linear forest $(C_0 \backslash P_0) \cup M_-^i$, which all belong to $K(Q_1^i)$ and form a perfect matching on $K(A_1)$, where $A_1 := V(Q_1^i) \cap V(M_-^i)$. Applying Theorem 5 inductively to $K(Q_1^i)$, we obtain that $M_1^i \cup P_1$ can be extended to a cycle $C_1$. Observe that $C := (C_0 \backslash P_0) \cup M_-^i \cup (C_1 \backslash P_1)$ is a single cycle in $K(Q_d)$ that extends $M$ and avoids $z$, as desired.

**Case 2:** $|M_-^i| \geq 5$ is odd. We give an outline of the construction steps in this case. Several details need to be filled in subsequently, and we also need to verify that all assumptions required to apply the various theorems are indeed satisfied.

The points with missing information in the outline are labeled $\boxed{1}$–$\boxed{4}$, and the missing details can be found in [13].

The cycle we construct that extends $M$ uses one additional edge between $Q_0^i$ and $Q_1^i$ in addition to the edges $M_-^i$. For this purpose we carefully choose a vertex $u$ in $Q_0^i$ different from $z$ such that $uu^i \notin M$ $\boxed{1}$, and we take $uu^i$ as the edge to be included in the cycle. Depending on whether and how its end vertices $u$ and $u^i$ are covered by $M$, this creates different conditions in $Q_0^i$ and $Q_1^i$ for the induction step. Specifically, there are three possible cases for $u$, namely $u \notin V(M)$, $u^M \in V(Q_0^i)$, or $u^M \in V(Q_1^i)$, and similarly three cases for $u^i$. Because of our assumption that $M$ is (H)-maximal, the case $u, u^i \notin V(M)$ cannot occur, as Lemma 15 would yield that $|M_-^i|$ is even. Consequently, for every $u \in V(Q_0^i)$ with $u \neq z$ we have that $u \notin V(M)$ implies that $u^i \in V(M)$.

We create a modified matching $N$ from $M$ as follows: If $u \in V(M)$ we remove the edge $uu^M$, and if $u^i \in V(M)$ we remove the edge $u^i u^{iM}$. Furthermore, we add the edge $vw$ where $v := u$ if $u \notin V(M)$ and $v := u^M$ if $u \in V(M)$, and similarly $w := u^i$ if $u^i \notin V(M)$ and $w := u^{iM}$ if $u^i \in V(M)$. We will construct a cycle $C$ that extends $N$ and avoids $z$ and $u$ if $u \in V(M)$ as well as $u^i$ if $u^i \in V(M)$. Written compactly, the cycle $C$ avoids $z$ and $\{u, u^i\} \setminus \{v, w\}$. From $C$, the desired cycle that extends $M$ and avoids $z$ can be obtained by straightforward modifications. Specifically, we remove the edge $vw$ from $C$, add the edge $uu^i$ and the edge $uu^M$ if $u \in V(M)$ as well as $u^i u^{iM}$ if $u^i \in V(M)$.

The construction of $C$ proceeds as follows: We carefully choose a perfect matching $P_0$ on $K(B_0)$, where $B_0 := V(Q_0^i) \cap V(N_-^i)$ $\boxed{2}$. Having chosen $u$ and $P_0$, we apply Theorem 8 inductively to $K(Q_0^i)$ and obtain that $N_0^i \cup P_0$ can be extended to a cycle $C_0$ that avoids $z$. If $u \in V(M)$, then we argue that $C_0$ also avoids $u$ $\boxed{3}$. Let $P_1$ be the shortcut edges of the linear forest $(C_0 \setminus P_0) \cup N_-^i$. If $u^i \notin V(M)$, then we apply Theorem 5 and obtain that $N_1^i \cup P_1$ can be extended to a cycle $C_1$. Otherwise, we apply Theorem 8 and obtain that $N_1^i \cup P_1$ can be extended to a cycle $C_1$ that avoids $u^i$. As mentioned before, we need to argue that the assumptions of the those theorems are met in the subcubes $Q_0^i$ and $Q_1^i$ for the matchings $N_0^i \cup P_0$ and $N_1^i \cup P_1$, and this is where our choices of $u$ and $P_0$ become crucial $\boxed{4}$. Observe that $C := (C_0 \setminus P_0) \cup N_-^i \cup (C_1 \setminus P_1)$ is a single cycle that extends $N$ and avoids $z$ and $u$ if $u \in V(M)$ as well as $u^i$ if $u^i \in V(M)$. $\square$

## 5   Open Questions

We conclude this paper with some interesting directions for future investigations.

The construction described in this paper translates straightforwardly to an algorithm that computes a cycle $C$ that extends a given matching $M$ of $K(Q_d)$ in polynomial time (polynomial in the size of the graph $Q_d$ or $K(Q_d)$). Can this be improved, along the lines described in [11], where part of the output can be computed already while knowing only part of the input?

Our Theorems 6 and 7 provide a lower bound for the length of a cycle $C$ that extends a given matching $M$. The reason for this is our approach of proving Theorem 8, which starts by adding edges to $M$ so that it becomes (H)-maximal.

Consequently, even for very small matchings $M$, the cycle $C$ might be very long. Can we instead prove a theorem about a cycle $C$ extending a given matching $M$, such that the length of $C$ is relatively short compared to the size of $M$, i.e., such that $|C| \leq f(|M|)$ for some reasonable function $f$? More generally, given a matching $M$ of $Q_d$ or $K(Q_d)$, for which integers $\ell$ is there a cycle $C$ with $|C| = \ell$ that extends $M$? Rephrased as a decision problem: Given $M$ and $\ell$, does $C$ exist?

Complementing Theorem 4, Dvořák and Fink [6] constructed a matching of $B(Q_d)$ that cannot be extended to a cycle factor. Is there a matching of $K(Q_d)$ or $B(Q_d)$ that is extendable to a cycle factor but not to a Hamilton cycle?

# References

1. Alahmadi, A., Aldred, R.E.L., Alkenani, A., Hijazi, R., Solé, P., Thomassen, C.: Extending a perfect matching to a Hamiltonian cycle. Discrete Math. Theor. Comput. Sci. **17**(1), 241–254 (2015)
2. Castañeda, N., Gotchev, I.S.: Embedded paths and cycles in faulty hypercubes. J. Comb. Optim. **20**(3), 224–248 (2010). https://doi.org/10.1007/s10878-008-9205-6
3. Chan, M.Y., Lee, S.J.: On the existence of Hamiltonian circuits in faulty hypercubes. SIAM J. Discrete Math. **4**(4), 511–527 (1991). https://doi.org/10.1137/0404045
4. Dimitrov, D., Dvořák, T., Gregor, P., Škrekovski, R.: Gray codes avoiding matchings. Discrete Math. Theor. Comput. Sci. **11**(2), 123–147 (2009)
5. Dvořák, T.: Hamiltonian cycles with prescribed edges in hypercubes. SIAM J. Discrete Math. **19**(1), 135–144 (2005). https://doi.org/10.1137/S0895480103432805
6. Dvořák, T., Fink, J.: Gray codes extending quadratic matchings. J. Graph Theory **90**(2), 123–136 (2019). https://doi.org/10.1002/jgt.22371
7. Dvořák, T., Gregor, P.: Hamiltonian paths with prescribed edges in hypercubes. Discrete Math. **307**(16), 1982–1998 (2007). https://doi.org/10.1016/j.disc.2005.12.045
8. Felsner, S., Trotter, W.T.: Colorings of diagrams of interval orders and $\alpha$-sequences of sets, vol. 144, pp. 23–31 (1995). https://doi.org/10.1016/0012-365X(94)00283-O. Combinatorics of ordered sets (Oberwolfach, 1991)
9. Fink, J.: Perfect matchings extend to Hamilton cycles in hypercubes. J. Combin. Theory Ser. B **97**(6), 1074–1076 (2007). https://doi.org/10.1016/j.jctb.2007.02.007
10. Fink, J.: Matchings extend into 2-factors in hypercubes. Combinatorica **39**(1), 77–84 (2019). https://doi.org/10.1007/s00493-017-3731-8
11. Fink, J.: Two algorithms extending a perfect matching of the hypercube into a Hamiltonian cycle. Eur. J. Combin. **88**, 103111 (2020). https://doi.org/10.1016/j.ejc.2020.103111
12. Fink, J., Gregor, P.: Long cycles in hypercubes with optimal number of faulty vertices. J. Comb. Optim. **24**(3), 240–265 (2012). https://doi.org/10.1007/s10878-011-9379-1
13. Fink, J., Mütze, T.: Matchings in hypercubes extend to long cycles (2024). Full preprint of the present article available at https://arxiv.org/abs/2401.01769
14. Gregor, P.: Perfect matchings extending on subcubes to Hamiltonian cycles of hypercubes. Discrete Math. **309**(6), 1711–1713 (2009). https://doi.org/10.1016/j.disc.2008.02.013

15. Gregor, P., Novotný, T., Škrekovski, R.: Extending perfect matchings to Gray codes with prescribed ends. Electron. J. Combin. **25**(2), Paper No. 2.56, 18 p (2018). https://doi.org/10.37236/6928

16. Johnson, J.R.: Long cycles in the middle two layers of the discrete cube. J. Combin. Theory Ser. A **105**(2), 255–271 (2004). https://doi.org/10.1016/j.jcta.2003.11.004

17. Kreweras, G.: Matchings and Hamiltonian cycles on hypercubes. Bull. Inst. Combin. Appl. **16**, 87–91 (1996)

18. Latifi, S., Zheng, S.Q., Bagherzadeh, N.: Optimal ring embedding in hypercubes with faulty links. In: Digest of Papers: FTCS-22, The Twenty-Second Annual International Symposium on Fault-Tolerant Computing, Boston, Massachusetts, USA, 8–10 July 1992, pp. 178–184. IEEE Computer Society (1992). https://doi.org/10.1109/FTCS.1992.243602

19. Liu, J.J., Wang, Y.L.: Hamiltonian cycles in hypercubes with faulty edges. Inform. Sci. **256**, 225–233 (2014). https://doi.org/10.1016/j.ins.2013.09.012

20. Mütze, T.: Proof of the middle levels conjecture. Proc. Lond. Math. Soc. (3) **112**(4), 677–713 (2016). https://doi.org/10.1112/plms/pdw004

21. Ruskey, F., Savage, C.: Hamilton cycles that extend transposition matchings in Cayley graphs of $S_n$. SIAM J. Discrete Math. **6**(1), 152–166 (1993). https://doi.org/10.1137/0406012

22. Savage, C.D., Winkler, P.: Monotone Gray codes and the middle levels problem. J. Combin. Theory Ser. A **70**(2), 230–248 (1995). https://doi.org/10.1016/0097-3165(95)90091-8

23. Simmons, G.J.: Almost all $n$-dimensional rectangular lattices are Hamilton-laceable. Congr. Numer. **XXI**, 649–661 (1978). Proceedings of the Ninth Southeastern Conference on Combinatorics, Graph Theory and Computing (Boca Raton, FL, 1978)

24. Vandenbussche, J., West, D.B.: Extensions to 2-factors in bipartite graphs. Electron. J. Combin. **20**(3), Paper 11, 10 p (2013). https://doi.org/10.37236/3594

# Weighted Group Search on the Disk and Improved LP-Based Lower Bounds for Priority Evacuation

Konstantinos Georgiou[✉] and Xin Wang

Department of Mathematics, Toronto Metropolitan University, Toronto, ON, Canada
{konstantinos,x85wang}@torontomu.ca

**Abstract.** We consider *weighted group search on a disk*, which is a search-type problem involving 2 mobile agents with unit-speed. The two agents start collocated and their goal is to reach a (hidden) target at an unknown location and a known distance of exactly 1 (i.e., the search domain is the unit disk). The agents operate in the so-called *wireless* model that allows them instantaneous knowledge of each others findings. The termination cost of agents' trajectories is the worst-case *arithmetic weighted average*, which we quantify by parameter $w$, of the times it takes each agent to reach the target, hence the name of the problem.

Our work follows a long line of research in search and evacuation, but quite importantly it is a variation and extension of two well-studied problems, respectively. The known variant is that where the search domain is the line, and for which an optimal solution is known. Our problem is also the extension of the so-called *priority evacuation*, which we obtain by setting the weight parameter $w$ to 0. For the latter problem the best upper/lower bound gap known is significant.

Our contributions for weighted group search on a disk are threefold. *First*, we derive upper bounds for the entire spectrum of weighted averages $w$. Our algorithms are obtained as a adaptations of known techniques, however the analysis is much more technical. *Second*, our main contribution is the derivation of lower bounds for all weighted averages. This follows from a *novel framework* for proving lower bounds for combinatorial search problems based on linear programming and inspired by metric embedding relaxations. *Third*, we apply our framework to the priority evacuation problem, improving the previously best lower bound known from 4.38962 to 4.56798, thus reducing the upper/lower bound gap from 0.42892 to 0.25056.

**Keywords:** Disk · Search · Evacuation · Combinatorial Search · Mobile Agents · Linear Programming · Lower Bounds

## 1 Introduction

Autonomous mobile agent searching over geometric domains such as lines, disks, circles, triangles and polygons has been the subject of extensive research over the

Research supported in part by NSERC.

This is an extended abstract. Due to space limitations, many proofs are omitted.

A. A. Rescigno and U. Vaccaro (Eds.): IWOCA 2024, LNCS 14764, pp. 28–42, 2024.
https://doi.org/10.1007/978-3-031-63021-7_3

last decades. In these problems, a fleet of agents is tasked with finding a hidden item in the geometric domain while complying with searchers' specifications. One of the most significant parameters that distinguish these variations is how the cost of the solution is quantified, effectively identifying the set of feasible solutions and changing the computational boundaries of the underlying problem. Indeed, in traditional search problems, e.g. *search* and rescue scenarios, one is concerned just with the finding of the hidden item, hence quantifying the cost as the time it takes the first agent to reach the target. In the other extreme, e.g. in *evacuating* scenarios, one is concerned with minimizing the time the last agent reaches the hidden item. These objective variations are combined with other fundamental specifications, such as the communication model and possibility of faultiness.

Despite the growing number of problems in the field, the number of problems for which matching upper and lower bounds are known is also increasing. Empirically, problems that admit full symmetry, involving the agents' specifications, domain, communication model, etc., are those that admit strong or matching upper and lower bounds. However, even a small deviation from symmetry makes any lower bounds to the problems particularly difficult to tackle. Examples include the face-to-face communication model, where knowledge is shared in an asymmetric way during the execution of the search, different agent speeds, or abstract domains such as triangles. Recently, asymmetry was also introduced as part of the way the search cost is quantified by considering the objective of only a *distinguished agent* reaching the hidden item, a problem known as *priority evacuation*. Negative results for this and similar problems are rare, and usually weak and challenging.

The main contribution of our work is a framework that provides lower bounds for combinatorial search problems. More generally, our starting point is the study of a generalization of the priority evacuation problem on the disk, which we obtain by considering *asymmetric* cost functions. Aside from the upper bounds we derive for these problems, our primary contribution is a framework for proving lower bounds (for these as well as for more general problems) that is based on linear programming and metric embedding relaxations. We demonstrate the usefulness of the framework by establishing lower bounds for our weighted evacuation problems, and in particular by improving the previously best lower bound known for the priority evacuation problem.

## 1.1 Related Work

The study of search-type problems dates back to the 1960's and has resulted to a rich theory that has been summarized in books [1,2] and surveys [21,30,37], among others. Originally, the problem pertained to the identification of a hidden object within a search domain, and hence the objective was to minimize the time the object was found [7]. Already with one searcher, the problem has seen interesting, and surprisingly challenging variations touching on the searcher's specs and the search domain, some of which we briefly discuss next. More recently, and with the emergence of fleet-robotics, search-type problems were considered with multiple searchers, see e.g. [15].

A basic example of a search space is the so-called linear search problem that has been considered with one [5] or multiple searchers [11]. The same problem has also been considered with the objective of minimizing the weighted average of the searchers' termination times [36] (that we also use in this work). Other 1-dimensional settings include variations such as searching rays [10] or graphs [4], or also search for multiple objects [9, 14].

Searching two-dimensional domains has become a more dominant topic in the last decade or so. Considered domains include polygons [29], the quite popular disk [15], the plane [28], regular polygons [19] the equilateral triangle and square [13,26], arbitrary triangles [32], and $\ell_p$ unit disks [35]. Many of these problems have also seen variations pertaining to the communication model or more generally to the searchers' specifications, e.g. searching the disk in the face-to-face model [27], with different searchers' speeds [6], and with different searchers' communication capabilities [23,31]. Last but not least, search problems have also been considered under faultiness settings, see for example [8,16,22,24,25,34,39]

A number of search problems have also been considered with less standard objectives. For example [12] considered a multi-objective search-type problem, [3] studied search problems under a broad competitive algorithmic lens, [38] considered information/cost trade-offs, [17,18] considered time/energy trade-offs, and [33] introduced search-and-fetch problems in two dimensions. More closely related to our work, is the so-called priority evacuation objective introduced in [19,20] where search termination is called when a distinguished searcher reaches the target.

## 1.2  Discussion on New Results

In this study, we explore a natural extension of the priority evacuation problem on the disk introduced in [19,20]. Our specific problem, termed the *weighted group search* on the disk, is parameterized by $w \in [0,1]$. Parameter $w$ represents a designated arithmetic weighted average (i.e. $\frac{x+wy}{1+w}$, when considering the weighted average of $x,y$) that defines the objective function, hence the name of the problem. The same objective was previously considered with the line as the search domain and solved optimally in [36]. Moreover, our weighted group search problem with $w = 0$ corresponds exactly to the previously studied priority evacuation problem on the disk for which the best lower and upper bound known are 4.38962 and 4.81854, respectively, creating a notable gap that remains an open problem.

Our results are derivations of upper and lower bounds for the weighted group search problem on the disk for all $w \in [0,1]$. Our upper bounds are obtained by a family of mobile agent trajectories (algorithms) that are adapting with $w$. The derived algorithm for $w = 0$ is exactly the best algorithm known for the previously studied priority evacuation problem (with performance 4.81854), and therefore our positive results can be understood as generalizations, still technical to analyze, of known algorithmic techniques. The main contribution of this work is the introduction of a novel framework for deriving lower bounds for general search problems. We discuss the key technical and conceptual ideas in the following section. Here we emphasize the significance of the obtained results in an area where finding strong lower bounds is notoriously challenging and rare. Indeed, we design a framework for proving lower bounds that is applicable to general objective functions, and to general discrete search domains. Our framework utilizes Linear Programming relaxations that are inspired by metric embedding relaxations of Euclidean 2-dimensional metric spaces induced by optimal search strategies. Indeed, we apply our framework to the weighted group search problem on the disk and we find lower bounds for all $w \in [0,1]$, resulting in upper/lower bound gaps that are diminishing with $w$. When $w = 1$ the bound is optimal. However, the punchline of our new methodology pertains to the lower bound obtained for $w = 0$, corresponding to previously studied priority evacuation. For this problem we improve the previously best lower bound known to 4.56798, reducing this way the upper/lower bound gap from 0.42892 to 0.25056.

### 1.3 Key Technical and Conceptual Ideas

Our approach to establishing *upper bounds* for weighted group search involves a generalization of the algorithm presented in [19]. We achieve this by specifically parameterizing the algorithm's behavior and performance according to the parameter $w$ of the cost function. Unsurprisingly, when $w = 0$, our algorithm aligns with the one detailed in [19]. The inherent challenge in analyzing our family of algorithms, which vary with $w$, lies in the fact that they generate diverse trajectories for the agents, where critical domain points are visited with varying order. These points are possible locations of the hidden item, which have been explored already, but whose close neighborhood has not been explored yet, and hence a adversarial placement of the hidden item arbitrarily close to those points is a possibility. Such points are necessarily introduced by strong search algorithms, as in these algorithms an agent may abandon the search of the disk in order to move in its interior, potentially in order to expedite evacuation should the hidden item be located by the other agent. Hence, identifying the optimal choices within that family (as a function of $w$) becomes more technical than the case $w = 0$. Consequently, quantifying the solution cost for all inputs necessitates a meticulous analysis, which at a high level considers various searched space configurations. For this reason, the contribution of this section is primarily only technical in nature.

The main contribution of this work lies in introducing a novel framework for proving *lower bounds* applicable to various combinatorial search-type problems. As an illustration of our methodology, we derive lower bounds for the weighted group search problem across all values of $w$. Specifically, we establish a matching lower bound for $w = 1$ and, notably, an improved lower bound for $w = 0$ addressing the priority evacuation problem. The challenge inherent in proving lower bounds for the weighted group search (as well as the priority evacuation or similar problems) stems from the contiguous search domain and the asymmetry of the objective function. Any lower bound argument necessitates the consideration of snapshots of an algorithm at certain time-stamps, identifying a finite collection of potential input placements for which the algorithm must perform well. Additionally, due to the online nature of the problem, the algorithm must operate identically until the input is discovered. The critical question revolves around how these finite points are processed by the two agents. A key observation is that the time-stamps, combined with potential input placements and agent trajectories, induce a finite 2-dimensional $\ell_2$ metric space. By conditioning on the order in which input placements are visited, one can model the optimal algorithm's performance using a Non-Linear Program (NLP). While solving all NLPs (corresponding to all permutations of placement visitations) identifies the optimal algorithm for the specific finite search domain, the challenge lies in the difficulty of solving these NLPs and providing proper certificates of optimality. In that direction, a second key observation is that one can relax the induced 2-dimensional $\ell_2$ metric spaces to abstract metric spaces, leading to Linear Programs (LPs). These LPs can be solved efficiently, with accompanying certificates of optimality. Importantly, the optimal values of these LPs serve as lower bounds for the NLPs, consequently establishing lower bounds for the original problems. Our framework relies among others on symbolic (not numerical), computer-assisted calculations, which we did not exhaust when deriving the lower bounds, suggesting that further improvements are possible by bypassing current computational limitations.

## 2   Preliminaries

### 2.1   Problem Definition, Notation and Some Observations

We consider a class of search-type problems $\text{WS}_{\mathcal{D}}^{f}$ (for Wireless Search) with 2 agents in the wireless model, which we define next. In these problems, 2 unit speed agents are initially collocated at the origin of the Euclidean plane, i.e. $\mathbb{R}^2$ equipped with the $\ell_2$ metric. The 2 agents have distinct (known) identities that we call $A_0, A_1$. Given a known geometric object $\mathcal{D} \subseteq \mathbb{R}^2$ and a known cost function $f : \mathbb{R}^2 \mapsto \mathbb{R}$ (non-decreasing in both coordinates), a solution to the problem is given by trajectories $\tau_i : \mathbb{R}_{\geq 0} \mapsto \mathbb{R}^2$, $i = 0, 1$, that induce movements for the 2 agents of speed at most 1. The two agents operate in the *wireless model* and they need to find a hidden target, in the following mathematical sense. For each fixed *target* $I \in \mathcal{D}$, we allow *both* trajectories $\tau_i$ to depend on $I$ *only after* $I$ is visited by *any* agent, i.e. when for some $t \geq 0$ and $i \in \{0, 1\}$ we have that $\tau_i(t) = I$. Equivalently, two executions of the trajectories are identical for different inputs up to the moment the first of these inputs is hit by a trajectory.

For each $I \in \mathcal{D}$, we also denote the *termination time* $T_i(I)$ of agent $A_i$ as the first time that agent $A_i$ visits target $I$. The agent who is the first that visits the target is referred to as the *finder*. We emphasize that each $T_i(I)$ depends on all $\tau_0, \tau_1$ and $I$ (while trajectories $\tau_j$ may only depend on $I$ after the target is visited). The objective of $\text{WS}_{\mathcal{D}}^{f}$ is to determine trajectories $\tau_0, \tau_1$ so as to minimize the *search cost* $\sup_{I \in \mathcal{D}} \frac{f(T_0(I), T_1(I))}{f(\|I\|_2, \|I\|_2)}$. When $f$ is the max function, the problem is known as *evacuation*, and when $f$ is the projection function, then it is known as *priority evacuation*. For this reason, also refer to the objective also as the *evacuation cost*. In this work we consider the following search domains: (a) $\mathcal{D} = \text{DISK}$, the unit radius disk, and (b) $\mathcal{D} = n\text{-GON}$, the vertices of a regular $n$-gon inscribed in the unit radius disk.

Some important observations are in place. *First*, the underlying search space (i.e. the space where agents' movements take place) is the 2-dim Euclidean space.

Searching in other $\ell_p$ metrics has been considered in [35]. For convenience, we think of the search domains embedded on the Cartesian plane so that the unit disks are centered at the origin. The fact that the underlying search space is a metric space, and in particular the Euclidean metric space will be essential in our lower bound arguments. *Second*, for a given $I \in \mathcal{D}$, let agent $A_i$ be the finder of the target, and suppose that this happens at time $t$, i.e. $I = \tau_i(t)$. Since the other agent can have her trajectory depend on this finding (wireless model), we may assume that $A_{1-i}$ moves directly to the target, i.e. that

$$T_{1-i}(I) = t + \|\tau_i(t) - \tau_{1-i}(t)\|_2 = t + \|I - \tau_{1-i}(t)\|_2. \tag{1}$$

*Third*, for the search domains we consider, i.e. the disk or the vertices of $n$-gons, all possible targets are at distance 1 from the origin. Agents that knew in advance the position of any target $I$ (i.e. if trajectories $\tau_i(t)$ could depend on $I$ for all $t \geq 0$) would need time 1 to reach the target, inducing cost $f(\|I\|_2), \|I\|_2)) = f(1,1)$, independent of the target $I$. Hence, when quantifying the performance of a search trajectory as per its search cost, we perform both worst case and competitive analysis. In particular, both performance quantifications admit the same optimal trajectories, and the corresponding optimal search costs are off by constant multiplicative factor $f(1,1)$ that depends only on $f$, hence they are equal when $f(1,1) = 1$.

## 2.2    Past Results, Search Domains and Cost Functions

A number of past results can be described in the framework of $\text{WS}_{\mathcal{D}}^f$ problems as we demonstrate next. The discussion focuses on the wireless model. For other communication models, one has to adjust the definition of feasible trajectories.

Typical search problems, where search is complete when the first agent reaches the target, are associated by definition with cost function $f(x,y) = \min\{x,y\}$. When multiple agents are involved in search, and one quantifies the performance by the time the last agent reaches the target, then one uses $f(x,y) = \|(x,y)\|_\infty = \max\{x,y\}$. The case where the performance is determined by the termination time of a designated agent, then one needs the projection function $f(x,y) = \text{proj}_2(x,y) = y$ (or $\text{proj}_1$). Finally, one may also consider a weighted average of the termination times of the two agents, by using cost function $f(x,y) = g_w(x,y) := wx + y$, where without loss of generality $w \in [0,1]$ (note that one may use a scaling factor $w+1$ without affecting optimizers). With these definitions in mind, we have that $g_0 = \text{proj}_2$ and $g_1 = \|\cdot\|_1$.

There are four notable examples of $\text{WS}_{\mathcal{D}}^f$ problems that were considered before. Our new findings build upon ideas found in these results:

- $\text{WS}_{\mathbb{R}\setminus[-1,1]}^{g_w}$, also known as weighted group search on a line, was considered in [36] where the search space $\mathcal{D}$ was the one dimensional real line excluding all points at most 1 away from the origin. The problem has been solved optimally. Our work is the first to consider the same cost function but on the disk.
- $\text{WS}_{\text{DISK}}^{\|\cdot\|_\infty}$ is the classic evacuation problem on the disk, solved optimally in [15], where the cost function is simply described as the time that the last agent reaches the target.

- $WS_{DISK}^{proj_2}$ is the priority evacuation problem (with 1 servant and 1 distinguished searcher, the queen) considered in [19]. In our current work, first we improve the lower bound for the problem, as well as we generalize it by considering cost functions $g_w$, where $w \in [0,1]$, and the same search domain. Recall that $g_0 = proj_2$.
- $WS_{6\text{-}GON}^{proj_2}$ was considered in [19] and solved optimally. Our methodology implies the same lower bound. Moreover, we are the first to study search domain $n$-GON, with $n > 6$, where in particular we derive lower bounds for $WS_{n\text{-}GON}^{g_w}$, $w \in [0,1]$, for $n = 7,8,9$.

The motivation for studying search domain $n$-GON relates to the fact that a lower bound to the problem is also a lower bound to searching domain DISK (for the same cost function), an idea introduced in [15], and later used in [19]. The following lemma is explicit in these works. Even though it was previously used only for the $\|\cdot\|_\infty$ and $proj_2$ cost functions, it holds more generally for any non-decreasing cost function $f$. Here we generalize the statement, since we will need it when studying the weighted group search problem. We emphasize that the lemma establishes lower bounds to algorithms addressing the search of the disk, conditioning on their performance of searching $n$-gons.

**Lemma 1.** *Let $t_0$, $t_1$ be the termination time lower bounds of agents $A_0, A_1$, respectively, for some input to $WS_{n\text{-}GON}^f$. Then, no algorithm for $WS_{DISK}^f$ with these termination times of the agents has evacuation cost better than $\frac{f(t_0 + \pi/n, t_1 + \pi/n)}{f(1,1)}$.*

Specifically all cost-functions $f \in \{proj_2, \|\cdot\|_\infty, g_w\}$ have the property that $\frac{f(x+a, y+a)}{f(1,1)} = \frac{f(x,y)}{f(1,1)} + a$. Now in Lemma 1, let $c_n$ be the derived lower bound for $WS_{n\text{-}GON}^f$ induced by target $I$. The same target $I$ is an eligible adversarial choice for $WS_{DISK}^f$, hence for the latter problem, the target is reached at least $\pi/n$ time later. This means, the following quantity is a valid lower bound to $WS_{DISK}^f$.

$$\frac{f\left(T_0(I) + \frac{\pi}{n}, T_1(I) + \frac{\pi}{n}\right)}{f(1,1)} = \frac{f(T_0(I), T_1(I))}{f(1,1)} + \frac{\pi}{n} = c_n + \frac{\pi}{n} \tag{2}$$

In fact the best lower bound of $\frac{\pi}{6} + 3 + \frac{\sqrt{3}}{2}$ for $WS_{DISK}^{proj_2}$ is due to the provable lower bound of $3 + \frac{\sqrt{3}}{2}$ for $WS_{6\text{-}GON}^{proj_2}$, see [19]. In this work we strengthen this result by deriving *new lower bounds* for $WS_{n\text{-}GON}^{proj_2}$, $n > 6$.

## 3  Improved Framework for Proving Lower Bounds

In this section we leverage the existing framework for showing lower bounds for $WS_{n\text{-}GON}^f$, and in light of Lemma 1, that would imply adjusted lower bounds for $WS_{DISK}^f$, too. In what follows $[n]$ denotes the set $\{0, 1, \ldots, n\}$, so for example $[1] = \{0,1\}$. Moreover, we denote the set of all permutations of $[n]$ by $\mathscr{R}_n$.

Our first task is to provide a systematic way in order to find the optimal solution to $WS_{n\text{-}GON}^f$. At a high level that would be accomplished by solving $\frac{(n-1)! 2^{n-1}}{n}$

many Non-Linear Programs. Recall that we think of the $n$-GON embedded on the Cartesian Euclidean space, inscribed in a unit-radius circle and centered at the origin. For this reason its vertices are identified by points $(\cos(2i\pi/n), \sin(2i\pi/n))$, $i = 0, \ldots, n-1$. Now, for a permutation $\rho \in \mathscr{R}_n$ (corresponding to the $n$ vertices of $n$-GON), we slightly abuse notation and we write $\rho_i$ to denote both the $i$-th element of the permutation, as well as the corresponding point $(\cos(2\rho_i\pi/n), \sin(2\rho_i\pi/n))$ on the plane.

The main idea of our formulation is that search strategies for $\mathrm{WS}^f_{n\text{-GON}}$ can be classified with respect to the order that vertices are visited, given by some permutation $\rho$ over $[n]$, and the identities of the agents that visit these vertices, given by a binary string $b \in [1]^n$. In particular, this means vertex $\rho_i$, i.e. the $i$-th visited vertex, is visited no later than vertex (target) $\rho_j$, when $i < j$, and that target $\rho_i$ is visited by agent $A_{b_i}$. We call such a search strategy a $(\rho, b)$-algorithm. Note that without loss of generality (due to symmetry), we may assume that the first vertex to be visited is vertex 0, and that the second visited vertex is one among $1, \ldots, \lfloor(n-1)/2\rfloor + 1$. This gives rise to at most $\frac{(n-1)!2^{n-1}}{n}$ classes of search algorithms. Next we show how to find the optimal $(\rho, b)$-algorithm, for each fixed $\rho, b$, by solving a Non-Linear Program.

**Lemma 2.** *For each $n \in \mathbb{N}$, permutation $\rho$ of $[n]$, binary string $b \in [1]^n$ and $t_0 \in \mathbb{R}_{\geq 0}$, consider the Non-Linear Program[1]*

$$\min \max_{i \in [n]} \{f(c_i^0, c_i^1)\} \qquad (\mathrm{NLP}^f_n(\rho, b, t_0))$$

$$s.t.: \ t_n \geq t_{n-1} \geq \ldots \geq t_1 \geq t_0$$

$$t_{i+1} - t_i \geq \left\| L_{i+1}^j - L_i^j \right\|_2, \ i \in [n-1], j \in [1]$$

$$c_i^j \geq t_i + \left\| L_i^j - L_i^{b_i} \right\|_2, \ i \in [n], j \in [1]$$

*in variables $\{c_i^j, t_i, x_i, y_i\}_{j \in [1], i \in [n]}$, where in particular we used abbreviations $L_i^{b_i} = \rho_i$ and $L_i^{1-b_i} = (x_i, y_i)$.*

*Then, for every convex cost function $f$, $(\mathrm{NLP}^f_n(\rho, b, t_0))$ admits a unique (hence global) minimum, and the optimizer for $t_0 = 1$ corresponds to the optimal $(\rho, b)$-algorithm for $\mathrm{WS}^f_{n\text{-GON}}$.*

We slightly abuse notation, and we denote by $\mathrm{NLP}^f_n(\rho, b, t_0)$ also the optimal value of the same Non-Linear Program. The following corollary follows immediately by our definitions, and Lemma 2.

**Corollary 1.** *The optimal search cost for $\mathrm{WS}^f_{n\text{-GON}}$ equals $\frac{1}{f(1,1)} \min_{\rho \in \mathscr{R}_n, b \in [1]^n}$ $\mathrm{NLP}^f_n(\rho, b, 1)$.*

Note that any solution to $(\mathrm{NLP}^f_n(\rho, b, t_0))$ is associated with an embedding of points $L_i^j$ in the $(\mathbb{R}^2, \ell_2)$ metric space, where in particular $d(L_i^j, L_{i'}^{j'}) = \left\| L_i^j - L_{i'}^{j'} \right\|_2$.

---

[1] For the intended meaning of the variables, the reader may consult the proof of the lemma.

This metric space satisfies the triangle inequality. Therefore, requiring that distances $d_{j,i,j',i'} := d(L_i^j, L_{i'}^{j'})$ satisfy the triangle inequality, but not necessarily that the space is embeddable to $\mathbb{R}^2$ (or even that it is $\ell_2$), gives rise to a natural relaxation to the problem. This idea is materialized in the next lemma.

**Lemma 3.** *For each $n \in \mathbb{N}$, permutation $\rho$ of $[n]$, $b \in [1]^n$ and $t_0 \in \mathbb{R}_{\geq 0}$, consider the Non-Linear Program (Relaxation)*

$$\min_{i \in [n]} \max \{f(c_i^0, c_i^1)\} \qquad (\mathrm{REL}_n^f(\rho, b, t_0))$$

$$\begin{aligned} s.t.: \ & t_n \geq t_{n-1} \geq \ldots \geq t_1 \geq t_0 \\ & t_{i+1} - t_i \geq d_{j,i+1,j,i}, \ i \in [n-1], j \in [1] \\ & c_i^j \geq t_i + d_{j,i,b_i,i}, \ i \in [n], j \in [1] \\ & d_{b_i,i,b_{i'},i'} = \|\rho_i - \rho_{i'}\|_2, \ i,i' \in [n] \\ & \left( \{L_i^j\}_{j \in [1], i \in [n]}, d(L_i^j, L_{i'}^{j'})) = d_{j,i,j',i'} \right) \text{ is a metric} \end{aligned}$$

*in variables $\{c_i^j, t_i, d_{j,i,j',i'}\}_{j,j' \in [1], i,i' \in [n]}$. Then, $(\mathrm{REL}_n^f(\rho, b, t_0))$ admits a unique (global) minimum, that we denote by $\mathrm{REL}_n^f(\rho, b, t_0)$, and in particular $\mathrm{NLP}_n^f(\rho, b, t_0) \geq \mathrm{REL}_n^f(\rho, b, t_0)$.*

The significance of $(\mathrm{REL}_n^f(\rho, b, t_0))$ over $(\mathrm{NLP}_n^f(\rho, b, t_0))$ is that the former can be solved much faster, especially when $f$ is a linear function (e.g. when $f \in \{\mathrm{proj}_2, \|\cdot\|_\infty, g_w\}$), in which case the resulting relaxation is a Linear Program (a basic trick introduces a new linear variable, and a collection of inequality constraints that simulate that the optimal solution simulates the max function). Because $(\mathrm{REL}_n^f(\rho, b, t_0))$ is easy to solve, it will be used to provide search cost lower bounds to $\mathrm{WS}_{\mathrm{DISK}}^f$, as the next lemma suggests.

**Lemma 4.** *For every $n \in \mathbb{N}$, no algorithm to $\mathrm{WS}_{\mathrm{DISK}}^f$ has search cost better than*

$$\mathscr{L}_n^f := \frac{1}{f(1,1)} \min_{\rho \in \mathscr{R}_n, b \in [1]^n} \mathrm{REL}_n^f\left(\rho, b, 1 + \frac{\pi}{n}\right).$$

Lemma 4 implies that for each $n$, one can solve $(n-1)!2^{n-1}/n$ Convex Programs (or more specifically Linear Programs, if $f$ is linear), in order to obtain lower bound $\mathscr{L}_n^f$ for $\mathrm{WS}_{\mathrm{DISK}}^f$. We explore this idea in Sect. 4 specifically for $f = \mathrm{proj}_2$, and more generally in Sect. 5 for cost functions $f = g_w$, $w \in [0,1]$.

## 4   Implied (and Improved) Lower Bounds for Priority Evacuation

In this section we present and prove one of our main contributions, which is an improved lower bound to the Priority Evacuation Problem $\mathrm{WS}_{\mathrm{DISK}}^{\mathrm{proj}_2}$.

**Theorem 1.** *No algorithm for* $\mathrm{WS}_{\mathrm{DISK}}^{proj_2}$ *has evacuation cost less than* $1 + \pi/9 + \sqrt{3}/2 + \cos(\pi/18) + 4\sin(\pi/9)$.

As a reminder, the previously best lower bound known for the problem was 4.38962. Note that not only the improvement is significant, but most importantly, the provided framework allows for even further improvements if one utilizes computational resources more efficiently, e.g. use specialized software for solving LP's instead of MATHEMATICA that was used in this project.

The proof of Theorem 1 follows immediately from Lemma 4 once we present new lower bounds for $\mathrm{WS}_{n\text{-GON}}^{proj_2}$, $n > 6$. Indeed, for $f = proj_2$, we have that $f(1,1) = 1$ and the cost function is linear. For this reason, due to Lemma 4, and as per the calculations (2), no algorithm for $\mathrm{WS}_{\mathrm{DISK}}^{proj_2}$ has evacuation cost less than $\frac{1}{f(1,1)} \min_{\rho \in \mathcal{R}_n, b \in [1]^n} \mathrm{REL}_n^{proj_2}\left(\rho, b, 1 + \frac{\pi}{n}\right) = \frac{\pi}{n} + \min_{\rho \in \mathcal{R}_n, b \in [1]^n} \mathrm{REL}_n^{proj_2}(\rho, b, 1)$, where in particular $\mathrm{REL}_n^{proj_2}(\rho, b, 1)$ is a lower bound to the optimal $(\rho, b)$-algorithm for $\mathrm{WS}_{n\text{-GON}}^{proj_2}$. For each $n = 3, \ldots, 9$, we solve $\mathrm{REL}_n^{proj_2}(\rho, b, 1)$ for all permutations $\rho \in \mathcal{R}_n$ and all binary strings $b \in [1]^n$, and we report the smallest values in Table 1. The calculations were computer-assisted but also symbolic (non-numerical).[2]

**Table 1.** Lower bounds to $\mathrm{WS}_{n\text{-GON}}^{proj_2}$, for $n = 3, \ldots, 9$ are given in the first column as the solutions to the Linear Programs $\mathrm{REL}_n^{proj_2}(\rho, b, 1)$. The second column is the numerical value of the lower bound, while the last two columns contain the corresponding permutation $\rho \in \mathcal{R}_n$ and binary string $b \in [1]^n$.

| $n$ | $min_{\rho \in \mathcal{R}_n, b \in [1]^n} \mathrm{REL}_n^{proj_2}(\rho, b, 1)$ | Num | $b$ | $\rho$ |
|---|---|---|---|---|
| 3 | $1 + \sqrt{3}$ | 2.73205 | {0, 1, 0} | (1,2,3) |
| 4 | $1 + 3/\sqrt{2}$ | 3.12132 | {1, 0, 0, 1} | (1,2,3,4) |
| 5 | $1 + \sqrt{25 + 2\sqrt{5}}/2$ | 3.71441 | {0, 1, 0, 1, 1} | (1, 3, 2, 4, 5) |
| 6 | $3 + \sqrt{3}/2$ | 3.86603 | {1, 0, 0, 1, 0, 1} | (1, 2, 3, 5, 4, 6) |
| 7 | $1 + \cos(3\pi/14) + 5\sin(\pi/7)$ | 3.95125 | {1, 0, 0, 1, 0, 1, 0} | (1, 2, 3, 7, 4, 6, 5) |
| 8 | $1 + \sqrt{2}/2 + 6\sin(\pi/8)$ | 4.00321 | {1, 0, 0, 1, 0, 1, 0, 1} | (1, 2, 3, 8, 4, 6, 5, 7) |
| 9 | $1 + \sqrt{3}/2 + \cos(\pi/18) + 4\sin(\pi/9)$ | 4.21891 | {1, 0, 0, 1, 0, 1, 0, 1, 0} | (1, 2, 3, 9, 4, 8, 5, 7, 6) |

From Table 1, and for $n = 9$, we derive a new lower bound to $\mathrm{WS}_{\mathrm{DISK}}^{proj_2}$, which is

$$\frac{\pi}{9} + \min_{\rho \in \mathcal{R}_9, b \in [1]^9} \mathrm{REL}_9^{proj_2}(\rho, b, 1) = \frac{\pi}{9} + 1 + \sqrt{3}/2 + \cos(\pi/18) + 4\sin(\pi/9) \approx 4.56798,$$

as promised. The lower bounds for $n = 3, 4, 5$ are only derived for comparison (and to illustrate how easily our methodology can derive new lower bounds to

---

[2] Calculations were performed symbolically with MATHEMATICA. The solution to any LP comes with a proof of optimality.

$\text{WS}_{n\text{-GON}}^{\text{proj}_2}$). We also emphasize that the reported lower bound for $n = 6$ agrees with the one reported and proven in [19]. In particular, this implies that for $n = 6$, and due to the matching upper bound for $\text{WS}_{6\text{-GON}}^{\text{proj}_2}$ in [19], we have that relaxation $\text{REL}_6^{\text{proj}_2}(\rho, b, 1)$ to $\text{NLP}_6^{\text{proj}_2}(\rho, b, 1)$ is exact, at least for the optimizers $\rho, b$ on Table 1. This is no longer the case for $n = 7, 8, 9$, since the solution found for $\text{REL}_7^{\text{proj}_2}(\rho, b, 1)$ is not embeddable in $(\mathbb{R}^2, \ell_2)$. For this reason, it is no surprise that quantities $\frac{\pi}{n} + \min_{\rho \in \mathscr{R}_n, b \in [1]^n} \text{REL}_n^{\text{proj}_2}(\rho, b, 1)$ are not increasing in $n$, and only the lower bounds for $n = 7, 9$ improve upon the one derived for $n = 6$. Finally, the computations for $n = 10$ are more intense, but unfortunately they do not give rise to further improvement (without exhausting all permutations and binary strings, a solution was found that gives a weaker lower bound that the case $n = 9$). Finally, the case $n = 11$ cannot be treated exhaustively since the number of configurations (approximately $7 \cdot 10^8$) becomes forbidden, especially for the size of the linear programs that is growing too, with $\Theta(n^3)$ constraints and $\Theta(n^2)$ variables.

## 5  Weighted Group Search on the Disk

In this section we derive upper and lower bounds for $\text{WS}_{\text{DISK}}^{g_w}$, stemming from the cost function $g_w(x, y) = wx + y$, $w \in [0, 1]$. Our upper and lower bounds are quantified by concrete formulas. Our lower bounds are derived by utilizing Lemma 4.

We emphasize that for $w = 0$ we have that $g_w = g_0 = \text{proj}_2$. In particular, the upper bound is also the one reported in [19] for $\text{WS}_{\text{DISK}}^{\text{proj}_2}$, while the depicted lower bound is an improvement of the one derived in [19], but lower than the one we proved for the special problem $\text{WS}_{\text{DISK}}^{g_0}$ in Sect. 4. This is because for the lower bounds to $\text{WS}_{\text{DISK}}^{g_w}$ we relied on the lower bounds we managed to prove for $\text{WS}_{7\text{-GON}}^{g_w}$. Indeed, dealing with lower bounds to $\text{WS}_{9\text{-GON}}^{g_w}$ was computationally too demanding, taking into consideration that parameter $w$ also ranges in $[0, 1]$.

### 5.1  Upper Bounds to Weighted Group Search on the Disk

The upper bound results are quantified in the following lemma.

**Lemma 5.** *For each $w \in [0, 0.2]$, let $\alpha = \alpha_w$ and $\beta = \beta_w$ be the solutions to non-linear system[3] $\gamma_1 = \gamma_2 = \gamma_3$, where $\gamma_1 = \alpha + \frac{2\sin(\alpha)}{w+1}$, $\gamma_2 = 2\pi - \alpha - \beta + \frac{2\sin\left(\alpha + \frac{\beta}{2} + \sin\left(\frac{\beta}{2}\right)\right)}{w+1}$, and $\gamma_3 = 2\sin\left(\frac{\beta}{2}\right) + \frac{\alpha + \beta + w(-\alpha - \beta + 2\pi)}{w+1}$. Then for each $w \in [0, 1]$, $\text{WS}_{\text{DISK}}^{g_w}$ admits a solution with search cost at most $1 + d_w + \frac{2\sin(d_w)}{w+1}$, where $d_w = \alpha_w$ if $w \le w_0$ and $d_w = \arccos\left(-\frac{w+1}{2}\right)$ otherwise. Moreover, $w_0$ is defined by equation $\alpha_w = \arccos\left(-\frac{w+1}{2}\right)$, with $w_0 \approx 0.0456911$.*

---

[3] $\alpha_w, \beta_w$ are to be invoked only for smaller values of $w$. Also $\gamma_1 = \gamma_2 = \gamma_3$, admits a solution also for higher values, but not for all $w \in [0, 1]$. The value of 0.2 was chosen only for aesthetic reasons.

As we will see next, the threshold of $w \leq 0.0456911$ corresponds to the critical ratio of the two weights $1/0.0456911 \approx 21.8861$ that indicates when the agent with the higher weight has incentive to deviate from the search in order to expedite her evacuation. When the agents' weight ratio is less than $21.8861$, then a plain vanilla algorithm is the best we can report, which however is optimal when the weight ratio is 1, i.e. when $w = 1$, see Lemma 7.

First we present the so-called $(a, b)$-detour Algorithm, introduced in [19], specifically for $\mathrm{WS}_{\mathrm{DISK}}^{\mathrm{proj}_2}$. For this, we present the two agent trajectories $\tau_0, \tau_1 : \mathbb{R}_{\geq 0} \mapsto \mathbb{R}^2$, where the functions depend on parameters $a, b$. Both trajectories will be piecewise movements along arcs and chords of the unit radius disk. The description of the trajectories is given under the assumption that no target is reported or found. Should the target be located by the agent, then the agent halts and transmits the finding message to her peer. Should the finding message reach an agent, then the agent moves along the shortest path towards the target. We emphasize that the algorithm is applicable to the search domain DISK, and therefore provides some solution to $\mathrm{WS}_{\mathrm{DISK}}^f$, for any cost function $f$. For the exposition, we also use abbreviation $c(t) := (\cos(t), \sin(t))$, which is the parametric equation of the unit-radius disk. Next we give formal description of the search algorithm. All movements are always at unit speed.

**Definition 1 (The $(a, b)$-Detour Algorithm for $\mathrm{WS}_{\mathrm{DISK}}^f$).** *Trajectory $\tau_0 : \mathbb{R}_{\geq 0} \mapsto \mathbb{R}^2$ of agent $A_0$, starting from origin $O$: Move to point $L = c(-a)$; Search clockwise up to point $E = c(b)$; Move to $D = c(0)$ along chord $ED$.*
*Trajectory $\tau_1 : \mathbb{R}_{\geq 0} \mapsto \mathbb{R}^2$ of agent $A_1$, starting from origin $O$: Move to point $L = c(-a)$; Search counter-clockwise up to point $D = c(0)$; Move to $E = c(b)$ along chord $ED$; Search clockwise up to $D = c(0)$.*

The following lemma effectively proves our upper bound claim of Lemma 5.

**Lemma 6.** *For each $w \in [0, 1]$, let*

$$a_w = \begin{cases} \alpha_w & w \leq w_0 \\ \pi & w > w_0 \end{cases}, \quad b_w = \begin{cases} \beta_w & w \leq w_0 \\ 0 & w > w_0 \end{cases},$$

*where $\alpha_w, \beta_w$ and $w_0 \approx 0.0456911$ are as in Lemma 5. Then, the $(a_w, b_w)$-Detour Algorithm for $\mathrm{WS}_{\mathrm{DISK}}^{g_w}$ has evacuation cost $1 + d_w + \frac{2\sin(d_w)}{w+1}$, where $d_w = \alpha_w$ if $w \leq w_0$ and $d_w = \arccos\left(-\frac{w+1}{2}\right)$ otherwise.*

## 5.2 Lower Bounds to Weighted Group Search on the Disk

In this section we give the details of how we obtained the lower bound values to $\mathrm{WS}_{\mathrm{DISK}}^{g_w}$.

**Theorem 2.** *For all $w \in [0, 1]$, no algorithm for $\mathrm{WS}_{\mathrm{DISK}}^{g_w}$ has evacuation cost less than $\max\left\{1 + \pi, 1 + \pi/7 + \frac{1}{w+1}\cos(3\pi/14) + 5\sin(\pi/7)\right\} \approx \max\{4.14159, 3.61822 + \frac{0.781831}{w+1}\}$.*

We start with a weak lower bound.

**Lemma 7.** *No algorithm for* $\mathrm{WS}_{\mathrm{DISK}}^{gw}$ *has evacuation cost less than* $1 + \pi$.

Notably, the result is tight for $w = 1$. Indeed, by Lemma 6 the upper bound for $\mathrm{WS}_{\mathrm{DISK}}^{g1}$ uses the $(\pi, 0)$-Detour Algorithm with performance $1 + d_1 + 2\sin(d_1)2 = 1 + \pi$ (note that $d_1 = \arccos\left(-\frac{1+1}{2}\right) = \pi$).

For small values of $w$, we obtain better bounds by deriving lower bounds to $\mathrm{WS}_{n\text{-GON}}^{gw}$ for $n = 7$. Unlike how we approached $\mathrm{WS}_{\mathrm{DISK}}^{\mathrm{proj}_2}$ (which is the same as $\mathrm{WS}_{\mathrm{DISK}}^{g0}$), this time we need to solve $\mathrm{REL}_n^{gw}\left(\rho, b, 1 + \frac{\pi}{n}\right)$ not only for all permutations $\rho$ and binary strings $b$, but also for enough many values of $w \in [0, 1]$. In this direction, we use $n = 7$ and compute the lower bounds to $\mathrm{WS}_{\mathrm{DISK}}^{gw}$ from $w = 0$ up to 1 with step size 0.001. The following is a strict generalization of the results in Table 1 pertaining to $\mathrm{WS}_{7\text{-GON}}^{gw}$ with $n = 7$, and was obtained by computer assisted calculations for determining the minimum value of $\mathrm{REL}_7^{gw}\left(\rho, b, 1\right)$ over all permutations $\rho$ and binary strings $b$.

**Lemma 8.** *No algorithm for* $\mathrm{WS}_{7\text{-GON}}^{gw}$ *has evacuation cost less than* $1 + \frac{1}{w+1}\cos\left(\frac{3\pi}{14}\right) + 5\sin\left(\frac{\pi}{7}\right)$.

Strictly speaking the bound of Lemma 8 equals $\min_{\rho \in \mathcal{R}_7, b \in [1]^7} \mathrm{REL}_7^{gw}\left(\rho, b, 1\right)$ only for $w \leq 0.8$ (for higher values, this technique gives a slightly better lower bound), however, this is already subsumed by the bound of Lemma 7. We conclude that for each $w$, and using Lemma 4 and the calculations (2), no algorithm for $\mathrm{WS}_{\mathrm{DISK}}^{gw}$ has evacuation cost less than

$$\frac{1}{g_w(1,1)} \min_{\rho \in \mathcal{R}_7, b \in [1]^7} \mathrm{REL}_7^{gw}\left(\rho, b, 1 + \frac{\pi}{7}\right) = \frac{\pi}{7} + \min_{\rho \in \mathcal{R}_7, b \in [1]^7} \mathrm{REL}_7^{gw}\left(\rho, b, 1\right),$$

where in particular $\mathrm{REL}_7^{gw}\left(\rho, b, 1\right)$ is a lower bound to the optimal $(\rho, b)$-algorithm for $\mathrm{WS}_{7\text{-GON}}^{gw}$. To resume, we solve $\mathrm{REL}_7^{gw}\left(\rho, b, 1\right)$ for all permutations $\rho \in \mathcal{R}_7$ and all binary strings $b \in [1]^n$, and we report the smallest values in Lemma 8. Overall, this implies that no algorithm for $\mathrm{WS}_{\mathrm{DISK}}^{gw}$ has search cost less than $1 + \pi/7 + \frac{1}{w+1}\cos(3\pi/14) + 5\sin(\pi/7)$. This observation, together with Lemma 7 give the proof of Theorem 2. Lastly we note that the transition value of $w$ for which the lower bound of Lemma 7 is weaker equals $\frac{7\cos\left(\frac{3\pi}{14}\right)}{6\pi - 35\sin\left(\frac{\pi}{7}\right)} - 1 \approx 0.493827$.

## References

1. Alpern, S., Fokkink, R., Gasieniec, L., Lindelauf, R., Subrahmanian, V.: Search Theory: A Game Theoretic Perspective. Springer, New York (2013). https://doi.org/10.1007/978-1-4614-6825-7
2. Alpern, S., Gal, S.: The Theory of Search Games and Rendezvous. Kluwer, Dordrecht (2003)
3. Angelopoulos, S., Dürr, C., Jin, S.: Best-of-two-worlds analysis of online search. In: Niedermeier, R., Paul, C. (eds.) 36th International Symposium on Theoretical Aspects of Computer Science, STACS 2019, 13–16 March 2019, Berlin, Germany. LIPIcs, vol. 126, pp. 7:1–7:17. Schloss Dagstuhl - Leibniz-Zentrum für Informatik (2019)

4. Angelopoulos, S., Dürr, C., Lidbetter, T.: The expanding search ratio of a graph. Discret. Appl. Math. **260**, 51–65 (2019)

5. Baeza Yates, R., Culberson, J., Rawlins, G.: Searching in the plane. Inf. Comput. **106**(2), 234–252 (1993)

6. Bampas, E., et al.: Linear search by a pair of distinct-speed robots. Algorithmica **81**(1), 317–342 (2019)

7. Beck, A.: On the linear search problem. Israel J. Math. **2**(4), 221–228 (1964)

8. Bonato, A., Georgiou, K., MacRury, C., Prałat, P.: Algorithms for p-faulty search on a half-line. Algorithmica **85**, 1–30 (2022)

9. Borowiecki, P., Das, S., Dereniowski, D., Kuszner, Ł: Distributed evacuation in graphs with multiple exits. In: Suomela, J. (ed.) SIROCCO 2016. LNCS, vol. 9988, pp. 228–241. Springer, Cham (2016). https://doi.org/10.1007/978-3-319-48314-6_15

10. Brandt, S., Foerster, K.-T., Richner, B., Wattenhofer, R.: Wireless evacuation on m rays with k searchers. Theor. Comput. Sci. **811**, 56–69 (2020)

11. Chrobak, M., Gąsieniec, L., Gorry, T., Martin, R.: Group search on the line. In: Italiano, G.F., Margaria-Steffen, T., Pokorný, J., Quisquater, J.-J., Wattenhofer, R. (eds.) SOFSEM 2015. LNCS, vol. 8939, pp. 164–176. Springer, Heidelberg (2015). https://doi.org/10.1007/978-3-662-46078-8_14

12. Chuangpishit, H., Georgiou, K., Sharma, P.: A multi-objective optimization problem on evacuating 2 robots from the disk in the face-to-face model; trade-offs between worst-case and average-case analysis. Information **11**(11), 506 (2020)

13. Chuangpishit, H., Mehrabi, S., Narayanan, L., Opatrny, J.: Evacuating equilateral triangles and squares in the face-to-face model. Comput. Geom. **89**, 101624 (2020)

14. Czyzowicz, J., Dobrev, S., Georgiou, K., Kranakis, E., MacQuarrie, F.: Evacuating two robots from multiple unknown exits in a circle. In: ICDCN, pp. 28:1–28:8. ACM (2016)

15. Czyzowicz, J., Gąsieniec, L., Gorry, T., Kranakis, E., Martin, R., Pajak, D.: Evacuating robots via unknown exit in a disk. In: Kuhn, F. (ed.) DISC 2014. LNCS, vol. 8784, pp. 122–136. Springer, Heidelberg (2014). https://doi.org/10.1007/978-3-662-45174-8_9

16. Czyzowicz, J., et al.: Evacuation from a disc in the presence of a faulty robot. In: Das, S., Tixeuil, S. (eds.) SIROCCO 2017. LNCS, vol. 10641, pp. 158–173. Springer, Cham (2017). https://doi.org/10.1007/978-3-319-72050-0_10

17. Czyzowicz, J., et al.: Energy consumption of group search on a line. In: Baier, C., Chatzigiannakis, I., Flocchini, P., Leonardi, S. (eds.) 46th International Colloquium on Automata, Languages, and Programming (ICALP 2019). Leibniz International Proceedings in Informatics (LIPIcs), vol. 132, pp. 137:1–137:15, Dagstuhl, Germany. Schloss Dagstuhl – Leibniz-Zentrum für Informatik (2019)

18. Czyzowicz, J., et al.: Time-energy tradeoffs for evacuation by two robots in the wireless model. Theoret. Comput. Sci. **852**, 61–72 (2021)

19. Czyzowicz, J., et al.: Priority evacuation from a disk: the case of n= 1, 2, 3. Theoret. Comput. Sci. **806**, 595–616 (2020)

20. Czyzowicz, J., et al.: Priority evacuation from a disk: The case of $n \geq 4$. Theoret. Comput. Sci. **846**, 91–102 (2020)

21. Czyzowicz, J., Georgiou, K., Kranakis, E.: Group search and evacuation. In: Flocchini, P., Prencipe, G., Santoro, N. (eds.) Distributed Computing by Mobile Entities. LNCS, vol. 11340, pp. 335–370. Springer, Cham (2019). https://doi.org/10.1007/978-3-030-11072-7_14

22. Czyzowicz, J., et al.: Search on a line by byzantine robots. Int. J. Found. Comput. Sci. **32**(04), 369–387 (2021)

23. Czyzowicz, J., et al.: Group evacuation on a line by agents with different communication abilities. In: ISAAC 2021, pp. 57:1–57:24 (2021)

24. Czyzowicz, J., Killick, R., Kranakis, E., Stachowiak, G.: Search and evacuation with a near majority of faulty agents. In: SIAM Conference on Applied and Computational Discrete Algorithms (ACDA 2021), pp. 217–227. SIAM (2021)

25. Czyzowicz, J., Kranakis, E., Krizanc, D., Narayanan, L., Opatrny, J.: Search on a line with faulty robots. Distrib. Comput. **32**(6), 493–504 (2019)

26. Czyzowicz, J., Kranakis, E., Krizanc, D., Narayanan, L., Opatrny, J., Shende, S.: Wireless autonomous robot evacuation from equilateral triangles and squares. In: Papavassiliou, S., Ruehrup, S. (eds.) ADHOC-NOW 2015. LNCS, vol. 9143, pp. 181–194. Springer, Cham (2015). https://doi.org/10.1007/978-3-319-19662-6_13

27. Disser, Y., Schmitt, S.: Evacuating two robots from a disk: a second cut. In: Censor-Hillel, K., Flammini, M. (eds.) SIROCCO 2019. LNCS, vol. 11639, pp. 200–214. Springer, Cham (2019). https://doi.org/10.1007/978-3-030-24922-9_14

28. Feinerman, O., Korman, A.: The ants problem. Distrib. Comput. **30**(3), 149–168 (2017)

29. Fekete, S., Gray, C., Kröller, A.: Evacuation of rectilinear polygons. In: Wu, W., Daescu, O. (eds.) COCOA 2010. LNCS, vol. 6508, pp. 21–30. Springer, Heidelberg (2010). https://doi.org/10.1007/978-3-642-17458-2_3

30. Flocchini, P., Prencipe, G., Santoro, N. (eds.): Distributed Computing by Mobile Entities, Current Research in Moving and Computing. LNCS, vol. 11340. Springer, Cham (2019). https://doi.org/10.1007/978-3-030-11072-7

31. Georgiou, K., Giachoudis, N., Kranakis, E.: Evacuation from a disk for robots with asymmetric communication. In: 33rd International Symposium on Algorithms and Computation (ISAAC 2022). Schloss Dagstuhl-Leibniz-Zentrum für Informatik (2022)

32. Georgiou, K., Jang, W.: Triangle evacuation of 2 agents in the wireless model. In: Erlebach, T., Segal, M. (eds.) ALGOSENSORS 2022. LNCS, vol. 13707, pp. 77–90. Springer, Cham (2022). https://doi.org/10.1007/978-3-031-22050-0_6

33. Georgiou, K., Karakostas, G., Kranakis, E.: Search-and-fetch with 2 robots on a disk: wireless and face-to-face communication models. Discrete Math. Theor. Comput. Sci. **21** (2019)

34. Georgiou, K., Kranakis, E., Leonardos, N., Pagourtzis, A., Papaioannou, I.: Optimal circle search despite the presence of faulty robots. In: Dressler, F., Scheideler, C. (eds.) ALGOSENSORS 2019. LNCS, vol. 11931, pp. 192–205. Springer, Cham (2019). https://doi.org/10.1007/978-3-030-34405-4_11

35. Georgiou, K., Leizerovich, S., Lucier, J., Kundu, S.: Evacuating from $\ell_p$ unit disks in the wireless model. Theoret. Comput. Sci. **944**, 113675 (2023)

36. Georgiou, K., Lucier, J.: Weighted group search on a line & implications to the priority evacuation problem. Theoret. Comput. Sci. **939**, 1–17 (2023)

37. Hohzaki, R.: Search games: literature and survey. J. Oper. Res. Soc. Japan **59**(1), 1–34 (2016)

38. Miller, A., Pelc, A.: Tradeoffs between cost and information for rendezvous and treasure hunt. J. Parallel Distributed Comput. **83**, 159–167 (2015)

39. Sun, X., Sun, Y., Zhang, J.: Better upper bounds for searching on a line with byzantine robots. In: Du, D.-Z., Wang, J. (eds.) Complexity and Approximation. LNCS, vol. 12000, pp. 151–171. Springer, Cham (2020). https://doi.org/10.1007/978-3-030-41672-0_9

# Simple Random Sampling of Binary Forests with Fixed Number of Nodes and Trees

Stoyan Dimitrov$^{(\boxtimes)}$ 

Rutgers University, Piscataway, NJ 08854, USA
tostoyan@gmail.com

**Abstract.** We generalize the classical algorithm of Rémy for random sampling of full binary trees with given number of leaves. As a result, we give a simple linear time algorithm for random generation of full binary forests with given number of trees and leaves. The algorithm is obtained from an elegant bijection that we construct in order to give a direct proof of the well-known fact that these forests are counted by the $k$-th fold self-convolution of the Catalan numbers. Via some well-known bijections, the given algorithm can be used to sample random objects from several other classes enumerated by self-convolutions of the Catalan numbers, e.g., binary forests with given number of trees, lists of given number of balanced strings and others.

**Keywords:** random sampling · binary trees · binary forests · bijection · Catalan numbers

## 1 Introduction

Binary trees are a family of objects of fundamental importance in both mathematics and computer science. They are used in efficient implementations of different algorithms and data structures (e.g., *dictionaries* and *priority queues*), as well as when solving large number of tasks, such as searching (*binary search trees*), sorting (*binary heaps*), representing code structure (*syntax trees*) and others. In the computer science literature, a binary tree is a rooted tree, such that each internal node has exactly two children (possibly empty), distinguished as the left and the right child, respectively. Another class of trees for which every node has either 0 or 2 children (but not 1) are the so-called *full binary trees*. A node with 0 children will be called a *leaf*. The binary trees with $n = 3$ nodes are shown in Fig. 1a and the full binary trees with 4 leaves are shown in Fig. 1b below.

It is well-known that the binary trees with $n$ nodes and the full binary trees with $n + 1$ leaves are both counted by the $n$-th Catalan number

$$C_n = \frac{1}{n+1}\binom{2n}{n}. \tag{1}$$

© The Author(s), under exclusive license to Springer Nature Switzerland AG 2024
A. A. Rescigno and U. Vaccaro (Eds.): IWOCA 2024, LNCS 14764, pp. 43–54, 2024.
https://doi.org/10.1007/978-3-031-63021-7_4

The Catalan numbers count numerous other sets of important objects in combinatorics and computer science. The book [14] lists more than 200 such sets of objects that we will call *Catalan objects*. There exist several widely known bijections between sets of Catalan objects (see [7]). For instance, looking at Fig. 1a and Fig. 1b, we can see that if we add edges to a given binary tree, so that all of its initial nodes have exactly two children, we get a full binary tree. Conversely, removing all leaves of a full binary tree (together with the edges adjacent to them) results in a binary tree. The image of the described map between binary trees and full binary trees can be clearly obtained in linear time and linear space, for every possible input.

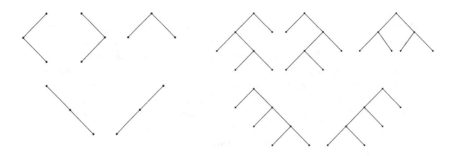

(a) All binary trees with 3 nodes.          (b) All full binary trees with 4 leaves.

**Fig. 1. .**

This paper generalizes the classical approach for random sampling of full binary trees, introduced by Rémy [10], where he shows how this can be done with a very simple algorithm in linear time and linear space. Our algorithm also runs in linear time and space and can generate a uniformly random forest of full binary trees with known total number of trees and leaves in them. Note that such an approach can be used to generate other lists of random Catalan objects of certain total size since mapping a full binary tree to a Catalan object in another class can be often done in linear time, as we showed that this is true for the standard mapping to binary trees. Since we have that map, our algorithm can be used to generate a uniformly random forest of binary trees with fixed number of trees and given number of nodes in them.

### 1.1   Problem, Notation and Motivation

Let us denote the set of full binary trees with $n+1$ leaves (and $n$ internal nodes) by $\mathcal{B}_n$. A *forest* of full binary trees with $k$ trees and $n$ leaves is a list $[T_1, \ldots, T_k]$ of $k$ full binary trees with total number of $n$ leaves, where $k$ and $n$ are some fixed numbers and $n \geq k$ since every tree has at least one leaf. The number of full binary trees with $n$ leaves is known, but do we know the number of full binary forests with $k$ trees and $n$ leaves? In fact, as explained in [8], the original problem that Eugène Catalan solved did not have the Catalan numbers as an answer, but the more general $k$-*fold self-convolution* of the Catalan numbers, which enumerates the forests we want. If $\{a_i\}_{n=0}^{\infty}$ and $\{b_n\}_{n=0}^{\infty}$ are two sequences, their convolution is the sequence $\{a * b\}_{n=0}^{\infty}$, such that

$$(a * b)_{n+1} = \sum_{i=0}^{n} a_i b_{n-i}.$$

The convolution $a * a$ is usually called *two-fold self-convolution* of $a$. In general, the $k$-fold self-convolution of the sequence $a$ is the sequence $a^{(k)}$ for which

$$a_{n+1}^{(k)} = \sum_{\substack{n_1+\cdots+n_k=n \\ n_i \geq 0}} a_{n_1} a_{n_2} \cdots a_{n_k}. \tag{2}$$

Note that if $\{a_n\}_{n=0}^{\infty}$ is a sequence with generating function $A(x)$, the $k$-th fold self-convolution of $a$ is a sequence with generating function $A^{k+1}(x)$. Note also that Eq. 1 implies $C_n^{(2)} = C_{n+1} = \frac{1}{n+2}\binom{2n+2}{n+1} = \frac{1}{n+1}\binom{2n+2}{n}$. For arbitrary $k$, Catalan himself (see [8]) showed that

$$C_n^{(k)} = \frac{k}{2n+k}\binom{2n+k}{n}. \tag{3}$$

We can see that the number of full binary forests with $k$ trees and $n$ leaves in total is

$$\sum_{\substack{n_1+\cdots+n_k=n-k \\ n_i \geq 0}} C_{n_1} C_{n_2} \cdots C_{n_k}, \tag{4}$$

since each of the $k$ trees must have at least one leaf and for $m \geq 0$, $C_m$ is the number of full binary trees with $m + 1$ leaves. Equation (2) tells us that the expression in Eq. 4 is equal to $C_{n-k}^{(k)}$. An example for a full binary forest (an ordered list of trees) is shown in Fig. 2 below. We will think about a full binary forest as a single ordered tree by adding an additional vertex (the root of that ordered tree), which will be connected to the roots of all trees in the forest. We denote the corresponding edges with dotted lines since this root will not be counted as a proper node of the forest. However, it will be used substantially in the random forest generation algorithm that we want to come up with (Algorithm 2 in Sect. 3).

Our algorithm for uniform sampling of full binary forests will use the recursive form of Eq. 3. In particular, having a bijective proof of that recursive form (Eq. 1 in Sect. 2), will allow us to generate a random forest of the wanted kind by adding the vertices one by one. Furthermore, to determine where to add a new vertex, we just need to generate a random pair of vertices among those already added in the tree. This can be achieved by simple generation of a pair of random integers in a given interval. For more details, see Algorithm 2 in Sect. 3. We should also note that Eq. (3) counts the number of lattice paths starting at the

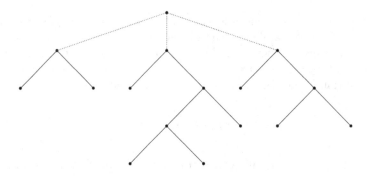

**Fig. 2.** A full binary forest with three trees and nine leaves.

point $(0,0)$ and comprised of $n + (k-1)$ Up steps (that is, $(1,1)$ steps) and $n$ Down steps ($(1,-1)$ steps), such that the path does not go below the $x$-axis [8]. Finally, the numbers $C_n^{(k)}$ must not be confused with the Fuss-Catalan numbers $C_{n,s} = \frac{1}{sn+1}\binom{(s+1)n}{n}$, which count the number of subdivisions of an $n$-gon into $k$-gons [14, Appendix B.4].

Efficient random sampling of binary trees or other Catalan objects, as well as forests of such objects, is needed in different cases. For instance, when we test functionality of data structures, one wants to check that functionality for a random instance of the structure. When using Monte Carlo algorithms to estimate certain quantity, we also need to generate a number of objects uniformly at random, and these objects are often Catalan. Random generation is also needed to conjecture or test hypothesis related to probability questions on Catalan objects.

### 1.2   Previous Work on Generation of Random Trees and Forests

How can we generate a random binary tree, full binary tree or another random Catalan object? Knott [5] was the first one to give an important recursive algorithm for generation of random binary trees running in $\mathcal{O}(n \log n)$ time. This algorithm utilizes the natural idea of using the recurrence

$$C_{n+1} = \sum_{i=0}^{n} C_i C_{n-i}, \tag{5}$$

to define a bijective function $\mathrm{rank}(T)$ which assigns a unique value between 1 and $C_n$ to every binary tree $T$ with $n$ nodes. Then, to generate a random such tree, it suffices to generate a random integer $x$ in $[1, n]$ and return $\mathrm{rank}^{-1}(x)$ (the so-called "unranking"). For a non-empty tree $T$, the particular rank function defined in [5], is the following.

$$\mathrm{rank}(T) := [(\mathrm{rank}(l(T)) - 1) \times C_{|r(T)|} + \mathrm{rank}(r(T))] + \sum_{j < |l(T)|} G_{j,|T|}, \tag{6}$$

where $|T|$ denotes the number of nodes of $T$, $G_{jn}$ denotes the number of binary trees with $n$ nodes, whose left subtree has $j$ nodes, and $l(T)$ (respectively, $r(T)$) denotes the size of the left (respectively, right) subtree of $T$. This recursive function is the only one for which $\text{rank}(T_1) < \text{rank}(T_2)$, if and only if one of the following natural conditions holds:

1) $|T_1| < |T_2|$.
2) $|T_1| = |T_2|$ and $l(T_1) < l(T_2)$.
3) $|T_1| = |T_2|$, $l(T_1) = l(T_2)$ and $r(T_1) < r(T_2)$.

Equation (6) shows that the time required to compute the given $\text{rank}(T)$, for a binary tree $T$ with $|T| = n$ is upper bounded by a function $T(n)$ satisfying $T(n) = 2T(\lceil \frac{n}{2} \rceil) + n$. Therefore, the master theorem [3, Section 4.3] gives that computing $\text{rank}(T)$ has worst-case time complexity $\mathcal{O}(n \log n)$. Two works also constructing bijective mappings, which give slight improvements to the algorithm of Knott are [11] and [13]. A separate group of previous works (for instance, [15] and [16]) establish injective maps from the set $\mathcal{B}_n$ to certain sets of words that can be generated in an easier way. Afterwards, in his classical paper from 1985 [10], Rémy gave the first linear time algorithm for sampling of random full binary trees that is also quite simple and widely used.

**Rémy's Algorithm.** The algorithm of Rémy is devised from a proof of the equation below that is equivalent to Eq. (1).

$$(n + 2)C_{n+1} = 2(2n + 1)C_n. \tag{7}$$

On the right, $2n + 1$ counts the number of all nodes in a full binary tree with $n + 1$ leaves since every such tree has $n$ internal nodes. On the left, we count the number of full binary trees with $n + 2$ leaves, where one of the leaves is marked. Noticing that, Rémy gave a simple bijection

$$f : (T, \ell) \rightarrow (T', v, d), \tag{8}$$

where $T \in \mathcal{B}_{n+1}$, $\ell$ is a leaf of $T$, $T' \in \mathcal{B}_n$, $v$ is a node of $T'$ and $d \in \{L, R\}$ stands for "direction". The action of $f$ is illustrated with the example in Fig. 3 below.

Having $T$ and its leaf $\ell$, the parent $p$ of $\ell$ must have another child $c$, since $T$ is a full binary tree. Delete $\ell$ and merge the vertices $p$ and $c$ into a single vertex $v$ (the edge between $p$ and $c$ is deleted, too). You obtain a tree $T' \in \mathcal{B}_n$ in which the subtree rooted at $v$ is the same as the one rooted at $c$ in $T$. We also have direction $d =$ 'L' (respectively, 'R'), if $c$ was a left (respectively, right) child of $p$ in $T$. Rémy's random tree generation algorithm is based on the map $f^{-1}$, that is, the inverse of $f$. We describe it below with Algorithm 1, assuming that we want to generate a random full binary tree $T \in \mathcal{B}_n$.

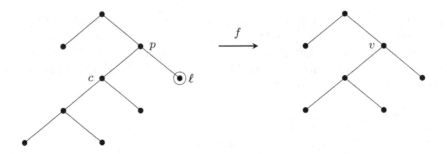

**Fig. 3.** Example of the action of the bijection $f$ defined by Rémy [10]

---

### Algorithm 1

---

Let $T'$ be the full binary tree having a single vertex, i.e., the only tree in $\mathcal{B}_0$.
For $j = 0, \ldots, n-1$, perform the steps below:

1. Given $T' \in \mathcal{B}_j$, choose one of its $2j + 1$ nodes $v$ at random.
2. Choose a direction $d$ ($L$ or $R$) at random.
3. Replace $v$ with a new node having $v$ as his child in direction $d$ and a new leaf vertex in the other direction. We get a tree $T \in \mathcal{B}_{j+1}$. Set $T' = T$.

---

Figure 4 gives an example of a sequence of trees we can get when following the described algorithm. The last tree in any such sequence is some $T \in \mathcal{B}_n$. The claim is that the probability for $T$ to be each of the trees in $\mathcal{B}_n$ is the same. In fact, the sequence of trees given in Fig. 4 is a walk in a Markov chain with its states being the full binary trees in $\mathcal{B} := \cup_{i=1}^{\infty} \mathcal{B}_i$. Because $f$ is a bijection, one can see the following: for each state corresponding to a tree in $\mathcal{B}_n$, we have exactly $2 \cdot (2n + 1)$ outgoing edges to states corresponding to trees in $\mathcal{B}_{n+1}$. At the same time, each state corresponding to a tree in $\mathcal{B}_{n+1}$ has exactly $n + 2$ ingoing edges (one such edge for each leaf) coming from trees in $\mathcal{B}_n$. Now, using

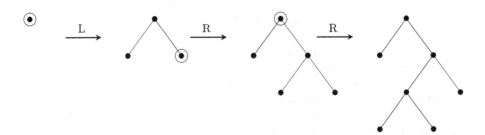

**Fig. 4.** Example of a path in the Markov chain of full binary trees corresponding to the algorithm of Rémy. The random node $v$ from step 1 of the algorithm is circled on each step. The direction $d$ is written over the transition arrows. We obtain a random such tree with four nodes.

induction one can prove that such a random walk in the described Markov chain would give a full binary tree with an uniform distribution. An implementation of Rémy's algorithm is discussed in Sect. 3.

**Other Works.** Sprugnoli [12] described a way to achieve the desired random generation in time $\mathcal{O}(n)$ and space $\mathcal{O}(\sqrt{n})$, based on another simple recursive idea that uses a closed form expression for $\sum_{j<k} G_{j,n}$ as a function of $k$, and some numerical methods approximating the inverse of this function. In a more recent work, Bacher, Bodini and Jacquot [2] improved the number of random bits used in the algorithm of Rémy from $n \log n$ to roughly $2n$ bits, but their algorithm is much more complicated for implementation. In another recent article, Lescanne [9] generalized the approach of Rémy to Motzkin and Schröder trees relying on previously found bijections for equations satisfied by their respective counting sequences.

We should note that in a work published in 1997, Alonso, Rémy and Schott [1] described a general algorithm for linear time generation of wide class of forests at random, which includes forests with given number of trees and leaves (see their Sect. 5.2). However, the method they use is different from ours as it is based on an injective map with a set of words. Our algorithm is simpler and easy to implement, as it is a direct generalization of the popular algorithm of Rémy from his original paper [10]. Another proof inspired by the elegant argument of Rémy is described by Foata and Zeilberger [4].

## 2  Bijection for Binary Forests

First, it is important to note that the approach of Rémy cannot be adapted easily to generation of random full binary forests. In particular, if one wants to generate a forest with $R$ trees, one might be tempted to first generate a random number $r \in [1, R]$ and then to use the algorithm of Rémy in the tree with number $r$. However, it is easily seen that this procedure does not generate a forest with uniform distribution. For instance, for a given $R$ and $n$, when one uses this procedure, the distribution of the number of nodes in the first tree would be binomial with parameters $n$ and $\frac{1}{R}$. This is not true for a uniformly generated forest with $R$ trees and $n$ leaves. Other simple approaches, e.g., randomly choosing a set of $R$ numbers that sum up to $n$, and running Remy's algorithm separately for each set, does not work either. One way to see this is by noticing that for a given number $n_1 = m$, the number of forests with $m$ nodes in the first tree, generated by this method for the given $n$ and $R$, will be $C_{m-1}$ times the number of solutions of $n_2 + n_3 + \ldots + n_R = n - m$, where $n_i \geq 1$ for every $i \in [2, R]$. Thus, we will have $C_{m-1}\binom{n-m-1}{R-2}$ such forests in total. On the other hand, Eq. 3 gives that the true number of forests that a uniform sampler must generate, with $m$ nodes in the first tree and for the given $n$ and $R$, is
$$C_{n-m-(R-1)}^{(R-1)} = \frac{R-1}{2(n-m)-(R-1)}\binom{2(n-m)-(R-1)}{(n-m)}.$$

The purpose of this section is to give a bijective proof for the number of full binary forests with given number of trees and leaves, which leads to a random sampling algorithm for such forests. For a fixed number $R \geq 2$, let $B_R(n)$ be the set of lists $[T_1, \ldots, T_R]$ comprised of $R$ full binary trees with total number of $n$ leaves. As mentioned in Sect. 1.1, we will think about every such list as a single ordered tree $T$ with the roots of the $R$ binary trees in the list being the children of the root of $T$. Let us also denote $b_R(n) := |B_R(n)|$. As pointed out in Sect. 1.1, we have $b_R(n) = C_{n-R}^{(R)} = \frac{R}{2n-R}\binom{2n-R}{n}$. From here, one can see that

$$\frac{C_{n-R}^{(R)}}{C_{n-1-R}^{(R)}} = \frac{(2n-R-1)(2n-R-2)}{n(n-R)}.$$

Therefore, to give a new proof of the fact that $b_R(n) = \frac{R}{2n-R}\binom{2n-R}{n}$, it suffices to prove the theorem below, since $b_R(R) = 1$, for every $R \geq 2$.

**Theorem 1.** *For any $R \geq 2$ and $n \geq R$,*

$$n(n-R)b_R(n) = (2n-R-1)(2n-R-2)b_R(n-1). \tag{9}$$

*Proof.* In the left-hand side, $n$ is the total number of leaves in the $R$ subtrees of the root of every tree $L$ in $B_R(n)$. At the same time, $n - R$ is the total number of internal nodes (non-leaves) for the $R$ subtrees since each of them has number of internal nodes one less than the number of leaves in it. Let us give an interpretation for the first two factors on the right! Every forest in $B_R(n-1)$ has $(n-1) + (n-1-R) + 1 = 2n - R - 1$ vertices. Thus $(2n-R-1)(2n-R-2)$ is the number of ordered pairs of different vertices $(v_1, v_2)$ among the vertices of a forest $L$ in $B_R(n-1)$.

It suffices to describe a bijective map

$$g : (T, \ell, i) \to (T', v_1, v_2),$$

where $T \in B_R(n)$ and $(\ell, i)$ is an ordered pair of a leaf and an internal node in one of the subtrees of the root of $T$, respectively, while $(v_1, v_2)$ is an ordered pair of different vertices of $T' \in B_R(n-1)$.

Denote the left and the right subtrees of the internal node $i$ by $U_L$ and $U_R$, respectively. Intuitively, the idea of the map $g$ is to move one of the subtrees $U_L$ or $U_R$, so that $\ell$ becomes its root, while $i$ becomes the root of the other subtree. In addition, $i$ and $\ell$ become $v_1$ and $v_2$ in some order, with a small exception addressing the case when the two vertices coincide. Formally, apply the following operation that will be called the *subtree flip*:

Delete the two edges between $i$ and its children. If $l$ is before (respectively, after) $i$ in the inorder traversal of the vertices of $T$, move the subtree $U_R$ (respectively, $U_L$), so that its root becomes $\ell$. Also, move the subtree $U_L$ (respectively, $U_R$), so that its new root is $i$.

We consider a few cases in order to determine $v_1$ and $v_2$ in the resulting tree.

Case 1. $\ell$ and $i$ are in different subtrees of the root of $T$.

Let the vertex at the place of $\ell$ in the new tree be $v_1$ and the vertex at the place of $i$ in this new tree be $v_2$. You can look at Fig. 5 for illustration. The case when $\ell$ is after $i$ in the inorder traversal of $T$ is analogous.

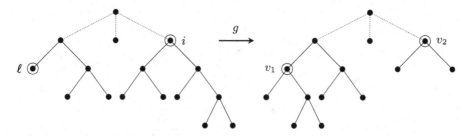

**Fig. 5.** Example of the subtree flip operation, applied when $R = 3$, $\ell$ and $i$ are in different subtrees of the root of $T$ (Case 1), and when $\ell$ is before $i$ in the inorder traversal of $T$.

Case 2. $\ell$ and $i$ are in the same subtrees of the root of $T$.

Note that in this case $i$ must be a predecessor of $\ell$. Thus, if we keep assigning $v_1$ to $\ell$ and $v_2$ to $i$, we will not have a bijection, as $v_2$ will be always a predecessor of $v_1$ and we must have the same chance for each vertex to be $v_1$ or $v_2$. Therefore, we do the following: if $\ell$ was in the left (respectively, right) subtree of $i$, then let $v_1$ (respectively, $v_2$) be the vertex at the place of $\ell$ in the new tree.

Case 2.1 $\ell$ and $i$ are different vertices after the flip (initially $\ell$ was <u>not</u> a child of $i$). Look at Fig. 6 given below. The case when $\ell$ is in the right subtree of $i$ is analogous.

Case 2.2 $\ell$ and $i$ become the same vertex after the flip ($\ell$ was a child of $i$). Then let $v_1$ (respectively, $v_2$) be the root of $T$, if $\ell$ is a left child of $i$ (respectively, right child). Also, let $v_2$ (respectively, $v_1$) be the vertex at the place of $i$. Look at Fig. 7 for illustration. The case when $\ell$ is a right child of $i$ is analogous.

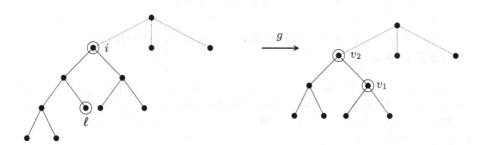

**Fig. 6.** Example of the action of the map $g$ when $R = 3$, $\ell$ is in the left subtree of $i$, but $\ell$ is not a child of $i$ (Case 2.1).

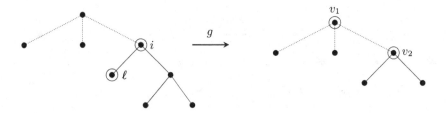

**Fig. 7.** Example of the action of the map $g$ when $R = 3$ and when $l$ is a left child of $i$.

To show that $g$ is a bijection, we will describe its inverse: Given a triple $(T', v_1, v_2)$, where $T' \in B_R(n-1)$ and $v_1, v_2$ are a pair of different vertices, we will obtain a triple $(T, \ell, i)$, where $T \in B_R(n)$, $\ell$ is a leaf of $T$ and $i$ is an internal node for some of the $R$ subtrees of the root of $T$.

First, if $v_1$ or $v_2$ is the root of $T'$, then $\ell$ must be a child of $i$ ($\ell$ and $i$ should certainly be in the same subtree of $T$) and thus we were in Case 2.2 when applying the map $g$. If $v_1$ (respectively, $v_2$) is the root of $T$, then $i$ must be $v_2$ (respectively, $v_1$) and $\ell$ must be left (respectively, right) child of $i$. Now, since know $i$ and $\ell$, we are ready to apply the inverse of the subtree flip. Add two edges connecting $v_2$ (respectively, $v_1$) with two children, where the current subtree with root $v_2$ becomes the right (respectively, left) subtree of $v_2$. The left (respectively, right) subtree of $v_2$ is a single vertex, i.e., we must have a leaf there.

If neither of $v_1$ or $v_2$ is the root of $T'$, then $\ell$ must not be a child of $i$. If $v_1$ and $v_2$ are in the same subtree of the root of $T'$, then we were in Case 2.1 when applying the map $g$. Otherwise, we were in Case 1. We can identify $i$ and $\ell$ in both cases. In Case 2.1, the node closest to the root among $v_1$ and $v_2$ corresponds to $i$. The label of that node, being $v_1$ or $v_2$ determines whether $\ell$ came from the left of the right. In Case 1, we know that $v_1$ corresponds to the leaf $\ell$ and $v_2$ to $i$. What remains is to apply the inverse of the subtree flip.

## 3   Linear Time Algorithm for Random Sampling of Forests

We can generate a random forest in $B_R(n)$ by utilizing the bijective map $g^{-1}$ for the map $g$ given in Sect. 2. Algorithm 2 given below gives a formal description of this procedure. While describing the algorithm, we are referring to the cases in the proof of Theorem 1.

In [6, Chapter 7.2.1.6] Knuth described a simple linear time and space implementation of the algorithm of Rémy using only one array. Assume we have a full binary tree with $n$ leaves and $n - 1$ internal vertices. To represent the tree, we can create an array $A[0..(2n-2)]$ keeping a label in $\{0, 1, \ldots, 2n-2\}$ for each node in the following way: let the odd labels be for internal nodes and even values for the leaves. The first element $A[0]$ contains the label of the root. In addition, for every $k \in [0, n-1]$, $A[2k+1]$ and $A[2k+2]$ are the labels of the

---

**Algorithm 2.** (Generation of a random full binary forest with $R$ trees and $n$ leaves)

Let $F'$ be the forest in $B_R(R)$ comprised of $R$ trees each having a single vertex.
For $j = R, \ldots, n$, perform the steps below:

1. Given $F' \in B_R(j)$, choose an ordered pair $(v_1, v_2)$ of different nodes in $F'$ at random (including the artificially added root for $F'$).
2. If $v_1$ or $v_2$ is the root of $F'$, then perform $g^{-1}$ according to Case 2.2.
   Otherwise, if $v_1$ and $v_2$ are in the same subtree of $F'$, perform $g^{-1}$ according to Case 2.1.
   Otherwise, if $v_1$ and $v_2$ are in different subtrees of $F'$, perform $g^{-1}$ according to Case 1.
3. Ignore the marked nodes $i$ and $\ell$ in the obtained forest. We obtain a forest $F \in B_R(j+1)$. Set $F' = F$.

Return $F$.

---

left and respectively the right child of the internal node with label $2k + 1$. An example of such an array is shown below (Table 1).

**Table 1.** Example of an array used in the algorithm of Rémy to represent a full binary tree.

| indices | 0 | 1 | 2 | 3 | 4 | 5 | 6 |
|---------|---|---|---|---|---|---|---|
| values  | 1 | 3 | 0 | 5 | 2 | 4 | 6 |

Having such an array of size $2n - 1$, the insertion of a new internal vertex and a leaf, as in the random generation procedure described in Sect. 1.2 is straightforward. We just need to generate a random node $x \in [0, 2n - 2]$ in the current tree and a direction - either 'left' or 'right'. Then, replace the label $x$ in the array by the label $2n - 1$ of the new internal node. Finally, let $A[2n - 1] = x$ and $A[2n] = 2n$, if the direction was 'left' and $A[2n - 1] = 2n$ and $A[2n] = x$ in case the direction was 'right'.

One can implement Algorithm 2 in linear time and space, by storing two dynamic lists, $L$ and $L_{\text{children}}$, representing the full binary forest $T = [T_1, \ldots, T_R]$. The list $L$ can contain the labels of the current nodes in $T$ and $L_{\text{children}}$ can store the children of the nodes in $L$. Initially, $L$ must contain the root of $T$ labeled 0 and the roots of $T_1, T_2, \ldots, T_R$ labeled by $1, 2, \ldots, R$, respectively. $L_{\text{children}}$ will be initially empty. Then, while running Algorithm 2, and for every value of $j$ in it, we add one new node to $L$ and two new edges to $L_{\text{children}}$. Once we assign a label to a node, we keep that label until the end of the algorithm's execution. The children of $L[i]$ can be stored by convention at the entries $L_{\text{children}}[2i]$ and $L_{\text{children}}[2i + 1]$ (possibly both being NULL, if $L[i]$ is a leaf). At the end, we must have $2n - R$ nodes in $L$. Swapping the places of two subtrees within a forest can be always done with a few operations and thus performing $g^{-1}$ takes a constant time regardless of the case that we are in step 2 of Algorithm 2. Therefore, we have an implementation of Algorithm 2 running in linear time and linear space. Note, however, that we will need at least $2n$ generations of a random number during the execution of Algorithm 2, as we generate a pair of random nodes of the forest at each step.

**Acknowledgments.** The author is grateful to Doron Zeilberger, who is the main reason for this work to exist, as he suggested to look for a proof of Eq. (9) similar to the one in the paper of Rémy [10].

# References

1. Alonso, L., Rémy, J.L., Schott, R.: A linear-time algorithm for the generation of trees. Algorithmica **17**, 162–182 (1997)
2. Bacher, A., Bodini, O., Jacquot, A.: Efficient random sampling of binary and unary-binary trees via holonomic equations. Theoret. Comput. Sci. **695**, 42–53 (2017)
3. Cormen, T.H., Leiserson, C.E., Rivest, R.L., Stein, C.: Introduction to Algorithms. MIT Press, Cambridge (2022)
4. Foata, D., Zeilberger, D.: A classic proof of a recurrence for a very classical sequence. arXiv preprint math/9805015 (1998)
5. Knott, G.D.: A numbering system for binary trees. Commun. ACM **20**(2), 113–115 (1977)
6. Knuth, D.E.: Art of Computer Programming, Volume 4, Fascicle 4, The: Generating All Trees–History of Combinatorial Generation. Addison-Wesley Professional (2013)
7. Tri Lai, Bijection between Catalan objects. https://www.math.unl.edu/~tlai3/Catalan%20Objects.pdf
8. Larcombe, P.J., French, D.R.: The Catalan number kappa-fold self-convolution identity: the original formulation. J. Comb. Math. Comb. Comput. **46**, 191–204 (2003)
9. Lescanne, P.: Holonomic equations and efficient random generation of binary trees (2022)
10. Rémy, J.L.: Un procédé itératif de dénombrement d'arbres binaires et son application à leur génération aléatoire. RAIRO. Informatique théorique **19**(2), 179–195 (1985)
11. Rotem, D.: On a correspondence between binary trees and a certain type of permutation. Inf. Process. Lett. **4**(3), 58–61 (1975)
12. Sprugnoli, R.: The generation of binary trees as a numerical problem. J. ACM (JACM) **39**(2), 317–327 (1992)
13. Solomon, M., Finkel, R.A.: A note on enumerating binary trees. J. ACM (JACM) **27**(1), 3–5 (1980)
14. Stanley, R.P.: Catalan Numbers. Cambridge University Press, Cambridge (2015)
15. Zaks, S.: Lexicographic generation of ordered trees. Theoret. Comput. Sci. **10**(1), 63–82 (1980)
16. Zerling, D.: Generating binary trees using rotations. J. ACM (JACM) **32**(3), 694–701 (1985)

# Maximizing Minimum Cycle Bases Intersection

Ylène Aboulfath[1], Dimitri Watel[2,3(✉)], Marc-Antoine Weisser[4],
Thierry Mautor[1], and Dominique Barth[1]

[1] DAVID, Université Versailles Saint-Quentin-En-Yvelines, Versailles, France
{ylene.aboulfath,thierry.mautor,dominique.barth}@uvsq.fr
[2] SAMOVAR, Evry, France
[3] ENSIIE, Evry, France
dimitri.watel@ensiie.fr
[4] LISN, CentraleSupélec, Gif-sur-Yvette, France
marc-antoine.weisser@centralesupelec.fr

**Abstract.** We address a specific case of the matroid intersection problem: given a set of graphs sharing the same set of vertices, select a minimum cycle basis for each graph to maximize the size of their intersection. We provide a comprehensive complexity analysis of this problem, which finds applications in chemoinformatics. We establish a complete partition of subcases based on intrinsic parameters: the number of graphs, the maximum degree of the graphs, and the size of the longest cycle in the minimum cycle bases. Additionally, we present results concerning the approximability and parameterized complexity of the problem.

**Keywords:** Minimum cycle basis · Matroids intersection problem · Complexity

## 1 Introduction

In chemoinformatics and bioinformatics, a molecular dynamics trajectory represents evolution of the 3D positions of atoms constituting a molecule, thus forming a sequence of molecular graphs derived from these positions at discrete time intervals. These graphs share the same vertex set (atoms) but vary in their edges, particularly those representing hydrogen bonds, which may appear or disappear over time, unlike covalent bonds, which are persistent. A research objective is to characterize the evolution of molecular structure during the trajectory [4,12].

The structure of a molecule is intricately linked to the interactions among elementary cycles within its associated graph. It appears that it is mainly the induced cycles of the molecule, therefore often the smallest cycles, which impact its structure. Therefore, it is commonly represented by a minimum cycle basis of the graph [6,7,9,10,15]. Cycle bases are a concise representation of cycles, and finding a minimum cycle basis (*i.e.* minimizing the total weight of cycles in the base) can be done in polynomial time [3,8]. Thus, given a sequence of molecular graphs modeling the trajectory, to evaluate the conservation of the molecular structure during the trajectory [1], we seek to obtain a minimum cycle basis for each graph such that they have the most cycles in common.

In this context, we define the problem, referred to as MAX-MCBI, as follows: given a set of $k$ graphs $\{G_1, G_2, \ldots, G_k\}$ with the same vertex set, find for each graph a minimum cycle basis such that the size of their overall intersection is maximum. Note that MAX-MCBI is a special case of the matroid intersection problem (MI) wherein, given $k$ matroids with the same ground set $C$, we search for one independent set in each matroid such that the size of their intersection is maximum. This is primarily because the set of cycles in an undirected graph forms a vector space. MI is NP-Complete when $k = 3$ [16] but polynomial in $|C|$ when $k = 2$ [5] and $\frac{1}{k}$-approximable [11]. Transferring this positive results to MAX-MCBI can be achieved once we address the challenge posed by the potentially exponential number of cycles in a graph.

**Our contributions.** In this paper, we exploit the distinctive features of our specific instance of MI to establish a NP-Hard/Polynomial partition of subcases based on intrinsic parameters. Additionally, we investigate the parameterized complexity and approximability of the problem MAX-MCBI and its decision version MCBI. In the decision version, given $k$ graphs and a non-negative integer $K$, the objective is to determine whether there exists a minimum cycle basis for each graph such that the size of their intersection is greater than $K$ or not.

We distinguish four intrinsic parameters: the number of graphs $k$; the maximum size $\gamma$ of the cycles in a minimum cycle basis of any graph $G_i$; the maximum degree $\Delta$ in any graph $G_i$; and the decision integer $K$. The first parameter $k$ directly arises from the complexity of MI which is polynomial for $k = 2$ and $NP-complete$ otherwise. Parameters $\gamma$ and $\Delta$ arise from our application. Those are classical parameters studied in molecular contexts. Finally, $K$ is a classical parameter in parameterized complexity.

The results are summarized in Table 1. Note that the case $\Delta = 2$ is trivially polynomial as each graph is a set of disjoint cycles. Considering the parameterized complexity and the approximability, few questions remain open.

Section 2 and 3 are dedicated to formal definitions and the proof that MAX-MCBI is indeed a subproblem of MI. As a result we get the $\frac{1}{k}$- approximation algorithm, the polynomial case when $k = 2$ and the belonging of MCBI to XP with respect to $K$. In Sect. 4, we prove the hardness results. Finally Sect. 5 is dedicated to the cases where $\gamma = 3$ or where $\gamma = 4$ and $\Delta = 3$.

## 2 Formal Definitions

We adopt a standard definition of a cycle in a graph $G$ as any subgraph in which each vertex has an even degree[1] [8]. The sum of two cycles, denoted as $c_1 \oplus c_2$,

---

[1] It's important to note that a cycle can then be composed of elementary cycles, which may seem counter-intuitive at first.

**Table 1.** Summary of the results of this paper. The hardness results hold true only when the parameters in the *blank cells* are not fixed and remain valid even when the parameters in the *non-blank cells* are fixed to the specified values (or higher). The polynomial results hold true only when the parameters in the *non-blank cells* are fixed to the specified value (or lower) and remain valid regardless of the values of the parameters in the *blank cells*. Additionally, parameters in the parameterized results are indicated with a cross.

| k | Δ | γ | K | MCBI | MAX-MCBI | |
|---|---|---|---|---|---|---|
| 3 | 4 / 3 | 4 / 5 | | NP-Complete | NP-Hard | |
| | 4 / 3 | 4 / 5 | - | - | $\frac{1}{k}$ Inapprox. | Theorem 3 |
| | 4 / 3 | 4 / 5 | × | W[1]-Hard | - | |
| | 2 | | | P | P | Section 1 |
| | | 3 | | P | P | Theorem 4 |
| | 3 | 4 | | P | P | Theorem 5 |
| 2 | | | | P | P | |
| | | | - | - | $\frac{1}{k}$ Approx. | Theorem 2 |
| | | | × | XP | | |

is the subgraph containing the set of edges present in one and only one of the two cycles. In this paper, when $D$ represents a set of cycles, we denote the sum of cycles as $\bigoplus D = \bigoplus_{d \in D} d$. Note also that if $\bigoplus D = c$ then $\bigoplus D \oplus c = 0$. This general definition of cycles with the sum operation $\oplus$ defines a vector space in the field $\mathbb{Z}/2\mathbb{Z}$. A cycle basis of a graph $G$ is a linearly independent set $B$ of cycles that spans the cycle space of $G$. The terms *span* and *linearly independent* refer to the classical linear algebra definitions.

- $B$ is linearly independent if, for all $B' \subseteq B$, $\bigoplus B' \neq 0$.
- $B$ spans a cycle $d$ if there exists $B' \subseteq B$ such that $\bigoplus B' = d$. If $B$ is a basis, then the subset $B'$ is unique for each cycle $d$.

We define $\lambda_B : B \times C \to \{0, 1\}$ as the function such that if $B'$ is the subset of $B$ with $\bigoplus B' = d$, then $c \in B'$ if and only if $\lambda_B(c, d) = 1$.

The weight of a cycle is defined as its number of edges, denoted for a cycle $c$ by $\omega(c)$. The weight of a cycle basis $B$ is given by $\sum_{c \in B} \omega(c)$. Therefore, a minimum cycle basis is a cycle basis that minimizes its weight. The set of minimum cycle bases of a graph $G$ is denoted by $\mathcal{MCB}(G)$. Polynomial time algorithms for finding a minimum cycle basis have been proposed [3,8].

*Problem 1 (*MAX-MCBI*).* Given a set of $k$ graphs $G_1, G_2, \ldots, G_k$, find a subset of cycles $B$ of $\bigcap_{i=1}^{k} G_i$ such that, for all $i \in [\![1; k]\!]$, there exists $B_i \in \mathcal{MCB}(G_i)$ with $B \subseteq B_i$, and maximizing $|B|$.

The decision problem MCBI associated with MAX-MCBI is, given an integer $K$, to determine if there exists a solution with $|B| \geq K$.

The rest of this section proves that the search for $B$ can be performed within a polynomial-size subset of cycles. We begin by presenting common lemmas on minimum cycle bases. Their proofs are provided in [2].

**Lemma 1.** *Given $B$ a cycle basis of a graph $G$ with two cycles $c_1 \in B$ and $c_2 \notin B$, if $\lambda_B(c_1, c_2) = 1$ then $(B \setminus \{c_1\}) \cup \{c_2\}$ is a cycle basis of $G$.*

**Lemma 2.** *$B \in \mathcal{MCB}(G)$ if and only if $B$ is a cycle basis and for $c_1, c_2$ with $c_1 \in B$ and $c_2 \notin B$ such that $\lambda_B(c_1, c_2) = 1$, we have $\omega(c_1) \leq \omega(c_2)$.*

**Lemma 3.** *If $B_1, B_2 \in \mathcal{MCB}(G)$, for every $c_1 \in B_1 \backslash B_2$, there exists $c_2 \in B_2 \backslash B_1$ such that $(B_1 \backslash \{c_1\}) \cup \{c_2\} \in \mathcal{MCB}(G)$.*

Given $G$, let $\mathcal{M}(G) = (C, I)$ be the couple where $C$ is the set of cycles of $G$ and $I$ are the subsets $D$ of $C$ such that there exists $B \in \mathcal{MCB}(G)$ with $D \subseteq B$.

**Lemma 4.** *$\mathcal{M}(G)$ is a matroid.*

*Proof.* Lemma 3 proves the basis exchange axiom. As $\mathcal{MCB}(G)$ is not empty (it possibly contains the empty set of $G$ is a tree), then $\mathcal{M}(G)$ satisfies the basis axioms and is then a matroid.    □

Note that MAX-MCBI can be simply rewritten as the search for a maximum-size set of cycles that are independent in each matroid $\mathcal{M}(G_i)$ using the matroid terminology for *independent*. In order to avoid confusion with the *linear independency*, whenever referring to linear algebra independence, we will explicitly use that terminology. This proves that MAX-MCBI is indeed a subproblem of the matroid intersection problem (MI). However, we cannot use algorithms dedicated to MI to prove any polynomial complexity result as, currently, the ground sets of our matroids have exponential size. We address this in the next section; however, we introduce a polynomial time independency oracle for $\mathcal{M}(G)$.

**Lemma 5.** *Given a subset $D$ of cycles of $G$, we can check in polynomial time if $D$ is independent in $\mathcal{M}(G)$.*

*Proof.* This can be achieved by running a modified version of the Horton algorithm [8]. Given a graph $G = (V, E)$, the Horton algorithm generates a minimum cycle basis by enumerating a list $L$ of $O(|V||E|)$ cycles, then sorting $L$ from the smallest cycles to the largest and finally using a greedy polynomial-time procedure to build a minimum cycle basis. To adapt this algorithm, we introduce a modification. Before the sorting step, we replace $L$ with $D \cup L$ and during the sorting process, in case of a tie, cycles from $D$ are given priority. The set $D$ is independent if and only if all cycles of $D$ are in the resulting basis.    □

---

**Algorithm 1.** Building the list of candidate cycles containing an optimal solution of MAX-MCBI

1: **function** CANDIDATESLIST($G_1, G_2, \ldots, G_k$)
2:     $L \leftarrow \emptyset$
3:     $G = (V, E) \leftarrow \bigcap_{i=1}^{k} G_i$
4:     **for** $u \in V$, $(v, w) \in E$ **do** add to $L$ the cycle consisting in the edge $(v, w)$, a shortest path from $u$ to $v$ and a shortest path from $u$ to $w$ in $G$, if such a cycle is elementary.
5:     **for** $(u, v) \in E$, $(w, x) \in E$ **do** add to $L$ two cycles consisting in $(u, v)$, $(w, x)$, one with the shortest paths from $u$ to $w$ and from $v$ to $x$ in $G$, and a second with the shortest paths from $u$ to $x$ and from $v$ to $w$ in $G$, if such a cycle is elementary.
6:     **return** $L$

---

## 3   Case with $k = 2$, Approximability and Parameterized Complexity

With Theorem 1, we prove we may only focus on a polynomial subset of cycles.

**Theorem 1.** *Given an instance $\{G_1, G_2, \ldots, G_k\}$ of* MAX-MCBI *and let $L$ be the list returned by the function* CANDIDATESLIST$(G_1, G_2, \ldots, G_k)$ *described in Algorithm 1, there exists an optimal solution $B^*$ such that $B^* \subseteq L$.*

*Proof.* Let $B^*$ be any optimal solution and $B_i$ a minimum cycle basis of $G_i$ containing $B^*$ for every $i \in [\![1; k]\!]$.

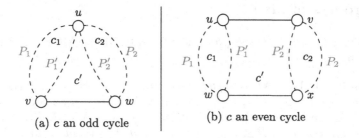

(a) $c$ an odd cycle          (b) $c$ an even cycle

**Fig. 1.** Example of cycles such that $c_1 \oplus c' \oplus c_2 = c$.

Let us suppose that there exists a cycle $c \in B^* \backslash L$. Hereinafter, we prove that $c$ can be replaced by a cycle $c' \in L \backslash B^*$ such that $B^* \backslash \{c\} \cup \{c'\}$ is still optimal.

Assuming $c$ is an odd cycle containing an edge $(v, w)$. Let $P_1$ and $P_2$ be the two paths included in $c$ connecting respectively $v$ and $w$ to the same node $u$ such that $|P_1| = |P_2|$. At line 4 of Algorithm 1, when the loop enumerates $u$ and $(v, w)$, we obtain a cycle $c'$ that is added to $L$ if it is elementary. As depicted by Fig. 1a, there exist two (possibly non-elementary) cycles $c_1$ and $c_2$ such that $c = c_1 \oplus c' \oplus c_2$. As $P_1'$ and $P_2'$ are shortest paths and as $|P_1| = |P_2| = (|c| - 1)/2$, we have $\omega(c_1) < \omega(c), \omega(c') \leq \omega(c)$ and $\omega(c_2) < \omega(c)$. If $c'$ is not elementary, then $c$ is the sum of strictly smaller cycles, this contradicts Lemma 2. Consequently, $c'$ is elementary and is added to $L$. For the same reason, $\omega(c') = \omega(c)$.

For all $i \in [\![1;k]\!]$, let $B_i$ be a minimum cycle basis of $G_i$ containing $B^*$. By Lemma 2, every $d$ such that $\lambda_{B_i}(d, c_1) = 1$ has a weight $\omega(d) \leq \omega(c_1)$. Similarly for $c_2$. Thus those cycles cannot be $c'$ and $c$. As $c' = \bigoplus\{d : \lambda_{B_i}(d, c_1)\} \oplus \bigoplus\{d : \lambda_{B_i}(d, c_2)\} \oplus c$, then, $c' \notin B_i$, otherwise $B_i$ is not linearly independent.

Consequently, $\omega(c) = \omega(c')$ and $\lambda_{B_i}(c, c') = 1$ for all basis $B_i$. We can then replace $c$ by $c'$ in $B_i$ by Lemma 1. The same property occurs for even cycles (see Fig. 1b). This operation can be repeated until $B^* \subseteq L$. As the size of $B^*$ is unchanged, we get another optimal solution. □

MAX-MCBI can now be seen as a subproblem of the matroid intersection problem, MI. We deduce the following theorem.

**Theorem 2.** MCBI *and* MAX-MCBI *are polynomial when* $k = 2$, MAX-MCBI *is* $\frac{1}{k}$-*approximable and, finally* MCBI *is XP with respect to* $K$.

*Proof.* MAX-MCBI consists in solving MI in $\{\mathcal{M}(G_i), i \in [\![1;k]\!]\}$. By Theorem 1, we can restrict each matroid to $L$, the cycles output by Algorithm 1. Let $\mathcal{M}(G_i)_{|L}$ be the resulting restricted matroid. By Lemma 5 and Theorem 1 $\mathcal{M}(G_i)_{|L}$ is a matroid with a polynomial size ground set and a polynomial time independence oracle. In that case MI is polynomial when $k = 2$ [5] and is $\frac{1}{k}$-approximable [11]. Finally if $K$ is fixed, we can simply enumerate every subset of $L$ of size $K$ and check if that subset is independent in all the matroids. □

## 4  Hardness of MCBI

This section gives the hardness proofs for MAX-MCBI and MCBI. The latter is in NP as by Lemma 5 we can check independence in polynomial time.

**Theorem 3.** MCBI *is NP-Complete even if* $k = 3$. *In addition,* MCBI *is W[1]-Hard with respect to* $K$. *Moreover, unless* $P = NP$, *for every* $\varepsilon > 0$, *there is no polynomial approximation algorithm with ratio* $\frac{1}{k^{1-\varepsilon}}$ *for* MAX-MCBI. *All those results remain true even if* $\Delta = 3$ *and* $\gamma = 5$ *or if* $\Delta = 4$ *and* $\gamma = 4$.

*Proof.* We provide a reduction from the *Maximum Independent Set* in a graph that consists, given a graph $H = (V, E)$ and an integer $K'$, in the search for an independent set of $H$ of size at least $K'$, that is a subset in which no pair of nodes is linked by an edge in $E$. In order to avoid confusions with the word *independent*, we then use the *Stable set* terminology instead of Independent set.

The reduction does not depend on the value of $\gamma$ except for a useful procedure described hereinafter and in Fig. 2. Let $c_1 = (u_1, u_2, \ldots, u_l)$ and $c_2 = (v_1, v_2, \ldots, v_l)$ be two disjoint cycles where $l \in \{4, 5\}$. The procedure $CONN(c_1, c_2)$ connects the nodes of the two cycles. We add the edges $(u_i, v_i)$ for $i \in [\![1;l]\!]$ and, if $l = 4$, we add the edges $(u_1, v_2), (u_2, v_3), (u_3, v_4)$ and $(u_4, v_1)$.

Note that, by Lemma 2, after using the procedure, no more minimum cycle basis containing $c_1$ and $c_2$ at the same time.

We now describe a general reduction with more than 3 graphs and then show how to reduce $k$. Let then $(H = (V, E), K')$ be an instance of the maximum Stable set problem. We create an instance of MCBI as follows. We set $l$ either to 4 or 5. We set $K = K'$. For each edge $e \in E$, we build a graph $G_e$. For each node $v \in V$, we add a cycle $c_v$ of size $l$ to all the graphs. All the cycles are disjoint. Finally, if $e = (u, v)$, then, in the graph $G_e$, we connect $c_u$ and $c_v$ using the procedure $CONN(c_u, c_v)$. Every other cycle of $G_e$ is not connected to the rest of the graph and is its own connected component.

Note that $l = \gamma$ and if $\gamma = 4$, then $\Delta = 4$ and if $\gamma = 5$ then $\Delta = 3$. Note also that only the cycles $\{c_v | v \in V\}$ belong to the intersection of the graphs $\{G_e | e \in E\}$. A feasible solution contains only those cycles.

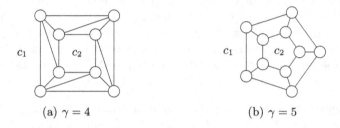

(a) $\gamma = 4$        (b) $\gamma = 5$

**Fig. 2.** Illustration of the procedure $CONN(c_1, c_2)$

Given an edge $e = (u, w) \in E$, the graph $G_e$ contains exactly two minimum cycles bases. If $\gamma = 4$ (resp. 5), let $D$ be the set of triangles (resp. squares) connecting $c_u$ and $c_w$. Then $\mathcal{MCB}(G_e) = \{D \cup \{c_v | v \in V \setminus \{w\}\}, D \cup \{c_v | v \in V \setminus \{u\}\}\}$. As a consequence, for every subset of nodes $V' \subset V$, the set $\{c_v | v \in V'\}$ is independent in $G_e$ if and only if $u \notin V'$ or $w \notin V'$. Thus, there exists a feasible set of cycles of size $K$ if and only if there exists a stable set of size $K$ in $H$. As the Maximum Stable Set is NP-Complete and W[1]-Hard with respect to $K'$, this reduction proves the NP-Completeness of MCBI and the W[1]-Hardness with respect to $K$.

Now note that, if $e = (u, v)$ and $f = (u', v')$ are two non-incident edges, then instead of $G_e$ and $G_f$ we can add to the instance a graph $G_{(e,f)}$ containing the union of $G_e$ and $G_f$. As the two edges are not incident, we have that a feasible set of cycles cannot contain $c_u$ and $c_v$ at the same time and the same for $c_{u'}$ and $c_{v'}$. Note that this transformation does not change the values of $\gamma$ and $\Delta$. We can extend this idea to any matching $M$ of $H$. By the Vizing Theorem [13], using a polynomial greedy algorithm, edges of $H$ can be covered with $|V|$ disjoint matchings. This reduces the number $k$ of graphs from $|E|$ to $|V|$. By [17], unless $P = NP$, for all $\varepsilon > 0$, there is no polynomial approximation with ratio $\frac{1}{|V|^{1-\varepsilon}}$ for the Maximum Stable Set problem. This proves the inapproximability result.

The Maximum Stable Set remains NP-Complete even if $H$ has degree at most 3 and if any path linking two nodes with degree 3 contains at least 3 edges [14]. Such a graph is always 3-edge-colorable. This proves the NP-Completeness of MCBI when $k = 3$. □

## 5   Cases $\gamma = 3$, and $\gamma = 4$, $\Delta = 3$

We describe two useful lemmas to prove that we can focus on the cycles size by size independently. In this section, we call $L$ the list returned by Algorithm 1, and we denote by $T(G)$ and $S(G)$ the triangles and squares of a graph $G$.

**Lemma 6.** *If* $B_1, B_2 \in \mathcal{MCB}(G)$, *for every* $l \in \mathbb{N}$, $\{c \in B_1 | \omega(c) \neq l\} \cup \{c \in B_2 | \omega(c) = l\} \in \mathcal{MCB}(G)$.

*Proof.* Using Lemma 3, for every $c \in B_1 \backslash B_2$, there is some $d \in B_2 \backslash B_1$ such that $B_1 \backslash \{c\} \cup \{d\} \in \mathcal{MCB}(G)$. Note that $\omega(c) = \omega(d)$. We can then swap the cycles until all the cycles of $B$ with size $l$ are in $B_2$. After the exchanges, $|\{c \in B | \omega(c) = l\}| = |\{c \in B_2 | \omega(c) = l\}|$, otherwise, the basis exchange axiom would be false for $B$ and $B_2$. Then $B = \{c \in B_1 | \omega(c) \neq l\} \cup \{c \in B_2 | \omega(c) = l\}$. □

Using Lemma 6 we see that all the subset of cycles of same size in two bases are interchangeable. This means that, from two feasible solutions $B_3$ and $B_4$ respectively maximizing the number of triangles and squares, we can build a feasible solution maximizing both of them. We define with $\mathcal{M}(G, l)$ the matroid $\mathcal{M}(G)$ retricted to the cycles of size $l$ in $L$. Maximizing the triangles (respectively squares) consists in finding a maximum set of cycles that are independent in $\mathcal{M}(G, 3)$ (resp. $\mathcal{M}(G, 4)$). The following lemma gives a characterization of the independent sets. Due to lack of space, the proof may be found in [2].

**Lemma 7.** *$B$ is independent in* $\mathcal{M}(G, l)$ *if and only if, for all* $D \subseteq B$, $\bigoplus D \notin span(cycle\ c | \omega(c) \leq l - 1)$.

**Theorem 4.** *Given an instance of* MAX-MCBI, *one can find a feasible solution maximizing the number of triangles in polynomial time. As a consequence,* MCBI *and* MAX-MCBI *are polynomial when* $\gamma = 3$.

*Proof.* Because we work with simple graphs, there are no cycles of size 2. By Lemma 7, any set of linearly independent triangles of $L$ is independent in $\mathcal{M}(G_i, 3)$ for all $i$. We can then start with an empty solution $B$ and add each cycle $c \in T(G) \cap L$ to $B$ if $c \notin span(B)$. □

Now we consider the case for $\gamma = 4$ and $\Delta = 3$. An instance where $\gamma = 4$ has triangles and squares in $L$. Contrary to the previous case, we can have two squares that are linearly independent but not together independent in $\mathcal{M}(G_i, 4)$ for all $i$. For instance one of the squares may contain a diagonal in $G_i$ or the sum of the squares may belong to the span of the triangles like in Fig. 2a.

Interestingly, the algorithm we use is almost the same as for $\gamma = 3$. It is described in the proof of Theorem 5.

We consider that $L$ do not contain any square that is spanned by triangles of $T(G)$ as such a square cannot belong to any independent set of $\mathcal{M}(G_i, 4)$. Such cycles can be removed from $L$ in polynomial time.

In the following lemmas, we use the matroid terminology of *circuit*. By Lemma 7, a circuit of $\mathcal{M}(G, 4)$ is a subset $C \subseteq S(G)$ such that $\bigoplus C \in span(T(G))$ but, for all subsets $C'$ of $C$, we have $\bigoplus C' \notin span(T(G))$. We now caracterize the circuits of the graphs when $\Delta = 3$.

**Lemma 8.** *Let $G \in \{G_i, i \leq k\}$ with $\Delta = 3$ and $s_1 = (a_1, b_1, c_1, d_1) \in L$, $s_2 = (a_2, b_2, c_2, d_2) \in L$, with $s_1 \neq s_2$ and $t_3 = (a_3, b_3, c_3) \in T(G)$. Figure 3 gives the possible intersections of $s_1$ and $s_2$, and of $s_1$ and $t_3$, up to an isomorphism.*

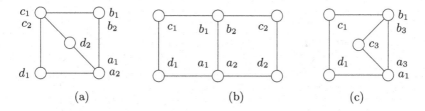

**Fig. 3.** Possible cases of intersection of squares and of triangles.

*Proof.* Recall that we removed from $L$ the squares generated by triangles of $T(G)$: they do not contain a chord. For the squares intersection, if $s_1$ and $s_2$ have only one common node, that node has degree 4 and 4 common nodes imply two diagonals. Thus they have three common nodes (Fig. 3a) or two common nodes without diagonal (Fig. 3b). Similarly, if $s_1$ and $t_3$ have 1 or 3 common nodes, there is a contradiction. Without diagonal, this leads to Fig. 3c.    □

**Lemma 9.** *Let $G \in \{G_i, i \leq k\}$ with $\Delta = 3$ and $C \subseteq L$ be a circuit of $\mathcal{M}(G_i, 4)$. Then $\bigcup C$ is a connected subgraph of $G$.*

*Proof.* As $C$ is a circuit of $\mathcal{M}(G_i, 4)$, by Lemma 7, there exists $T \subseteq T(G)$ such that $\bigoplus T = \bigoplus C$. We assume that $T$ is minimal, meaning that for $T' \subsetneq T$, $\bigoplus T \neq 0$. We now show that $\bigcup C \cup \bigcup T$ is a connected graph. Indeed, otherwise, let assume there exists a connected component $G'$ of $\bigcup C \cup \bigcup T$, and let $C'$ and $T'$ be respectively the proper subets of $C$ and $T$ in $G'$. No cycle of $C \backslash C'$ and $T \backslash T'$ intersects the edges of $\bigcup C' \cup \bigcup T'$. As $\bigoplus T = \bigoplus C$ then $\bigoplus T' = \bigoplus C'$. Consequently either $C'$ is empty in which case $\bigoplus T' = 0$ or $C'$ is a dependent proper subset of $C$. The first case contradicts the minimality of $T$ and the second one contradicts the fact that $C$ is a circuit. The only left possibility is that $T' = C' = \emptyset$ meaning that $\bigcup C \cup \bigcup T$ is connected.

Now we assume that $\bigcup C$ is disconnected. In $\bigcup C \cup \bigcup T$, any two connected components of $\bigcup C$ are connected by a chain $P$ included in $\bigcup T$. Let $s_1 = (a_1, b_1, c_1, d_1)$ and $s_2 = (a_2, b_2, c_2, d_2)$ be two squares of $C$ respectively in the first and second component linked by $P$. Let $w_1, w_2, \ldots, w_q$ be the nodes of $P$, with $w_1 \in s_1$ and $w_q \in s_2$. Without loss of generality, we state that $w_1 = a_1$ and $w_q = a_2$. Note that we cannot have $q = 1$ as the squares do not intersect.

Each edge $(w_i, w_{i+1})$ belongs to a triangle $t_i$. As $w_1 = a_1$ has 3 neighbors, $b_1, d_1$ and $w_2$, then $t_1$ is either $(w_1, w_2, b_1)$ or $(w_1, w_2, d_1)$. We assume, wlog, that it is $(w_1, w_2, b_1)$. As a result, $q \neq 2$, otherwise, $w_2 = a_2$ has four neighbors $(a_1, b_1, b_2$ and $d_2)$. Thus $q \geq 3$. As $w_2$ has already 3 neighbors, $w_3, a_1$ and $b_1$, the nodes of $t_2$ are $w_2, w_3$ and either $a_1$ or $b_1$. This means that $w_3$ is connected to that node, and then the degree of $a_1$ or $b_1$ is 4. We then have a contradiction. $\square$

**Lemma 10.** *Let $G \in \{G_i, i \leq k\}$ with $\Delta = 3$. Then Fig. 4 gives the possible circuits $C$ of $G$ such that $\bigoplus C \neq 0$.*

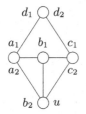
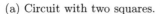
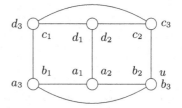

(a) Circuit with two squares.             (b) Circuit with three squares.

**Fig. 4.** Possible circuits containing squares in a graph $G$ with degree at most 3. Each square contains four node $(a_i, b_i, c_i, d_i)$.

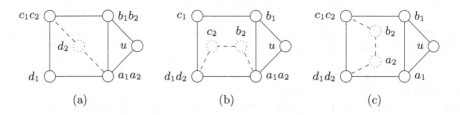

(a)                          (b)                          (c)

**Fig. 5.** Illustrations for the proof of Lemma 10

*Proof.* As $\bigoplus C \neq 0$, then, there exists $T \subseteq T(G)$ with $T \neq \emptyset$ and $\bigoplus C = \bigoplus T$. There exists at least one triangle $t$ that intersects a square $s_1$ of $C$. By Lemma 8, that intersection is the graph depicted by Fig. 3c. We set $(a_1, b_1, c_1, d_1)$

and $t = (a_1, b_1, u)$. As we removed from $L$ the squares generated by triangles, $|C| > 1$. By Lemma 9, there exists another square $s_2$ that intersects $s_1$. Let $s_2 = (a_2, b_2, c_2, d_2)$. In the following, we rename the nodes of $s_2$ so that if $a_1$ (resp. $b_1, c_1, d_1$) is in $s_1 \cap s_2$, then $a_1 = a_2$ (resp. $b_1 = b_2$, $c_1 = c_2$, $d_1 = d_2$).

- If $s_1$ and $s_2$ have three common nodes as in Fig. 3a, then:
  - $s_1 \cap s_2 = \{a_1 = a_2, b_1 = b_2, c_1 = c_2\}$ and $d_1 \neq d_2$ (see Fig. 5a). Then $u, b_1, d_1$ and $d_2$ are neighbors of $a_1$, then two of those nodes are equal. Otherwise the degree of $a_1$ is 4. By hypothesis, $d_1 \neq d_2$. We cannot have $d_1 = b_1$ as $s_1$ would not be a square, similarly $b_1 \neq u$. And $d_1 \neq u$ otherwise $s_1$ would contain a diagonal. The only possibilities are $d_2 = u$ or $d_2 = b_1$. In the first case $s_2$ contains a diagonal $(b_2, d_2)$. In the second case, $d_2 = b_2$. Thus there is a contradiction.
  - Or $s_1 \cap s_2 = \{a_1 = a_2, c_1 = c_2, d_1 = d_2\}$ and $b_1 \neq b_2$. Similarly we can deduce that $b_2 = u$. We obtain the graph of Fig. 4a.
  - The two other intersections are symmetrical cases.
- If $s_1$ and $s_2$ have two common nodes as in Fig. 3b, then:
  - $s_1 \cap s_2 = \{a_1 = a_2, b_1 = b_2\}$ and $c_1 \neq c_2, d_1 \neq d_2$. Then $c_2 = u$ and $d_2 = u$, otherwise the degree of $a_1$ or $b_1$ is 4. But $c_2$ cannot equal $d_2$.
  - Or $s_1 \cap s_2 = \{a_1 = a_2, d_1 = d_2\}$ and $b_1 \neq b_2, c_1 \neq c_2$ (see Fig. 5b). Then $b_2 = u$ otherwise the degree of $a_1$ is 4, and $c_2 \neq u$.
    The edge $(b_2 = u, c_2)$ necessarily belongs either to a triangle or to another square of $C$, otherwise, we cannot have $\bigoplus T + \bigoplus C = 0$. Assuming it belongs to a triangle $t'$. As $u$ has 3 neighbors, $a_1, b_1$ and $c_2$, the third node of $t'$ is necessarily $a_1$ or $b_1$. This implies that $a_1$ or $b_1$ are linked to $c_2$. This is a contradiction as that nodes has then 4 neighbors. Consequently, the edge $(b_2 = u, c_2)$ belongs to a third square $s_3 = (a_3, b_3 = b_2 = u, c_3 = c_2, d_3)$. Again, as $u$ has already 3 neighbors, we have that $a_3 \in \{a_1 = a_2, b_1\}$. If $a_3 = a_1 = a_2$ then $d_3$ must be an existing neighbor of $a_1$, that is $b_1$ or $d_1 = d_2$. It cannot be $d_2$ otherwise $s_2 = s_3$. Then $b_1 = d_2$, $b_1$ has four neighbors and there is a contradiction. If, on the other hand, $a_3 = b_1$ then $d_3$ must be an existing neighbor of $b_1$, that is $a_1$ or $c_1$. In the first case, $a_1$ has four neighbors. In the second case, we obtain the graph of Fig. 4b.
  - The case $s_1 \cap s_2 = \{b_1, c_1\}$ is symmetrical.
  - Or $s_1 \cap s_2 = \{c_1 = c_2, d_1 = d_2\}$ and $a_1 \neq a_2, b_1 \neq b_2$ (see Fig. 5c). The node $a_2$ cannot equal $a_1$ or $b_1$. It is then either $u$ or an additional node. The same property occurs for $b_2$. If $a_2 = u$ then $u$ has four neighbors $a_1, b_1, d_1 = d_2$ and $b_2$. As $b_2$ cannot equal any of the three first nodes, there is a contradiction. Similarly, there is a contradiction if $b_2 = u$.
    The edge $(a_1, d_1)$ necessarily belongs to either a triangle or another square of $C$, otherwise, we cannot have $\bigoplus T + \bigoplus C = 0$. It cannot belong to a triangle as the two nodes have three neighbors and no common neighbor. There is then a third square $s_3 = (a_3 = a_1, b_3, c_3, d_3 = d_1)$ containing that edge. Then $b_3$ is a neighbor of $a_1$ and $c_3$ is a neighbor of $d_3$. As a consequence $b_3 = u$ and $c_3 = a_2$. Thus $u$ and $a_2$ are neighbors.

Similarly, the edge $(b_1, c_1)$ belongs to a fourth square and $u$ and $b_2$ are neighbors, and then $u$ has four neighbors. This is a contradiction.

Note finally that no square can be added to extend the circuits of Fig. 4 as, by definition, $C$ is a minimal dependent set of cycles.                                    □

**Theorem 5.** MCBI *and* MAX-MCBI *are polynomial when* $\gamma = 4$ *and* $\Delta = 3$.

*Proof.* We use the following algorithm. First, we compute the squares of the list $L$ with Algorithm 1. We then remove from $L$ any square that is generated by $T(G_i)$ for some $i \in [\![1; k]\!]$. Then, we initialize an empty solution $B$ and, for each cycle $c \in L$, add $c$ to $B$ if $c \notin \bigcup_{i=1}^{k} span(B \cup T(G_i))$. We finally return $B$.

Let $B^*$ be an optimal solution and $B$ be the solution resulting from the algorithm. Let $s_{(j)}$ be the $j$-th added cycle of $L$ and let also $B_{(j)}$ be the set $B$ at the beginning of the j-th iteration of the algorithm (before adding $s_{(j)}$). Let $\alpha$ be the first index where $s_{(\alpha)} \in B^* \backslash B$ or $s_{(\alpha)} \in B \backslash B^*$. We assume that, among all the optimal solutions, $B^*$ is the solution maximizing the index $\alpha$.

If $s_{(\alpha)} \in B^* \backslash B$, then at the $\alpha$-th iteration of the algorithm, $s_{(\alpha)}$ is not added to $B$, meaning that $s_{(\alpha)} \in span(B \cup T(G_i))$ for some $i$, that is $\{s_{(i)}\} \cup B_{(i)}$ is not independent in $\mathcal{M}(G_i, 4)$. Note that, by definition of $\alpha$, $B_{(\alpha)} \subseteq B^*$. And because $s_{(\alpha)} \in B^*$ then $\{s_{(\alpha)}\} \cup B_{(\alpha)} \subseteq B^*$, which is a contradiction.

If, on the other hand, $s_{(\alpha)} \in B \backslash B^*$, then $B^* \cup \{s_{(\alpha)}\}$ contains a circuit $C$ in $\mathcal{M}(G_{i_C}, 4)$, with $s_{(\alpha)} \in C$, for some $i_C \in [\![1; k]\!]$. We first make the assumption that $C$ is dependent in all the graphs. Let $s \in C \cap B^*$.

-  Either the set $B(s) = B^* \backslash \{s\} \cup \{s_{(\alpha)}\}$ is feasible.
-  Or for some $i \in [\![1; k]\!]$, $B^* \backslash \{s\} \cup \{s_{(\alpha)}\}$ is dependent in $\mathcal{M}(G_i, 4)$. In that last case, there exists a circuit $C'$ of $\mathcal{M}(G_i, 4)$ with $C' \subset B(s)$ and $s_{(\alpha)} \in C'$. As $s \in C$ and $s \notin C'$ then $C \neq C'$ ; and thus $(C \cup C') \backslash (C \cap C')$ is dependent in $\mathcal{M}(G_i, 4)$. However $(C \cup C') \backslash (C \cap C') \subset B^*$ and this is a contradiction.

Therefore $B(s)$ is feasible for every $s \in C \cap B^*$. Note that $B(s)$ is also optimal. Let $j_1 < j_2 < \cdots < j_{|C|}$ be the $|C|$ indices of the squares of $C$ in $L$ (one of them is $\alpha$). By definition of $\alpha$, for every $j \leq \alpha$, $s_{(j)} \in B$. As $B$ is independent, $C \nsubseteq B$: there exists a square $s_{(j)}$, with $j \in [\![\alpha + 1; j_{|C|}]\!]$ such that $s_{(j)} \notin B$. The set $B(s_{(j)})$ is optimal and then contradicts the maximization of $\alpha$ by $B^*$.

Consequently, $C$ is not dependent in all the graphs. Let $T$ be the set of triangles of $G_{i_C}$ such that $\bigoplus C = \bigoplus T$. There exists a graph $G_i$ such that $\bigcup T \nsubseteq G_i$. Indeed, if we assume the contrary, we get that $C$ is dependent in all graphs. As a consequence, first $T \neq \emptyset$ and at least one edge in $\bigcup T$ is not part of any square in $C$. This implies that $C$ is not the circuit depicted by Fig. 4b.

By Lemma 10, $C$ is the circuit depicted by Fig. 4a. It contains $s_{(\alpha)}$ and another square $s_\epsilon$. Let $s_{(\alpha)} = (a_1, b_1, c_1, d_1)$ and $s_\epsilon = (a_2 = a_1, b_2 \neq b_1, c_2 = c_1, d_2 = d_1)$. We now prove that, in each graph $G \in \{G_1, G_2, \ldots, G_k\}$, if $C$ is not a circuit in $\mathcal{M}(G, 4)$ then $= B^* \cup \{s_{(\alpha)}\}$ is independent in that matroid. Indeed, we have that $(b_1, b_2) \notin G$. Also, if $B^* \cup \{s_{(\alpha)}\}$ is not independent, then there is another circuit $C' \subseteq B^* \cup \{s_{(\alpha)}\}$ in $\mathcal{M}(G, 4)$ containing $s_{(\alpha)}$. This circuit $C'$ is again depicted by Fig. 4a, similar to $C$. Let $s_{(\lambda)} = (a_3, b_3, c_3, d_3)$ be the second square of $C'$. The four possible cases are considered.

- $b_1 = b_3, c_1 = c_3, d_1 = d_3$ and $(a_1, a_3) \in G$. Note that $a_3 \neq b_2$ as the edge $(b_1 = b_3, b_2)$ is not in $G$. However in that case $a_1$ has four neighbors.
- The case $a_1 = a_3, b_1 = b_3, d_1 = d_3$ and $(c_1, c_3) \in G$ is symmetrical.
- $a_1 = a_3, c_1 = c_3, d_1 = d_3$ and $(b_1, b_3) \in G$. Note that $b_3 \neq b_2$ as $(b_1, b_2) \notin G$. However in that case $a_1$ has four neighbors.
- $a_1 = a_3, b_1 = b_3, c_1 = c_3$ and $(d_1, d_3) \in G$. If $d_3 = b_2$, then $(d_3 = b_2, d_1 = d_2)$ is a diagonal of $s_{(\varepsilon)}$. However, if $d_3 \neq b_2$ then $a_1$ has four neighbors.

The existence of $C'$ is a contradiction, $B^* \cup \{s_{(\alpha)}\}$ is independent in $\mathcal{M}(G, 4)$.

As a consequence, $B^* \backslash \{s_{(\varepsilon)}\} \cup \{s_{(\alpha)}\}$ is feasible and optimal. However, as $B$ is independent and contains $s_{(\alpha)}$, then $s_{(\varepsilon)} \notin B$, this means that $\varepsilon > \alpha$. The optimality of $B^* \backslash \{s_{(\varepsilon)}\} \cup \{s_{(\alpha)}\}$ contradicts the fact that $B^*$ maximises the value of $\alpha$. As a conclusion no such square $s_{(\alpha)}$ exists and $B$ is optimal.    $\square$

## 6    Concluding Remarks

This paper introduces the problem of maximizing the intersection of minimum cycle bases of graphs and studies the complexity based on four natural parameters. Several questions remain open regarding the minimum value of $k$ for which inapproximability and W[1]-hardness hold. Furthermore, from a chemical perspective, one could argue that the difference between the sets of edges of the graphs (the edit distance) may be small relative to $k$, $\Delta$, and $\gamma$. This observation may lead to the discovery of new tractable algorithms for the problem.

## References

1. Aboulfath, Y., Bougueroua, S., Cimas, A., Barth, D., Gaigeot, M.P.: Time-resolved graphs of polymorphic cycles for h-bonded network identification in flexible biomolecules. J. Chem. Theory Comput. **20**, 1019–1035 (2024)
2. Aboulfath, Y., Watel, D., Weisser, M.A., Mautor, T., Barth, D.: Maximizing Minimum Cycle Bases Intersection (2024). https://hal.science/hal-04559959
3. Amaldi, E., Iuliano, C., Jurkiewicz, T., Mehlhorn, K., Rizzi, R.: Breaking the $O(m^2 n)$ barrier for minimum cycle bases. In: Fiat, A., Sanders, P. (eds.) ESA 2009. LNCS, vol. 5757, pp. 301–312. Springer, Heidelberg (2009). https://doi.org/10.1007/978-3-642-04128-0_28
4. Bougueroua, S., et al.: Graph theory for automatic structural recognition in molecular dynamics simulations. J. Chem. Phys. **149**(18), 184102 (2018). https://doi.org/10.1063/1.5045818
5. Edmonds, J.: Submodular functions, matroids, and certain polyhedra. In: Jünger, M., Reinelt, G., Rinaldi, G. (eds.) Combinatorial Optimization — Eureka, You Shrink! LNCS, vol. 2570, pp. 11–26. Springer, Heidelberg (2003). https://doi.org/10.1007/3-540-36478-1_2

6. Gaüzère, B., Brun, L., Villemin, D.: Relevant cycle hypergraph representation for molecules. In: Kropatsch, W.G., Artner, N.M., Haxhimusa, Y., Jiang, X. (eds.) GbRPR 2013. LNCS, vol. 7877, pp. 111–120. Springer, Heidelberg (2013). https://doi.org/10.1007/978-3-642-38221-5_12

7. Gleiss, P.M., Stadler, P.F., Wagner, A., Fell, D.A.: Relevant cycles in chemical reaction networks. Adv. Complex Syst. **4**(02n03), 207–226 (2001). https://doi.org/10.1142/s0219525901000140

8. Horton, J.D.: A polynomial-time algorithm to find the shortest cycle basis of a graph. SIAM J. Comput. **16**(2), 358–366 (1987)

9. Ilemo, S.N., Barth, D., David, O., Quessette, F., Weisser, M.A., Watel, D.: Improving graphs of cycles approach to structural similarity of molecules. PLOS ONE **14**(12), e0226680 (2019). https://doi.org/10.1371/journal.pone.0226680

10. Kavitha, T., et al.: Cycle bases in graphs characterization, algorithms, complexity, and applications. Comput. Sci. Rev. **3**(4), 199–243 (2009). https://doi.org/10.1016/j.cosrev.2009.08.001

11. Korte, B., Hausmann, D.: An analysis of the greedy heuristic for independence systems. Ann. Disc. Math. **2**, 65–74 (1978)

12. Likhachev, I.V., Balabaev, N., Galzitskaya, O.: Available instruments for analyzing molecular dynamics trajectories. Open Biochem. J. **10**(1), 1–11 (2016)

13. Misra, J., Gries, D.: A constructive proof of Vizing's theorem. Inf. Process. Lett. **41**(3), 131–133 (1992). https://doi.org/10.1016/0020-0190(92)90041-s

14. Murphy, O.J.: Computing independent sets in graphs with large girth. Disc. Appl. Math. **35**(2), 167–170 (1992)

15. Vismara, P.: Union of all the minimum cycle bases of a graph. Electron. J. Comb. **4**(1), R9 (1997). https://doi.org/10.37236/1294

16. Welsh, D.J.A.: Matroid Theory, p. 131. Courier Dover Publications, Mineola (1976)

17. Zuckerman, D.: Linear degree extractors and the inapproximability of max clique and chromatic number. In: Proceedings of the Thirty-Eighth Annual ACM Symposium on Theory of Computing, STOC 2006, pp. 681-690. Association for Computing Machinery, New York (2006). https://doi.org/10.1145/1132516.1132612

# Improving Online Bin Covering
# with Little Advice

Andrej Brodnik[1,4], Bengt J. Nilsson[2(✉)], and Gordana Vujović[1,3]

[1] University of Ljubljana, Ljubljana, Slovenia
andrej.brodnik@fri.uni-lj.si
[2] Malmö University, Malmö, Sweden
bengt.nilsson.TS@mau.se
[3] Complexity Science Hub, Vienna, Austria
vujovic@csh.ac.at
[4] University of Primorska, Koper, Slovenia

**Abstract.** The online bin covering problem is: given an input sequence of items find a placement of the items in the maximum number of bins such that the sum of the items' sizes in each bin is at least 1. Boyar *et al.* [3] present a strategy that with $O(\log \log n)$ bits of advice, where $n$ is the length of the input sequence, achieves a competitive ratio of $8/15 \approx 0.5333\dots$. We show that with a strengthened analysis and some minor improvements, the same strategy achieves the significantly improved competitive ratio of $135/242 \approx 0.5578\dots$, still using $O(\log \log n)$ bits of advice.

**Keywords:** Bin covering · Online computation · Competitive analysis · Advice complexity

## 1 Introduction

In the bin covering problem ([2]), we are given a set of items of sizes in the range $]0, 1[$ and the goal to cover a maximum number of bins where a bin is covered if the sum of sizes of items placed in it is at least 1. It has been shown that the problem is NP-hard [1].

The bin covering problem has applications in various situations in business and industry, from packing medical relief boxes to emergency areas so that each box contains at least a minimum net weight, to such complex problems as distribution of data/items to a maximum number of processors/bins, where the processor can run a task if at least some minimal amount of data is available.

In the online version, items are delivered successively (one-by-one) and each item has to be packed, either in an existing bin or a new bin, before the next item arrives. The first online strategy proposed for the problem is *Dual Next Fit* (DNF) and its competitive ratio of $1/2$ is proved by Assmann *et al.* [1]. Later Csirik and Totik [6] prove that no pure online strategy can achieve a better competitive ratio.

A. A. Rescigno and U. Vaccaro (Eds.): IWOCA 2024, LNCS 14764, pp. 69–81, 2024.
https://doi.org/10.1007/978-3-031-63021-7_6

Boyar *et al.* [3] look at bin covering using extra advice provided by an oracle through an advice tape that the strategy can read. If the input sequence consists of $n$ items, they show that with $o(\log \log n)$ bits of advice, no strategy can have better competitive ratio than $1/2$. In addition, they show that a linear number of bits of advice is necessary to achieve competitive ratio greater than $15/16$. Finally, they provide a strategy using $O(\log \log n)$ bits of advice and competitive ratio $8/15 - O(1/\log n)$. We thus differentiate between pure strategies and advice-based strategies.

Brodnik *et al.* [4], show a $2/3$-competitive strategy using $O(\zeta + \log n)$ advice, where $\zeta$ is the number of bits used to encode any item value in $\sigma$. However, this result does not follow the model used by Boyar *et al,* in that they make assumptions on the encoding of the input items.

*Our Results.* We present an improved analysis of the online strategy by Boyar *et al.* [3] and show that with additional modifications it is possible to achieve a competitive ratio of $135/242 - O(1/\log n)$ using $O(\log \log n)$ bits of advice.

## 2   Preliminaries

The *online bin covering problem* we consider is, given an input sequence $\sigma = (v_1, v_2, \ldots)$ of $|\sigma| = n$ values $v_i$ with $v_i \in \, ]0, 1[$, find the *maximum* number of unit sized bins that can be covered online with items from the input sequence $\sigma$. A bin $B$ is covered if $\sum_{v \in B} v \geq 1$.

We measure the quality of an online maximization strategy A by its *asymptotic competitive ratio*, the maximum bound $R$ such that

$$|A(\sigma)| \geq R \cdot |\mathsf{OPT}(\sigma)| - C, \tag{1}$$

for every possible input sequence $\sigma$, where $|A(\sigma)|$ is the number of bins covered by A, $\mathsf{OPT}(\sigma)$ is a solution for which $|\mathsf{OPT}(\sigma)|$ is maximal, and $C$ is some constant. It is useful to think of an online problem as a game played between the strategy and an *adversary*, that has full knowledge of the strategy and its decisions as the input items arrive. The adversary's objective is to produce the worst possible input sequence for the strategy.

The *dual next fit strategy* (DNF), maintains one active bin, into which it packs items until it is covered. It then opens a new empty active bin and continues the process. For the one-dimensional case DNF has a competitive ratio $1/2$, which is tight [1,6].

However, if input items are bounded by some value $v_i \leq \alpha < 1$, it is easy to show that DNF has a competitive ratio as stated in the following inequality from Brodnik *et al* [4].

$$|\mathsf{DNF}(\sigma_\alpha)| > \frac{1}{1 + \alpha} |\mathsf{OPT}(\sigma_\alpha)| - \frac{1}{1 + \alpha}. \tag{2}$$

The strategy described in this paper assumes the *advice-on-tape* model [5] in which an *oracle* has knowledge about both the strategy and the input sequence produced by the adversary, and writes a sequence of advice bits on an unbounded *advice tape* that the strategy can read at any time, before or while the requests are released by the adversary. The number of bits *read* by the strategy defines the *advice complexity* of the strategy. Since the length of the advice is not explicitly given, the oracle needs some mechanism for the strategy to infer how many bits of advice it should read. This can be done with a self-delimiting encoding that extends the length of a bit string only by an additive lower order term [7, Section 3.2].

*Quantity Approximations Using Fixed Point Numerical Values.* Let $v$ have a binary fixed point representation $v = \sum_{i=-\infty}^{q} v_i \cdot 2^i$, where $q + 1 \geq 1$ is the number of bits of the integer part of $v$ and $v_i$ is its $i^{\text{th}}$ bit. We define $v^{-b}$ ($v^{+b}$) to be the approximate value of $v$ obtained by taking the $b$ most significant bits of $v$ and replacing the remaining bits by zeroes (ones), i.e., $v^{-b} = \sum_{i=q-b+1}^{q} v_i \cdot 2^i$, ($v^{+b} = 2^{q-b+1} + \sum_{i=q-b+1}^{q} v_i \cdot 2^i$), with $b \geq 1$. Thus,

$$v - v^{-b} = \left( \sum_{i=-\infty}^{q-b} v_i \cdot 2^i \right) \leq 2^{-b+1} \cdot 2^q \leq \tau v, \tag{3}$$

for some value $2^{-b+1} \leq \tau$. Hence, choosing $b \geq \log 1/\tau$, for sufficiently small $\tau$, guarantees that $v^{-b} \geq (1 - \tau)v$.

Furthermore, $v^{-b}$ can be represented by $O(1/\tau + \log \log v)$ bits by providing the $b = 1/\tau$ most significant bits of $v$ and the length of the binary description of $v$, which has $O(\log \log v)$ size. The difference $v^{+b} - v$ can be similarly bounded.

# 3  Description of the Previous Strategies with Improvements

The Dual Harmonic strategy ($\mathsf{DH}_k$) has been used as the building block for the advice based strategies previously published [3,4]. The strategy partitions the items by sizes into $k$ groups, $]0, 1/k[$, $[1/k, 1/(k-1)[, \ldots, [1/3, 1/2[$, $[1/2, 1[$, and packs items in each group according to DNF. The items of size $]0, 1/k[$ are called *small items* and the items of size $[1/t, 1/(t-1)[$ are called *t-items*, for integers $t \leq k$. Evidently, since $\mathsf{DH}_k$ is a pure strategy, it is at best $1/2$-competitive using the same argument as in Csirik and Totik [6].

The partitioning of items into $k$ groups, as above, also facilitates in structuring the optimal solution. For a given input sequence $\sigma$ and an integer $k \geq 2$, a fixed optimal covering $\mathsf{OPT}(\sigma)$ can be partitioned into groups, $\mathcal{G}_{t_1 t_2 \cdots t_j}$, where the index $t_1 t_2 \cdots t_j$, with $2 \leq t_1 \leq t_2 \leq \cdots \leq t_j \leq k$, denotes that each bin in group $\mathcal{G}_{t_1 t_2 \cdots t_j}$ contains one $t_1$-item, one $t_2$-item, etc. Note that the multiplicity of an index denotes the number of times an item type occurs in the bin. If these items do not fill the bin, then it also contains the necessary amount of small items to do so. We further denote the group of bins that are only covered by small items by $\mathcal{G}_S$. For example, for $k = 2$, the small items have size less than $1/2$ and $\mathsf{OPT}(\sigma) = \mathcal{G}_{22} \cup \mathcal{G}_2 \cup \mathcal{G}_S$, while for $k = 3$, the small items have size less than $1/3$ and $\mathsf{OPT}(\sigma) = \mathcal{G}_{22} \cup \mathcal{G}_2 \cup \mathcal{G}_{233} \cup \mathcal{G}_{23} \cup \mathcal{G}_{333} \cup \mathcal{G}_{33} \cup \mathcal{G}_3 \cup \mathcal{G}_S$.

Brodnik et al. [4] show that a modification of strategy $\mathsf{DH}_4$ to include advice gives an asymptotic competitive ratio of $2/3$. The idea of the strategy is that the oracle provides two values, $\lfloor |\mathcal{G}_2|/3 \rfloor \le m \le \lfloor |\mathcal{G}_2|/2 \rfloor$ and $x_m$, the size of the $m^{\text{th}}$ largest item in the input sequence, to the strategy. The items of size at least $x_m$ are defined to be *good items*. The strategy then opens $m$ reserved bins, places 2-items of size at least $x_m$ into the reserved bins unless each reserved bin already contains a 2-item, and the remaining 2-items are packed two-by-two in separate bins. The small items are first used to pack the reserved bins up to a level of at least $1 - x_m$ and once this is done, the remaining small items are packed using DNF into separate bins. The remaining items are packed using DNF into separate bins according to their group as specified by $\mathsf{DH}_4$.

The advice size for the strategy is $O(\zeta + \log n)$ bits of advice, where $\zeta$ is the number of bits required to represent the value $x_m$, they assume input items are rational. Since $m \le n = |\sigma|$, $m$ can be represented with $O(\log n)$ bits and approximated with $O(\log \log n)$ bits (giving the competitive ratio $2/3 - O(1/\log n)$). We remark that if the binary representation of $x_m$ includes a 0 early, say within the first $n$ or $\log n$ bits, then the oracle can truncate this binary representation a few bits further on and use a short representation as an approximation of $x_m$, giving that $\zeta \in O(\log n)$ or $O(\log \log n)$ depending on the case. This would then lead to the strategy having advice size $O(\log n)$ or $O(\log \log n)$. Hence, for $\zeta$ to be very large, $x_m$ must be very close to 1.

Boyar et al. [3] consider a different computational model, where the input items are not necessarily assumed to be rational values of limited size. In fact, they use this assumption to obtain that $\Omega(n)$ bits of advice are necessary to achieve a competitive ratio above $15/16$. They also modify $\mathsf{DH}_2$ to use $O(\log \log n)$ advice bits, considering only two item types, 2-items and small items, of size $< 1/2$, and achieve an asymptotic competitive ratio of $8/15$.

We give an overview of their strategy without providing a detailed analysis. They define a covering to be $(\alpha, \epsilon)$-*desirable*, for $0 < \alpha \le 1$ and an error limitation parameter $\epsilon$, if it obeys the following three properties:

**Property I.** the covering has at least $\lfloor \alpha \cdot |\mathcal{G}_2| \rfloor$ bins that each contain one 2-item and small items,

**Property II.** the 2-items not placed according to Property I are placed pairwise (possibly together with small items) in bins,

**Property III.** the small items not placed together with 2-items according to Property II cover $2|\mathcal{G}_S|/3 - O(\epsilon|\mathcal{G}_2|)$ bins.

They prove that an $(\alpha, \epsilon)$-desirable covering, for any $\beta \ge 1$, where $|\mathcal{G}_{22}| + |\mathcal{G}_2| = \beta |\mathcal{G}_2|$, the number of covered bins is at least

$$\min \left\{ \frac{\alpha + 2\beta - 1}{2\beta}, \frac{2}{3} \right\} \cdot |\mathsf{OPT}(\sigma)| - O(\epsilon|\mathcal{G}_2|). \tag{4}$$

They claim a strategy such that if $\beta \geq 15/14$, they can apply a lemma similar to Lemma 1 below and claim a competitive ratio of 8/15. If $\beta < 15/14$ their main strategy yields an $(\alpha, \epsilon)$-desirable covering, for $\alpha = (7 - 6\beta)/15 - O(\epsilon)$ and $\epsilon > 1/2^{b/2} > 1/\log n$. To achieve this, they define $m = \lfloor |\mathcal{G}_2|/3 \rfloor$ and the $m^{-b}$ largest items in the input sequence as *good items*. The subsequence of 2-items in the input is then split into three parts, the first two having sizes $m^{-b}$ and the last part $2|\mathcal{G}_{22}| + |\mathcal{G}_2| - 2m^{-b}$. The strategy obtains a $b$ bit approximation $m^{-b}$ of the value $m$ from the oracle and opens $m^{-b}$ reserved bins. The oracle additionally provides the strategy with information on which of the three subsequences to place in the reserved bins. If it is one of the first two, the $\lfloor \alpha \cdot |\mathcal{G}_2| \rfloor$ largest 2-items are guaranteed to be good and the remaining ones are packed two-by-two using subsequent 2-items. If the oracle signals the last subsequence, slightly more than half of the 2-items in this sequence are placed in the reserved bins (since $\beta$ is small, these fit in the reserved bins). Again, the $\lfloor \alpha \cdot |\mathcal{G}_2| \rfloor$ largest 2-items in the reserved bins are guaranteed to be good and the remaining 2-items are packed two-by-two using subsequent 2-items from the sequence. Analyzing the different cases gives that at least $\lfloor \alpha \cdot |\mathcal{G}_2| \rfloor$ bins for $\alpha = (7 - 6\beta)/15 - O(\epsilon)$ among the reserved bins are covered with only one 2-item and small items.

The small items are packed in the reserved bins using a *dual worst fit* (DWF) strategy using no more than the total amount of small items that an optimal solution uses in the bins of $\mathcal{G}_2$ that contain non-good 2-items. The particulars are similar to our presentation in Sect. 4.2 guaranteeing an $(\alpha, \epsilon)$-desirable packing. Setting $b = O(\log \log n)$, Expression (4) gives a competitive ratio of $8/15 - O(1/\log n)$ using $O(b) = O(\log \log n)$ bits of advice.

A first possible improvement to the Boyar *et al.* strategy is to note that we can choose $m$ differently, as long as $2m^{-b}$ bins from $\mathcal{G}_2$ contain enough small items to pack the $m^{-b}$ reserved bins to a sufficient level to guarantee that the resulting packing is $(\alpha, \epsilon)$-desirable, for some $\alpha$. The good items are hence defined to be the $|\mathcal{G}_2| - 2m^{-b}$ largest ones in the input sequence. Analyzing this setup gives the optimal values of $m$ to be $m = \lfloor 2|\mathcal{G}_2|/9 + 2|\mathcal{G}_{22}|/3 \rfloor$ and $m = \lfloor |\mathcal{G}_2|/5 + |\mathcal{G}_{22}|/2 \rfloor$, both giving the competitive ratio $5/9 - O(1/\log n)$. The next section presents some additional modifications that improve the competitive ratio even further.

## 4    Modification and Analysis of the Boyar *et al.* Strategy

In the remainder of this exposition, we improve on the analysis of the strategy by Boyar *et al.* [3] and show that with some fundamental changes, we significantly improve the competitive ratio, still using advice of the same order of magnitude.

Let $T_2$ be the total number of 2-items in the input sequence $\sigma$, i.e., $T_2 = 2|\mathcal{G}_{22}| + |\mathcal{G}_2|$, since we, like Boyar *et al* [3], only use two item types. Further, define the parameter $\beta$ through $|\mathcal{G}_{22}| + |\mathcal{G}_2| = \beta |\mathcal{G}_2|$, with $\beta \geq 1$. This leads to the following immediate result.

**Lemma 1.** (Corresponding to Lemma 1 in [3]) *When $|\mathcal{G}_2| = 0$ or $\beta \geq 121/107$, the strategy* DH$_2$ *has competitive ratio at least 135/242.*

*Proof.* Boyar *et al.* [3] prove that if $|\mathcal{G}_2| = 0$, then the strategy has competitive ratio 2/3 and otherwise the competitive ratio of the strategy is $\min\{(2\beta - 1)/(2\beta), 2/3\}$. If $\beta \geq 121/107$, the result follows.                                    □

The oracle can signal with one bit to the strategy that $\beta \geq 121/107$ and tell it to use the pure $\mathsf{DH}_2$ strategy guaranteeing a competitive ratio of at least $135/242$. Therefore, we assume from now on that $|\mathcal{G}_2| > 0$ and that $\beta < 121/107$.

Let us define an $(\alpha, \alpha', \rho, \epsilon)$-*viable* covering for parameters $0 \leq \alpha, \alpha' < 1$, a 2-item size ratio $0 \leq \rho \leq 1$, and an error limitation parameter $\epsilon > 0$ that will be assigned a value in Theorem 1, if it obeys the following three properties:

**Property I.** the covering splits the subsequence of 2-items into two parts. The first $(1 - \rho)T_2$ 2-items, such that the covering of these has at least $\lfloor \alpha \cdot |\mathcal{G}_2| \rfloor$ bins, each containing one 2-item and small items, and the last subsequence of $\rho T_2$ 2-items, for which the covering has at least $\lfloor \alpha' \cdot |\mathcal{G}_2| \rfloor$ bins that each contain one 2-item and small items,

**Property II.** the 2-items from the first subsequence of $(1 - \rho)T_2$ 2-items not placed according to Property I are placed pairwise (possibly together with small items) in bins, (the last $\rho T_2$ 2-items are not placed pairwise in bins and these bins may therefore not necessarily be covered),

**Property III.** the small items not placed together with 2-items according to Properties I and II cover $2|\mathcal{G}_S|/3 - O(\epsilon|\mathcal{G}_2|)$ bins.

We prove the following lemma.

**Lemma 2.** *For any input sequence $\sigma$ and any $\beta \geq 1$, an $(\alpha, \alpha', \rho, \epsilon)$-viable covering, has at least the following number of covered bins*

$$\min\left\{\frac{(1 - \rho)(2\beta - 1) + \alpha + 2\alpha'}{2\beta}, \frac{2}{3}\right\} \cdot |\mathsf{OPT}(\sigma)| - O(\epsilon|\mathcal{G}_2|).$$

*Proof.* Let $a$ be the number of covered bins having exactly one 2-item from the first $(1 - \rho)T_2$ 2-items in $\sigma$ and let $a'$ be the number of covered bins having exactly one 2-item from the last $\rho T_2$ 2-items. The first $(1 - \rho)T_2$ are packed two-by-two except for the $a$ covered bins containing exactly one 2-item. Together with Property III, we obtain

$$\frac{(1 - \rho)T_2 - a}{2} + a + a' + \frac{2|\mathcal{G}_S|}{3} - O(\epsilon|\mathcal{G}_2|) = \frac{(1 - \rho)T_2 + a}{2} + a' + \frac{2|\mathcal{G}_S|}{3} - O(\epsilon|\mathcal{G}_2|)$$

$$\geq \frac{(1 - \rho)(2|\mathcal{G}_{22}| + |\mathcal{G}_2|) + \alpha|\mathcal{G}_2|}{2} + \alpha'|\mathcal{G}_2| + \frac{2|\mathcal{G}_S|}{3} - O(\epsilon|\mathcal{G}_2|)$$

$$= \frac{(1 - \rho)(2\beta - 1) + \alpha + 2\alpha'}{2}|\mathcal{G}_2| + \frac{2}{3}|\mathcal{G}_S| - O(\epsilon|\mathcal{G}_2|)$$

$$\geq \min\left\{\frac{(1 - \rho)(2\beta - 1) + \alpha + 2\alpha'}{2\beta}, \frac{2}{3}\right\} \cdot |\mathsf{OPT}(\sigma)| - O(\epsilon|\mathcal{G}_2|)$$

bins, since $I + J \geq \min\{I, J\}$ and $|\mathsf{OPT}(\sigma)| = |\mathcal{G}_{22}| + |\mathcal{G}_2| + |\mathcal{G}_S| = \beta|\mathcal{G}_2| + |\mathcal{G}_S|$.   □

We denote the modified strategy that we present by $\mathsf{DH}_2^b$, where $b$ is the number of bits used to approximate advice values. We fix $\epsilon > 1/2^{b/2}$, use $b$ bit fixed point approximations as described in Sect. 2 and will specify the value of $b$ later.

Let $m_R \overset{\text{def}}{=} \left\lfloor (1-\epsilon)^2 (27|\mathcal{G}_2|/121 + 2|\mathcal{G}_{22}|/3) \right\rfloor$ and $m \overset{\text{def}}{=} \lceil (1+\epsilon)m_R \rceil$. Define $n_g \overset{\text{def}}{=} |\mathcal{G}_2| - 2m$ and let the $n_g$ largest items of the input sequence $\sigma$ be the *good 2-items*. The oracle transmits the advice value $m_R^{-b}$ and the strategy opens $m_R^{-b}$ bins called *reserved bins*. It repeats the following steps for each item $v$.

**if** $v$ **is small** place it according to the presentation in Sect. 4.2,
**else** ($v$ is a 2-item), either place it in one of the reserved bins that does not already contain a 2-item according to the presentation in Sect. 4.1, ensuring that at least $\lfloor \alpha \cdot |\mathcal{G}_2| \rfloor$ of them are good 2-items; or place it in a reserved bin that contains a 2-item that is not good; or place it pairwise with another 2-item in a new bin.

## 4.1 Placing the 2-Items

We show the following result.

**Lemma 3.** *For $\beta < 121/107$, there exist threshold values $\delta^T > 0$, $\alpha_L^T > 0$, and size ratios $\rho > 0$ and $\rho' > 0$, such that the strategy $\mathsf{DH}_2^b$ using $O(b)$ bits of advice produces an*

1. *$(\alpha, \epsilon)$-desirable covering, if $\delta \leq \delta^T$,*
2. *$(\alpha_L^T, \delta, \rho, \epsilon)$-viable covering, if $\delta > \delta^T$ and $\alpha_L \geq \alpha_L^T$,*
3. *$(0, \alpha - \alpha_L^T + \delta, \rho', \epsilon)$-viable covering, if $\delta > \delta^T$ and $\alpha_L < \alpha_L^T$,*

*when $\alpha \leq 685/1452 - \beta/3 - \delta/4 - \epsilon/4$, for any $\epsilon > 1/2^{b/2}$.*

*Proof.* We only prove the appropriate properties I and II for the three cases stated in the Lemma and defer the proof of Property III to Lemma 5 in Sect. 4.2.

We implicitly subdivide the sequence of 2-items into four consecutive subsequences. The first three subsequences contain exactly $m_R^{-b}$ 2-items each and the last subsequence consists of the remaining 2-items. The oracle can signal using no more than two bits of advice which of the four subsequences the strategy should use to place in the reserved bins. If the strategy chooses the last subsequence the oracle can, with two additional bits of advice, identify whether the strategy should produce an $(\alpha, \epsilon)$-desirable covering or one of two possible $(\alpha, \alpha', \rho, \epsilon)$-viable coverings. We have four cases.

**Case 1.** If the oracle signals that one of the first three subsequences contains at least $\lfloor \alpha \cdot |\mathcal{G}_2| \rfloor^{-b} - 1$ good 2-items, using two bits and the approximate value $\lfloor \alpha \cdot |\mathcal{G}_2| \rfloor^{-b}$, then the strategy places the 2-items in previous subsequences two-by-two in separate bins, thus guaranteeing that these are covered. The 2-items that the oracle designates are placed in the reserved bins and once this is done, the strategy declares the largest $\lfloor \alpha \cdot |\mathcal{G}_2| \rfloor^{-b} - 1$ of them to be good. It then packs the other non-good 2-items in the reserved bins with subsequent 2-items, and places any further 2-items in the input sequence two-by-two in separate bins. This gives an $(\alpha, \epsilon)$-desirable covering.

**Case 2.** If the oracle signals the last subsequence, then each of the previous subsequences contains fewer than $\lfloor \alpha \cdot |\mathcal{G}_2| \rfloor^{-b} - 1$ good 2-items and the strategy places the 2-items in these subsequences two-by-two in separate bins. There are $Z = T_2 - 3m_R^{-b}$ 2-items left in the last subsequence, which in turn is further partitioned into three *chunks*, that consist of $X_L^{-b}$, $X_R^{-b}$, and $Y$ 2-items respectively. The relationships between these values are: $X_L^{-b} + X_R^{-b} + Y = Z$, $X_L^{-b} + X_R^{-b} - Y \le \lfloor \alpha \cdot |\mathcal{G}_2| \rfloor^{-b}$, $X_L^{-b} + X_R^{-b} \le m_R^{-b}$, $X_L^{-b} + Y \le m_R^{-b}$, and $X_R^{-b} + Y \le m_R^{-b}$. The initial two chunks contain at least $\lfloor \alpha \cdot |\mathcal{G}_2| \rfloor^{-b}$ good items, separated into $\lfloor \alpha_L \cdot |\mathcal{G}_2| \rfloor^{-b}$ good items in the first chunk and at least $\lfloor \alpha \cdot |\mathcal{G}_2| \rfloor^{-b} - \lfloor \alpha_L \cdot |\mathcal{G}_2| \rfloor^{-b}$ in the second chunk. The third chunk contains $\lfloor \delta \cdot |\mathcal{G}_2| \rfloor$ good 2-items.

Let $\delta^T$ be a threshold value to be determined later. If $\delta \le \delta^T$, the oracle signals the strategy to perform the steps of the next case.

**Case 2a.** The oracle transmits the values $X_L^{-b}$, $X_R^{-b}$, and the value $\lfloor \alpha \cdot |\mathcal{G}_2| \rfloor^{-b}$ to the strategy on the advice tape. The strategy places the first $X_L^{-b} + X_R^{-b}$ 2-items (all the items in the first two chunks) in the reserved bins, and declares the largest $\lfloor \alpha \cdot |\mathcal{G}_2| \rfloor^{-b} - 1$ of them to be good. Use the remaining $Y$ 2-items in the sequence (the last chunk) to pack together with the 2-items in the reserved bins that were not declared good. The 2-items in the first two chunks fit in the reserved bins since $X_L^{-b} + X_R^{-b} \le m_R^{-b}$ and we show that they contain at least $\lfloor \alpha \cdot |\mathcal{G}_2| \rfloor^{-b} - 1$ good 2-items.

Let $g$ be the number of good 2-items placed alone in the reserved bins. The final $Y$ 2-items in the input sequence contain $\lfloor \delta \cdot |\mathcal{G}_2| \rfloor$ good items, hence

$$
\begin{aligned}
g &\ge n_g - 3\left(\lfloor \alpha \cdot |\mathcal{G}_2| \rfloor^{-b} - 2\right) - \lfloor \delta \cdot |\mathcal{G}_2| \rfloor \ge |\mathcal{G}_2| - 2m - 3 \cdot \lfloor \alpha \cdot |\mathcal{G}_2| \rfloor - \lfloor \delta \cdot |\mathcal{G}_2| \rfloor + 6 \\
&\ge |\mathcal{G}_2| - 2(1 + \epsilon)m_R - 3 \cdot \alpha \cdot |\mathcal{G}_2| - \delta \cdot |\mathcal{G}_2| + 2 \\
&\ge |\mathcal{G}_2| - 2(1 + \epsilon)\left[(1 - \epsilon)^2 \left(27|\mathcal{G}_2|/121 + 2|\mathcal{G}_{22}|/3\right)\right] - 3\alpha \cdot |\mathcal{G}_2| - \delta \cdot |\mathcal{G}_2| + 2 \\
&\ge 67|\mathcal{G}_2|/121 - 4(\beta - 1) \cdot |\mathcal{G}_2|/3 - 3\alpha \cdot |\mathcal{G}_2| - \delta \cdot |\mathcal{G}_2| - \epsilon \cdot |\mathcal{G}_2| - 1 \\
&\ge \alpha \cdot |\mathcal{G}_2| - 1 \ge \lfloor \alpha \cdot |\mathcal{G}_2| \rfloor - 1 \ge \lfloor \alpha \cdot |\mathcal{G}_2| \rfloor^{-b} - 1,
\end{aligned}
$$

if $\alpha \le 685/1452 - \beta/3 - \delta/4 - \epsilon/4$. Thus, we obtain an $(\alpha, \epsilon)$-desirable covering.

**Case 2b.** If $\delta > \delta^T$, we consider the threshold value $\alpha_L^T$ also to be determined later. If $\alpha_L \ge \alpha_L^T$, the oracle signals this fact to the strategy in addition to the values $X_L^{-b}$ and $\lfloor \alpha_L^T \cdot |\mathcal{G}_2| \rfloor^{-b}$. The strategy places the 2-items in the first chunk in the reserved bins and declares $\lfloor \alpha_L^T \cdot |\mathcal{G}_2| \rfloor^{-b}$ of them to be good. It uses the 2-items in the second chunk to pack the declared non-good 2-items in the reserved bins two-by-two and packs the remaining 2-items in the second chunk two-by-two into separate bins. The 2-items of the third chunk are then placed in the remaining reserved bins. Thus, the strategy produces a $(\alpha_L^T, \delta, \rho, \epsilon)$-viable covering, with $\rho = Y/T_2$.

**Case 2c.** If $\delta > \delta^T$ and $\alpha_L < \alpha_L^T$, then the oracle signals this fact to the strategy in addition to the value $X_L^{-b}$. The strategy packs 2-items in the first chunk of the last group two-by-two in separate (not the reserved) bins and places the 2-items of the last two chunks in the reserved bins. Hence, the strategy gives an $(0, \alpha_R + \delta, \rho', \epsilon)$-viable covering, with $\rho' = (X_R^{-b} + Y)/T_2$. □

## 4.2   Placing the Small Items

As before, we let $\epsilon > 1/2^{b/2}$, with $b$ being the number of bits used to approximate integers given as advice, and let $m_R \stackrel{\text{def}}{=} \left\lfloor (1 - \epsilon)^2 (27|\mathcal{G}_2|/121 + 2|\mathcal{G}_{22}|/3) \right\rfloor$ be the number of reserved bins that the oracle proposes to the strategy. The strategy hence transmits the approximate value $m_R^{-b}$ to $\mathsf{DH}_2^b$. Next, let $m \stackrel{\text{def}}{=} \left\lceil (1 + \epsilon)m_R \right\rceil$ and let $1 - d$ be the size of the $(|\mathcal{G}_2| - 2m)^{\text{th}}$ input item in descending sorted order of size, i.e., the size of the smallest good item. By a simple exchange argument, we may assume that the bins in $\mathcal{G}_2$ contain the $|\mathcal{G}_2|$ largest 2-items in the input sequence $\sigma$, since any pair of 2-items will cover a bin in $\mathcal{G}_{22}$.

Using the terminology in [3] we define small items of size at least $d^{+b}$ as *black items*, and those of size less than $d^{+b}$ as *white items*. Now, let $n_B$ be the number of bins in $\mathcal{G}_2$ (of the fixed optimal solution) that contain black items.

**Placing Black Items.** Since $\mathcal{G}_2$ contains $n_B$ bins having black items it is possible to cover $m_{RB} \stackrel{\text{def}}{=} \min\{m_R^{-b}, n_B\}$ reserved bins with exactly one good item (of size $\geq 1 - d$) and one black item (of size $\geq d^{+b}$). Our objective is therefore to fill these reserved bins with the smallest black items. Again by an exchange argument, we can choose the smallest $n_B$ such items to be in bins of $\mathcal{G}_2$ and the remaining ones to be in bins of $\mathcal{G}_S$ without decreasing the size of the solution.

Let $s_B$ be the size of the $m_{RB}^{\text{th}}$ smallest black item in the input sequence, we let $x_B$ be the number of items among the $m_{RB}$ smallest black items that have size at most $s_B^{-b}$ and let $e_B$ be the number of items among the $m_{RB}$ smallest black items that have size in the range $]s_B^{-b}, s_B^{+b}]$, so $m_{RB} = x_B + e_B$. Let $m_B \stackrel{\text{def}}{=} x_B + e_B^{-b}$, following the presentation in [3].

The $\mathsf{DH}_2^b$ strategy receives the advice $d^{+b}$, $m_B^{-b}$, $s_B^{-b}$, and $e_B^{-b}$ from the oracle to deal with the black items. Out of the $m_R^{-b}$ reserved bins, the strategy allocates $m_B^{-b}$ of them as *black reserved bins*, each containing one black item. When a black item arrives of size at most $s_B^{-b}$ it is placed in a black reserved bin. When a black item in the range $]s_B^{-b}, s_B^{+b}]$ arrives it is also placed in a black reserved bin unless $e_B^{-b}$ such items have already been placed in the black reserved bins. We have two cases to take care of.

If $s_B \geq \epsilon$, then $s_B^{+b} < s_B + 1/2^b \leq s_B + \epsilon^2 \leq (1 + \epsilon)s_B$. Hence, at most $e_B^{-b}$ black items are used that have size $s_B^{+b} \leq (1 + \epsilon)s_B$, i.e., a factor of $(1 + \epsilon)$ larger than the $m_{RB}^{\text{th}}$ smallest black item.

If $s_B < \epsilon$, on the other hand, then the strategy packs a total amount of at most $s_B^{+b}(x_B + e_B^{-b}) < (\epsilon + \epsilon^2)m_{RB} \leq 2\epsilon \cdot m_{RB} \leq 2\epsilon \cdot |\mathcal{G}_2|$ black items.

Once the black reserved bins have been filled with a black item, the remaining black items that arrive in the input sequence are placed together with other small items using DNF into separate bins.

**Placing White Items.** Let $m_W = m_R^{-b} - m_B^{-b}$ be the number of remaining reserved bins, we call them *white reserved bins*. The strategy places white items as they arrive into these bins using the Dual Worst Fit strategy (DWF) until each remaining white reserved bin contains white items to a level of at least $d^{+b}$. When a white item arrives, the substrategy DWF places the item in the white reserved bin that contains the smallest sum of white items. Once each white reserved bin contains white items to the amount of $d^{+b}$, the remaining white items that arrive in the input sequence are placed together with other small items using DNF into separate bins as in the case with the black items.

We prove the following results as in [3].

**Lemma 4.** (Corresponding to Lemma 3 in [3]) *A reserved bin that includes a good item will be covered when* $\mathsf{DH}_2^b$ *terminates.*

*Proof.* By definition, a good item has size at least $1 - d$ so it is sufficient to show that each reserved bin contains small items to the amount of at least $d$. This is clearly true for the black reserved bins, since $d^{+b} \geq d$.

Consider the bins in $\mathcal{G}_2$ of the optimal solution. Of these, $|\mathcal{G}_2| - n_g = 2m$ bins contain a 2-item of size at most $1 - d$ and at least $2m - m_{RB}$ contain only white small items. The bins in $\mathcal{G}_2$ therefore contain white items to the amount of $S_W \geq d(2m - m_{RB})$.

The DWF strategy stops placing white items when each white reserved bin contains at least $d^{+b}$ amount of white items or there are no more white items in the input sequence. Assume for a contradiction that all the white items in the input sequence are placed in the white reserved bins, but that not all the white reserved bins contain at least $d$ amount of white items, i.e., the input sequence reaches the end without having placed at least $d$ amount of white items in each of the white reserved bins.

Let $a_1, \ldots, a_l$ be the sizes of the white items that have size at least $d$. Note that $l < m_W$, otherwise all white reserved bins have at least $d$ amount of white items, a contradiction. The remaining white items have size less than $d$. The white reserved bin that contains the item of size $a_i$ when DWF has processed all the white items contains less than $d + a_i$ amount of white items since DWF always places white items in the bin containing the least amount of white items, for $1 \leq i \leq l$. For the same reason, any other white reserved bin contains less than $2d$ amount of white items. This gives us the following bound on $S_W$,

$$S_W < \sum_{i=1}^{l}(a_i + d) + \sum_{i=l+1}^{m_W} 2d = 2m_W \cdot d + \sum_{i=1}^{l}(a_i - d).$$

On the other hand, since those bins in $\mathcal{G}_2$ that do not contain good items must each have at least $d$ amount of small items in them. Discounting those bins that contain black items, there are $2m - m_{RB} \geq 2m_W$ such bins in $\mathcal{G}_2$, giving us a lower bound on the amount of white items in bins of $\mathcal{G}_2$. We obtain

$$S_W \geq d(2m_W - l) + \sum_{i=1}^{l} a_i = 2m_W \cdot d + \sum_{i=1}^{l}(a_i - d),$$

giving us a contradiction. Hence, every white reserved bin contains white items to the amount of at least $d$ if the DWF strategy reaches the end of the input sequence before terminating. Thus, the DWF strategy guarantees that a white reserved bin contains at least $d$ and less than $2d^{+b}$ amount of white items. We have, if $d \geq \epsilon$, that

$$
\begin{aligned}
2d^{+b} m_W = 2d^{+b}\left(m_R^{-b} - m_B^{-b}\right) &\leq 2d^{+b}\left(m_R^{-b} - (1-\epsilon)m_B\right) \\
= 2d^{+b}\left(m_R^{-b} - (1-\epsilon)(x_B + e_B^{-b})\right) &\leq 2d^{+b}\left(m_R^{-b} - (1-\epsilon)^2(x_B + e_B)\right) \\
\leq 2d^{+b}\left(m_R - (1-\epsilon)^2 m_{RB}\right) &\leq 2(d+1/2^b)\left(m_R - (1-\epsilon)^2 m_{RB}\right) \\
< 2(d+\epsilon^2)\left(m_R - (1-\epsilon)^2 m_{RB}\right) &\leq 2d(1+\epsilon)\left(m_R - (1-\epsilon)^2 m_{RB}\right) \\
\leq d\left(2(1+\epsilon)m_R - 2(1+\epsilon)(1-\epsilon)^2 m_{RB}\right) &\leq d\left(2m - m_{RB}\right) \leq S_W,
\end{aligned}
$$

which holds for $1/2^{b/2} < \epsilon$ sufficiently small. Hence, there is sufficient amount of white items among those in bins of $\mathcal{G}_2$ to place among the reserved white bins and reach $d^{+b}$ white items in each such bin, thus guaranteeing that the DWF strategy terminates in this case.

If $d < \epsilon$, then following the same reasoning as above, we have

$$
\begin{aligned}
2d^{+b} m_W &\leq 2(d+1/2^b)m_R < 2(d+\epsilon^2)m_R < 2(\epsilon+\epsilon^2)m_R \leq 3\epsilon m_R \\
&\leq 3\epsilon(1-\epsilon)^2(27|\mathcal{G}_2|/121 + 2|\mathcal{G}_{22}|/3) < 12055 \cdot \epsilon|\mathcal{G}_2|/12947 \leq \epsilon \cdot |\mathcal{G}_2|,
\end{aligned}
$$

since $\beta < 121/107$ and $1/2^{b/2} < \epsilon$ sufficiently small, concluding the proof. □

**Lemma 5.** (Corresponding to Lemma 4 in [3]) *The strategy* $DH_2^b$ *covers at least* $2|\mathcal{G}_S|/3 - O(\epsilon|\mathcal{G}_2|)$ *small bins.*

*Proof.* From the proof of the previous lemma, if $d \geq \epsilon$, we can view all the white items placed in the reserved bins by the strategy to come from bins in $\mathcal{G}_2$ in the optimal solution and, if $d < \epsilon$, that the sum of all the white items in bins of $\mathcal{G}_2$ is at most $\epsilon \cdot |\mathcal{G}_2|$. In either case, the bins in $\mathcal{G}_S$ have not been reduced by more than $\epsilon \cdot |\mathcal{G}_2|$ amount of small items to place in the white reserved bins.

As for the black items, the strategy uses $m_{RB}$ of the smallest black items to place in the black reserved bins, except possibly in $e_B^{-b}$ cases when such an item lies in the range $]s_B^{-b}, s_B^{+b}]$, where the correct black item would have size just above $s_B^{-b}$ but the strategy chose an item of size no more than $s_B^{+b}$. The amount of small items we overuse to fill the black reserved bins is at most

$$1/2^b \cdot e_B^{-b} < \epsilon^2 \cdot e_B^{-b} \leq \epsilon \cdot m_{RB} \leq \epsilon \cdot |\mathcal{G}_2|.$$

Again, the bins in $\mathcal{G}_S$ have not been reduced by more than $\epsilon \cdot |\mathcal{G}_2|$ amount of small items to place in the black reserved bins.

Hence, the amount of small items that remains for the strategy to use is at least $|\mathcal{G}_S| - 2\epsilon \cdot |\mathcal{G}_2|$ and by Inequality (2), these small items cover at least $2|\mathcal{G}_S|/3 - O(\epsilon|\mathcal{G}_2|)$ bins, proving the lemma. □

We can now prove our main result.

**Theorem 1.** *The modified strategy* $\mathrm{DH}_2^{\lg\lg n}$, *using* $O(\log\log n)$ *bits of advice, achieves an asymptotic competitive ratio of at least* $135/242 - O(1/\log n) \approx 0.5578\ldots - O(1/\log n)$.

*Proof.* If $\beta \geq 121/107$ the result follows from Lemma 1 and the oracle can signal this case using one bit. If $1 \leq \beta < 121/107$, the oracle provides a constant number of integer values, all smaller than $n$ and the threshold value $d^{+b}$. In addition, the oracle provides a constant number of selection bits of advice to help the strategy decide between the appropriate strategy cases.

From Lemma 3, we know that there exist parameters $\delta^T$, $\alpha_L^T$, $\rho$, and $\rho'$ such that the strategy, with appropriate advice will produce desirable or viable coverings as claimed for $\alpha$ sufficiently small. Let $\delta^T = 1/11$, $\alpha_L^T = 5\alpha/14$, $\rho = 13/121$ and $\rho' = 26/121$.

If $\delta \leq \delta^T = 1/11$, then the strategy produces an $(\alpha, \epsilon)$-desirable covering in Cases 1 and 2a of the proof of Lemma 3. Choosing $\alpha = 685/1452 - \beta/3 - \delta/4 - \epsilon/4$, as in the prerequisite of the lemma, Expression (4) gives a competitive ratio of

$$\frac{\alpha + 2\beta - 1}{2\beta} = \frac{\frac{685}{1452} - \frac{\beta}{3} - \frac{\delta}{4} - \frac{\epsilon}{4} + 2\beta - 1}{2\beta} \geq \frac{5}{6} - \frac{100}{363\beta} - \epsilon \geq \frac{135}{242} - O\left(\frac{1}{\log n}\right).$$

If $\delta > \delta^T = 1/11$ and $\alpha_L \geq \alpha_L^T = 5\alpha/14$, then the strategy produces an $(\alpha_L^T, \delta, \rho, \epsilon)$-viable covering in Case 2b of the proof of Lemma 3. With $\alpha_L^T = 5\alpha/14 = 3425/20328 - 5\beta/42 - 5\delta/56 - 5\epsilon/56$, Lemma 2 gives a competitive ratio of

$$\frac{(1-\rho)(2\beta - 1) + \alpha_L^T + 2\delta}{2\beta} \geq \frac{8467}{10164} - \frac{2797}{10164\beta} - \epsilon \geq \frac{135}{242} - O\left(\frac{1}{\log n}\right).$$

If $\delta > \delta^T = 1/11$ and $\alpha_L < \alpha_L^T = 5\alpha/14$, then the strategy produces an $(0, \alpha_R + \delta, \rho', \epsilon)$-viable covering in Case 2b of the proof of Lemma 3. Again, with $\alpha_L^T = 5\alpha/14$, Lemma 2 gives a competitive ratio of

$$\frac{(1-\rho')(2\beta - 1) + 2(\alpha_R + \delta)}{2\beta} \geq \frac{(1-\rho')(2\beta - 1) + 2(\alpha - \alpha_L^T + \delta)}{2\beta} \geq$$

$$\geq \frac{967}{1694} - \frac{1}{77\beta} - \epsilon \geq \frac{135}{242} - O\left(\frac{1}{\log n}\right).$$

By setting $b = 2\log\log n$ and $\epsilon = 2/2^{b/2} = 2/\log n$, the strategy can approximate any integer of size up to $n = |\sigma|$, proving the claim. □

## 5   Conclusions

We provide a deeper analysis and improvements to the strategy for online bin covering by Boyar *et al.* [3] that uses $O(\log \log n)$ bits of advice and achieves a significantly better competitive ratio of $135/242 - O(1/\log n)$, where $n$ is the length of the input sequence.

For future work, improving the competitive ratio further using $O(\log \log n)$ advice bits is an interesting challenge.

The authors would like to thank the anonymous reviewers for their comments and constructive criticism of an earlier draft of this manuscript.

## References

1. Assmann, S.A., Johnson, D.S., Kleitman, D.J., Leung, J.Y.T.: On a dual version of the one-dimensional bin packing problem. J. Algor. **5**(4), 502–525 (1984)
2. Assmann, S.F.: Problems in discrete applied mathematics. PhD thesis, Massachusetts Institute of Technology (1983)
3. Boyar, J., Favrholdt, L.M., Kamali, S., Larsen, K.S.: Online bin covering with advice. Algorithmica **83**(3), 795–821 (2021)
4. Brodnik, A., Nilsson, B.J., Vujovic, G.: Online bin covering with exact advice. Informatica Int. J. Comput. Inf. (2024)
5. Böckenhauer, H.-J., Komm, D., Královič, R., Královič, R., Mömke, T.: Online algorithms with advice: the tape model. Inf. Comput. **254**, 59–83 (2017)
6. Csirik, J., Totik, V.: Online algorithms for a dual version of bin packing. Disc. Appl. Math. **21**(2), 163–167 (1988)
7. Komm, D.: An Introduction to Online Computation—Determinism, Randomization, Advice. Springer, Texts in Theoretical Computer Science (2016). ISBN 978-3-319-42747-8

# An Improved Bound for Equitable Proper Labellings

Julien Bensmail[1] and Clara Marcille[2(✉)]

[1] Université Côte d'Azur, CNRS, Inria, I3S, Nice, France
[2] Univ. Bordeaux, CNRS, Bordeaux INP, LaBRI, 33400 Talence, France
`clara.marcille@gmail.com`

**Abstract.** For every graph $G$ with size $m$ and no connected component isomorphic to $K_2$, we prove that, for $L = (1, 1, 2, 2, \ldots, \lfloor m/2 \rfloor + 2, \lfloor m/2 \rfloor + 2)$, we can assign labels of $L$ to the edges of $G$ in an injective way so that no two adjacent vertices of $G$ are incident to the same sum of labels. This implies that every such graph with size $m$ can be labelled in an equitable and proper way with labels from $\{1, \ldots, \lfloor m/2 \rfloor + 2\}$, which improves on a result proved by Haslegrave, and Szabo Lyngsie and Zhong, implying this can be achieved with labels from $\{1, \ldots, m\}$.

**Keywords:** proper labelling · equitable labelling · 1-2-3 Conjecture

## 1 Introduction

Let $G$ be a graph. For a set $S \subset \mathbb{R}$, an *S-labelling* $\ell : E(G) \to S$ of $G$ is an assignment of *labels* from $S$ to the edges. In case $S = \{1, \ldots, k\}$ for some $k \geq 1$, we say $\ell$ is a *k-labelling* of $G$. For every vertex $u$ of $G$, we denote by $\sigma_\ell(u)$ (or $\sigma(u)$ if there are no ambiguities) its *sum* by $\ell$, being the sum of labels assigned to the edges incident to $u$, that is, $\sigma(u) = \sum_{v \in N(u)} \ell(uv)$. Now, $\ell$ is said *proper* if we have $\sigma(u) \neq \sigma(v)$ for every edge $uv$ of $G$ (that is, the resulting $\sigma_\ell$ is a proper vertex-colouring). Last, we say $G$ is *nice* if $G$ has no connected component isomorphic to $K_2$, in which case we set $\chi_\Sigma(G)$ as the smallest $k \geq 1$ such that $G$ admits proper $k$-labellings. It can be checked that, indeed, $\chi_\Sigma(G)$ is well defined if and only if $G$ is nice.

Proper labellings and the parameter $\chi_\Sigma$ have been mostly investigated in the context of the so-called 1-2-3 Conjecture, raised by Karoński, Łuczak, and Thomason [5] in 2004:

*Conjecture 1 (Karoński, Łuczak, Thomason [5]).* If $G$ is a nice graph, then $\chi_\Sigma(G) \leq 3$.

Several aspects behind the 1-2-3 Conjecture have been investigated in literature to date, including approaching results, algorithmic results, peculiar behaviours, and variants. Definitely the most appealing and significant result in this context is a full proof of the conjecture proposed recently by Keusch [6].

A. A. Rescigno and U. Vaccaro (Eds.): IWOCA 2024, LNCS 14764, pp. 82–97, 2024.
https://doi.org/10.1007/978-3-031-63021-7_7

This apart, most of the most interesting related results can be found *e.g.* in the survey [9] by Seamone.

Despite Keusch's proof of the 1-2-3 Conjecture, there are many more or less closely related open, interesting questions and problems in the field. One of these deals with a concept of equitability for proper labellings, first considered in [1] by Baudon, Pilśniak, Przybyło, Senhaji, Sopena, and Woźniak, and studied further in [2,3]. This concept revolves around the following notions. Let $G$ be a graph, and $\ell$ be a labelling of $G$. For any $l \in \mathbb{R}$, we denote by $\mathrm{nb}(l, \ell) \geq 0$ the number of edges of $G$ to which label $l$ is assigned by $\ell$. We say that $\ell$ is *equitable* if, for any two labels $l$ and $l'$ assigned by $\ell$, we have $|\mathrm{nb}(l, \ell) - \mathrm{nb}(l', \ell)| \leq 1$, or, in other words, if any two label values are assigned about the same number of times by $\ell$. Assuming $G$ is nice, we denote by $\overline{\chi_\Sigma}(G)$ the smallest $k \geq 1$ such that $G$ admits equitable proper $k$-labellings.

In the very first work on the topic [1], the authors investigated the parameter $\overline{\chi_\Sigma}(G)$ for particular classes of nice graphs $G$, without raising a particular conjecture. It is later in [2] that the authors raised an "Equitable 1-2-3 Conjecture", reading as follows:

*Conjecture 2 (Bensmail, Fioravantes, Mc Inerney, Nisse [2]).* If $G$ is a nice graph different from $K_4$, then $\overline{\chi_\Sigma}(G) \leq 3$.

Proper labellings are objects that are rather hard to comprehend in general, so, unsurprisingly, even harder to comprehend equitable proper labellings are. To date, it is known from [1,2] that the Equitable 1-2-3 Conjecture holds for several easy, common classes of nice graphs. In [2], the authors observed that there are infinitely many graphs $G$ with $2 = \chi_\Sigma(G) < \overline{\chi_\Sigma}(G) = 3$, and they notably proved that determining whether $\overline{\chi_\Sigma}(G) = 2$ holds for a given graph $G$ with $\chi_\Sigma(G) = 2$ is NP-complete.

In this work, we are mainly interested in upper bounds on the parameter $\overline{\chi_\Sigma}$. To date, it is not even clear that there is an absolute constant $k \geq 1$ such that $\overline{\chi_\Sigma}(G) \leq k$ holds for every nice graph $G$. Actually, even establishing non-constant bounds is not that clear. Through greedy arguments, in [1] the authors observed that $\overline{\chi_\Sigma}(G) \leq 2^{|E(G)|}$ holds for every nice graph $G$, thereby establishing that it makes sense investigating upper bounds on $\overline{\chi_\Sigma}$, and also that the notion of nice graphs remains relevant in the context of equitable proper labellings. Later on, in [3], Bensmail, Senhaji, and Szabo Lyngsie, through the study of a combination of the 1-2-3 Conjecture and of the so-called Antimagic Labelling Conjecture[1], asked whether all nice graphs $G$ admit a proper $S$-labelling $\ell$ for $S = \{1, \ldots, |E(G)|\}$ and the extra property that $\mathrm{nb}(l, \ell) = 1$ for all $l \in S$ – the point being that, if this was true, then it would imply that $\overline{\chi_\Sigma}(G) \leq |E(G)|$ holds for every nice graph $G$. This was later proved independently by Haslegrave in [4], and by Szabo Lyngsie and Zhong in [7], through rather different approaches (probabilistic tools for the former proof, constructive ones for the latter one).

---

[1] Raised by Hartsfield and Ringel in [8], asking whether, in general, for every graph $G$ there is a $\{1, \ldots, |E(G)|\}$-labelling $\ell$ where $\mathrm{nb}(l, \ell) = 1$ for every assigned label $l$, and no two vertices $u$ and $v$ verify $\sigma(u) = \sigma(v)$.

This is where the investigations on Conjecture 2 stand to date, the best general upper bound on $\overline{\chi_{\Sigma}}(G)$ we know of to date being $|E(G)|$ for every nice graph $G$.

Our main goal in this work is to improve upon this upper bound, which we do by about a factor 2. Namely, our main result reads as follows:

**Theorem 1.** *If $G$ is a nice graph, then $\overline{\chi_{\Sigma}}(G) \leq \left\lfloor \frac{|E(G)|}{2} \right\rfloor + 2$.*

Although our upper bound in Theorem 1 is still not constant, we believe the way we prove it remains of interest, as we mostly build upon the proof of Szabo Lyngsie and Zhong from [7], which we enhance with a new approach and different, dedicated arguments. Our proof of Theorem 1 can be found in Sect. 2. Afterwards, we finish off with a concluding discussion in Sect. 3, in which we explain why it might be difficult to improve upon Theorem 1 further, and come up with other, different questions and problems of independent interest for further work on the topic.

## 2   Proof of Theorem 1

Our proof of Theorem 1 relies mainly on a peculiar point in the definition of an equitable labelling $\ell$, being that the equitability constraint (*i.e.*, that $|\mathrm{nb}(l, \ell) - \mathrm{nb}(l', \ell)| \leq 1$ holds) is only required to hold for pairs of label values $l$ and $l'$ that are actually assigned by $\ell$. What we mean by that is that having $\mathrm{nb}(l, \ell) = 0$ and $\mathrm{nb}(l', \ell) \geq 2$ is not regarded as an objection to $\ell$ being equitable. Hence, if $\ell$ is an $S$-labelling for some set $S$ of labels and $\mathrm{nb}(l, \ell) \leq 2$ holds for all $l \in S$, then $\ell$ is considered equitable.

Before we get to the actual proof of Theorem 1, we need some preparation first. To make the proof more legible, and due to the ideas developed in the previous paragraph, we work with labellings in a slightly different way. Let $G$ be a graph, and $L = (l_1, \ldots, l_q)$ be a sequence of labels from $\mathbb{R}$ (that is, a given label value may appear more than once in $L$). Throughout, we assume every sequence of labels is ordered increasingly, that is, $l_1 \leq \cdots \leq l_q$ in the present case. For any $l \in \mathbb{R}$, we denote by $\mathrm{nb}(l, L) \geq 0$ the number of times label $l$ appears in $L$, and by $\mathrm{mult}(L) \geq 1$ the largest value of $\mathrm{nb}(l, L)$ over all $l \in \mathbb{R}$. Assuming $|L| \geq |E(G)|$, an $L$-*strict-labelling* $\ell$ of $G$ is an assignment of labels from $L$ to the edges of $G$ that complies with the elements of $L$ and their number of occurrences; that is, every label value $l$ appearing in $L$ must be assigned at most $\mathrm{nb}(l, L)$ times. In some sense, $L$ is a pool of labels in which one has to pick elements when assigning labels (so that once a label is assigned, it is no longer part of the pool). For $L'$ being a subsequence of $L$, we denote by $L - L'$ the sequence obtained from $L$ by removing all elements in $L'$.

As mentioned earlier, to prove Theorem 1 our goal is to show that every nice graph $G$ admits a proper $L$-strict-labelling $\ell$, where $L = (1, 1, 2, 2, \ldots, k + 1, k + 1, k + 2, k + 2)$ and $k = \left\lfloor \frac{|E(G)|}{2} \right\rfloor$. Note that $|L| \in \{|E(G)| + 3, |E(G)| + 4\}$ depending on whether $|E(G)|$ is odd or even. Before getting to the formal details of the proof, let us outline its rough ideas first. Essentially, we build $\ell$ by induction, starting initially with all edges of $G$ being unlabelled. We then start by picking some vertex $u$ with degree $d = d(u)$, and $d$ labels $l_{i_1}, \ldots, l_{i_d}$ from $L$, before then assigning these labels $l_{i_1}, \ldots, l_{i_d}$ to the edges incident to $u$, and then proceeding by induction on $G - u$ and $L - (l_{i_1}, \ldots, l_{i_d})$. Of course, these very general ideas suffer several issues. In particular, in $G - u$, the edges incident to $u$ are no longer present, which means that, for any neighbour $v$ of $u$ in $G$, 1) the adjacency between $u$ and $v$ is present in $G$ only, not in $G - u$, so the fact that $u$ and $v$ are distinguished by $\ell$ must rely solely on how we labelled the edges incident to $u$, and 2) when labelling the edges of $G - u$, we have to take into account that, for all neighbours of $u$ in $G$, their eventual sums also involve labels assigned to the edges incident to $u$, which are not present in $G - u$.

- To deal with the latter problem above, we will actually deal with weighted graphs, where a *weighted graph* $(H, c)$ is a graph $H$ given together with some function $c : V(H) \to \mathbb{R}$ modelling a sum contribution for every vertex. Essentially, above, in our initial graph $G$ all contributions will be equal to 0, modelling the fact that all edges are actually present in $G$, and initially unlabelled (which is similar, in terms of sums, to having all edges being assigned label 0 – any graph can actually be seen as a weighted graph with all contributions 0). Then, during the inductive step, whenever considering, in some remaining graph, a vertex $u$ with $d$ neighbours $v_1, \ldots, v_d$ and $d$ labels $l_{i_1}, \ldots, l_{i_d}$ remaining from the initial $L$, assuming we assign label $l_{i_j}$ to $uv_j$ for every $j \in \{1, \ldots, d\}$ before pursuing with $G - u$, we will modify the contribution (by $c$) of each $v_j$ by $l_{i_j}$ to take into account that, in the more global picture, edge $uv_j$ is assigned label $l_{i_j}$, thereby contributing $l_{i_j}$ to the full sum of $v_j$.
- To deal with the former problem, we need to guarantee that the edges incident to $u$ are labelled so that, taking the sum contributions into account, no matter how the remaining edges of $G - u$ are labelled, we cannot get any conflict[2] between $u$ and its neighbours. To guarantee this, we will employ the notion of smallest vertex. In a weighted graph $(H, c)$, the *smallest possible sum* (w.r.t. a sequence $L$ of labels) of a vertex $u$ is $\text{small}(u) = c(u) + x$, where $x$ is the sum of the $d(u)$ smallest elements of $L$. In some sense, the smallest possible sum of $u$ is the smallest sum we can achieve for $u$ through assigning labels of $L$ to all the edges incident to $u$, taking into account the contribution $c(u)$. Now, a *smallest vertex* of $(H, c)$ is a vertex with minimum smallest possible sum. Note that a smallest vertex is not necessarily unique (consider *e.g.* the case of a regular graph with all contributions being 0). However, a convenient property is that if, when building an $L$-strict-labelling of a given weighted graph, we start by considering a smallest vertex $u$ and assigning the smallest labels of $L$ to the edges incident to $u$, and then label the other edges arbitrarily, then in most cases we are sure $u$ cannot get involved in conflicts, as formalised in the upcoming lemma. There and further, we say a sequence $L = (l_1, l_2, \ldots)$ of labels with $l_1 \le l_2 \le \ldots$ is *shifted* if $l_1 < l_2$. Also, to be clear, in a weighted

---

[2] In a weighted graph $(H, c)$, two vertices $u$ and $v$ are *conflicting* by a labelling $\ell$ if $\sigma_\ell(u) + c(u) = \sigma_\ell(v) + c(v)$.

graph $(H, c)$ with a labelling $\ell$, for any vertex $v$ when writing $\sigma_\ell(v)$ (or $\sigma(v)$) we mean the sum of labels assigned to the edges incident to $v$ (*i.e.*, we do not take $c(v)$ into account).

**Lemma 1.** *Let $(G, c)$ be a weighted graph with a smallest vertex $u$ having $d$ neighbours $v_1, \ldots, v_d$, and $L = (l_1, l_2, \ldots)$ be a sequence of labels with $|L| \geq |E(G)|$ and $l_1 \leq l_2 \leq \ldots$, where $l_1, l_2, \cdots \in \mathbb{R}^*$. If $\mathrm{mult}(L) \leq 2$ and $u$ is a smallest vertex of maximum degree, then, for every $L$-strict-labelling $\ell$ of $(G, c)$ such that the edges incident to $u$ are assigned the smallest $d$ labels $l_1, \ldots, l_d$ of $L$, all conflicts involving $u$ must also include some smallest vertex $v_i$ where either:*

- *$d(v_i) = 1$, and $\ell(uv_i) = l_1$ (if $L$ is shifted) or $\ell(uv_i) \in \{l_1, l_2\}$ (otherwise); or*
- *$d(v_i) = d(u) = 2$, $L$ is shifted, $\ell(uv_i) = l_1$, and the second edge incident to $v_i$ is assigned the second smallest label value in $L$ ($l_2$ if $l_2 < l_3$, or $l_2$ or $l_3$ if $l_2 = l_3$).*

*Proof.* Note that, by definition, upon assigning labels $l_1, \ldots, l_d$ to the edges incident to $u$, since $u$ is a smallest vertex it cannot be that $u$ gets in conflict with an adjacent vertex that is not a smallest vertex itself. We now analyse two cases, depending on whether $L$ is shifted or not. Assume first $L$ is not shifted.

- Assume first $d(u) \geq 3$. Since $L$ is not shifted, $l_1 = l_2 = l$, and $l_2 < l_3$ since $\mathrm{mult}(L) \leq 2$. Consider any $v_i$. If $d(v_i) \geq 2$, then $v_i$ is incident to at most one edge assigned a label with value $l$ while two edges have been assigned this label value, meaning $\sigma(v_i) + c(v_i) > \mathrm{small}(v_i) \geq \mathrm{small}(u) = \sigma(u) + c(u)$ (in particular, this inequality still holds if $v_i$ is not incident to an edge assigned a label with value $l$); thus, $u$ and $v_i$ cannot be in conflict. Now, if $d(v_i) = 1$, then, since $\mathrm{small}(v_i) \geq \mathrm{small}(u)$, so that we get $\sigma(v_i) + c(v_i) = \sigma(u) + c(u)$ it must be that $v_i$ is a smallest vertex and $uv_i$ was assigned label $l_1$ or $l_2$.
- Assume now $d(u) = 2$. Since $L$ is not shifted, then $l_1 = l_2 = l$ and label value $l$ was assigned to $uv_1$ and $uv_2$ only. Focus on any $v_i \in \{v_1, v_2\}$. If $d(v_i) \geq 2$, then only one edge incident to $v_i$ is assigned a label with value $l$, while $L$ contains two occurrences of it; thus, $\sigma(v_i) + c(v_i) > \mathrm{small}(v_i) \geq \mathrm{small}(u) = \sigma(u) = \sigma(u) + c(u)$, and $u$ and $v_i$ cannot be in conflict. Now, if $d(v_i) = 1$, then, since $\sigma(u) + c(u) = \mathrm{small}(u)$, so that $\sigma(v_i) + c(v_i) = \sigma(u) + c(u)$ it must be that $v_i$ is a smallest vertex.
- Assume last $d(u) = 1$. Since $u$ is a smallest vertex of maximum degree, then, since $\ell(uv_1) = l_1$, so that there is a conflict between $u$ and $v_1$ it must be that $v_1$ is a smallest vertex, and, thus, $d(v_1) = 1$.

Assume second that $L$ is shifted. Then $l_1 < l_2$, and, if $d \geq 3$, then $l_2 \leq l_3$ and $l_1 < l_3$ (as $\mathrm{mult}(L) \leq 2$).

- First assume $d(u) \geq 3$. Focus on any $v_i$. If $d(v_i) \geq 2$, then, again, since $L$ contains only one occurrence of label value $l_1$, either $v_i$ is not incident to the unique edge assigned label $l_1$ (if $\ell(uv_i) \neq l_1$) or $v_i$ is not incident to any edge assigned label $l_2$ or $l_3$ (if $\ell(uv_i) = l_1$). Thus, in both cases $\sigma(v_i) + c(v_i) > \mathrm{small}(v_i) \geq \mathrm{small}(u) = \sigma(u) + c(u)$. Now, if $d(v_i) = 1$, then, since $\sigma(u) + c(u) = \mathrm{small}(u)$, the only way to have a conflict between $u$ and $v_i$ is that we have $\mathrm{small}(u) = \mathrm{small}(v_i) = \sigma(v_i) + c(v_i)$, which requires $v_i$ be a smallest vertex and $uv_i$ be assigned label $l_1$.

- Assume second that $d(u) = 2$. We here know that, w.l.o.g., $\ell(uv_1) = l_1$ and $\ell(uv_2) = l_2$.
  - Regarding $v_1$, if $d(v_1) = 1$, then we have $\sigma(v_1) + c(v_1) = \text{small}(v_1)$, so we can only have a conflict with $u$ if $v_1$ is a smallest vertex. If $d(v_1) = 2$, then, since $\text{small}(v_1) \geq \text{small}(u) = \sigma(u) + c(u)$, so that we get a conflict between $u$ and $v_1$ it must be (apart from $v_1$ be a smallest vertex) that the second edge incident to $v_1$ is assigned a label with value $l_2$, which, since $\ell(uv_2) = l_2$, is only possible (since $\text{mult}(L) \leq 2$) if $l_2 = l_3$ and the second edge incident to $v_1$ is assigned label $l_3$. Now, if $d(v_1) \geq 3$, then, necessarily, either $l_2 \neq l_3$ and thus there is no edge incident to $v_1$ assigned a label with value $l_2$ (as only $uv_2$ verifies this), or $l_2 = l_3 = l$ and there is at most one edge incident to $v_1$ assigned a label with value $l$ (possibly its second incident edge) while two edges of $(G, c)$ are assigned a label with value $l$ (one of which is $uv_2$). Thus we have $\sigma(v_1) + c(v_1) > \text{small}(v_1) \geq \text{small}(u) = \sigma(u) + c(u)$.
  - Regarding $v_2$, we cannot have $\sigma(u) + c(u) = \sigma(v_2) + c(v_2)$ since $L$ contains label $l_1$, which unique label value was assigned to $uv_1$ only. So $u$ and $v_2$ cannot be in conflict here, regardless of $d(v_2)$.
- Assume third $d(u) = 1$. Recall that, in order to have a conflict between $u$ and $v_1$, since $u$ is a smallest vertex and $\ell(uv_1) = l_1$, it must be that $v_1$ is a smallest vertex; since $u$ is a smallest vertex of maximum degree, then in that case we deduce $d(v_1) = 1$.

This concludes the proof.      □

Last, we need a notion of niceness for weighted graphs. We say a weighted graph $(G, c)$ is *nice* if it has no connected component $uv$ isomorphic to $K_2$ with $c(u) = c(v)$. Note that, clearly, any weighted graph that is not nice admits no proper $L$-strict-labelling for any sequence $L$ of labels.

We are now ready to prove our main result, from which Theorem 1 follows as a corollary.

**Theorem 2.** *Every nice weighted graph $(G, c)$ admits a proper $L$-strict-labelling for every sequence $L$ of labels with $|L| \geq |E(G)| + 3$ and $\text{mult}(L) \leq 2$.*

*Proof.* The proof is by induction on the order $n$ of $G$. If $n = 1$, then there is nothing to prove. If $n = 2$, then either $G$ has no edge and again there is nothing to prove, or $G$ consists of a single edge $uv$ with $c(u) \neq c(v)$ (as otherwise $(G, c)$ would not be nice), in which case, through assigning any label $l \in L$ to $uv$ we get $\sigma(u) + c(u) = l + c(u) \neq l + c(v) = \sigma(v) + c(v)$, thus what is desired.

We now proceed with the general case. That is, we now suppose that the claim holds for all weighted graphs on up to $n - 1$ vertices, and we prove the claim for $(G, c)$. By the induction hypothesis, we can suppose $(G, c)$ is connected. Indeed, if $(G, c)$ is the disjoint union of two nice weighted graphs $(G_1, c)$ and $(G_2, c)$ (where, abusing the notation, $c$ is here restricted to $G_1$ and $G_2$), then, by induction, we can get a proper $L$-strict-labelling $\ell_1$ of $(G_1, c)$, thus assigning $|E(G_1)|$ labels from $L$; what remains of $L$ is then a sequence $L'$ of labels with $|L'| \geq |E(G)| + 3 - |E(G_1)| = |E(G_2)| + 3$ with $\text{mult}(L') \leq 2$, so, by induction, we can get a proper $L'$-strict-labelling $\ell_2$ of $(G_2, c)$, which, with $\ell_1$, forms what is desired for $(G, c)$.

From now on, we thus assume $(G, c)$ is connected. To lead the rest of the proof, we essentially look at the vertex degrees, and show we can employ induction properly in case certain degree configurations are present. As an illustration, we first deal with a specific case that will later simplify the proof a lot.

*Remark 1.* The result holds if $(G, c)$ has a vertex of degree 2 adjacent to a vertex of degree 1.

*Proof of the Remark.* Assume $v$ is a vertex of degree 2 of $G$, and let $u$ and $w$ be its two neighbours, where $d(u) = 1$. If $G$ is the path $uvwx$ of length 3 (that is, $d(w) = 2$ and $w$, besides $v$, is also adjacent to a vertex $x$ of degree 1), then $|E(G)| = 3$ and $|L| \geq |E(G)| + 3 = 6$, so, since $\text{mult}(L) \leq 2$, there are at least three pairwise distinct labels $l_{i_1}, l_{i_2}, l_{i_3}$ in $L$. We here assign one $l_i$ of these labels to $vw$ so that $c(v) + l_i \neq c(u)$ and $c(w) + l_i \neq c(x)$, which guarantees $u$ and $v$, and similarly $x$ and $w$, cannot eventually get in conflict regardless how we label $uv$ and $wx$. Next, we assign another $l_j$ of these labels to $uv$ so that $L - (l_i, l_j)$ contains at least two distinct labels. It then suffices to assign any remaining label $l_k$ to $wx$ so that $v$ and $w$ are distinguished to obtain a proper $L$-strict-labelling of $(G, c)$.

Under the assumption that $d(u) = 1$, $d(v) = 2$, and $G$ is not the path of length 3, note that $(G - v, c)$ must be nice. Also, we have $|E(G)| \geq 2$, and thus $|L| \geq |E(G)| + 3 = 5$, meaning that $L$ must contain three pairwise distinct labels $l_{i_1}, l_{i_2}, l_{i_3}$. Let $l_i$ be any of these three labels such that $c(u) \neq c(v) + l_i$. Set now $G' = G - v$, $L' = L - (l_i)$, and define $c' : V(G') \to \mathbb{R}$ as $c'(w) = c(w) + l_i$ and $c'(x) = c(x)$ for all $x \in V(G') \setminus \{w\}$. By the induction hypothesis, $(G', c')$ admits a proper $L'$-strict-labelling $\ell'$, since $\text{mult}(L') \leq 2$ and $|L'| = |L| - 1 \geq |E(G)| + 2 = |E(G')| + 4$. Now, in $(G, c)$, start from $\ell'$, assign label $l_i$ to $vw$, and last, since only $uv$ remains to be labelled, at least four labels of $L$ have not been assigned yet, at least two of which must have distinct values since $\text{mult}(L) \leq 2$, so we assign one $l_j$ to $uv$ so that, denoting $\ell$ the resulting labelling of $(G, c)$, we get $\sigma_\ell(v) + c(v) \neq \sigma_\ell(w) + c(w)$. Note that $w$ does not get involved into a conflict since we get $\sigma_\ell(w) + c(w) = \sigma_{\ell'}(w) + c'(w)$, $\sigma_\ell(w) = \sigma_{\ell'}(w) + l_i$, and $c'(w) = c(w) + l_i$. Likewise, by the choice of $l_i$, we cannot have a conflict between $u$ and $v$, and, by the choice of $l_j$, we cannot have one between $v$ and $w$. Meanwhile, all vertices of $G$ different from $u$, $v$, and $w$ have their sums not altered by these modifications. Thus, $\ell$ is a proper $L$-strict-labelling of $(G, c)$.                                  ◇

Back to the proof of Theorem 1, we focus on $u$, a smallest vertex of $(G, c)$. Among all possible choices as $u$, we choose one with maximum degree $d(u)$. We consider first when $d(u)$ is small, *i.e.*, at most 2. Recall that, throughout what follows, the elements $l_1, l_2, \ldots$ of $L$ are ordered increasingly, *i.e.*, $l_1 \leq l_2 \leq \ldots$.

- Assume first that $d(u) = 1$, and let $v$ denote the unique neighbour of $u$ in $G$. If $d(v) = 1$, then actually $|V(G)| = 2$, a case we have covered already. Thus, assume $d(v) \geq 2$, and even $d(v) \geq 3$ due to Remark 1. Then note that $G' = G - u$ is necessarily nice, and thus so is $(G', c')$ for any $c'$. Note also that since $u$ is a smallest vertex of $(G, c)$ with maximum degree, then $\text{small}(u) < \text{small}(v)$. Now set $L' = L - (l_1)$, and define $c' : V(G') \to \mathbb{R}$ as $c'(v) = c(v) + l_1$ and $c'(w) = c(w)$ for all $w \in V(G') \setminus \{v\}$. Since $|L'| = |L| - 1 \geq |E(G)| + 2 = |E(G')| + 3$, by the induction hypothesis $(G', c')$ admits a proper $L'$-strict-labelling $\ell'$. Now let $\ell$ be the $L$-strict-labelling of $(G, c)$ obtained from $\ell'$ by assigning label $l_1$ to $uv$. As a result, note that $c'(v) + \sigma_{\ell'}(v) = c(v) + \sigma_\ell(v)$ since $\sigma_\ell(v) = \sigma_{\ell'}(v) + l_1$. Also, since we assigned label $l_1$ to $uv$, we get $c(u) + \sigma_\ell(u) = \text{small}(u) < \text{small}(v) = c(v) + \sigma_\ell(v)$, while assigning label $l_1$ to $uv$ changed the sums of $u$ and $v$ only. Thus, $\ell$ is a proper $L$-strict-labelling of $(G, c)$.

- Assume now that $d(u) = 2$, and let $v$ and $w$ be the two neighbours of $u$ in $G$. By Remark 1, we can assume $d(v), d(w) \geq 2$. Since neither $v$ nor $w$ can be a vertex of degree 2 adjacent to a vertex of degree 1 by Remark 1, if $G - u$ is not nice then it must be that $vw$ is an edge, and actually that $(G, c)$ is a weighted triangle. In this case, $|E(G)| = 3$ and $|L| \geq |E(G)| + 3 = 6$, so $L$ contains three pairwise distinct labels $l_{i_1}, l_{i_2}, l_{i_3}$ since $\text{mult}(L) \leq 2$. Note that, w.l.o.g., we can also assume $c(u) \leq c(v) \leq c(w)$. In this case, assuming $l_{i_1} < l_{i_2} < l_{i_3}$, we assign label $l_{i_1}$ to $uv$, label $l_{i_2}$ to $uw$, and label $l_{i_3}$ to $vw$. This guarantees $c(u) + l_{i_1} + l_{i_2} < c(v) + l_{i_1} + l_{i_3} < c(w) + l_{i_2} + l_{i_3}$. Thus, the sums of $u$, $v$, and $w$ are distinct, and we get a proper $L$-strict-labelling of $(G, c)$.

So, now, we can assume $G - u$ is nice. We consider two cases, depending on whether $L$ is shifted.

- If $L$ is not shifted, then let $l = l_1 = l_2$ be the value of the two smallest labels in $L$ ($l_1$ and $l_2$), which are thus equal. We here set $G' = G - u$, $L' = L - (l_1, l_2)$, and define $c' : V(G') \to \mathbb{R}$ as $c'(v) = c(v) + l$, $c'(w) = c(w) + l$, and $c'(x) = c(x)$ for all $x \in V(G') \setminus \{v, w\}$. Since $G'$ is nice, so is $(G', c')$, so by the induction hypothesis there is a proper $L'$-strict-labelling $\ell'$ of $(G', c')$. We extend $\ell'$ to some labelling $\ell$ of $(G, c)$ by setting $\ell(uv) = l_1$ and $\ell(uw) = l_2$. As a result, we get $\sigma_\ell(u) + c(u) = \text{small}(u)$. Meanwhile, note that both $v$ and $w$ are incident to only one edge assigned a label with value $l$, while this label value was assigned twice, and, recall, we have $d(v), d(w) \geq 2$. Thus, $\sigma_\ell(v) + c(v) > \text{small}(v)$ and $\sigma_\ell(w) + c(w) > \text{small}(w)$, while by our choice of $u$ we have $\text{small}(v), \text{small}(w) \geq \text{small}(u)$. Thus $u$ cannot be involved in conflicts (recall Lemma 1), while for all other vertices $x$ (in $V(G') \setminus \{v, w\}$) we have $\sigma_\ell(x) + c(x) = \sigma_{\ell'}(x) + c'(x)$, implying $\ell$ is thus a desired proper $L$-strict-labelling of $(G, c)$.

- If $L$ is shifted, then $l_1$ and $l_2$ are two smallest label values of $L$, where $l_1 < l_2$ (that is, there is only one occurrence of label value $l_1$ in $L$). Essentially we would here like to proceed just as in the previous case, considering $G' = G - u$ and $L' = L - (l_1, l_2)$, and defining $c' : V(G') \to \mathbb{R}$ as earlier, assuming we would then, say, assign label $l_1$ to $uv$ and label $l_2$ to $uw$. We need to be careful however, as in some cases such arguments do not apply; precisely, in some cases we might end up, for the resulting $\ell$ in $(G, c)$, with $\sigma_\ell(v) + c(v) = \text{small}(v) = \sigma_\ell(u) + c(u)$. It can be checked, however, that these arguments apply right away when $d(v), d(w) \geq 3$. Likewise, they apply as is when $d(v) \geq 3$ and $d(w) = 2$, and, free to rename $v$ and $w$, when $d(v) = 2$ and $d(w) \geq 3$. It can be checked also that we cannot have any conflict between $u$ and $v$ when $d(v) = 2$ and $\text{small}(v) > \text{small}(u)$, and, again free to rename $v$ and $w$, when $d(w) = 2$ and $\text{small}(w) > \text{small}(u)$. This follows from arguments alike those used to prove Lemma 1: as long as we assign labels $l_1$ and $l_2$ to the edges incident to $u$, any conflict involving $u$ must be with a vertex whose incident edges have been assigned the smallest label values (recall we might have $l_2 = l_3$ since $\text{mult}(L) \leq 2$, in which case $l_3$ is also one of the two smallest label values).

So, the very last case to consider is when $d(v) = d(w) = 2$, $\text{small}(v) = \text{small}(w) = \text{small}(u)$, and thus $c(v) = c(w) = c(u)$. If considering $v$ instead of $u$ does not lead to a more favourable situation, then it means that $v$ as well is adjacent to two vertices of degree 2 being smallest vertices (one of which is $u$, and the other one could be $w$). Of course, the same goes for $w$. Through repeatedly considering adjacent vertices of degree 2 of $(G, c)$ this way, either at some point we reach a smallest vertex of degree 2 whose neighbours allow for previous arguments to apply, or we determine that $(G, c)$ is actually a weighted cycle with all vertices being of smallest possible sum $\text{small}(u)$. In this case, we can obtain a proper $L$-strict-labelling of $(G, c)$ e.g. in the following way. Set $G = v_0 \ldots v_{n-1} v_0$. For any two adjacent vertices $v_i$ and $v_{i+1}$ (where indexes are modulo $n$ throughout), since $c(v_i) = c(v_{i+1})$, through labelling $(G, c)$ by a labelling $\ell$, to guarantee that $\sigma_\ell(v_i) + c(v_i) \neq \sigma_\ell(v_{i+1}) + c(v_{i+1})$ it suffices to guarantee that $\sigma_\ell(v_i) \neq \sigma_\ell(v_{i+1})$; and for that we need only to have $\ell(v_{i-1} v_i) \neq \ell(v_{i+1} v_{i+2})$. If $G$ is a cycle of length at least 4, then we obtain a proper $L$-strict-labelling of $(G, c)$ when considering the edges following the ordering $(v_0 v_1, v_1 v_2, \ldots, v_{n-1} v_0)$ and assigning to them labels of $L$ in increasing order. Indeed, this guarantees every two edges at distance 2 get assigned distinct labels, since $\text{mult}(L) \leq 2$. Now if $G$ is a triangle $uvwu$, then we can reuse arguments we introduced at the very beginning of the current case (to deal with cases where $G - u$ is not nice).

So, we can now assume that $d(u) = d \geq 3$. Set $G' = G - u$; note that if some connected component of $G'$ is just an edge $vw$, then, due to Remark 1, we deduce that $uv$ and $uw$ must be edges of $G$ as well. So it might be that $G'$ is not nice, but this must be caused by triangles attached at $u$ in $G$.

Recall that $l_1, \ldots, l_d$ denote the smallest $d$ labels of $L$, where $l_1 \leq \cdots \leq l_d$. To build a proper $L$-strict-labelling $\ell$ of $(G, c)$, our goal now is, assuming $(G', c')$ is nice for some $c'$, to manage to invoke the induction hypothesis just like we did before. More precisely, in general, this will be achieved as follows. Let $uv_1, \ldots, uv_d$ denote the $d$ edges incident to $u$, and let $\phi$ be a permutation of $\{1, \ldots, d\}$. An *attempt* (w.r.t. $\phi$) will consist in considering $L' = L - (l_1, \ldots, l_d)$ and $(G', c')$ the weighted graph where, recall, $G' = G - u$, and we have $c' : V(G') \to \mathbb{R}$ where $c'(v_i) = c(v_i) + l_{\phi(i)}$ for every $i \in \{1, \ldots, d\}$ and $c'(w) = c(w)$ for all $w \in V(G') \setminus \{v_1, \ldots, v_d\}$. Assuming $(G', c')$ is nice, we then deduce a proper $L$-strict-labelling $\ell$ of $(G, c)$ from a proper $L'$-strict-labelling $\ell'$ of $(G', c')$ (obtained by induction), through simply assigning label $l_{\phi(i)}$ to every edge $uv_i$. In many cases, there is actually a $\phi$ guaranteeing a successful attempt (*i.e.*, such that the eventual $\ell$ is proper), see below. However, there are two reasons why an attempt might fail.

– Nothing guarantees $(G', c')$ is nice, implying the induction hypothesis cannot be invoked. However, as pointed out above, this can only occur if $G' = G - u$ contains an isolated edge $v_i v_j$ (that is, $u$ is adjacent, in $G$, to two adjacent vertices of degree 2) such that $c'(v_i) = c'(v_j)$.
– To guarantee $u$ cannot eventually get involved in conflicts by $\ell$, then, by Lemma 1, since, through an attempt, we are assigning the smallest $d$ labels of $L$ to the edges incident to $u$, and $d(u) \geq 3$, we must make sure $v_{\phi(1)}$ is not a vertex of degree 1 being a smallest vertex of $(G, c)$. Likewise, if $L$ is not shifted, then by definition $l_2$ is also the smallest label value in $L$ (that is, $l_1 = l_2$), and so, here as well, we must make sure $v_{\phi(2)}$ is not a vertex of degree 1 being a smallest vertex of $(G, c)$.

In particular, if we rename $v_1, \ldots, v_d$ as $w_1, \ldots, w_a$ for some $a \geq 0$, $x_1, y_1, \ldots, x_b, y_b$ for some $b \geq 0$, $x'_1, y'_1, \ldots, x'_{b'}, y'_{b'}$ for some $b' \geq 0$, and $z_1, \ldots, z_c$ for some $c \geq 0$, where $a + 2b + 2b' + c = d$ and

– every $w_i$ verifies $d(w_i) = 1$ and $\mathrm{small}(w_i) = \mathrm{small}(u)$;
– every pair $\{x_i, y_i\}$ forms an isolated edge $x_i y_i$ in $G'$ and $c(x_i) < c(y_i)$;
– every pair $\{x'_i, y'_i\}$ forms an isolated edge $x'_i y'_i$ in $G'$ and $c(x'_i) = c(y'_i)$;
– every $z_i$ meets none of the previous conditions[3],

then a permutation leading to a successful attempt can be obtained in most situations. In most cases, this will be done through considering another permutation $\phi'$ obtained from some original $\phi$ by *swapping* two values $\phi(i)$ and $\phi(j)$ by $\phi$ to get a different permutation $\phi'$, which means that $\phi'(i) = \phi(j)$, $\phi'(j) = \phi(i)$, and $\phi'(k) = \phi(k)$ for all $k \in \{1, \ldots, d\} \smallsetminus \{i, j\}$, and repeating this operation as long as necessary. Throughout what follows, when dealing with any of the next configurations, we implicitly assume that none of the previous ones applies.

– $a = 0$. In this case, so that an attempt through any initial permutation $\phi$ is successful, we just need to guarantee, for some distinct $\alpha, \beta \in \{1, \ldots, d\}$, that there is no $i \in \{1, \ldots, b\}$ (if $b \geq 1$) such that $(x_i, y_i) = (v_\alpha, v_\beta)$ and $c'(x_i) = c'(y_i)$, and no $i \in \{1, \ldots, b'\}$ (if $b \geq 1$) such that $(x'_i, y'_i) = (v_\alpha, v_\beta)$ and $l_{\phi(\alpha)} = l_{\phi(\beta)}$. Note that the former configuration, since we assumed $c(v_\alpha) = c(x_i) < c(y_i) = c(v_\beta)$, is avoided through simply having $l_{\phi(\alpha)} \leq l_{\phi(\beta)}$; so, in case, through $\phi$, we have $c'(v_\alpha) = c'(v_\beta)$, then we can get rid of this by simply considering the permutation $\phi'$ obtained from $\phi$ by swapping $\phi(\alpha)$ and $\phi(\beta)$.
  • Assume first $b' = 0$. If we are not done with $\phi$, then there exists a pair of vertices $\{x_i, y_i\}$ such that, for some distinct $\alpha, \beta \in \{1, \ldots, d\}$, we have $(x_i, y_i) = (v_\alpha, v_\beta)$ and $\phi(\alpha) > \phi(\beta)$. As explained earlier, this issue is no

---

[3] Note that each $z_i$ is either a vertex of degree at least 3, a vertex of degree 2 not part of a triangle attached at $u$ (thus not part of any pair $\{x_i, y_i\}$ or $\{x'_i, y'_i\}$), or a non-smallest vertex of degree 1.

longer present through the permutation $\phi'$ obtained from $\phi$ by swapping $\phi(\alpha)$ and $\phi(\beta)$. After repeating this for all such conflicting pairs $\{x_i, y_i\}$, eventually we get a permutation such that the associated $L$-strict-labelling of $(G, c)$ is proper.

- Assume now $b = 0$. Again, if we are not done with $\phi$, then $b' > 0$ (as otherwise the previous case would apply), and there is a pair of vertices $\{x_i', y_i'\}$ such that for some distinct $\alpha, \beta \in \{1, \dots, d\}$, we have $(x_i', y_i') = (v_\alpha, v_\beta)$ and $\phi(\alpha) = \phi(\beta)$. Since $\text{mult}(L) \le 2$ and $d(u) \ge 3$, there exists a vertex $v_\gamma \not\in \{v_\alpha, v_\beta\}$ such that $\phi(\alpha) \ne \phi(\gamma)$. If $v_\gamma$ does not belong to another pair $\{x_j', y_j'\}$ (where $i \ne j$), then we can freely swap $\phi(\alpha)$ and $\phi(\gamma)$ so that $\{x_i', y_i'\}$ is no longer a conflicting pair (while new ones are not created) by the resulting permutation. Otherwise, if $v_\gamma$ belongs to some pair $\{x_j', y_j'\}$ with $(x_j', y_j') = (v_{\alpha'}, v_{\beta'})$ for some $1 \le j \le b'$ and distinct $\alpha', \beta' \in \{1, \dots, d\}$ with $\{\alpha, \beta\} \cap \{\alpha', \beta'\} = \emptyset$, then we consider the permutation $\phi'$ obtained by swapping $\phi(\alpha)$ and $\phi(\alpha')$. Note that this cannot yield a $\phi'$ where $\phi'(\alpha) = \phi'(\beta)$ or $\phi'(\alpha') = \phi'(\beta')$ since $\text{mult}(L) \le 2$ and $\phi(\alpha) = \phi(\beta)$. By then repeating these arguments for every conflicting pair $\{x_k', y_k'\}$, eventually we end up with a desired permutation.
- Assume last that $b, b' \ge 1$. We start, if needed, by applying the exact same arguments as in the previous case to obtain, from $\phi$, some permutation $\phi'$ so that for every pair $\{x_i', y_i'\}$ such that $(x_i', y_i') = (v_\alpha, v_\beta)$ for some distinct $\alpha, \beta \in \{1, \dots, d\}$, we have $\phi'(\alpha) \ne \phi'(\beta)$. We then apply, if needed, the exact same arguments (on conflicting pairs $\{x_i, y_i\}$, if any) as in the first case above to reach a permutation such that the associated $L$-strict-labelling of $(G, c)$ is proper.

- $b \ge 1$.

  In this case, assuming w.l.o.g. that $(v_1, v_2) = (x_1, y_1)$, we consider any permutation $\phi$ where $\phi(1) = 1$ and $\phi(2) = 2$. This way, note that there is no $v_i$ of degree 1 such that $uv_i$ is assigned label $l_1$ or $l_2$, the smallest two label values of $L$. By remarks above, $u$ cannot be involved in conflicts as long as we do not swap $\phi(1)$ nor $\phi(2)$ with a $\phi(\gamma)$ such that $\gamma \in \{1, \dots, d\}$ and $v_\gamma$ is a $w_i$. First off, in case there is a conflicting pair $(x_i', y_i') = (v_\alpha, v_\beta)$, then we get rid of it by swapping $\phi(1)$ and $\phi(\alpha)$. Then, in case there are more conflicting pairs of the form $\{x_j', y_j'\}$, then we get rid of them by swapping elements as in an earlier case. So we can assume we get to some point where, by $\phi$, there are no conflicting pairs of the form $\{x_i', y_i'\}$, and there is no $v_i$ of degree 1 such that $uv_i$ is assigned label $l_1$ or $l_2$. From here, for every conflicting pair $(x_i, y_i) = (v_\alpha, v_\beta)$, just as in a previous case we can get rid of it by just swapping the labels assigned to $uv_\alpha$ and $uv_\beta$, to eventually reach a permutation such that the associated $L$-strict-labelling of $(G, c)$ is proper. In particular, note that, through swapping labels, labels $l_1$ and $l_2$ remain assigned to edges $uv_i$ with $v_i$ being an $x_i$, a $y_i$, an $x_i'$, or a $y_i'$.

- $b' \ge 1$ and $c \ge 1$.

  Here, assume w.l.o.g. that $z_1 = v_1$, $x_1' = v_2$, and $y_1' = v_3$.

  - If $L$ is shifted, then recall that $l_1 < l_2$, so $L$ contains only one label, $l_1$, with smallest value. We here consider any permutation $\phi$ where $\phi(2) = 1$.

This guarantees $u$ cannot be involved in conflicts with $v_i$'s of degree 1. Likewise, this guarantees we cannot get any conflict between $v_2$ and $v_3$. Now, if some conflicts remain, then they must involve vertices in some pair $\{x_i', y_i'\} \neq \{x_1', y_1'\}$ (recall we can assume $b = 0$ since the previous case does not apply), in which case, since $\{x_1', y_1'\}$ is a pair whose vertices are not in conflict, we can again repeatedly modify $\phi$ by swapping elements to reach another permutation $\phi'$ whose associated $L$-strict-labelling of $(G, c)$ is proper.

- If $L$ is not shifted, then recall that $l_1 = l_2$. Here, we consider any permutation $\phi$ where $\phi(1) = 1$ and $\phi(2) = 2$. Note that this guarantees that $v_2$ and $v_3$ cannot be in conflict. To attain a permutation $\phi'$ from $\phi$ such that the associated $L$-strict-labelling of $(G, c)$ is proper, we can then, if necessary, swap elements as in the last case. In particular, since labels $l_1$ and $l_2$ get assigned to edges not incident to smallest vertices of degree 1, eventually vertex $u$ cannot be involved in any conflict.

In both cases, we are thus done as well.

- $c \geq 2$.

Assuming w.l.o.g. that $z_1 = v_1$ and $z_2 = v_2$, we are here done through considering any permutation $\phi$ with $\phi(1) = 1$ and $\phi(2) = 2$. This guarantees the smallest two labels of $L$ get assigned to edges going to $z_i$'s, which, recall, cannot be in conflict with $u$ by associated $L$-strict-labellings of $(G, c)$ assigning the smallest $d$ labels to edges incident to $u$. Again, this also guarantees $u$ cannot be involved in conflicts at all. Recall also that, since none of the previous cases applies, we may assume $b = b' = 0$, so here we do not have to consider possible conflicting pairs $\{x_i, y_i\}$ and $\{x_i', y_i'\}$.

- $b' \geq 2$.

Since none of the previous cases applies, we have $b = c = 0$. In this case, we can essentially be done just as in the previous one, assuming $v_1 = x_1'$, $v_2 = y_1'$, $v_3 = x_2'$, and $v_4 = y_2'$, by considering any initial permutation $\phi$ with $\phi(1) = 1$, $\phi(2) = 3$, $\phi(3) = 2$, and $\phi(4) = 4$. This guarantees $x_1'$ and $y_1'$ cannot be in conflict, and similarly for $x_2'$ and $y_2'$ (in particular, since $\text{mult}(L) \leq 2$, we have $l_1 < l_3$ and $l_2 < l_4$). All conflicts, if any, must now involve pairs $\{x_i', y_i'\}$ with $i \geq 3$, and we can again get rid of any such conflict by swapping the labels assigned to $ux_i'$ and $uy_i'$.

Thus, since $d(u) \geq 3$, the last situations we have to consider are when

- $a \geq 3$, and $b = b' = c = 0$;
- $a \geq 1$, $b' = 1$, and $b = c = 0$;
- $a \geq 2$, $b = b' = 0$, and $c = 1$.

In the first and second cases, note that the whole structure of $(G, c)$ is discovered (in that all edges are incident either to $u$ or to neighbours of $u$ only), so, assuming we label all edges of $(G, c)$ at once (i.e., induction is not invoked), it is no longer necessary, by a proper $L$-strict-labelling of $(G, c)$, to guarantee that $u$ is of smallest possible sum, and, thus, that the smallest $d$ labels of $L$ are assigned to the edges incident to $u$.

To conclude, we consider each of the three possible remaining cases separately.

- $a \geq 3$, and $b = b' = c = 0$.

  Start from the $L$-strict-labelling $\ell$ of $(G, c)$ where $\ell(uw_i) = l_i$ for every $i \in \{1, \ldots, d\}$ (recall $d = a$ here). Since $w_1$ is a smallest vertex and $\ell(uw_1) = l_1$, note that we have $\sigma_\ell(u) + c(u) = l_1 + \cdots + l_d + c(u) = x = l_1 + c(w_1)$, and thus $c(w_1) = x - l_1$. Likewise, if, for every $i \in \{2, \ldots, d\}$, we consider the labelling obtained from $\ell$ by swapping $l_1$ and $l_i$, then we deduce that $c(w_i) = x - l_1$. Since $|L| \geq |E(G)| + 3$, the three largest label values $l_{d+1}, l_{d+2}, l_{d+3}$ of $L$ have not been assigned, where $l_{d+1} \leq l_{d+2} \leq l_{d+3}$ and thus $l_{d+1} < l_{d+3}$ since $\mathrm{mult}(L) \leq 2$, and $l_2 < l_{d+2}$ since $d \geq 3$. Now consider the $L$-strict-labelling $\ell'$ of $(G, c)$ obtained from the initial $\ell$ by assigning label $l_{d+2}$ to $uw_1$ and label $l_{d+3}$ to $uw_2$. As a result, note that we now have $\sigma_{\ell'}(u) + c(u) = x + l_{d+2} + l_{d+3} - l_1 - l_2$. Meanwhile, $\sigma_{\ell'}(w_1) + c(w_1) = l_{d+2} + c(w_1) = x + l_{d+2} - l_1$; since $l_{d+3} - l_2 > 1$, we thus have that $u$ and $w_1$ are not in conflict. Likewise, $\sigma_{\ell'}(w_2) + c(w_2) = l_{d+3} + c(w_2) = x + l_{d+3} - l_2$; since $l_{d+2} - l_1 > 1$, again $u$ and $w_2$ cannot be in conflict. Now, for every $i \in \{3, \ldots, d\}$, we have $\sigma_{\ell'}(w_i) + c(w_i) = \sigma_\ell(w_i) + c(w_i) \leq l_d + c(w_i) = x + l_d - l_1$; since $l_{d+2} - l_1 > l_d - l_1$, and $l_{d+3} - l_2 > 0$, we have that $u$ and $w_i$ cannot be in conflict. Thus, $\ell'$ is proper.

- $a \geq 1$, $b' = 1$, and $b = c = 0$.

  Recall that, here, $u$ is adjacent to $a$ vertices $w_1, \ldots, w_a$ of degree 1 where $\mathrm{small}(u) = \mathrm{small}(w_i)$ for every $i \in \{1, \ldots, a\}$, and two vertices $x'_1$ and $y'_1$ such that $x'_1 y'_1$ is an edge and $c(x'_1) = c(y'_1)$. Also, by the choice of $u$ we have $\mathrm{small}(u) \leq \mathrm{small}(x'_1), \mathrm{small}(y'_1)$.

  - If $L$ is shifted, then $l_1 < l_2$, so we here assign labels $l_1, \ldots, l_d$ to the edges incident to $u$ so that label $l_1$ is assigned to $ux'_1$ and label $l_2$ is assigned to $uy'_1$ (and the other labels are assigned arbitrarily), thus so that $ux'_1$ and $uy'_1$ are assigned distinct labels and the smallest label value of $L$ is not assigned to any edge $uw_i$; by then assigning any remaining label to $x'_1 y'_1$ we obtain a proper $L$-strict-labelling of $(G, c)$. In particular, $u$ can be in conflict with neither the $w_i$'s nor with $x'_1$ and $y'_1$ by Lemma 1 (recall $d(u) \geq 3$), and similarly for $x'_1$ and $y'_1$ since $c(x'_1) = c(y'_1)$ and we assigned distinct labels to $ux'_1$ and $uy'_1$ (while $x'_1$ and $y'_1$ are adjacent vertices of degree 2).

  - Thus consider now when $L$ is not shifted, i.e., $l_1 = l_2$, and further assume that $a \geq 2$. We start similarly as in the previous case, assigning label $l_1$ to $ux'_1$ and label $l_3$ to $uy'_1$, where $l_1 < l_3$ since $\mathrm{mult}(L) \leq 2$. We then assign the remaining labels $l_2, l_4, \ldots, l_d$ to the $uw_i$'s arbitrarily, and call $\ell$ the resulting labelling. Then, as in a previous case, assuming $l_2$ was assigned to some $uw_i$, we have $\sigma_\ell(u) + c(u) = l_1 + \cdots + l_d + c(u) = x = l_2 + c(w_i)$, and thus $c(w_i) = x - l_2$, where recall $l_2 = l_1$. Again, free to swapping label $l_2$ (assigned to $uw_i$) and the label assigned to any other edge $uw_j$ (with $j \neq i$), we can further assume $c(w_i) = x - l_2$ for all $i \in \{1, \ldots, a\}$. Now, similarly as earlier, if $a \geq 2$, then we can replace labels $l_2$ and $l_4$ with labels $l_{d+3}$ and $l_{d+4}$ (since $x'_1 y'_1$ is not incident to $u$, note that $|L| \geq d + 4$; then, $l_3 < l_{d+3}$) to make sure $u$ is not in conflict with the $w_i$'s. It then remains to label $x_1' y_1'$. Recall that, since we assigned distinct labels ($l_1$ and $l_3$) to $ux_1'$ and $uy_1'$ and $c(x_1') = c(y_1')$, eventually we cannot get a conflict between $x_1'$ and $y_1'$ whatever label we assign to $x_1' y_1'$. So it all falls down to assigning a label to $x_1' y_1'$ so that $u$ is in conflict with neither $x_1'$ nor $y_1'$. Note that there are at least four labels of $L$ that have not assigned, namely $l_2$, $l_4$, $l_{d+1}$, and $l_{d+2}$. Since $\mathrm{mult}(L) \le 2$, we have $l_2 < l_4 < l_{d+2}$. So there must be a label in $\{l_2, l_4, l_{d+2}\}$ we can assign to $x_1' y_1'$ so that $u$ is in conflict with neither $x_1'$ nor $y_1'$. Eventually, this process thus results in a proper $L$-strict-labelling of $(G, c)$.

- A very last case to consider is when $a=1$ and $L$ is not shifted (as, in the last case we just considered, it was necessary that $a \ge 2$). In that case, we start by setting $\ell(uw_1) = l_1$, $\ell(ux_1') = l_2$, and $\ell(uy_1') = l_3$, where, recall, $l_1 = l_2$ and $l_2 < l_3$. Since for now we necessarily have a conflict between $u$ and $w_1$ (since $w_1$ is a smallest vertex of degree 1), we deduce $c(w_1) + l_1 = c(u) + l_1 + l_2 + l_3$. We here change the label of $uy_1'$ to $l_6$ (since $|E(G)| = 4$, recall $|L| \ge 7$), which gets rid of the conflict between $u$ and $w_1$ since $l_3 < l_6$. Also, $l_2 < l_6$ so we cannot get a conflict between $x_1'$ and $y_1'$ whatever label we assign to $x_1' y_1'$. So it remains to label $x_1' y_1'$ so that $u$ gets in conflict with neither $x_1'$ nor $y_1'$, and for that any label in $\{l_3, l_4, l_5, l_7\}$ is available. Since $\mathrm{mult}(L) \le 2$ we deduce that $l_3, l_5, l_7$ are pairwise distinct. So one of these labels can be assigned to $x_1' y_1'$ to get a proper $L$-strict-labelling of $(G, c)$.

$- a \ge 2$, $b = b' = 0$, and $c = 1$.

Recall that $u$ is here adjacent to $w_1, \dots, w_a$ and $z_1$, where the $w_i$'s are smallest vertices of degree 1, and $z_1$ is neither a smallest vertex of degree 1 nor part of a triangle attached at $u$. Thus, contrarily to the previous two cases, since $c=1$ recall that the structure of $G$ is not quite revealed at this point (unless $z_1$ is a non-smallest vertex of degree 1, in which case the upcoming arguments still apply).

- If $L$ is shifted, then recall $l_1 < l_2$. We here consider $(G', c')$ and $L'$, where $G' = G - \{u, w_1, \dots, w_a\}$ (which is nice; possibly $G'$ is empty if $z_1$ has degree 1 in $G$) and $L' = L - (l_1, \dots, l_d)$, and $c'$ is defined as $c'(z_1) = c(z_1) + l_1$ and $c'(v) = c(v)$ for all $v \in V(G') \setminus \{z_1\}$. By the induction hypothesis, $(G', c')$ admits a proper $L'$-strict-labelling $\ell'$ (possibly a trivial one, if $G'$ is empty). To extend it to a proper $L$-strict-labelling $\ell$ of $(G, c)$, we first set $\ell(uz_1) = l_1$, so that $\sigma_\ell(z_1) + c(z_1) = \sigma_{\ell'}(z_1) + c'(z_1)$, guaranteeing $z_1$, vertex $u$ apart, cannot be involved in conflicts. It then suffices to assign, by $\ell$, labels $l_2, \dots, l_d$ to the other edges incident to $u$ arbitrarily. This way we get $\sigma_\ell(u) = c(u) = \mathrm{small}(u)$, so, by Lemma 1, since $l_1 < l_2$, we know $u$ cannot be involved in conflicts. Thus, $\ell$ is proper.

- Last, if $L$ is not shifted, i.e., $l_1 = l_2$, then we proceed as follows. We start off similarly as in the previous case until we get $\ell$. This time, however, we know there is a conflict between $u$ and the unique $w_i$ such that $\ell(uw_i) = l_2$. As in previous cases, this implies that if we set $x = l_1 + \dots + l_d + c(u)$, then $c(w_i) = x - l_2$, and through swapping the label assigned to $uw_i$ and the label assigned to any $uw_j$ with $j \ne i$, we also deduce that, actually, $c(w_j) = x - l_2$

for all $j \in \{1, \ldots, a\}$. Now, assuming w.l.o.g. that $\ell(uw_1) = l_2$ and $\ell(uw_2) = l_3$ (recall that $w_2$ exists since $a \geq 2$), we consider the $L$-strict-labelling $\ell'$ of $(G, c)$ obtained from $\ell$ by assigning label $l_{d+2}$ to $uw_1$ and label $l_{d+3}$ to $uw_2$. By previous arguments, since $c(w_i) = x - l_2$ for all $i \in \{1, \ldots, a\}$, we know $u$ cannot be in conflict with none of the $w_i$'s (in particular, $l_3 < l_{d+2}$ since $\text{mult}(L) \leq 2$). If $u$ is not in conflict with $z_1$, then we are done. Otherwise, since $\text{mult}(L) \leq 2$, observe that it must be that $l_{d+1} + l_{d+2} \neq l_{d+2} + l_{d+3}$ (as equality would imply $l_{d+1} = l_{d+3}$); so by changing the labels assigned to $uw_1$ and $uw_2$ to $l_{d+1}$ and $l_{d+2}$, respectively, we can alter the sum of $u$ to guarantee $z_1$ cannot be involved in conflicts, and, again, since we altered both $\ell(uw_1)$ and $\ell(uw_2)$ (since, because $d \geq 3$, $l_2 < l_{d+1}$ and $l_3 < l_{d+2}$), also $u$ cannot be involved in conflicts with the $w_i$'s. So we again end up with a desired proper $L$-strict-labelling of $(G, c)$.

This concludes the whole proof.                                      □
As mentioned earlier, Theorem 1 now follows directly from Theorem 2.

## 3   Conclusion

In this work, through Theorem 1 we essentially improved the best known general upper bound on $\overline{\chi_\Sigma}$ by about a factor 2. To achieve this, we mainly considered proper labellings assigning any label value at most twice, and proved that every nice graph can actually be labelled this way.

We note that improving Theorem 2 by a bit, that is, even only to sequences of labels of size at least $|E(G)| + 1$, is not quite clear, as, analysing the cases we considered in our proof, one can come up with examples of weighted graphs $(G, c)$ that cannot be labelled as desired. An obvious example is *e.g.* when $(G, c)$ is a weighted star with center $u$ and an odd number $d \geq 3$ of leaves $v_1, \ldots, v_d$, $L$ is a sequence of $|E(G)| + 1$ labels with $\text{mult}(L) \leq 2$ and all label values of $L$ appear exactly twice in $L$, and all vertices are smallest vertices. Indeed, as seen in the proof of Theorem 2, whatever $|L| - 1$ labels we assign to $uv_1, \ldots, uv_d$ it is not possible to have $u$ being not involved in one conflict. It might be, however, that if we exclude such bad weighted stars, then we can improve Theorem 2 further down.

Generalising the approach we considered, one could also legitimately wonder about the more general question of labelling graphs in a proper way but so that every assigned label value is assigned to at most $k$ edges, for some fixed $k \geq 1$. For $k = 1$, this is exactly what was considered in [3], and the best result we could hope for was proved in [4,7], being that for any sequence of labels $L$ with $\text{mult}(L) \leq 1$ containing at least $|E(G)|$ pairwise distinct labels there is a proper $L$-strict-labelling of every nice graph $G$. What we investigated through Theorem 2 is essentially the case $k = 2$ of these considerations; although our result, being that for any sequence of labels $L$ with $\text{mult}(L) \leq 2$ containing at least $\left\lfloor \frac{|E(G)|}{2} \right\rfloor + 2$ pairwise distinct labels there is a proper $L$-strict-labelling of every nice graph $G$, can maybe be improved, as mentioned above some pathological cases might arise when restricting $L$ even by a bit. More generally speaking, one could wonder how these considerations behave as $k$ grows larger.

Note that considering larger values of $k$ this way would not bring anything new regarding Conjecture 2 and Theorem 1, as, if we are allowed to assign any label value to at least three edges, then, through a labelling $\ell$, we might end up with $\mathrm{nb}(l, \ell) = 1$ and $\mathrm{nb}(l', \ell) = 3$ for two label values $l$ and $l'$, which would thus not be considered equitable. Thus, towards Conjecture 2, and towards results better than Theorem 1, one has to consider other approaches. From a more general perspective, we are still far from a constant upper bound on the parameter $\overline{\chi_\Sigma}$, and it is probable that new ideas are needed.

# References

1. Baudon, O., Pilśniak, M., Przybyło, J., Senhaji, M., Sopena, É., Woźniak, M.: Equitable neighbour-sum-distinguishing edge and total colourings. Disc. Appl. Math. **222**, 40–53 (2017)
2. Bensmail, J., Fioravantes, F., Mc Inerney, F., Nisse, N.: Further results on an equitable 1-2-3 conjecture. Disc. Appl. Math. **297**, 1–20 (2021)
3. Bensmail, J., Senhaji, M., Szabo Lyngsie, K.: On a combination of the 1-2-3 conjecture and the antimagic labelling conjecture. Disc. Math. Theor. Comput. Sci. **19**(1) (2017)
4. Haslegrave, J.: Proof of a local antimagic conjecture. Disc. Math. Theor. Comput. Sci. **20**(1) (2018)
5. Karoński, M., Łuczak, T., Thomason, A.: Edge weights and vertex colours. J. Comb. Theory Ser. B **91**, 151–157 (2004)
6. Keusch, R.: A solution to the 1-2-3 conjecture. J. Comb. Theory Ser. B **166**, 183–202 (2024)
7. Lyngsie, K.S., Zhong, L.: A generalized version of a local antimagic labelling conjecture. Graphs Combinat. **34**, 1363–1369 (2018)
8. Hartsfield, N., Ringel, G.: Pearls in Graph Theory. Academic Press, San Diego (1990)
9. Seamone, B.: The 1-2-3 Conjecture and related problems: a survey. Preprint (2012). http://arxiv.org/abs/1211.5122

# Approximate Realizations
# for Outerplanaric Degree Sequences

Amotz Bar-Noy[1], Toni Böhnlein[2], David Peleg[3], Yingli Ran[3(✉)],
and Dror Rawitz[4]

[1] City University of New York (CUNY), New York, USA
[2] Huawei, Zurich, Switzerland
[3] Weizmann Institute of Science, Rehovot, Israel
yingli.ran@weizmann.ac.il
[4] Bar Ilan University, Ramat-Gan, Israel

**Abstract.** We study the question of whether a sequence $d = (d_1, d_2, \ldots, d_n)$ of positive integers is the degree sequence of some outerplanar (a.k.a. 1-page book embeddable) graph $G$. If so, $G$ is an outerplanar realization of $d$ and $d$ is an outerplanaric sequence. The case where $\sum d \leq 2n - 2$ is easy, as $d$ has a realization by a forest (which is trivially an outerplanar graph). In this paper, we consider the family $\mathcal{D}$ of all sequences $d$ of even sum $2n \leq \sum d \leq 4n - 6 - 2\omega_1$, where $\omega_x$ is the number of $x$'s in $d$. (The second inequality is a necessary condition for a sequence $d$ with $\sum d \geq 2n$ to be outerplanaric.) We partition $\mathcal{D}$ into two disjoint subfamilies, $\mathcal{D} = \mathcal{D}_{NOP} \cup \mathcal{D}_{2PBE}$, such that every sequence in $\mathcal{D}_{NOP}$ is provably non-outerplanaric, and every sequence in $\mathcal{D}_{2PBE}$ is given a realizing graph $G$ enjoying a 2-page book embedding (and moreover, one of the pages is also bipartite).

## 1 Introduction

**Background.** Our study is concerned with degree sequences of graphs. The degree of a vertex in a graph $G$ is the number of its incident edges, and the sequence of vertex degrees of $G$ is denoted by $\deg(G)$. Studied for more than 60 years, the DEGREE REALIZATION problem requires to decide, given a sequence $d$ of positive integers, whether $d$ has a *realizing graph*, namely, a graph $G$ such that $\deg(G) = d$, and to find such a graph if exists. A sequence $d$ that has a realizing graph is called *graphic*. One of the first characterizations of all graphic degree sequences was given by Erdös and Gallai [6]. About the same time, Havel and Hakimi [10,11] described an algorithm that, given a sequence $d$, generates a realization if $d$ is graphic or verifies that $d$ is not graphic.

The realization problem has also been investigated for various well known classes of graphs. In particular, realization by planar graphs was considered in several papers, but currently a complete solution seems out of reach. The survey by Rao [17] listed a number of partial results and studies on related questions. In

---

This work was supported by US-Israel BSF grant 2022205.

this paper, we address the question whether a given sequence has an *outerplanar* realization. Outerplanar graphs are planar graphs that can be embedded in the plane so that edges do not intersect each other and additionally, each vertex lies on the outer face (i.e., no vertex is surrounded by edges) [20].

A sequence $d \in \mathcal{D}$ is *outerplanaric* if it has a realizing outerplanar graph. Similarly, a sequence $d \in \mathcal{D}$ is *maximal outerplanaric (MOP)* if it has a realizing *maximal* outerplanar graph (namely, an outerplanar graph with the property that adding any edge to it violates outerplanarity).

Throughout, we denote by $d = (d_1, d_2, \ldots, d_n)$ a nonincreasing sequence of $n$ positive integers and denote its *volume* by $\sum d = \sum_{i=1}^{n} d_i$. We use the shorthand $a^k$ to denote a subsequence of $k$ consecutive integers $a$, and denote the multiplicity of degree $x$ in a sequence $d$ by $\omega_x$. By the well-known Handshaking Lemma, $\sum d$ is even for every graphic sequence $d$. If $\sum d \leq 2n-2$ and $\sum d$ is even, then $d$ has a forest realization with $(2n - \sum_i d_i)/2$ components [9]. Therefore, we focus on $d$ with $\sum d \geq 2n$. We show later (Lemma 4) that if $d$ is outerplanaric but not forestic (i.e., $\sum d > 2n - 2$), then $\sum d \leq 4n - 6 - 2\omega_1$, and moreover, $\sum d = 4n - 6$ if and only if $d$ is maximal outerplanaric. Consequently, we focus on the family of nonincreasing sequences of $n$ positive integers

$$\mathcal{D} = \left\{ d = (d_1, \ldots, d_n) \mid \sum d \text{ is even and } 2n \leq \sum d \leq 4n - 6 - 2\omega_1 \right\}.$$

Let $\mathcal{D}_{\leq 4} \subseteq \mathcal{D}$ be the subclass of sequences whose largest degree satisfies $d_1 \leq 4$.

While we are unable to fully characterize the sequences of $\mathcal{D}$ w.r.t. outerplanarity, we obtain a classification result that can be thought of as an "approximate" solution to the outerplanarity problem. To do that, we employ the concept of *book embeddings* which is used in graph theory to measure how far a graph is *away* from being outerplanar. Given a graph $G = (V, E)$, let $E = E_1 \cup \ldots \cup E_p$ be a partition of its edges such that each subgraph $G_i = (V, E_i)$ is outerplanar. For a book embedding of $G$, think of a book in which the pages (half-planes) are filled by outerplanar embeddings of the $G_i$'s such that the vertices are embedded on the spine of the book and in the same location on each page. This constraint is equivalent to requiring that the vertices appear in the same order along the cyclic order (defined later as the vertex ordering on the outer face) of each of the outerplanar embeddings of the $G_i$'s. The *book thickness* or *pagenumber* [2] is the minimal number of pages for which a graph has a valid book embedding. Note that a graph is outerplanar if and only if it has pagenumber 1, and it is known that the pagenumber of planar graphs is at most 4 [22].

**Our Contribution.** We give a classification of $\mathcal{D}$ into two sub-families, $\mathcal{D} = \mathcal{D}_{NOP} \cup \mathcal{D}_{2PBE}$, such that

- every sequence in $\mathcal{D}_{NOP}$ is provably non-outerplanaric, and
- every sequence in $\mathcal{D}_{2PBE}$ is given (in polynomial time) a realizing graph $G = (V, E)$, $E = E_1 \cup E_2$, enjoying a 2-page book embedding with pages $G_1$ and $G_2$ (hereafter referred to as a *2PBE realization*). Moreover, $G_2$ is bipartite (hereafter, an *OP+bi realization*), and in some cases, $G_2$ consists of at most one or two edges (hereafter, an *OP+1* or *OP+2* realization).

Note that every *outerplanaric* sequence in $D$ falls in $\mathcal{D}_{2PBE}$. Thus, the above result can be interpreted as yielding a polynomial time algorithm providing an "approximation" for outerplanaric sequences, in the sense that if the given sequence $d$ is outerplanaric (i.e., has a 1-page realization), then the algorithm will produce a 2-page realization for it. We remark that for many of the outerplanaric sequences in $\mathcal{D}$, we are in fact able to find an outerplanar realization, but as this refined classification is incomplete, it is omitted here.

Figure 1 gives an overview of the algorithm.

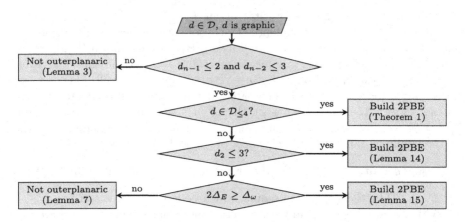

**Fig. 1.** The resulting flowchart. $\Delta_E = (4n - 6 - \sum d)/2$ **and** $\Delta_\omega = 3\omega_1 + 2\omega_2 + \omega_3 - n$.

**Additional Related Work.** For the OUTERPLANAR DEGREE REALIZATION problem, a full characterization of *forcibly* outerplanar graphic sequences (namely, sequences each of whose realizations is outerplanar) was given in [5]. A characterization of the degree sequence of maximal outerplanar graphs having exactly two 2-degree nodes was provided in [16]. This was also mentioned in [3]. A characterization of the degree sequences of maximal outerplanar graphs with at most four vertices of degree 2 was given in [13].

Sufficient conditions for the realization of 2-trees were given in [14]. A full characterization of the degree sequences of 2-trees was presented in [3]. A graph $G$ is a 2-tree if $G = K_3$ or $G$ has a vertex $v$ with degree 2, whose neighbors are adjacent and $G[V \setminus \{v\}]$ is a 2-tree. This definition implies, in particular, that 2-trees have $\sum d = 4n - 6$. By Theorem 1 of [3], if a sequence $d$ is maximal outerplanaric, i.e., it is outerplanaric and satisfies $\sum d = 4n - 6$, then $d$ can be realized by a 2-tree. In [18] it was shown that every 2-tree has a 2-page book embedding. Note that the class of 2-trees is not hereditary, i.e., subgraphs of 2-trees are not necessarily 2-trees. Therefore, the result of [3] does not apply to non-maximal degree sequences.

For the PLANAR DEGREE REALIZATION problem, regular planar graphic sequences were classified in [12], and planar bipartite biregular degree sequences were studied in [1]. In [19], Schmeichel and Hakimi determined which graphic sequences with $d_1 - d_n = 1$ are planaric, and presented similar results for $d_1 - d_n = 2$, with a small number of unsolved cases. Some of the sequences left unsolved in [19] were later resolved in [7,8]. Some additional studies on special cases of the planaric degree realization problem were listed in [17].

## 2 Preliminaries

**Definitions and Notation.** Given two $n$-integer sequences $d$ and $d'$, denote their componentwise addition and difference, respectively, by $d \oplus d' = (d_1 + d'_1, \ldots, d_n + d'_n)$ and $d \ominus d' = (d_1 - d'_1, \ldots, d_n - d'_n)$. For a nonincreasing sequence $d$ of $n$ nonnegative integers, let $\mathsf{pos}(d)$ denote the prefix consisting of the positive integers of $d$. Given two sequences $d'$ and $d''$ of length $n'$ and $n''$ respectively, let $d' \circ d'' = (d'_1, \ldots, d'_{n'}, d''_1, \ldots, d''_{n''})$.

Consider a simple (undirected) graph $G = (V, E)$. A (directed) *circuit* in $G$ is an *ordered set* of vertices $C = \{v_0, v_1, \ldots, v_{k-1}\}$ (with possible repetitions) such that $(v_i, v_{(i+1) \bmod k}) \in E$ for every $i = 0, \ldots, k - 1$. The directed edge set of $C$ is $E(C) = \{\langle v_i, v_{(i+1) \bmod k} \rangle \mid i = 0, \ldots, k - 1\}$ such that no directed edge may appear on it more than once. If the vertices of $C$ are distinct then it is called a *cycle*. A *triangle* is a cycle consisting of three edges.

For a graph $G = (V, E)$, one may define an accompanying embedding in the plane, namely, a mapping of every vertex to a point in two-dimensional space, with a curve connecting the two endpoints $u, w$ of every edge $(u, w) \in E$. An *outerplanar* graph $G$ is a graph that has a planar embedding in which all the vertices occur on the outer face. (This embedding can be transformed into a 1-page book embedding of $G$, with all vertices placed on the *spine* of the book.) We refer to the order in which the vertices occur on the outer face as the *cyclic-order* of $G$, and denote it by $C_{order}(G)$. A *maximal outerplanar* (MOP) graph $G$ is an outerplanar graph such that adding any new edge to it results in a non-outerplanar graph. It is folklore that for a 2-connected outerplanar graph, the cyclic order corresponds to a Hamiltonian cycle in $G$.

Hereafter, whenever we discuss a given outerplanar (resp., 2PBE) graph, we implicitly assume that some outerplanar (resp., 2PBE) embedding for it is given as well, and when constructing an outerplanar (resp., 2PBE) graph, we also give an outerplanar (resp., 2PBE) embedding for it.

Given an outer-planar embedding, an *external* edge is an edge residing on the outer face. That is, $e = (u, w)$ is external if $u$ and $w$ are neighbors on a circuit which is part of the outer face. All other edges of $G$ are *internal*. An *outer* (respectively, *inner*) *cycle* in $G$ is a cycle all of whose edges are external (resp., internal). Note that the outer face may be comprised of several outer-cycles.

**Degree Sequences of Trees and Forests.** Consider a nonincreasing sequence $d = (d_1, \ldots, d_n)$ of positive integers such that $\sum d$ is an even number. It is known that if $\sum d \leq 2n - 2$ then $d$ is graphic, and moreover, $d$ can be realized by an acyclic graph (forest). In this case, $d$ is called a *forestic* sequence. If $\sum d = 2n - 2$, then $d$ can be realized only by trees, and $d$ is called a *treeic* sequence. A vertex of degree one is called a *leaf*.

We make use of a special type of realizations of treeic sequences by *caterpillar graphs*. In a caterpillar graph $G = (V, E)$, all the non-leaves vertices are arranged on a path, called the *spine* of the caterpillar. Formally, the spine of a caterpillar graph is an ordered sequence $S = (x_1, \ldots, x_s) \subseteq V$, $s \geq 1$, in which $(x_i, x_{i+1}) \in E$ for $i = 1, \ldots, s - 1$ (see Fig. 2 for an example).

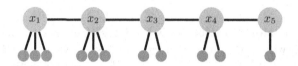

**Fig. 2.** A caterpillar graph with degree sequence $(5, 4^3, 2, 1^{11})$.

The other ingredient in the construction of a realization to a forestic sequence is a collection of disjoint edges, hereafter referred to informally as a *matching*. (Some proofs are deferred to the full paper due to space constraints.)

**Observation 1.** *A forestic sequence* $d = (d_1, d_2, \ldots, d_n)$ *of positive integers can be realized by a union of a caterpillar graph and a matching.*

**Some Known Results.** We next summarize known conditions for outerplanarity.

**Lemma 2** [21]. *If* $d = (d_1, \ldots, d_n)$ *is an outerplanaric degree sequence where* $n \geq 2$, *then* $\sum d \leq 4n - 6$, *with equality if and only if* $d$ *is maximal outerplanaric.*

**Lemma 3** [20]. *If* $d = (d_1, \ldots, d_n)$ *is outerplanaric, then (i)* $d_{n-1} \leq 2$, *and (ii)* $d_{n-2} \leq 3$. *Moreover,* $d_{n-1} = d_n = 2$ *in a maximal outerplanaric sequence* $d$.

For the case $\sum d > 2n - 2$, the next lemma improves Lemma 2 if the multiplicity $\omega_1$ in $d$ is taken into account (the proof is deferred to the full paper).

**Lemma 4.** *If* $d$ *is outerplanaric and* $\sum d > 2n - 2$, *then* $\sum d \leq 4n - 6 - 2\omega_1$.

We now recall some facts on maximal outerplanaric sequences (or MOP's).

**Fact 1** [15]. *A graph is a MOP if and only if it can be constructed from a triangle by a finite number of applications of repeatedly applying the following operation: to the graph already constructed, add a new vertex in the exterior face and join it to two adjacent vertices on that face.*

We rely on the construction method of Lemma 1 to prove the following simple but useful facts. Parts (1) and (2) are known (cf. [4, 21]. We could not find a reference for Part (3); The proof will be given in the full paper for completeness.

**Lemma 5.** *Consider a maximal outerplanar graph* $G$ *with* $n \geq 4$ *vertices.*

*(1) Every two vertices of degree 2 in* $G$ *are nonadjacent.*
*(2) there is a outerplanar embedding of* $G$ *such that the boundary of the outer face is a Hamiltonian cycle* $C$ *and each inner face is a triangle.*
*(3) Removing the edges of* $C$ *from* $G$ *and discarding the resulting degree-0 vertices yields a connected outerplanar graph.*

## 3  Necessary and Sufficient Conditions for Outerplanarity

This section presents some necessary and sufficient conditions for outerplanarity, used to prove the results of the main section concerning 2-page book embedding. The next lemma provides a necessary condition for maximal outerplanarity.

**Lemma 6.** *If $d$ is a maximal outerplanaric sequence where $d_1 > 3$ then if $d_2 > 3$ or $\omega_2 > 2$ then $2\omega_2 + \omega_3 \leq n$ otherwise $2\omega_2 + \omega_3 \leq n + 1$.*

Note that this lemma does not hold for non-maximal outerplanaric sequences. For example, the outerplanaric sequence $(4, 2^4)$ is a counterexample for which $2\omega_2 + \omega_3 = 8 = n + 3$. We next define for a sequence $d$ the functions *edge deficit* and *low degree surplus* that play a key role in our analysis.

$$\Delta_E(d) = (4n - 6 - \sum d)/2, \qquad \Delta_\omega(d) = 3\omega_1 + 2\omega_2 + \omega_3 - n.$$

Note that $\Delta_E(d) \geq 0$ for any outerplanaric sequence. If $\Delta_E(d) > 0$ and $G$ is a nonmaximal outerplanar realization of $d$, then $\Delta_E(d)$ is the number of edges that need to be added to $G$ to generate a maximal outerplanar graph. If $d$ is clear from the context we write simply $\Delta_E, \Delta_\omega$ instead of $\Delta_E(d), \Delta_\omega(d)$. The next lemma shows a connection between $\Delta_E$ and $\Delta_\omega$.

**Lemma 7.** *Let $d$ be an outerplanaric sequence with $d_2 > 3$. Then $2\Delta_E \geq \Delta_\omega$.*

*Proof.* Let $G$ be an outerplanar realization for $d$. Add $\Delta_E$ edges to $G$ to transform it into a maximal outerplanar graph $G'$ (without adding vertices). For $0 \leq i \leq \Delta_E$, let $G_i$ be $G$ after adding $i$ edges, with degree sequence $d^i$. By definition, $G_0 = G$ and $G_{\Delta_E} = G'$. Define $f(G_i) = 2\Delta_E(d^i) - \Delta_\omega(d^i)$ for $0 \leq i \leq \Delta_E$. We show:

$$f(G_i) \geq f(G_{i+1}) \text{ for any } 0 \leq i \leq \Delta_E - 1. \tag{1}$$

Let $d' = \deg(G')$. Then $\sum d' = 4n - 6$ and $\omega'_1 = 0$. Lemma 6 implies that $2\omega'_2 + \omega'_3 \leq n$. Thus, $\Delta_E(d') = 0$ and $\Delta_\omega(d') = 3\omega'_1 + 2\omega'_2 + \omega'_3 - n \leq 0$. Hence,

$$f(G_{\Delta_E}) = 2\Delta_E(d') - \Delta_\omega(d') \geq 0. \tag{2}$$

| $\deg_{G_i}(u), \deg_{G_i}(v)$ | $\delta$ |
|---|---|
| $(1,1) \rightarrow (2,2)$ | $-2$ |
| $(1,2) \rightarrow (2,3)$ | $-2$ |
| $(1,3) \rightarrow (2,4+)$ | $-2$ |
| $(1,4+) \rightarrow (2,5+)$ | $-1$ |
| $(2,2) \rightarrow (3,3)$ | $-2$ |

| $\deg_{G_i}(u), \deg_{G_i}(v)$ | $\delta$ |
|---|---|
| $(2,3) \rightarrow (3,4+)$ | $-2$ |
| $(2,4+) \rightarrow (3,4+)$ | $-1$ |
| $(3,3) \rightarrow (4+,4+)$ | $-2$ |
| $(3,4+) \rightarrow (4+,4+)$ | $-1$ |
| $(4+,4+) \rightarrow (4+,4+)$ | $0$ |

We now compute how adding a single edge $(u, v)$ to $G_i$ decreases the value of $3\omega_1(G_i) + 2\omega_2(G_i) + \omega_3(G_i)$. There are ten possibilities for the changes in the degrees of $u$ and $v$ ($\deg_{G_i}(u)$ and $\deg_{G_i}(v)$) that cause $3\omega_1(G_i)+2\omega_2(G_i)+\omega_3(G_i)$ to change by $\delta$. We list them in the above table where we denote degrees of 4 or more by 4+. It follows that adding a single edge decreases $3\omega_1(G_i) + 2\omega_2(G_i) + \omega_3(G_i)$ by at most 2 which implies that $\Delta_\omega(d^i)$ is also decreased by at most 2. Also, $2\Delta_E(d^i)$ is decreased by exactly 2 when adding a single edge to $G_{i-1}$. Therefore, Inequality (1) holds. Combining this with inequality (2) implies that $f(G_0) \geq 0$. The lemma follows because $f(G_0) = 2\Delta_E - \Delta_\omega$.    □

We start with a sequence $d$ satisfying $d_n = 2$ and $\omega_2 = 2$. To prove our next result, we generate a reduced sequence $d'$ of $d$ where each degree is decreased by 2. We verify that $d'$ is forestic and realize it by a graph $G'$ constructed as the union of a caterpillar graph and a matching, following Observation 1. Then, a realization $G$ of $d$ is found by adding a Hamiltonian cycle to $G'$. The Hamiltonian cycle is the outer-cycle of $G$. It follows that in this case, there is a realization of $d$ where each vertex appears exactly once on the outer-cycle.

**Lemma 8.** *Let $d \in \mathcal{D}$. If $d_n = 2$ and $\omega_2 = 2$, then $d$ is outerplanaric, and has a realization with an outerplanar embedding that has a Hamiltonian outer-cycle.*

*Proof.* Let $d$ be as in the lemma. First, we construct the sequence $d' = (d'_1, \ldots, d'_{n'})$ by subtracting 2 from each degree of $d$, i.e., let $d'_i = d_i - 2$ for $i \in \{1, \ldots, n-2\}$. Note that $n' = n - 2$ and that $\sum d'$ is even. Moreover, note that $d'$ is forestic, since $\sum d \leq 4n - 6$. Hence,

$$\sum d' = \sum d - 2n \leq 2n - 6 = 2(n' + 2) - 6 = 2n' - 2.$$

By Observation 1, $d'$ can be realized by a graph $G' = (V', E')$ composed of the union of a caterpillar graph $T'$ and a matching $M'$. Let $S = (x_1, \ldots, x_s)$ be the vertices on the spine of $T'$, and let $X_i = \{\ell_{i,1}, \ldots, \ell_{i,k_i}\} \subseteq V'$ be the leaves adjacent to the spine vertex $x_i$, for $i \in \{1, \ldots, s\}$. Note that vertex $x_i$ has degree $k_i + 1$ if $i \in \{1, s\}$, otherwise it has degree $k_i + 2$. Assume $M' = \{(x'_1, x''_1), \ldots, (x'_t, x''_t)\}$. To find an outerplanar realization of $d$, we add a set of edges to $G'$ that form a Hamiltonian cycle (including two additional vertices of degree 2). We explain our construction in four steps for which we assume that $s$ is odd. If $s$ is even, an analogogous construction applies.

(1) Add the edges along the following two paths $P_1 = (x_1, x_3, \ldots, x_{s-2}, x_s)$ and $P_2 = (x_{s-1}, x_{s-3}, \ldots, x_4, x_2)$ connecting spine vertices in odd and even positions, respectively.

(2) Add two additional vertices $x_0, x_{s+1}$ and use them to form a cycle $C$ together with $P_1$ and $P_2$. Let $C = (x_0, x_1, x_3, \ldots, x_s, x_{s+1}, x_{s-1}, x_{s-3}, \ldots, x_4, x_2, x_0)$.

(3) The leaves of a vertex in $P_1$ are inserted into $C$ between two vertices of $P_2$. I.e., the leaves of $x_i$, for $i = 1, \ldots, s$, are inserted between $x_{i-1}$ and $x_{i+1}$.

(4) Insert vertices $x'_1, \ldots, x'_t$ on the segment between $\ell_{s,k_s}$ and $x_{s+1}$ in clockwise order. Similarly, insert vertices $x''_1, \ldots, x''_t$ on the segment between $x_{s+1}$ and $x_s$ in a counter-clockwise order. Connect $x'_i$ and $x''_i$ for $1 \leq i \leq t$.

**Fig. 3.** $G$ as constructed in Lemma 8. The vertices $x_1, \ldots, x_6$ and the leaves in yellow (labels omitted) together with the black edges form the caterpillar graph $G'$. Paths $P_1$ and $P_2$ are depicted in green and blue, respectively. Together, they form the cycle $C$. (Color figure online)

Figure 3 illustrates the construction of $G$. Following the four steps, we have

$$C = (x_0, x_1, \ell_{2,1}, \ell_{2,2}, \ldots, \ell_{2,k_2}, x_3, \ell_{4,1}, \ell_{4,2}, \ldots, \ell_{4,k_2}, x_5, \ldots, x_{s-1}, \ell_{s,1}, \ell_{s,2}, \ldots, \ell_{s,k_s},$$
$$x'_1, \ldots, x'_t, x_{s+1}, x''_t, \ldots, x''_1, x_s, \ell_{s-1,k_s-1}, \ell_{s-1,k_s-1-1}, \ldots, x_2, \ell_{1,k_1}, \ell_{1,k_1-1}, \ldots, \ell_{1,1}, x_0).$$

Formally, let $V = V' \cup \{x_0, x_{s+1}\}$. The set $E'$ contains the path connecting the spine vertices, $\{(x_i, x_{i+1}) \mid i = 1, 2, \ldots, s-1\}$, and the edges connecting each spine vertex to its leaves. The set $L$ contains edges among the leaves while the sets $A_1$ and $A_2$ cntain edges connecting spine and leaf vertices

$$L = \{(\ell_{i,j}, \ell_{i,j+1}) \mid i = 1, \ldots, s, \quad j = 1, \ldots, k_i - 1\},$$
$$A_2 = \{(x_i, \ell_{i+1,1}) \cup (\ell_{i+1,k_{i+1}}, x_{i+2}) \mid i = 1, \ldots, s \text{ and } k_{i+1} > 0\},$$
$$A_1 = \{(x_i, x_{i+2}) \mid i = 1, \ldots, s \text{ and } k_{i+1} = 0\}$$

Note that $A_2, A_1, L, E', M'$ are pairwise disjoint. Let $E = E' \cup L \cup A_2 \cup A_1 \cup M'$, and let $G = (V, E)$. Notice that Fig. 3 depicts an outerplanar embedding of $G$. By the construction of $G$, every vertex in $V$ appears on the Hamiltonian outer-cycle. No two edges in $E(G)$ cross each other. Therefore, $G$ is an outerplanar graph. By the definition of $d'$, two vertices of degree 2 are deleted, which correspond to $x_0, x_{s+1}$. Every vertex in $V \setminus \{x_0, x_{s+1}\}$ has its degree decreased by 2. Recall that $T' \cup M'$ realizes $d'$. After adding $C$ to $T'$ and inserting $M'$ in step (4), every vertex $i \in V \setminus \{x_0, x_{s+1}\}$ has its degree increased by 2, making it equal to $d_i$. Combining this with the fact that $x_0$ and $x_{s+1}$ have degree 2 in $G$, we have $\deg(G) = d$. The lemma follows. $\qquad\square$

**Corollary 1.** $d \in \mathcal{D}$ is outerplanaric If $\sum d \leq 4n - 2w_2 - 2$ and $d_{n-1} = d_n = 2$.

## 4   Approximate Realizations by Book Embeddings

This section concerns finding approximate realizations. Our main result is a classification of $\mathcal{D}$ into two sub-families, $\mathcal{D} = \mathcal{D}_{NOP} \cup \mathcal{D}_{2PBE}$, such that

- every sequence in $\mathcal{D}_{NOP}$ is provably non-outerplanaric, and
- every sequence in $\mathcal{D}_{2PBE}$ is given a realizing gaph enjoying a 2-page book embedding (hereafter referred to as a *2PBE realization*).

Note that if a graph has pagenumber 2, then it is planar (the converse it false).

We first analyze the subclass $\mathcal{D}_{\leq 4} \subseteq \mathcal{D}$ of sequences with $d_1 \leq 4$. We show that every $d \in \mathcal{D}_{\leq 4}$ enjoys a 2PBE realization.

**Lemma 9.** *For a sequence $d \in \mathcal{D}_{\leq 4}$ s.t. $d_n = d_{n-1} = 2$, (a) if $\omega_3 > 0$ then $d$ is outerplanaric, and (b) if $d_1 \leq n - 1$ then $d$ has an OP+2 realization.*

**Lemma 10.** *Every $d \in \mathcal{D}$ s.t. $d_n \geq 2$ and $d_{n-1} \geq 3$ has an OP+1 realization.*

**Lemma 11.** *Every $d \in \mathcal{D}_{\leq 4}$ s.t. $d_n = 1$ and $d_1 \leq n-1$ has an OP+2 realization.*

**Theorem 1.** *Let $d \in \mathcal{D}_{\leq 4}$ and $d_1 \leq n - 1$. Then $d$ has an OP+2 realization.*

Next, we analyze the subclass $\mathcal{D}_{\geq 5} = \mathcal{D} \setminus \mathcal{D}_{\leq 4}$ of sequences with $d_1 \geq 5$. We first consider the special case of *maximal-volume* sequences in $\mathcal{D}_{\geq 5}$, namely, ones whose volume equals exactly $4n - 6$. By Lemma 4, $\omega_1 = 0$. Combining this with Lemma 3, $\omega_2 \geq 2$. Since Lemma 8 has already considered the case for $\omega_2 = 2$, it remains to consider sequences $d$ where $\omega_2 > 2$. This class of sequences turns out to be the most challenging. Their characterization is covered by the next key lemma, whose proof is by far the most complex part of our analysis.

**Lemma 12.** *Let $d \in \mathcal{D}_{\geq 5}$. If (1) $\sum d = 4n - 6$, (2) $2\omega_2 + \omega_3 \leq n+1$, (3) $\omega_2 > 2$, and (4) $d_n = 2$, Then $d$ has an OP+bi realization.*

*Proof.* Let $d$ be as in the lemma. To prove the claim, we define sequences $d', d''$, of lengths $n', n''$ respectively, such that $d = d' \oplus d''$.

Intuitively, it would appear that if property (1) of the lemma holds, then $d'$ and $d''$ can be defined such that $\sum d' = 2n' - 4$ and $\sum d'' = 2n'' - 2$, so both $d'$ and $d''$ can be realized by a forest and embedded on its own page, implying a 2-page book embedding. Proving this intuition is complicated by the need to ensure two properties: (a) that no edge $(u, w)$ occurs in *both* pages, and (b) that the vertices occur in the same order on the outer face of both pages.

Instead, we select the sequence $d''$ (ignoring the 0-degrees) so that it meets the preconditions of Lemma 8, i.e., $d''$ satisfies (i) $\omega_2'' = 2$, (ii) $d_{n''}'' = 2$, and (iii) $d'' \in \mathcal{D}$, namely, $\sum d'' \leq 4n'' - 6$ and $\sum d''$ is even. Then, Lemma 8 provides an outerplanar realization $G''$ of $d''$ and its embedding fills the first page of the book. The embedding of $G''$ imposes an ordering on the vertices, as they appear along the spine of the book. For the second page, we construct an outerplanar realization $G'$ of $d'$ such that the vertices follow the same ordering along the book spine. To obtain $d''$, we apply the following two transformations to $d$.

**(TR1)** remove $q := \omega_2 - 2$ degree-2 entries from $d$ to meet requirement (i).
**(TR2)** decrease the total volume of the remaining sequence by $2q$ to satisfy requirement (iii), while carefully avoiding the creation of new degrees 1 or 2.

Denote the number of degrees that are 5 or more by $\ell := \sum_{i \geq 5} \omega_i = n - \omega_2 - \omega_3 - \omega_4$. The remaining proof, describing the construction of $G'$ for $d'$, is split into two cases, depending on the relation between $q$ and $\ell$.

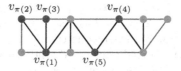

**Fig. 4.** The graph realizing $\hat{d}$. Blue circles: vertices in $B$. Green circles: vertices in $V(\hat{G}) \setminus B$. (Color figure online)

*Case A: $q < \ell$.* In this case, a path sequence $d' = (2^{n'-2}, 1^2)$ is a suitable choice, and the construction of $G'$ is simplified considerably.

We define the sequence $d'$ to have length $n' = n$. It contains vertices of degree 0 and we do *not* consider it in non-increasing order, to satisfy the constraint $d = d' \oplus d''$. Once $d'$ is defined, $d''$ is implicitly defined as well, as $d'' = d \ominus d'$. It can be regarded as having length $n'' = n$ where $d''_i = 0$ if $i > n - q$.

$$d'_i = \begin{cases} 1, & \text{if } i = 1, \\ 2, & \text{if } i = 2, \ldots, q, \\ 1, & \text{if } i = q + 1, \\ 0, & \text{if } i = q + 2, \ldots, n - q, \\ 2, & \text{if } i = n - q + 1, \ldots, n. \end{cases} \qquad d''_i = \begin{cases} d_i - 1, & \text{if } i = 1, \\ d_i - 2, & \text{if } i = 2, \ldots, q, \\ d_i - 1, & \text{if } i = q + 1, \\ d_i, & \text{if } i = q + 2, \ldots, n - q, \\ 0, & \text{if } i = n - q + 1, \ldots, n. \end{cases}$$

We use the sequence $d = (6^2, 5^4, 4, 3^2, 2^6)$ as a running example for Case A. In this case, $n = 15$, $q = 4$, and $\ell = 6$. It is not hard to verify that $d$ satisfies the conditions of this lemma and the condition of Case A. We have that $d' = (1, 2^3, 1, 0^2, 2^4)$, $d'' = (5, 4, 3^2, 4, 5, 4, 3^2, 2^2, 0^4)$.

For the next step, consider $\hat{d} = (d''_1, d''_2, \ldots, d''_{n-q})$, which is $d''$ without the trailing 0 degrees. In the above example, $\hat{d} = (5^2, 4^3, 3^4, 2^2)$.

**Claim.** Sequence $\hat{d}$ satisfies the preconditions of Lemma 8.

By Lemma 8, $\hat{d}$ can be realized as the outerplanar graph $\hat{G}$. (For our example, this graph is given in Fig. 4.) We add $q$ isolated vertices to $V(\hat{G})$ to generate an outerplanar realization $G'' = (V, E'')$ of $d''$. Let $V = \{v_1, v_2, \ldots, v_n\}$ such that $\deg_{G''}(v_i) = d''_i$, for $i = 1, 2, \ldots, n$. Hence, the set $R := \{v_{n-q+1}, \ldots, v_n\}$ contains the isolated vertices of degree 0. Consider an outerplanar embedding of $\hat{G}$, and note that the isolated vertices in $R$ can be placed in arbitrary positions on the cyclic order to yield a planar embedding of $G''$.

Let $B := \{v_1, v_2, \ldots, v_{q+1}\}$ be the set of vertices whose degrees were reduced in $d''$ compared to the original $d$ and whose original degree is at least 5. Let $\pi$ be the permutation by which the vertices of $B$ appear in the cyclic-order defined by the planar embedding of $G''$. Without loss of generality, let $\pi(1) = 1$. Hence, there is an embedding of $G''$ into the page of a book where the vertices in $B$ are arranged along the spine in the order $(v_{\pi(1)}, v_{\pi(2)}, \ldots, v_{\pi(q+1)})$. The vertices in $R$ can be placed arbitrarily along the spine. In our example $\hat{G}$, $B = \{v_1, \ldots, v_5\}$, and its vertices are arranged in the cyclic order $v_{\pi_1}, \ldots, v_{\pi_5}$, as in Fig. 4.

To finish the proof of case A, construct a realization $G' = (V, E')$ of $d'$ such that $\deg_{G'}(v_i) = d'_i$, for $i = 1, 2, \ldots, n$. The vertices $v_{q+2}, \ldots, v_{n-q}$ have vertices of degree 0 in $G'$ and are ignored when constructing $G'$. The vertices in $R$ must have degree 2 in $G'$. In the embeddings of $G''$ and $G'$, we place vertex $v_{n-q+i}$ between $v_{\pi(i)}$ and $v_{\pi(i+1)}$ along the spine, for $i = 1, \ldots, q$. Let index $j$ be such that $d''_{\pi(j)} = d_{q+1} - 1$. For every $i = 1, \ldots, \pi(j) - 2, \pi(j), \pi(j) + 1, \ldots, q$, add to $G'$ the two edges $(v_{\pi(i)}, v_{n-q+i})$ and $(v_{n-q+i}, v_{\pi(i+1)})$. Finally, add the edges $(v_{\pi(j)-1}, v_{n-q+j})$ and $(v_{n-q+j}, v_{q+1})$.

By construction, $V(G'') = V(G')$. By definition of $d'$ and $d''$, $\deg_{G'}(v_i) + \deg_{G''}(v_i) = d_i$ for $i = 1, 2, \ldots, n$. The vertices in $B \cup R$ form a path in $G'$ where $v_1$ and $v_{\pi(j)}$ are the two ends (with degree 1). Finally, observe that $E'' \cap E' = \emptyset$ since each edge of $G'$ is incident to a vertex in $R$, which has vertices of degree 0 in $G''$. Hence, $G''$ and $G'$ form a valid book embedding of a graph that realizes $d$. Notice that $G'$ is bipartite outerplanar and $G''$ is an outerplanar graph.

Notice that in our example $v_{\pi(1)} = v_1$, $v_{\pi(2)} = v_3$, $v_{\pi(3)} = v_4$, $v_{\pi(4)} = v_2$, $v_{\pi(5)} = v_5$. Then in $G'$, $d'(v_{\pi(1)}) = 1$, $d'(v_{\pi(2)}) = d'(v_{\pi(3)}) = d'(v_{\pi(4)}) = 2$, $d'(v_{\pi(5)}) = 1$, $R = \{u_1, \ldots, u_q\}$. Figure 5 shows a valid embedding of $G'$ into the page of a book. This completes the proof of Case A.

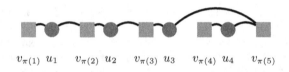

$$v_{\pi(1)} \ u_1 \qquad v_{\pi(2)} \ u_2 \qquad v_{\pi(3)} \ u_3 \qquad v_{\pi(4)} \ u_4 \qquad v_{\pi(5)}$$

**Fig. 5.** $G'$ ($j = 4$) constructed in Lemma 12-A's proof. The blue rectangles are B's vertices. (Color figure online)

*Case B:* This case is significantly more complicated than case A, and its detailed and rather lengthy analysis is deferred to the full paper.                           □

Lemma 12 considers maximal-volume sequences. The following lemma mainly focuses on non-maximal-volume sequences. Recall that $\Delta_E(d) = (4n - 6 - \sum d)/2$ and $\Delta_\omega(d) = 3\omega_1 + 2\omega_2 + \omega_3 - n$.

**Lemma 13.** *Let $d \in \mathcal{D}_{\geq 5}$ and $d_1 \leq n - 1$. If (1) $2\Delta_E \geq \Delta_\omega$, (2) $\omega_2 > 2$ (3) $d_2 > 3$, and (4) $d_n = 2$, Then $d$ has an OP+bi realization.*

**Lemma 14.** *Every graphic $d \in \mathcal{D}_{\geq 5}$ with $d_2 \leq 3$ has an OP+bi realization.*

**Corollary 2.** *Let $d \in \mathcal{D}_{\geq 5}$ and $d_1 \leq n - 1$. If $2\Delta_E \geq \Delta_\omega$ and $d_n = d_{n-1} = 2$, then $d$ has an OP+bi realization.*

**Lemma 15.** *Let $d \in \mathcal{D}_{\geq 5}$ and $d_1 \leq n-1$. If $2\Delta_E \geq \Delta_\omega$ and $d$ satisfies $d_{n-1} \leq 2$ and $d_{n-2} \leq 3$, then $d$ has an OP+bi realization.*

Combining Lemmas 3, 7 and 15, we conclude the following.

**Corollary 3.** *There is a polynomial time algorithm that, given an outerplanaric sequence $d \in \mathcal{D}_{\geq 5}$ with $d_2 > 3$, returns an OP+bi realization for d.*

Combining Theoream 1, Lemma 14 and Corollary 3, we conclude with the following.

**Theorem 2.** *There is a polynomial time algorithm that given a sequence $d \in \mathcal{D}$ either returns "d is not outerplanaric" or returns a 2PBE realization for d.*

**Corollary 4.** *There is a poly-time algorithm constructing a 2PBE realization for any sequence in a superclass[1] $\mathcal{S}$ of the class $\mathcal{OP}$ of all outerplanaric sequences.*

# References

1. Adams, P., Nikolayevsky, Y.: Planar bipartite biregular degree sequences. Disc. Math. **342**, 433–440 (2019)
2. Bernhart, F., Kainen, P.C.: The book thickness of a graph. J. Comb. Th. B **27**, 320–331 (1979)
3. Bose, P., Dujmović, V., Krizanc, D., Langerman, S., Morin, P., Wood, D.R., Wuhrer, S.: A characterization of the degree sequences of 2-trees. JGT **58**, 191–209 (2008)
4. Campos, C.N., Wakabayashi, Y.: On dominating sets of maximal outerplanar graphs. Disc. Appl. Math. **161**(3), 330–335 (2013)
5. Choudum, S.A.: Characterization of forcibly outerplanar graphic sequences. In: Combinatorics and Graph Theory, pp. 203–211 (1981)
6. Erdös, P., Gallai, T.: Graphs with prescribed degrees of vertices [Hungarian]. Mat. Lapok (N.S.) **11**, 264–274 (1960)
7. Fanelli, S.: On a conjecture on maximal planar sequences. JGT **4**, 371–375 (1980)
8. Fanelli, S.: An unresolved conjecture on nonmaximal planar graphical sequences. Disc. Math. **36**(1), 109–112 (1981)
9. Gupta, G., Joshi, P., Tripathi, A.: Graphic sequences of trees and a problem of Frobenius. Czechoslovak Math. J. **57**, 49–52 (2007)
10. Hakimi, S.L.: On realizability of a set of integers as degrees of the vertices of a linear graph -I. SIAM J. Appl. Math. **10**(3), 496–506 (1962)
11. Havel, V.: A remark on the existence of finite graphs [in Czech]. Casopis Pest. Mat. **80**, 477–480 (1955)
12. Hawkins, A., Hill, A., Reeve, J., Tyrrell, J.: On certain polyhedra. Math. Gaz. **50**(372), 140–144 (1966)
13. Li, Z., Zuo, Y.: On the degree sequences of maximal outerplanar graphs. Ars Comb. **140**, 237–250 (2018)
14. Lotker, Z., Majumdar, D., Narayanaswamy, N.S., Weber, I.: Sequences characterizing $k$-trees. In: Chen, D.Z., Lee, D.T. (eds.) COCOON 2006. LNCS, vol. 4112, pp. 216–225. Springer, Heidelberg (2006). https://doi.org/10.1007/11809678_24
15. Proskurowski, A.: Minimum dominating cycles in 2-trees. Int. J. Comput. Inf. Sci. **8**, 405–417 (1979)

---

[1] $\mathcal{S}$ strictly contains $\mathcal{OP}$. For example, $(4^2, 2^4)$ is not outerplanaric but satisfies the conditions of Lemma 9(b), hence our algorithm constructs a 2PBE realization for it.

16. Rafid, M.T., Ibrat, R., Rahman, M.S.: Generating scale-free outerplanar networks. In: Hsieh, S.Y., Hung, L.J., Klasing, R., Lee, C.W., Peng, S.L. (eds.) ICS 2022, vol. 1723, pp. 156–166. Springer, Heidelberg (2022). https://doi.org/10.1007/978-981-19-9582-8_14

17. Rao, S.B.: A survey of the theory of potentially p-graphic and forcibly p-graphic degree sequences. In: Combinatorics and Graph Theory, pp. 417–440 (1981)

18. Rengarajan, S., Veni Madhavan, C.E.: Stack and queue number of 2-trees. In: Du, D.-Z., Li, M. (eds.) COCOON 1995. LNCS, vol. 959, pp. 203–212. Springer, Heidelberg (1995). https://doi.org/10.1007/BFb0030834

19. Schmeichel, E.F., Hakimi, S.L.: On planar graphical degree sequences. SIAM J. Appl. Math. **32**, 598–609 (1977)

20. Syslo, M.M.: Characterizations of outerplanar graphs. Disc. Math. **26**, 47–53 (1979)

21. West, D.: Introduction to Graph Theory. Prentice Hall, Upper Saddle river (2001)

22. Yannakakis, M.: Four pages are necessary and sufficient for planar graphs. In: 18th ACM Symposium on Theory of Computing (STOC), pp. 104–108 (1986)

# Hypergraph Dualization with **FPT**-delay Parameterized by the Degeneracy and Dimension

Valentin Bartier[1], Oscar Defrain[2]([✉]), and Fionn Mc Inerney[3]

[1] LIRIS, CNRS, Université Claude Bernard Lyon 1, Lyon, France
[2] Aix-Marseille Université, CNRS, LIS, 13009 Marseille, France
oscar.defrain@lis-lab.fr
[3] Algorithms and Complexity Group, TU Wien, Vienna, Austria

**Abstract.** At STOC 2002, Eiter, Gottlob, and Makino presented a technique called *ordered generation* that yields an $n^{O(d)}$-delay algorithm listing all minimal transversals of an $n$-vertex hypergraph of degeneracy $d$. Recently at IWOCA 2019, Conte, Kanté, Marino, and Uno asked whether this XP-delay algorithm parameterized by $d$ could be made FPT-delay for a weaker notion of degeneracy, or even parameterized by the maximum degree $\Delta$, i.e., whether it can be turned into an algorithm with delay $f(\Delta) \cdot n^{O(1)}$ for some computable function $f$. Moreover, and as a first step toward answering that question, they note that they could not achieve these time bounds even for the particular case of minimal dominating sets enumeration. In this paper, using ordered generation, we show that an FPT-delay algorithm can be devised for minimal transversals enumeration parameterized by the degeneracy and dimension, giving a positive and more general answer to the latter question.

**Keywords:** algorithmic enumeration · hypergraph dualization · minimal transversals · minimal dominating sets · FPT-delay enumeration

## 1 Introduction

Graphs and hypergraphs are ubiquitous in discrete mathematics and theoretical computer science. Formally, a *hypergraph* $\mathcal{H}$ consists of a family $E(\mathcal{H})$ of subsets called *edges* (or *hyperedges*) over a set $V(\mathcal{H})$ of elements called *vertices*, and it is referred to as a *graph* when each of its edges is of size precisely two. A *transversal* in a hypergraph is a set of vertices that intersects every edge. The problem of enumerating the family $Tr(\mathcal{H})$ of all (inclusion-wise) minimal transversals of a given hypergraph $\mathcal{H}$, usually denoted by TRANS-ENUM and also known as

This work was supported by the ANR project DISTANCIA (ANR-17-CE40-0015) and the Austrian Science Fund (FWF, project Y1329).

*hypergraph dualization*, is encountered in many areas of computer science including logic, database theory, and artificial intelligence [16,19,27,31,38]. The best-known algorithm to date for the problem is due to Fredman and Khachiyan [23], and runs in time $N^{o(\log N)}$, where $N = |V(\mathcal{H})| + |E(\mathcal{H})| + |Tr(\mathcal{H})|$. The question of the existence of an output-polynomial-time algorithm has been open for over 30 years, and is arguably one of the most important open problems in algorithmic enumeration [19,28]. An enumeration algorithm is said to run in *output-polynomial* time if it outputs all solutions and stops in a time that is bounded by a polynomial in the combined sizes of the input and the output. It is said to run with *polynomial delay* if the run times before the first output, after the last output,[1] and between two consecutive outputs are bounded by a polynomial depending on the size of the input only. Clearly, any polynomial-delay algorithm defines an output-polynomial-time algorithm. We refer the reader to [28,41] for more details on the usual complexity measures of enumeration algorithms, and to [19] for a survey on minimal transversals enumeration.

In 2002, an important step toward the characterization of tractable cases of TRANS-ENUM was made by Eiter, Gottlob, and Makino [17]. The problem, studied under its equivalent formulation of monotone Boolean dualization, was shown to admit a polynomial-delay algorithm whenever the hypergraph has bounded degeneracy, for an appropriate notion of hypergraph degeneracy (see Sect. 2 for definitions of graph and hypergraph degeneracies) that generalizes graph degeneracy. Specifically, the authors proved the following result.

**Theorem 1** [17,18]. *There is an algorithm that, given an n-vertex hypergraph of weak degeneracy d, generates its minimal transversals with $n^{O(d)}$-delay.*

In particular, the above theorem captures $\alpha$-acyclic hypergraphs and several other classes of hypergraphs related to restricted cases of conjunctive normal form (CNF) formulas. Another consequence of Theorem 1 is a polynomial-delay algorithm for the intimately related problem of enumerating minimal dominating sets in graphs, usually denoted by DOM-ENUM, when the graph has bounded degeneracy. A *minimal dominating set* in a graph is a (inclusion-wise) minimal subset of vertices intersecting each closed neighborhood. As the family of closed neighborhoods of a graph defines a hypergraph, DOM-ENUM naturally reduces to TRANS-ENUM. However, quite surprisingly, and despite the fact that not all hypergraphs are hypergraphs of closed neighborhoods of a graph, the two problems were actually shown to be equivalent in [29]. This led to a significant interest in DOM-ENUM, and the characterization of its complexity status in various graph classes has since been explored in the literature [4,24,25,30].

The theory of parameterized complexity provides a framework to analyze the run time of an algorithm as a function of the input size and some parameters of the input such as degeneracy. It has proved to be very useful in finding tractable instances of NP-hard problems, as well as for devising practical algorithms. In this context, a problem is considered tractable for a fixed parameter $k$ if it is

---

[1] Some authors equivalently choose to ignore the postprocessing by assuming that the last solution is kept in memory and output just before halting, see, e.g., [8].

in FPT *when parameterized by* $k$, that is, if it admits an algorithm running in time $f(k) \cdot n^{O(1)}$ for some computable function $f$, where $n$ is the size of the input. It is important to note that the combinatorial explosion of the problem can be restrained to the parameter $k$ here. There are also weaker notions of tractability in this context, with a problem being in XP *when parameterized by* $k$ if it admits an algorithm running in time $n^{f(k)}$. Studying the tractability of important problems within the paradigm of parameterized complexity has led to a fruitful line of research these last decades, and we refer the reader to [13] for an overview of techniques and results in this field.

The study of algorithmic enumeration from a parameterized complexity perspective is quite novel. It may be traced back to works such as [14, 21], where the authors devise algorithms listing all solutions of size $k$ in time $f(k) \cdot n^{O(1)}$, where $n$ is the size of the input. Note, however, that these kinds of results hint at a relatively low number of solutions (typically not superpolynomial in $n$ for fixed $k$), while this is not the case for many enumeration problems including the one that we consider in this paper. New parameterized paradigms for enumeration problems of this second type were later developed in [12, 26], and a growing interest for FPT-enumeration has since been developing as witnessed by the recent Dagstuhl Seminar on output-sensitive, input-sensitive, parameterized, and approximative enumeration [22], and the recent progress on this topic [10, 11, 26, 35, 36]. Among the different complexity measures one could want to satisfy, the notion of FPT-delay naturally arises. An algorithm A for an enumeration problem parameterized by $k$ runs with FPT-*delay* if there exists a computable function $f$ such that A has delay $f(k) \cdot n^{O(1)}$, where $n$ in the size of the input. An XP-delay algorithm for an enumeration problem parameterized by $k$ is analogously defined with the delay being $n^{f(k)}$ instead. Note that the algorithm given by Theorem 1 is XP-delay, but not FPT-delay. The next question naturally arises within the context of parameterized complexity.

*Question 1.* Can the minimal transversals of an $n$-vertex hypergraph of weak degeneracy $d$ be listed with delay $f(d) \cdot n^{O(1)}$ for some computable function $f$?

To the best of our knowledge, no progress has been made on this question to date. At IWOCA 2019, Conte, Kanté, Marino, and Uno formulated the following weakening by considering a weaker notion of degeneracy $d$ and the maximum degree $\Delta$ as an additional parameter. Specifically, they asked the following.

*Question 2* [10]. Can the minimal transversals of an $n$-vertex hypergraph be listed with delay $2^d \cdot \Delta^{f(d)} \cdot n^{O(1)}$ for some computable function $f$, where $d$ is the strong degeneracy and $\Delta$ is the maximum degree of the hypergraph?

As $d \leq \Delta$ (see Sect. 2 for the definition of strong degeneracy) such a time bound is FPT parameterized by the maximum degree only. This formulation, however, is not arbitrary. It is chosen to match the complexity of an FPT-delay algorithm the authors in [10] give for another related problem—namely the enumeration of maximal irredundant sets in hypergraphs—as well as giving a finer analysis of the dependence on $d$ and $\Delta$. A set of vertices is *irredundant* if removing any of its vertices decreases the number of edges it intersects. It is easy to see that minimal transversals are maximal irredundant sets. However, the converse is not true, and the two problems of listing minimal transversals and maximal irredundant sets behave differently. Indeed, while TRANS-ENUM is known to admit an output quasi-polynomial-time algorithm [23], it was announced in [7] that the enumeration of maximal irredundant sets is impossible in output-polynomial time unless P = NP. As put in [10], answering Question 2 would help to precise the links between the two problems, as maximal irredundancy is commonly believed to be more difficult than minimal transversality. In the same paper, the authors state that their algorithm for maximal irredundancy does not apply to minimal transversality, and that they were also unable to turn it into one for graph domination with the same delay. The next question naturally follows, and may be regarded as a weakening of Question 2 that is implicit in [10].

*Question 3* [10]. Can the minimal dominating sets of an $n$-vertex graph be enumerated with delay $2^d \cdot \Delta^{f(d)} \cdot n^{O(1)}$ for some computable function $f$, where $d$ is the degeneracy and $\Delta$ is the maximum degree of the graph?

In this paper, we give a positive answer to Question 3, suggesting that minimal domination and maximal irredundancy are of relatively comparable tractability as far as the degeneracy is concerned. In fact, we prove a stronger result. That is, we design an FPT-delay algorithm for TRANS-ENUM parameterized by the degeneracy and the maximum size of a hyperedge, usually referred to as *dimension*, even for the notion of hypergraph degeneracy considered in [17,18], which is less restrictive than the one considered in [10], a point that is discussed in Sect. 2. More formally, we prove the following result.

**Theorem 2.** *The minimal transversals of an $n$-vertex hypergraph of weak degeneracy $d$ and dimension $k$ can be enumerated with delay $k^d \cdot n^{O(1)}$.*

The FPT-delay algorithm we obtain is based on the ordered generation technique originally presented by Eiter, Gottlob, and Makino in [17], combined with elementary arguments on the number of minimal subsets hitting the neighborhood of a well-chosen vertex of bounded "backdegree". As a corollary of Theorem 2, we obtain an algorithm solving DOM-ENUM running within the time bounds required in Question 3. We then discuss the limitations of the technique and show that it cannot be directly used to get a positive answer to Question 1 unless FPT = W[1] (see Sect. 6 for more details and references).

Additionally, we believe that our presentation is of valuable interest as far as the ordered generation technique and its limitations in the design of parameterized algorithms are concerned. Furthermore, we hope that it will raise interest in making headway in the recent and promising line of research concerning the study of enumeration problems in the parameterized setting.

*Organization of the Paper.* In the next section, we introduce the problem and related notation, and discuss the aforementioned variants of hypergraph degeneracy considered in [10,18]. In Sect. 3, we present the ordered generation technique applied to hypergraphs, and show how it amounts to generating extensions of a partial solution while preserving FPT-delay. In Sect. 4, we show how to generate extensions with FPT-delay parameterized by the weak degeneracy and dimension, proving Theorem 2. The answer to Question 3 is given in Sect. 5 as a corollary. Future research directions and the limitations of the presented techniques are discussed in Sect. 6.

## 2    Preliminaries

*Graphs and Hypergraphs.* For a hypergraph $\mathcal{H}$, its vertex set is denoted by $V(\mathcal{H})$, and its edge set is denoted by $E(\mathcal{H})$. If every edge has size precisely two, then it is referred to as a graph and is usually denoted by $G$. Given a vertex $v \in V(\mathcal{H})$, the set of edges that contain $v$ (also called edges *incident* to $v$) is denoted by inc($v$). The *degree* of $v$ in $\mathcal{H}$, denoted by $\deg_{\mathcal{H}}(v)$, is the number of edges in inc($v$). The maximum degree of a vertex in $\mathcal{H}$ is denoted by $\Delta(\mathcal{H})$ (or simply $\Delta$). The *dimension* of $\mathcal{H}$ is the maximum size of an edge in $\mathcal{H}$. A *transversal* of $\mathcal{H}$ is a subset $T \subseteq V(\mathcal{H})$ such that $E \cap T \neq \emptyset$ for all $E \in E(\mathcal{H})$. It is *minimal* if it is minimal by inclusion, or equivalently, if $T \setminus \{v\}$ is not a transversal for any $v \in T$. The set of minimal transversals of $\mathcal{H}$ is denoted by $Tr(\mathcal{H})$.

In this paper, we are interested in the following problem known to admit an XP-delay algorithm [18] parameterized by the degeneracy, but for which the existence of an FPT-delay algorithm parameterized by the same parameter (even in addition to the maximum degree) is open [10].

MINIMAL TRANSVERSALS ENUMERATION (TRANS-ENUM)
**Input:**    A hypergraph $\mathcal{H}$.
**Output:** The set $Tr(\mathcal{H})$.

We introduce notation that is convenient for minimal transversality. Let $S \subseteq V(\mathcal{H})$ and let $v$ be a vertex in $S$. A *private edge of $v$ with respect to $S$* is an edge $E \in E(\mathcal{H})$ such that $E \cap S = \{v\}$. The set of private edges of $v$ with respect to $S$ are denoted by priv($S, v$). It is well-known that a transversal in a hypergraph is minimal if and only if each vertex it contains has a private edge.

The *degeneracy* of a graph is a well-known parameter that has received considerable attention as a measure of sparsity in a graph [37], as well as a suitable parameter to devise tractable graph algorithms [1,18,39,43]. For any graph $G$, it corresponds to the least integer $d$ such that every subgraph of $G$ has a vertex of degree at most $d$, and it can be computed in time $O(|V(G)| + |E(G)|)$ [34]. Several generalizations to hypergraphs have been proposed in the literature [2,15,18,33]. Among these generalizations, two of them naturally arise depending on the notion of subhypergraph we consider, and were previously considered in the context of hypergraph dualization [10,18]. We review them here.

Let $\mathcal{H}$ be a hypergraph and $S$ a subset of vertices of $\mathcal{H}$. The hypergraph *induced* by $S$ is the hypergraph

$$\mathcal{H}[S] := (S, \{E \in E(\mathcal{H}) : E \subseteq S\}).$$

The *trace* of $S$ on $\mathcal{H}$ is the hypergraph

$$\mathcal{H}_{|S} := (S, \{E \cap S : E \in E(\mathcal{H}), \; E \cap S \neq \emptyset\}).$$

Several edges in $\mathcal{H}$ may have the same intersection with $S$, but this is only counted once in $\mathcal{H}_{|S}$, i.e., the trace is a set and not a multiset. An example of these notions of subhypergraph is in Fig. 1. The two following definitions generalize graph degeneracy (with the strong degeneracy being at most one more than the graph degeneracy), and differ on whether we consider the subhypergraph to be the hypergraph induced by the first $i$ vertices or its trace on these vertices.

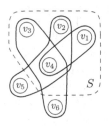

**Fig. 1.** A hypergraph $\mathcal{H}$ on 6 vertices and 3 edges. The hypergraph induced by $S :=$ $\{v_1, v_2, v_3, v_4\}$ has a single edge $\{v_1, v_4\}$, while the trace of $S$ on $\mathcal{H}$ contains three edges $\{v_1, v_4\}$, $\{v_2, v_4\}$, and $\{v_3, v_4\}$.

**Definition 1 (Weak degeneracy, [18]).** *The weak degeneracy of $\mathcal{H}$ is the smallest integer $d$ such that there exists an ordering $v_1, \ldots, v_n$ of its vertices satisfying for every $1 \leq i \leq n$:*

$$\deg_{\mathcal{H}[\{v_1, \ldots, v_i\}]}(v_i) \leq d.$$

**Definition 2 (Strong degeneracy, [10,33]).** *The strong degeneracy of $\mathcal{H}$ is the smallest integer $d$ such that there exists an ordering $v_1, \ldots, v_n$ of its vertices satisfying for every $1 \leq i \leq n$:*

$$\deg_{\mathcal{H}_{|\{v_1, \ldots, v_i\}}}(v_i) \leq d.$$

The next inequality follows from the fact that, for any subset $S$ of $V(\mathcal{H})$, an edge $E$ that belongs to $\mathcal{H}[S]$ also belongs to $\mathcal{H}_{|S}$ by definition.

*Remark 1.* A hypergraph's weak degeneracy is at most its strong degeneracy.

Note that the strong degeneracy of a hypergraph can be arbitrarily larger than its weak degeneracy. To see this, consider the hypergraph $\mathcal{H}$ formed by taking a clique on $k$ vertices $v_1, \ldots, v_k$ (a graph), and adding one new vertex $u$ in each edge, so that each hyperedge in $\mathcal{H}$ has size 3. Let $U$ be the set of vertices $u$ added in the construction, and consider any ordering of $V(\mathcal{H})$ such that the elements $v_1, \ldots, v_k$ appear before those in $U$. The weak degeneracy of $\mathcal{H}$ is 1 since each vertex $u$ has degree 1 and the hypergraph induced by $v_1, \ldots, v_k$ has no edge. On the other hand, the strong degeneracy of $\mathcal{H}$ is at least $k$ since the trace of $\mathcal{H}$ on $v_1, \ldots, v_k$ is a clique whose vertices all have degree $k$.

Also, note that while there exist $n$-vertex hypergraphs with $O(2^n)$ edges, any hypergraph that has (weak or strong) degeneracy $d$ contains $O(dn)$ edges. This explains why time bounds such as those expected in Questions 2 and 3, and obtained in Theorem 2 and subsequent lemmas in this paper are reasonable.

For both notions of degeneracy, an *elimination ordering* is an order $v_1, \ldots, v_n$ as in their definitions. Recall that Question 2 is stated for strong degeneracy [10]. In this paper, we show Theorem 2 to hold for weak degeneracy, which implies the same result for strong degeneracy. Note that the weak (strong, respectively) degeneracy of a hypergraph $\mathcal{H}$ along with an elimination ordering witnessing that degeneracy can be computed in time $O(|V(\mathcal{H})| + \sum_{E \in E(\mathcal{H})} |E|)$ [34, 40].

*Graph Domination.* If a hypergraph is a graph, then its transversals are usually referred to as vertex covers in the literature. This is not to be confused with domination, that we define now. Let $G$ be a graph. The *closed neighborhood* of a vertex $v$ of $G$ is the set $N[v] := \{v\} \cup \{u : \{u, v\} \in E(G)\}$. A *dominating set* in $G$ is a subset $D \subseteq V(G)$ intersecting every closed neighborhood in the graph. It is called *minimal* if it is minimal by inclusion, and the set of minimal dominating sets of $G$ is denoted by $D(G)$. In the minimal dominating sets enumeration problem, denoted DOM-ENUM, the task is to generate $D(G)$ given $G$.

As stated in the introduction, TRANS-ENUM and DOM-ENUM are polynomially equivalent, one direction being a direct consequence of the easy observation that $D(G) = Tr(\mathcal{N}(G))$ for any graph $G$, where $\mathcal{N}(G)$ denotes the *hypergraph of closed neighborhoods* of $G$ whose vertex set is $V(G)$ and whose edge set is $\{N[v] : v \in V(G)\}$ (see, e.g., [3]). The other direction was exhibited by Kanté, Limouzy, Mary, and Nourine in [29] even when restricted to co-bipartite graphs. However, the construction in [29] does not preserve the degeneracy. Hence, the existence of an FPT-delay algorithm for DOM-ENUM parameterized by the degeneracy (and more) would not directly imply one for TRANS-ENUM.

## 3   Ordered Generation

Our algorithm is based on the *ordered generation* technique introduced in [17] for monotone Boolean dualization, and later adapted to the context of graph domination in [4, 5]. Our presentation of the algorithm is based on these works.

Let $\mathcal{H}$ be a hypergraph and $v_1, \ldots, v_n$ an ordering of its vertex set. For readability, given any $i \in \{1, \ldots, n\}$, we set $V_i := \{v_1, \ldots, v_i\}$, $\mathcal{H}_i := \mathcal{H}[V_i]$, and denote by $\mathsf{inc}_i(u)$, for any $u \in V(\mathcal{H}_i)$, the incident edges of $u$ in $\mathcal{H}_i$. By convention, we set $V_0 := \emptyset$, $\mathcal{H}_0 := (\emptyset, \emptyset)$, and $Tr(\mathcal{H}_0) := \{\emptyset\}$. Further, for all $0 \leq i \leq n$, $S \subseteq V(\mathcal{H})$, and $v \in S$, we denote by $\mathsf{priv}_i(S, v)$ the private edges of $v$ with respect to $S$ in $\mathcal{H}_i$. Let us now consider some integer $i \in \{0, \ldots, n-1\}$. For $T \in Tr(\mathcal{H}_{i+1})$, we call the *parent of $T$ with respect to $i + 1$*, denoted by $\mathsf{parent}(T, i + 1)$, the set obtained by applying the following greedy procedure:

*While there exists a vertex $v \in T$ such that $\mathsf{priv}_i(T, v) = \emptyset$,*
*remove from $T$ a vertex of smallest index with that property.*

Observe that $\mathsf{parent}(T, i + 1)$ is uniquely defined by this routine. Given a minimal transversal $T^*$ of $\mathcal{H}_i$, we call the *children of $T^*$ with respect to $i$*, denoted by $\mathsf{children}(T^*, i)$, the family of sets $T \in Tr(\mathcal{H}_{i+1})$ such that $\mathsf{parent}(T, i+1) = T^*$.

**Lemma 1.** *Let $0 \leq i \leq n - 1$, $T \in Tr(\mathcal{H}_{i+1})$. Then, $\mathsf{parent}(T, i + 1) \in Tr(\mathcal{H}_i)$.*

*Proof.* By definition, every edge of $\mathcal{H}_i$ is an edge of $\mathcal{H}_{i+1}$. Hence, $T$ is a transversal of $\mathcal{H}_i$ as well. Now, as the greedy procedure above only removes vertices $v$ that have no private edge in $\mathcal{H}_i$, the obtained set is a transversal of $\mathcal{H}_i$ at each step, and ends up minimal by construction. □

**Lemma 2.** *Let $0 \leq i \leq n - 1$ and $T^* \in Tr(\mathcal{H}_i)$. Then, either:*

- $T^* \in Tr(\mathcal{H}_{i+1})$ *and* $\mathsf{parent}(T^*, i + 1) = T^*$; *or*
- $T^* \cup \{v_{i+1}\} \in Tr(\mathcal{H}_{i+1})$ *and* $\mathsf{parent}(T^* \cup \{v_{i+1}\}, i + 1) = T^*$.

*Proof.* As $T^* \in Tr(\mathcal{H}_i)$, we have $\mathsf{priv}_i(T^*, v) \neq \emptyset$ for every $v \in T^*$, and since every edge of $\mathcal{H}_i$ is an edge of $\mathcal{H}_{i+1}$, we derive that $\mathsf{priv}_{i+1}(T^*, v) \neq \emptyset$ for every $v \in T^*$. Hence, if $T^*$ is a transversal of $\mathcal{H}_{i+1}$, then we obtain $T^* \in Tr(\mathcal{H}_{i+1})$, and thus, $\mathsf{parent}(T^*, i + 1) = T^*$ by definition, yielding the first item of the lemma.

Let us thus assume that $T^*$ is not a transversal of $\mathcal{H}_{i+1}$. Then, there is an edge $E$ in $\mathcal{H}_{i+1}$ that is not hit by $T^*$. As $T^*$ is a transversal of $\mathcal{H}_i$, that edge must contain the vertex $v_{i+1}$. We derive that $T := T^* \cup \{v_{i+1}\}$ is a transversal of $\mathcal{H}_{i+1}$, which furthermore satisfies $\mathsf{priv}_{i+1}(T, v_{i+1}) \neq \emptyset$ as $E$ is not hit by $T^*$. Let us show that $\mathsf{priv}_{i+1}(T, v) \neq \emptyset$ holds for any other vertex $v \in T$. Consider one such vertex $v \neq v_{i+1}$ and $E \in \mathsf{priv}_i(T^*, v)$. Such an edge exists as $T^* \in Tr(\mathcal{H}_i)$. If $E \notin \mathsf{priv}_{i+1}(T, v)$, then we have that $v_{i+1} \in E$, which contradicts the fact that $E$ belongs to $\mathcal{H}_i$. Hence, $E \in \mathsf{priv}_{i+1}(T, v)$, and the lemma follows. □

The *parent* relation defines a rooted tree, called the *solution tree*, with vertex set $\{(T, i) : T \in Tr(\mathcal{H}_i), \ 0 \leq i \leq n\}$, leaves $\{(T, n) \mid T \in Tr(\mathcal{H})\}$, and root $(\emptyset, 0)$. Our generation algorithm consists in a traversal of that tree, starting from its root, that outputs every leaf. Lemma 1 ensures that each node $(T, i)$, $0 < i \leq n$, can be obtained from its parent. Lemma 2 ensures that each node $(T, i)$, $0 \leq i < n$, has at least one child, i.e., that every branch of the tree leads to a different minimal transversal of $\mathcal{H}$. In what follows, the set $T$ of an internal node $(T, i)$ is referred to as a *partial solution*; see Fig. 2 for a representation of the tree. The complexity of the algorithm depends on the delay needed to generate the children of an internal node, which is formalized in the next theorem.

**Theorem 3.** *Let $f, s \colon \mathbb{N}^2 \to \mathbb{Z}^+$ be two functions. Suppose there is an algorithm that, given a hypergraph $\mathcal{H}$ on vertices $v_1, \ldots, v_n$ and $m$ edges, an integer $i \in \{0, \ldots, n - 1\}$, and $T^* \in Tr(\mathcal{H}_i)$, enumerates $\mathsf{children}(T^*, i)$ with delay $f(n, m)$ and using $s(n, m)$ space. Then, there is an algorithm that enumerates the set $Tr(\mathcal{H})$ with delay $O(n \cdot f(n, m))$ and total space $O(n \cdot s(n, m))$.*

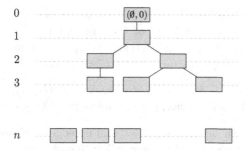

**Fig. 2.** A representation of the solution tree defined by the parent relation.

*Proof.* We assume the existence of an algorithm A of the first type, and describe an algorithm B listing $Tr(\mathcal{H})$ with the desired delay. The algorithm first chooses an ordering $v_1, \ldots, v_n$ of the vertex set of $\mathcal{H}$ in $O(n)$ time. Then, it runs a DFS of the solution tree defined by the parent relation above, starting from its root $(\emptyset, 0)$ and only outputting sets $T$ corresponding to visited leaves $(T, n)$. Recall that these sets are exactly the minimal transversals of $\mathcal{H}$. First, the correctness of B follows from Lemmas 1 and 2. Let us now consider the time and space complexity of such an algorithm. At first, the root $(\emptyset, 0)$ is constructed in $O(1)$ time and space. Then, for each internal node $(T^*, i)$ of the tree, the set children$(T^*, i)$ is generated with delay $f(n, m)$ and space $s(n, m)$ by A. We furthermore note that when visiting a node of the solution tree, we do not need to generate all its children. We can instead compute one child that has not been processed yet, and continue the DFS on that child. Thus, when we visit some node $(T, i)$, we only need to store the data of the $i$ paused executions of A enumerating the children of the ancestors of $(T, i)$ including the root $(\emptyset, 0)$, together with the data of the current instance of A listing children$(T, i)$. Hence, the maximum delay of B between two consecutive outputs is bounded by twice the height of the solution tree times the delay of A, and the same goes for the space complexity. Since the height of the tree is $n$, we get the desired bounds of $O(n \cdot f(n, m))$ and $O(n \cdot s(n, m))$ for the delay and space complexities of B, respectively.     □

Note that only a multiplicative factor of $f(n, m)$ is added to the input size in the delay of the algorithm given by Theorem 3. Thus, an algorithm generating children with FPT-delay $f(k) \cdot \text{poly}(n, m)$, for some parameter $k$ and computable function $f$, yields an FPT-delay algorithm for TRANS-ENUM parameterized by $k$.

## 4   Parameterized Children Generation

In hypergraphs, we show that the children of an internal node of the solution tree, as defined in Sect. 3, may be enumerated with FPT-delay parameterized by the weak degeneracy and the dimension. By Theorem 3, this leads to an algorithm listing minimal transversals in hypergraphs with FPT-delay parameterized by the weak degeneracy and the dimension, i.e., Theorem 2.

Let $\mathcal{H}$ be a hypergraph of weak degeneracy $d$ and dimension $k$, and let $v_1, \ldots, v_n$ be an elimination ordering of $\mathcal{H}$ witnessing the degeneracy. By definition, for any $i \in \{0, \ldots, n-1\}$ and $E \in \mathsf{inc}_{i+1}(v_{i+1})$, we have $|\mathsf{inc}_{i+1}(v_{i+1})| \leq d$ and $|E| \leq k$. We fix $i \in \{0, \ldots, n-1\}$ and $T^* \in Tr(\mathcal{H}_i)$ in what follows.

**Lemma 3.** *For any $X \subseteq V_{i+1} \setminus T^*$, if $T^* \cup X \in Tr(\mathcal{H}_{i+1})$, then every $x \in X$ has a distinct private edge in $\mathsf{inc}_{i+1}(v_{i+1})$, and thus, $|X| \leq d$.*

*Proof.* This holds since $|\mathsf{inc}_{i+1}(v_{i+1})| \leq d$ and no edge in $\mathcal{H}_i$ may be the private edge of $x \in X$ with respect to $T^* \cup X$ as all such edges are hit by $T^*$. $\qquad \square$

**Lemma 4.** *The set $\mathsf{children}(T^*, i)$ can be generated in $k^d \cdot n^{O(1)}$ time.*

*Proof.* Recall from Lemma 2 that either $\mathsf{children}(T^*, i) = T^*$ or $T^* \cup \{v_{i+1}\}$ belongs to $\mathsf{children}(T^*, i)$. Clearly, these two situations can be sorted out in $n^{O(1)}$ time. In the first case, we have generated the children within the claimed time. Let us thus assume that $\mathsf{children}(T^*, i) \neq T^*$ and $T^* \cup \{v_{i+1}\}$ has already been produced as a first child in $n^{O(1)}$ time.

We now focus on the generation of sets $X \subseteq V_{i+1} \setminus T^*$ with $v_{i+1} \notin X$ such that $T^* \cup X \in Tr(\mathcal{H}_{i+1})$. For each edge $E$ in $\mathsf{inc}_{i+1}(v_{i+1})$, we guess at most one element $x_E \neq v_{i+1}$ in $E$ to be the element of $X$ that has $E$ as its private edge. As $\mathsf{inc}_{i+1}(v_{i+1})$ has at most $d$ edges and each of these edges has size at most $k$ and contains $v_{i+1}$, we get at most $(k + 1 - 1)^d$ choices for $X$. For every such set $X$, we check whether the obtained set $T := T^* \cup X$ is a minimal transversal of $\mathcal{H}_{i+1}$ satisfying $T^* = \mathsf{parent}(T, i + 1)$ in $n^{O(1)}$ following the definition. This takes $k^d \cdot n^{O(1)}$ time in total. By Lemma 3, this procedure defines an exhaustive search for the children of the node $(T^*, i)$ in the solution tree. $\qquad \square$

Theorem 2 then follows by Theorem 3 and Lemma 4. In Lemma 4, we do not need to store all such sets $X$ as they can be generated following a lexicographic order on the vertices. So, this approach only needs space that is polynomial in $n$.

## 5    Consequences for Minimal Domination

Let $G$ be a graph of degeneracy $d$ and maximum degree $\Delta$. We show that, as a corollary of Theorem 2, we obtain an FPT-delay algorithm listing the minimal dominating sets of $G$ within the time bounds of Question 3.

In what follows, let $v_1, \ldots, v_n$ be an elimination ordering of $G$ witnessing the degeneracy, and recall that it can be computed in time $O(|V(G)| + |E(G)|)$ [34]. Further, recall that $\mathcal{N}(G)$ denotes the hypergraph of closed neighborhoods of $G$ (see Sect. 2). Note that $v_1, \ldots, v_n$ witnesses that the weak degeneracy of $\mathcal{N}(G)$ is at most $d+1$. Indeed, for any $i \in \{1, \ldots, n\}$, the number of hyperedges in $\mathsf{inc}_i(v_i)$ is bounded by the number of neighbors of $v_i$ in $\{v_1, \ldots, v_{i-1}\}$ plus possibly one if the neighborhood of $v_i$ is contained in $\{v_1, \ldots, v_{i-1}\}$. We furthermore have the following relation between the dimension of $\mathcal{N}(G)$ and $\Delta$.

**Observation 4.** *The dimension of $\mathcal{N}(G)$ is at most $\Delta(G) + 1$.*

*Proof.* Let $E$ be a hyperedge in $\mathcal{N}(G)$. Then, $E = N[u]$ for some $u \in V(G)$, and hence, $|E|$ is at most the degree of $u$ plus one, which is bounded by $\Delta(G) + 1$. □

As a corollary of Theorem 2 and Observation 4, we obtain a $(\Delta + 1)^{d+1} \cdot n^{O(1)}$-time algorithm for DOM-ENUM, and thus, a $\Delta^{f(d)} \cdot n^{O(1)}$-time algorithm for DOM-ENUM for a computable function $f$. This answers Question 3 in the affirmative.

It should be mentioned that using the Input Restricted Problem technique (IRP for short) from [9], more direct arguments yield an FPT algorithm parameterized by $\Delta$. This is of interest if no particular care on the dependence in the degeneracy is required. In a nutshell, this technique amounts to reducing the enumeration to a restricted instance consisting of a given solution $S$ and a vertex $v$ not in $S$, in the same (but even more restricted) spirit as in Sect. 4. Specifically, the authors show that if such an enumeration can be solved in polynomial time (in the size of the input only), then the general problem may be solved with polynomial delay. It can moreover be seen that such a technique preserves FPT delay, a point not considered in [9]. Then, using Observation 4 and noting that the degree of $\mathcal{N}(G)$ is bounded by $\Delta + 1$ as well, the IRP can be solved in $2^{\Delta^2} \cdot n^{O(1)}$ time, and using [9, Theorems 5.8 and 5.9], we deduce an FPT-delay algorithm parameterized by $\Delta$ for DOM-ENUM. Note that more work is to be done in order to get the dependence in the degeneracy, or for such a technique to be applied to hypergraphs of bounded degeneracy and dimension, namely to obtain Theorem 2. In order to get such time bounds, the algorithm of [9] must be guided by the degeneracy order. The interested reader will notice that this guided enumeration is in essence what is proposed by the ordered generation technique, earlier introduced by Eiter et al. in [17] for hypergraph dualization specifically, while the work of [9] aimed at being more general by considering hereditary properties, with both approaches building up on ideas from [42].

## 6    Discussion and Limitations

We exhibited an FPT-delay algorithm for TRANS-ENUM parameterized by the weak degeneracy and dimension, yielding an FPT-delay algorithm for DOM-ENUM parameterized by the (degeneracy and) maximum degree, answering a question in [10]. In light of the existence of an FPT-delay algorithm for DOM-ENUM parameterized by the maximum degree, it is tempting to ask whether it admits an FPT-delay algorithm for other structural parameters. However, for well-studied parameters like the clique number in graphs [4] or poset dimension in comparability graphs [6], even the existence of an XP-delay algorithm for DOM-ENUM remains open. A natural strengthening of Question 3 and weakening of Question 1, would be to get rid of the maximum degree:

*Question 4.* Can the minimal dominating sets of an $n$-vertex $d$-degenerate graph be enumerated with delay $f(d) \cdot n^{O(1)}$ for some computable function $f$?

However, it is valuable to note that the techniques presented here seem to fail in the context of hypergraphs, as an FPT-delay algorithm listing children parameterized by the weak degeneracy (for any elimination ordering witnessing the weak degeneracy) yields an FPT-delay algorithm for MULTICOLORED INDEPENDENT SET parameterized by the number of colors, a notorious $W[1]$-hard problem [13]. Recall that, in MULTICOLORED INDEPENDENT SET, the vertices of a graph $G$ are colored with colors $1, \ldots, k$, and we have to find an independent set containing exactly one vertex from each color. Formally, we prove the following.

**Theorem 5.** *Let $G$ be an instance of MULTICOLORED INDEPENDENT SET with colors $1, \ldots, k$ with $k, |E(G)| \geq 2$. Then, there exists an $n$-vertex hypergraph $\mathcal{H}$ of weak degeneracy $k + 1$ such that, for any elimination ordering $v_1, \ldots, v_n$ witnessing the weak degeneracy, there exists a minimal transversal $T^* \in Tr(\mathcal{H}_{n-1})$ whose children are all multicolored independent sets of $G$ except for one.*

*Proof.* We construct an instance of children generation as follows. We start with the vertices of $G$ and add the vertices $w, u_1, \ldots, u_{k+2}$. For each color $1 \leq \ell \leq k$ and integer $1 \leq j \leq k+2$, we create a hyperedge $E_{\ell,j}$ consisting of the vertices of color $\ell$ together with $u_j$. For each edge $xy \in E(G)$, we add a vertex $z_{xy}$ and two hyperedges $A_{xy} := \{x, z_{xy}\}$ and $B_{xy} := \{y, z_{xy}\}$. Let $Z := \{z_{xy} : xy \in E(G)\}$. We add a hyperedge containing $w$, $u_1$, and all the vertices of $Z$. Lastly, for all $2 \leq j \leq k + 2$, we add a hyperedge $\{w, u_j\}$ and a hyperedge containing $u_j$ and all the vertices of $Z$. This completes the construction of the hypergraph $\mathcal{H}$.

We first show that $\mathcal{H}$ has weak degeneracy $k + 1$. The lower bound comes from the minimum degree which is $k + 1$. The upper bound is witnessed by the degeneracy ordering that first removes $u_1, w, u_2, \ldots, u_{k+2}$ in that order, then the $z_{xy}$'s, and then all of the remaining (isolated) vertices in an arbitrary order.

Let $v_1, \ldots, v_n$ be any elimination ordering of $\mathcal{H}$ witnessing the weak degeneracy. Since $u_1$ is the only vertex of minimum degree $k + 1$ in $\mathcal{H}$, it must be that $v_n = u_1$. Thus, $\mathcal{H}_n = \mathcal{H}$ and the hyperedges belonging to $\mathrm{inc}_n(v_n)$ are $E_{\ell,1}$ for all $1 \leq \ell \leq k$, and the one containing $w$, $u_1$, and all the vertices of $Z$.

Let $T^* := Z \cup \{u_2, \ldots, u_{k+2}\}$. By construction, $T^* \in Tr(\mathcal{H}_{n-1})$, $A_{xy}$ and $B_{xy}$ are the only private edges of each $z_{xy} \in T^*$, and $\{w, u_j\}$ is the only private edge of each $u_j \in T^*$. Consider the generation of the children of $T^*$. For $X \subseteq V_n \setminus T^*$ to be such that $T^* \cup X \in Tr(\mathcal{H}_n)$, $X$ must hit $E_{\ell,1}$ for all $1 \leq \ell \leq k$ (the hyperedge containing $w$, $u_1$, and all the vertices of $Z$ is already hit by vertices in $T^*$). Further, each such hyperedge must be hit exactly once since $u_1$ hits every such hyperedge and any two other vertices contained in a same such hyperedge have the same trace on $\mathrm{inc}_n(v_n)$. Moreover, $X$ should not contain the two endpoints of any $xy \in E(G)$, as otherwise the corresponding vertex $z_{xy} \in T^*$ would lose its private edges. Hence, among all such possible candidate sets $X$, one consists of the singleton $\{v_n\}$, and the others are multicolored independent sets of $G$. $\square$

Another natural strengthening of Theorem 2 would be to relax the degeneracy, and require to be FPT-delay in the dimension only. We note however that even the existence of an XP-delay algorithm in that context seems open [32].

Finally, for the open questions stated above, an intermediate step would be to aim at FPT-*total-time algorithms*, that is, algorithms producing all the solutions in time $f(k) \cdot N^{O(1)}$ for $k$ the parameter, $f$ some computable function, and $N$ the size of the input plus the output. If successful, it would be interesting to know if these algorithms can be made FPT-*incremental*, that is, if they can produce the $i^{\text{th}}$ solution in $f(k) \cdot (i + n)^{O(1)}$ time, where $n$ is the size of the input. In that direction, FPT-incremental time was obtained in [20] for the maximum degree.

**Acknowledgements.** We would like to thank the anonymous reviewers for their careful reading and for providing simpler arguments in Sect. 4.

# References

1. Alon, N., Gutner, S.: Linear time algorithms for finding a dominating set of fixed size in degenerated graphs. Algorithmica **54**(4), 544–556 (2009)
2. Bar-Noy, A., Cheilaris, P., Olonetsky, S., Smorodinsky, S.: Online conflict-free colouring for hypergraphs. Comb. Probab. Comput. **19**(4), 493–516 (2010)
3. Berge, C.: Hypergraphs: combinatorics of finite sets, North-Holland Mathematical Library, vol. 45. Elsevier (1984)
4. Bonamy, M., Defrain, O., Heinrich, M., Pilipczuk, M., Raymond, J.F.: Enumerating minimal dominating sets in $K_t$-free graphs and variants. ACM Trans. Algorithms (TALG) **16**(3), 39:1–39:23 (2020)
5. Bonamy, M., Defrain, O., Heinrich, M., Raymond, J.F.: Enumerating minimal dominating sets in triangle-free graphs. In: 36th International Symposium on Theoretical Aspects of Computer Science, STACS 2019, Leibniz International Proceedings in Informatics (LIPIcs), vol. 126, pp. 16:1–16:12. Schloss Dagstuhl–Leibniz-Zentrum fuer Informatik (2019)
6. Bonamy, M., Defrain, O., Micek, P., Nourine, L.: Enumerating minimal dominating sets in the (in)comparability graphs of bounded dimension posets. arXiv preprint arXiv:2004.07214 (2020)
7. Boros, E., Makino, K.: Generating maximal irredundant and minimal redundant subfamilies of a given hypergraph. In: WEPA 2016 & Boolean Seminar 2017 (2016). http://clp.mff.cuni.cz/booleanseminar/presentations/endre.pdf
8. Capelli, F., Strozecki, Y.: Geometric amortization of enumeration algorithms. In: 40th International Symposium on Theoretical Aspects of Computer Science (STACS 2023). Schloss Dagstuhl – Leibniz-Zentrum für Informatik (2023)
9. Cohen, S., Kimelfeld, B., Sagiv, Y.: Generating all maximal induced subgraphs for hereditary and connected-hereditary graph properties. J. Comput. Syst. Sci. **74**(7), 1147–1159 (2008)
10. Conte, A., Kanté, M.M., Marino, A., Uno, T.: Maximal irredundant set enumeration in bounded-degeneracy and bounded-degree hypergraphs. In: Colbourn, C.J., Grossi, R., Pisanti, N. (eds.) IWOCA 2019. LNCS, vol. 11638, pp. 148–159. Springer, Cham (2019). https://doi.org/10.1007/978-3-030-25005-8_13
11. Creignou, N., Ktari, R., Meier, A., Müller, J.S., Olive, F., Vollmer, H.: Parameterised enumeration for modification problems. Algorithms **12**(9), 189 (2019)
12. Creignou, N., Meier, A., Müller, J.S., Schmidt, J., Vollmer, H.: Paradigms for parameterized enumeration. Theor. Comput. Syst. **60**, 737–758 (2017)
13. Cygan, M., Fomin, F.V., Kowalik, Ł, Lokshtanov, D., Marx, D., Pilipczuk, M., Pilipczuk, M., Saurabh, S.: Parameterized Algorithms. Springer, Cham (2015). https://doi.org/10.1007/978-3-319-21275-3

14. Damaschke, P.: Parameterized enumeration, transversals, and imperfect phylogeny reconstruction. Theoret. Comput. Sci. **351**(3), 337–350 (2006)
15. Dutta, K., Ghosh, A.: On subgraphs of bounded degeneracy in hypergraphs. In: Heggernes, P. (ed.) WG 2016. LNCS, vol. 9941, pp. 295–306. Springer, Heidelberg (2016). https://doi.org/10.1007/978-3-662-53536-3_25
16. Eiter, T., Gottlob, G.: Identifying the minimal transversals of a hypergraph and related problems. SIAM J. Comput. **24**(6), 1278–1304 (1995)
17. Eiter, T., Gottlob, G., Makino, K.: New results on monotone dualization and generating hypergraph transversals. In: Proceedings of the Thirty-Fourth Annual ACM Symposium on Theory of Computing, pp. 14–22 (2002)
18. Eiter, T., Gottlob, G., Makino, K.: New results on monotone dualization and generating hypergraph transversals. SIAM J. Comput. **32**(2), 514–537 (2003)
19. Eiter, T., Makino, K., Gottlob, G.: Computational aspects of monotone dualization: a brief survey. Discret. Appl. Math. **156**(11), 2035–2049 (2008)
20. Elbassioni, K., Hagen, M., Rauf, I.: Some fixed-parameter tractable classes of hypergraph duality and related problems. In: Grohe, M., Niedermeier, R. (eds.) IWPEC 2008. LNCS, vol. 5018, pp. 91–102. Springer, Heidelberg (2008). https://doi.org/10.1007/978-3-540-79723-4_10
21. Fernau, H.: On parameterized enumeration. In: Computing and Combinatorics: 8th Annual International Conference, COCOON 2002, pp. 564–573. Springer, Cham (2002)
22. Fernau, H., Golovach, P.A., Sagot, M.F.: Algorithmic enumeration: output-sensitive, input-sensitive, parameterized, approximative (dagstuhl seminar 18421). In: Dagstuhl Reports, vol. 8, Schloss Dagstuhl-Leibniz-Zentrum fuer Informatik (2019)
23. Fredman, M.L., Khachiyan, L.: On the complexity of dualization of monotone disjunctive normal forms. J. Algorithms **21**(3), 618–628 (1996)
24. Golovach, P.A., Heggernes, P., Kanté, M.M., Kratsch, D., Sæther, S.H., Villanger, Y.: Output-polynomial enumeration on graphs of bounded (local) linear MIM-width. Algorithmica **80**(2), 714–741 (2018)
25. Golovach, P.A., Heggernes, P., Kanté, M.M., Kratsch, D., Villanger, Y.: Enumerating minimal dominating sets in chordal bipartite graphs. Discret. Appl. Math. **199**, 30–36 (2016)
26. Golovach, P.A., Komusiewicz, C., Kratsch, D., Le, V.B.: Refined notions of parameterized enumeration kernels with applications to matching cut enumeration. J. Comput. Syst. Sci. **123**, 76–102 (2022)
27. Gunopulos, D., Mannila, H., Khardon, R., Toivonen, H.: Data mining, hypergraph transversals, and machine learning. In: PODS, pp. 209–216. ACM (1997)
28. Johnson, D.S., Yannakakis, M., Papadimitriou, C.H.: On generating all maximal independent sets. Inf. Process. Lett. **27**(3), 119–123 (1988)
29. Kanté, M.M., Limouzy, V., Mary, A., Nourine, L.: On the enumeration of minimal dominating sets and related notions. SIAM J. Discret. Math. **28**(4), 1916–1929 (2014)
30. Kanté, M.M., Limouzy, V., Mary, A., Nourine, L., Uno, T.: A polynomial delay algorithm for enumerating minimal dominating sets in chordal graphs. In: Mayr, E.W. (ed.) WG 2015. LNCS, vol. 9224, pp. 138–153. Springer, Heidelberg (2016). https://doi.org/10.1007/978-3-662-53174-7_11
31. Kavvadias, D., Papadimitriou, C.H., Sideri, M.: On horn envelopes and hypergraph transversals. In: Ng, K.W., Raghavan, P., Balasubramanian, N.V., Chin, F.Y.L. (eds.) ISAAC 1993. LNCS, vol. 762, pp. 399–405. Springer, Heidelberg (1993). https://doi.org/10.1007/3-540-57568-5_271

32. Khachiyan, L., Boros, E., Elbassioni, K., Gurvich, V.: On the dualization of hypergraphs with bounded edge-intersections and other related classes of hypergraphs. Theoret. Comput. Sci. **382**(2), 139–150 (2007)
33. Kostochka, A.V., Zhu, X.: Adapted list coloring of graphs and hypergraphs. SIAM J. Discret. Math. **22**(1), 398–408 (2008)
34. Matula, D.W., Beck, L.L.: Smallest-last ordering and clustering and graph coloring algorithms. J. ACM **30**(3), 417–427 (1983)
35. Meier, A.: Enumeration in incremental FPT-time. arXiv preprint arXiv:1804.07799 (2018)
36. Meier, A.: Incremental FPT delay. Algorithms **13**(5), 122 (2020)
37. Nešetřil, J., Ossona de Mendez, P.: Sparsity: graphs, structures, and algorithms, vol. 28. Springer, New York (2012). https://doi.org/10.1007/978-3-642-27875-4
38. Nourine, L., Petit, J.M.: Extending set-based dualization: application to pattern mining. In: Proceedings of the 20th European Conference on Artificial Intelligence, pp. 630–635. IOS Press (2012)
39. Philip, G., Raman, V., Sikdar, S.: Polynomial kernels for dominating set in graphs of bounded degeneracy and beyond. ACM Trans. Algorithms (TALG) **9**(1), 1–23 (2012)
40. Shahrokhi, F.: Bounding the trace function of a hypergraph with applications. In: Hoffman, F. (eds.) Combinatorics, Graph Theory and Computing. SEIC-CGTC 2020. Springer Proceedings in Mathematics Statistics, vol. 388, pp. 117–126. Springer, Cham (2022). https://doi.org/10.1007/978-3-031-05375-7_8
41. Strozecki, Y.: Enumeration complexity. Bull. EATCS **1**(129) (2019)
42. Tsukiyama, S., Ide, M., Ariyoshi, H., Shirakawa, I.: A new algorithm for generating all the maximal independent sets. SIAM J. Comput. **6**(3), 505–517 (1977)
43. Wasa, K., Arimura, H., Uno, T.: Efficient enumeration of induced subtrees in a k-degenerate graph. In: Ahn, H.-K., Shin, C.-S. (eds.) ISAAC 2014. LNCS, vol. 8889, pp. 94–102. Springer, Cham (2014). https://doi.org/10.1007/978-3-319-13075-0_8

# Star-Forest Decompositions of Complete Graphs

Todor Antić[1]([✉]), Jelena Glišić[1]([✉]), and Milan Milivojčević[1,2]([✉])

[1] Faculty of Mathematics and Physics, Charles University, Prague, Czech Republic
todor@kam.mff.cuni.cz, {jglisic,milivojcevic}@matfyz.cz
[2] Faculty of Mathematics, Natural Sciences and Information Technologies, University
of Primorska, Koper, Slovenia

**Abstract.** We deal with the problem of decomposing a complete geometric graph into plane star-forests. In particular, we disprove a recent
conjecture by Pach, Saghafian and Schnider by constructing for each $n$
a complete geometric graph on $n$ vertices which can be decomposed into
$\lceil \frac{n}{2} \rceil + 1$ plane star-forests. Additionally we prove that for even $n$, every
decomposition of a complete abstract graph on $n$ vertices into $\frac{n}{2} + 1$ star-
forests is composed of a perfect matching and $\frac{n}{2}$ star-forests with two
edge-balanced components, which we call broken double stars.

## 1 Introduction

A classic question asked in graph theory is the following: "Given a graph $G$, what
is the minimal number of subgraphs with property $P$ that the edges of $G$ can be
partitioned into?" Historically, this question was asked for abstract graphs and
property $P$ was replaced with forests, trees, complete bipartite graphs and many
more [2,11,14]. Similar questions can be asked about graphs drawn in the plane
or on any other surface. Here we want to decompose a complete graph on the
surface into subgraphs that have a certain geometric property in addition to the
property $P$. Answering such questions is a similar, but separate research direction
that has been pursued by many authors in discrete geometry and graph drawing
communities. A *geometric* graph is a graph drawn in the plane, with vertices
represented by points in general position and edges as straight line segments
between them. Recently, there has been a lot of work done on decomposing
geometric graphs into planar subgraphs of a special kind, such as trees, stars,
double stars etc. [7,15]. This paper will be concerned with decomposing complete
geometric graphs into plane star-forests. A *star* is a connected graph on $k$ vertices
with one vertex of degree $k - 1$ (center) and $k - 1$ vertices of degree 1. Our
definition allows a graph that has two vertices and a single edge to be a star but
it is not clear what the center should be. In this case we define the center to be

T. Antić and J. Glišić—The author is supported by project 23-04949X of the Czech
Science Foundation (GAČR).
T. Antić and J. Glišić—The research was conducted during a scholarship provided by
Višegrad Fund.

A. A. Rescigno and U. Vaccaro (Eds.): IWOCA 2024, LNCS 14764, pp. 126–137, 2024.
https://doi.org/10.1007/978-3-031-63021-7_10

one of the endpoints of the edge, and this choice is arbitrary. The definition also allows that a single vertex is a star and in this case this vertex is also the center of said star. A *star-forest* is a forest whose every connected component is a star. It is easy to observe that a complete graph $K_n$ can be decomposed into $n-1$ stars. Furthermore, $K_n$ cannot be decomposed into less than $n-1$ stars [3]. In the same paper, Akiyama and Kano proved that $K_n$ can be decomposed into at most $\lceil \frac{n}{2} \rceil + 1$ star-forests and that this bound is tight. Besides complete graphs, star-forest decompositions have been studied in the case of general abstract graphs as well . The number of star-forsets needed to decompose a graph $G$ is called the *star arboricity* of $G$ and has been intensively studied [4,5,10]. The story is different for complete geometric graphs. To the best of our knowledge, the first mention of star-forest decompositions for geometric graphs was made by Dujmović and Wood in [8]. They asked if one can decompose a complete geometric graph on $n-1$ vertices, whose vertices form a convex polygon into less than $n-1$ star-forests. Recently, this question was answeered in the negative by Pach, Saghafian and Schnider [12]. They showed that a complete geometric graph whose vertices form a convex polygon cannot be decomposed into fewer than $n-1$ plane star-forests. In the same paper, the authors posed the following question.

*Question 1.* What is the minimal number of plane star-forests that a complete geometric graph can be decomposed into?

Based on their findings they made the following conjecture:

*Conjecture 1* [12]. Let $n \geq 1$. There is no complete geometric graph with $n$ vertices that can be decomposed into fewer than $\lceil 3n/4 \rceil$ plane star-forests.

## Contribution

The main aim of this note is to answer this conjecture in the negative. The authors in [12] give a special configuration of $n = 4k$ points and construct a simple decomposition into $3n/4$ plane star-forests. Motivated by this example, we first describe a method generalizing it. This is the content of Theorem 3. Then we provide a point set on $n = 6$ points which can be decomposed into $\frac{2n}{3} = 4$ star-forests, disproving the conjecture. We then improve the bound further by constructing complete geometric graphs which can be decomposed into $\frac{n}{2} + 1$ plane star-forests, which is best possible. This is the content of Theorem 5. Attacking this problem raised some further questions regarding decompositions of abstract complete graphs into star-forests. Mainly, our computations [6] have shown us that for $n \in \{6,8\}$, the decomposition of $K_n$ into $\frac{n}{2} + 1$ star-forests is unique in a certain sense. In each fitting decomposition one star-forest was a perfect matching on $\frac{n}{2}$ vertices while the other $\frac{n}{2}$ star-forests were edge balanced, spanning and had centers at endpoints of an edge of the perfect matching. We call such a decomposition a broken double stars decomposition. We prove that for even $n$, every decomposition of the abstract complete graph $K_n$ into $\frac{n}{2} + 1$ star-forests is a broken double stars decomposition. This is the content of Theorem 1.

## 2    Decompositions of Complete Graphs into Star-Forests

In this section our goal is to define a unique construction of star-forest decompositions of the complete graph. These star-forests will be a byproduct of a decomposition of $K_n$ into special trees, called *double stars*. A double star is a graph composed of two vertex-disjoint stars, whose centers are joined by an edge. For an even $n = 2k$, the double star decomposition of $K_n$ is obtained in the following way. Let $M$ be a perfect matching in $K_n$. Then, for each edge $e \in M$ we create a double star by connecting each endpoint of $e$ to $\frac{n-2}{2}$ vertices of $K_n$ in such a way that we do not obtain a cycle. This results in a decomposition of the edge set into $\frac{n}{2}$ double stars. From this we can define a decomposition of $K_n$ into star-forests in the most natural way. One forest is a perfect matching on $k$ edges, and each of the other $k$ forests is composed of two stars with $k - 1$ edges each, whose centers are the endpoints of an edge of the perfect matching. This construction was described in [3] and again in [12]. For a visual explanation see Fig. 1. We will call any such decomposition of $K_n$ a *broken double stars* decomposition. We now formally state the main result of this section.

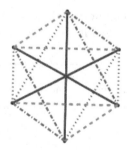

**Fig. 1.** A broken double stars decomposition of $K_6$

**Theorem 1.** *Let $n = 2k$ be an even integer. Then any decomposition of $K_n$ into $k + 1$ star-forests is a broken double stars decomposition.*

But before proving this we need a couple of building blocks. We first note that we can assume that a decomposition of $K_{2k}$ into $k+1$ star-forests cannot contain a star forest with a single component. This is because removing the center of this component and all of the adjacent edges would result in a decomposition of $K_{2k-1}$ into $k < \lceil \frac{2k-1}{2} \rceil + 1$ star forests, which is impossible by the result of Akiyama and Kano [3].

**Lemma 1.** *Let $n = 2k$ be an even integer and let $F_0, \ldots, F_k$ be a decomposition of $K_n$ into $k + 1$ star-forests. If $F_0$ is a perfect matching then every other star-forest is spanning and consists of two components.*

*Proof.* Assume that we have such a decomposition and that one of the star-forests has at least three components. Then it can have at most $2k - 3$ edges. Then the number of edges covered by the decomposition is at most $k + 2k - 3 + (k - 1)(2k - 2) < k(2k - 1) = \binom{2k}{2}$.    □

**Lemma 2.** *Let $n = 2k$ be an even integer and let $\mathcal{F}$ be a decomposition of $K_n$ into $k + 1$ star-forests. If some $F \in \mathcal{F}$ is a perfect matching then this decomposition is a broken double stars decomposition.*

*Proof.* Let us label the vertices of $K_n$ as $\{v_1, v_2, \ldots, v_n\}$. Assume that we have a decomposition of $K_n$ into star-forests such that one of the forests is a perfect matching. Assume without loss of generality that the edges of the perfect matching are given by $\{v_i, v_{i+k}\}$ for $1 \leq i \leq k$. Now we will prove that for each forest $F' \neq F$ there exists $i$ such that the centers of the stars in star-forest $F'$ are exactly $v_i$ and $v_{i+k}$. Assume for contradiction that there is a star-forest $F^*$ in the decomposition whose stars have centers $v_i$ and $v_j$ for some $j \neq i \pm k$. Then consider the star-forest with center in $v_{i+k}$ and let $u$ be the center of the other star in this forest. By Lemma 1, we know that $F^*$ is spanning. Thus $\{v_j, v_{i+k}\} \in E(F^*)$ and either $\{v_i, u\} \in E(F^*)$ or $\{v_j, u\} \in E(F^*)$. In both cases, since $\{v_i, v_{i+k}\}$ is an edge in the perfect matching it means that the forest with centers $v_{i+k}$ and $u$ cannot be spanning since it cannot reach one of the centers of $F^*$. Lastly, we need to prove that inside of each star-forest, stars have equal number of edges. For this we will assume that the forest with centers in $v_i, v_{i+k}$ does not have stars with equal number of edges, without loss of generality the star with center $v_i$ has fewer than $k$ edges. But then there exists $j$ such that neither $v_j$ nor $v_{j+k}$ is connected to $v_i$ in this forest. So both of the edges $\{v_i, v_j\}$ and $\{v_i, v_{j+k}\}$ need to be used by the star-forest with centers $v_j, v_{j+k}$, which is impossible.    □

The following observation has been checked computationally.

**Observation 2.** *Any decomposition of $K_6$ into 4 star-forests is a broken double stars decomposition.*

Finally we proceed with a proof of Theorem 1.

*Proof (of Theorem 1).* We will proceed by induction on $k$. For the base case $k = 3$ the claim holds by Observation 2. Now assume that it holds for $k - 1$. Suppose that we have a decomposition of $K_{2k}$ into $k + 1$ star-forests. We know that we cannot have a star-forest with a single component which is not a single vertex. Further, if each star-forest has at least three components, we cannot cover all of the edges since each component can have at most $2k - 3$ edges and $(k + 1)(2k - 3) < \binom{2k}{2}$. Therefore there is at least one star-forest with two components. Consider the graph obtained by removing the centers (call them $c, d$) of such a star-forest from $K_{2k}$. We are then left with a decomposition of $K_{2k-2}$ into $k$ star-forests. By the inductive hypothesis, this must be a broken double star decomposition. One of the edges removed was the one between the two centers, call it $e$. Assume that $e$ extends a forest with centers $a, b$, forming a forest with centers $a, b, c$ for example. Then, none of the edges between $c, d$ and $a, b$ can be in the same forest as $e$. So assume that two of these edges, $\{a, c\}$ and $\{d, b\}$ belong to another forest with centers $a', b'$. Then, without loss of generality, the edges $\{a, a'\}, \{b, b'\}$ belong to the forest with centers $a, b$ and the edges $\{a, b'\}, \{b, a'\}$ must belong to a forest different than the one with centers $a', b'$, contradicting the inductive assumption. Therefore, $e$ must extend the perfect matching in the decomposition of the smaller graph and the result follows by Lemma 2. □

## 3    Decomposing Complete Geometric Graphs into Plane Star-Forests

Firstly, we will describe a blowup method for generating infinite families of complete geometric graphs on $n$ vertices which can be decomposed into $cn$ star-forests. This blowup method was most likely already known to authors of [12], but they did not explicitly state it. We then use this method to construct a counterexample for Conjecture 1. Afterwards, we construct a complete geometric graph on $2k$ vertices which can be decomposed into $k+1$ plane star-forests. From now on, we will write $GP$ for a complete geometric graph whose underlying point set is $P \subseteq \mathbb{R}^2$.

**Theorem 3.** *Let $c \in (1/2, 1)$ be a constant. If there is a complete geometric graph on $n_0$ points which can be partitioned into $cn_0$ plane star-forests, in such a way that each vertex is a center of at least one tree, then for each integer $k \geq 1$, there exists a complete geometric graph on $kn_0$ points that can be partitioned into $ckn_0$ plane star-forests.*

*Proof.* Let $S$ be the underlying point set of the original complete geometric graph and let $k > 1$ be an integer. Label the points in $S$ by $a_1, \ldots, a_{n_0}$. Now, replace each $a_i$ by a set $A_i = \{a_i^1, \ldots, a_i^k\}$ of $k$ points in general position in such a way that if we choose $b_1, \ldots, b_{n_0}$ where $b_i \in A_i$, we obtain a point set of the same order type as $S$. Call the new point set $S^k$. Now if $F_1, \ldots, F_{cn_0}$ is the decomposition of $GS$ into plane star-forests, from this, we will obtain the decomposition of $GS^k$ into $c(kn_0)$ plane star-forests. Let $a_j$ be the center of a star in $F_i$. We will construct $k$ new stars with centers in $a_j^1, \ldots, a_j^k$. Start with $a_j^1$, add to it all of the edges of the form $\{a_j^1, a_j^l\}$ that were not already used (in the case of $a_j^1$, none were used). Now for each edge of the form $\{a_j, a_{j'}\}$ in $F_i$, add all of the edges from $a_j^1$ to the vertices in $A_{j'}$. Continue doing this for each vertex $a_j^l$, where $l \in \{1, 2, \ldots, k\}$. We do this for each star in $F_i$ and for each forest in the original decomposition. The result of this process is $cn_0$ families of star-forests, each of size $k$. And the planarity of the star-forests follows from the definition of the point set $S^k$. To see this, assume that a tree in the new decomposition has an intersection. Then the intersection is between edges whose 4 vertices are in different $A_i$'s. But if this was the case, then a choice of transversal that includes this 4 vertices would induce a crossing inside the original decomposition of $GS$. □

We note that the assumption that each point is a center of at least one forest is crucial as otherwise the star-forests constructed in the proof do not cover all of the edges.

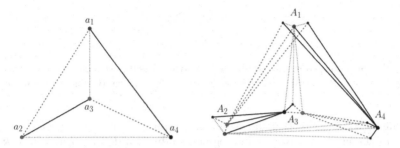

**Fig. 2.** A complete geometric graph on 4 vertices decomposed into three plane star-forests and the corresponding graph on 12 vertices with the decomposition into 9 star-forests (only 4 are drawn for readability). Each vertex of the point set on the left has been used as a center of one tree and colored accordingly.

While Theorem 3 gives us a nice way of constructing infinitely many complete geometric graphs that can be partitioned into few plane star-forests, we still need concrete small examples to be able to produce the infinitudes. One example was given by the authors in [12] and can be found in Fig. 2. This example motivated Conjecture 1. We proceed in a similar fashion.

**Lemma 3.** *There exists a configuration of 6 points in the plane which can be partitioned into 4 plane star-forests in such a way that each point is a center of at least one star.*

*Proof.* We consider a configuration of 6 points which is crossing-minimal according to [13]. We decompose the graph into 4 star-forests as in Fig. 3. The graph has thus been decomposed into three 2-component star-forests colored blue, red and black and one 3-component forest colored purple. □

Now, using the point set of $n_0 = 6$ elements from the above lemma, which can be decomposed into $2n_0/3 = 4$ star-forests, we obtain as an easy corollary a family of point sets on $n = 6k$ points which can be decomposed into $2n/3$ star-forests, thus disproving Conjecture 1. We state this formally below.

**Corollary 1.** *For every $n$ divisible by 6, there exists a geometric graph on $n$ vertices which can be decomposed into $2n/3$ plane star-forests.*

For every $k \in \mathbb{N}$ we construct a point set on $2k$ points that can be decomposed into $k + 1$ plane star-forests. By Theorem 1, one star-forest in the decomposition will be a perfect matching. So our approach will be to first construct the perfect matching as an arrangement of line segments in the plane and then use it to construct the other star-forests. We will say that the arrangement of $k$ line

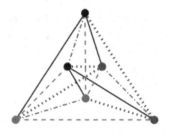

**Fig. 3.** A complete geometric graph on 6 vertices decomposed into four star-forests, vertices are colored the same as trees whose centers they are. (Color figure online)

segments is *SF-extendable* if the geometric graph with underlying point set consisting of the endpoints of the line segments admits a decomposition into $k+1$ plane star-forests, one of which is the perfect matching given by the arrangement. We say that two line segments are in a *stabbing* position if the convex hull of their endpoints is a triangle. If $s = ab$ and $l = cd$ are two line segments in a stabbing position and the convex hull of $\{a, b, c, d\}$ contains $c$ or $d$ in the interior, we say that $l$ *stabs* $s$.

We now provide a necessary condition for an arrangement to be SF-extendable.

**Lemma 4.** *Let $\mathcal{L}$ be an arrangement of line segments in the plane. If $\mathcal{L}$ is SF-extendable then every pair of segments from $\mathcal{L}$ is in a stabbing position.*

*Proof.* Assume on the contrary that there are two line segments $ab$ and $cd$ which are not in a stabbing position. Then the convex hull of $\{a, b, c, d\}$ forms a convex quadrangle. Assume that the cyclic ordering of vertices along the convex hull is $(a, b, c, d)$. The two diagonals of this quadrangle intersect. Assume that $\mathcal{L}$ is SF-extendable. If the star-forest with centers in $a$ and $b$ contains the edges $\{a, c\}$ and $\{b, d\}$, then the star-forest with centers in $c$ and $d$ is not planar, contradicting the assumption that $\mathcal{L}$ is SF-extendable. If the star-forest with centers in $a$ and $b$ contains the edges $\{a, d\}$ and $\{b, c\}$ then it is not planar, again contradicting the assumption that $\mathcal{L}$ is SF-extendable. $\qquad\square$

Let $L_1$ be a segment of length 1 with center at the origin. Call its left endpoint $a_1$ and the right endpoint $b_1$. Let $a_1, a_2, \ldots, a_k, b_1$ be vertices of some convex $(k+1)$-gon $P$ such that $L_1$ is an edge of $P$ and each $a_i$ for $i > 1$ is placed inside of the top left quadrant of the plane. We will now construct line segments $L_2, \ldots L_k$ with endpoints $a_i, b_i$ respectively. For each $i > 1$ place each $b_i$ in the intersection of interiors of all of the triangles $(a_l, b_l, a_j)$ where $l < j \leq i$ and the top right quadrant of the plane. We call such a line arrangement a $k$-staircase. See Fig. 4 for examples of a 3-staircase and a 4-staircase and their extension into a star-forest decomposition. It is not completely obvious that this construction is always feasible. So we prove it.

**Lemma 5.** *For any $k \in \mathbf{N}$ and any convex $(k+1)$-gon $P$ whose vertices are positioned as described above, there exists a $k$-staircase whose convex hull is $P$.*

*Proof.* The essence of this proof will be to prove that at each step we can place $b_i$ as described above. So we write $S_i$ for the intersection of all the triangles $(a_l, b_l, a_j)$ for $l < j \leq i$ and the upper right quadrant of the plane. It is clear that the interior of $S_2$ is nonempty. Now assume that the interior of $S_i$ is nonempty for some $i > 2$. Then $S_{i+1} = S_i \cap \bigcap_{j < i}(a_j, b_j, a_{i+1})$. And since by construction $b_i$ is in the interior of $S_i$ then there exists an open ball $B$ of radius $\epsilon > 0$ centered at $b_i$ such that $B \subset S_i$. Further, $B$ clearly has to intersect the triangles $(a_j, b_j, a_{i+1})$ for every $j \leq i$ since each of these triangles either contains $b_i$ in the interior or as a vertex in case of $j = i$. Thus it is clear that the interior of $S_{i+1}$ is nonempty as well and the result follows.                                                   □

**Theorem 4.** *For each $k \geq 1$, there exists an SF-extendable arrangement of $k$ line segments.*

*Proof.* We will show that a $k$-staircase is SF-extendable. The star-forest with centers in $a_i, b_i$ has the edge set

$$\{\{a_i, a_j\} : j > i\} \cup \{\{a_i, b_k\} : k < i\} \cup \{\{b_i, b_j\} : j > i\} \cup \{\{b_i, a_k\} : k < i\}.$$

It is clear that every edge will be covered by this decomposition. Now we will check planarity of the forest with centers $a_i, b_i$. Edges of the form $\{a_i, a_j\}$ for $j > i$ cannot cross edges of the form $\{b_i, b_j\}$ since points $a_i$ are separated from points $b_i$ by the $y$-axis. Edges of the form $\{a_i, a_j\}$ cannot cross edges of the form $\{b_i, a_k\}$ where $k < i \leq j$ since $a_j, a_k$ are in different half-planes determined by a line through $a_i, b_i$. The other cases are similar.                          □

**Theorem 5.** *For each $n$ there exists a complete geometric graph on $n$ points which can be decomposed into $\lceil \frac{n}{2} \rceil + 1$ plane star-forests.*

*Proof.* If $n = 2k$ is even, the complete geometric graph is given by the $k$-staircase. In the case $n = 2k - 1$, take a single point away from the $k$-staircase and the resulting complete geometric graph gives the result.                                         □

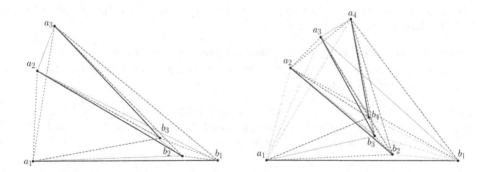

**Fig. 4.** Point sets in 3-staircase and 4-staircase configurations.

The $k$-staircase produces a point set with a fairly large convex hull. So a natural question is to ask if this is in fact necessary or can we find point sets with a smaller convex hull. The following construction shows that we can in fact do this. We will again construct an arrangement of line segments. Start with a segment of length 1 centered at the origin, and call its endpoints $a_1, b_1$ as before. We place points $a_2, \ldots, a_k$ on a concave chain and place $b_2, \ldots, b_k$ so that they obey the same conditions as in the construction of the $k$-staircase. This way we obtain an arrangement of line segments $L_1, \ldots, L_k$ which is SF-extendable which we will call $k$-comet. To prove this fact we can use the same star-forest decomposition as we did in the proof of Theorem 4. However, where $k$-staircase defines a point set on $2k$ points with convex hull of size $k + 1$, the $k$-comet defines a point set on $2k$ points with convex hull of size 3, see Fig. 5.

In fact, it is not hard to construct such a point set with a convex hull of any size between 3 and $k + 1$. First note that the $k$-comet is "obtained" from a $k$-staircase by turning a convex polygonal line into a concave one. We say "obtained" since one also needs to adjust the positions of $b_i$'s in the construction. Now we can do the same thing but turn only an initial segment of the convex polygonal line into a concave segment and obtain a smaller convex hull. Even further, one can make the following observation.

**Observation 6.** *Let $k \geq 2$. If $G$ is a complete geometric graph on $n = 2k$ vertices which can be decomposed into $k + 1$ plane star-forests, then the size of the convex hull of $V(G)$ is at most $k + 1$.*

*Proof.* By Theorem 1, we know that one of the star-forests will be a perfect matching. If $V(G)$ has a convex hull of size at least $k + 2$, then the perfect matching needs to use at least 2 edges of the convex hull. But then the perfect matching cannot be SF-extendable by Lemma 4 since the edges of the convex hull are never in a stabbing position. □

Based on this observation we make the following conjecture.

*Conjecture 2.* Let $n \geq 3$ be odd, $G$ a complete geometric graph on $n$ vertices which can be decomposed into $\lceil \frac{n}{2} \rceil + 1$ plane star-forests. Then the convex hull of $V(G)$ has size at most $\lceil \frac{n}{2} \rceil + 1$.

What is not clear is if one can construct point sets with convex hull of size $k + 1$ that are not a $k$-staircase. Our computations found no such point sets. So we conjecture the following.

*Conjecture 3.* If a complete geometric graph $G$ on $2k$ points admits a decomposition into $k + 1$ plane star-forests and the size of the convex hull of $V(G)$ is $k + 1$ then $V(G)$ can be described as a $k$-staircase.

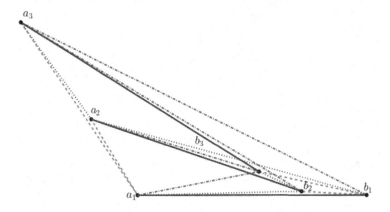

**Fig. 5.** Point set in 3-comet configuration.

# 4    Computing Plane Star-Forest Decompositions on Small Point Sets

Using a simple computer search, we managed to find all point sets on 6 points that can be decomposed into 4 plane star-forests and all point sets on 8 points that can be decomposed into 5 plane star-forests. Out of the 16 order types on 6 points, which can be found on [1], we have found decompositions which satisfy the requirements from Theorem 3 for 6 of them. Those point sets and corresponding partitions can be seen in Fig. 6. Out of 3315 order types on 8 points, we have found such decompositions for 411 of them. The code is available at [6]. Of course, for even bigger point sets, a completely different approach would be needed, as the number of different order types for $n \geq 12$ becomes too large to handle.

# 5    Further Research and Open Questions

While our construction is optimal in the sense that it minimizes the size of minimal decomposition into plane star-forests, it is still unclear what characterizes complete geometric graphs that admit such a decomposition. We have made no progress towards solving this problem but computational results show that these point sets can be quite diverse. However, a possibly easier question to answer is the following.

*Question 2.* What are sufficient conditions for an arrangement of line segments to be SF-extendable?

Lemma 4 gives us a necessary condition, but it is not hard to see that this is not sufficient to guarantee SF-extendability. For example, Felsner et. al constructed in [9], a point set on points called the exploding double chain. This point set has other interesting properties, but for us it is important since its subsets can be embedded with pairwise stabbing segments, see Fig. 7. However it does

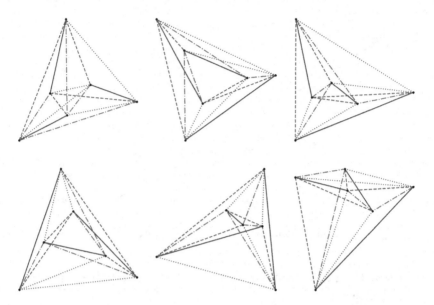

**Fig. 6.** Star-forest decompositions of the point sets on 6 points that admit them.

not admit a decomposition into plane star-forests already for $n = 6$, as can be seen by observing that it is not present in Fig. 6. We also note that there is an interesting variation of the original problem that we have not explored yet, but where our approach from Theorem 3 can also be used. We define a *k-star-forest* to be a star-forest with at most $k$ components. Authors in [12] proposed the following conjecture:

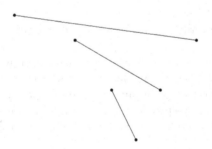

**Fig. 7.** Exploding double chain on 6 points

*Conjecture 4.* [12] The number of plane $k$-star-forests needed to decompose a complete geometric graph is at least $\frac{(k+1)n}{2k}$.

Our example does not show anything regarding Conjecture 4. But, it is not hard to see that the construction from Theorem 3 preserves the maximal number of components among all forests. Thus, we believe a similar approach could be used to attack this conjecture.

**Acknowledgments.** We thank Pavel Valtr and Jan Kynčl who proposed the problem to us and gave a lot of useful comments and suggestions during the combinatorial problems seminar at Charles University.

# References

1. Aichholzer, O.: Enumerating order types for small point sets with applications. http://www.ist.tugraz.at/staff/aichholzer/research/rp/triangulations/order types/
2. Akbari, S., Jensen, T.R., Siggers, M.H.: Decompositions of graphs into trees, forests, and regular subgraphs. Discret. Math. **338**(8), 1322–1327 (2015)
3. Akiyama, J., Kano, M.: Path factors of a graph. In: Graphs and applications (Boulder, Colo., 1982), pp. 1–21. Wiley-Intersci. Publ., Wiley, New York (1985)
4. Algor, I., Alon, N.: The star arboricity of graphs. Discrete Math. **75**(1-3), 11–22 (1989). https://doi.org/10.1016/0012-365X(89)90073-3, graph theory and combinatorics (Cambridge, 1988)
5. Alon, N., McDiarmid, C., Reed, B.: Star arboricity. Combinatorica **12**(4), 375–380 (1992). https://doi.org/10.1007/BF01305230
6. Antić, T., Glišić, J., Milivojčević, M.: https://github.com/milivojcevic6/Star-Forest-Decompositions
7. Bose, P., Hurtado, F., Rivera-Campo, E., Wood, D.R.: Partitions of complete geometric graphs into plane trees. Comput. Geom. **34**(2), 116–125 (2006). https://doi.org/10.1016/J.COMGEO.2005.08.006
8. Dujmović, V., Wood, D.R.: Graph treewidth and geometric thickness parameters. Discrete Comput. Geom. **37**(4), 641–670 (2007). https://doi.org/10.1007/s00454-007-1318-7
9. Felsner, S., Schrezenmaier, H., Schröder, F., Steiner, R.: Linear size universal point sets for classes of planar graphs. In: 39th International Symposium on Computational Geometry, LIPIcs. Leibniz International Proceedings in Informatics, vol. 258, pp. Art. No. 31, 16. Schloss Dagstuhl. Leibniz-Zent. Inform., Wadern (2023). https://doi.org/10.4230/lipics.socg.2023.31
10. Karisani, N., Mahmoodian, E.S., Sobhani, N.K.: On the star arboricity of hypercubes. Australas. J. Combin. **59**, 282–292 (2014)
11. Lonc, Z.: Decompositions of graphs into trees. J. Graph Theory **13**(4), 393–403 (1989)
12. Pach, J., Saghafian, M., Schnider, P.: Decomposition of geometric graphs into star-forests. In: Bekos, M.A., Chimani, M. (eds.) Graph Drawing and Network Visualization. GD 2023. LNCS, vol. 14465, pp. 339–346. Springer, Cham (2023). https://doi.org/10.1007/978-3-031-49272-3_23
13. Pilz, A., Welzl, E.: Order on order types. Discret. Comput. Geom. **59**(4), 886–922 (2018). https://doi.org/10.1007/S00454-017-9912-9
14. Shyu, T.: Decomposition of complete graphs into paths and stars. Discret. Math. **310**(15–16), 2164–2169 (2010)
15. Trao, H.M., Chia, G.L., Ali, N.A., Kilicman, A.: On edge-partitioning of complete geometric graphs into plane trees (2019). https://arxiv.org/abs/1906.05598

# Convex-Geometric $k$-Planar Graphs Are Convex-Geometric $(k + 1)$-Quasiplanar

Todor Antić[(⊠)]

Faculty of Mathematics and Physics, Department of Applied Mathematics, Charles
University, Prague, Czech Republic
todor@kam.mff.cuni.cz

**Abstract.** A drawing of a graph is $k$-planar if every edge has at most $k$
crossings with other edges of the graph and it is $k$-quasiplanar if it has no
set of $k$ pairwise crossing edges. We say that a graph drawing is simple
if two edges intersect at most once. In 2020, Angelini et al. proved that
all simple $k$-planar graphs are simple $(k + 1)$-quasiplanar, which was the
first non-trivial relationship between these two classes. We say that a
graph drawing is convex-geometric if its vertices are drawn as points on
a circle and its edges are drawn as straight line segments between them.
In this paper we prove that, for $k \geq 2$, every convex-geometric $k$-planar
graph is convex-geometric $(k + 1)$-quasiplanar.

## 1 Introduction

A *topological* graph is a graph drawn in the plane with vertices as points and
edges as simple curves between the points. We say that a topological graph is
*simple* if each pair of edges crosses at most once. A topological graph is called
a *geometric graph* if its edges are drawn as straight-line segments between the
vertices. A geometric graph is called *convex-geometric* graph if its vertices are
in convex position. For the problem we are considering the actual position of
the vertices on the convex hull is not relevant and we only care about the cyclic
ordering of the vertices. Therefore we assume that the vertices lie on a common
circle in the plane.

We say that a graph drawing is *$k$-quasiplanar* if there is no set of $k$ pairwise crossing edges. We say that an abstract graph is (convex-geometric) $k$-
quasiplanar if it admits a (convex-geometric) $k$-quasiplanar drawing.

Another interesting class of graphs are so-called *$k$-planar* graphs. A graph
drawing is said to be $k$-planar if every edge has at most $k$ crossings with other
edges in the graph. Moreover, an abstract graph is said to be (convex-geometric)
$k$-planar if it admits a (convex-geometric) $k$-planar drawing. Notice that both
$k$-planar and $k$-quasiplanar graphs can be defined by forbidden configurations;
for an example see Fig. 1. Also note the following connection between these two
classes of graphs:

**Observation 1** *If $G$ is a $k$-planar graph drawing then $G$ is $(k + 2)$-quasiplanar.*

A. A. Rescigno and U. Vaccaro (Eds.): IWOCA 2024, LNCS 14764, pp. 138–150, 2024.
https://doi.org/10.1007/978-3-031-63021-7_11

To see that this is true, notice that in a set of $k + 2$ pairwise intersecting edges, each edge crosses $k + 1$ other edges, so the condition for $k$-planarity is not satisfied.

**Fig. 1.** Forbidden configurations in 3-planar (left) and 3-quasiplanar graphs (right).

Further, every $k$-planar ($k$-quasiplanar) graph is $(k + 1)$-planar ($(k + 1)$-quasiplanar) so they form a natural hirearchy. It is natural to ask if there exists a function $f$ such that every $k$-planar graph is $f(k)$-quasiplanar. For a long time the best bound for $f$ was the obvious one from Observation 1. The first improvement was made by Angelini et al. [3], who proved that every simple $k$-planar graph is simple $(k + 1)$-quasiplanar. For now, there is no better bound for this problem nor an analogous result for non-simple drawings.

Both $k$-planar and $k$-quasiplanar graphs are a part of the family of so-called beyond-planar graphs. The study of beyond-planar graphs has emerged as an area of interest in graph drawing community recently. This is in part due to practical applications in data visualisation, since it has been shown that topological and geometrical properties of graph drawings impact human understanding of the graph. For more information about beyond-planar graphs see the survey by Didimo, Liotta and Montecchiani [6].

Convex-geometric, $k$-quasiplanar graphs have been a particular area of interest and have proven to be easier to work with than general geometric and topological counterparts. For example, the following conjecture is still open for general $k$-quasiplanar graphs:

*Conjecture 1.* Every $k$-quasiplanar graph on $n$ vertices has at most $O_k(n)$ edges.

Conjecture 1 is solved for $k = 4$ [1,2] and is still open for $k > 4$ [7]. While it is known to be true for convex-geometric $k$-quasiplanar graphs [5]. For more information about convex-geometric graphs see [8].

## 1.1 Contribution

In this paper we focus on the class of convex-geometric graphs. Inspired by the main result from the paper by Angelini et al. [3], we use a similar approach and prove that convex-geometric $k$-planar graphs are convex-geometric $(k + 1)$-quasiplanar for all sufficiently large $k$. It is important to note that we do not guarantee that the resulting $(k + 1)$-quasiplanar drawing is still $k$-planar. Our proof relies on the fact that crossings in convex-geometric graphs are completely determined by the ordering of the vertices of the graph along the circle. This allows us to define a very natural flipping operation which "untangles" sets of $k + 1$ crossing edges. For simplicity, we call a set of $k + 1$ pairwise crossing edges a $(k+1)$-*crossing*. By *untangling* of a $(k+1)$-crossing we understand the process of redrawing the graph in such a way that at least two edges of a given $(k + 1)$-crossing do not cross anymore. We also give partial results that could be used with some further case analysis to improve our main result.

### 1.2   Organization of the Paper

The paper will be organized as follows: In Sect. 2, we introduce the notation and most of the definitions that we use throughout the paper. In Sect. 3, we discuss the structure of $(k + 1)$-crossings in convex-geometric graphs. In Sect. 4, we introduce the flipping operation for the untangling of $(k + 1)$-crossings. In Sect. 5, we construct a special ordering on the set of $(k+1)$-crossings of a convex-geometric $k$-planar graph. In Sect. 6, we introduce the global flipping operation and prove the main result of the paper. In Sect. 7, we give some results regarding $k$-crossings in convex-geometric $k$-planar graphs and strategy for showing that such graphs are convex-geometric $k$-quasiplanar.

## 2   Notation

Throughout the paper we usually abuse notation and write $G$ for both the drawing of the graph and the underlying abstract graph. In cases where distinction is important it will be specified. Further we assume that all the vertices are drawn on the circle of radius 1 centered at $(0, 0)$. We call a set of $h$ pairwise crossing edges an $h$-*crossing*. We will often refer to an $h$-crossing $X$ as a set of pairwise crossing edges or as the graph (drawing) containing the edges and their points interchangeably.

   If $X$ is a given $h$-crossing, we label its vertices in the following way. We pick a vertex arbitrarily and label it $v_1^X$, then we move clockwise from $v_1^X$ and label next $h - 1$ vertices by $v_2^X, \ldots, v_h^X$. Then we continue moving clockwise and label the next $h$ vertices as $u_1^X, \ldots, u_h^X$. Since $X$ is a $h$-crossing, it is not hard to see that two vertices $a, b$ of $X$ form an edge if and only if there are $h - 1$ vertices of $X$ in between them when traversing the circle from $a$ to $b$. In particular, this means that $\{u_i^X, v_i^X\}$ is always an edge of $X$ and we label it $e_i^X$. We omit the superscript $X$ when it is clear which crossing we are talking about. For an example, see Fig. 2.

   Now let $G$ be a convex-geometric drawing of a graph and $a, b$ two vertices of $G$. Assume that $a$ comes before $b$ when traversing the circle in a clockwise direction starting from the point $(0, 1)$. We define the *closed interval* between $a$

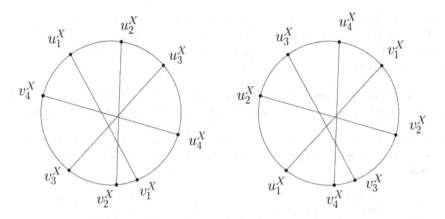

**Fig. 2.** Two labellings of a given 4-crossing, starting from different vertices.

and $b$ to be the set of all vertices between $a$ and $b$ starting from $a$ and moving clockwise along the circle and ending at $b$. We denote it by $[a, b]$. If we do not want to include the endpoints we write $(a, b)$ and call it the *open interval* between $a$ and $b$. If we want to include $a$ but not $b$ we write $[a, b)$ and $(a, b]$ in the opposite case.

## 3    (k+1)-Crossings in Convex-Geometric K-Planar Graphs

Now we describe the structure of $(k+1)$-crossings in a convex-geometric $k$-planar graph $G$.

**Lemma 1.** *Let* $X_1, X_2, ..., X_m$ *be* $(k + 1 - i)$-*crossings in a* $k$-*planar graph* $G$ *where* $1 \leq i \leq k$. *Then either* $\bigcap_{j=1}^{m} |E(X_j)| \geq \lceil k + \frac{m-mi}{m-1} - 1 \rceil$ *or* $\bigcap_{j=1}^{m} |E(X_j)| = 0$.

*Proof.* We restate this in terms of intersection graph of $E(G)$. Then we have $m$ cliques of size $(k + 1 - i)$ in a graph of maximum degree $k$. Assume now that they have $f$ common vertices. Then degree of any such vertex $v$ is at most $k$. So we calculate:

$$\deg(v) = m(k + 1 - i - f) + f - 1 \leq k$$

$$k(m - 1) + m(1 - i) - f(m - 1) - 1 \leq 0$$

$$k + \frac{m - mi}{m - 1} - 1 \leq f.$$

And the result follows.                                                      □

An easy corollary of Lemma 1 is that $(k + 1)$-crossings are edge disjoint. Further, we can say something about the possible number of shared vertices between $(k + 1)$-crossings. Before that, we need the following lemmas:

**Lemma 2.** *Let $X$ be a $(k+1)$-crossing in a convex-geometric $k$-planar graph $G$. Let $v_i$ be a vertex of $X$, where without loss of generality $k+1 \neq i \neq 1$. If $e$ is an edge in $E(G) \setminus E(X)$ with endpoint $v_i$ then the other endpoint of $e$ is in one of the intervals $(v_i, v_{i+1}]$ or $[v_{i-1}, v_i)$.*

*Proof.* If $e$ has the other endpoint anywhere else in $G$, then it would cross the edge $e_{i-1}$ or the edge $e_{i+1}$, contradicting the assumption that $G$ is $k$-planar. □

**Lemma 3.** *Let $X$ be a $(k+1)$-crossing in a convex-geometric $k$-planar graph $G$. Then there is no edge in $E(G) \setminus E(X)$ between vertices located in disjoint open intervals between vertices of $X$.*

**Lemma 4.** *If $X, Y$ are two $(k+1)$-crossings in a convex-geometric $k$-planar graph $G$, then convex hulls of $V(X)$ and $V(Y)$ are interior-disjoint.*

*Proof.* Denote the convex hulls from the statement by $C_X$ and $C_Y$. If $\mathrm{Int}(C_X) \cap \mathrm{Int}(C_Y) \neq \emptyset$ that means that there are points $a_X, b_X \in V(X)$ and an edge $e$ of $C(Y)$ such that $a_X$ and $b_X$ are in different halfplanes determined by $e$. However, this would imply that there is an edge in $E(X)$ with endpoints in different open intervals determined by vertices of $Y$, contradicting Lemma 3. □

Now we can state the result we have been building towards:

**Theorem 2.** *Let $X, Y, Z$ be $(k+1)$-crossings in a convex-geometric $k$-planar graph $G$. Then:*

1. *$X$ and $Y$ share at most two vertices, and*
2. *$X, Y$ and $Z$ share at most one vertex.*

*Proof.* Assume for contradiction that $|V(Y) \cap V(X)| = 3$. Then we contradict Lemma 4 since both convex hulls would contain any convex combination of the three points in the intersection. Moreover, by Lemma 3, this means that $Y$ is contained in an interval between two vertices of $X$ which are consecutive along the circle. For the second part of the theorem, assume $V(Y) \cap V(X) \cap V(Z) = \{a, b\}$. Denote by $C_X$, $C_Y$ and $C_Z$ the convex hulls of $V(X)$, $V(Y)$ and $V(Z)$ respectively. Therefore, the line segment between $a$ and $b$ is in all of $C_X$, $C_Y$ and $C_Z$. Without loss of generality, this means that $C_X$ and $C_Y$ are not interior-disjoint, contradicting Lemma 4 again. □

**Observation 3** *Let $G$ be a convex-geometric $k$-planar graph and $X, Y$ two $(k+1)$-crossings in $G$ such that $X \cap Y = \{v_i^X, v_{i+1}^X\}$. Then the edge $\{v_i^X, v_{i+1}^X\}$ (if it exists) is not a part of $Y$.*

*Proof.* Assume on contrary that $\{v_i, v_{i+1}\}$ is an edge of $Y$, then it is crossed by all the other edges in $Y$. As the crossings in the convex-geometric drawing are completely determined by the ordering of vertices this would imply that $Y$ contains vertices both inside and outside of the interval $[v_i^X, v_{i+1}^X]$, which we know is impossible by Lemma 3. □

## 4  The Flipping Operation

In this section and the next one we describe our main methods for untangling $(k + 1)$-crossings. We begin by describing the local operation and then we show how to apply it globally.

Assume for now that $G$ is a convex-geometric $k$-planar graph and that $X$ is a $(k + 1)$-crossing in $G$. We then pick two consecutive vertices $a, b$ in $X$. Then we change the position of vertices $a$ and $b$ and we invert completely the interval $(a, b)$ not to disturb any of the crossings inside the interval nor create new ones. We denote this operation by $\mathrm{Flip}(a, b)$. For greater clarity, consider Fig. 3.

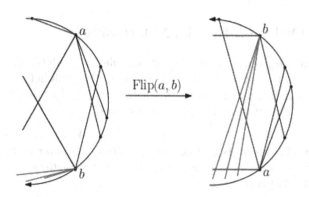

**Fig. 3.** The flip operation.

It is obvious that $\mathrm{Flip}(a, b)$ untangles the $(k + 1)$-crossing $X$. However, it is not completely clear that it creates no new $(k + 1)$-crossings. This is the topic of the following lemma.

**Lemma 5.** *Let $G$ be a convex-geometric $k$-planar graph and $X$ a $(k+1)$-crossing in $G$. Let $G'$ be the graph obtained from $G$ by $\mathrm{Flip}(v_i, v_{i+1})$. If $e$ and $e'$ are edges which cross in $G'$ but not in $G$ then (up to renaming), one of the following holds:*

1. *$e$ is the edge $e_i$ and $e'$ is an edge connecting $v_{i+1}$ to a vertex in the interval $(v_{i+1}, v_{i+2}]$.*
2. *$e$ is the edge $e_{i+1}$ and $e'$ is an edge connecting $v_i$ to a vertex in the interval $[v_{i-1}, v_i)$.*
3. *$e$ is an edge connecting $v_i$ to a vertex in the interval $[v_{i-1}, v_i)$ and $e'$ is an edge connecting $v_{i+1}$ to a vertex in the interval $(v_{i+1}, v_{i+2}]$.*

*Proof.* The flipping operation by definition only affects the interval $[v_i, v_{i+1}]$ and the two intervals neighbouring it. Further, we flip the entire interval $[v_i, v_{i+1}]$ so none of the edges in it are affected. All the vertices in the intervals $[v_{i-1}, v_i)$ and $(v_{i+1}, v_{i+2}]$ remain in their original positions so none of the edges inside those intervals are altered. By Lemma 2, there are no edges between the intervals $[v_{i-1}, v_i), (v_i, v_{i+1})$ and $(v_{i+1}, v_{i+2}]$ so the only edges that could be altered are the ones from the statement of the lemma. □

**Corollary 1.** *Let $G$ be a convex-geometric $k$-planar graph and $X$ a $(k+1)$-crossing in $G$. Let $G'$ be the graph obtained from $G$ by $Flip(v_i, v_{i+1})$. Then $G'$ contains no 3-crossings which were not present in $G$.*

*Proof.* Assume that there is a new 3-crossing $X$ in $G'$, denote by $a, b, c$ the edges of $X$. By Lemma 5, we know that in order to form a new 3-crossing, two of the edges need to intersect in $G$, assume that $b, c$ intersect in $G$. If $a, b$ and $a, c$ form configurations from part 1. of Lemma 5, that would mean that $a = e_i$ and that $b, c$ share an endpoint, contradicting our assumption that $b$ and $c$ intersect. Similar logic leads to contradiction in the case when $a, b$ and $a, c$ form configurations from part 2 of Lemma 5. □

## 5    Linear Order on the (k+1)-Crossings

Assume that we are given a convex-geometric $k$-planar graph $G$. We now describe a partial order on the set of $(k+1)$-crossings in $G$ that will help us perform the flipping operation globally. This ordering will be essential to prove our theorem. We denote by $\chi$ the set of all $(k+1)$-crossings in $G$.

**Definition 1.** *Let $X \in \chi' \subseteq \chi$ be a $(k+1)$ crossing. We call $X$ a leaf-crossing if it has an interval between its vertices which contains all other vertices of $X$ and no vertices of any other $(k+1)$-crossing. We call the vertices of $X$ that define this interval the leaf vertices of $X$*

It is easy to observe that a leaf always exists for any nonempty $\chi' \subseteq \chi$. Then we can define our linear ordering as follows:

1. Let $\chi_0 = \chi$. Let $X_1, ..., X_{l_0}$ be the leaves in $\chi_0$. They will be $l_0$ smallest elements in our linear order, ordered arbitrarily among themselves.
2. For $i \geq 1$, let $X_1^i, ... X_{l_{i-1}}^i$ be the leaves in $\chi_{i-1}$ and define $\chi_i = \chi_{i-1} \setminus \bigcup_{j=0}^{l_{i-1}} X_j^i$. Order the elements $X_1, ..., X_{l-1}$ arbitrarily and add them to the linear order up to now.

Now we prove an important property of this ordering.

**Lemma 6.** *Let $G$ be a convex-geometric $k$-planar graph with a fixed drawing. If $X_n$ is the $n^{th}$ smallest element in the ordering on $\chi$, then:*

$$|V(X_n) \cap \bigcup_{i=1}^{n-1} V(X_i)| \leq 2.$$

*Proof.* Assume that the statement is not true, i.e. that

$$|V(X_n) \cap \bigcup_{i=1}^{n-1} V(X_i)| \geq 3.$$

That would mean that $V(X_n)$ occupies two intervals between the vertices of $\bigcup_{i=1}^{n-1} V(X_i)$, but this cannot happen by the proof of Theorem 2. □

# 6   The Global Flip

In this section, we develop a method to untangle all $(k+1)$-crossings without creating new ones. We assume that we have a convex-geometric $k$-planar graph $G$ with $k \geq 2$ and denote the set of all $(k+1)$-crossings by $\chi$. We fix the ordering from Sect. 5.

Now we explain how to globalize the flipping operation. We start from $X_1$ and perform $\text{Flip}(a^{X_1}, b^{X_1})$, where $a^{X_1}, b^{X_1}$ are the two consecutive vertices of $X_1$ such that there are $k$ vertices of $X_1$ between $a^{X_1}$ and the closest leaf vertex of $X_1$ and there are $k$ vertices of $X_1$ between $b^{X_1}$ and the closest leaf vertex of $X_1$. Then we proceed to the next $(k+1)$-crossing in the ordering and repeat the action. We repeat this process until we have no more $(k+1)$-crossings left.

It is not immediately clear that this works. Firstly, we proved that if $G$ is $k$-planar then a single flip creates no new $(k+1)$-crossing, but it is very easy to see that flipping operation destroys the $k$-planarity (in fact it can make the local crossing number almost arbitrarily big) so we cannot apply Lemma 5 to any subsequent flips. In general, we write $G_i$ for the graph drawing obtained in the $i^{th}$ step of the global flipping and $G_0$ for the original graph drawing.

When talking about a specific interval in $G$, for example $[a, b]$, it may happen that the order of the vertices inside $[a, b]$ is changed during the flipping process or even that new vertices have been added to the interval which were not in the interval originally. In order to avoid confusion, we write $[a, b]_0$ to denote the set of vertices that are in the interval $[a, b]$ inside the original drawing of $G$.

The following observation is a consequence of Lemma 6 and the definition of the ordering.

**Observation 4** *Let $G$ be as above. Then for every $n > 0$, $[v_1^{X_{n+1}}, u_k^{X_{n+1}}]_0$ has the same order along the circle in $G_n$ as in $G$, up to reversal.*

*Proof.* If $[v_1^{X_{n+1}}, u_{k+1}^{X_{n+1}}]_0$ has different ordering in $G_n$, that would mean that during the flipping process, a vertex of $X_{n+1}$ was flipped. However, by definition of the global flip we know that at each step of the flipping we perform $\text{Flip}(a^{X_i}, b^{X_i})$ where $a^{X_i}, b^{X_i}$ are the vertices which are furthest away from the leaf vertices of $X_i$. It is then impossible that $a^{X_i}, b^{X_i}$ belong to another $(k+1)$-crossing $X_j$ for $j > i$.  □

Now this corollary is easy to check.

**Corollary 2.** *For every $n > 0$, drawing of the subgraph induced by $V(X_{n+1})$ is $k$-planar inside $G_n$.*

In essence, what Observation 4 tells us is that each new flip affects a disjoint interval from all the previous ones, and that is enough to prove our main result:

**Theorem 5.** *For every $k \geq 2$, every convex-geometric $k$-planar graph is convex-geometric $(k+1)$-quasiplanar.*

*Proof.* We already know that each flip untangles a $(k + 1)$-crossing inside of which it was performed. If we also show that it creates no new $(k + 1)$ crossings, we are done. We prove even more. We prove the following: "For each $n > 0$, if $S$ is a set of edges that pairwise cross in $G_n$ but not in $G_{n-1}$, then $|S| \leq 2$". We proceed by induction on $n$, where the base case $n = 1$ is the content of Lemma 5. We can now assume that the statement holds for all numbers $1, \ldots, n$. Then we look at $G_{n+1}$, we know that it has been obtained by operation $\text{Flip}(a^{X_{n+1}}, b^{X_{n+1}})$ in $G_n$. But by Observation 4 we know that this is the same as performing this operation in $G_0$ so by a logic analogous to the proof of the Lemma 5, we can complete the proof.                                                                                      □

It is not hard to see that the convex-geometric $(k + 1)$-quasiplanar drawing obtained by the flipping process is usually not going to be convex-geometric $k$-planar. We also provide the following construction of a $k$-planar graph which does not admit a convex-geometric $k$-planar, $(k + 1)$-quasiplanar drawing.

For each $k \geq 2$, let $G_k$ be a convex geometric graph constructed as follows. We start with a $(k + 1)$-crossing $X_k$. We abandon our previous convention and label the vertices of $X_k$ by $v_1, v_2, \ldots, v_{2k+2}$, starting from an arbitrary vertex and labeling them as we traverse the circle clockwise. Label the edges starting from $v_1, \ldots, v_{k+1}$ by $e_1, \ldots, e_{k+1}$ respectively. Finally, we add $k + 1$ vertices inside each interval $[v_i, v_{i+1}]$ and connect each vertex to both endpoints of the interval. We label these vertices $a_i^1, \ldots a_i^{k+1}$. We label them by traversing the interval from $v_i$ to $v_{i+1}$ and labelling the $j^{th}$ vertex we see by $a_i^j$. This finishes the construction of $G_k$. See Fig. 4 for an illustration. The proof of the following observation is omitted due to length constraints and will be available in the full version.

**Observation 6** *For each $k \geq 2$, graph $G_k$ cannot be redrawn so that it is convex-geometric $k$-planar and $(k + 1)$-quasiplanar.*

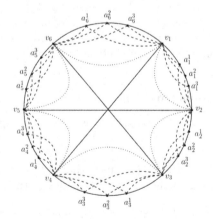

**Fig. 4.** Graph $G_2$, some edges are drawn as curves for the purpose of readability.

# 7    Convex-Geometric K-Quasiplanarity

Starting from now, we will be talking about *maximal $k$-crossings*. We say that a $k$-crossing is maximal if it is not contained in any $(k + 1)$-crossing. We now generalize our previous work to prove that for a sufficiently big $k$, every convex-geometric $k$-planar graph is convex-geometric $(k + 1)$-quasiplanar. We believe that this illustrates a template on how to improve this result further, but better analysis would be needed. All of the proofs from this section are very similar to the ones in the previous sections, with slight differences in the details which we leave out due to length constrains.

From now on, we focus on the case where $G$ is convex-geometric $k$-planar.

**Lemma 7.** *Let $X, Y$ be two edge-disjoint maximal $(k + 1 - i)$-crossings in $G$. Then $V(X) \cap V(Y) \leq 2 + 2i$. Further, all except $2i$ of the vertices of $Y$ are contained in a single $X$ interval.*

*Proof.* Assume that $|V(X) \cap V(Y)| \geq 2i + 3$. That means that there is an edge $e \in V(X)$ such that there are at least $i + 1$ vertices of $Y$ in each half-circle determined by $e$. This further means that at least $i + 1$ edges of $Y$ need to cross $e$, which is impossible by $k$-planarity of $X$.    □

From now onwards, we focus on maximal $k$-crossings in convex-geometric $k$-planar graphs. Next observations all have very similar proofs as to the corresponding statements for $(k + 1)$-crossings in convex-geometric $k$-planar graphs.

**Observation 7** *Let $X, Y, Z$ be three edge-disjoint $k$-crossings in $G$, then $|V(X) \cap V(Y) \cap V(Z)| \leq 1$.*

**Observation 8** *Let $X, Y$ be two $k$-crossings in $G$ sharing $k - 1$ edges. Then $|V(X) \cap V(Y)| \in \{2k - 2, 2k - 1\}$.*

We want to describe a variation of the flip operation that can be done to untangle $k$-crossings. We first note that the original flip operation can fail miserably if we are flipping inside an interval which contains a smaller $(k - 2)$-crossing. This is because this $(k - 2)$-crossing may be a part of two maximal $(k - 1)$-crossings as is seen in Fig. 5. Thus flipping this interval would cause the $(k - 1)$-crossings to become a single $(k + 1)$-crossing.

However, all is not lost, the following observations tell us that if this happens it is in a sense limited. The proofs can be obtained by induction and counting the crossings at appropriate edges.

**Observation 9** *Let $X$ be a $k$-crossing in $G$ and let $a, b$ be two consecutive vertices of $X$. Then there is at most two edges connecting $(a, b)$ to $(b, a)$.*

**Observation 10** *Let $X$ be a $k$-crossing in $G$ and $a, b$ two consecutive vertices of $X$. Assume that there is exactly 2 non-crossing edges connecting $(a, b)$ to $(b, a)$, called $e_1, e_2$. Then $e_1, e_2$ can intersect all edges of at most one $(k - 1)$-crossing or two $(k - 2)$-crossings fully contained in $(a, b)$. Further if $x_1, x_2$ are endpoints of $e_1, e_2$ which are inside $(a, b)$ then there is no other vertices of the $h$-crossings whose edges $e_1$ and $e_2$ intersect in $(x_1, x_2)$.*

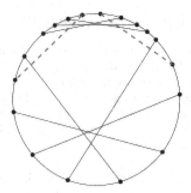

**Fig. 5.** A 4-crossing with two edges going inside an interval and crossing all of the edges of a 3-crossing contained inside said interval.

Now we call interval $(a, b)$ between consecutive edges of a $k$-crossing *regular* if there is no $(k-1)$-crossing inside of it whose edges are all crossed by two different edges connecting $(a, b)$ to $(b, a)$. Otherwise we call it *critical*. The proof of the following Lemma is essentially the same as the proof of Lemma 5.

**Lemma 8.** *Let $X$ be a maximal $k$-crossing in a convex-geometric $k$-planar graph $G$. Let $G'$ be the drawing obtained by performing $Flip(a, b)$ on a regular interval $(a, b)$ between consecutive vertices of $X$. Then if $Y$ is an $h$-crossing in $G'$ which did not appear in $G$, $h \leq 3$.*

If we want to perform Flip operation inside of a maximal $k$-crossing $X$, ideally we would like to perform it on a regular interval. However, as one might imagine, this is not always possible. In the case we first perform the flip on the critical crossing, then if $x_1, x_2$ are as in Observation 10, we perform the appropriate flip to put $x_1, x_2$ in their original places. However, it can happen that the interval $(x_1, x_2)$ was critical in the newly created crossing, in this case we just repeat the same procedure inside of it. Since our graphs are finite, we are guaranteed to only need a finite amount of flips. If the original critical interval was $(a, b)$, we call this operation Multiflip$(a, b)$. With this we can prove the following.

**Lemma 9.** *Let $X$ be a $k$-crossing in $G$ and $(a, b)$ a critical interval between two consecutive vertices of $X$. If $G'$ is a the drawing obtained by Multiflip$(a, b)$, then there is no $h$-crossings in $G'$ which are not in $G$ where $h > k - 1$. Further if $Y$ is an $h$-crossing in $G'$ which was not present in $G$ and $h > 3$ then all but two vertices of $Y$ are contained in the interval $(b, a)$ in $G'$.*

Further, we can redefine what a *leaf* is. We say that a $(k + 1)$-crossing $X$ is a leaf if there is an interval between two vertices of $X$ that contains all vertices of $X$ and no vertices of any other $(k + 1)$-crossing nor any other $k$-crossing. We say that a maximal $k$-crossing $Y$ is a leaf if there is an interval between two vertices of $Y$ that contains all but possibly two vertices of $Y$ and no vertices of any other $(k + 1)$-crossing nor any other $k$-crossing, except possibly vertices of a crossing that shares $k - 1$ of its edges. With this, one can describe how to make a global flipping operation on the set of all $k$-crossings and $(k + 1)$-crossings as in the previous chapter. This, and a detailed proof of the following theorem can be found in the full version.

**Theorem 11.** *There exists a $k > 0$ such that for every convex-geometric $k$-planar graph $G$ there is a convex-geometric $k$-quasiplanar drawing of $G$.*

## 8 Open Questions

The original question that was asked when we started working on this problem was the following:

*Question 1.* Is there a function $f : \mathbb{N} \to \mathbb{N}$, $f(k) = o(k)$ such that every simple/convex-geometric/geometric $k$-planar graph is $f(k)$-quasiplanar?

This seemed to be a hard problem to tackle. Relationships between crossings of different size are hard to describe. The proof in [3] heavily depended on the knowledge that a $k$-planar drawing of a graph is automatically $(k + 2)$-quasiplanar. The problem also seems to be fairly hard in the more general case of geometric graphs, where the structure of the graph is not as combinatorial in nature as convex-geometric drawings and not so flexible as in the case of topological drawings [3], so similar ideas are hard to apply. Still, a possibly easier question that can still be interesting is the following.

*Question 2.* Is it true that every simple/geometric/convex-geometric $k_c$-planar graph is simple/geometric/convex-geometric $(k - c)$-quasiplanar for every $c \geq 0$ and sufficiently big $k_c$?

**Ackgnowledgments.** This work is supported by project 23-04949X of the Czech Science Foundation (GAČR). The research was conducted during a scholarship provided by Višegrad fund. I would like to thank Jank Kynčl and Pavel Valtr for suggesting this problem and all of the helpful conversations.

## References

1. Ackerman, E.: On the maximum number of edges in topological graphs with no four pairwise crossing edges. Discrete Comput. Geom. **41**(3), 365–375 (2009)
2. Ackerman, E.: Quasi-planar Graphs. In: Hong, S.-H., Tokuyama, T. (eds.) Beyond Planar Graphs, pp. 31–45. Springer, Singapore (2020). https://doi.org/10.1007/978-981-15-6533-5_3
3. Angelini, P., et al.: Simple $k$-planar graphs are simple $(k+1)$-quasiplanar. J. Combin. Theory Ser. B **142**, 1–35 (2020)
4. Angelini, P., Da Lozzo, G., Förster, H., Schneck, T.: 2-layer $k$-planar graphs: density, crossing lemma, relationships, and pathwidth. In: Graph drawing and network visualization, vol. 12590 of LNCS, pp. 403–419. Springer, Cham [2020] (2020)

5. Capoyleas, V., Pach, J.: A Turán-type theorem on chords of a convex polygon. J. Combin. Theory Ser. B **56**(1), 9–15 (1992)
6. Didimo, W., Liotta, G., Montecchiani, F.: A survey on graph drawing beyond planarity. ACM Comput. Surv. **52**, 1–37 (2019)
7. Fox, J., Pach, J., Suk, A.: Quasiplanar graphs, string graphs, and the erdős-gallai problem. In: Angelini, P., von Hanxleden, R. (eds.) Graph Drawing and Network Visualization, GD 2022, LNCS, vol. 13764, pp. 219–231. Springer, Cham (2023). https://doi.org/10.1007/978-3-031-22203-0_16
8. Pach, J.: Geometric graph theory. In: Handbook of discrete and computational geometry, pp. 257–279. CRC Press, Boca Raton, FL (2018)

# Detecting $K_{2,3}$ as an Induced Minor

Clément Dallard[1]([✉]) [iD], Maël Dumas[2], Claire Hilaire[3] [iD], Martin Milanič[3] [iD],
Anthony Perez[2], and Nicolas Trotignon[4] [iD]

[1] Department of Informatics, University of Fribourg, Fribourg, Switzerland
clement.dallard@unifr.ch
[2] Université d'Orléans, INSA Centre Val de Loire, LIFO EA 4022, Orléans, France
{mael.dumas,anthony.perez}@univ-orleans.fr
[3] FAMNIT and IAM, University of Primorska, Koper, Slovenia
{claire.hilaire,martin.milanic}@upr.si
[4] LIP, CNRS, École Normale Supérieure de Lyon, Lyon, France
nicolas.trotignon@ens-lyon.fr

**Abstract.** We consider a natural generalization of chordal graphs, in which every minimal separator induces a subgraph with independence number at most 2. Such graphs can be equivalently defined as graphs that do not contain the complete bipartite graph $K_{2,3}$ as an induced minor, that is, graphs from which $K_{2,3}$ cannot be obtained by a sequence of edge contractions and vertex deletions.

We develop a polynomial-time algorithm for recognizing these graphs. Our algorithm relies on a characterization of $K_{2,3}$-induced minor-free graphs in terms of excluding particular induced subgraphs, called Truemper configurations.

**Keywords:** induced minor · minimal separator · Truemper configurations

## 1 Introduction

Chordal graphs have been extensively studied in the literature, mainly owing to their well-understood structure, their multiple characterizations and linear-time recognition algorithms (see, e.g. [1]). Among these characterizations, one states that chordal graphs are exactly the graphs in which every minimal separator forms a clique, that is, has independence number at most 1. In this paper, we investigate the complexity of recognizing graphs whose minimal separators have independence number at most 2. More specifically, we design a polynomial-time algorithm that decides whether each minimal separator of a given graph induces a subgraph with independence number at most 2.

Another way to generalize the above characterization of chordal graphs is to require that each minimal separator is a union of two cliques (rather than having independence number at most 2). Graphs that satisfy this property are called *bisimplicial* and were recently studied by Milanič, Penev, Pivač, and

Vuškovič [18]. One of their results (namely [18, Corollary 3.8]), combined with [11, Theorem 6.3], implies that the class of bisimplicial graphs forms a generalization of the class of 1-perfectly orientable graphs (see, e.g., [11,12]), which in turn forms a common generalization of the classes of chordal graphs and circular-arc graphs.

All these graph classes have the nice property that they are closed under vertex deletions and edge contractions; in other words, they are closed under induced minors. This is also the case for several geometrically defined graph classes such as the class of planar graphs, and for intersection graph classes. More precisely, for any graph class $\mathcal{G}$, the class of graphs defined as intersection graphs of connected subgraphs of graphs in $\mathcal{G}$ is closed under induced minors. Among others, this framework captures the classes of chordal graphs and circular-arc graphs, corresponding respectively to the case when $\mathcal{G}$ is the class of all trees and all cycles. The connection between induced minors and the property that each minimal separator of a graph induces a graph with independence number at most 2 comes from the following result of Dallard, Milanič, and Štorgel [5, Lemma 3.2]: For any positive integer $q$, a graph $G$ does not contain the complete bipartite graph $K_{2,q}$ as an induced minor if and only if each minimal separator of $G$ induces a graph with independence number less than $q$. This implies, in particular, that each minimal separator of a graph $G$ induces a graph with independence number at most 2 if and only if $G$ does not contain $K_{2,3}$ as an induced minor. As shown in [18], the aforementioned class of bisimplicial graphs is characterized by an infinite family of forbidden induced minors, that is, graphs that do not belong to the class but for which deleting any vertex or contracting any edge results in a graph in the class. The smallest forbidden induced minor of bisimplicial graphs happens to be $K_{2,3}$.

Every induced-minor closed class can be characterized by a family $\mathcal{F}$ of forbidden induced minors. The set $\mathcal{F}$ may be finite or infinite (as in the case of bisimplicial graphs).[1] However, contrary to case of the induced subgraph relation, finiteness of the family of forbidden induced minors does not imply polynomial-time recognition (unless P = NP). Indeed, as shown by Fellows, Kratochvíl, Middendorf, and Pfeiffer in 1995 [8], there exists a graph $H$ such that determining if a given graph $G$ contains $H$ as an induced minor is NP-complete. Recently, Korhonen and Lokshtanov showed that this can happen even if $H$ is a tree [13].

It is thus a natural question to determine for which graphs $H$ determining if a given graph $G$ contains $H$ as an induced minor can be done in polynomial time. This is the case for all graphs $H$ on at most four vertices. The problem is particularly simple if $H$ is a forest on at most four vertices or the disjoint union a vertex and a triangle. Otherwise, excluding $H$ always leads to a well structured class that is easy to recognize. When $H$ is the 4-cycle, that is, $K_{2,2}$, the class of graphs excluding $H$ as an induced minor is precisely the class of chordal graphs. For $H = K_4$, excluding $H$ as an induced minor is equivalent to excluding it as a minor, which leads to the class of graphs with treewidth at most two. For

---

[1] For a simpler example, consider the case when $\mathcal{F}$ consists of all complements of cycles.

the diamond, that is, the graph $H$ obtained from $K_4$ by removing one edge, excluding $H$ leads to graphs whose blocks are all complete graphs or chordless cycles (see [10, Theorem 4.6]). For the paw, that is the graph $H$ obtained from $K_4$ by removing two adjacent edges, a direct algorithm is easy. In this latter case, we note that excluding $H$ as an induced minor is equivalent to excluding $H$ as an induced topological minor, and this leads to a very simple class [3].

**Results and Organization of the Paper.** Our main result is a polynomial-time algorithm to decide whether an input graph contains $K_{2,3}$ as an induced minor. Our method relies on the study of the induced subgraphs that must be present whenever the input graph contains $K_{2,3}$ as an induced minor.

In Sect. 3, we prove that all of them are particular instances of graphs known as *Truemper configurations*. Fortunately, the detection of Truemper configurations is a well-studied question (see [6] for the most recent survey). For instance, detecting $K_{2,3}$ as an induced topological minor can be equivalently stated as detecting a Truemper configuration called the *theta*, a problem that can be solved in polynomial time [4,14]. It turns out that by relying on several such previous works, we can easily detect three of the four Truemper configurations identified in Sect. 3 (namely the 3-path-configurations known as the *long prisms*, *pyramids*, and *thetas*, the definitions are given later). This is all done in Sect. 4.

In Sect. 5, we give an algorithm to detect the last configuration identified in Sect. 3 (namely the *broken wheel*) provided that the graph does not contain any of the 3-path-configurations detected in Sect. 4, thus completing our algorithm. Our approach relies on what is called a *shortest path detector*, a method originally invented for the detection of pyramids in [2].

Observe that detecting $K_{2,3}$ as an induced minor might look easy, in particular because the topological version (that is, detecting a theta) is known already. We might indeed have overlooked a simpler method to solve it. But several NP-completeness results for questions similar to the ones that we address here (namely the detection of prisms and broken wheel, see below for the details) suggest that a too naive attempt is likely to fail.

All our algorithms are written for the decision version of the problem they address, but writing an algorithm that actually outputs the structure whose existence is decided is straightforward. We omit it for the sake of readability.

We conclude the paper with some open questions, see Sect. 6.

Due to lack of space, proofs of results marked by ♦ are omitted.

## 2   Preliminaries

All graphs considered in this paper are finite, simple, and undirected, with $n$ vertices and $m$ edges. Given a graph $G$ and a set $S \subseteq V(G)$, the *subgraph of $G$ induced by $S$* is the graph denoted by $G[S]$ with vertex set $S$ in which two vertices are adjacent if and only if they are adjacent in $G$. A *subdivision* of a graph $G$ is any graph obtained from $G$ by a sequence of edge subdivisions. Given two graphs $G$ and $H$, the graph $H$ is an *induced subgraph* of $G$ if $H$ can be obtained from $G$

by a sequence of vertex deletions, an *induced minor* of $G$ if $H$ can be obtained from $G$ by a sequence of vertex deletions or edge contractions, and an *induced topological minor* of $G$ if some subdivision of $H$ is an induced subgraph of $G$. Note that $H$ is an induced minor of $G$ if and only if $G$ admits an *induced minor model* of $H$, that is, a set $\{X_v \colon v \in V(H)\}$ of pairwise disjoint nonempty subsets of $V(G)$ indexed by the vertices of $H$, each inducing a connected subgraph of $G$ and such that for any two distinct vertices $u, v \in V(H)$, there is an edge in $G$ between a vertex in $G[X_u]$ and a vertex in $G[X_v]$ if and only if $u$ and $v$ are adjacent in $H$. Given a graph $G$ and a set $S \subseteq V(G)$, we denote by $N(S)$ the set of vertices in $V(G) \setminus S$ having a neighbor in $S$ and by $N[S]$ the set $S \cup N(S)$. For $S = \{v\}$ for some $v \in V(G)$, we write $N(v)$ for $N(\{v\})$ and similarly for $N[v]$. Given two graphs $G$ and $H$, we say that $G$ is $H$-*free* if $G$ contains no induced subgraph isomorphic to $H$.

A *path* in a graph $G$ is a sequence $v_1 \ldots v_k$ of pairwise distinct vertices of $G$ such that $v_i$ is adjacent to $v_{i+1}$, for every integer $i$ such that $0 \leq i \leq k - 1$. We also view the empty set as an empty path. Given a nonempty path $P = v_1 \ldots v_k$, we define the *length* of $P$ as $k - 1$, its *endpoints* as the vertices $v_1$ and $v_k$, and its *interior* as the subpath $v_2 \ldots v_{k-1}$ (which is empty if $k \leq 2$). A *chord* of a path in $G$ is an edge in $G$ whose endpoints are non-consecutive vertices of the path. A path is *chordless* if it has no chord. Given a path $P = v_1 \ldots v_k$ and two vertices $a$ and $b$ on $P$ such that $a = v_i$ and $b = v_j$ where $1 \leq i \leq j \leq k$, we denote by $aPb$ the subpath of $P$ from $a$ to $b$, that is, the path $v_i \ldots v_j$. Given an integer $k \geq 3$, a *cycle of length* $k$ in a graph $G$ is a sequence $v_1 \ldots v_k v_1$ of vertices of $G$ such that the vertices $v_1, \ldots, v_k$ are pairwise distinct, $v_i$ is adjacent to $v_{i+1}$ for all $i \in \{1, \ldots, k - 1\}$, and $v_k$ is adjacent to $v_1$. A *chord* of a cycle in $G$ is an edge in $G$ whose endpoints are non-consecutive vertices of the cycle. A cycle is *chordless* if it has no chord. For convenience, and when there is no ambiguity, we will not distinguish between a path and its vertex set (and the same for a cycle), nor between a set of vertices and the subgraph it induces. For example, for a vertex $v$ in a path $P$, we write $P \setminus v$ to denote the subgraph induced by $V(P) \setminus \{v\}$, etc.

An *independent set* in a graph $G$ is a set of pairwise nonadjacent vertices. A *clique* in $G$ is a set of pairwise adjacent vertices. A *triangle* is a clique of size three. The graph $K_{p,q}$ is a complete bipartite graph with parts of size $p$ and $q$, respectively.

## 3    Reducing to Truemper Configurations

In this section, we show that detecting $K_{2,3}$ as an induced minor is in fact equivalent to detecting whether some specific graphs are contained as induced subgraphs. First, we give some definitions.

A *prism* is a graph made of three vertex-disjoint chordless paths $P_1 = a_1 \ldots b_1$, $P_2 = a_2 \ldots b_2$, $P_3 = a_3 \ldots b_3$ of length at least 1, such that $\{a_1, a_2, a_3\}$ and $\{b_1, b_2, b_3\}$ are triangles and no edges exist between the paths except those of the two triangles. A prism is *long* if at least one of its three paths has length at least 2.

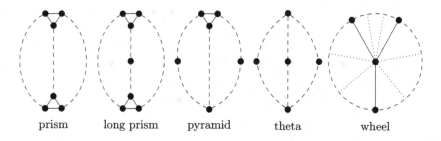

prism       long prism       pyramid       theta       wheel

**Fig. 1.** A schematic representation of a prism, a long prism, a pyramid, a theta, and a wheel. Dashed edges represent paths of length at least 1. Dotted edges may be present or not.

A *pyramid* is a graph made of three chordless paths $P_1 = a \ldots b_1$, $P_2 = a \ldots b_2$, $P_3 = a \ldots b_3$ of length at least 1, two of which have length at least 2, vertex-disjoint except at $a$, and such that $\{b_1, b_2, b_3\}$ is a triangle and no edges exist between the paths except those of the triangle and the three edges incident to $a$.

A *theta* is a graph made of three internally vertex-disjoint chordless paths $P_1 = a \ldots b$, $P_2 = a \ldots b$, $P_3 = a \ldots b$ of length at least 2 and such that no edges exist between the paths except the three edges incident to $a$ and the three edges incident to $b$. Prisms, pyramids, and thetas are referred to as *3-path-configurations*.

A *hole* in a graph is a chordless cycle of length at least 4. Observe that the lengths of the paths in the definitions of prisms, pyramids, and thetas are designed so that the union of any two of the paths induces a hole. A *wheel* $W = (H, x)$ is a graph formed by a hole $H$ (called the *rim*) together with a vertex $x$ (called the *center*) that has at least three neighbors in the hole. Prism, pyramids, thetas, and wheels are referred to as *Truemper configurations*; see Fig. 1.

A *sector* of a wheel $(H, x)$ is a path $P$ that is contained in $H$, whose endpoints are neighbors of $x$ and whose internal vertices are not. Note that $H$ is edgewise partitioned into the sectors of $W$. Also, every wheel has at least three sectors. A wheel is *broken* if at least two of its sectors have length at least 2.

We denote by $\mathcal{S}$ the class of subdivisions of the graph $K_{1,3}$; see Fig. 2. The vertex with degree 3 in a graph $G \in \mathcal{S}$ is called the *apex* of $G$. The class $\mathcal{T}$ is the class of graphs that can be obtained from three paths of length at least one by selecting one endpoint of each path and adding three edges between those endpoints so as to create a triangle. The class $\mathcal{M}$ is the class of graphs $H$ that consist of a path $P$ and a vertex $a$, called the *center* of $H$, such that $a$ is nonadjacent to the endpoints of $P$ and $a$ has at least two neighbors in $P$. Given a graph $H \in \mathcal{S} \cup \mathcal{T} \cup \mathcal{M}$, the *extremities* of $H$ are the vertices of degree one as well as the center of $H$ in case $H \in \mathcal{M}$. Observe that any $H \in \mathcal{S} \cup \mathcal{T} \cup \mathcal{M}$ has exactly three extremities.

The key ingredient in our proof of Theorem 1 is the following sufficient condition for the presence of an induced subgraph from $\mathcal{S} \cup \mathcal{T} \cup \mathcal{M}$ (which is also used for proving Lemma 2).

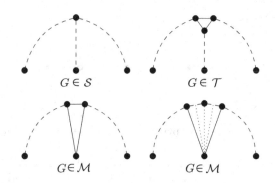

**Fig. 2.** A schematic representation of graphs in $\mathcal{S}$, $\mathcal{T}$, and $\mathcal{M}$. Dashed edges represent paths of length at least 1. Dotted edges may be present or not.

♦ **Lemma 1.** *Let $G$ be a graph and $I$ be an independent set in $G$ with $|I| = 3$. If there exists a component $C$ of $G \setminus I$ such that $I \subseteq N(C)$, then there exists an induced subgraph $H$ of $G[N[C]]$ such that $H \in \mathcal{S} \cup \mathcal{T} \cup \mathcal{M}$ and $I$ is exactly the set of extremities of $H$.*

♦ **Theorem 1.** *A graph contains $K_{2,3}$ as an induced minor if and only if it contains a long prism, a pyramid, a theta, or a broken wheel as an induced subgraph.*

*Proof Sketch.* It is easy to check that a long prism, a pyramid, a theta, or a broken wheel all contain $K_{2,3}$ as an induced minor.

Conversely, assume that a graph $G$ contains $K_{2,3}$ as an induced minor. Let us denote the vertices of $K_{2,3}$ as $\{u, v, a, b, c\}$ so that $\{u, v\}$ and $\{a, b, c\}$ form a bipartition of $K_{2,3}$. Let $\mathcal{H} = \{X_u, X_v, X_a, X_b, X_c\}$ be an induced minor model of $K_{2,3}$ in $G$ minimizing $|X_u \cup X_v \cup X_a \cup X_b \cup X_c|$, and subject to that, minimizing $|X_a \cup X_b \cup X_c|$.

It easy to check under these assumptions that $X_a$, $X_b$ and $X_c$ each consists of one vertex. So $I = X_a \cup X_b \cup X_c$ is a stable set of size 3. The idea is then to apply Lemma 1 to $X_u \cup I$ and $X_v \cup I$. The result then follows from a straightforward case by case analysis. □

## 4    Detecting 3-Path-Configurations

Detecting a pyramid in polynomial time was first done by Chudnovsky and Seymour [2]. A faster method was later discovered by Lai, Lu, and Thorup [14]. The $\widetilde{O}(\cdot)$ notation suppresses polylogarithmic factors.

**Theorem 2.** *Determining whether a given graph contains a pyramid as an induced subgraph can be done in time $\widetilde{O}(n^5)$.*

Detecting a theta in polynomial time was first done by Chudnovsky and Seymour [4]. A faster method was later discovered by Lai, Lu, and Thorup [14].

**Theorem 3.** *Determining whether a given graph contains a theta as an induced subgraph can be done in time $\widetilde{O}(n^6)$.*

We now explain how a long prism can be detected in a graph with no pyramid. Note that detecting a prism is known to be an NP-complete problem [17]. Our algorithm is inspired from the algorithm in [17] to detect a pyramid or a prism. Adapting it to detect a long prism in a pyramid-free graph is straightforward.

♦ **Lemma 2.** *Determining whether a given pyramid-free graph contains a long prism as an induced subgraph can be done in time $\mathcal{O}(n^6(n+m))$.*

## 5 Detecting a Broken Wheel

We give here a polynomial-time algorithm to detect a broken wheel in a graph with no long prism, no pyramid, and no theta. Note that detecting a broken wheel in general is an NP-complete problem, even when restricted to bipartite graphs. The proof can be found in [7]. Note that this article addresses the question of the detection of wheels (broken or not), however, the proof (of [7, Theorem 4]) that detecting a wheel is NP-complete proves without a single modification that detecting a broken wheel is NP-complete.

Let $W = (H, x)$ be a broken wheel. Since $W$ is broken, it contains at least two sectors of length at least 2, say $P = a \ldots b$ and $R = c \ldots d$. We orient $H$ clockwise and assume that $a, b, c,$ and $d$ appear clockwise in this order along $H$. We denote by $Q$ the path of $H$ from $b$ to $c$ that does not contain $a$ and $d$. We denote by $S$ the path of $H$ from $d$ to $a$ that does not contain $b$ and $c$. Note that

$$\text{possibly } b = c \text{ or } d = a, \text{ but not both,} \tag{1}$$

since $W$ is a wheel by assumption. Note also that $a \neq b$, $c \neq d$, $ab \notin E(G)$, and $cd \notin E(G)$.

For every vertex $v$ in $H$, we denote by $v^+$ its neighbor in $H$ in the clockwise direction, and by $v^-$ its neighbor in the counter-clockwise direction. Note that possibly $a^+ = b^-$, $a^+b^- \in E(G)$, $c^+ = d^-$, or $c^+d^- \in E(G)$.

We say that the 13-tuple $F = (x, a, b, c, d, a^+, b^+, c^+, d^+, a^-, b^-, c^-, d^-)$ is a *frame* for $W$. Every broken wheel has at least one frame. In what follows, once a frame is fixed, we use the notation for the paths $P$, $Q$, $R$, $S$ without recalling it; see Fig. 3.

If $W$ is a broken wheel in a graph $G$ with frame $F$, we say that $(W, F)$ is *optimal* if:

- no broken wheel of $G$ has fewer vertices than $W$, and
- among all broken wheels with the same number of vertices as $W$ and their frames, the sum of the lengths of $S$ and $Q$ is minimal.

The *cube* is the graph with vertex set $\{v_1, v_2, v_3, v_4, v_5, v_6, x, y\}$ such that $v_1v_2v_3v_4v_5v_6v_1$ is a hole, $N(x) = \{v_1, v_3, v_5\}$, and $N(y) = \{v_2, v_4, v_6\}$. Observe

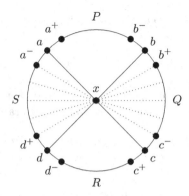

**Fig. 3.** A frame for a broken wheel. Dotted edges may be present or not.

that removing any vertex from the cube yields a broken wheel on seven vertices. Hence, every graph that does not contain a broken wheel is cube-free.

Given a graph $G$, a wheel $W = (H, x)$ contained in $G$ as an induced subgraph, and a vertex $y \in G \setminus W$, we denote by $N_H(y)$ the set of vertices in $H$ that are adjacent to $y$.

We start with three lemmas analyzing the structure around an optimal pair $(W, F)$ of a broken wheel and a frame.

**Lemma 3.** *Let $G$ be a (long prism, pyramid, theta, cube)-free graph. Let $W = (H, x)$ be a broken wheel in $G$ with frame $F = (x, a, b, c, d, a^+, b^+, c^+, d^+, a^-, b^-, c^-, d^-)$ such that $(W, F)$ is optimal. Let $y$ be a vertex of $G \setminus W$ that is not adjacent to $a$, $b$, $c$, or $d$. Then $N_H(y)$ induces a (possibly empty) path of length at most 2. In particular, $N_H(y)$ is contained in the interior of one of $P$, $Q$, $R$, or $S$.*

*Proof.* Assume first that $y$ has neighbors in at most one path among $P$, $Q$, $R$, and $S$. If $N_H(y)$ is not contained in some path of $H$ of length at most 2, it can be used to replace some subpath of $H$ to obtain a broken wheel with fewer vertices than $W$, a contradiction (note that this is correct if $xy \in E(G)$ and if $xy \notin E(G)$). Hence, $N_H(y)$ is contained in some path $T$ of $H$ of length at most 2, and it must be equal to $T$, because otherwise $T$ has length 2 and $y$ is non adjacent to its middle vertex, and $y$ and $H$ form a theta. So, the conclusion of the lemma holds. Hence, from now on, we assume that $y$ has neighbors in at least two paths among $P$, $Q$, $R$, and $S$.

If $N_H(y) \subseteq Q \cup S$, then $y$ has neighbors in both $Q$ and $S$. Since $G$ is theta-free, $y$ has at least three neighbors in $H$. It follows that $y$ is the center of a broken wheel with rim $H$, with the same number of vertices as $W$ and that has a frame contradicting the minimality of the sum of the lengths of $Q$ and $S$ (since $y$ is not adjacent to the endpoints of $Q$ and $S$). Hence, $y$ has neighbors in the interior of at least one of $P$ or $R$, say in $P$ up to symmetry.

If $y$ is adjacent to $x$, then $P$, $x$, and $y$ induce a theta or a broken wheel with center $y$ and with fewer vertices than $W$, a contradiction. So, $y$ is not adjacent to $x$.

Assume that $y$ has neighbors in $S$. Let $a'$ be the neighbor of $y$ in $P$ closest to $a$ along $P$. Let $b'$ be the neighbor of $y$ in $P$ closest to $b$ along $P$. Let $d'$ be the neighbor of $y$ in $S$ closest to $d$ along $S$. Let $T$ be a shortest path from $x$ to $d'$ in the subgraph of $G$ induced by the vertices of the path $xdSd'$. If $d'a \notin E(G)$, the paths $xaPa'$, $xbPb'$, and $xTd'y$ induce a theta or a pyramid (namely, if $a' = b'$ they form a theta with apices $a'$ and $x$, if $a'b' \in E(G)$ they form a pyramid, and otherwise they form a theta with apices $x$ and $y$). Hence, $d'a \in E(G)$. It follows that $d' = a^-$ is the unique neighbor of $y$ in $S$.

We claim that $a' = b'$ and $a'a \in E(G)$. Indeed, if $a' = b'$ and $a'a \notin E(G)$, then the three paths $aPa'$, $axbPa'$, and $ad'ya'$ form a theta with apices $a$ and $a'$; if $a'$ and $b'$ are adjacent, then the three paths $aPa'$, $axbPb'$, and $ad'y$ form a pyramid with apex $a$ and triangle $\{a', b', y\}$; and if $a'$ and $b'$ are distinct and nonadjacent, then the three paths $aPa'y$, $axbPb'y$, and $ad'y$ form a theta with apices $a$ and $y$. It follows that $a' = a^+$ is the unique neighbor of $y$ in $P$. Therefore, the hole $xTa^-ya^+Pbx$ is the rim of a broken wheel $W'$ centered at $a$. To avoid that the wheel $W'$ has fewer vertices than $W$, we infer that $b = c$, $R$ has length exactly 2, and $T = xdSa^-$ (which implies $xa^- \notin E(G)$). So, $Q = b$ and $R = bb^+d$. Vertex $y$ must be adjacent to $b^+$, since otherwise $H$ and $y$ form a theta with apices $a^-$ and $a^+$ (recall that $y$ is not adjacent to $a, b, c$ and $d$). We have $a^-d \in E(G)$, for otherwise the three paths $dSa^-$, $dxaa^-$, and $db^+ya^-$ form a theta with apices $d$ and $a^-$. By a symmetric argument, we must have $a^+b \in E(G)$. Hence, $\{x, y, a, b, d, a^-, a^+, b^+\}$ induces a cube, a contradiction.

We proved that $y$ cannot have neighbors in $S$. Symmetrically, it cannot have neighbors in $Q$. Hence, $y$ has neighbors in both $P$ and $R$ and no neighbor in $Q$ and $S$. Since $G$ contains no theta, $y$ must have at least two neighbors in either $P$ or $R$, say in $R$ up to symmetry. Up to symmetry, we may assume by (1) that $a \neq d$ (so possibly $b = c$). Let $a'$ be the neighbor of $y$ in $P$ closest to $a$ along $P$. Let $d'$ be the neighbor of $y$ in $R$ closest to $d$ along $R$. Let $c'$ be the neighbor of $y$ in $R$ closest to $c$ along $R$. Note that $c' \neq d'$. If $ca' \notin E(G)$, then the three paths $yd'Rdx$, $yc'Rcx$, and $ya'Pax$ induce a long prism, a pyramid, or a theta (namely, they induce a long prism if $c'd' \in E(G)$ and $ad \in E(G)$, a pyramid if exactly one of $c'd' \in E(G)$ and $ad \in E(G)$ happens, and a theta otherwise). So, $ca' \in E(G)$. Hence, $b = c$ and $a'$ is the unique neighbor of $y$ in $P$. Hence, since $c' \neq d'$, the three paths $yd'Rdxb$, $yc'Rb$, and $ya'b$ induce a pyramid with triangle $\{y, c', d'\}$ or a theta with apices $y$ and $b$, a contradiction. $\qquad\square$

**Lemma 4.** *Let $G$ be a (long prism, pyramid, theta, cube)-free graph. Let $W = (H, x)$ be a broken wheel of $G$ with frame $F = (x, a, b, c, d, a^+, b^+, c^+, d^+, a^-, b^-, c^-, d^-)$ such that $(W, F)$ is optimal. Let $P' = a^+ \ldots b^-$ be a shortest path from $a^+$ to $b^-$ in $G \setminus ((N[a] \cup N[b] \cup N[c] \cup N[d] \cup N[x]) \setminus \{a^+, b^-\})$. Then the induced subgraph of $G$ obtained from $W$ by removing the internal vertices of $P$ and adding the vertices of $P'$ instead is a broken wheel with frame $F$, and $(W', F)$ is optimal. A similar result holds for $R$.*

*Proof.* If no vertex of $P'$ has a neighbor in the interior of $Q$, $R$, or $S$ (in which case $P'$ is vertex-disjoint from $Q \cup R \cup S$), then the conclusion of the lemma holds. Hence, suppose for a contradiction that some vertex $v$ of $P'$ has a neighbor in the interior of $Q$, $R$, or $S$, and choose $v$ closest to $a^+$ along $P'$. Note that $v$ is in the interior of $P'$, hence, $v \notin W$ and $v$ is not adjacent to any of $a$, $b$, $c$, $d$, and $x$. Let $w$ be the vertex closest to $v$ along $vP'a^+$ that has some neighbor in $P$. Note that $w$ exists because $a^+ \in P$; moreover, $w$ is in the interior of $P'$; hence, $w \notin W$ and $w$ is not adjacent to any of $a$, $b$, $c$, $d$, and $x$. Since $v$ has a neighbor in the interior of $Q$, $R$, or $S$, Lemma 3 implies that $N_H(v)$ induces a nonempty path of length at most 2 that is contained in the interior of either $Q$, $R$, or $S$. Similarly, $N_H(w)$ induces a nonempty path of length at most 2 that is contained in the interior of $P$. Note that $w \neq v$. We set $T = vP'w$. Now, we focus on $T$ and $H$. By construction, $T$ is disjoint from $H$ and no vertex in the interior of $T$ has a neighbor in $H$. If $vw \notin E(G)$ or if one of $|N_H(v)|$ or $|N_H(w)|$ has size at most 2, then $T \cup H$ induces a long prism, a pyramid, or a theta. Hence, $vw \in E(G)$, and both $N_H(v)$ and $N_H(w)$ induce a path of length 2. It follows that the subgraph of $G$ induced by $H \cup T \cup \{x\}$ contains a theta with apices $v$ and $x$, a contradiction. $\qquad\square$

The next lemma might look identical to the previous one, but there is an important difference: the vertices of the shortest path are allowed to be adjacent to $x$.

♦ **Lemma 5.** *Let $G$ be a (long prism, pyramid, theta, cube)-free graph. Let $W = (H, x)$ be a broken wheel of $G$ with frame $F = (x, a, b, c, d, a^+, b^+, c^+, d^+, a^-, b^-, c^-, d^-)$ such that $(W, F)$ is optimal. Let $Q' = b^+ \ldots c^-$ be a shortest path from $b^+$ to $c^-$ in $G \setminus ((N[a] \cup N[b] \cup N[c] \cup N[d]) \setminus \{b^+, c^-\})$. Then the induced subgraph of $G$ obtained from $W$ by removing the internal vertices of $Q$ and adding the vertices of $Q'$ instead is a broken wheel with frame $F$, and $(W', F)$ is optimal. A similar result holds for $S$.*

Now we have everything ready to prove the main result of this section.

**Lemma 6.** *Determining whether a given (long prism, pyramid, theta)-free graph contains a broken wheel as an induced subgraph can be done in time $\mathcal{O}(n^{13}(n + m))$.*

*Proof.* Consider the following algorithm whose input is any (long prism, pyramid, theta)-free graph $G$.

1. Enumerate all 7-tuples of vertices $G$, and if one of them induces a broken wheel, output "$G$ contains a broken wheel" and stop.
2. Enumerate all 13-tuples $F = (x, a, b, c, d, a^+, b^+, c^+, d^+, a^-, b^-, c^-, d^-)$ of vertices of $G$. For each of them:
   (i) Compute a shortest path $P'$ from $a^+$ to $b^-$ in $G \setminus ((N[a] \cup N[b] \cup N[c] \cup N[d] \cup N[x]) \setminus \{a^+, b^-\})$. If no such path exists, set $P' = \emptyset$.

(ii) If $b = c$, set $Q' = \emptyset$. Otherwise, compute a shortest path $Q'$ from $b^+$ to $c^-$ in $G \setminus ((N[a] \cup N[b] \cup N[c] \cup N[d]) \setminus \{b^+, c^-\})$. If no such path exists, set $Q' = \emptyset$.

(iii) Compute a shortest path $R'$ from $c^+$ to $d^-$ in $G \setminus ((N[a] \cup N[b] \cup N[c] \cup N[d] \cup N[x]) \setminus \{c^+, d^-\})$. If no such path exists, set $R' = \emptyset$.

(iv) If $d = a$, set $S' = \emptyset$. Otherwise, compute a shortest path $S'$ from $d^+$ to $a^-$ in $G \setminus ((N[a] \cup N[b] \cup N[c] \cup N[d]) \setminus \{d^+, a^-\})$. If no such path exists, set $S' = \emptyset$.

(v) If $P' \cup Q' \cup R' \cup S' \cup F$ induces a broken wheel centered at $x$, output "$G$ contains a broken wheel" and stop.

3. Output "$G$ contains no broken wheel".

If the algorithm outputs that $G$ contains a broken wheel in Step 2 (v), it obviously does. Conversely, assume that $G$ contains a broken wheel, and let us check that the algorithm gives the correct answer.

If $G$ contains a broken wheel on 7 vertices, then it is detected in Step 1. So we may assume that $G$ contains no broken wheel on 7 vertices, which implies that $G$ contains no cube and that the algorithm continues with Step 2.

Since $G$ contains a broken wheel, it contains a broken wheel $W$ with a frame $F$ such that $(W, F)$ is optimal. We denote by $P$, $Q$, $R$, and $S$ the paths of the broken wheel as above. At some point in Step 2, the algorithm considers $F$. By Lemma 4 applied to $W$, the paths $P'$ (defined in Step 2 (i)), $Q$, $R$, and $S$ together with $F$ form a broken wheel $W_1$. By Lemma 5 applied to $W_1$ if $b \neq c$ (and trivially otherwise), the paths $P'$, $Q'$ (defined in Step 2 (ii)), $R$, and $S$ together with $F$ form a broken wheel $W_2$. By Lemma 4 applied a second time to $W_2$, the paths $P'$, $Q'$, $R'$ (defined in Step 2 (iii)), and $S$ together with $F$ form a broken wheel $W_3$. By Lemma 5 applied a second time to $W_3$ if $d \neq a$ (and trivially otherwise), the paths $P'$, $Q'$, $R'$, and $S'$ (defined in Step 2 (iv)) together with $F$ form a broken wheel $W_4$. Hence, the algorithm actually finds a broken wheel, and therefore gives the correct answer.

The algorithm relies on a brute force enumeration of 13-tuples of vertices that can be implemented in time $\mathcal{O}(n^{13})$, and a search of the graph for computing shortest paths $P'$, $Q'$, $R'$, $S'$, as well as testing if $P'$, $Q'$, $R'$, $S'$, and $F$ induce a broken wheel centered at $x$ can be implemented in time $\mathcal{O}(n + m)$.  □

## 6    Conclusion

We can now state our main theorem.

**Theorem 4.** *Determining whether a given graph contains $K_{2,3}$ as an induced minor can be done in time $\mathcal{O}(n^{13}(n + m))$.*

*Proof.* By Theorem 1, it is enough to decide whether $G$ contains a long prism, a pyramid a theta, or a broken wheel as an induced subgraph. Detecting a pyramid can be performed in time $\widetilde{O}(n^5)$ by Theorem 2. Detecting a theta can be performed in time $\widetilde{O}(n^6)$ by Theorem 3. Hence, it can be assumed that the input graph contains no pyramid and no theta, so that detecting a long prism can be performed in time $\mathcal{O}(n^6(n+m))$ by Lemma 2. Hence, it can be assumed that the input graph contains no pyramid, no theta, and no long prism, so that detecting a broken wheel can be performed in time $\mathcal{O}(n^{13}(n+m))$ by Lemma 6.    □

## Open Questions

We wonder what is the complexity of detecting a broken wheel in a theta-free graph. Detecting a theta or a wheel (possibly not broken) in an input graph can be performed in polynomial time. However, the algorithm given in [19] relies on a complicated structural description of the class of (theta, wheel)-free graphs. We wonder whether a simpler approach, based on computing shortest paths as we do here, might work. Symmetrically, our polynomial-time algorithm suggests that a full structural description of graphs that do not contain $K_{2,3}$ as an induced minor may be worth investigating.

In particular, we propose the following.

*Conjecture 1.* There exists a polynomial $p$ and an integer $k$ such that if $G$ has no clique cutset and does not contain $K_{2,3}$ as an induced minor, then either $G$ has at most $p(|V(G)|)$ minimal separators or $G$ has no hole of length at least $k$.

If Conjecture 1 is true, it would give an alternative polynomial-time recognition algorithm for determining whether a given graph contains $K_{2,3}$ as an induced minor.

Let us mention that it is easy to settle the complexity status of the most classical optimization problems when restricted to graphs that do not contain $K_{2,3}$ as an induced minor. The maximum independent set and $k$-coloring (for any fixed $k$) problems are polynomial-time solvable because the tree-independence number is bounded (see [5]). Computing the chromatic number is NP-hard, because it is NP-hard for the subclass of circular-arc graphs (see [9]). Many problems such as the maximum clique and coloring the complement are NP-hard because the class contains the complement of all triangle-free graphs.

Finally, note that the complexity of recognizing bisimplicial graphs, that is, graphs in which each minimal separator is a union of two cliques (see [18]) remains open. In this regard, our main result implies that the problem is solvable in polynomial time when restricted to the class of perfect graphs, since in a perfect graph, a set of vertices is a union of two cliques if and only if it induces a graph with independence number at most 2 (see [15,16]).

**Acknowledgments.** The authors are grateful to Nevena Pivač for helpful discussions. This work is supported in part by the *Slovenian Research and Innovation Agency* (I0-0035, research program P1-0285 and research projects J1-3001, J1-3002, J1-3003, J1-4008, J1-4084, N1-0102, and N1-0160, and BI-FR/22-23-PROTEUS-01), by the research program *CogniCom* (0013103) at the University of Primorska, by the French *Fédération de Recherche ICVL* (Informatique Centre-Val de Loire), by the LABEX MILYON (ANR-10-LABX-0070) of Université de Lyon, within the program Investissements d'Avenir (ANR-11-IDEX-0007) operated by the French National

Research Agency (ANR), by Agence Nationale de la Recherche (France) under research grant ANR DIGRAPHS ANR-19-CE48-0013-01 and by H2020-MSCA-RISE project CoSP-GA No. 823748.

# References

1. Blair, J.R.S., Peyton, B.: An introduction to chordal graphs and clique trees. In: George, A., Gilbert, J.R., Liu, J.W.H. (eds.) Graph Theory and Sparse Matrix Computation. The IMA Volumes in Mathematics and its Applications, vol. 56, pp. 1–29. Springer, New York (1993). https://doi.org/10.1007/978-1-4613-8369-7_1
2. Chudnovsky, M., Cornuéjols, G., Liu, X., Seymour, P., Vušković, K.: Recognizing Berge graphs. Combinatorica **25**(2), 143–186 (2005). https://doi.org/10.1007/s00493-005-0012-8
3. Chudnovsky, M., Penev, I., Scott, A., Trotignon, N.: Excluding induced subdivisions of the bull and related graphs. J. Graph Theory **71**(1), 49–68 (2012). https://doi.org/10.1002/JGT.20631
4. Chudnovsky, M., Seymour, P.: The three-in-a-tree problem. Combinatorica **30**(4), 387–417 (2010). https://doi.org/10.1007/s00493-010-2334-4
5. Dallard, C., Milanič, M., Štorgel, K.: Treewidth versus clique number. III. Tree-independence number of graphs with a forbidden structure. J. Comb. Theory Ser. B **167**, 338–391 (2024). https://doi.org/10.1016/j.jctb.2024.03.005
6. Diot, E., Radovanović, M., Trotignon, N., Vušković, K.: The (theta, wheel)-free graphs Part I: Only-prism and only-pyramid graphs. J. Combin. Theory Ser. B **143**, 123–147 (2020). https://doi.org/10.1016/j.jctb.2017.12.004
7. Diot, E., Tavenas, S., Trotignon, N.: Detecting wheels. Appl. Anal. Discrete Math. **8**(1), 111–122 (2014). https://doi.org/10.2298/AADM131128023D
8. Fellows, M.R., Kratochvíl, J., Middendorf, M., Pfeiffer, F.: The complexity of induced minors and related problems. Algorithmica **13**(3), 266–282 (1995). https://doi.org/10.1007/BF01190507
9. Garey, M.R., Johnson, D.S., Miller, G.L., Papadimitriou, C.H.: The complexity of coloring circular arcs and chords. SIAM J. Algebraic Discret. Methods **1**(2), 216–227 (1980). https://doi.org/10.1137/0601025
10. Hartinger, T.R.: New Characterizations in Structural Graph Theory: 1-Perfectly Orientable Graphs, Graph Products, and the Price of Connectivity. Ph.D. thesis, University of Primorska (2017). https://www.famnit.upr.si/sl/studij/zakljucna_dela/download/532
11. Hartinger, T.R., Milanič, M.: Partial characterizations of 1-perfectly orientable graphs. J. Graph Theory **85**(2), 378–394 (2017). https://doi.org/10.1002/jgt.22067
12. Kammer, F., Tholey, T.: Approximation algorithms for intersection graphs. Algorithmica **68**(2), 312–336 (2014). https://doi.org/10.1007/s00453-012-9671-1
13. Korhonen, T., Lokshtanov, D.: Induced-minor-free graphs: Separator theorem, subexponential algorithms, and improved hardness of recognition. In: Woodruff, D.P. (ed.) Proceedings of the 2024 ACM-SIAM Symposium on Discrete Algorithms, SODA 2024, Alexandria, VA, USA, 7–10 January 2024, pp. 5249–5275. SIAM (2024). https://doi.org/10.1137/1.9781611977912.188
14. Lai, K.Y., Lu, H.I., Thorup, M.: Three-in-a-tree in near linear time. In: STOC 2020—Proceedings of the 52nd Annual ACM SIGACT Symposium on Theory of Computing, pp. 1279–1292. ACM, New York (2020). https://doi.org/10.1145/3357713.3384235

15. Lovász, L.: A characterization of perfect graphs. J. Comb. Theory Ser. B **13**, 95–98 (1972). https://doi.org/10.1016/0095-8956(72)90045-7

16. Lovász, L.: Normal hypergraphs and the perfect graph conjecture. Discrete Math. **2**, 253–267 (1972). https://doi.org/10.1016/0012-365X(72)90006-4

17. Maffray, F., Trotignon, N.: Algorithms for perfectly contractile graphs. SIAM J. Discrete Math. **19**(3), 553–574 (2005). https://doi.org/10.1137/S0895480104442522

18. Milanič, M., Penev, I., Pivač, N., Vušković, K.: Bisimplicial separators. J. Graph Theory (2024, to appear). https://doi.org/10.1002/jgt.23098

19. Radovanović, M., Trotignon, N., Vušković, K.: The (theta, wheel)-free graphs Part II: structure theorem. J. Combin. Theory Ser. B **143**, 148–184 (2020). https://doi.org/10.1016/j.jctb.2019.07.004

# Computing Maximal Palindromes
# in Non-standard Matching Models

Mitsuru Funakoshi[1] , Takuya Mieno[2(✉)] , Yuto Nakashima[1] ,
Shunsuke Inenaga[1] , Hideo Bannai[3] , and Masayuki Takeda[1]

[1] Department of Informatics, Kyushu University, Fukuoka, Japan
{nakashima.yuto.003,inenaga.shunsuke.380}@m.kyushu-u.ac.jp
[2] Department of Computer and Network Engineering, University of
Electro-Communications, Chofu, Japan
tmieno@uec.ac.jp
[3] M&D Data Science Center, Tokyo Medical and Dental University, Tokyo, Japan
hdbn.dsc@tmd.ac.jp

**Abstract.** *Palindromes* are popular and important objects in textual
data processing, bioinformatics, and combinatorics on words. Let $S = XaY$ be a string where $X$ and $Y$ are of the same length, and $a$ is either
a single character or the empty string. Then, there exist two alternative
definitions for palindromes: $S$ is said to be a palindrome if $S$ is equal
to its reversal $S^R$ (Reversal-based definition); or if its right-arm $Y$ is
equal to the reversal of its left-arm $X^R$ (Symmetry-based definition). It
is clear that if the "equality" ($\approx$) used in both definitions is exact character matching ($=$), then the two definitions are the same. However, if
we apply other string-equality criteria $\approx$, including the *complementary-
matching model* for biological sequences, the *Cartesian-tree model* [Park
et al., TCS 2020], the *parameterized model* [Baker, JCSS 1996], the *order-
preserving model* [Kim et al., TCS 2014], and the *palindromic-structure
model* [I et al., TCS 2013], then are the reversal-based palindromes and
the symmetry-based palindromes the same? To the best of our knowledge, no previous work has considered or answered this natural question. In this paper, we first provide answers to this question, and then
present efficient algorithms for computing all *maximal palindromes under
the non-standard matching models* in a given string. After confirming
that Gusfield's offline suffix-tree-based algorithm for computing maximal symmetry-based palindromes can be readily extended to the aforementioned matching models, we show how to extend Manacher's online
algorithm for computing maximal reversal-based palindromes in linear
time for all the aforementioned matching models.

**Keywords:** Palindromes · Complementary matching · Watson-Crick
matching · Cartesian-tree matching · Parameterized matching ·
Order-preserving matching · Palindromic matching

M. Funakoshi—Current affiliation: NTT Communication Science Laboratories, Japan.

A. A. Rescigno and U. Vaccaro (Eds.): IWOCA 2024, LNCS 14764, pp. 165–179, 2024.
https://doi.org/10.1007/978-3-031-63021-7_13

# 1    Introduction

Finding characteristic patterns in strings, such as tandem repeat, unique factor, and palindrome, is a fundamental task in string data processing. Of course, what kind of characteristic string we want depends on the application domain. For example, when strings represent DNA/RNA sequences, tandem repeats can help find genetic diseases [9], unique factors may be helpful in preprocessing PCR primer design [25], and (gapped) palindromes may represent second structures of RNA called hairpin [9]. Let us consider the case when a string represents a stock chart, which is a numerical sequence. In this case, one is less interested in their exact values (i.e., exact prices) than in finding price fluctuations from the stock chart. Motivated by discovering such fluctuations, Park et al. [24] introduced a notion of *Cartesian-tree match*. The *Cartesian-tree* of a numeric string $S$ of length $n$ is an ordered binary tree recursively defined below: the root is the leftmost occurrence $i$ of the smallest value in $S$, the left child of the root is the Cartesian-tree of $S[1..i-1]$, and the right child of the root is the Cartesian-tree of $S[i+1..n]$. We say that two strings of equal length Cartesian-tree match if their Cartesian-trees are isomorphic. Cartesian-tree matching can capture shapes of stock chart patterns such as *head-and-shoulder* [24]. Also, some significant stock chart patterns exhibit *palindrome-like* structures (e.g., head-and-shoulder and double-top). Hence, enumerating such palindrome-like structures in a stock chart is expected to improve the efficiency of chart pattern searches. The first motivation of our research is to explore efficient methods to compute such palindrome-like structures by applying Cartesian-tree matching.

First, let us revisit the definition of palindromes in the context of the standard matching model, that is, the "equality" ($\approx$) between two strings is the exact matching on a character-by-character basis ($=$). There are two common definitions for palindromes: Let $S = XaY$ be a string, where $X$ and $Y$ are the same length, and $a$ is either a single character or the empty string. Then, $S$ is said to be a palindrome if:

**Reverse-based definition:** $S$ is equal to its reversal $S^R$;
**Symmetry-based definition:** $Y$ is equal to the reversal of $X$.

It is clear that if we assume the standard matching model, then the two definitions are the same, meaning that a string $S$ is a reverse palindrome iff $S$ is a symmetric palindrome. Observe this with examples such as racecar and noon. However, the two definitions above differ if we apply the Cartesian-tree model. For instance, the Cartesian-tree of $S_1 = 223$ is the tree with no branches that slopes down to the right while the Cartesian-tree of $S_1^R = 322$ is shaped like a *logical AND symbol* ($\wedge$) because the leftmost smallest element in $S_1$ is the second one. Thus, $S_1$ is not a reverse palindrome. On the other hand, $S_1 = 223$ is a symmetric palindrome since the Cartesian-trees of its right-half 3 and the reversal of its left-half $(2)^R = 2$ are both singletons. In addition, the two definitions are also different if we apply the *complementary-matching model* where A matches T, and G matches C, which is a standard matching model for DNA sequences. For instance, a string $S_2 = $ AGCT is not a reverse palindrome since $S_2[3] = $ T and

$S_2^R[3] = \text{T}$ do not complementary-match, but is a symmetric palindrome since $(\text{AG})^R = \text{GA}$ and CT complementary-match. Now we raise the following question: If we apply other string-equality criteria $\approx$, including the *parameterized model* [3], the *order-preserving model* [16], and the *palindromic-structure model* [12], then are the reverse palindromes and the symmetric palindromes the same? This question is interesting because, while the symmetry-based palindrome definition requires only $X^R \approx Y$, the reversal-based palindrome definition requires more strict matching under $\approx$ so that $S \approx S^R \Leftrightarrow XaY \approx (XaY)^R \Leftrightarrow XaY \approx Y^RaX^R$. Thus, these two types of palindromes can be quite different for some matching criterion $\approx$. To our knowledge, no previous work has considered or answered this natural question. In this paper, we first provide quick answers to this question and present efficient algorithms for computing such palindromes in a given string under the non-standard matching models.

One of the well-studied topics regarding palindromes is *maximal palindromes*, which are substring palindromes whose left-right extensions are not palindromes. It is interesting and important to find all maximal palindromes in a given string $T$ because any substring palindrome of $T$ can be obtained by removing an equal number of characters from the left and the right of some maximal palindrome. Hence, by computing all maximal palindromes of a string, we obtain a compact representation of all palindromes in the string. Manacher [20] showed an elegant algorithm for finding all maximal palindromes in a given string $T$. Manacher's algorithm works in an online manner (processes the characters of $T$ from left to right) and runs in $O(n)$ time and space for general (unordered) alphabets, where $n$ is the length of the input string $T$. Later, Gusfield [9] showed another famous algorithm for computing all maximal palindromes. Gusfield's algorithm uses the suffix tree [30] built on a concatenation of $T$ and its reversal $T^R$ that is enhanced with an LCA (lowest common ancestor) data structure [26]. Gusfield's method is offline (processes all the characters of $T$ together) and works in $O(n)$ time and space for linearly-sortable alphabets, including integer alphabets of size $\text{poly}(n)$. We remark that Gusfield's algorithm uses only the symmetry-based definition for computing maximal palindromes. That is, computing the longest common prefix of $T[c..n]$ and $(T[1..c-1])^R$ for each integer position $c$ in $T$ gives us the maximal palindrome of even length centered at $c - 0.5$. Maximal palindromes of odd lengths can be computed analogously. On the other hand, Manacher's algorithm (which will be briefly recalled in a subsequent section) uses both reversal-based and symmetry-based definitions for computing maximal palindromes.

In this paper, we propose a new framework for formalizing palindromes under the non-standard matching models, which we call *Substring Consistent Symmetric and Two-Transitive Relations* (*SCSTTRs*) that include the complementary-matching model and *Substring Consistent Equivalence Relations* (*SCERs*) [21]. We note that SCERs include the Cartesian-tree model [24], the parameterized model [3], the order-preserving model [16], and the palindromic-structure model [12]. As far as we are aware, the existing algorithms are designed only for computing standard maximal palindromes based on exact character matching, and it is not clear how they can be adapted for the aforementioned palindromes

**Table 1.** Time complexities of algorithms for computing each type of palindromes, where $n$ denotes the length of the input string, $\sigma$ is the (static) alphabet size, and $\pi$ is the size of the parameterized alphabet for parameterized matching. The time complexities are valid under some assumptions for the alphabet, which are designated in parentheses. Each of the algorithms uses $O(n)$ space.

| Matching | Type | |
|---|---|---|
| | Symmetric Palindromes | Reverse Palindromes |
| Exact | $O(n)$ time [9] (linearly sortable) | $O(n)$ time [20] (general unordered) |
| Complementary | $O(n)$ time (linearly sortable) | $O(n)$ time (general unordered) |
| Parameterized | $O(n \log(\sigma + \pi))$ time (linearly sortable) | $O(n)$ time (linearly sortable) |
| Order-Preserving | $O(n \log \log^2 n / \log \log \log n)$ time (linearly sortable) | $O(n)$ time (general unordered) |
| Cartesian-Tree | $O(n \log n)$ time (general ordered) | $O(n)$ time (linearly sortable) |
| Palindromic-Structure | $O(n \min\{\sqrt{\log n}, \log \sigma / \log \log \sigma\})$ time (general unordered) | $O(n)$ time (general unordered) |

under the non-standard matching models. Our first claim is that Gusfield's framework is easily extensible for palindromes under the non-standard matching models: if one has a suffix-tree-like data structure that is built on a suitable encoding for a given matching criterion $\approx$ enhanced with the LCA data structure, then one can readily compute all maximal symmetric palindromes under the non-standard matching models in $O(n)$ time. Thus, the construction time for the suffix-tree-like data structure dominates the total time complexity (see Table 1). On the other hand, extending Manacher's algorithm for palindromes under the non-standard matching models is much more involved. The main contribution of this paper is to design Manacher-style algorithms that compute all maximal reverse palindromes in $O(n)$ time under *all* the non-standard matching models considered in this paper (see Table 1). We remark that a straightforward implementation of the extension of Manacher's algorithm requires quadratic time. We then reduce the time complexity to (near) linear by exploiting and utilizing combinatorial properties of the non-standard matching models (in particular, the Cartesian-tree model, the parameterized model, and the palindromic-structure model require non-trivial elaborations).

*Related Work.* The Cartesian-tree matching problem is introduced by Park et al. [24]. They showed a linear-time algorithm to solve the problem and also proposed an $O(n)$-space index for the problem, which can be constructed in $O(n \log n)$ time. Kim and Cho [17] developed a compact $3n + o(n)$-*bits* index. Nishimoto et al., [22] proposed another $O(n)$-space index, which can be constructed in $O(n \log \sigma)$ time where $\sigma$ is the alphabet size. A probabilistic method for multiple pattern Cartesian-tree matching is also studied [27]. Many other variants of the Cartesian-tree matching problem are studied, including matching on indeterminate strings [7], subsequence matching [23], approximate matching [1], and the longest common factor computing [6].

As for other or more general settings, Kikuchi et al. [15] considered the problem of finding *covers* under SCERs. Kociumaka et al. [18] introduced *squares* (tandem repeats) in the non-standard matching models, including order-preserving matching. Gawrychowski et al. [8] presented a worst-case optimal time algorithm to compute the distinct order-preserving matching squares in a string. We note that our results are not readily obtained by the results above and require new techniques due to the different natures between covers/squares and palindromes.

## 2 Preliminaries

### 2.1 Notations

Let $\Sigma$ be an *alphabet*. An element of $\Sigma^*$ is called a *string*. The length of a string $T$ is denoted by $|T|$. The empty string $\varepsilon$ is the string of length 0, namely, $|\varepsilon| = 0$. For a string $T = xyz$, $x$, $y$ and $z$ are called a *prefix*, *substring*, and *suffix* of $T$, respectively. For a string $T$ and an integer $1 \leq i \leq |T|$, $T[i]$ denotes the $i$-th character of $T$, and for two integers $1 \leq i \leq j \leq |T|$, $T[i..j]$ denotes the substring of $T$ that begins at position $i$ and ends at position $j$. For convenience, let $T[i..j] = \varepsilon$ when $i > j$. The reversal of a string $T$ is denoted by $T^R$, i.e., $T^R = T[|T|] \cdots T[1]$. For two strings $X$ and $Y$, $lcp(X, Y)$ denotes the length of the longest common prefix of $X$ and $Y$. Namely, $lcp(X, Y) = \max\{\ell \mid X[1..\ell] = Y[1..\ell]\}$. A *rightward longest common extension* (*rightward LCE*) query on a string $T$ is to compute $lcp(T[i..|T|], T[j..|T|])$ for given two positions $1 \leq i < j \leq |T|$. Similarly, a *leftward LCE* query is to compute $lcp(T[1..i]^R, T[1..j]^R)$. Then, an *outward LCE* (resp. *inward LCE*) query is, given two positions $1 \leq i < j \leq |T|$, to compute $lcp(T[1..i]^R, T[j..|T|])$ (resp. $lcp(T[i..|T|], T[1..j]^R)$).

A string $T$ is called a *palindrome* if $T = T^R$. We remark that the empty string $\varepsilon$ is also considered to be a palindrome. A non-empty substring palindrome $T[i..j]$ is said to be a *maximal palindrome* in $T$ if $T[i-1] \neq T[j+1]$, $i = 1$, or $j = |T|$. For any non-empty substring palindrome $T[i..j]$ in $T$, $\frac{i+j}{2}$ is called its *center*. It is clear that for each center $c = 1, 1.5, \ldots, n-0.5, n$, we can identify the maximal palindrome $T[i..j]$ whose center is $c$ (namely, $c = \frac{i+j}{2}$). Thus, there are exactly $2n - 1$ maximal palindromes in a string of length $n$ (including empty strings, which occur at non-integer centers $c$ when $T[c-0.5] \neq T[c+0.5]$). Manacher [20] showed an online algorithm that computes all maximal palindromes in a string $T$ of length $n$ in $O(n)$ time. An alternative offline approach is to use outward LCE queries for $2n-1$ pairs of positions in $T$. Using the suffix tree [30] for string $T\$T^R\#$ enhanced with a lowest common ancestor data structure [4,10,26], each outward LCE query can be answered in $O(1)$ time where \$ and \# are special characters that do not appear in $T$. Preprocessing for this approach takes $O(n)$ time and space [5,9] when alphabet $\Sigma$ is linearly-sortable.

## 2.2   Substring Consistent Symmetric and Two-Transitive Relation

A *Substring Consistent Equivalence Relation (SCER)* is an equivalence relation $\approx_{\text{SCER}}$ such that $X \approx_{\text{SCER}} Y$ for two strings $X, Y$ of equal length means that $X[i..j] \approx_{\text{SCER}} Y[i..j]$ holds for all $1 \le i \le j \le |X| = |Y|$. Matsuoka et al. [21] defined the notion and considered the pattern matching problems with SCER. Afterward, several problems over SCER have been considered (e.g., [11,13,15]).

In this paper, we treat a more general relation; Substring Consistent Symmetric and Two-Transitive Relation (SCSTTR). An SCSTTR is a relation $\approx_{\text{SCSTTR}}$ that satisfies the following conditions for strings $W, X, Y, Z$ of equal length:

1. Satisfies the symmetric law; namely, if $X \approx_{\text{SCSTTR}} Y$, then $Y \approx_{\text{SCSTTR}} X$ also holds.
2. Satisfies the two-transitive law; namely, if $W \approx_{\text{SCSTTR}} X$, $X \approx_{\text{SCSTTR}} Y$, and $Y \approx_{\text{SCSTTR}} Z$, then $W \approx_{\text{SCSTTR}} Z$ also holds.
3. Satisfies the substring consistency; namely, if $X \approx_{\text{SCSTTR}} Y$, then $X[i..j] \approx_{\text{SCSTTR}} Y[i..j]$ holds for all $1 \le i \le j \le |X| = |Y|$.

We remark that any SCER is also an SCSTTR. Our aim for introducing SCSTTRs is to deal with the complementary-matching model including the Watson-Crick (WK) model [29] for $\Sigma = \{\mathsf{A}, \mathsf{C}, \mathsf{G}, \mathsf{T}\}$ that is *not* an SCER. For each of the SCSTTR matching models below, encodings with which the matching can be reduced to exact matching are known.

**Complementary Matching** [14,19]. Let $f$ be a function on $\Sigma$ that has the two following properties: (1) $f(uv) = f(u)f(v)$ for all strings $u, v \in \Sigma^*$, and (2) $f^2$ equals the identity. Two strings $X$ and $Y$ are said to complementary-match if there is a function $f$ satisfying $X = f(Y)$. For example, if $f(\mathsf{A}) = \mathsf{T}$, $f(\mathsf{C}) = \mathsf{G}$, $f(\mathsf{T}) = \mathsf{A}$, and $f(\mathsf{G}) = \mathsf{C}$ on $\Sigma = \{\mathsf{A}, \mathsf{T}, \mathsf{G}, \mathsf{C}\}$, then $X = \mathsf{AGCTAT}$ and $Y = \mathsf{TCGATA}$ complementary-match. Reversal-based palindromes under complementary-matching have been studied as $\theta$-palindromes [14,19][1]. In this paper, we consider both the reverse and symmetric definitions and call them Theta rev-palindromes and Theta sym-palindromes, respectively.

**Cartesian-tree Matching** [24]. The Cartesian-tree $\mathsf{CT}(T)$ of a string $T$ is the ordered binary tree recursively defined as follows [28]:

1. If $T$ is the empty string, $\mathsf{CT}(T)$ is empty.
2. If $T$ is not empty, let $T[i]$ be the leftmost occurrence of the smallest character in $T$. Then, the root of $\mathsf{CT}(T)$ is $T[i]$, the left-side subtree of $\mathsf{CT}(T)$ is $\mathsf{CT}(T[1..i-1])$, and the right-side subtree of $\mathsf{CT}(T)$ is $\mathsf{CT}(T[i+1..|T|])$.

We say that two strings $X$ and $Y$ Cartesian-tree match and denote it by $X \approx_{\text{ct}} Y$ if $\mathsf{CT}(X)$ and $\mathsf{CT}(Y)$ are isomorphic as unlabeled ordered trees. For example, $X \approx_{\text{ct}} Y$ holds for two strings $X = \mathsf{cabdcf}$ and $Y = \mathsf{eaacbc}$ (See Fig. 1).

---

[1] $\theta$-palindromes are defined by the function $\theta$ that is not morphic but is antimorphic in the literature, while we use the morphic function $f$ since they can be treated as the same in our arguments.

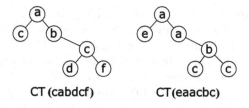

**Fig. 1.** Illustration for Cartesian-trees of $X = $ cabdcf and $Y = $ eaacbc. Since they are isomorphic except their node labels, $X \approx_{ct} Y$ holds.

The *parent distance encoding* $\mathsf{PD}_T$ of a string $T$ is an integer sequence of length $|T|$ such that:

$$\mathsf{PD}_T[i] = \begin{cases} i - \max_{1 \leq j < i}\{j : T[j] \preccurlyeq T[i]\} & \text{if such } j \text{ exists;} \\ 0 & \text{otherwise.} \end{cases}$$

For any two strings $X$ and $Y$, $X \approx_{ct} Y$ iff $\mathsf{PD}_X = \mathsf{PD}_Y$ [24]. For example, again consider two string $X = $ cabdcf and $Y = $ eaacbc, which satisfy $X \approx_{ct} Y$, then $\mathsf{PD}_{X=\text{cabdcf}} = 001121 = \mathsf{PD}_{Y=\text{eaacbc}}$. We note that if $S$ is a prefix of $T$, then the number of 0s in $\mathsf{PD}_T$ is at least the number of 0s in $\mathsf{PD}_S$.

**Parameterized Matching [2].** Let $\Sigma$ and $\Pi$ be disjoint sets of characters, respectively, called a static and parameterized alphabet. Two strings $X$ and $Y$ are said to parameterized match if there is a renaming bijection over the alphabet $\Pi$ that transforms $X$ into $Y$. We write $X \approx_{\text{para}} Y$ iff strings $X$ and $Y$ parameterized match.

**Order-preserving Matching [16].** For ordered alphabets, two strings $X$ and $Y$ are said to order-preserving match if the relative orders of $X$ correspond to those of $Y$. Namely, $X[i] \preccurlyeq X[j] \Leftrightarrow Y[i] \preccurlyeq Y[j]$ for all $1 \leq i, j \leq |X| = |Y|$. We write $X \approx_{\text{op}} Y$ iff strings $X$ and $Y$ order-preserving match.

**Palindromic-structure Matching [12].** Two strings $X$ and $Y$ are said to palindromic-structure match if the length of the maximal palindrome at each center position in $X$ is equal to that of $Y$. We write $X \approx_{\text{pal}} Y$ iff strings $X$ and $Y$ palindromic-structure match.

## 2.3 Our Problems

In this paper, we consider the following two definitions of palindromes:

**Definition 1 (SCSTTR symmetry-based palindromes).** *A string $P$ is called an SCSTTR sym-palindrome if $(P[1..\lfloor |P|/2 \rfloor])^R \approx_{\text{SCSTTR}} P[\lceil |P|/2 \rceil..|P|]$ holds.*

**Definition 2 (SCSTTR reversal-based palindromes).** *A string $P$ is called an SCSTTR rev-palindrome if $P \approx_{\text{SCSTTR}} P^R$ holds.*

As for palindromes under the exact matching, the above two definitions are the same. However, as for palindromes under the non-standard matching models, these definitions are not equivalent. For example, ATTGAAT is not a WK rev-palindrome but is a WK sym-palindrome. Also, CACB is not a parameterized rev-palindrome but is a parameterized sym-palindrome. This paper considers problems for computing maximal SCSTTR (complementary, parameterized, order-preserving, Cartesian-tree, and palindromic-structure) palindromes for the two definitions. Firstly, we notice that SCSTTR rev-palindromes have symmetricity:

**Lemma 1.** *Let $P$ be an SCSTTR rev-palindrome. If a substring $P[i..j]$ of $P$ is an SCSTTR rev-palindrome, then the substring at the symmetrical position is also an SCSTTR rev-palindrome, namely, $P[|P| - j + 1..|P| - i + 1]$ is an SCSTTR rev-palindrome.*

*Proof.* Since $P$ is an SCSTTR rev-palindrome, $P[i..j] \approx_{\text{SCSTTR}} P[|P| - j + 1..|P| - i + 1]^R$ and $P[i..j]^R \approx_{\text{SCSTTR}} P[|P| - j + 1..|P| - i + 1]$ hold. Also, since $P[i..j]$ is an SCSTTR rev-palindrome, $P[i..j] \approx_{\text{SCSTTR}} P[i..j]^R$ holds. Combining these three equations under SCSTTR, we obtain $P[|P| - j + 1..|P| - i + 1] \approx_{\text{SCSTTR}} P[|P| - j + 1..|P| - i + 1]^R$ from the symmetric law and two-transitive law. Therefore, $P[|P| - j + 1..|P| - i + 1]$ is an SCSTTR rev-palindrome. □

This lemma allows us to design a Manacher-like algorithm for computing maximal SCSTTR rev-palindromes in Sect. 4.

Next, we notice that the Cartesian-tree matching is not closed under reversal; namely, $X \approx_{\text{ct}} Y$ does not imply $X^R \approx_{\text{ct}} Y^R$. For instance, aaaa $\approx_{\text{ct}}$ abcd but aaaa $\not\approx_{\text{ct}}$ dcba. Hence, there are cases where the Cartesian-tree sym-palindrome obtained by extending outward direction is not equal to that obtained by extending inward direction for the same center position $c$. Therefore, we introduce the two following variants for the Cartesian-tree sym-palindromes:

**Definition 3 (outward Cartesian-tree sym-palindromes).** *A string $P$ is an outward Cartesian-tree sym-palindrome if $(P[1..\lfloor |P|/2 \rfloor])^R \approx_{\text{ct}} P[\lceil |P|/2 \rceil..|P|]$.*

**Definition 4 (inward Cartesian-tree sym-palindromes).** *A string $P$ is an inward Cartesian-tree sym-palindrome if $P[1..\lfloor |P|/2 \rfloor] \approx_{\text{ct}} (P[\lceil |P|/2 \rceil..|P|])^R$.*

## 3   Algorithms for Computing Maximal SCSTTR Symmetry-Based Palindromes

This section considers algorithms for computing maximal SCSTTR sym-palindromes. The main idea is the same as Gusfield's algorithm; to use the outward LCE query on the matching model in SCSTTR for each center position. Then, the complexity of the algorithm can be written as $O(nq + t_{\text{SCSTTR}})$ time and $O(n + s_{\text{SCSTTR}})$ space, where $q$ is the outward LCE query time, $t_{\text{SCSTTR}}$ is the construction time of the data structure for the LCE query on the matching model, and $s_{\text{SCSTTR}}$ is the space of the data structure. Then, we obtain the following results for several matching models in SCSTTR.

**Theorem 1.** *All maximal SCSTTR sym-palindromes can be computed with $O(n)$ space for the following matching models:*

1. *For linearly sortable alphabets, all maximal Theta sym-palindromes can be computed in $O(n)$ time.*
2. *For ordered alphabets, all outward/inward maximal Cartesian-tree sym-palindromes can be computed in $O(n \log n)$ time.*
3. *For linearly sortable alphabets, all maximal parameterized sym-palindromes can be computed in $O(n \log(\sigma + \pi))$ time.*
4. *For linearly sortable alphabets, all maximal order-preserving sym-palindromes can be computed in $O(n \log \log^2 n / \log \log \log n)$ time.*
5. *For general unordered alphabets, all maximal palindromic-structure sym-palindromes can be computed in $O(n \min\{\sqrt{\log n}, \log \sigma / \log \log \sigma\})$ time.*

These results can be obtained by using SCSTTR suffix trees [3,5,9,12,16,24]. All the details and proofs can be found in the full version of this paper.

## 4    Algorithms for Computing Maximal SCSTTR Reversal-Based Palindromes

This section considers algorithms for computing maximal SCSTTR rev-palindromes. If we were to use the SCSTTR suffix tree and outward LCE queries as in the previous section, how to choose the starting positions of outward LCE queries is unclear. Therefore, a naïve approach would require $O(n)$ inward LCE queries for each center position, and the total complexity will be $O(n^2 + t_{\mathrm{SCSTTR}})$ time and $O(n + s_{\mathrm{SCSTTR}})$ space. By combining inward LCE queries and binary search, we can further achieve $O(n \log n + t_{\mathrm{SCSTTR}})$ time and $O(n + s_{\mathrm{SCSTTR}})$ space with this approach. This section shows $O(n)$-time algorithms without constructing SCSTTR suffix trees:

**Theorem 2.** *There exist $O(n)$-time algorithms which compute (1) all maximal Theta rev-palindromes for general unordered alphabets, (2) all maximal Cartesian-tree rev-palindromes for linearly sortable alphabets, (3) all maximal parameterized rev-palindromes for linearly sortable alphabets, (4) all maximal order-preserving rev-palindromes for general unordered alphabets, and (5) all maximal palindromic-structure rev-palindromes for general unordered alphabets.*

Proofs for latter three matching models are provided in the full version.

### 4.1    Outline of Computing SCSTTR Reversal-Based Palindromes

At first, we show a framework for computing maximal SCSTTR rev-palindromes, which is a generalization of Manacher's algorithm [20]. For the sake of simplicity, we denote SCSTTR rev-palindromes by just *palindromes* in the description of the framework below. In the following, we describe how to compute all odd-lengthed maximal palindromes. Even-lengthed maximal palindromes can be obtained analogously.

We consider finding the odd-lengthed maximal palindromes in ascending order of the center position. Let MPal[$c$] denote the interval corresponding to the odd-lengthed maximal palindrome centered at position $c$. The length of MPal[1] is always one. Assuming that all odd-lengthed maximal palindromes whose center is at most $c-1$ have been computed and let MPal[$c'$] = [$b, e$] be the longest interval whose ending position $e$ is largest among {MPal[1], ..., MPal[$c-1$]}. Further, let $P' = T[b..e]$. Also, let $d$ be the distance between $c'$ and $c$; namely, $d = c - c'$. If $T[i-1..j+1]$ is a palindrome, we say palindrome $T[i..j]$ can be extended. Also, we call computing the length of MPal[$(i+j)/2$] computing the extension of $T[i..j]$.

Now, we consider how to compute MPal[$c$]. If $e < c$, then $c = e+1$ holds. We then compute MPal[$c$] by computing the extension of $T[e+1..e+1]$. Otherwise (i.e., if $c \le e$ holds), we compute MPal[$c$] as in the following two cases according to the relationship between $P'$ and MPal[$c'-d$].

1. If the starting position of MPal[$c'-d$] is larger than $b$, then |MPal[$c$]| = |MPal[$c'-d$]| holds (by Lemma 1). Thus we *copy* MPal[$c'-d$] to MPal[$c$] with considering symmetry.
2. Otherwise, the ending position of MPal[$c$] is at least $e$. Then we compute the extension of $T[2c-e..e]$, and $P'$ is updated to MPal[$c$] if $T[2c-e..e]$ can be extended.

Note that, in the exact matching model, this framework is identical to that of Manacher's algorithm. The framework includes two non-trivial operations; copying and extension-computing. How many times both operations are called can be analyzed in the same manner as for Manacher's algorithm, and is $O(n)$. Also, each copying operation can be done in constant time. Thus, if we can perform each extension-computing in (amortized) constant time, the total time complexity becomes $O(n)$. The following sections focus only on how to compute an extension in (amortized) constant time under each of the two matching models; the complementary-matching model and the Cartesian-tree matching model.

### 4.2 Maximal Theta Reversal-Based Palindromes

In the complementary-matching model, the correspondence of matching characters, although different, is predetermined, and naïve character comparisons can be done in constant time. Thus, the following result can be obtained immediately.

**Proposition 1.** *For general unordered alphabets, all maximal Theta rev-palindromes can be computed in $O(n)$ time with $O(n)$ space.*

### 4.3 Maximal Cartesian-Tree Reversal-Based Palindromes

Here, we consider how to compute the extension of a maximal Cartesian-tree rev-palindrome $T[i..j]$. We consider the difference between $PD_{T[i..j]}$ and $PD_{T[i-1..j+1]}$. There are three types of positions of $PD_{T[i-1..j+1]}$ in which values differ from $PD_{T[i..j]}$; the first and last positions (which do not exist in $PD_{T[i..j]}$), and each position $k$ such that $\mathsf{PD}_{T[i..j]}[k] = 0$ and $T[i-1] \preccurlyeq T[k]$ with $i \leq k \leq j$ (i.e., $\mathsf{PD}_{T[i-1..j+1]}[k+1] \neq 0$). Let $m_{[i..j]}$ be the leftmost occurrence position of the smallest value in $T[i..j]$. Then, $k'$ such that $\mathsf{PD}_{T[i..j]}[k'] = 0$ is always to the left or equal to $m_{[i..j]}$. However, since the number of such positions is not always constant, the time required for computing the extension from $T[i..j]$ to $T[i-1..j+1]$ can be $\omega(1)$. For example, when $T[i-1..j+1] = \text{becaebdaefc}$, $\mathsf{PD}_{T[i..j]} = [\underline{0}, \underline{0}, 0, 1, 2, 1, 4, 1, 1]$ and $\mathsf{PD}_{T[i-1..j+1]} = [0, \underline{1}, \underline{2}, 0, 1, 2, 1, 4, 1, 1, 3]$ hold. There are two differences between $\mathsf{PD}_{T[i..j]}$ and $\mathsf{PD}_{T[i-1..j+1]}[2..10]$ (underlined). We will consider the total number of updates of such positions through the entire algorithm.

Let $c$ be the center of $T[i..j]$ and $T[b..e]$ be the maximal Cartesian-tree rev-palindrome centered at $c$. Also, let $Z_c$ be the number of 0's in $\mathsf{PD}_{T[i..j]}$, $E_c$ be the length of the extension of $T[i..j]$ (i.e., $E_c = i - b$), and $U_c$ be the number of updates of positions of 0 from $\mathsf{PD}_{T[i..j]}$ to $\mathsf{PD}_{T[b..e]}$. Let $T[i'..j']$ be the next palindrome of center $c'$ which we have to compute the extension of, after computing the extension of $T[i..j]$. Namely, $T[i'..j']$ is the longest proper suffix palindrome of $T[b..e]$ if $|T[b..e]| > 1$. Otherwise, $T[i'..j'] = T[e+1..e+1]$ holds.

As the first step to considering the total sum $\sum U_c$ over the entire algorithm, we consider the simpler case such that each character is distinct from the other in the string. We call this case *permutation-Cartesian-tree matching*.

In this case, $c = m_{[i..j]}$ holds since $T[m_{[i..j]}]$ is the only smallest value in $T[i..j]$ and $m_{[i..j]}$ is the root of $\mathsf{CT}(T[i..j])$ and $\mathsf{CT}(T[i..j]^R)$. Also, $m_{[i..j]} = m_{[b..e]}$ holds. Now we consider the starting position of the next palindrome $T[i'..j']$.

**Lemma 2.** *Let $T[i'..j']$ be the next palindrome. Then, $i' > m_{[i..j]}$ holds.*

*Proof.* The next palindrome $T[i'..j']$ is either $T[e+1..e+1]$ or the longest proper suffix palindrome of $T[b..e]$. Since the former case is obvious, we consider only the latter in the following. For the sake of contradiction, we assume that $i' \leq m_{[i..j]}$. Then $T[m_{[i'..j']}] < T[m_{[i..j]}]$ holds. Since $m_{[i'..j']} \in [b..e]$, this leads to a contradiction with $m_{[i..j]} = m_{[b..e]}$. ∎

From Lemma 2, 0s in $\mathsf{PD}_{T[i..j]}$ and 0s in $\mathsf{PD}_{T[i'..j']}$ all correspond to different positions in $T$. Hence, $\sum Z_c \leq n$ holds. Also, $U_c \leq Z_c + E_c$ and $\sum E_c \leq n$ holds. Therefore, we obtain $\sum U_c \leq \sum Z_c + \sum E_c \leq 2n$. In the preprocessing, we compute $\mathsf{PD}_T$ and $\mathsf{PD}_{T^R}$ in $O(n)$ time [24] and create a link from each position $p$ to $q$ with $\max_{1 \leq q < p}\{q : T[q] \preccurlyeq T[p]\}$. Then, when we compute the extension of $T[i..j]$, each 0's position to be updated can be obtained in $O(1)$ time by traversing back the links from $i - 1$. Thus, the extension from $T[i..j]$ to $T[b..e]$ can be computed in $O(E_c + U_c)$ time by updating 0s and checking the equalities of values of all updated positions between $\mathsf{PD}_{T[i-\ell..j+\ell]}$ and $\mathsf{PD}_{T[i-\ell..j+\ell]^R}$ for each $1 \leq \ell \leq E_c$, since we can access each element of $\mathsf{PD}_{T[i-\ell..j+\ell]}$ and $\mathsf{PD}_{T[i-\ell..j+\ell]^R}$ in constant time if we have $\mathsf{PD}_T$ and $\mathsf{PD}_{T^R}$ [24]. Therefore, we obtain the following result:

**Lemma 3.** *For linearly sortable alphabets, all maximal permutation-Cartesian-tree rev-palindromes can be computed in $O(n)$ time with $O(n)$ space.*

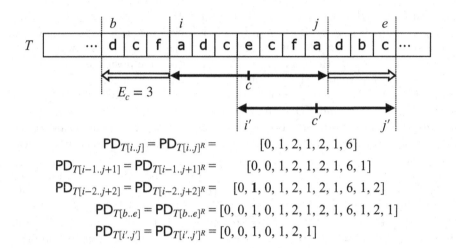

**Fig. 2.** A concrete example for Cartesian-tree palindromes. In this example, $Z_c = 1$, $E_c = 3$, and $U_c = 1$ hold since when $\mathrm{PD}_{T[i-1..j+1]}$ grows to $\mathrm{PD}_{T[i-2..j+2]}$, a single value (shown in bold font) is updated from 0 to non-zero.

Now we consider the sum $\sum U_c$ in the general case. As for the relationship between $Z_c$ and $Z_{c'}$, we show the following lemma:

**Lemma 4.** *For the center $c$ of $T[i..j]$ and the center $c'$ of the next palindrome $T[i'..j']$, $Z_{c'} \leq Z_c + E_c - U_c$ holds.*

*Proof.* The next palindrome $T[i'..j']$ is either $T[e+1..e+1]$ or the longest proper suffix palindrome of $T[b..e]$. Since the former case is obvious, we consider only the latter case. By definitions, the number of 0s in $\mathrm{PD}_{T[b..e]}$ is equal to $Z_c + E_c - U_c$. Also, since $T[b..e]$ is a Cartesian-tree palindrome, the number of 0s in $\mathrm{PD}_{T[b..e]^R}$ is also equal to $Z_c + E_c - U_c$. Similarly, since $T[i'..j']$ is a Cartesian-tree palindrome, the number of 0s in $\mathrm{PD}_{T[i'..j']^R}$ is equal to $Z_{c'}$. Further, the number of 0s in $\mathrm{PD}_{T[b..e]^R}$ is at least $Z_{c'}$ since $T[i'..j']^R$ is a prefix of $T[b..e]^R$ (see also Fig. 2). Therefore $Z_{c'} \leq Z_c + E_c - U_c$ holds.

From Lemma 4, $U_c \leq Z_c - Z_{c'} + E_c$ holds. Therefore, the total number of updates $\sum U_c \leq \sum(Z_c - Z_{c'}) + \sum E_c \leq 2n$. Hence, we obtain the following result:

**Proposition 2.** *For linearly sortable alphabets, all maximal Cartesian-tree rev-palindromes can be computed in $O(n)$ time with $O(n)$ space.*

**Acknowledgments.** This work was supported by JSPS KAKENHI Grant Numbers JP20J21147 (MF), JP23H04381(TM), JP21K17705 (YN), JP23H04386 (YN), JP22H03551 (SI), JP20H04141 (HB), and JP18H04098 (MT).

# References

1. Auvray, B., David, J., Groult, R., Lecroq, T.: Approximate Cartesian tree matching: an approach using swaps. In: Nardini, F.M., Pisanti, N., Venturini, R. (eds.) String Processing and Information Retrieval, SPIRE 2023, LNCS, vol. 14240, pp. 49–61. Springer, Cham (2023). https://doi.org/10.1007/978-3-031-43980-3_5
2. Baker, B.S.: A theory of parameterized pattern matching: algorithms and applications. In: 25th Annual ACM Symposium on Theory of Computing, ACM 1993, pp. 71–80. ACM (1993). https://doi.org/10.1145/167088.167115
3. Baker, B.S.: Parameterized pattern matching: algorithms and applications. J. Comput. Syst. Sci. **52**(1), 28–42 (1996). https://doi.org/10.1006/jcss.1996.0003
4. Bender, M.A., Farach-Colton, M.: The LCA problem revisited. In: Gonnet, G.H., Viola, A. (eds.) LATIN 2000. LNCS, vol. 1776, pp. 88–94. Springer, Heidelberg (2000). https://doi.org/10.1007/10719839_9
5. Farach-Colton, M., Ferragina, P., Muthukrishnan, S.: On the sorting-complexity of suffix tree construction. J. ACM **47**(6), 987–1011 (2000). https://doi.org/10.1145/355541.355547
6. Faro, S., Lecroq, T., Park, K., Scafiti, S.: On the longest common Cartesian substring problem. Comput. J. **66**(4), 907–923 (2023). https://doi.org/10.1093/COMJNL/BXAB204
7. Gawrychowski, P., Ghazawi, S., Landau, G.M.: On indeterminate strings matching. In: 31st Annual Symposium on Combinatorial Pattern Matching, CPM 2020, June 17-19, 2020, Copenhagen, Denmark. LIPIcs, vol. 161, pp. 14:1–14:14. Schloss Dagstuhl - Leibniz-Zentrum für Informatik (2020). https://doi.org/10.4230/LIPICS.CPM.2020.14
8. Gawrychowski, P., Ghazawi, S., Landau, G.M.: Order-preserving squares in strings. In: 34th Annual Symposium on Combinatorial Pattern Matching, CPM 2023, 26–28 June 2023, Marne-la-Vallée, France. LIPIcs, vol. 259, pp. 13:1–13:19. Schloss Dagstuhl - Leibniz-Zentrum für Informatik (2023). https://doi.org/10.4230/LIPICS.CPM.2023.13
9. Gusfield, D.: Algorithms on Strings, Trees, and Sequences - Computer Science and Computational Biology. Cambridge University Press, Cambridge (1997). https://doi.org/10.1017/cbo9780511574931
10. Harel, D., Tarjan, R.E.: Fast algorithms for finding nearest common ancestors. SIAM J. Comput. **13**(2), 338–355 (1984). https://doi.org/10.1137/0213024
11. Hendrian, D.: Generalized dictionary matching under substring consistent equivalence relations. In: Rahman, M.S., Sadakane, K., Sung, W.-K. (eds.) WALCOM 2020. LNCS, vol. 12049, pp. 120–132. Springer, Cham (2020). https://doi.org/10.1007/978-3-030-39881-1_11
12. Palindrome pattern matching: I, T., Inenaga, S., Takeda, M. Theor. Comput. Sci. **483**, 162–170 (2013). https://doi.org/10.1016/j.tcs.2012.01.047
13. Jargalsaikhan, D., Hendrian, D., Yoshinaka, R., Shinohara, A.: Parallel algorithm for pattern matching problems under substring consistent equivalence relations. In: 33rd Annual Symposium on Combinatorial Pattern Matching, CPM 2022. LIPIcs, vol. 223, pp. 28:1–28:21. Schloss Dagstuhl - Leibniz-Zentrum für Informatik (2022). https://doi.org/10.4230/LIPIcs.CPM.2022.28
14. Kari, L., Mahalingam, K.: Watson-Crick conjugate and commutative words. In: Garzon, M.H., Yan, H. (eds.) DNA 2007. LNCS, vol. 4848, pp. 273–283. Springer, Heidelberg (2008). https://doi.org/10.1007/978-3-540-77962-9_29

15. Kikuchi, N., Hendrian, D., Yoshinaka, R., Shinohara, A.: Computing covers under substring consistent equivalence relations. In: Boucher, C., Thankachan, S.V. (eds.) SPIRE 2020. LNCS, vol. 12303, pp. 131–146. Springer, Cham (2020). https://doi.org/10.1007/978-3-030-59212-7_10

16. Kim, J., Eades, P., Fleischer, R., Hong, S., Iliopoulos, C.S., Park, K., Puglisi, S.J., Tokuyama, T.: Order-preserving matching. Theor. Comput. Sci. **525**, 68–79 (2014). https://doi.org/10.1016/j.tcs.2013.10.006

17. Kim, S., Cho, H.: A compact index for Cartesian tree matching. In: 32nd Annual Symposium on Combinatorial Pattern Matching, CPM 2021, July 5-7, 2021, Wrocław, Poland. LIPIcs, vol. 191, pp. 18:1–18:19. Schloss Dagstuhl - Leibniz-Zentrum für Informatik (2021). https://doi.org/10.4230/LIPICS.CPM.2021.18

18. Kociumaka, T., Radoszewski, J., Rytter, W., Walen, T.: Maximum number of distinct and nonequivalent nonstandard squares in a word. Theor. Comput. Sci. **648**, 84–95 (2016). https://doi.org/10.1016/j.tcs.2016.08.010

19. de Luca, A., Luca, A.D.: Pseudopalindrome closure operators in free monoids. Theor. Comput. Sci. **362**(1–3), 282–300 (2006). https://doi.org/10.1016/j.tcs.2006.07.009

20. Manacher, G.K.: A new linear-time "on-line" algorithm for finding the smallest initial palindrome of a string. J. ACM **22**(3), 346–351 (1975). https://doi.org/10.1145/321892.321896

21. Matsuoka, Y., Aoki, T., Inenaga, S., Bannai, H., Takeda, M.: Generalized pattern matching and periodicity under substring consistent equivalence relations. Theor. Comput. Sci. **656**, 225–233 (2016). https://doi.org/10.1016/j.tcs.2016.02.017

22. Nishimoto, A., Fujisato, N., Nakashima, Y., Inenaga, S.: Position heaps for Cartesian-tree matching on strings and tries. In: Lecroq, T., Touzet, H. (eds.) SPIRE 2021. LNCS, vol. 12944, pp. 241–254. Springer, Cham (2021). https://doi.org/10.1007/978-3-030-86692-1_20

23. Oizumi, T., Kai, T., Mieno, T., Inenaga, S., Arimura, H.: Cartesian tree subsequence matching. In: 33rd Annual Symposium on Combinatorial Pattern Matching, CPM 2022, 27-29 June 2022, Prague, Czech Republic. LIPIcs, vol. 223, pp. 14:1–14:18. Schloss Dagstuhl - Leibniz-Zentrum für Informatik (2022). https://doi.org/10.4230/LIPICS.CPM.2022.14

24. Park, S.G., Bataa, M., Amir, A., Landau, G.M., Park, K.: Finding patterns and periods in Cartesian tree matching. Theor. Comput. Sci. **845**, 181–197 (2020). https://doi.org/10.1016/j.tcs.2020.09.014

25. Pei, J., Wu, W.C., Yeh, M.: On shortest unique substring queries. In: 29th IEEE International Conference on Data Engineering, ICDE 2013, Brisbane, Australia, 8–12 April 2013, pp. 937–948. IEEE Computer Society (2013). https://doi.org/10.1109/ICDE.2013.6544887

26. Schieber, B., Vishkin, U.: On finding lowest common ancestors: simplification and parallelization. SIAM J. Comput. **17**(6), 1253–1262 (1988). https://doi.org/10.1137/0217079

27. Song, S., Gu, G., Ryu, C., Faro, S., Lecroq, T., Park, K.: Fast algorithms for single and multiple pattern Cartesian tree matching. Theor. Comput. Sci. **849**, 47–63 (2021). https://doi.org/10.1016/J.TCS.2020.10.009

28. Vuillemin, J.: A unifying look at data structures. Commun. ACM **23**(4), 229–239 (1980). https://doi.org/10.1145/358841.358852

29. Watson, J.D., Crick, F.H.: Molecular structure of nucleic acids: a structure for deoxyribose nucleic acid. Nature **171**, 737–738 (1953). https://doi.org/10.1038/171737a0
30. Weiner, P.: Linear pattern matching algorithms. In: 14th Annual Symposium on Switching and Automata Theory, SWAT 1973, pp. 1–11. IEEE Computer Society (1973). https://doi.org/10.1109/SWAT.1973.13

# On the Structure of Hamiltonian Graphs with Small Independence Number

Nikola Jedličková$^{(\boxtimes)}$ and Jan Kratochvíl

Department of Applied Mathematics, Faculty of Mathematics and Physics,
Charles University, Prague, Czech Republic
{jedlickova,honza}@kam.mff.cuni.cz

**Abstract.** A Hamiltonian path (cycle) in a graph is a path (cycle, respectively) which passes through all of its vertices. The problems of deciding the existence of a Hamiltonian cycle (path) in an input graph are well known to be NP-complete, and restricted classes of graphs which allow for their polynomial-time solutions are intensively investigated. Until very recently the complexity was open even for graphs of independence number at most 3. A so far unpublished result of Jedličková and Kratochvíl [arXiv:2309.09228] shows that for every integer $k$, the problems of deciding the existence of a Hamiltonian path and cycle are polynomial-time solvable in graphs of independence number bounded by $k$. As a companion structural result, in this paper, we determine explicit obstacles for the existence of a Hamiltonian path for small values of $k$, namely for graphs of independence number 2, 3, and 4. Identifying these obstacles in an input graph yields alternative polynomial-time algorithms for deciding the existence of a Hamiltonian path with no large hidden multiplicative constants.

**Keywords:** Graph · Hamiltonian path · Hamiltonian cycle · Path Cover · Independence number · Polynomial-time algorithm

## 1 Introduction

A cycle in a graph is *Hamiltonian* if it contains all vertices of the graph. A graph is called *Hamiltonian* if it contains a Hamiltonian cycle. The notion of Hamiltonian graphs is intensively studied in graph theory. Many sufficient conditions for Hamiltonicity of graphs are known, but a simple necessary and sufficient condition is not known (and not likely to exist, as the problem itself is NP-complete). On the other hand, many open questions and conjectures are around. The more than 50 years old conjecture of Barnette [1] states that every cubic planar bipartite 3-connected graph is Hamiltonian, to mention at least one.

A *Hamiltonian path* in a graph is a path that contains all vertices of the graph. Obviously, every Hamiltonian graph contains a Hamiltonian path, and a graph is Hamiltonian if it contains a Hamiltonian path connecting a pair of

Supported by GAUK 370122.

adjacent vertices. A graph is called *Hamiltonian connected* if every two distinct vertices are connected by a Hamiltonian path. Chvátal and Erdős stated and proved elegant sufficient conditions for Hamiltonian connectedness and for the existence of Hamiltonian paths in terms of comparing the vertex connectivity and independence number of the graph under consideration, cf. Proposition 1. For an excellent survey on Hamiltonian graphs, see [10] and a more recent one [11].

From the computational complexity point of view, all these problems are hard. Karp [13] proved already in 1972 that deciding the existence of Hamiltonian paths and cycles in an input graph are NP-complete problems. In order to study the borderline between easy (polynomial-time solvable) and hard (NP-complete) variants of the problems, researchers have intensively studied the complexity of Hamiltonian-related problems in special graph classes. Deciding the existence of a Hamiltonian cycle remains NP-complete on planar graphs [8], circle graphs [4] and for several other generalizations of interval graphs [2], for chordal bipartite graphs [15] and for split (and therefore also for chordal) graphs [9], but is solvable in linear time in interval graphs [14] and in convex bipartite graphs [15]. The existence of a Hamiltonian path can be decided in polynomial time for cocomparability graphs [6] and for circular arc graphs [5].

Many of the above mentioned graph classes are hereditary, i.e., closed in the induced subgraph order, and as such can be described by collections of forbidden induced subgraphs. This has led to carefully examining $H$-free graphs, i.e., graph classes with a single forbidden induced subgraph. For Hamiltonian-type problems, consider three-vertex graphs $H$. For $H = K_3$, both HAMILTONIAN CYCLE and HAMILTONIAN PATH (the problems of deciding the existence of a Hamiltonian cycle or Hamiltonian path, respectively) are NP-complete, since they are NP-complete on bipartite graphs [15]. For $H = 3K_1$, the edgeless graph on three vertices, both HAMILTONIAN CYCLE and HAMILTONIAN PATH are polynomial time solvable [7]. The remaining two graphs, $P_3$ and $K_1 + K_2$, are induced subgraphs of $P_4$, thus the corresponding class of $H$-free graphs is a subclass of cographs, and as such a subclass of cocomparability graphs, in which the HAMILTONIAN PATH and HAMILTONIAN CYCLE problems are solvable in polynomial time [6]. However, a complete characterization of graphs $H$ for which HAMILTONIAN PATH or HAMILTONIAN CYCLE problems are solvable in polynomial time (and for which they are NP-complete) when restricted to $H$-free input graphs is not in sight.

### Our Results

Recently, there has been a substantial progress at least for the case when $H$ is an edgeless graph, i.e., $H = kK_1$ for some $k$. It is shown in [12] that HAMILTONIAN PATH and HAMILTONIAN CYCLE are polynomial-time solvable in $kK_1$-free graphs, for every $k$, i.e., that these problems are in the class XP when parameterized by the independence number of the input graph. The goal of this paper is to provide a companion structural description of feasible instances for small values of $k$, namely $k = 2, 3, 4$. The connectivity level of the input graph plays a role in the case distinction of the characterization, and theorems of Chvátal and Erdős are handy here (cf. Sect. 2). It is relatively easy to describe the feasible instances of HAMILTONIAN PATH with prescribed end-vertices in a $3K_1$-free graph, cf. Theorem 1. All the conditions can be checked quickly, and thus the obstacles provide not only a structural result, but also a practical algorithm. For $4K_1$-free and $5K_1$-free graphs, the conditions become more complicated (cf. Theorems 4, 7). In this conference version, we show in detail how the most general Theorem 7 follows from the previous theorems, whose proofs will appear in the journal version of the paper.

## 2    Preliminaries

We consider simple undirected graphs without loops or multiple edges. The vertex set of a graph $G$ is denoted by $V(G)$, its edge set by $E(G)$. Edges are considered as two-element sets of vertices, thus we write $u \in e$ to express that a vertex $u$ is incident with an edge $e$. For the sake of brevity, we write $uv$ instead of $\{u, v\}$ for the edge containing vertices $u$ and $v$. We say that $u$ is *adjacent to* $v$ if $uv \in E(G)$. The subgraph of $G$ induced by vertices $A \subseteq V(G)$ will be denoted by $G[A]$. The *independence number* of a graph $G$, denoted by $\alpha(G)$, is the order of the largest edgeless induced subgraph of $G$. With the standard notion of $K_k$ being the complete graph with $k$ vertices and $G + H$ being the disjoint union of graphs $G$ and $H$, $\alpha(G)$ is equal to the largest $k$ such that $kK_1$ is an induced subgraph of $G$. A graph is called *H-free* if it contains no induced subgraph isomorphic to $H$.

A *path* in a graph $G$ is a sequence of distinct vertices such that any two consecutive ones are adjacent. A *cycle* is formed by a path connecting two adjacent vertices. The path (cycle) is *Hamiltonian* if it contains all vertices of the graph. The *path cover number* of $G$, denoted by $\mathrm{pc}(G)$, is the smallest number of vertex disjoint paths that contain all vertices of $G$.

A graph is connected if any two vertices are connected by a path. A *vertex cut* in a connected graph $G$ is a set $A \subset V(G)$ of vertices such that the graph $G - A = G[V(G) \backslash A]$ is disconnected. The *vertex connectivity* $c_v(G)$ of a graph $G$ is the order of a minimum cut in $G$, or $|V(G)| - 1$ if $G$ is a complete graph. Since we will not consider edge connectivity, we will often omit the adjective when talking about the connectivity measure, we always have vertex connectivity in mind.

Although we consider undirected graphs, when we talk about a path in a graph, the path itself is considered traversed in the direction from its starting vertex to the ending one. Formally, when we say that a path $P$ connects a vertex $x$ to a vertex $y$, then by $P^{-1}$ we denote the same path, but traversed from $y$ to $x$. This is important when creating a longer path by concatenating shorter ones.

We build upon the following results of Chvátal and Erdős.

**Proposition 1** [3]. *Let $G$ be an s-connected graph.*

1. *If $\alpha(G) < s + 2$, then $G$ has a Hamiltonian path.*
2. *If $\alpha(G) < s + 1$, then $G$ has a Hamiltonian cycle.*
3. *If $\alpha(G) < s$, then $G$ is Hamiltonian connected (i.e. every pair of vertices is joined by a Hamiltonian path).*

# 3   Hamiltonian Paths in Graphs with Small Independence Number

In this section we explicitly describe the situations when Hamiltonian paths do not exist for $kK_1$-free graphs for small cases of $k$, namely $k = 3, 4, 5$.

For the description, we will use the following notation. A path cover of size 2 with one path starting in a vertex $u$ and the other one starting in a vertex $v$ will be denoted by $PC(u, v)$. A path $P$ with endpoints $u, v$ will be denoted by $P_{u,v}$. Furthermore, $c(G)$ will denote the number of connected components of $G$. Note that a vertex cut of size 1 is called an *articulation point*. The set of articulation points in a graph $G$ will be denoted by $Art(G)$.

## 3.1   $3K_1$-Free Graphs

It follows from Proposition 1.1 that each connected $3K_1$-free graph has a Hamiltonian path. We will establish when a $3K_1$-free graph has a Hamiltonian path, resp. path cover starting in a specified vertex (resp. vertices).

**Theorem 1.** *Let $G$ be a connected $3K_1$-free graph and let $u, v$ be distinct vertices of $G$. Then $G$ has a Hamiltonian path $P_{u,v}$ if and only if*

*(a) none of the vertices $u$ and $v$ is an articulation point in $G$, and*
*(b) for every articulation point $x$, the vertices $u$ and $v$ belong to different components of $G - \{x\}$, and*
*(c) $\{u, v\}$ is not a minimum vertex cut in $G$.*

*Moreover, this happens if and only if $G$ satisfies (a), (b) and*

*(c') $\{u, v\}$ is not a vertex cut of size two in $G$.*

**Theorem 2.** *Let $G$ be a connected $3K_1$-free graph. Then $G$ has a Hamiltonian path starting at $u \in V(G)$ if and only if $u$ is not an articulation point in $G$.*

**Theorem 3.** *Let $G$ be a connected $3K_1$-free graph. Then $G$ has a path cover $PC(u, v)$ for every two distinct vertices $u, v \in V(G)$.*

## 3.2   $4K_1$-Free Graphs

**Theorem 4.** *Let $G$ be a connected $4K_1$-free graph. Then $G$ has a Hamiltonian path if and only if the following conditions are satisfied.*

*(a) For every articulation point $x$, the graph $G - \{x\}$ has exactly 2 components.*
*(b) There are no 3 articulation points inducing a triangle in $G$.*

*Moreover, if $G$ does not have a Hamiltonian path, then $G$ has a path cover of size 2 (Fig. 1).*

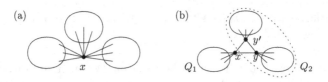

**Fig. 1.** An illustration of violation of the conditions in Theorem 4.

**Theorem 5.** *Let $G$ be a connected $4K_1$-free graph. The graph $G$ has a Hamiltonian path starting in a vertex $u$ if and only if the following conditions are satisfied.*

(a) *The graph $G$ has a Hamiltonian path (i.e. the conditions of Theorem 4 are satisfied).*

(b) *The vertex $u$ is not an articulation point of $G$.*

(c) *Let $x \neq u$ be an articulation point of $G$ and let $Q_u$ be the component of $G - \{x\}$ containing $u$. If $G[Q_u]$ is not 2-connected, then for every articulation point $y \neq u$ of $G[Q_u]$, there exists a vertex $v \in Q_u$ adjacent to $x$ such that $u, v$ are in different components of $G[Q_u] - \{y\}$.*

(d) *Let $x \neq u$ be an articulation point of $G$ and let $Q_u$ be the component of $G - \{x\}$ containing $u$. If $G[Q_u]$ is 2-connected, then there exists $v \in Q_u$ adjacent to $x$ such that $v \neq u$ and $\{u, v\}$ does not form a minimum vertex cut in $G[Q_u]$.*

(e) *There is no vertex $x$ such that $c(G - \{u, x\}) \geq 3$.*

(f) *Let $G$ be 2-connected and let $\{x, y\}$ be a minimum vertex cut of $G$ such that $c(G - \{x, y\}) = 2$. Let $Q_1$ be a component of $G - \{x, y\}$ which is a clique and let $Q_2$ be the other component of $G - \{x, y\}$. If all neighbours of $x$ in $Q_2$ are articulation points of $G[Q_2]$, then $u \neq y$.*

Note that if we allowed $y = u$ in condition (c) and (c) were violated with $y = u$, then condition (e) would be violated (Fig. 2).

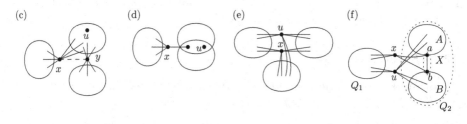

**Fig. 2.** An illustration of violation of conditions (c), (d), (e) and (f) in Theorem 5. Dashed edge can be present or not.

**Theorem 6.** *Let $G$ be a $4K_1$-free graph. If $G$ is connected, then $G$ has a path cover $PC(u, v)$ if and only if the following conditions are satisfied.*

(a) For every articulation point $x$ such that $G - \{x\}$ has at least 3 components, $u, v$ are different from $x$ and belong to different components of $G - \{x\}$.

(b) For any three articulation points $x, y, z$ inducing a triangle in $G$,
    - if both $u, v$ are different from $x, y, z$, then $u, v$ are not in the same component of $G - \{x, y, z\}$,
    - if exactly one of $u, v$ is in $\{x, y, z\}$, say $u$, then $v$ is not in the component of $G - \{x, y, z\}$ adjacent to $u$.

(c) The pair $\{u, v\}$ does not form a vertex cut (not necessarily minimum) in $G$ such that $c(G - \{u, v\}) \geq 3$.

(d) Let $x \neq u$ be an articulation point of $G$ such that $c(G - \{x\}) = 2$, let $Q_1$ be the component of $G - \{x\}$ containing $u$, and let $Q_2$ be the other component of $G - \{x\}$. Suppose that $G[Q_1]$ contains an articulation point $y \neq u$ and denote the component of $G[Q_1] - \{y\}$ containing $u$ by $J_1$ and the other component by $J_2$. If $v \in J_1$, $|J_1| > 2$, and $x$ is adjacent only to vertices from $J_1 \cup \{y\}$ in $G[Q_1]$, then there exists $x' \in J_1, x' \neq u, v$ adjacent to $x$ or $y$.

If $G$ is not connected, then there is a $PC(u, v)$ if and only if $G$ has exactly 2 components, $u, v$ are in different components and none of $u, v$ is an articulation point in its component (Fig. 3).

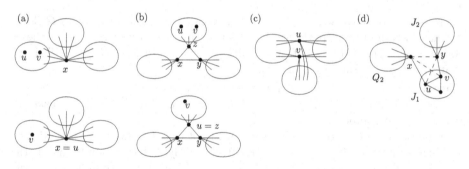

**Fig. 3.** An illustration of violations of conditions (a), (b), (c) and (d) in Theorem 6. Dashed edges can be present or not.

## 3.3   $5K_1$-Free Graphs

In this section we present necessary and sufficient conditions for a connected $5K_1$-free graph to contain a Hamiltonian path.

**Theorem 7.** *Let $G$ be a connected $5K_1$-free graph. The graph $G$ has a Hamiltonian path if and only if the following conditions are satisfied.*

(a) For every articulation point $x$, $c(G - \{x\}) \leq 2$.

(b) There are no 3 articulation points inducing a triangle in $G$.

(c) *If G is not 2-connected, and x is an articulation point of G such that one of the components of $G - \{x\}$, denoted by $Q_1$, is a clique and the other one, denoted by $Q_2$, induces a $4K_1$-free graph, then there exists a vertex $u \in Q_2$ adjacent to x such that $G[Q_2]$ has a Hamiltonian path starting in u.*

(d) *If G is 2-connected, but not 3-connected, then for every minimum vertex cut $\{x, y\}$ in G, $c(G - \{x, y\}) \leq 3$.*

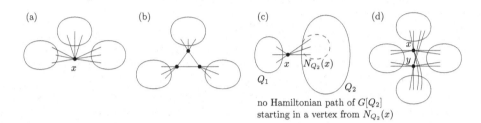

no Hamiltonian path of $G[Q_2]$
starting in a vertex from $N_{Q_2}(x)$

**Fig. 4.** An illustration of violations of the conditions (a), (b), (c) and (d) in Theorem 7.

*Proof.* We will use notation $N_Q(x) = \{y : y \in Q \wedge xy \in E(G)\}$ for a vertex $x \in V(G)$ and a set $Q$ of vertices of $G$.

It is easy to observe that if any of the conditions is violated, then $G$ does not have a Hamiltonian path. See Fig. 4 for an illustration of violations of these conditions. Suppose that all of the conditions (a)–(d) are satisfied. We will show that then $G$ has a Hamiltonian path.

**Case 1.** *Suppose that G is connected, but not 2-connected.*

Let $x$ be an articulation point of $G$. By condition (a), $G - \{x\}$ has exactly two connected components.

First suppose that one of the components of $G - \{x\}$ (say $Q_1$) is a clique, and hence the other one (say $Q_2$) is $4K_1$-free. Clearly, $G$ has a Hamiltonian path if and only if there exists $u \in N_{Q_2}(x)$ such that there exists a Hamiltonian path of $G[Q_2]$ starting in $u$, and this is guaranteed by condition (c) of this theorem.

If both $Q_1$, $Q_2$ induce $3K_1$-free graphs, then $G$ has a Hamiltonian path if and only if $x$ has a neighbour $x_1 \in Q_1$ and a neighbour $x_2 \in Q_2$ such that $x_1$ is not an articulation point in $G[Q_1]$ and $x_2$ is not an articulation point in $G[Q_2]$ (because, by Theorem 2, in these, and only in these cases $G[Q_i]$ has a Hamiltonian path, for $i = 1, 2$). Observe that a connected $3K_1$-free graph has at most two articulation points, and if it has two of them, then they are adjacent. If $x$ is adjacent only to one articulation point in $G[Q_1]$ and no other vertices of this component, then condition (a) of this theorem is violated. If $x$ is adjacent to two articulation points in $G[Q_1]$ and no other vertices of this component, then condition (b) is violated. Similarly for $G[Q_2]$).

**Case 2.** *Suppose that G is 2-connected, but not 3-connected.*

Condition (d) of this theorem implies that for every vertex cut $\{x, y\}$ of size 2 in $G$, $c(G - \{x, y\}) \leq 3$. If there is a vertex cut $\{x, y\}$ of size 2 in $G$ such that $c(G - \{x,y\}) = 3$, then two components of $G - \{x,y\}$, say $Q_1, Q_2$, are cliques and the third component, say $Q_3$, induces a $3K_1$-free graph. In this case, $G$ has a Hamiltonian path if and only if there is a vertex in $Q_3$ which is not an articulation point of $G[Q_3]$ and which is adjacent to $x$ or $y$. Such a vertex always exists since $G$ is 2-connected.

Suppose from now on that for every vertex cut $\{x,y\}$ of size 2 in $G$, $c(G - \{x,y\}) = 2$. Fix a vertex cut $\{x,y\}$ in $G$ and let $Q_1, Q_2$ be the two components of $G - \{x,y\}$.

**Case 2a.** *Both $Q_1, Q_2$ induce $3K_1$-free graphs.*

Recall that $3K_1$-free graphs have path covers of size two starting at any pair of vertices by Theorem 3.

If at least one of $G[Q_1]$ and $G[Q_2]$ has a Hamiltonian path starting in a neighbour of $x$ and ending in a neighbour of $y$, then $G$ has a Hamiltonian path. If this is not the case, then for each $i = 1, 2$ and for every choice of $x' \in N_{Q_i}(x)$, $y' \in N_{Q_i}(y)$, $x' \neq y'$, at least one of the conditions (a)–(c) of Theorem 1 is violated. We will show that even if this is the case, $G$ contains a Hamiltonian path anyway.

*Claim 1.* If for some $i = 1, 2$, all neighbours of at least one of the vertices $x, y$ are articulation points of $G[Q_i]$, then $G$ has a Hamiltonian path. (Note that in this case all choices of $(x', y')$ such that $x' \in N_{Q_i}(x)$, $y' \in N_{Q_i}(y)$ violate condition (a) of Theorem 1.)

*Proof of the Claim.* Suppose without loss of generality that $N_{Q_1}(x) \subseteq Art(G[Q_1])$. Since $x$ has at least one neighbour in $Q_1$, this means that $G[Q_1]$ is not 2-connected, and since $G[Q_1]$ is $3K_1$-free, $|Art(G[Q_1])| \leq 2$. Let $u$ be an articulation point of $G[Q_1]$ adjacent to $x$. Denote the connected components of $G[Q_1 - \{u\}]$ by $J_1, J_2$. Since $G[Q_1]$ is $3K_1$-free, $J_1$ and $J_2$ are cliques, $u$ is adjacent to all vertices of one component, say $J_1$, and $u$ has at least one neighbour $v \in J_2$. If $v$ is the only neighbour of $u$ in $J_2$ and $|J_2| > 1$, then $v$ is also an articulation point of $G[Q_1]$ and there are no other articulation points. Thus, $N_{Q_1}(x) \subseteq \{u, v\}$ if $v$ is an articulation point of $G[Q_1]$, and $N_{Q_1}(x) = \{u\}$ otherwise. Since $G$ is 2-connected, $y$ has neighbours both in $J_1$ and $J_2$.

It follows that $G[Q_1 \cup \{x,y\}]$ has a Hamiltonian path starting in a vertex of $J_2$, going through all vertices of $G[J_2]$ to a neighbour of $y$, proceeding to $y$, then traversing all vertices of $G[J_1]$, proceeding to $u$ and ending in $x$. If $G[Q_2]$ has a Hamiltonian path starting at a neighbour of $x$, then we are done.

Suppose for a contradiction that $G[Q_2]$ does not have a Hamiltonian path starting at a neighbour of $x$. By Theorem 2, we have that $N_{Q_2}(x) \subseteq Art(G[Q_2])$. Then $x$ together with one vertex from each component of $G - (\{x,y\} \cup N_{Q_1}(x) \cup N_{Q_2}(x))$ create a copy of $5K_1$ in $G$, a contradiction. $\square$

*Claim 2.* Suppose that for some $i = 1, 2$, $G[Q_i]$ is not 2-connected and that neither $N_{Q_i}(x)$ nor $N_{Q_i}(y)$ is a subset of $Art(G[Q_i])$. Then $x$ and $y$ have neighbours in different components of $G[Q_i] - Art(G[Q_i])$.

*Proof of the Claim.* We know that every vertex of $Q_i$ is adjacent to at most one articulation point of $G[Q_i] - Art(G[Q_i])$, this $G[Q_i] - Art(G[Q_i])$ has two components, and the vertices of each component are adjacent to the same articulation point (if they are adjacent to any articulation point at all). If both $N_{Q_i}(x)$ and $N_{Q_i}(y)$ were disjoint with the same component of $G[Q_i] - Art(G[Q_i])$, then the articulation point adjacent to this component would be an articulation point of entire $G$, contradicting the 2-connectivity of $G$.    □

**Claim 3.** Suppose that both $G[Q_1]$ and $G[Q_2]$ are 2-connected. Then $G$ has a Hamiltonian path.

*Proof of the Claim.* Consider $Q_1$. Since $G[Q_1]$ is 2-connected, conditions (a) and (b) of Theorem 1 are satisfied for any choice of $a \in N_{Q_1}(x), b \in N_{Q_1}(y)$. If, for some choice of $a \neq b$, condition (c) is also satisfied, we construct a Hamiltonian path in $G$ as it was described at the beginning of Case 2a. Hence suppose that for any pair of distinct vertices $a \in N_{Q_1}(x), b \in N_{Q_1}(y)$, $\{a, b\}$ is a minimum vertex cut in $G[Q_2]$. If $|N_{Q_1}(x) \cup N_{Q_1}(y)| = 2$, then $N_{Q_1}(x) \cup N_{Q_1}(y)$ is a vertex cut of size 2 in $G[Q_1]$, and then it is also a vertex cut of size 2 in $G$ and after removing it, we get at least 3 components. This would violate condition (d) of this theorem.

Suppose that $|N_{Q_1}(x) \cup N_{Q_1}(y)| > 2$ and suppose that $x$ has at least 2 neighbours in $G[Q_1]$. It follows from observations done in Case 2c of Theorem 5 that $G[Q_1 \cup \{x\}]$ has a Hamiltonian cycle. Thus, if $G[Q_2]$ has a Hamiltonian path starting in a neighbour of $y$, then we are done. If this not the case, by Theorem 2, we have that $y$ is adjacent only to articulation points of $G[Q_2]$ which contradicts our assumption that $G[Q_2]$ is also 2-connected.    □

Let us now summarize the Case 2a. Suppose first that at least one of the components $Q_1, Q_2$ induces a graph which is not 2-connected, let it be $Q_1$. If all neighbours of $x$ or $y$ in $Q_1$ are articulation points of this components, then $G$ contains a Hamiltonian path by Claim 1. If neither $N_{Q_1}(x)$ nor $N_{Q_1}(y)$ is a subset of $Art(G[Q_1])$, then by Claim 2 we can choose a neighbour $a$ of $x$ and a neighbour $b$ of $y$, both in $Q_1$, in such a way that they belong to different components of $G[Q_1] - Art(G[Q_1])$. It follows that for any articulation point $u \in Art(G[Q_1])$, $a$ and $b$ belong to different components of $G[Q_1] - \{u\}$ and hence condition (b) of Theorem 1 is satisfied. Condition (a) is satisfied because $\{a, b\} \cap Art(G[Q_1]) = \emptyset$ and condition (c) is satisfied because $G[Q_1]$ is not 2-connected. Thus a Hamiltonian path from $a$ to $b$ in $G[Q_1]$ exists, and it can be extended to a Hamiltonian path in $G$ by a path cover $PC(a', b')$ in $G[Q_2]$, with $a' \in N_{Q_2}(x)$, $b' \in N_{Q_2}(b')$, $a' \neq b'$. Finally, if both $G[Q_1]$ and $G[Q_2]$ are 2-connected, $G$ has a Hamiltonian path by Claim 3.

**Case 2b.** *One of the components, say $Q_1$, is a clique and the other one, say $Q_2$, induces a $4K_1$-free graph.*

Observe that if there exits a path cover of size 2 in $G[Q_2]$ with one path starting at a vertex in $N_{Q_2}(x)$ and the other one starting in a vertex in $N_{Q_2}(y)$, then $G$ has a Hamiltonian path. Such a path cover does not exist if and only if at least one condition of Theorem 6 is violated for $G[Q_2]$. Several cases can occur.

*Case A. There is an articulation point $z$ of $G[Q_2]$, such that $G[Q_2] - \{z\}$ has at least 3 components.*

Let $z$ be such an articulation point of $G[Q_2]$, i.e., $G[Q_2] - \{z\}$ has at least 3 components, let us denote them $J_1, J_2, J_3$. Since $G[Q_2]$ is $4K1$-free, all of $G[J_1], G[J_2], G[J_3]$ induce cliques. First suppose that (a) of Theorem 6 is violated for any pair of distinct vertices, one from $N_{Q_2}(x)$ and second one from $N_{Q_2}(y)$. If $N_{Q_2}(x)$ (or $N_{Q_2}(y)$) contains only $z$, then $c(G - \{y, z\}) \geq 3$ (or $c(G - \{x, z\}) \geq 3$) which would violate condition (d) of this theorem. Also, $N_{Q_2}(x) \cup N_{Q_2}(y)$ cannot be a subset of $(G[J_i] \cup \{z\})$ for any $i = 1, 2, 3$, since $G$ is 2-connected. Thus, this is not possible.

Assume that $x$ and $y$ have neighbours in different components of $G[Q_2] - \{z\}$, say $x$ has a neighbour $u$ in $J_1$ and $y$ has a neighbour $v$ in $J_2$ (thus, (a) of Theorem 6 is satisfied for $u, v$). It follows from the structure of $G[Q_2]$ that $G[Q_2]$ does not have 3 articulation points inducing a triangle, and hence condition (b) of Theorem 6 must be satisfied for $u, v$.

Suppose that (c) of Theorem 6 is violated for $u, v$. This means that $\{u, v\}$ is a vertex cut in $G[Q_2]$ such that $c(G[Q_2] - \{u, v\}) \geq 3$. This is possible only if $u$ is the only neighbour of $z$ in $J_1$ and $v$ is the only neighbour of $z$ in $J_2$ and $|J_1|, |J_2| > 1$. Since $G[Q_2]$ is $4K_1$-free, $z$ is adjacent to all vertices in $J_3$. Since $G$ is 2-connected, then $x$ or $y$ must have a neighbour in $J_3$, say $x$ has a neighbour $u'$ in $J_3$. Also note that $u'$ is not an articulation point of $G[Q_2]$.

Now consider the pair $u', v$. We can obtain a path cover $PC(u', v)$ by taking one path starting at $u'$, then traversing all vertices of $J_3$, $z$, $u$ and all vertices of $J_1$. The second path starts in $v$ and traverses all vertices of $J_2$.

*Case B. There exist three articulation points $a, b, c$ inducing a triangle in $G[Q_2]$.*

Let $a, b, c$ be three articulation points inducing a triangle in $G[Q_2]$. Denote the components of $G[Q_2] - \{a, b, c\}$ by $J_a, J_b, J_c$ correspondingly. Since, $G[Q_2]$ is $4K_1$-free, $a$ is adjacent to all vertices in $J_a$, $b$ is adjacent to all vertices in $J_b$ and $c$ is adjacent to all vertices in $J_c$.

First suppose that (b) of Theorem 6 is violated for any pair of distinct vertices, one from $N_{Q_2}(x)$ and second one from $N_{Q_2}(y)$. Since $G$ is 2-connected, then $x$ or $y$ has a neighbour in $G[Q_2]$ which is not an articulation point and $N_{Q_2}(x) \cup N_{Q_2}(y)$ cannot be a subset of vertices $\{a, b, c\}$ together with vertices from one component of $G[Q_2] - \{a, b, c\}$. Thus, this is not possible.

It follows that at least one of $x, y$ has a neighbour which is not articulation point of $G[Q_2]$, say $x$ has a neighbour $u$ in $J_a$, and $y$ has a neighbour in $G[Q_2 - (J_a \cup \{a\})]$, say $y$ is adjacent to $v$ in $J_b \cup \{b\}$. We can obtain one path starting in $x$, going through $u$, all vertices of $J_a$, $a$, $c$ and all vertices of $J_c$. The second path starts at $y$, goes through $b$ and all vertices of $J_b$.

*Case C. For a pair of distinct vertices $u \in N_{Q_2}(x)$ and $v \in N_{Q_2}(y)$, $\{u, v\}$ is a vertex cut (not necessarily minimal) in $G[Q_2]$ such that $c(G[Q_2 - \{u, v\}]) \geq 3$.*

First suppose that $|N_{Q_2}(x) \cup N_{Q_2}(y)| = 2$. Since $\{u, v\} = |N_{Q_2}(x) \cup N_{Q_2}(y)|$, $\{u, v\}$ is also a vertex cut of size 2 in $G$ and $c(G - \{u, v\}) \geq 4$ which would violate condition (d) of this theorem.

Hence suppose that $|N_{Q_2}(x) \cup N_{Q_2}(y)| > 2$. We will distinguish two cases.

*Case C.1. The pair $\{u, v\}$ is a minimal vertex cut in $G[Q_2]$.*

Since $G[Q_2]$ is $4K_1$-free graph, $G[Q_2 - \{u, v\}]$ has exactly 3 components, denote them $J_1, J_2, J_3$, and they are cliques. Each of $u, v$ has at least one neighbour in each component.

Suppose that $G[Q_2]$ is not 2-connected. There is an articulation point $z$ of $G[Q_2]$. From the minimality of the vertex cut $\{u, v\}$ we have that $z \neq u, v$. Since $J_1, J_2, J_3$ are cliques, the only possibility is that $z$ is the only neighbour of $u$ and $v$ in one of the cliques, say $J_1$, and the size of $J_1$ is at least 2. Also note that $z$ is adjacent to all other vertices of $J_1$. The graph $G$ is 2-connected, thus $x$ or $y$, say $x$, has a neighbour $x'$ in $G[J_1 - \{z\}]$. We obtain one path starting at vertex $x'$, traversing through all vertices of $J_1$, continuing to $z$, $u$ and finally traversing all vertices of $J_2$. The second one starts in $v$ and goes through all vertices of $J_3$.

Suppose that $G[Q_2]$ is 2-connected. Since $|N_{Q_2}(x) \cup N_{Q_2}(y)| > 2$, at least one of $x, y$ has a neighbour in $G[Q_2]$ other than $u$ and $v$, say $x$ has a neighbour $u'$ in $J_1$. The pair $\{u, v\}$ is a minimal vertex cut in $G[Q_2]$ and $G[Q_2]$ is 2-connected, thus, $u$ and $v$ have together two distinct neighbours in each component. We obtain one path starting at vertex $u'$ and traversing all vertices of $J_1$. The second one starts at $v$, goes through all vertices of $J_2$, continues to $u$ and finally traverses all vertices of $J_3$.

*Case C.2. The vertex cut $\{u, v\}$ in $G[Q_2]$ is not minimal.*

Thus, $G[Q_2]$ is not 2-connected. If there is an articulation point $z$ of $G[Q_2]$ such that $c(G[Q_2 - \{z\}]) = 3$ (possibly $z = u$ or $z = v$), then Case A applies.

Suppose that for each articulation point of $G[Q_2]$, $G[Q_2]$ has 2 components after removing it. Since $\{u, v\}$ is not a minimal vertex cut in $G[Q_2]$, at least one of $u, v$ is an articulation point of $G[Q_2]$, say $u$. The graph $G[Q_2 - \{u\}]$ has 2 components, denote the one containing $v$ by $J_1$ and the other one by $J_2$. The vertex $v$ must be an articulation point in $G[J_1]$ and thus, $G[J_2]$ is a clique. Denote the components of $G[J_1 - \{v\}]$ by $R_1, R_2$. Let us summarise the situation. The graph $G[Q_2 - \{u, v\}]$ has 3 components $J_2, R_1, R_2$, all of them are cliques, $u$ is adjacent to $J_2$ and $J_1 = R_1 \cup \{v\} \cup R_2$. The vertex $v$ is adjacent to $R_1$ and $R_2$.

If $v$ is the only neighbour of $u$ in $G[J_1]$, then $c(G[Q_2 - \{v\}]) = 3$ and again Case A applies. Thus, we can assume without loss of generality that $u$ has a neighbour $u'$ in $R_1$. Since $G$ is 2-connected, there exists a vertex $w$ in $G[J_2]$ which is not the only neighbour (if $|J_2| > 1$) of $u$ in $G[J_2]$ such that $w$ is adjacent to $x$ or $y$.

If $w$ is a neighbour of $x$, then we obtain one path starting at $w$, then going through all vertices of $J_2$, continuing to $u$, $u'$ and finally traversing all vertices of $R_1$ (if $|J_2| = 1$, the path simply starts in $w$ being the only neighbour of $u$ in $J_2$, continues to $u$, $u'$ and traverses $R_1$). The second path starts at $v$ and goes through all vertices of $R_2$.

Suppose that $w$ is a neighbour of $y$. If all neighbours of $u$ in $J_1$ are articulation points of $G[J_1]$, then at least one of the Cases A or B applies. Otherwise, we obtain one path starting at $w$ and traversing all vertices of $J_2$. The second path starts at $u$ and traverses all vertices of $J_1$ (the existence of such a path is guaranteed by Theorem 2).

*Case D. There is a pair of distinct vertices $u \in N_{Q_2}(x)$ and $v \in N_{Q_2}(y)$, such that condition (d) of Theorem 6 is violated for $u, v$.*

Thus, we have the following situation. There is an articulation point $w$ of $G[Q_2]$ such that $c(G[Q_2] - \{w\}) = 2$, with $J_1, J_2$ being the components of $G[Q_2] - \{w\}$. Further $G[J_1]$ contains an articulation point $z$, with $R_1, R_2$ being the components of $G[J_1] - \{z\}$ such that $u, v \in R_1$, $|R_1| > 2$, $N_{J_1}(w) \subseteq \{u, v, z\}$ and $N_{R_1}(z) \subseteq \{u, v\}$. Observe that $R_1, R_2, J_2$ are cliques.

Since $G$ is 2-connected, $w$ is not an articulation point of $G$, and thus there exists a vertex $w'$ in $J_2$ such that $w'$ is adjacent to $x$ or $y$. This $w'$ also cannot be an articulation point of entire $G$, and thus there are two possibilities – either $|J_2| = 1$ and $\{w'\} = N_{J_2}(w) = N_{J_2}(x) \cup N_{J_2}(y)$, or $|N_{J_2}(w) \cup N_{J_2}(x) \cup N_{J_2}(y)| > 1$. In the latter case, $w'$ could have been chosen so that $w$ has a neighbour in $J_2$ other than $w'$, and we further assume it had been chosen in this way.

Suppose that $w'$ is adjacent to $x$. If $v$ is not an articulation point of $J_1$, then we take one path starting in $w'$, traversing all vertices of $J_2$ and ending in $w$. As the other path we take a Hamiltonian path of $G[J_1]$ starting at $v$, which exists according to Theorem 2. Hence suppose that $v$ is an articulation point of $G[J_1]$, which means that $v$ is the only neighbour of $z$ in $R_1$. If $w$ is adjacent to $u$, we obtain one path starting in $w'$, traversing all vertices of $J_2$, continuing to $w, u$ and all vertices of $R_1 \setminus \{v\}$, and the other path starting in $v$, continuing to $z$ and all vertices of $R_2$. If $w$ is not adjacent to $u$, i.e., when $N_{J_1}(w) \subseteq \{v, z\}$, one of the Cases A or B applies – Case B when $N_{J_1}(w) = \{v, z\}$, Case A with $v$ being the articulation point when $N_{J_1}(w) = \{v\}$ and likewise Case A with $z$ being the articulation point when $N_{J_1}(w) = \{z\}$ (needless to say, $N_{J_1}(w) \neq \emptyset$ since $Q_2$ is a connected component of $G - \{x, y\}$).

Analogously, if $w'$ is adjacent to $y$, we distinguish two cases depending on whether $u$ is an articulation point of $G[J_1]$ or not, because the roles of the pairs $x, u$ and $y, v$ are symmetric for $Q_2$.

*Case E. None of the Cases A–D occurs.*

Since $G$ is 2-connected, $\{x, y\}$ is a minimal cut in $G$, and therefore $|N_{Q_2}(x) \cup N_{Q_2}(y)| \geq 2$. Choose different vertices $u, v$ such that $u \in N_{Q_2}(x)$ and $v \in N_{Q_2}(y)$. Since none of the Cases A–D occurs, $G[Q_2]$ has a path cover of size 2 with the paths starting in vertices $u$ and $v$ according to Theorem 3. Thus, $G$ has a Hamiltonian path.

**Case 3.** *$G$ is at least 3-connected.*

In this case, $G$ has a Hamiltonian path by Proposition 1.1.

# 4  Conclusion

We characterized the structure of graphs having a Hamiltonian path and path cover of size two with prescribed endpoints for the class of $3K_1$-free and $4K_1$-free graphs. The characterization was then applied to characterize $5K_1$-free graphs with a Hamiltonian path and gives an explicit algorithm for checking the existence of a Hamiltonian path in a $5K_1$-free graph.

All the proofs were constructive and such that we can also construct a Hamiltonian path in polynomial time, if one exists. We leave structural description of $kK_1$-free graphs having a Hamiltonian path, cycle or being Hamiltonian connected for higher $k$ as a future research.

# References

1. Barnette, D.: Conjecture 5. Recent progress in combinatorics. In: Proceedings of the Third Waterloo Conference on Combinatorics. Academic Press, New York (1969)
2. Bertossi, A.A., Bonuccelli, M.A.: Hamiltonian circuits in interval graph generalizations. Inf. Process. Lett. **23**(4), 195–200 (1986)
3. Chvátal, V., Erdös, P.: A note on Hamiltonian circuits. Discrete Math. **2**(2), 111–113 (1972)
4. Damaschke, P.: The Hamiltonian circuit problem for circle graphs is NP-complete. Inf. Process. Lett. **32**(1), 1–2 (1989)
5. Damaschke, P.: Paths in interval graphs and circular-arc graphs. Discrete Math. **112**(1–3), 49–64 (1993)
6. Damaschke, P., Deogun, J.S., Kratsch, D., Steiner, G.: Finding Hamiltonian paths in cocomparability graphs using the bump number algorithm. Order **8**(4), 383–391 (1991)
7. Duffus, D., Gould, R., Jacobson, M.: Forbidden subgraphs and the Hamiltonian theme. In: I4th International Conference on the Theory and Applications of Graphs, Kalamazoo, pp. 297–316. Wiley (1981)
8. Garey, M.R., Johnson, D.S., Tarjan, R.E.: The planar Hamiltonian circuit problem is NP-complete. SIAM J. Comput. **5**(4), 704–714 (1976)
9. Golumbic, M.C.: Algorithmic Graph Theory and Perfect Graphs. Elsevier (2004)
10. Gould, R.J.: Advances on the Hamiltonian problem - a survey. Graphs Comb. **19**(1), 7–52 (2003)
11. Gould, R.J.: Recent advances on the Hamiltonian problem: survey III. Graphs Comb. **30**(1), 1–46 (2014)
12. Jedličková, N., Kratochvíl, J.: Hamiltonian path and Hamiltonian cycle are solvable in polynomial time in graphs of bounded independence number. arXiv preprint arXiv:2309.09228 (2023)
13. Karp, R.M.: Reducibility among combinatorial problems. In: Miller, R.E., Thatcher, J.W., Bohlinger, J.D. (eds.) Complexity of Computer Computations. IRSS, pp. 85–103. Springer, Boston (1972). https://doi.org/10.1007/978-1-4684-2001-2_9
14. Keil, J.M.: Finding Hamiltonian circuits in interval graphs. Inf. Process. Lett. **20**(4), 201–206 (1985)
15. Müller, H.: Hamiltonian circuits in chordal bipartite graphs. Discrete Math. **156**(1–3), 291–298 (1996)

# Resolving Unresolved Resolved and Unresolved Triplets Consistency Problems

Daniel J. Harvey[1], Jesper Jansson[1], Mikołaj Marciniak[1(✉)], and Yukihiro Murakami[2]

[1] Graduate School of Informatics, Kyoto University, Kyoto, Japan
Daniel.Harvey87@gmail.com, jj@i.kyoto-u.ac.jp, marciniak@int.pl
[2] Delft University of Technology, Delft, The Netherlands
y.murakami@tudelft.nl

**Abstract.** The $R^{+-}F^{+-}$ CONSISTENCY problem is a basic problem related to the construction of phylogenetic trees. Its input is two sets $R^+$ and $R^-$ of resolved triplets and two sets $F^+$ and $F^-$ of unresolved triplets (also known as *fan triplets*). The objective of the problem is to determine if there exists a phylogenetic tree that includes all elements in $R^+ \cup F^+$ and excludes all elements in $R^- \cup F^-$ as embedded subtrees, and to construct such a tree if one exists. Jansson *et al.* [Journal of Computational Biology, 2018] cataloged the computational complexity of the problem under various restrictions, with four notable exceptions in which the output tree is required to be ternary, i.e., has degree at most three. Here, we resolve these four remaining cases by proving that for ternary trees: (i) $F^+$ CONSISTENCY as well as $R^+F^+$ CONSISTENCY are solvable in polynomial time; and (ii) $F^{+-}$ CONSISTENCY and $R^+F^{+-}$ CONSISTENCY are NP-hard. To obtain (i), we develop a novel way of expressing the triplets CONSISTENCY problem for ternary trees as a system of equations whose nontrivial solutions can be used to partition the leaf labels into subsets that label subtrees of the output tree. Result (ii) is obtained after observing some new equivalences between resolved triplets and fan triplets consistent with a given phylogenetic tree.

**Keywords:** phylogenetic tree · rooted triplets consistency · tree algorithm · computational complexity

## 1 Introduction

Phylogenetic trees are a key tool in evolutionary biology, used to describe evolutionary history and relationships between different species. They are also used in other fields; for example, in linguistics to model relationships between languages. An important challenge associated with phylogenetic trees is their reconstruction from different types of data [6,16]. Finding a phylogenetic tree that best describes the given data can be a difficult and, for large datasets, time-consuming

© The Author(s), under exclusive license to Springer Nature Switzerland AG 2024
A. A. Rescigno and U. Vaccaro (Eds.): IWOCA 2024, LNCS 14764, pp. 193–205, 2024.
https://doi.org/10.1007/978-3-031-63021-7_15

task. In many cases, however, a compromise between computational efficiency and accuracy may be achieved by the *supertrees* technique [2,3]: First, using a computationally intensive method such as maximum likelihood [4,6], create a set of phylogenetic trees for small (e.g., 3-element), overlapping subsets of the leaf labels that each have a high probability of being correct. Next, using a combinatorial algorithm, merge all the small trees into a single tree.

A rooted phylogenetic tree with exactly three leaves can be either binary, in which case it is called a *resolved triplet*, or not, in which case it is called an *unresolved triplet* or a *fan triplet*. (This article will use the term fan triplet.) In the context of merging trees into a supertree, the following problem is fundamental: Determine if there exists a tree corresponding to a given collection of required and forbidden sets of resolved and fan triplets, and if so, construct such a tree. This problem has been investigated by many researchers [5,7–9,11–15,17] under different assumptions. In particular, a classic algorithm by Aho *et al.* [1] named BUILD that solves the problem for the case of required resolved triplets in polynomial time has been well studied and extended in various ways.

Jansson *et al.* [10] surveyed and cataloged the computational complexity of the many variants of the problem and obtained several new results in the process, but left four borderline variants open. The main goal of the present paper is to resolve these remaining variants, thereby completely characterizing the computational complexity of the problem when the output tree is required to have degree at most $D$ for any integer $D \geq 2$, for all possible 15 nonempty combinations of the different types of inputs.

## 2    Preliminaries

Recall that a tree is a simple connected graph without cycles. A tree is *rooted* if one of its vertices has been designated as the root. The edges of a rooted tree can be assigned a natural orientation forming directed paths from the root to each leaf. If there exists a path from a vertex $u$ to a vertex $v$, we say that $u$ is an *ancestor* of $v$ and that $v$ is a *descendant* of $u$. When such a path is of length one (that is, there exists an edge from $u$ to $v$) then we also say that $u$ is the *parent* of $v$ and that $v$ is the *child* of $u$.

A *phylogenetic tree* is a rooted tree in which each leaf is given a unique label and each internal vertex has at least two children. Additionally, we will also treat the degenerate cases of a tree with a single vertex and the empty tree as phylogenetic trees. For simplicity, we will refer to phylogenetic trees as trees and identify each leaf with its label. For any (phylogenetic) tree $T$, we will denote the set of its leaves (labels) by $L_T$. For any two leaves $u, v \in L_T$, we denote by $\mathrm{lca}^T(u, v)$ their lowest common ancestor, that is, the vertex $w$ that is an ancestor of both $u$ and $v$ such that no child of $w$ is also an ancestor to both $u$ and $v$.

Here we let the *degree of a vertex* be the number of its children, and the *degree of a tree* be the maximum degree of a vertex taken over all vertices of the tree. A *binary tree* is a tree in which the degree of each vertex is at most 2, and a *ternary tree* is a tree in which the degree of each vertex is at most 3.

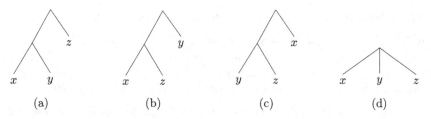

**Fig. 1.** All possible rooted triplets with the leaf set $\{x, y, z\}$ – resolved triplets (a) $xy|z$, (b) $xz|y$, and (c) $yz|x$ and fan triplet (d) $x|y|z$.

A *rooted triplet*, or *triplet* for short, is a tree with exactly three leaves. Let $t$ be a rooted triplet and suppose that $L_t = \{x, y, z\}$. If $t$ is a binary tree, then $t$ is called a *resolved triplet* and we write $xy|z$, where $\mathrm{lca}^t(x, y)$ is a proper descendant of $\mathrm{lca}^t(x, z) = \mathrm{lca}^t(y, z)$. Otherwise, the triplet $t$ is called a *fan triplet* and we write $x|y|z$. Note that there exist only four triplets (see Fig. 1) with a fixed set of leaves $\{x, y, z\}$, namely $xy|z, xz|y, yz|x$, and $x|y|z$.

Given a tree $T$ and three distinct leaves $\{x, y, z\} \subseteq L_T$, the resolved triplet $xy|z$ is *consistent with* $T$ if and only if $\mathrm{lca}^T(x, y)$ is a proper descendant of $\mathrm{lca}^T(x, z) = \mathrm{lca}^T(y, z)$. Likewise, the fan triplet $x|y|z$ is *consistent with* $T$ if and only if $\mathrm{lca}^T(x, y) = \mathrm{lca}^T(x, z) = \mathrm{lca}^T(y, z)$. (Equivalently, we may say in each case that the tree is consistent with the triplet.) In other words, the rooted triplet consistent with $T$ describes the relative position of its three leaves in $T$.

We will say that a tree $T$ is *over* a set of leaves $L$ if each leaf of $T$ belongs to $L$. Let $T$ be a tree. We define the *restriction* of the tree $T$ into the set $L' \subseteq L_T$, denoted by $T|_{L'}$, to be the tree $T'$ with $L_{T'} = L'$ such that each triplet $t$ consistent with the tree $T'$ is also consistent with the tree $T$. Less formally, $T|_{L'}$ can be obtained by "removing" all leaves $x \notin L'$ not belonging to the set $L'$. In the special case where the set $L'$ has only 3 elements, a tree $t = T|_{L'}$ is a rooted triplet over $L'$ consistent with the tree $T$. Let $t_T$ denote the set of all resolved and fan triplets consistent with $T$, i.e., $t_T = \{T|_{\{x,y,z\}} : \{x, y, z\} \subseteq L_T\}$. Using the new notation, we can note that $T' = T|_{L'}$ is the restriction of tree $T$ into the set of leaves $L' \subseteq L_T$ if and only if $t_{T'} \subseteq t_T$.

To illustrate, let

$$F = \{1|2|3, 1|4|5, 1|6|7, 2|4|7, 2|5|6, 3|4|6\}$$

be a set of fan triplets over the leaf set $\{1, 2, \ldots, 7\}$. There is no ternary tree $T$ with $F \subseteq t_T$ (and we *strongly recommend* the reader to try to prove it), but if the last fan triplet $3|4|6 \in F$ is replaced by $3|4|7$ then the situation changes; more precisely, the resulting set is consistent with a ternary tree with three maximal proper subtrees leaf-labeled by $\{4, 6\}$, $\{1, 2, 3\}$, and $\{5, 7\}$, respectively. It may seem to the reader that the only way to distinguish between such cases is through an optimized brute-force check of all possibilities. However, we will demonstrate how to perform this task in polynomial time. We hope that after reading this paper, the reader will appreciate the usefulness of the newly developed theory.

The general problem considered in this paper is as follows.

**The $\mathcal{R}^{+-}\mathcal{F}^{+-}$ Consistency Problem.** *Given two sets $R^+$ and $R^-$ of resolved triplets and two sets $F^+$ and $F^-$ of fan triplets over a set of leaves $L$, either return a tree with $L_T = L$ such that $R^+ \bigcup F^+ \subseteq t_T$ and $(R^- \bigcup F^-) \bigcap t_T = \emptyset$ if such a tree exists, or return the answer null if no such tree exists.*

In other words, the output tree must contain all triplets from the sets $R^+$ and $F^+$, and must not contain any triplets from the sets $R^-$ or $F^-$. Different variants of the problem arise when some of the given sets are forced to be empty. In such cases, we name the problem by omitting the symbols from $\mathcal{F}, \mathcal{R}, +, -$ corresponding to the empty sets. For example, the $\mathcal{R}^+$ Consistency problem is the variant of the $\mathcal{R}^{+-}\mathcal{F}^{+-}$ Consistency problem where $R^- = F^+ = F^- = \emptyset$.

Tables 1 to 4 below (all entries can be found in [10]), show the computational complexity of the 15 variants of the $\mathcal{R}^{+-}\mathcal{F}^{+-}$ Consistency problem when the output tree has no degree limitations and when the output tree is required to have outdegree at most $D$ for $D = 2$, $D = 3$, and any fixed integer $D \geq 4$, respectively. In the tables, P means that the corresponding problem is polynomial-time solvable, while NP-h means that the corresponding problem is NP-hard. For example, the $\mathcal{R}^-\mathcal{F}^-$ Consistency problem is NP-hard. Furthermore, a question mark indicates that the computational complexity was unknown.

As can be seen in Tables 1 to 4, the problem was almost fully solved except four variants in the borderline case of $D = 3$. For the remaining four variants, i.e., $\mathcal{F}^+$, $\mathcal{R}^+\mathcal{F}^+$, $\mathcal{F}^{+-}$, and $\mathcal{R}^+\mathcal{F}^{+-}$, the computational complexity remained open. In the rest of the paper, we resolve the last four remaining variants by either finding a polynomial-time solution or an NP-hardness proof for each one, thereby completely characterizing the computational complexity of the problem when the output tree is required to have degree at most $D$ for any integer $D \geq 2$, for all possible 15 nonempty combinations of the different types of inputs.

**Table 1.** Unbounded degree case

|  | $\emptyset$ | $\mathcal{F}^+$ | $\mathcal{F}^-$ | $\mathcal{F}^{+-}$ |
|---|---|---|---|---|
| $\emptyset$ | – | P | P | NP-h |
| $\mathcal{R}^+$ | P | P | P | NP-h |
| $\mathcal{R}^-$ | P | P | NP-h | NP-h |
| $\mathcal{R}^{+-}$ | P | P | NP-h | NP-h |

**Table 2.** Binary case

|  | $\emptyset$ | $\mathcal{F}^+$ | $\mathcal{F}^-$ | $\mathcal{F}^{+-}$ |
|---|---|---|---|---|
| $\emptyset$ | – | P | P | P |
| $\mathcal{R}^+$ | P | P | P | P |
| $\mathcal{R}^-$ | NP-h | NP-h | NP-h | NP-h |
| $\mathcal{R}^{+-}$ | NP-h | NP-h | NP-h | NP-h |

**Table 3.** Ternary case

|  | $\emptyset$ | $\mathcal{F}^+$ | $\mathcal{F}^-$ | $\mathcal{F}^{+-}$ |
|---|---|---|---|---|
| $\emptyset$ | – | ? | P | ? |
| $\mathcal{R}^+$ | P | ? | P | ? |
| $\mathcal{R}^-$ | NP-h | NP-h | NP-h | NP-h |
| $\mathcal{R}^{+-}$ | NP-h | NP-h | NP-h | NP-h |

**Table 4.** $D$-bounded degree case, $D \geq 4$

|  | $\emptyset$ | $\mathcal{F}^+$ | $\mathcal{F}^-$ | $\mathcal{F}^{+-}$ |
|---|---|---|---|---|
| $\emptyset$ | – | NP-h | P | NP-h |
| $\mathcal{R}^+$ | P | NP-h | P | NP-h |
| $\mathcal{R}^-$ | NP-h | NP-h | NP-h | NP-h |
| $\mathcal{R}^{+-}$ | NP-h | NP-h | NP-h | NP-h |

## 3   The $\mathcal{F}^+$ CONSISTENCY Problem for a Ternary Tree

This section focuses on the following variant of the CONSISTENCY problem.

**The Ternary $\mathcal{F}^+$ CONSISTENCY Problem.** *For a given set of fan triplets $F = \{a_1|b_1|c_1, \ldots, a_n|b_n|c_n\}$ over a set of leaves $L$, return a ternary tree $T$ with $L_T = L$ such that $F \subseteq t_T$ if such a tree exists, or return the answer null if it does not.*

Below, we will prove the following theorem.

**Theorem 1.** *The ternary $\mathcal{F}^+$ CONSISTENCY problem is solvable in polynomial time.*

For each fan triplet $a|b|c \in F$, we identify it with an integer domain modulo 3 equation $w_a + w_b + w_c \equiv^3 0$. The set $F$ can thus be identified with the following system of integer domain equations:

$$\begin{cases} w_{a_1} + w_{b_1} + w_{c_1} \equiv^3 0, \\ \quad \vdots \\ w_{a_n} + w_{b_n} + w_{c_n} \equiv^3 0. \end{cases}$$

The above system of equations can be represented as a matrix equation $M(F,L)\overrightarrow{w} \equiv^3 \overrightarrow{0}$, where $\overrightarrow{w} = (w_j)_{j \in L}$ and the matrix $M(F,L)$ has $|F| = n$ rows and $|L|$ columns, with entries from the set $\{0,1,2\}$ and initially equal to 0 or 1. Specifically, $M(F,L)_{i,j} = 1$ if the leaf $j$ appears in the $i$-th equation, that is, if $j \in \{a_i, b_i, c_i\}$; otherwise, the matrix element is equal to 0.

The solution to a system of equations will be called *trivial* if all the variables are equal. Otherwise, if at least two variables are different, a solution will be called *nontrivial*. Consider the following simple yet crucial observation.

**Observation 1.** *The equality $w_a + w_b + w_c \equiv^3 0$ holds if and only if either all variables are distinct $\{w_a, w_b, w_c\} = \{0,1,2\}$ or all are equal $w_a = w_b = w_c$.*

*Proof.* The three variables $w_a, w_b$, and $w_c$ can take one of three values: 0, 1, or 2. If the three variables are not all different, then at least two of them are equal; without loss of generality, let $w_a = w_b$. Then

$$w_c \equiv^3 -w_a - w_b \equiv^3 -2w_a \equiv^3 w_a.$$

Hence if the three variables are not all different, then they are all equal.  □

Let $T$ be a ternary tree consistent with a set of fan triplets $F$. The root vertex has at most 3 children and each child forms a separate subtree. Each triplet $a|b|c \in F$ can have all leaves in different subtrees or all in the same subtree. This property corresponds to the solution of the equation $w_a + w_b + w_c \equiv^3 0$. Intuitively, the value of the variable $w_j$ corresponds to the index (0, 1, or 2) of the subtree in which the leaf $j$ is located. More generally, we can formulate the following lemma.

**Lemma 1.** *Let $F \neq \emptyset$ be a set of fan triplets over a set of leaves $L$. There exists a ternary tree $T$ with the leaf set $L_T = L$ such that $T$ is consistent with $F$ if and only if the equation $M(F, L)\overrightarrow{w} \equiv^3 \overrightarrow{0}$ has a nontrivial solution $\overrightarrow{w} = (w_j)_{j \in L}$ and there exist ternary trees $T_0, T_1$, and $T_2$ such that for each $i \in \{0, 1, 2\}$, the tree $T_i$ with the set of leaves $L_{T_i} = L_i$ is consistent with $F_i$, where*

$$L_i = \{j : w_j = i\},$$
$$F_i = \{a|b|c \in F : a, b, c \in L_i\}.$$

*Proof.* Let $F \neq \emptyset$ be a set of fan triplets over $L$, and let $T$ be a ternary tree with the leaf set $L_T = L$ such that $T$ is consistent with $F$. Let $T_0, T_1$, and $T_2$ denote the subtrees of the root vertex in the tree $T$. For each leaf $d$ located in the $i$-th subtree, we put $w_d = i$. Note that each triplet $a|b|c \in F$ has all leaves located either in different subtrees, in which case $\{w_a, w_b, w_c\} = \{0, 1, 2\}$, or all located in the same subtree, in which case $w_a = w_b = w_c$. In both cases, by Observation 1, we have the equality $w_a + w_b + w_c \equiv^3 0$. We have an identical equality for all triplets in $F$, and it follows that $\overrightarrow{w}$ is a solution to the equation $M(F, L)\overrightarrow{w} \equiv^3 \overrightarrow{0}$. Since $F \neq \emptyset$, the root vertex of $T$ must have at least two children. Therefore, at least two of the trees $T_0, T_1$, and $T_2$ are non-empty, and thus the obtained solution is nontrivial. Since $T$ is consistent with $F$, then trees $T_0, T_1, T_2$ are consistent consecutively with the sets $F_0, F_1, F_2$ by their definition.

Let $F \neq \emptyset$ be a set of fan triplets. Suppose we have given a nontrivial solution to the matrix equation $M(F, L)\overrightarrow{w} \equiv^3 \overrightarrow{0}$ and let $T_0, T_1, T_2$ be ternary trees such that each $T_i$ is consistent with $F_i$ and $L_{T_i} = L_i$, where $L_i, F_i$ for $i \in \{0, 1, 2\}$ are defined as in the lemma statement. Since we have a nontrivial solution, at least two of the sets $L_i$ are non-empty, meaning that at least two of the trees $T_i$ are non-empty. We create a tree $T$ by taking a root vertex and adding the trees $T_0, T_1, T_2$ as subtrees of the root. Note that for each triplet $a|b|c \in F$ the equality $w_a + w_b + w_c \equiv^3 0$ holds and by Observation 1 we obtain that $\{w_a, w_b, w_c\} = \{0, 1, 2\}$ or $w_a = w_b = w_c$. In the first case, the leaves $a, b$, and $c$ are located in distinct subtrees of $T$. Then $\text{lca}^T(a, b) = \text{lca}^T(b, c) = \text{lca}^T(a, c)$ is the root vertex of the tree $T$, and thus the triplet $a|b|c$ is consistent with $T$. In the second case, the triplet $a|b|c$ is consistent with some tree $T_i$ and therefore also with $T$. Therefore, all triplets in $F$ are consistent with $T$. $\qquad\square$

On the implementation side, we will represent each inner vertex of a ternary tree as a list of (two or three) pointers to its children, and a tree as a pointer to the root vertex. We now introduce the following auxiliary functions.

- SOLVE($F$, $L$) – finds a nontrivial solution to a system of equations identified with a set of triplets $F$ in variables from the set $L$ and divides the variables into three sets according to their values. In the first step, we create the matrix $M(F, L)$. The equation we wish to solve is $M(F, L)\overrightarrow{w} \equiv^3 \overrightarrow{0}$. Using Gauss–Jordan row elimination, we reduce the matrix to reduced row echelon form. In this way, the variables were divided into two subsets, that is, the dependent variables corresponding to the columns with leading ones and the independent variables corresponding to the remaining columns. Any solution can be obtained by fixing any values of the independent variables and calculating from them the values of the dependent variables. When there are at least two independent variables, we find a nontrivial solution by fixing at least two different values to the independent variables. Otherwise, when there is at most one independent variable, we find all solutions by considering all possible values for that potential variable. When there exists a nontrivial solution $\vec{w} = (w_j)_j \in L$, the function finds and returns a triple of sets $(L_0, L_1, L_2)$, where $L_i = \{j : w_j = i\}$. Otherwise, the function returns null.

- DIVIDE($F$, $L$) – returns all triplets from the set $F$ where all leaves of the triplet belong to the set $L$.

- NEWVERTEX() – returns a pointer to the newly created vertex.

- NEWEDGE($T$, $V$) – creates at vertex $T$ an edge to vertex $V$.

- ADDMISSINGVERTICES($T$, $L$) – adds the missing vertices from the set $L$ to the tree $T$ in an arbitrary way such that the modified tree still remains a ternary tree. (Later, we will provide an example of such an addition.)

The first function is dominated by Gauss–Jordan row elimination. Since $2 \equiv^3 -1$, we will never have to use division, and thus SOLVE($F$, $L$) can be implemented in $O(|L|^2|F|)$ time using a brute-force approach. All other auxiliary functions can be implemented in linear time.

Now, using Lemma 1 and the divide-and-conquer method we present an algorithm for the ternary $\mathcal{F}^+$ CONSISTENCY problem. When $|F| > 0$, we apply Lemma 1 and divide the original problem into two or three smaller instances. Otherwise, in the base case, when $|F| = 0$, we return any tree as the answer (for example, by creating an empty tree and adding missing leaves to it). The pseudocode of our algorithm is summarized in FANTERNARYBUILD.

On a high level, our new FANTERNARYBUILD algorithm, just like Aho *et al.*'s BUILD algorithm from [1], uses the natural divide-and-conquer strategy on the set of leaves. Both algorithms construct a tree by partitioning the set of leaves into blocks containing leaves belonging to the same subtree. They then recursively find a solution for each block and finally attach the recursively found subtrees as children to a new root vertex. If, at any point during execution, the current set of leaves contains more than one leaf but cannot be divided into more than one block, the algorithms stop and return null.

The crucial aspect is the method of partitioning the leaf set. The FANTERNARYBUILD algorithm creates a system of modulo 3 equations defined by the input triplets, solves it, and then assigns all leaves whose variables have the same value to one block of the partition. In comparison, BUILD creates an undirected graph whose vertices represent the leaves of the constructed tree, with edges defined by the input triplets. It then computes the connected components of the graph and assigns the leaves in each connected component to one block of the partition.

---

**Algorithm** FANTERNARYBUILD

---

**Input:** a set of leaves $L$, and a set of fan triplets $F$ over $L$.
**Output:** a tree $T$ consistent with the set $F$ or null if no such tree exists.

1: **function** FANTERNARYBUILD($F$, $L$)
2:     $T = \emptyset$
3:     **if** $|F| > 0$ **then**
4:         $(L_0, L_1, L_2) = $ SOLVE($F, L$)
5:         **if** $(L_0, L_1, L_2) = $ null **then** return null
6:         T=NEWVERTEX()
7:         **for** $i = 0, 1, 2$ **do**
8:             $F_i = $ DIVIDE($F, L_i$)
9:             $T_i = $ FANTERNARYBUILD($F_i, L_i$)
10:             **if** $T_i = $null **then** return null
11:             **if** $|L_i| > 0$ **then** NEWEDGE($T$, $T_i$)
12:         ADDMISSINGVERTICES($T$, $L$)
13:     return $T$

---

The computational complexity of FANTERNARYBUILD is $O(|L|^3 |F|)$, since the output tree has at most $|L| - 1$ inner vertices, and $O(|L|^2 |F|)$ operations are performed at each inner vertex. Thus the ternary $\mathcal{F}^+$ CONSISTENCY problem has a polynomial-time solution, proving Theorem 1.

Equipped with the FANTERNARYBUILD algorithm, the reader can directly show that the set $F$ in the example in Sect. 2 has no solution and also easily find a ternary tree that is consistent with the modified $F$.

## 4    The Remaining Ternary CONSISTENCY Problems

To determine the exact boundary between NP-hard versions of the ternary CONSISTENCY problem and those with a polynomial time solution, we now study the remaining variants. In what follows, we will often need to modify trees by adding or removing leaves. We will start with the simple observation that adding and removing leaves can be done in constant time and does not affect the other triplets unrelated to the added or removed leaves.

**Observation 2.** *Let $T$ be a ternary tree over a set of leaves $L$, and let $t$ be a triplet over $\{a, b, x\}$, where $a, b \in L_T$ are leaves and $x \notin L_T$ is not a leaf of $T$. We can add the leaf $x$ to the tree $T$ in constant time, thus obtaining a new ternary tree $T'$ such that $t \in t_{T'}$.*

*Similarly, for each leaf $x$ in $L_{T'}$ of a given ternary tree $T'$, we can remove it in constant time, thus obtaining a new ternary tree $T$ over $L_{T'} \setminus \{x\}$.*

*Moreover, in both addition and removal, the triplets not containing leaf $x$ will remain unchanged, that is, $t_T \subseteq t_{T'}$.*

*Proof.* We will first consider adding a new leaf $x$ to a tree $T$. Let $u = \text{lca}^T(a, b)$ be the lowest common ancestor of the vertices $a, b \in L_T$. Consider the first case when $t = ab|x$. If $u$ is the root of the tree $T$, then we create a new vertex $w$ as the new root of the tree $T$ and then add $x$ and the previous root as a children of $w$. Otherwise, we subdivide the edge from $u$ to its parent to create a new vertex $w$, and add $x$ as a child of $w$. Now consider the second case when $t = ax|b$. Let $c$ be the child of $u$ that is an ancestor of $a$. Subdivide the edge from $u$ to $c$ to create a new vertex $w$, and again add $x$ as a child of $w$. The case $t = bx|a$ can be realized by symmetry. Finally consider the last case when $t = a|b|x$. If $u$ has degree 2, then add $x$ as a child of $u$. Otherwise, let $c$ be the child of the vertex $u$ which is neither an ancestor of $a$ nor $b$. Subdivide the edge from $u$ to $c$ to create a new vertex $w$, and add $x$ as a child of $w$.

Now consider the removal of a leaf. Let $w$ be the parent of a leaf $x$. We remove $x$ and its incident edge. If after removing $x$, the vertex $w$ has degree 1, we need to remove the vertex $w$. If $w$ is the root, then we remove $w$ and its incident edge, setting a unique child of $w$ as the new root. Otherwise, we remove the vertex $w$ by replacing the two-edge path through $w$ with a single edge.

In both adding and removing a leaf $x$, the relative positions of the other leaves remain unchanged. Thus, all triplets not containing $x$ will remain unchanged. □

*Remark.* The most well-known algorithmic definition of tree restriction uses the above method for leaf removal. However, since each tree is uniquely defined by the set of its all triplets, the definitions are equivalent.

We will formulate two lemmas. The first lemma shows an equivalence between consistency with a resolved triplet and a related fan triplet.

**Lemma 2.** *Let $T$ be a ternary tree such that $a, b, c, x \in L_T$ and $T$ is consistent with fan triplet $a|c|x$. The tree $T$ is consistent with the resolved triplet $p = ab|c$ if and only if the fan triplet $q = b|c|x$ is consistent with $T$.*

*Proof.* First, we will prove the theorem in the base case when $L_T = \{a, b, c, x\}$. Let $L_0$, $L_1$, and $L_2$ be the sets of leaves of the three connected components of $T$ minus the root and its edges. As we already know from Sect. 3, the sets $L_0$, $L_1$, and $L_2$ correspond to a nontrivial solution of the system of equations corresponding to the set of fan triplets. Then, the variables $w_a$, $w_b$ and $w_c$ satisfy $w_a + w_c + w_x \equiv^3 0$, since $a|c|x$ is consistent with $T$.

If $T$ is consistent with $b|c|x$, then $w_b \equiv^3 -w_c - w_x \equiv^3 w_a$ by Observation 1.

If $T$ is consistent with $ab|c$, then also $w_a \equiv^3 w_b$.

In both cases, since the solution is nontrivial, by Observation 1 we obtain $\{w_a = w_b, w_c, w_x\} = \{0, 1, 2\}$. Thus $T$ is consistent with both $ab|c$ and $b|c|x$.

Now we will prove the theorem for any tree. Let $T$ be a tree consistent with the triplet $a|c|x$. Let $T' = T|_{\{x,a,b,c\}}$ be a restriction of $T$. It follows that $t_{T'} \subseteq t_T$. If the triplet $p$ is consistent with $T$, then $p$ is also consistent with $T'$ because $t_{T'}$ consists of a triplet over the set of leaves $\{a, b, c\}$. Using the result for the base case, we obtain $q \in t_{T'} \subseteq t_T$. Symmetrically, if $q \in t_T$, then $q \in t_{T'}$, and thus $p \in t_{T'} \subseteq t_T$. □

The following lemma is a generalization of the preceding one and allows for the conversion of any triplet into another chosen unrelated triplet.

**Lemma 3.** *Let $T$ be a ternary tree such that $a, b, c, x, y, z \in L_T$ and $T$ is consistent with the set of fan triplets $F = \{a|c|x, a|b|y, a|x|z\}$. Define the triplets*

$$p_1 = ab|c, \quad p_2 = ac|b, \quad p_3 = bc|a, \quad p_4 = a|b|c;$$
$$q_1 = x|y|z, \quad q_2 = xz|y, \quad q_3 = xy|z, \quad q_4 = yz|x.$$

*Then, for any index $j$, the tree $T$ is consistent with the triplet $p_j$ if and only if the triplet $q_j$ is consistent with the tree $T$.*

*Proof.* First, we will present a proof in the base case when $L_T = \{x, y, z, a, b, c\}$. Let $L_0$, $L_1$, and $L_2$ be the sets of leaves of the three connected components of $T$ minus the root and its edges. As we already know from Sect. 3, the sets $L_0$, $L_1$, and $L_2$ correspond to a nontrivial solution to the equation $M(F, L)\overrightarrow{w} \equiv^3 \overrightarrow{0}$. By subtracting the row corresponding to $a|c|x$ from the row corresponding to $a|x|z$, we obtain the system of equations

$$\begin{cases} w_x \equiv^3 2w_a + 2w_c, \\ w_y \equiv^3 2w_a + 2w_b, \\ w_z \equiv^3 w_c. \end{cases}$$

Each solution can be determined by fixing the values of the independent variables $w_a, w_b, w_c$ and calculating values the dependent variables $w_x, w_y, w_z$. Thus, there exist $3^3 = 27$ solutions. Consider the following four nontrivial solutions of the above equation system:

| | | | |
|---|---|---|---|
| 1. | $L_0 = \{x\}$, | $L_1 = \{a, b, y\}$, | $L_2 = \{c, z\}$; |
| 2. | $L_0 = \{y\}$, | $L_1 = \{a, c, x, z\}$, | $L_2 = \{b\}$; |
| 3. | $L_0 = \{x, y\}$, | $L_1 = \{a\}$, | $L_2 = \{b, c, z\}$; |
| 4. | $L_0 = \{c, y, z\}$, | $L_1 = \{a\}$, | $L_2 = \{b, x\}$. |

It can be easily checked that the $j$-th solution provides consistency with the triplets $p_j$ and $q_j$. For each of these solutions we can construct $3! - 1 = 5$ additional solutions of the system by permuting the indices. Additionally, by setting $w_a = w_b = w_c$ we obtain all 3 trivial solutions. This gives a total of $4*6+3 = 27$ solutions, and hence there are no other solutions. Thus the lemma holds in this base case.

Now we will prove the theorem for any tree. Let $T$ be a tree consistent with $F$. Let $T' = T|_{\{x,y,z,a,b,c\}}$ be a restriction of $T$. It follows that $t_{T'} \subseteq t_T$. Let $p_j \in t_T$ be a triplet consistent with $T$. Thus $p_j$ is also consistent with $T'$ because $t_{T'}$ consists of a triplet over the set of leaves $\{a, b, c\}$. Using the result for the base case, we obtain $q_j \in t_{T'} \subseteq t_T$. Symmetrically, if $q_j \in t_T$, then $q_j \in t_{T'}$, and thus $p_j \in t_{T'} \subseteq t_T$. □

Using Lemmas 2 and 3, we prove two theorems concerning the computational complexity of two ternary CONSISTENCY problems.

**Theorem 2.** *The ternary* $\mathcal{R}^+\mathcal{F}^+$ CONSISTENCY *problem is solvable in polynomial time.*

*Proof.* Let $F^+$ be the set of fan triplets over $L$ and $R^+$ be the set of resolved triplets over $L$. For each triplet $ab|c \in R^+$, we create a new additional leaf label $x = x(a, b, c) \notin L$. We convert the sets of resolved triplets in $R^+$ and fan triplets in $F^+$ into a new set of fan triplets

$$F = F^+ \cup \{a|c|x, b|c|x : ab|c \in R^+\}$$

over an extended set of leaves $L_F = L \cup \{x : ab|c \in R^+\}$.

We will prove that the ternary $\mathcal{R}^+\mathcal{F}^+$ CONSISTENCY problem for sets of triplets $R^+$ and $F^+$ over $L$ is equivalent to the ternary $\mathcal{F}^+$ CONSISTENCY problem for a set of triplets $F$ over $L_F$.

Let $T$ be a tree over $L$ that is consistent with $F^+$ and $R^+$. For each triplet $ab|c \in R^+$, we define a new leaf $x = x(a, b, c)$, and using Observation 2, we add the new leaf $x$ to the tree $T$ so that the resulting tree $T'$ is consistent with $a|c|x$. By Lemma 2 the tree $T'$ is consistent with the triplet $b|c|x$. We repeat this for all other elements of $R^+$ to obtain a tree that is consistent with $F$.

Conversely, let $T'$ be a tree over $L_F$ that is consistent with $F$. Obviously, the tree $T'$ is consistent with $F^+$ and by Lemma 2, the tree $T'$ is also consistent with $R^+$. Remove all leaves $x \in L_F \backslash L$ to obtain a tree $T$ over $L$. By Observation 2, $T$ remains consistent with $R^+$ and $F^+$.

By Theorem 1, the ternary $\mathcal{F}^+$ CONSISTENCY problem has a polynomial time solution. Since we can convert every instance of the ternary $\mathcal{R}^+\mathcal{F}^+$ CONSISTENCY problem into an equivalent instance of the ternary $\mathcal{F}^+$ CONSISTENCY in polynomial time, the ternary $\mathcal{R}^+\mathcal{F}^+$ CONSISTENCY problem is also in $P$. $\square$

**Theorem 3.** *The ternary* $\mathcal{F}^{+-}$ CONSISTENCY *problem is NP-hard.*

*Proof.* We give a polynomial-time reduction from the ternary $\mathcal{R}^-$ CONSISTENCY problem, which is known to be NP-hard. Let $R^-$ be a set of resolved triplets over $L$. For each triplet $ab|c \in R^-$, we create three new additional leaf labels $x = x(a, b, c)$, $y = y(a, b, c)$, and $z = z(a, b, c)$. We convert the set of resolved triplets $R^-$ into two new sets of fan triplets

$$F^+ = \{a|c|x, a|b|y, a|x|z : ab|c \in R^-\}$$
$$F^- = \{x|y|z : ab|c \in R^-\}$$

over an extended set of leaves $L_F = L \cup \{x, y, z : ab|c \in R^-\}$.

We will prove that the ternary $\mathcal{R}^-$ CONSISTENCY problem for a set of triplets $R^-$ over $L$ is equivalent to the ternary $\mathcal{F}^{+-}$ CONSISTENCY problem for sets of fan triplets $F^+$ and $F^-$ over $L_F$.

Let $T$ be a tree over $L$ that is not consistent with $R^-$. For each triplet $ab|c \in R^-$, we invoke Observation 2 to successively add leaves $x$, $y$, and $z$ to $T$ so that the resulting tree $T'$ is consistent with $a|c|x$, $a|b|y$, and $a|x|z$. By Lemma 3, since $T'$ is not consistent with $ab|c$, it is also not consistent with $x|y|z$. We repeat this for all other elements of $R^-$ and obtain a tree that is consistent with $F^+$ and not consistent with $F^-$.

Conversely, let $T'$ be a tree over $L_F$ that is consistent with $F^+$ and not consistent with $F^-$. By Lemma 3, $T'$ is not consistent with $R^-$. Remove all leaves in $L_F \backslash L$ to obtain a tree $T$. By Observation 2, $T$ is not consistent with the set $R^-$, since $T = T'|_L$ and $t_T \subseteq t_{T'}$.

Thus, the ternary $\mathcal{F}^{+-}$ CONSISTENCY problem is NP-hard, because the ternary $\mathcal{R}^-$ CONSISTENCY problem is NP-hard.                                    □

Since the ternary $\mathcal{R}^+\mathcal{F}^{+-}$ CONSISTENCY problem is a generalization of the ternary $\mathcal{F}^{+-}$ CONSISTENCY problem, the following result is immediate.

**Corollary 1.** *The ternary $\mathcal{R}^+\mathcal{F}^{+-}$ CONSISTENCY problem is NP-hard.*

## 5   Conclusion

Finally, we summarize our new findings and complete the classification.

**Proposition 1.** *The computational complexity of all 15 variants of the ternary $\mathcal{R}^{+-}\mathcal{F}^{+-}$ CONSISTENCY problem is as presented in Table 5.*

*Proof.* The variants $\mathcal{F}^+$ and $\mathcal{R}^+\mathcal{F}^+$ have polynomial-time solutions by Theorem 1 and its generalization Theorem 2. By [10, Corollary 1], the variant $\mathcal{R}^+\mathcal{F}^-$, and thereby also $\mathcal{R}^+$ and $\mathcal{F}^-$, have polynomial-time solutions, too. In contrast, using Theorem 3 and its consequent Corollary 1, we obtain that the variant $\mathcal{F}^{+-}$, and hence $\mathcal{R}^+\mathcal{F}^{+-}$, are NP-hard. All other variants are NP-hard since the variant $\mathcal{R}^-$ is NP-hard according to [10, Theorem 3].                                    □

**Table 5.** Completed version of Table 3, presenting the computational complexity of all variants of the ternary CONSISTENCY problem. As before, P means that the corresponding problem is polynomial-time solvable, while NP-h means that the problem is NP-hard.

|            | $\emptyset$ | $\mathcal{F}^+$ | $\mathcal{F}^-$ | $\mathcal{F}^{+-}$ |
|------------|------|------|------|------|
| $\emptyset$ | –    | P    | P    | NP-h |
| $\mathcal{R}^+$ | P    | P    | P    | NP-h |
| $\mathcal{R}^-$ | NP-h | NP-h | NP-h | NP-h |
| $\mathcal{R}^{+-}$ | NP-h | NP-h | NP-h | NP-h |

**Acknowledgments.** This work was partially funded by JSPS KAKENHI grant 22H03550. Mikołaj Marciniak was additionally supported by Narodowe Centrum Nauki, Grant Number 2017/26/A/ST1/00189 and Narodowe Centrum Badań i Rozwoju, Grant Number POWR.03.05.00-00-Z302/17-00.

# References

1. Aho, A.V., Sagiv, Y., Szymanski, T.G., Ullman, J.D.: Inferring a tree from lowest common ancestors with an application to the optimization of relational expressions. SIAM J. Comput. **10**(3), 405–421 (1981)
2. Bininda-Emonds, O.R.P., et al.: The delayed rise of present-day mammals. Nature **456**, 274 (2008)
3. Bininda-Emonds, O.R.P.: The evolution of supertrees. Trends Ecol. Evol. **19**(6), 315–322 (2004)
4. Chor, B., Hendy, M., Penny, D.: Analytic solutions for three taxon ML trees with variable rates across sites. Discrete Appl. Math. **155**(6), 750–758 (2007). Computational Molecular Biology Series, Issue V
5. Constantinescu, M., Sankoff, D.: An efficient algorithm for supertrees. J. Classif. **12**, 101–112 (1995)
6. Felsenstein, J.: Inferring Phylogenies. Sinauer (2003)
7. Guillemot, S., Jansson, J., Sung, W.-K.: Computing a smallest multi-labeled phylogenetic tree from rooted triplets. IEEE/ACM Trans. Comput. Biol. Bioinform. **8**, 1141–1147 (2009)
8. He, Y.-J., Huynh, T.N.D., Jansson, J., Sung, W.-K.: Inferring phylogenetic relationships avoiding forbidden rooted triplets. J. Bioinform. Comput. Biol. **4**, 59–74 (2006)
9. Huber, K., van Iersel, L., Moulton, V., Scornavacca, C., Wu, T.: Reconstructing phylogenetic level-1 networks from nondense binet and trinet sets. Algorithmica **77**, 173–200 (2014)
10. Jansson, J., Lingas, A., Rajaby, R., Sung, W.-K.: Determining the consistency of resolved triplets and fan triplets. J. Comput. Biol. **25**(7), 740–754 (2018). PMID: 29451395
11. Jansson, J., Nguyen, N.B., Sung, W.-K.: Algorithms for combining rooted triplets into a galled phylogenetic network. SIAM J. Comput. **35**, 1098–1121 (2006)
12. Ng, M.P., Wormald, N.C.: Reconstruction of rooted trees from subtrees. Discrete Appl. Math. **69**(1–2), 19–31 (1996)
13. Semple, C.: Reconstructing minimal rooted trees. Discrete Appl. Math. **127**(3), 489–503 (2003)
14. Semple, C., Daniel, P., Hordijk, W., Page, R., Steel, M.: Supertree algorithms for ancestral divergence dates and nested taxa. Bioinformatics **20**, 2355–2360 (2004)
15. Snir, S., Rao, S.: Using max cut to enhance rooted trees consistency. IEEE/ACM Trans. Comput. Biol. Bioinf. **3**, 323–333 (2006)
16. Sung, W.-K.: Algorithms in Bioinformatics: A Practical Introduction. Chapman & Hall/CRC (2009)
17. Willson, S.: Constructing rooted supertrees using distances. Bull. Math. Biol. **66**, 1755–1783 (2004)

# Complexity Framework for Forbidden Subgraphs IV: The Steiner Forest Problem

Hans L. Bodlaender[1], Matthew Johnson[2], Barnaby Martin[2],
Jelle J. Oostveen[1], Sukanya Pandey[1], Daniël Paulusma[2], Siani Smith[3],
and Erik Jan van Leeuwen[1(✉)]

[1] Department of Information and Computing Sciences, Utrecht University,
Utrecht, The Netherlands
{h.l.bodleander,j.j.oostveen,s.pandey1,e.j.vanleeuwen}@uu.nl
[2] Durham University, Durham, UK
{matthew.johnson2,barnaby.d.martin,daniel.paulusma}@durham.ac.uk
[3] University of Bristol and Heilbronn Institute for Mathematical Research,
Bristol, UK
siani.smith@bristol.ac.uk

**Abstract.** We study STEINER FOREST on $H$-subgraph-free graphs, that is, graphs that do not contain some fixed graph $H$ as a (not necessarily induced) subgraph. We are motivated by a recent framework that completely characterizes the complexity of many problems on $H$-subgraph-free graphs. However, in contrast to, e.g. the related STEINER TREE problem, STEINER FOREST falls outside this framework. Hence, the complexity of STEINER FOREST on $H$-subgraph-free graphs remained tantalizingly open. We make significant progress on this open problem: our main results are four novel polynomial-time algorithms for different excluded graphs $H$ that are central to further understand its complexity. Along the way, we study the complexity of STEINER FOREST for graphs with a small $c$-deletion set, that is, a small set $X$ of vertices such that each component of $G - X$ has size at most $c$. Using this parameter, we give two algorithms that we later employ as subroutines. First, we present a significantly faster parameterized algorithm for STEINER FOREST parameterized by $|X|$ when $c = 1$ (i.e. the vertex cover number), which by a recent result is best possible under ETH [Feldmann and Lampis, arXiv 2024]. Second, we prove that STEINER FOREST is polynomial-time solvable for graphs with a 2-deletion set of size at most 2. The latter result is tight, as the problem is NP-complete for graphs with a 3-deletion set of size 2.

**Keywords:** Steiner forest · forbidden subgraph · complexity dichotomy · vertex cover number · deletion set

J.J. Oostveen is supported by the NWO grant OCENW.KLEIN.114 (PACAN).

A. A. Rescigno and U. Vaccaro (Eds.): IWOCA 2024, LNCS 14764, pp. 206–217, 2024.
https://doi.org/10.1007/978-3-031-63021-7_16

# 1  Introduction

We consider the complexity of a classical graph problem, STEINER FOREST, restricted to graphs that do not contain some fixed graph $H$ as a subgraph. Such graphs are said to be $H$-*subgraph-free*, that is, they cannot be modified to $H$ by a sequence of edge deletions and vertex deletions. A graph $G$ is $H$-*free* if $G$ cannot be modified into $H$ by a sequence of vertex deletions only. Even though $H$-free graphs are more widely studied in the literature, $H$-subgraph-free graphs are also highly interesting, as was recently shown through the introduction of a large, general framework for the subgraph relation [15–17,19].

For a set of graphs $\mathcal{H}$, a graph $G$ is $\mathcal{H}$-*subgraph-free* if $G$ is $H$-subgraph-free for every $H \in \mathcal{H}$. In order to unify known classifications for INDEPENDENT SET [1], DOMINATING SET [1], LIST COLOURING [14], LONG PATH [1] and MAX-CUT [18] on $\mathcal{H}$-subgraph-free graphs (for finite $\mathcal{H}$), a systematic approach was developed in [15]. We first explain this framework.

For $k \geq 1$, the $k$-*subdivision* of an edge $e = uv$ of a graph replaces $e$ by a path of length $k + 1$ with endpoints $u$ and $v$ (and $k$ new vertices). The $k$-*subdivision* of a graph $G$ is the graph obtained from $G$ after $k$-subdividing each edge. For a graph class $\mathcal{G}$ and an integer $k$, let $\mathcal{G}^k$ consist of the $k$-subdivisions of the graphs in $\mathcal{G}$. A graph problem $\Pi$ is NP-complete *under edge subdivision of subcubic graphs* if for every integer $j \geq 1$, there is an integer $\ell \geq j$ such that: if $\Pi$ is NP-complete for the class $\mathcal{G}$ of subcubic graphs (graphs with maximum degree at most 3), then $\Pi$ is NP-complete for $\mathcal{G}^\ell$. Now, $\Pi$ is a *C123-problem* if:

C1. $\Pi$ is polynomial-time solvable for every graph class of bounded treewidth,
C2. $\Pi$ is NP-complete for the class of subcubic graphs, and
C3. $\Pi$ is NP-complete under edge subdivision of subcubic graphs.

A *subdivided* claw is a graph obtained from a *claw* (4-vertex star) by subdividing each of its three edges zero or more times. The *disjoint union* of two vertex-disjoint graphs $G_1$ and $G_2$ is the graph $(V(G_1) \cup V(G_2), E(G_1) \cup E(G_2))$. The set $\mathcal{S}$ consists of all graphs that are disjoint unions of subdivided claws and paths. We can now state the complexity classification of [15].

**Theorem 1** ([15]). *Let $\Pi$ be a C123-problem. For a finite set $\mathcal{H}$, the problem $\Pi$ on $\mathcal{H}$-subgraph-free graphs is polynomial-time solvable if $\mathcal{H}$ contains a graph from $\mathcal{S}$ (or equivalently, if the class of $\mathcal{H}$-subgraph-free graphs has bounded treewidth) and NP-complete otherwise.*

See the (long) table of problems in [15], for examples of C123-problems other than the ones above and for examples of problems that do not satisfy C2 or C3. There are also problems that only satisfy C2 and C3 but not C1. For example, SUBGRAPH ISOMORPHISM is NP-complete even for input pairs of path-width 1. A few years ago, Bodlaender et al. [6] settled the complexity of SUBGRAPH ISOMORPHISM for $H$-subgraph-free graphs apart from essentially two open cases. ($H = P_5$ and $H = 2P_5$). Hence, the following question is challenging:

*How do* C23-*problems, i.e., problems that satisfy C2 and C3 but not C1, behave for H-subgraph-free graphs? Can we still classify their computational complexity?*

We consider this question for STEINER FOREST. A *Steiner forest* of an undirected graph $G$, with a set $T = \{(s_1, t_1), \ldots, (s_p, t_p)\}$ of specified pairs of vertices called *terminals*, is a subgraph $F$ of $G$, such that $s_i$ and $t_i$, for every $i \in \{1, \ldots, p\}$, belong to the same connected component of $F$. This leads to the problem:

---

STEINER FOREST
    *Instance:* A graph $G$, a set $T$ of terminal pairs and an integer $k$.
    *Question:* Does $(G, T)$ have a Steiner forest $F$ with $|E(F)| \leq k$?

---

STEINER FOREST generalizes the C123-problem STEINER TREE [15], which is to decide if for a given integer $k$, a graph $G$ with some specified set $T$ of vertices has a tree $S$ with $|E(S)| \leq k$ containing every vertex of $T$: take all pairs of vertices of $T$ as terminal pairs to obtain an equivalent instance of STEINER FOREST.

For a constant $c$, a *c-deletion set* of a graph $G = (V, E)$ is a set $T \subseteq V$ such that each connected component of $G - T$ has size at most $c$. The *c-deletion set number* (or $(c + 1)$-*component order connectivity*) of $G$ is the size of a smallest $c$-deletion set (see also [3, 7–9, 11, 12]). The following theorem is a crucial result of Bateni, Hajiaghayi and Marx [4] and plays an important role in our paper:

**Theorem 2** ([4]). STEINER FOREST *is polynomial-time solvable for graphs of treewidth at most 2, but* NP-*complete for graphs of treewidth 3, tree-depth 4, and 3-deletion set number 2.*

This shows that STEINER FOREST does not satisfy C1, unlike STEINER TREE [2]. As STEINER TREE satisfies C2 and C3 [15], STEINER FOREST satisfies C2 and C3 and is a C23-problem, unlike STEINER TREE which is C123 [15]. This leaves the complexity of STEINER FOREST on $H$-subgraph-free graphs open.

**Our Results.** Let $K_{a,b}$ be the complete bipartite graph with $a$ vertices on one side and $b$ on the other. Let $S_{a,b,c}$ be the graph obtained from the claw $(K_{1,3})$ by subdividing its three edges $a - 1$, $b - 1$ and $c - 1$ times, respectively. Let $P_r$ be the path on $r$ vertices. For two graphs $H_1$ and $H_2$, we write $H_1 \subseteq H_2$ if $H_1$ is a subgraph of $H_2$, i.e., $V(H_1) \subseteq V(H_2)$ and $E(H_1) \subseteq E(H_2)$. We write $H_1 + H_2$ for the disjoint union of $H_1$ and $H_2$ and $sH_1$ to denote the disjoint union of $s$ copies of $H_1$. Our results on STEINER FOREST for $H$-subgraph-free graphs are:

**Theorem 3.** *For a graph $H$, STEINER FOREST on $H$-subgraph-free graphs is*

- *polynomial-time solvable if $H \subseteq 2K_{1,3} + P_3 + sP_2, 2P_4 + P_3 + sP_2, P_9 + sP_2$ or $S_{1,1,4} + sP_2$ for each $s \geq 0$, and*
- *NP-complete if $H \supseteq 3K_{1,3}, 2K_{1,3} + P_4, K_{1,3} + 2P_4, 3P_4$ or if $H \notin \mathcal{S}$.*

The gap between the easy and hard cases could be significantly reduced if we could resolve an intriguing open problem (see Sect. 5). As graphs of tree-depth 3 are $P_8$-subgraph-free, Theorems 2 and 3 yield the following dichotomy:

**Corollary 1.** *For a constant* $t$, STEINER FOREST *on graphs of tree-depth* $t$ *is polynomial-time solvable if* $t \leq 3$ *and* NP-*complete if* $t \geq 4$.

The NP-hardness part of Theorem 3 follows from the gadget from Theorem 2 and NP-completeness of STEINER FOREST when $H \notin \mathcal{S}$ [5], as shown in Sect. 2. For the polynomial part, proven in Sect. 4, we first make some useful observations in Sect. 3.

Gima et al. [13] showed that STEINER FOREST has an $n^{O(k)}$-time algorithm in the weighted case and an $(2k^2 2^k k^{2^k})^{O(k)} n^{O(1)}$-time algorithm in the unweighted case, where $k$ is the vertex cover number. We need their result as a subroutine for several of our algorithms, but we were able to significantly improve on it, as we show in the following result (proof omitted).

**Theorem 4.** STEINER FOREST *has a* $2^{O(k \log k)} n^{O(1)}$ *time algorithm, where* $k$ *is the vertex cover number of the input graph, even in the weighted case.*

Afterwards, Feldmann and Lampis [10] showed our $2^{O(k \log k)} n^{O(1)}$ algorithm is best possible under the Exponential Time Hypothesis. They also gave an alternative algorithm with the same runtime that relies on an algorithm of Bateni et al. [4]. In contrast, our algorithm is self-contained.

Another important subroutine for some of our algorithms is the polynomial-part of the following dichotomy. We prove this part in Lemma 3 in Sect. 3, while the NP-completeness part is taken from Theorem 2, which is due to [4].

**Theorem 5.** *For a constant* $c$, STEINER FOREST *on graphs with a* $c$-*deletion set of size at most* 2 *is polynomial-time solvable if* $c \leq 2$ *and* NP-*complete if* $c \geq 3$.

## 2    NP-Completeness Results

Bateni, Hajiaghayi and Marx [4] explicitly proved that STEINER FOREST is NP-complete for graphs of treewidth 3, see also Theorem 2. The additional properties in the lemma below can be easily verified from inspecting their gadget, which is displayed in Fig. 1.

**Lemma 1.** STEINER FOREST *is* NP-*complete for* $(3K_{1,3}, 2K_{1,3} + P_4, K_{1,3} + 2P_4, 3P_4)$-*subgraph-free graphs of tree-depth* 4 *with* 3-*deletion set number* 2.

We can now show the NP-completeness part of Theorem 3. It is known that STEINER TREE, and thus STEINER FOREST, is NP-complete for $H$-subgraph-free graphs if $H \notin \mathcal{S}$ [5]. The NP-completeness part of Theorem 3 now follows immediately from this observation and Lemma 1.

## 3    Polynomial Subroutines

A *minimum* Steiner forest for an instance $(G, T)$ is one with the smallest number of edges. Denote the number of edges of such a forest by $\mathsf{sf}(G, T)$. We assume

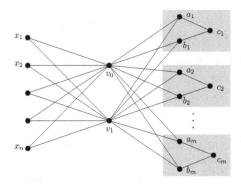

**Fig. 1.** The graph $G$ (gadget from [4]) used in the proof of Lemma 1. On the left there are vertices $x_1 \ldots x_n$ representing the boolean variables. On the right, each $P_3$ represents a clause of the given $R$-formula $\phi$.

that for any terminal pair $(s,t) \in T$, $s$ and $t$ are distinct (as any pair where $s = t$ can be removed without affecting the feasibility or the size of a minimum solution). A vertex $v$ is a *terminal vertex* if there is a pair $(s,t) \in T$ with $v = s$ or $v = t$. A *cut vertex* of a connected graph is a vertex whose removal yields a graph with at least two connected components. A graph with no cut vertices is 2-*connected*. A 2-*connected component* is a maximal subgraph that is 2-connected. A graph class is *hereditary* if it is closed under deleting vertices.

**Lemma 2.** *For every hereditary graph class $\mathcal{G}$, if* STEINER FOREST *is polynomial-time solvable for the subclass of* 2-*connected graphs of $\mathcal{G}$, then it is polynomial-time solvable for $\mathcal{G}$.*

*Proof (Sketch).* Assume $G$ is connected. If $G$ is not 2-connected, then let $G_1$ be a 2-connected component of $G$ with only one cut-vertex $v$ of $G$. Let $G_2 = G - (V(G_1) \setminus \{v\})$. For $i = 1, 2$, let $T_i$ be the set of all terminal pairs in $G_i$ with both terminals in $V(G_i)$, to which we add the pair $(s_j, v)$ for each $(s_j, t_j) \in T$ such that $s_j \in V(G_i)$ and $t_j \notin V(G_i)$. Similarly, if $(s_j, t_j) \in T$ such that $s_j \notin V(G_i)$ and $t_j \in V(G_i)$, then we add $(v, t_j)$ to $T_i$. Now apply the algorithm for 2-connected graphs to $(G_1, T_1)$, and recurse on $(G_2, T_2)$. $\qquad\square$

The *contraction* of an edge $e = (u,v)$ in a graph $G$ replaces $u$ and $v$ by a new vertex $w$ that is adjacent to all former neighbours of $u$ and $v$ in $G$. Our next result shows the polynomial part of Theorem 5.

**Lemma 3.** STEINER FOREST *is polynomial-time solvable for graphs with a 2-deletion set of size at most 2.*

*Proof.* Let $G$ be an $n$-vertex graph that, together with a set $T$ of terminal pairs, is an instance of STEINER FOREST. If $G$ has a 2-deletion set of size 1, then this vertex forms a cut vertex of $G$. Hence, we can apply Lemma 2 to reduce to the case of graphs of size 3, which is trivial. So, we assume that $G$ has a 2-deletion set $C$ of size 2, say $C = \{u, v\}$. By Lemma 2 we may assume that $G$ is 2-connected.

First suppose $u$ and $v$ are adjacent. If the edge $uv$ is part of the solution, we may contract $uv$ to form a 2-deletion set of size 1. We remember the size of the Steiner Forest found. If $uv$ is not part of the solution, we reduce to the case when $u$ and $v$ are not adjacent. From now on, we assume $u$ and $v$ are non-adjacent.

We first search for a minimum solution for $(G,T)$ over all solutions with one connected component. We compute a minimum solution over all solutions that contain $u$ but not $v$, and a minimum solution over all solutions that contain $v$ but not $u$. This takes polynomial time, as we reduce to the case of graphs of size 3 by applying Lemma 2. We now compute a minimum solution for $(G,T)$ over all solutions with one connected component that contain both $u$ and $v$. As $G$ is 2-connected, there exists a path from $u$ to $v$ in $G$. As $C$ is a 2-deletion set, every path from $u$ to $v$ has at most two inner vertices. We check each of the $O(n^2)$ paths from $u$ to $v$ (with at most two inner vertices). For each choice of $P$, we contract the path $P$ to a single vertex and apply Lemma 2 to reduce to the case of graphs of size 3 by applying Lemma 2.

It remains to find a minimum solution for $(G,T)$ over all solutions with two connected components, one containing $u$ and the other one $v$, and to check this minimum solution with the minimum solutions of the other types found above.

If some connected component $D$ of $G \setminus \{u,v\}$ does not contain any terminal vertex, we can safely delete $D$. Hence, we assume that every connected component of $G \setminus \{u,v\}$ contains at least one terminal vertex. We now show that we can restrict ourselves to solutions that contain at most one non-terminal vertex.

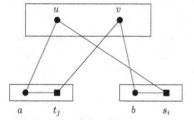 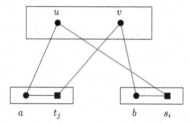

**Fig. 2.** Left: a solution containing two non-terminal vertices $a$ and $b$. Right: the conversion to an equivalent solution containing $b$ as the only non-terminal vertex.

Suppose that there exists an optimal solution $F$ with two non-terminal vertices $a$ and $b$ (Fig. 2). We may assume without loss of generality that $v$ is adjacent to $b$, and that $b$ is adjacent to terminal $s_i$. Hence, $F$ contains the edges $vb$ and $bs_i$. As $G$ is 2-connected, the edge $us_i$ must exist in $G$ as well. Consider another connected component of $G \setminus \{u,v\}$, say $at_j$, where $t_j$ is a terminal. First assume that $au \in E(G)$. As $G$ is 2-connected, we also have $vt_j \in E(G)$. As $a \in V(F)$, the edges $ua$ and $at_j$ must belong to $E(F)$. We now remove the edges $ua$ and $at_j$ from $F$ and add the edges $us_i$ and $vt_j$ to $F$. This yields a minimum Steiner Forest $F'$ with fewer non-terminal vertices. If $au \notin E(G)$, we have that $av \in E(G)$. In this case we can make a similar replacement and come to the same conclusion.

Due to the above, we may restrict ourselves to finding minimum solutions for $(G,T)$ with two connected components that contain at most one non-terminal vertex. Hence, we branch by considering all $O(n)$ options for the set of non-terminal vertices used in a minimum solution. We consider each of the $O(n)$ branches separately and as follows.

First, we remove all non-terminal vertices that we did not guess to be in the solution. For the guessed non-terminal vertex we do as follows. By construction, such a guessed non-terminal vertex $z$ is adjacent to $u$ or $v$, and we contract the edge $zu$ or $zv$, respectively. Afterwards, we apply Lemma 2 again such that we may assume that the resulting instance, which we denote by $(G, T)$ again, is 2-connected. Now every vertex of $V(G) \setminus \{u, v\}$ is a terminal vertex. Hence, every vertex of $V(G) \setminus \{u, v\}$ will be in the solution we are trying to construct.

For every connected component $D$ of $G - \{u, v\}$ we do as follows. As $C$ is a 2-deletion set, $D$ consists of at most two vertices. We only consider $D$ if $D$ has exactly two vertices, $x$ and $y$. If $(x, y) \in T$ and neither $x$ nor $y$ appears in any other terminal pair, then we add the edge $xy$ to the solution; remove the vertices $x$ and $y$ from $G$; and also remove the pair $(x, y)$ from $T$. This is indeed optimal, as edges from $x$ and $y$ to $u$ or $v$ would not be involved in connecting any other terminal pairs. So we may assume that $(x, y) \notin T$, or $x$ or $y$ appears in some other pair than $(x, y)$ in $T$.

We now apply the following operation on the component $D$, depending on its adjacency to $u$ and $v$. First suppose $x$ and $y$ are both adjacent to both $u$ and $v$. Then we may remove the edge $xy$ for the following reason. Recall that $x$ and $y$ are terminal vertices such that $(x, y) \notin T$, or $x$ or $y$ appears in some other pair than $(x, y)$ in $T$. Hence, we need to connect $x$ to either $u$ or $v$, and we also need to connect $y$ to either $u$ or $v$. For doing this, we do not need to use the edge $xy$. Now suppose one of $x, y$, say $x$, is adjacent to $u$ and $v$, whereas the other one, $y$, is adjacent to only one of $u$ and $v$, say to $u$. For the same reasons as before, we need to connect $x$ to either $u$ or $v$, and we also need to connect $y$ to either $u$ or $v$. If we use $xu$, then we must also use $yu$, and in that case we can replace $xu$ by $xy$. Hence, we may remove $xu$ from $G$.

So, afterwards, we reduced the instance in polynomial time to a new instance, which we will also denote by $(G, T)$, with the following properties. Every connected component of $G - \{u, v\}$ has at most two vertices. By 2-connectivity, for every connected component that contains exactly one vertex $z$ we have the edges $uz$ and $vz$. Moreover, for every connected component that contains exactly two vertices $x$ and $y$, we have the edges $ux$ and $vy$ but not $uy$ and not $vx$.

As every vertex in $V(G) \setminus \{u, v\}$ is a terminal vertex and we search for a minimum solution for $(G, T)$ with two connected components, we need one edge in the solution for each 1-vertex connected component of $G - \{u, v\}$ and two edges for each 2-vertex connected component. First, suppose one of $u, v$, say $u$, is not a terminal vertex. Then $G - \{v\}$ is a minimum solution, so we can stop (note that we already found this solution before).

Now suppose that both $u$ and $v$ are terminal vertices. We discard the branch if $u$ and $v$ represent terminals of the same pair (as then the solution must be connected). Else we do as follows. If $u$ represents $s_i$, then we connect the terminal vertex that represents $t_i$ to $u$. This can only be done in one way: if the terminal vertex representing $t_i$ is not adjacent to $u$, it contains a unique neighbour adjacent to $u$. Afterwards, we remove the terminal pairs that we connected in this way from $G$. We repeat these steps for $v$. If it was not possible to connect some terminal pair, then we discard the branch. Otherwise, we apply Lemma 2 again, such that the resulting instance, which we denote by $(G, T)$ again, is 2-connected. The other properties are maintained, and neither $u$ nor $v$ is a terminal vertex. Hence, we can take as solution for $(G, T)$ either $G - \{u\}$ or $G - \{v\}$. We now construct a solution for the original graph and terminal set, and out of all the solutions found take a minimum one.                                      □

For some of our polynomial-time results we need to extend Lemma 3 as follows (proof omitted).

**Lemma 4.** STEINER FOREST *is polynomial-time solvable for graphs $G$ with a set $X \subseteq E(G)$ of bounded size such that $G - X$ has a 2-deletion set $C$ of size at most 2 and each end-point of every edge of $X$ is either an isolated vertex of $G - X$ or a vertex of $C$.*

## 4  Polynomial Cases

To prove the polynomial part of Theorem 3, we start with a general lemma. To prove it we use Theorem 4; we omit the details.

**Lemma 5.** *For a graph $H$, if STEINER FOREST can be solved in polynomial time on the class of $H$-subgraph-free graphs, then STEINER FOREST can be solved in polynomial time on the class of $(H + P_2)$-subgraph-free graphs.*

We now consider specific graphs $H$. The first one is the case $H = 2K_{1,3}$. We omit the proof.

**Lemma 6.** STEINER FOREST *is polynomial-time solvable for $2K_{1,3}$-subgraph-free graphs.*

By using Lemma 4 (which generalized Lemma 3) we can extend Lemma 6 (again we omit the proof details).

**Lemma 7.** STEINER FOREST *is polynomial-time solvable for $(2K_{1,3} + P_3)$-subgraph-free graphs.*

We now consider the case $H = S_{1,1,4}$. We omit the proof.

**Lemma 8.** STEINER FOREST *is polynomial-time solvable for $S_{1,1,4}$-subgraph-free graphs.*

We use Lemma 8 to prove the case $H = P_9$. We also need Lemma 3 again.

**Lemma 9.** STEINER FOREST *is polynomial-time solvable for $P_9$-subgraph-free graphs.*

*Proof.* Let $G = (V, E)$ be a $P_9$-subgraph-free graph that is part of an instance of STEINER FOREST. By Lemma 2 we may assume that $G$ is 2-connected. Let $P = u_1 \cdots u_r$, for some $r \geq 2$, be a longest (not necessarily induced) path in $G$. As $G$ is $P_9$-subgraph-free, we have that $r \leq 8$. If $r \leq 5$, then $G$ is $P_6$-subgraph-free, and thus $S_{1,1,4}$-subgraph-free, and we can apply Lemma 8. Hence, $r \in \{6, 7, 8\}$.

**Case 1.** $r = 6$.
Then $G$ is $P_7$-subgraph-free. Suppose $G - V(P)$ has a connected component $D$ with more than one vertex. As $G$ is $P_7$-subgraph-free and $|V(D)| \geq 2$, no vertex of $D$ is adjacent to $u_1$, $u_2$, $u_5$ or $u_6$. As $G$ is connected, at least one of $u_3$ or $u_4$, say $u_3$, has a neighbour $v$ in $D$. Suppose $w \in V(D)$ is adjacent to $u_4$ (where $w = v$ is possible). Let $vQw$ be a path from $v$ to $w$ in $D$; note that $Q$ might be empty. Now the path $u_1u_2u_3vQwu_4u_5u_6$ has at least seven vertices, contradicting that $G$ is $P_7$-subgraph-free. Hence, no vertex of $D$ is adjacent to $u_4$. This means that $u_3$ is a cut-vertex of $G$, contradicting the 2-connectivity. We conclude that every connected component of $G - V(P)$ consists of one vertex. In other words, $\{u_1, \ldots, u_6\}$ is a vertex cover of $G$, and we can apply Theorem 4.

**Case 2.** $r = 7$.
Then $G$ is $P_8$-subgraph-free. Suppose $G - V(P)$ has a connected component $D$ with more than one vertex. As $G$ is $P_8$-subgraph-free and $|V(D)| \geq 2$, no vertex of $D$ is adjacent to $u_1$, $u_2$, $u_6$ or $u_7$. As $G$ is connected, at least one of $u_3$, $u_4$ or $u_5$ has a neighbour $v$ in $D$.

First assume that $u_3$ or $u_5$, say $u_3$, has a neighbour $v$ in $D$. As $|V(D)| \geq 2$, $v$ has a neighbour $w$ in $D$. If $w$ has a neighbour $x \neq v$ in $D$, then $xwvu_3u_4u_5u_6u_7$ is a path on eight vertices, contradicting that $G$ is $P_8$-subgraph-free. Hence, $v$ is the only neighbour of $w$ in $D$. As $G$ is 2-connected, $w$ has a neighbour on $P$. Recall that no vertex of $D$ is not adjacent to $u_1$, $u_2$, $u_6$ or $u_7$. If $w$ is adjacent to $u_4$, then $u_1u_2u_3vwu_4u_5u_6$ is a path on eight vertices. If $w$ is adjacent to $u_5$, then $u_1u_2u_3vwu_5u_6u_7$ is a path on eight vertices. Hence, $w$ must be adjacent to $u_3$ (and $u_3$ is the only neighbour of $w$ on $P$). We now find that $w$ is the only neighbour of $v$ in $D$, as otherwise, if $v$ has a neighbour $w' \neq w$ on $D$, then $w'vwu_3u_4u_5u_6u_7$ is a path on eight vertices. In other words, $V(D) = \{v, w\}$. By the same arguments, but now applied on $v$, we find that $u_3$ is the only neighbour of $v$ on $P$. Hence, $u_3$ is a cut-vertex of $G$, contradicting the 2-connectivity of $G$.

From the above we conclude that no vertex of $D$ is adjacent to $u_3$. By the same reason, no vertex of $D$ is adjacent to $u_5$. We find that $u_4$ disconnects $D$ from the rest of $G$, contradicting the 2-connectivity of $G$. We conclude that every connected component of $G - V(P)$ consists of one vertex. In other words, $\{u_1, \ldots, u_7\}$ is a vertex cover of $G$, and we can apply Theorem 4.

**Case 3.** $r = 8$.
Recall that $G$ is $P_9$-subgraph-free, so this is the last case to consider. If every connected component of $G - V(P)$ consists of one vertex, then $\{u_1, \ldots, u_8\}$ is a vertex cover of $G$, and we can apply Theorem 4. Now suppose that $G - V(P)$ has a connected component $D$ with more than one vertex. As $G$ is $P_9$-subgraph-free

and $|V(D)| \geq 2$, no vertex of $D$ is not adjacent to $u_1$, $u_2$, $u_7$ or $u_8$. As $G$ is connected, at least one of $u_3$, $u_4$, $u_5$ or $u_6$ has a neighbour in $D$.

First, suppose that neither $u_3$ nor $u_6$ has a neighbour in $D$. Then $u_4$ or $u_5$, say $u_4$, has a neighbour $v$ in $D$. As $G$ is 2-connected, there exists a path $vQu_5$ from $v$ to $u_5$ that does not contain $u_4$. As no vertex from $\{u_1, u_2, u_3, u_6, u_7, u_8\}$ has a neighbour in $D$, the vertices of $Q$ belong to $D$. Now, $u_1u_2u_3u_4vQu_5u_6u_7u_8$ is a path on nine vertices, contradicting that $G$ is $P_9$-subgraph-free.

Hence, at least one of $u_3$ or $u_6$, say $u_3$, has a neighbour $v$ in $D$. As $|V(D)| \geq 2$, we find that $v$ has a neighbour $w$ in $D$. If $w$ has a neighbour $x \neq v$ in $D$, then $xwvu_3u_4u_5u_6u_7u_8$ is a path on nine vertices, contradicting that $G$ is $P_9$-subgraph-free. Hence, $v$ is the only neighbour of $w$ in $D$. As $G$ is 2-connected, this means that $w$ has a neighbour on $P$. Recall that no vertex of $D$ is adjacent to $u_1$, $u_2$, $u_7$ or $u_8$. If $w$ is adjacent to $u_4$, then $u_1u_2u_3vwu_4u_5u_6u_7$ is a path on nine vertices. If $w$ is adjacent to $u_5$, then $u_1u_2u_3vwu_5u_6u_7u_8$ is a path on nine vertices. Hence, $w$ must be adjacent to either or both $u_3$ and $u_6$ (and $w$ has no other neighbours on $P$). We now find that $w$ is the only neighbour of $v$ in $D$ for the following reason. Suppose $v$ has a neighbour $w' \neq w$ on $D$. If $w$ is adjacent to $u_3$, then $w'vwu_3u_4u_5u_6u_7u_8$ is a path on nine vertices. If $w$ is adjacent to $u_6$, then $w'vwu_6u_5u_4u_3u_2u_1$ is a path on nine vertices. In other words, $V(D) = \{v, w\}$. By the same arguments, but now applied on $v$, we find that apart from $u_3$, it holds that $v$ may have only $u_6$ as a neighbour of $P$.

We have proven for any path $Q = q_1 \ldots q_8$ on eight vertices that every connected component of size at least 2 in $G - Q$ has size exactly 2, and moreover, $q_3$ and $q_6$ are the only vertices of $Q$ with neighbours in such a connected component. We may assume without loss of generality that $v$ is adjacent to $u_3$ and $w$ is adjacent to $u_6$, as otherwise one of $u_3, u_6, v, w$ is a cut-vertex of $G$, contradicting the 2-connectivity of $G$. By replacing $P$ with $P' = u_1u_2u_3vwu_6u_7u_8$, we find that $u_4$ and $u_5$ have no neighbours outside $\{u_3, u_6\}$. By replacing $P$ with $P' = u_5u_4u_3vwu_6u_7u_8$, we find that $u_2$ has no neighbours outside $\{u_3, u_6\}$ (just like $u_1$). By symmetry, $u_7$ has no neighbours outside $\{u_3, u_6\}$ (just like $u_8$). Hence, every connected component of $G - \{u_3, u_6\}$ has at most two vertices. Thus, $\{u_3, u_6\}$ is a 2-deletion set of size at most 2, and we can apply Lemma 3.    □

We use Lemma 9 to show the case $H = 2P_4 + P_3$ (proof details are omitted).

**Lemma 10.** STEINER FOREST *is polynomial-time solvable for* $(2P_4 + P_3)$-*subgraph-free graphs.*

We now prove the polynomial part of Theorem 3 by combining Lemma 5 with each of the other lemmas in this section.

# 5   Conclusions

Our aim was to increase our understanding of the complexity of STEINER FOREST and more generally, C23-problems, that is, graph problems that are not only NP-complete on subcubic graphs (C2) and under edge division of subcubic graphs

(C3), but also NP-complete for graphs of bounded treewidth (not C1). Therefore, we studied STEINER FOREST for $H$-subgraph-free graphs. We significantly narrowed the number of open cases, thereby proving a number of boundary cases (see Theorem 3). So far, we could not generalize Lemma 5 from $P_2$ to $P_3$:

**Open Problem 1.** *Let* $H$ *be a graph. Is* STEINER FOREST *polynomial-time solvable on* $(H + P_3)$-*subgraph-free graphs if it is so for* $H$-*subgraph-free graphs?*

An affirmative answer to this question would reduce the number of open cases in Theorem 3 to a finite number. However, this requires a polynomial-time algorithm for STEINER FOREST on graphs with a 2-deletion set of size $d$ for any constant $d$.

The question of whether such an algorithm exists turned out to be challenging. One important attempt that we made to solve this question was to reduce instances to highly structured instances. In particular, we were able to reduce it to the case where the vertices of the deletion set itself belong to different connected components of a minimum Steiner forest, and all vertices not in the deletion set are terminals. However, even solving such highly structured instances seems difficult. We managed to reduce it to a Constraint Satisfaction Problem (CSP). Interestingly, this CSP can be solved in polynomial time for 2-deletion sets of size 2 (cf. Lemma 3). The same CSP is NP-complete when we consider deletion sets of size 3. Unfortunately, our reduction is only one way, so this does not directly imply NP-completeness of STEINER FOREST in this case. Still, this hints that the problem might be NP-complete for $H = sP_3$ for some $s \geq 4$.

Finally, the C123 problem STEINER TREE is also classified with respect to the induced subgraph relation: it is polynomial-time solvable for $H$-free graphs if $H \subseteq_i sP_1 + P_4$ for some $s \geq 0$ and NP-complete otherwise [5]. The hardness part of this result immediately carries over to STEINER FOREST. However, we do not know the complexity of STEINER FOREST on $(sP_1 + P_4)$-free graphs.

**Acknowledgments.** We thank Daniel Lokshtanov for pointing out a possible relationship between STEINER FOREST and CSP, as discussed in Sect. 5. We also thank the anonymous reviewers of earlier versions of this paper for their helpful comments and suggestions.

# References

1. Alekseev, V.E., Korobitsyn, D.V.: Complexity of some problems on hereditary graph classes. Diskret. Mat. **4**, 34–40 (1992)
2. Arnborg, S., Lagergren, J., Seese, D.: Easy problems for tree-decomposable graphs. J. Algorithms **12**, 308–340 (1991)
3. Barefoot, C.A., Entringer, R., Swart, H.: Vulnerability in graphs - a comparative survey. J. Comb. Math. Comb. Comput. **1**, 13–22 (1987)
4. Bateni, M., Hajiaghayi, M.T., Marx, D.: Approximation schemes for steiner forest on planar graphs and graphs of bounded treewidth. J. ACM **58**, 21:1–21:37 (2011)
5. Bodlaender, H.L., Brettell, N., Johnson, M., Paesani, G., Paulusma, D., van Leeuwen, E.J.: Steiner trees for hereditary graph classes: a treewidth perspective. Theoret. Comput. Sci. **867**, 30–39 (2021)

6. Bodlaender, H.L., et al.: Subgraph Isomorphism on graph classes that exclude a substructure. Algorithmica **82**, 3566–3587 (2020)

7. Bulteau, L., Dabrowski, K.K., Köhler, N., Ordyniak, S., Paulusma, D.: An algorithmic framework for locally constrained homomorphisms. In: Bekos, M.A., Kaufmann, M. (eds.) WG 2022. LNCS, vol. 13453, pp. 114–128. Springer, Cham (2022). https://doi.org/10.1007/978-3-031-15914-5_9

8. Drange, P.G., Dregi, M.S., van 't Hof, P.: On the computational complexity of vertex integrity and component order connectivity. Algorithmica **76**, 1181–1202 (2016)

9. Dvorák, P., Eiben, E., Ganian, R., Knop, D., Ordyniak, S.: Solving integer linear programs with a small number of global variables and constraints. In: Proceedings of IJCAI 2017, pp. 607–613 (2017)

10. Feldmann, A.E., Lampis, M.: Parameterized algorithms for steiner forest in bounded width graphs. In: Proceedings of ICALP 2024. LIPIcs (2024, to appear)

11. Fujita, S., Furuya, M.: Safe number and integrity of graphs. Discret. Appl. Math. **247**, 398–406 (2018)

12. Fujita, S., MacGillivray, G., Sakuma, T.: Safe set problem on graphs. Discret. Appl. Math. **215**, 106–111 (2016)

13. Gima, T., Hanaka, T., Kiyomi, M., Kobayashi, Y., Otachi, Y.: Exploring the gap between treedepth and vertex cover through vertex integrity. Theoret. Comput. Sci. **918**, 60–76 (2022)

14. Golovach, P.A., Paulusma, D.: List coloring in the absence of two subgraphs. Discret. Appl. Math. **166**, 123–130 (2014)

15. Johnson, M., Martin, B., Oostveen, J.J., Pandey, S., Paulusma, D., Smith, S., van Leeuwen, E.J.: Complexity framework for forbidden subgraphs I: the framework. CoRR abs/2211.12887 (2022)

16. Johnson, M., Martin, B., Pandey, S., Paulusma, D., Smith, S., van Leeuwen, E.J.: Edge multiway cut and node multiway cut are hard for planar subcubic graphs. In: Proceedings of SWAT 2024. LIPIcs, vol. 294, pp. 29:1–29:17 (2024)

17. Johnson, M., Martin, B., Pandey, S., Paulusma, D., Smith, S., van Leeuwen, E.J.: Complexity framework for forbidden subgraphs III: when problems are polynomial on subcubic graphs. In: Proceedings of MFCS 2023, LIPIcs, vol. 272, pp. 57:1–57:15 (2023)

18. Kamiński, M.: Max-cut and containment relations in graphs. Theoret. Comput. Sci. **438**, 89–95 (2012)

19. Martin, B., Pandey, S., Paulusma, D., Siggers, M., Smith, S., van Leeuwen, E.J.: Complexity framework for forbidden subgraphs II: when hardness is not preserved under edge subdivision. CoRR abs/2211.14214 (2022)

# Minimizing Distances Between Vertices and Edges Through Tree $t$-Spanners

Fernanda Couto[2] , Luís Cunha[1(✉)] , Edmundo Pinto[1] ,
and Daniel Posner[2]

[1] Universidade Federal Fluminense, Niterói, Rio de Janeiro, Brazil
lfignacio@ic.uff.br, edmundo.pinto@bndes.gov.br
[2] Universidade Federal Rural do Rio de Janeiro, Nova Iguaçu, Rio de Janeiro, Brazil
{fernandavdc,posner}@ufrrj.br

**Abstract.** A tree $t$-spanner of a graph $G$ is a spanning tree $T$ of $G$ in which any two adjacent vertices of $G$ have distance at most $t$ in $T$. We say that $G$ is $t$-admissible if it admits a tree $t$-spanner, and $\sigma$ is the smallest $t$ for which $G$ is $t$-admissible. It is well-known that deciding whether $G$ has a tree $t$-spanner (the $t$-ADMISSIBILITY problem) is in P for $t \leq 2$, it is NP-complete for $t \geq 4$, and it is a long open problem to decide 3-ADMISSIBILITY. EDGE $t$-ADMISSIBILITY is a variation of the former problem, where the goal is decide whether the line graph of $G$ contains a tree $t$-spanner $T$, indicating that adjacent edges of $G$ have distance at most $t$ in $T$. It is known that EDGE $t$-ADMISSIBILITY is in P for $t \leq 3$, while it is NP-complete for $t \geq 8$. We investigate the complexity of dealing with minimizing distances at same time vertices and edges. This is the TOTAL ADMISSIBILITY problem. We prove that TOTAL $t$-ADMISSIBILITY is NP-complete, even for bipartite or planar graphs. Besides, total graphs include middle-graphs and almost-total graphs as subgraphs, which satisfy the operation we define as *clique-augmenting* of graphs $G$, denoted as $CA(G)$. We prove that deciding TREE $t$-ADMISSIBILITY for clique-augmenting graphs is NP-complete. We also prove that graphs $CA(G)$ can be classified into two types ($\sigma(CA(G)) = \sigma(G)$ or $\sigma(CA(G)) = \sigma(G)+1$) and it is NP-complete to decide whether $\sigma(CA(G)) = \sigma(G)$. Moreover, by showing special properties, we present several tractable cases to obtain $\sigma(CA(G))$.

**Keywords:** tree $t$-spanners · total graphs · clique-augmenting graphs

## 1 Introduction

Determining spanning trees is a classical problem on graphs with many applications in telecommunications networks in order to determine routing protocols of communications [1,2]. A *tree $t$-spanner* of a graph $G$ is a spanning tree $T$ of $G$ in which any two adjacent vertices of $G$ have distance at most $t$ in $T$. A graph $G$ having a tree $t$-spanner is called a *$t$-admissible* graph. The smallest $t$ for which a graph $G$ is $t$-admissible is the *stretch index of $G$* and is denoted by

$\sigma_T(G)$ (or simply $\sigma(G)$). The TREE $t$-ADMISSIBILITY problem (or $t$-admissibility for short) aims to decide whether a given graph $G$ has $\sigma(G) \leq t$. The problem of determining the stretch index has been studied by establishing bounds on $\sigma(G)$ or developing the computational complexity for several graph classes [3–5]. Cai and Corneil [3] proved that $t$-admissibility is NP-complete, for $t \geq 4$, whereas 2-admissible graphs can be recognized in polynomial-time. However, the characterization of 3-admissible graphs is still an open problem. In spite of that, many advances have been developed in order to characterize 3-admissibility for graph classes [6–8]. The characterization of 2-admissible graphs, deals with triconnected components (also considering $K_2$ and $K_3$ as triconnected components), and stated that a 2-connected graph $G$ is 2-admissible if and only if $G$ contains a spanning tree $T$ such that for each triconnected component $H$ of $G$, $T \cap H$ is a spanning star of $H$ [3].

*Graph Operations and Admissibility.* Graph operations are extensively studied when relating to $t$-admissibility in order to determine relationships between the stretch index before and after applying such operations.

Couto and Cunha [9] proved that any graph $G$ (distinct of a complete graph) can be transformed into a 4-admissible one by the union of $G$ and its complement graph $\overline{G}$, and adding a perfect matching between corresponding vertices of $G$ and $\overline{G}$. The resulted graph is the *complementary prism* of $G$, denoted as $G\overline{G}$. Deciding $t$-ADMISSIBILITY of those graphs is a restriction of deciding this problem for graphs with diameter at most $t + 1$, which is NP-complete [10] for $t \geq 4$. Because any $G\overline{G}$ graph has diameter at most 3, which is also at most 5, and deciding 4-ADMISSIBILITY can be done in polynomial time [9]. Another operation is obtaining a union of two copies of $G$ and adding a perfect matching between corresponding vertices of the two graphs $G$. The resulted graph $GG$ is the prism of $G$ (or the Cartesian product $G \times K_2$). Couto and Cunha [9] proved that for $GG$ graphs, $t$-admissibility is NP-complete, for $t \geq 4$. Gomez, Miyazawa and Wakabayashi [11] investigated spanning trees also restrict to prisms of graphs, and characterized in polynomial time those that admit a tree 3-spanner. Couto, Cunha and Posner [12] introduced EDGE TREE ADMISSIBILITY to investigate distances between edges of a graph, and proved NP-hardness for edge 8-admissibility and polynomial-time algorithm for edge 3-admissibility. Furthermore, they determined the stretch index for several graph classes, such as split graphs, its generalized graphs and graphs with few $P_4$'s.

*Total Tree Spanners.* The *distance between two edges* $e_1$ and $e_2$ of $G$, for $e_1, e_2 \in E(G)$, is the distance between their corresponding vertices in $L(G)$, for $L(G)$ being the line graph of $G$. The *total graph* of $G$ has $V(Tot(G)) = V(G) \cup E(G)$ and $E(Tot(G)) = E(G) \cup E(L(G)) \cup \{u\ uv \mid u \in V(G)\ and\ uv \in E(G)\}$. An equivalent way to define the total graph of a graph $G$ is by subdividing each edge of $G$, i.e. each edge $uv$ of $G$ is replaced by a path $u, uv, v$ where $uv$ is a new vertex of the graph, and for each pair of vertices at a distance 2 in this new graph, we add an edge between them.

The *total tree t-spanner* of a graph $G$ is a spanning tree $T$ of $\mathrm{Tot}(G)$ such that, for any two adjacent objects (vertices or edges) of $G$, their distance is at most $t$ in $T$. Hence, a total tree $t$-spanner of $G$ is a tree $t$-spanner of $\mathrm{Tot}(G)$. A graph $G$ that has a total tree $t$-spanner is called *total t-admissible*. The smallest $t$ for which $G$ is a total $t$-admissible graph is the *total stretch index of G*, denoted by $\sigma''_T(G)$ (or simply $\sigma''(G)$ or $\sigma''$). TOTAL $t$-ADMISSIBILITY aims to decide whether a graph $G$ has $\sigma''(G) \leq t$ (see Fig. 1 for an example).

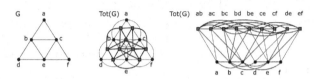

**Fig. 1.** On the left: A graph $G$; On the middle: One representation of $\mathrm{Tot}(G)$; On the right: Another representation of $\mathrm{Tot}(G)$ with the related total tree 4-spanner in red. (Color figure online)

*Contributions.* We analyze two perspectives of $t$-ADMISSIBILITY: 1) Total admissibility, generalizing studies on $t$-admissibility and edge $t$-admissibility; 2) Clique-augmenting graphs (including subgraphs of total graphs and other classes), for which we can distinguish into two types, where the stretch index changes or not.

In Sect. 2, we present a relation between the three parameters over spanning trees of a graph $G$, which are $\sigma(G), \sigma'(G)$ and $\sigma''(G)$. Section 3 is devoted to the total tree admissibility problem. We prove that $\sigma''(G)$ is upper bounded by $\sigma(G) + 2$, we prove that TOTAL TREE 18-ADMISSIBILITY is NP-complete, even for bipartite graphs or planar graphs (and adapted in Sect. 4 to prove that TOTAL TREE 17-ADMISSIBILITY is NP-complete), and since the problem is hard in general, it motivates us to present sufficient conditions for graphs to be total tree 4-admissible. As a byproduct, we determine polynomial-time solvable algorithms to determine the total stretch index for join graphs (such as cographs) and split graphs. Section 4 is devoted to clique-augmenting graphs, which generalizes the construction of middle-graphs and almost-total graphs, and we are able to relate the stretch indexes of clique-augmenting graphs and their root graphs. In Sect. 5, we present further work. Due to the space constrains, some results are presented with *sketch of proofs*, while some graph definitions and the complete proofs will appear in a full version of the paper.

Table 1 summarizes the known computational complexity results and the ones we have established in this work.

**Table 1.** Results of columns $G$ and $L(G)$ obtained in [3] and [12], resp. Other columns correspond to results obtained in this work. $Tot(G), M(G), AT(G)$ and $CA(G)$ are the total, middle, almost-total and clique-augmenting graphs of $G$, resp.

| $t$-admissibility | $G$ | $L(G)$ | $Tot(G)$ | $M(G)$ | $AT(G)$ | $CA(G)$ |
|---|---|---|---|---|---|---|
| Polynomial | $\leq 2$ | $\leq 3$ | $\leq 2$ | $\leq 2$ | $\leq 3$ | $\leq 2$ |
| NP-complete | $\geq 4$ | $\geq 8$ | $\geq 17$ | $\geq 17$ | $\geq 5$ | $\geq 5$ |

## 2    Bounds on Edge and Total Admissibility

When dealing with $\sigma(G), \sigma'(G)$ and $\sigma''(G)$, a natural attempt is to determine a relation between them. Since induced cycles in a graph $G$ correspond to cycles of the same length in $L(G)$, then $\sigma'(C_n) = \sigma(C_n) = n - 1$. Considering now $Tot(C_n)$, this graph is isomorphic to the 2-*cycle-power graph* $C_{2n}^2$, which is obtained from a $C_{2n}$ by adding edges between all vertices with distance at most 2 in $C_{2n}$. Couto and Cunha [7] determined the stretch index for any $p$-*cycle-power graph* [7], $\sigma(C_{2n}^2) = n$, and so, $\sigma''(C_n) = n$.

Trees are the graphs that have stretch index equal to 1, and the edge 1-admissible graphs are the ones whose their line graphs are trees. Since line graphs are claw-free, path graphs are the edge 1-admissible graphs. It is known that considering trees, if $G$ is a path graph, $\sigma'(G) = 1$, otherwise $\sigma'(G) = 2$ [12].

A total graph $Tot(G)$ is isomorphic to a tree if and only if $G$ is a trivial graph, because any edge $uv$ of $G$ forms a $C_3$ in $Tot(G)$ by the vertices $u, v, uv$, where $u, v \in V(G)$ and $uv \in L(G)$. Hence, we have that $K_1$ is the unique total 1-admissible graph. Given an arbitrary tree, we have Proposition 1.

**Proposition 1.** *If $G$ is a tree distinct of a trivial graph, then $\sigma''(G) = 2$.*

We now describe graphs $G$ where $\sigma$ is as far as we want from $\sigma'$ and $\sigma''$. It will imply that parameters $\sigma'$ and $\sigma''$ cannot bound $\sigma$ for a constant factor.

A graph $G$ is a $C_n$-*multi-cycle graph* if $G$ is obtained from the union of a cycle $C_n$ with vertices $\{v_1, \ldots, v_n\}$ and $n$ disjoint paths $P_{n-2}$ with vertices $\{p_{1,1}, \ldots, p_{1,n-2}\}$, $\{p_{2,1}, \ldots, p_{2,n-2}\}$, $\ldots$, $\{p_{n,1}, \ldots, p_{n,n-2}\}$, and adding the edges $\{p_{1,1}, v_1\}$, $\{p_{1,n}, v_2\}, \{p_{2,1}, v_2\}$, $\{p_{2,n}, v_3\}, \ldots, \{p_{n,1}, v_n\}, \{p_{n,n}, v_1\}$. Figure 2(a) depicts a multi-cycle graph $G$ obtained from a $C_6$.

In order to obtain a tree-spanner $T$ of $G$ (described in Fig. 2(b)) we must remove one edge per each cycle of $G$ in such a way that the resulted graph is connected. Optimum tree-spanners of $C_n$-multi-cycle graphs are obtained by removing one edge between degree-two vertices for each of the outer cycles and one of the edges of the inner cycle. Such removal implies two adjacent vertices in $G$ at a distance $2n - 3$ in $T$, and then $\sigma(G) = 2n - 3$.

We obtain an edge tree-spanner of $G$ with $\sigma'(G) = n + 1$ by connecting the edges of the inner cycle and using these edges to connect the edges of the outer cycles incident to them and the others edges of the outer cycles may be connected in anyway (described in Fig. 2(c)). The maximum distance between two edge-vertices which share a vertex in $G$ is $n + 1$.

Finally, we obtain a total tree-spanner of $G$ with $\sigma''(G) = n+2$ by including to the edge tree-spanner of $G$ with $\sigma'(G) = n+1$ the vertices of $G$ such that each vertex is adjacent to one of its endpoints in $G$, as described in Fig. 2(d). By increasing the value of the cycles of the multi-cycle graph class, we obtain graphs with $\sigma(G)$ far greater than $\sigma'(G)$ and $\sigma''(G)$.

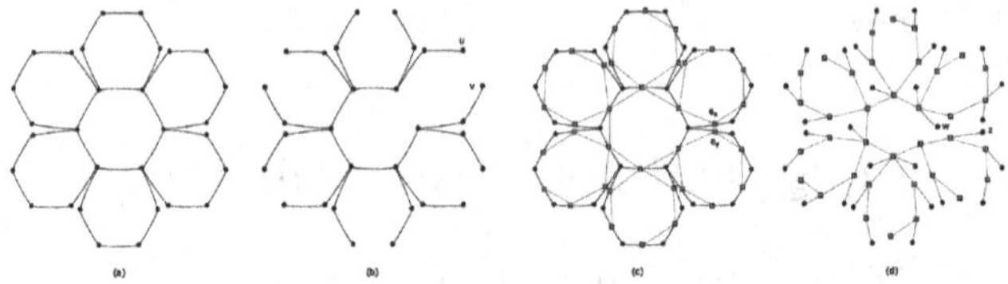

**Fig. 2.** Graph with $\sigma'$ and $\sigma''$ less than $\sigma$. (a) Graph using $C_6$'s. (b) tree-spanner with $\sigma = 9$. (c) edge tree-spanner with $\sigma' = 7$. (d) total tree-spanner with $\sigma'' = 8$.

## 3   Total Admissibility

First, in Theorem 1, we present an upper bound of $\sigma''$ in function of $\sigma(G)$, by constructing a total tree $t''$-spanner of $G$ with $t'' \leq \sigma(G) + 2$.

**Theorem 1.** *Given a graph $G$, $\sigma''(G) \leq \sigma(G) + 2$.*

*Sketch of Proof.* We consider that the solution tree $T''$ of $Tot(G)$ has the solution tree $T$ of $G$ as a subgraph. Considering the root of $T$ as the root of $T''$, we orient all the edges of $G$ from bottom to top when one vertex is at a lower depth than another and from left to right when it is at the same depth and we connect the vertices of $L(G)$ to head of oriented edges. This way, we have a solution $T''$ with the edges of $T$ plus the edges that connect the vertices of $L(G)$. As we take advantage of the edges of $T$, therefore the distance between vertices of $G$ is bounded by $\sigma(G)$. The distance between vertices of $G$ and $L(G)$ is bounded by $\sigma(G) + 1$, since a vertex $uv$ of $L(G)$ is connected to a vertex $u$ of $G$ that is at maximum $\sigma$ distance from another vertex $v$ adjacent to the vertex $uv$ of $L(G)$ in $Tot(G)$. Finally, the distances between two vertices of $L(G)$ are bounded to $\sigma(G) + 2$, because, due to the orientation of the edges, we were able to show that the vertices of $G$ connected respectively to 2 adjacent vertices of $L(G)$ in $Tot(G)$ has a distance bounded to $\sigma(G)$, so the distance between any 2 vertices of $L(G)$ is bounded by $\sigma + 2$.                                                    $\square$

### 3.1   Total $k$-Admissibility is **NP-Complete**

Next we present a polynomial-time transformation from the determination of a tree 4-spanner for any arbitrary graph $G$ [3] to the determination of a total tree 18-spanner for a transformed graph $H$ from $G$.

**Construction 31.** *Let $H$ be a graph obtained from a graph $G = (V, E)$ by replacing each edge $uv \in E(G)$ by two paths $u, c_{uv1}, c_{vu1}, v$ and $u, c_{uv2}, c_{uv3}, c_{uv4}, c_{uv5}, c_{uv6}, c_{uv7}, c_{muv}, c_{vu7}, c_{vu6}, c_{vu5}c_{vu4}, c_{vu3}, c_{vu2}, v.$*

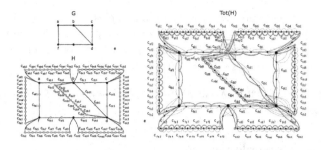

**Fig. 3.** (a) Graph $G$ and its tree 4-spanner in green. (b) Graph $H$ obtained from Construction 31 on $G$. (c) Total graph $Tot(H)$ and its solution tree 18-spanner (green marked edges) based on the tree 4-spanner of $G$ proved in Theorem 2. (Color figure online)

Figure 3 depicts a transformation of a graph $G$ into a graph $H$ described in Construction 31.

The key idea of the next theorem is that by replacing each edge of $G$ by a cycle of size 17 in $H$, we force each of these cycles to be connected as a path on a total tree 18-spanner of $H$. Afterwards, we verify that the remaining connections among those paths to construct the total tree 18-spanner of $H$ are directly related to the connections on a tree 4-spanner of $G$.

**Theorem 2.** TOTAL TREE 18-ADMISSIBILITY *is* NP-*complete.*

*Sketch of Proof.* We prove the claim: A graph $G$ has a tree 4-spanner if and only if the graph $H$ obtained from $G$ in Construction 31 has a total tree 18-spanner.

Having a tree 4-spanner $T$ of $G$, we build a tree $T''$ 18-spanner for $Tot(H)$ by the following way: each edge of $T$ corresponds to a 3-path in $T''$ associated to the power of cycle $C_{17}^2$, which is a gadget for each edge of $G$ and is 17-admissible; each non-edge of $T$ correspond to a 4-path that implies the existence of two adjacent vertices of $Tot(H)$ whose distance in $T''$ is at most 18, since a 4-path in $T$ implies in a 12-path in $T''$, plus 4 edges associated to a non-edge of $T'$. Moreover, we prove that is necessary to use at most more 2 edges $T''$, implying in a 18-path whose extremities are adjacent vertices of $Tot(H)$.

Now, given a tree 18-spanner of $Tot(H)$, we analyze the cases that guarantee a 18-path. It would be obtained by six 3-paths, corresponding to a tree 6-spanner of $G$. It is not possible, because we must use more than 18 edges in $Tot(H)$ to conclude a path. Similarly, it is not possible if such a 18-path was obtained by five 3-paths plus 3 edges, either. Thus, the unique way to have a 18-path in $Tot(H)$ is by four 3-paths plus 6 edges, implying $G$ be 4-admissible. □

Theorem 2 can be extended for any value of $\sigma(G) = k \geq 4$ and $\sigma''(G) = 15 + (k-3)3$ using cycles of size $14 + (k-3)3$. I.e, $\sigma(G) = 4$ and $\sigma''(G) = 18$, $\sigma(G) = 5$ and $\sigma''(G) = 21$, and so on.

It is NP-complete do decide whether $\sigma(G) = 5$ for bipartite graphs [6]. From graph $H$ obtained by Construction 31, replace each edge by cycles of size 20 and multiply by three the size of the original cycles of the bipartite graphs, it results on a bipartite graph. Besides, it is NP-complete to determine $\sigma(G)$ for planar graphs [8]. Furthermore, the transformation of Construction 31, applied to a planar graph $G$, results a planar graph $H$ as well.

**Corollary 1.** TOTAL TREE-ADMISSIBILITY *is* NP-*complete for bipartite graphs and for planar graphs.*

### 3.2   Total 4-Admissible Graphs

We have just proved that deciding TOTAL $t$-ADMISSIBILITY is a difficult task. Next, we present sufficient conditions for graphs to be total 4-admissible, which allow us to present polynomial-time algorithms to determine $\sigma''(G)$ for join graphs and split graphs.

**Lemma 1.** *If a graph $G$ has a vertex $v$ (or an edge $uv$) such that each edge of $G$ (each vertex of $G$) has a vertex in $N(v)$ as one of its endpoints (is adjacent to $u$ or to $v$), then $\sigma''(G) \leq 4$.*

The property of graphs being $t$-admissible is not hereditary, that is, a graph $G$ can be $t$-admissible, while an induced subgraph of $G$ can have an total stretch index greater than $t$. Despite of that, Lemma 2 states that if $G$ has a subgraph with a total stretch index equal to 4 then $\sigma''(G) \geq 4$.

**Lemma 2.** *If $G$ has a subgraph $H$ with $\sigma''(H) = 4$ then $\sigma''(G) \geq 4$.*

Complete graph $K_3$ with vertices $a, b, c$ has $\sigma''(K_3) = 3$ with total tree 3-spanning $T''$ where the root vertex is the edge-vertex $uv$, which is adjacent to all edge-vertices and vertices $a$ and $b$ in $T''$, and the remaining vertex $c$ is adjacent to the vertex-edge $ac$. For $n \geq 4$, we have the following Lemma 3.

**Lemma 3.** *The complete graph $K_n$ has $\sigma'' = 4$, for $n \geq 4$.*

**Theorem 3.** *There is a polynomial-time algorithm to determine $\sigma''(G)$ for split graphs $G = (S, K, E)$, where $S$ is an independent set and $K$ is a clique with maximum size.*

**Theorem 4.** *For a join graph $G$, $\sigma''(G)$ can be determined in polynomial-time.*

*Sketch of Proof.* By Lemma 1, $\sigma''(G) \leq 4$. We present characterizations of join graphs with stretch index equal $t$, for $t \in \{1, \cdots, 4\}$, yielding in a polynomial-time algorithm to determine $\sigma''(G)$.    □

## 4  Clique-Augmenting Graphs

Hamada and Yoshimura [13] defined the middle-graph. The *middle-graph* $M(G)$ of a graph $G$ is defined as follows. The vertex set $V(M(G))$ is $V(G) \cup E(G)$. Two vertices $v, w \in V(M(G))$ are adjacent in $M(G)$ if (i) $v, w \in E(G)$ and $v$ and $w$ associate to edges incident to a same vertex in $G$ or $v \in V(G)$ and $w \in E(G)$ and $w$ corresponds to an edge is incident to $v$ in $G$. The *almost-total graph* $AT(G)$ of a graph $G$ is defined as follows: The vertex set $V(AT(G))$ is $V(G) \cup E(G)$. Two vertices $v, w \in V(AT(G))$ are adjacent in $AT(G)$ if they are adjacent in $G$ or $v \in V(G)$, $w \in E(G)$ and $w$ corresponds to an edge is incident to $v$ in $G$. Another way to construct a middle-graph (an almost-total) of $G$ is from $Tot(G)$ we remove the edges of $G$ (of $L(G)$).

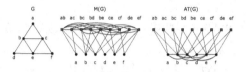

**Fig. 4.** On the left: graph $G$; on the middle: middle-graph of $G$, M($G$); on the right: almost-total graph of $G$, AT($G$).

Middle-graphs and almost-total graphs (as depicted in Fig. 4) are subgraphs of total graphs. Therefore, our following approach is to consider the admissibility for those graphs, once the property of being $t$-admissibility is not hereditary (i.e. subgraphs of a $t$-admissible graphs may not be $t$-admissible). When focusing on middle-graphs, there are graphs $G$ in which $\sigma(M(G)) = \sigma(L(G))$ and where $\sigma(M(G)) = \sigma(Tot(G))$.

Considering $G$ isomorphic to cycle graphs $C_n$, it is clear that $\sigma(G) = n - 1$, $\sigma'(G) = n - 1$, $\sigma''(G) = n$, and $\sigma(M(G)) = \sigma''(G) = n$. While considering $G$ isomorphic to $C_n$-multi-cycle graphs, we have that $\sigma(G) = 2n - 3, \sigma'(G) = n + 1, \sigma''(G) = n + 2$, and $\sigma(M(G)) = \sigma'(G) = n + 1$.

Hence, since $\sigma$ for middle-graphs (and for almost-total) is not always equal to $\sigma'$ nor to $\sigma''$, next we deal with those cases, which both of them can be viewed as subclasses of clique-augmenting graphs, defined below.

**Definition 1.** *A* clique-augmenting *of a graph* $G$ *is a graph* $H = CA(G)$ *obtained from* $G$ *and: i) choosing an arbitrary set of cliques of* $G$. *Let* $X_1, X_2, \ldots, X_k$ *be such cliques; ii) for each selected clique* $X_i$, *add a new vertex* $u_{X_i}$ *to the graph and all edges between* $u_{X_i}$ *and the vertices of* $X_i$, *i.e.* $N_H(u_{X_i}) = X_i$. *Hence,* $V(H) = V(G) \cup \{u_{X_1}, u_{X_2}, \ldots, u_{X_k}\}$ *and* $E(H) = E(G) \cup \{u_{X_i} v : v \in X_i, i = 1, 2, \ldots, k\}$.

**Lemma 4.** *Middle-graphs and almost-total-graphs are clique-augmenting graphs.*

As a direct consequence of Definition 1 we have Theorem 5.

**Theorem 5.** *A graph $G$ is a clique-augmenting graph if and only if $G$ has a simplicial vertex.*

Since we can check the existence of a simplicial vertex in polynomial-time [14], Theorem 5 yields in a polynomial-time characterization of clique-augmenting graphs. Furthermore, any chordal graph is a clique-augmenting graph, and clearly the converse is not true. In general, we have what follows:

**Theorem 6.** *Given a graph $G$, $\sigma(G) \leq \sigma(CA(G)) \leq \sigma(G) + 1$.*

In the following, we investigate clique-augmenting graphs and analyze them by the stretch index whether it is changed or not comparing to the stretch index of the original graph, the two types of Theorem 6. We analyse when cliques chosen of Definition 1 have size 2, 3 or arbitrary based on some constraints.

### 4.1   Clique-Augmenting Graphs on Cliques of Size 2

**Lemma 5.** *Let $H$ be any graph obtained in Definition 1 from a graph $G$, if $H$ is obtained by adding a vertex to all edges of $G$, then $\sigma(H) = \sigma(G) + 1$.*

By Lemma 5, if $H$ is obtained from $G$ of Definition 1 by adding a vertex to all edges of $G$, then $\sigma(H) = \sigma(G) + 1$. Since deciding $\sigma(G) \geq 4$ is NP-complete, then it follows that deciding $\sigma(CA(G)) \geq 5$ is NP-complete as well.

**Theorem 7.** *It is NP-complete to decide whether $\sigma(CA(G)) = \sigma(G)$.*

Note that almost-total-graphs can be obtained by adding a new vertex $w$ for each edge $uv$ and make it adjacent to both endpoints $u$ and $v$. From Lemma 5, this implies that $\sigma(AT(G)) = \sigma(G) + 1$. Moreover, 2-admissibility and 4-admissibility are polynomial-time solvable and NP-complete problems, resp.

**Corollary 2.** *Deciding the stretch index for almost-total graphs is in* P, *for $t \leq 3$ and* NP-*complete, for $t \geq 5$.*

One possible approach to settle the complexity $t$-admissibility for middle-graphs would be to deal with when $M(G) \approx AT(G')$, i.e. they are isomorphic.

**Theorem 8.** *Given graphs $G$ and $G'$, if $M(G) \approx AT(G')$ then $L(G)$ is $K_3$-free.*

From Theorem 8, $CA(L(G)) \approx M(G) \approx AT(L(G))$ so $L(G)$ is $K_3$-free. For $L(G)$ to be $K_3$-free, $G$ is a path or a cycle $C_n, n \geq 4$. Therefore, $\sigma(L(G))$ can be determined in polynomial time when $L(G)$ is $K_3$-free, because $\sigma'(C_n) = n - 1$ and $\sigma'(P_n) = 1$. This implies that $\sigma(AT(L(G))) = n$ or 2 for $L(G)$ $K_3$-free. If $G$ is a path, then $M(G) \not\approx AT(G')$, since $L(G)$ would also be a path and the construction of $M(G)$ adds degree vertices 2 adjacent to the endpoints of each edge of $L(G)$ plus 2 vertices of degree 1 at the beginning and end of the path of $L(G)$. Therefore, for $M(G) \approx AT(G')$, $G$ must be a cycle $C_n, n \geq 4$.

Determining $\sigma(AT(G))$ is NP-complete for arbitrary $G$ (Corollary 2). In [12] it is proved that the deciding $\sigma(L(G))$ is NP-complete even if $G$ is bipartite.

However, we conclude that $\sigma(L(G))$ can be determined in polynomial time when $L(G)$ is bipartite, since bipartite $L(G)$ implies that it is $K_{1,3}$-free (because it is a line graph and line graphs are graphs without claws) and $K_3$-free (because it is a bipartite graph, so it does not have odd cycles), therefore, these graphs have $\Delta \leq 2$. Therefore, we have the following Theorem 9.

**Theorem 9.** *Given a graph $G$, $\sigma(L(G))$ can be determined in polynomial time for $L(G)$ a bipartite graph.*

The complexity of deciding $\sigma(M(G))$ is settled in Sect. 4.3. From Lemma 4, determining stretch index for clique-augmenting graphs is NP-complete even if we add a new vertex adjacent to both endpoints of each edge of $G$. Now, we are interested in determining stretch index for clique-augmenting graphs even when there are edges without having new added vertices.

Note that instead of adding a new vertex adjacent to all edges of $G$, we add a new vertex adjacent to $u$ and $v$ in such a way that $d_T(u,v) = \sigma(G)$, for a solution tree $T$ of $G$, this is not sufficient to guarantee $\sigma(CA(G)) = \sigma(G)+1$. For example, if $G$ is a $C_4$ and we add a new vertex $w$ to only an edge $uv$ where $d_T(u,v) = \sigma(G) = 3$, we still keep $\sigma(CA(G)) = 3$.

Next, we deal with some of these cases where we do not add vertex for all edges and the obtained graph changes or not the stretch index.

**Proposition 2.** *Let $uv$ be an edge of $G$. If $d_T(u,v) < \sigma(G)$, for a solution $T$ of $G$, and $CA(G)$ is obtained from $G$ by adding a vertex $w$ such that $uw, vw \in E(G)$, then $\sigma(CA(G)) = \sigma(G)$.*

**Theorem 10.** *Let $H$ be obtained from $G$ in Definition 1 by adding a new vertex adjacent to all pairs of adjacent vertices of $G$ that belong to all paths of size $\sigma(G)$ of any solution tree of $G$ then $\sigma(H) = \sigma(G)+1$.*

Since a cycle graph of size $n$ has $n$ different solution trees by removing a unique edge of $G$ per tree, then all pair of adjacent vertices of $G$ belong to a path of size $\sigma(G)$ of any solution tree. Thus, as a consequence Theorem 10, if $G$ is a cycle it is necessary to add a vertex adjacent to all pairs of adjacent vertices of $G$ to construct a graph $H$ satisfying $\sigma(H) = \sigma(G)+1$.

A natural question is: is it possible to add vertices not corresponding to all edges of the graph and even though the stretch index increases? The answer is yes. Note that if we have a graph $G$ that is formed by a cycle $C_n$ and a pendant vertex (i.e. a vertex of degree 1) adjacent to a vertex of the cycle, we can add $n$ new vertices adjacent to each one of the $n$ edges of the $C_n$ in order to construct a graph $H$ with $\sigma(H) = n$, which is greater than $\sigma(G) = n-1$.

This construction satisfies Theorem 10 since the unique edge of $G$ we do not add a new vertex does not belong to any solution $\sigma(G)$-path of $G$. We can generalize the idea of increasing the stretch factor even without adding vertices with respect to all edges of the graph, as proved in Corollary 3.

**Corollary 3.** *Let $H = CA(G)$ be obtained by adding a vertex adjacent to each pair of adjacent vertices of $G$, except bridges, then $\sigma(H) = \sigma(G)+1$.*

In Corollary 3 we proved that it is not necessary to add vertex with respect to all edges of $G$ to increase the stretch index, but those are bridges. Hence, another question arises: is it possible to add vertices not corresponding to all edges of the graph and with respect to the edges we do not add new vertices there are edges belonging to cycles (these edges are not bridges) and even though the stretch index increases? The answer is still **yes**. For example, consider a *triangular-based prism graph* $G$ (graph obtained by the union of two triangles, each triangle is a base, and edges between any two correspondent vertices of the bases), that has $\sigma(G) = 3$, it is sufficient to add a vertex adjacent to each $u$ and $u'$, where $u$ and $u'$ are correspondent vertices of the two triangular bases, obtaining graph $H$. There are no bridges in this graph and the stretch index increases to $\sigma(H) = 4$. Moreover, we can increase the stretch index even by instead of adding those 3 new vertices, adding only 2 of such vertices.

One could ask whether for all prism graphs the stretch index increases when adding a vertex for each edge $uu'$, where $u$ and $u'$ are correspondent vertices of the two bases obtaining the resulted graph $H$, similar to the base case where the bases are triangles. However, the answer is **no**. Note that even considering each base as a $C_n$, for $n \geq 4$ in a prism graph $G$, this property of increasing the stretch index does not hold, because $\sigma(G) \geq 4$ and after the vertices addition as just mentioned, we form distances at most 4 in a solution tree of $H$.

## 4.2   Clique-Augmenting Planar Graphs on Cliques of Size 3

Fekete and Kremer [8] proved that determining the stretch index for planar graphs is NP-complete. A seminal result in [15] says that a graph $G$ has a planar line graph $L(G)$ if and only if $G$ is planar of maximum degree $\Delta(G) = 4$ and every vertex of degree 4 is a cut vertex. Based on that our interest is to investigate clique-augmenting graphs of planar graphs which result in planar graphs as well. Since to determine stretch indexes we can restrict the problem on biconnected graphs (i.e. without a cut vertex), once $\sigma(G) = \max_{B_i \in \mathcal{B}(G)}\{\sigma(G[B_i])\}$, for $\mathcal{B}(G)$ is the set of all maximum biconnected components of $G$, we can assume that $G$ is planar and $\Delta(G) = 3$ in order to consider the case where $L(G)$ is planar.

**Proposition 3.** *Every planar line graph $L(G)$ of a biconnected planar graph $G$ has $\Delta(L(G)) = 4$.*

Each $v \in V(G)$ with $d_G(v) = 3$ yields a $K_3$ in $L(G)$, so we have Proposition 4.

**Proposition 4.** *Every planar line graph $L(G)$ of a biconnected planar graph $G$ has the size a maximum clique $\omega(L(G)) \leq 3$.*

Based on Proposition 3 and Proposition 4, we obtain a clique-augmenting graph $CA(L(G))$ of a planar line graph $L(G)$ in the following way: Choose an arbitrary number of $K_3$'s in $L(G)$; Add a new vertex adjacent to each $K_3$ we have chosen in $L(G)$. As a direct consequence of this last construction, we have that $H = CA(L(G))$ is a planar graph. Moreover, we have Proposition 5.

**Proposition 5.** *The graph $H$ obtained from a planar line graph $L(G)$ by adding new vertices adjacent to each $K_3$ has $\Delta(H) = 8$.*

As consequence of Proposition 5, we ask about the stretch index of planar graphs with $\Delta = 8$. In order to solve this question we investigate two perspectives: What is known about the stretch index of planar graphs? What is known about the stretch index of bounded degree graphs?

For the first question, if a planar graph $G$ has bounded face degree, then $\sigma(G)$ can be determined in polynomial-time (Fekete and Kremer [8]). Since if $G$ has bounded face degree, its dual graph has bounded degree, and every planar graph is dual graph of some other graph (specifically, every planar graph is dual of its dual graph), it implies that $H = CA(L(G))$ satisfies Fekete and Kremer's condition. For the second one, the stretch index of graphs with bounded $\Delta$ can be determined in polynomial-time (Papoustsakis [10]). Thus, we have Theorem 11.

**Theorem 11.** *Given a graph $H = CA(L(G))$ obtained from a line planar graph $L(G)$ by adding new vertices adjacent to each $K_3$, we can determine $\sigma(H)$ in polynomial-time.*

### 4.3   Clique-Augmenting Line Graphs on Cliques of Arbitrary Sizes

From the graphs obtained by Construction 31, in a similar argument as the one given in Theorem 2, it can be proved that $G$ has a tree 4-spanner if and only if the middle-graph of $H$, $M(H)$, where $H$ is obtained from $G$ in Construction 31, has a tree 17-spanner. Hence, Theorem 12 follows straightforward.

**Theorem 12.** 17-ADMISSIBILITY *is* NP-*complete for middle-graphs.*

The construction obtained in Theorem 12 can also be adapted to obtain a reduction for TOTAL-17-ADMISSIBILITY, determined in Corollary 4.

**Corollary 4.** TOTAL TREE-17-ADMISSIBILITY *is* NP-*complete.*

Recall that $G$ is a line graph if its edges can be partitioned into maximal complete subgraphs such that no vertex lies in more than two of the subgraphs [16].

**Lemma 6.** *Let $G'$ be a line graph $L(G)$ with edge partition into maximal complete subgraphs. If $H = CA(G')$ is obtained from $G'$ by adding a new vertex with respect to any of such maximal clique of $G'$, then $H$ is a line graph.*

Analogous to Lemma 5, we have Lemma 7, when a line graph $G'$ is biconnected and distinct of a complete graph.

**Lemma 7.** *Let $G' = L(G)$ be a biconnected line graph distinct of a complete graph and let $C$ be an edge partition of $G'$ into maximal cliques. The graph $H$ obtained from $G'$ by adding each vertex adjacent to each maximal clique of $C$ has $\sigma(H) = \sigma(G') + 1$.*

Since it is known that deciding $\sigma'(G) = \sigma(L(G))$ is NP-complete for $t = 8$ [12], from Lemma 6 and Lemma 7 we have Theorem 13.

**Theorem 13.** 9-ADMISSIBILITY *for graphs H of Lemma 7 is* NP-*complete.*

Couto, Cunha and Posner [12] also showed that EDGE 3-ADMISSIBILITY is polynomial-time solvable. Hence, we have Theorem 14.

**Theorem 14.** *Deciding $\sigma(H) = 4$ and $\sigma(H) = 3$ is a polynomial-time solvable problem for H being graphs of Lemma 7.*

## 5    Further Work

Further investigation will be on characterizing when $\sigma(CA(G)) = \sigma(G)$, continuing the sufficient and necessary conditions we have presented. Besides, another approach will be on dealing with 3-ADMISSIBILITY of clique-augmenting graphs. So far, a great number of $CA(G)$ graphs that satisfy $\sigma \leq 3$ can be determined: If $\sigma(CA(G)) \in \{1, 2, 3\}$, then the polynomial-time algorithm for 2-ADMISSIBILITY is sufficient to determine $\sigma(CA(G)) = 3$ when $G$ is 2-admissible. Note that if $\sigma(G) = 2$, then $\sigma(CA(G)) \in \{2, 3\}$, thus when checking 2-ADMISSIBILITY for $CA(G)$, it determines $\sigma(CA(G)$ being equal to 2 if the answer is yes, or being equal to 3, otherwise. We left as open question to determine 3-ADMISSIBILITY for $CA(G))$ whose $\sigma(G) = 3$. This will indicate us to fully characterize chordal 3-admissible graphs, since these are clique-augmenting graphs.

## References

1. Bhatt, S., Chung, F., Leighton, T., Rosenberg, A.: Optimal simulations of tree machines. In: Foundations of Computer Science, pp. 274–282. IEEE (1986)
2. Peleg, D., Ullman, J.D.: An optimal synchronizer for the hypercube. SIAM J. Comput. **18**(4), 740–747 (1989)
3. Cai, L., Corneil, D.G.: Tree spanners. SIAM J. Discrete Math. **8**(3), 359–387 (1995)
4. Couto, F., Cunha, L.F.I.: Tree $t$-spanners of a graph: minimizing maximum distances efficiently. In: Kim, D., Uma, R.N., Zelikovsky, A. (eds.) COCOA 2018. LNCS, vol. 11346, pp. 46–61. Springer, Cham (2018). https://doi.org/10.1007/978-3-030-04651-4_4
5. Couto, F., Cunha, L.F.I.: Hardness and efficiency on minimizing maximum distances for graphs with few P4's and (k, ℓ)-graphs. Electron. Notes Theor. Comput. Sci. **346**, 355–367 (2019)
6. Brandstadt, A., Dragan, F.F., Le, H.-O., Uehara, R., et al.: Tree spanners for bipartite graphs and probe interval graphs. Algorithmica **47**(1), 27–51 (2007)
7. Couto, F., Cunha, L.F.I.: Hardness and efficiency on minimizing maximum distances in spanning trees. Theor. Comput. Sci. **838**, 168–179 (2020)
8. Fekete, S.P., Kremer, J.: Tree spanners in planar graphs. Discrete Appl. Math. **108**(1–2), 85–103 (2001)
9. Couto, F., Cunha, L.F.I.: Hardness and efficiency on t-admissibility for graph operations. Discrete Appl. Math. **304**, 342–348 (2021)

10. Papoutsakis, I.: Tree spanners of bounded degree graphs. Discrete Appl. Math. **236**, 395–407 (2018)
11. Gómez, R., Miyazawa, F.K., Wakabayashi, Y.: Tree 3-spanners on generalized prisms of graphs. In: Castañeda, A., Rodríguez-Henríquez, F. (eds.) LATIN 2022. LNCS, vol. 13568, pp. 557–573. Springer, Cham (2022). https://doi.org/10.1007/978-3-031-20624-5_34
12. Couto, F., Cunha, L., Posner, D.: Edge tree spanners. In: Gentile, C., Stecca, G., Ventura, P. (eds.) Graphs and Combinatorial Optimization: from Theory to Applications. ASS, vol. 5, pp. 195–207. Springer, Cham (2021). https://doi.org/10.1007/978-3-030-63072-0_16
13. Hamada, T., Yoshimura, I.: Traversability and connectivity of the middle graph of a graph. Discret. Math. **14**(3), 247–255 (1976)
14. Golumbic, M.C.: Algorithmic Graph Theory and Perfect Graphs. Elsevier, Amsterdam (2004)
15. Sedláček, J.: Some properties of interchange graphs (in theory of graphs and its applications, m. fiedler, editor), pp. 145–150 (1964)
16. Lehot, P.G.: An optimal algorithm to detect a line graph and output its root graph. J. ACM (JACM) **21**(4), 569–575 (1974)

# Enumerating Minimal Vertex Covers and Dominating Sets with Capacity and/or Connectivity Constraints

Yasuaki Kobayashi[1] , Kazuhiro Kurita[2(✉)] , Yasuko Matsui[3] ,
and Hirotaka Ono[2]

[1] Hokkaido University, Sapporo, Japan
koba@ist.hokudai.ac.jp
[2] Nagoya University, Nagoya, Japan
kurita@i.nagoya-u.ac.jp, ono@nagoya-u.jp
[3] Tokai University, Hiratsuka, Japan
yasuko@tokai.ac.jp

**Abstract.** In this paper, we consider the problems of enumerating minimal vertex covers and minimal dominating sets with capacity and/or connectivity constraints. We develop polynomial-delay enumeration algorithms for these problems on bounded-degree graphs. For the case of minimal connected vertex covers, our algorithm runs in polynomial delay even on the class of $d$-claw free graphs, which extends the result on bounded-degree graphs. To complement these algorithmic results, we show that the problems of enumerating minimal connected vertex covers and minimal capacitated vertex covers in bipartite graphs are at least as hard as enumerating minimal transversals in hypergraphs.

**Keywords:** Enumeration algorithms · Connected vertex cover · Connected dominating set · Capacitated vertex cover · Capacitated dominating set

## 1 Introduction

Enumerating minimal or maximal vertex subsets satisfying some graph properties has been widely studied for decades since it has various applications in many fields. For example, the problem of enumerating maximal cliques in graphs is an essential task in data mining, which is intensively studied from both theoretical and practical perspectives [6,9,10,20,31,33,34]. This enumeration problem is equivalent to that of enumerating minimal vertex covers as there is a one-to-one correspondence between the collection of all maximal cliques in $G$ and that of all minimal vertex covers of its complement $\overline{G}$.

Partially supported by JSPS KAKENHI Grant Numbers JP20H00595, JP20K04973, JP20H05964, JP20H05967, JP21K17812, JP21K19765, JP22H03549, JP22H00513, and JP23H03344 and JST ACT-X Grant Number JPMJAX2105.

A. A. Rescigno and U. Vaccaro (Eds.): IWOCA 2024, LNCS 14764, pp. 232–246, 2024.
https://doi.org/10.1007/978-3-031-63021-7_18

Another important enumeration problem is to enumerate minimal dominating sets in graphs. This enumeration problem is known to be "equivalent"[1] to dualizing monotone Boolean functions or enumerating minimal transversals in hypergraphs [25], which has many applications in a broad range of fields in computer science [3,13,14]. (See [16] for a survey).

As for the complexity of these enumeration problems, the current status of the problems of enumerating maximal cliques and enumerating minimal dominating sets of graphs is significantly different. By a classical result of Tsukiyama et al. [34], we can enumerate all maximal cliques (and hence minimal vertex covers) of a graph in polynomial delay. Here, an enumeration algorithm runs in *polynomial delay* if the time elapsed[2] between every pair of two consecutive outputs is upper bounded by a polynomial solely in the input size. In contrast to this, no output-polynomial time algorithm for enumerating minimal dominating sets of a graph is known so far, where an enumeration algorithm runs in *output-polynomial time* if the total running time is upper bounded by a polynomial in the combined size of the input and all outputs[3]. There are several results showing polynomial-delay or output-polynomial time algorithms for enumerating minimal dominating sets in several special classes of graphs [4,11,23,26,27,29]. The fastest known algorithm for general graphs is due to Fredman and Khachiyan [21], which runs in time $N^{o(\log N)}$, where $N$ is the number of vertices and hyperedges plus the number of minimal dominating sets of an input graph.

As variants of these two enumeration problems, we study the problems of enumerating minimal *connected* vertex covers and minimal *connected* dominating sets of graphs, which we call MINIMAL CONNECTED VERTEX COVER ENUMERATION and MINIMAL CONNECTED DOMINATING SET ENUMERATION, respectively. Although these two combinatorial objects are considered to be natural (as they are well studied in several areas [2,12,17,19,24,35], including input-sensitive enumeration algorithms [1,22]), enumeration algorithms (with an output-sensitive analysis [18]) for these variants are not much investigated in the literature. Kanté et al. [25] showed that the problem of enumerating minimal connected dominating sets in split graphs is at least as hard as that of enumerating minimal dominating sets in general graphs, while the problem without the connectivity requirement admits a polynomial-delay algorithm on this class of graphs. This indicates that the connected variant is also a challenging problem. Very recently, Kobayashi et al. [30] devised a polynomial-delay algorithm for enumerating minimal connected vertex covers with cardinality at most $t$ for a given subcubic graph and a threshold $t$ by exploiting a well-known relation between a connected vertex cover in a (sub)cubic graph and a "matching" in a certain matroid [35].

---

[1] This means that there is an output-polynomial time algorithm for enumerating minimal dominating sets in graphs if and only if there is an output-polynomial time algorithm for dualizing monotone Boolean functions.

[2] This also includes the running time of pre-processing and post-processing.

[3] Note that the output size can be exponential in the input size.

We also study another type of variants: the problems of enumerating minimal *capacitated* vertex covers and minimal *capacitated* dominating sets of graphs, which we call MINIMAL CAPACITATED VERTEX COVER ENUMERATION and MINIMAL CAPACITATED DOMINATING SET ENUMERATION, respectively. These two problems generalize the conventional minimal vertex cover and dominating set enumeration problems (see Sect. 2 for details).

In this paper, we tackle these enumeration problems by restricting our focus to *bounded-degree graphs*. We design polynomial-delay algorithms for these problems on graphs with maximum degree $\Delta = O(1)$ in Sects. 4 to 6. The polynomial-delay algorithm for MINIMAL CONNECTED VERTEX COVER ENUMERATION can be extended to that on $d$-claw free graphs with $d = O(1)$. For MINIMAL CONNECTED VERTEX COVER ENUMERATION and MINIMAL CAPACITATED VERTEX COVER ENUMERATION, we show, in Sects. 4 to 6, that these problems are at least as hard as enumerating minimal transversals in hypergraphs, even if the input graph is restricted to be bipartite. These results indicate that the connectivity and/or capacity requirements make the problems "harder" in a certain sense.

## 2    Preliminaries

Throughout this paper, we only consider undirected simple graphs (unless otherwise stated). Let $G$ be a graph. We use $n$ to denote the number of vertices in $G$. We denote by $V(G)$ and $E(G)$ the sets of vertices and edges in $G$, respectively. For $v \in V(G)$, the set of neighbors of $v$ in $G$ is denoted by $N(v)$. This notation is extended to vertex sets: $N(X) = \bigcup_{v \in X} N(v) \backslash X$ for $X \subseteq V(G)$. For $X \subseteq V(G)$, the subgraph of $G$ induced by $X$ is denoted by $G[X]$. We define $N[v] = N(v) \cup \{v\}$ and $N[X] = N(X) \cup X$ for $v \in V(G)$ and $X \subseteq V(G)$.

A set of vertices $C$ is called a *vertex cover* of $G$ if for every edge in $G$, at least one end vertex of it belongs to $C$. A vertex cover $C$ is said to be *connected* if $G[C]$ is connected. A *dominating set* of $G$ is a set of vertices $D$ such that $V(G) = N[D]$, that is, every vertex in $V(G) \backslash D$ has a neighbor in $D$. A connected dominating set of $G$ is defined analogously.

The "capacitated" variants of a vertex cover and dominating set are defined as follows. Let $c \colon V(G) \to \mathbb{N}$ be a capacity function. Notice that we allow a vertex $v$ such that $c(v) = 0$. A *capacitated vertex cover* of $(G, c)$ is a pair $(C, \alpha)$ of a vertex set $C \subseteq V(G)$ and a function $\alpha \colon E(G) \to C$ such that $\alpha(\{u, v\})$ is either $u$ or $v$ and $|\alpha^{-1}(v)| \leq c(v)$ for $v \in C$, where $\alpha^{-1}(v) = \{e \in E(G) : \alpha(e) = v\}$ is the set of edges mapped to $v$ under $\alpha$. In other words, a vertex $v \in C$ covers at most $c(v)$ edges. A *capacitated dominating set* of $(G, c)$ is a pair $(D, \beta)$ of a vertex set $D$ and a function $\beta \colon V(G) \backslash D \to D$ such that $\beta(v)$ is a neighbor of $v$ and $|\beta^{-1}(v)| \leq c(v)$ for $v \in D$, where $\beta^{-1}(v) = \{w \in V(G) \backslash D : \beta(w) = v\}$. We simply refer to a vertex set $X \subseteq V(G)$ as a capacitated vertex cover (resp. capacitated dominating set) of $(G, c)$ if there is a function $\alpha \colon E(G) \to X$ (resp. $\beta \colon V(G) \backslash X \to X$) such that $(X, \alpha)$ is a capacitated vertex cover (resp. $(X, \beta)$ is a capacitated dominating set) of $(G, c)$. Clearly, every capacitated vertex cover (resp. capacitated dominating set) of $(G, c)$ is a vertex cover (resp. dominating set) of $G$, and the converse holds when setting $c(v) = \Delta$ for $v \in V(G)$, where $\Delta$ is the maximum degree of a vertex in $G$.

Let $\mathcal{S} \subseteq 2^{V(G)}$ be a collection of vertex sets of $G$. We say that $\mathcal{S}$ is *monotone* if for $X, Y \subseteq V(G)$ with $X \subseteq Y$, $X \in \mathcal{S}$ implies $Y \in \mathcal{S}$. It is easy to see that the collections of capacitated vertex covers and capacitated dominating sets of $(G, c)$ are monotone, while the collection of connected vertex sets (i.e., vertex sets that induce connected subgraphs of $G$) is not monotone. The following proposition shows that the collections of connected vertex covers and connected dominating sets of $G$ are also monotone.

**Proposition 1.** *Let $G = (V, E)$ be a connected graph such that $|E| > 0$ and let $X$ be a connected vertex cover (resp. connected dominating set) of $G$. Then, for any $Y \subseteq V(G)$ with $X \subseteq Y$, $Y$ is a connected vertex cover (resp. connected dominating set) of $G$ as well.*

*Proof.* Let $X$ be a connected vertex cover of $G$. Since the empty set is not a vertex cover, we assume that $X$ has at least one vertex. As $G$ is connected and $X$ is a vertex cover of $G$, every vertex $v$ in $V(G) \backslash X$ has a neighbor in $X$. This implies that $G[X \cup \{v\}]$ is connected. The case of a connected dominating set is analogous. $\square$

We also show that one can decide in polynomial time whether a given vertex set $X$ is a capacitated vertex cover (and a capacitated dominating set) of $(G, c)$.

**Proposition 2.** *There are polynomial-time algorithms for checking whether a given vertex set $X$ is a capacitated vertex cover and is a capacitated dominating set of $(G, c)$, respectively.*

*Proof.* We first consider the case for capacitated vertex cover. We reduce this feasibility-checking problem to the bipartite matching problem as follows. The bipartite graph $H$ consists of two independent sets $V_E$ and $V$. The first set $V_E$ is defined as $V_E = \{w_e : e \in E(G)\}$, that is, $V_E$ contains a vertex $w_e$ for each edge $e$ in $G$. The second set $V$ is defined as $V = \{w_v^i : v \in V(G), 1 \leq i \leq c(v)\}$, that is, $V$ contains $c(v)$ vertices for each vertex $v$ in $V(G)$. For $v \in V(G)$, $1 \leq i \leq c(v)$, and $e \in E(G)$, we add an edge between $w_e$ and $w_v^i$ if $e$ is incident to $v$ in $G$. The graph constructed in this way is indeed bipartite, which we denote by $H$.

It is easy to observe that $(G, c)$ has a capacitated vertex cover $(X, \alpha)$ if and only if $H[V_E \cup \{w_v^i : v \in X, 1 \leq i \leq c(v)\}]$ has a matching $M$ saturating $V_E$, as the function $\alpha$ is straightforwardly defined from the matching and vice versa.

For capacitated dominating set, we modify the construction of $H$ as follows: Replace $V_E$ with $V' = \{w_v : v \in V\}$ and add edges between $w_u \in V'$ and $w_v^i \in V$ for all $1 \leq i \leq c(v)$ if and only if $\{u, v\} \in E(G)$. Similarly, $(G, c)$ has a capacitated dominating set $(X, \beta)$ if and only if $H[\{w_v : v \in V(G) \backslash X\} \cup \{w_v^i : v \in X, 1 \leq i \leq c(v)\}]$ has a matching saturating $\{w_v : v \in V(G) \backslash X\}$.

We can find a maximum cardinality bipartite matching in polynomial time, proving this proposition. $\square$

## 3    A Quick Tour of the Supergraph Technique

Before proceeding to our algorithms, we quickly review the *supergraph technique* (also known as $X - e + Y$ *method*), which is frequently used in designing enumeration algorithms [5,7,8,28,32]. The crux is summarized below in Theorem 1.

Let $S \subseteq 2^U$ be a collection of subsets of a finite set $U$. Here, we assume that every pair of distinct sets in $S$ is incomparable with respect to set inclusion, that is, for $X, Y \in S$, both $X \backslash Y$ and $Y \backslash X$ are nonempty unless $X = Y$. The basic idea to the supergraph technique is to define a strongly connected directed graph $\mathcal{D} = (S, \mathcal{A})$ on $S$ by appropriately defining an arc set $\mathcal{A}$. Given this directed graph, we can enumerate all sets in $S$ by solely traversing $\mathcal{D}$ from an arbitrary $S \in S$. To this end, we need to define the directed graph $\mathcal{D}$ so that it is strongly connected.

We first observe that, under the above incomparability on $S$, for $X, Y \in S$, $|X \backslash Y| = 0$ if and only if $X = Y$. We define a set of arcs of $\mathcal{D}$ in such a way that for any two $X, Y \in S$ with $X \neq Y$, $X$ has an outgoing arc to some $Z \in S$ such that $|X \backslash Y| > |Z \backslash Y|$. The (out-)neighborhood of $X$ defined in this way is denoted by $\mathcal{N}^+(X)$. By the above observation, this directed graph $\mathcal{D}$ is strongly connected, as we can inductively show that there is a directed path from $X$ to $Y$ in $\mathcal{D}$ via $Z \in \mathcal{N}^+(X)$. This idea is formalized as follows.

**Theorem 1 (e.g., [7,8,29]).** *Suppose that for $X, Y \in S$ with $X \neq Y$, the neighborhood $\mathcal{N}^+(X)$ of $X$ contains a set $Z \in S$ such that $|X \backslash Y| > |Z \backslash Y|$. Moreover, suppose that given $X \in S$, we can compute the neighborhood of $X$ in (total) time $T(n)$, where $n = |U|$. Then, given an arbitrary initial set $S \in S$, we can enumerate all sets in $S$ in delay $T(n) \cdot n^{O(1)}$.*

Due to Theorem 1, it suffices to define such a polynomial-time computable neighborhood $\mathcal{N}^+ \colon S \to 2^S$.

## 4    MINIMAL CONNECTED VERTEX COVER ENUMERATION

In order to enumerate all minimal connected vertex covers in a graph, it suffices to construct a directed graph discussed in the previous section. To be more precise, let $G$ be a connected graph with maximum degree $\Delta$ and let $S_{\mathrm{cvc}}$ be the collection of all minimal connected vertex covers of $G$. Let $X, Y \in S_{\mathrm{cvc}}$ with $X \neq Y$. By the minimality of $X$ and $Y$, there is a vertex $v \in X \backslash Y$. As $Y$ is a vertex cover of $G$ not including $v$, all the neighbors of $v$ are included in $Y$ (i.e., $N(v) \subseteq Y$). We define a vertex cover $X'$ of $G$ as $X' := (X \backslash \{v\}) \cup N(v)$. By the connectivity of $G[X]$, we have the following observation.

**Observation 1.** *Each component of $G[X']$ has at least one vertex of $N(v)$.*

This observation indicates that the number of connected components in $G[X']$ is upper bounded by $\Delta$.

**Lemma 1.** *Let $S \subseteq V(G)$ be an arbitrary vertex cover of $G$ and let $Y$ be a minimal connected vertex cover of $G$. Suppose that each component in $G[S]$ has at least one vertex of $Y$. Then, there is a vertex set $W \subseteq Y$ of at most $q - 1$ vertices such that $G[S \cup W]$ is connected, where $q$ is the number of connected components in $G[S]$.*

*Proof.* We prove the lemma by induction on the number of connected components $q$ in $G[S]$. If $G[S]$ is connected, we are done. Suppose otherwise. Let $C$ and $C'$ be distinct components in $G[S]$. Since both $C \cap Y$ and $C' \cap Y$ are nonempty, there is a path $P$ in $G[Y]$ connecting $C \cap Y$ and $C' \cap Y$. We choose $C$ and $C'$ to minimize the length of such a path $P$. Since $S$ is a vertex cover of $G$, the length of $P$ is exactly 2. Let $w$ be the (unique) internal vertex of $P$. Applying the induction to $S \cup \{w\}$ proves the lemma. $\square$

Thus, there is a set $W \subseteq Y$ of size at most $\Delta - 1$ such that $G[X' \cup W]$ is a connected vertex cover of $G$. We call such a vertex set $W$ a *valid augmentation* for $X'$. As $v \in X \backslash Y$, $N(v) \subseteq Y$, and $W \subseteq Y$, we have

$$|X \backslash Y| \geq |(X' \cup \{v\}) \backslash (N(v) \cup Y)| > |X' \backslash Y| = |(X' \cup W) \backslash Y| \geq |Z \backslash Y|,$$

where $Z$ is an arbitrary minimal connected vertex cover of $G$ with $Z \subseteq X' \cup W$.

Now, we formally define $\mathcal{N}^+(X)$ for $X \in \mathcal{S}_{\mathrm{cvc}}$. For a connected vertex cover $C$ of $G$, we let $\mu(C)$ be an arbitrary minimal connected vertex cover of $G$ with $\mu(C) \subseteq C$. By Proposition 1, $\mu(C)$ can be computed in polynomial time by greedily removing vertices from $C$. We define

$$\mathcal{N}^+(X) = \{\mu((X \backslash \{v\}) \cup N(v) \cup W) :$$
$$v \in X, W \text{ is a valid augmentation for } (X \backslash \{v\}) \cup N(v)\},$$

which can be computed in $n^{\Delta + O(1)}$ time by just enumerating all valid augmentations $W$ of size at most $\Delta - 1$. As discussed above, for each $Y \in \mathcal{S}_{\mathrm{cvc}}$ with $X \neq Y$, $\mathcal{N}^+(X)$ has a minimal connected vertex cover $Z \in \mathcal{S}_{\mathrm{cvc}}$ such that $|X \backslash Y| > |Z \backslash Y|$. By Theorem 1, we have the following.

**Theorem 2.** *There is an $n^{\Delta + O(1)}$-delay enumeration algorithm for* MINIMAL CONNECTED VERTEX COVER ENUMERATION.

The running time of the algorithm in Theorem 2 depends, in fact, on the number of connected components in $G[X']$ with $X' = (X \backslash \{v\}) \cup N(v)$ rather than the degree of $v$. We can extend this idea as follows. For $d \in \mathbb{N}$, a *d-claw* is a star graph with $d$ leaves. A graph $G$ is said to be *d-claw free* if it has no $d$-claw as an induced subgraph. Clearly, every graph with maximum degree $\Delta$ is $(\Delta + 1)$-claw free. Suppose that $G$ is $d$-claw free. As the class of $d$-claw free graphs is hereditary, every induced subgraph of $G$ is also $d$-claw free. A crucial observation to our extension is that for every vertex $v \in X$, $G[(X \backslash \{v\}) \cup N(v)]$ has at most $d - 1$ components as otherwise $G$ has an induced $d$-claw with center $v$. This observation shows that there always exists a valid augmentation for $(X \backslash \{v\}) \cup N(v)$ of size $d - 2$, which yields a polynomial-delay algorithm for $d = O(1)$.

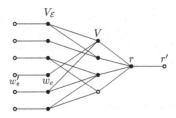

**Fig. 1.** The figure illustrates the constructed graph $G$. Filled circles indicate a minimal connected vertex cover of $G$.

**Theorem 3.** *There is an $n^{d+O(1)}$-delay enumeration algorithm for* MINIMAL CONNECTED VERTEX COVER ENUMERATION, *provided that the input graph $G$ is $d$-claw free.*

We complement these polynomial-delay enumeration algorithms by showing that MINIMAL CONNECTED VERTEX COVER ENUMERATION is "not so easy", even on bipartite graphs. For a hypergraph $\mathcal{H} = (V, \mathcal{E})$, a *transversal* of $\mathcal{H}$ is a vertex subset $S \subseteq V$ such that each hyperedge contains at least one vertex of $S$. Recall that no output-polynomial time algorithm for enumerating minimal transversals in hypergraphs is known so far.

**Theorem 4.** *If there is an output-polynomial time (or a polynomial-delay) algorithm for* MINIMAL CONNECTED VERTEX COVER ENUMERATION *on bipartite graphs, then there is an output-polynomial time (or a polynomial-delay) algorithm for enumerating minimal transversals in hypergraphs.*

*Proof.* The proof is done by constructing a bipartite graph $G$ from a hypergraph $\mathcal{H}$ and showing that there is a bijection between the collection of minimal connected vertex covers of $G$ and that of minimal transversals of $\mathcal{H}$.

Let $\mathcal{H} = (V, \mathcal{E})$ be a hypergraph. We construct a bipartite graph $G$ as follows. We start with the incidence bipartite graph of $\mathcal{H}$, that is, the vertex set of $G$ consists of two independent sets $V$ and $V_{\mathcal{E}} = \{w_e : e \in \mathcal{E}\}$, and two vertices $v \in V$ and $w_e \in V_{\mathcal{E}}$ are adjacent if and only if $v \in e$. From this bipartite graph, we add a pendant vertex $w'_e$ that is only adjacent to $w_e$ for each $w_e \in V_{\mathcal{E}}$ and a vertex $r$ that is adjacent to all vertices in $V$. To complete the construction of $G$, we add a pendant vertex $r'$ that is adjacent to $r$. See Fig. 1 for an illustration.

In the following, we show that there is a bijection between the collection of minimal connected vertex covers of $G$ and that of minimal transversals of $\mathcal{H}$.

Let $C$ be an arbitrary minimal connected vertex cover of $G$. We first observe that $w_e \in C$ and $w'_e \notin C$ for all $e \in \mathcal{E}$. To see this, fix a hyperedge $e \in \mathcal{E}$. As $C$ is a vertex cover of $G$, at least one of $w_e$ and $w'_e$ is contained in $C$. If both of them are contained in $C$, $C \setminus \{w'_e\}$ is still a connected vertex cover of $G$, contradicting to the minimality. Thus, exactly one of $w_e$ and $w'_e$ is contained in $C$. As $G[C]$ is connected, $w_e$ must be contained in $C$. Similarly, we have $r \in C$ and $r' \notin C$.

Let $S = C \cap V$. It is easy to see that $C \setminus S = V_{\mathcal{E}} \cup \{r\}$ is a vertex cover of $G$. Due to the connectivity of $C$, for every $e \in \mathcal{E}$, $C$ contains at least one vertex of $V$

that is adjacent to $w_e$. This implies that $S$ is a transversal of $\mathcal{H}$. If there is $v \in S$ such that $S\backslash\{v\}$ is a transversal of $\mathcal{H}$, $C\backslash\{v\}$ is a connected vertex cover of $G$ as every $w_e \in \mathcal{V}_\mathcal{E}$ has a neighbor in $S\backslash\{v\}$, contradicting the minimality of $C$. Thus, $S$ is a minimal transversal of $\mathcal{H}$. It is straightforward to verify that this relation is reversible, that is, for every minimal transversal $S$ of $\mathcal{H}$, $S \cup \mathcal{V}_\mathcal{E} \cup \{r\}$ is a minimal connected vertex cover of $G$.

From a hypergraph $\mathcal{H}$, we can construct the graph $G$ in polynomial time. Therefore, if there is an output-polynomial time algorithm for MINIMAL CONNECTED VERTEX COVER ENUMERATION, we can enumerate all minimal transversals of $\mathcal{H}$ in output-polynomial time as well. □

## 5   MINIMAL CONNECTED DOMINATING SET ENUMERATION

To enumerate all the minimal connected dominating sets, we use the same strategy in the previous section. Let $G$ be a connected graph with maximum degree $\Delta$ and let $\mathcal{S}_{\text{cds}}$ be the collection of all minimal connected dominating sets of $G$. Let $X, Y \in \mathcal{S}_{\text{cds}}$ be distinct minimal connected dominating sets of $G$. Then, there is a vertex $v \in X\backslash Y$.

**Lemma 2.** *There is a vertex set $W \subseteq Y$ of size at most $\Delta$ such that $(X\backslash\{v\}) \cup W$ is a dominating set of $G$. Moreover, $G[(X\backslash\{v\}) \cup W]$ has at most $\Delta$ components.*

*Proof.* Since $X$ is a dominating set of $G$, each vertex in $V(G)\backslash N[v]$ belongs to $X\backslash\{v\}$ or has a neighbor in $X\backslash\{v\}$. As $v \notin Y$, $Y$ contains at least one vertex $v'$ in $N(v)$. Moreover, for each $w \in N(v)\backslash(X \cup \{v'\})$, $Y$ contains at least one vertex $w'$ in $N[w]$. We let $W = \{v'\} \cup \{w' : w \in N(v)\backslash(X \cup \{v'\})\}$. Clearly, $W$ contains at most $|N(v)\backslash X| \leq \Delta$ vertices. Since each vertex in $N[v]$ belongs to $(X\backslash\{v\}) \cup W$ or has a neighbor in $(X\backslash\{v\}) \cup W$, $(X\backslash\{v\}) \cup W$ is a dominating set of $G$.

Since $G[X]$ is connected, $G[(X\backslash\{v\})]$ has at most $|N(v) \cap X|$ components. As $|W| \leq |N(v)\backslash X|$, $G[(X\backslash\{v\} \cup W]$ has at most $|N(v) \cap X| + |N(v)\backslash X| \leq \Delta$ components. □

Let $W$ be a set of vertices of size at most $\Delta$ discussed in Lemma 2 and let $X' = (X\backslash\{v\}) \cup W$ be a dominating set of $G$. Since $Y$ is a dominating set of $G$, the following observation holds.

**Observation 2.** *For each component $C$ of $G[X']$, $N[C] \cap Y \neq \emptyset$.*

**Lemma 3.** *Let $S \subseteq V$ be an arbitrary dominating set of $G$ and let $Y$ be a minimal connected dominating set of $G$. Then, there is a vertex set $U \subseteq Y$ of at most $2q - 2$ vertices such that $G[S \cup U]$ is connected, where $q$ is the number of connected components in $G[S]$.*

*Proof.* We prove the lemma by induction on the number of connected components $q$ in $G[S]$. If $G[S]$ is connected, we are done. Suppose otherwise. Let $C$ and $C'$ be distinct components in $G[S]$. As both $N[C] \cap Y$ and $N[C'] \cap Y$ are nonempty (by Observation 2), there is a path $P$ in $G[Y]$ connecting $N[C] \cap Y$ and $N[C'] \cap Y$. We choose $C$ and $C'$ to minimize the length of such a path $P$. The path $P$ contains at most two vertices of $V \backslash S$ as otherwise one of the internal vertices of $P$ has no neighbor in $S$, contradicting the fact that $S$ is a dominating set of $G$. Let $w, w'$ be these (possibly identical) two vertices of $V(P) \cap (V \backslash S)$. Applying the induction to $S \cup \{w, w'\}$ proves the lemma. □

Thus, there is a set $U \subseteq Y$ of size at most $2\Delta - 2$ such that $G[X' \cup U]$ is a connected dominating set of $G$. As $v \in X \backslash Y$ and $W \cup U \subseteq Y$, we have

$$|X \backslash Y| > |(X \backslash \{v\}) \backslash Y| = |((X \backslash \{v\}) \cup W) \backslash Y| = |(X' \cup U) \backslash Y| \geq |Z \backslash Y|,$$

where $Z$ is an arbitrary minimal connected dominating set of $G$ with $Z \subseteq X' \cup U$. Now, let us formally define $\mathcal{N}^+(X)$ for $X \in \mathcal{S}_{\mathrm{cds}}$. For connected dominating set $D$ of $G$, we let $\mu(D)$ be an arbitrary minimal connected dominating set of $G$ with $\mu(D) \subseteq D$. By Proposition 1, $\mu(D)$ can be computed in polynomial time. We then define

$$\mathcal{N}^+(X) = \{\mu((X \backslash \{v\}) \cup W^*) : v \in X, W^* \subseteq V(G), |W^*| \leq 3\Delta - 2,$$
$$G[(X \backslash \{v\}) \cup W^*] \text{ is a connected dominating set of } G\}.$$

By Lemma 2, for each $Y \in \mathcal{S}_{\mathrm{cds}}$, there is a set of vertices $W \subseteq Y$ with cardinality at most $\Delta$ such that $(X \backslash \{v\}) \cup W$ is a dominating set of $G$. Moreover, by Lemma 3, there is a set of vertices $U \subseteq Y$ with cardinality at most $2\Delta - 2$ such that $(X \backslash \{v\}) \cup W \cup U$ is a connected dominating set of $G$. Thus, for each $Y \in \mathcal{S}_{\mathrm{cds}}$ with $X \neq Y$, $\mathcal{N}^+(X)$ has a minimal connected dominating set $Z \in \mathcal{S}_{\mathrm{cds}}$ such that $|X \backslash Y| > |Z \backslash Y|$. By Theorem 1, we have the following.

**Theorem 5.** *There is an $n^{3\Delta + O(1)}$-delay enumeration algorithm for* MINIMAL CONNECTED DOMINATING SET ENUMERATION.

We would like to mention that a similar improvement for $d$-claw free graphs would not be straightforward. In fact, we can show that MINIMAL CONNECTED DOMINATING SET ENUMERATION is at least as hard as MINIMAL TRANSVERSAL ENUMERATION, even on claw-free graphs. A graph is said to be *co-bipartite* if its complement is bipartite. It is easy to observe that every co-bipartite graph has no independent set of size 3, and hence it is claw-free.

**Theorem 6.** *If there is an output-polynomial time (or a polynomial-delay) algorithm for* MINIMAL CONNECTED DOMINATING SET ENUMERATION *on co-bipartite graphs, then there is an output-polynomial time (or a polynomial-delay) algorithm for enumerating minimal transversals in hypergraphs.*

*Proof.* The reduction is identical to one in [25], which proves a similar "hardness" of enumerating minimal dominating sets in co-bipartite graphs.

Let $\mathcal{H} = (V, \mathcal{E})$ be a hypergraph. We assume that $\mathcal{E}$ is nonempty and $\emptyset \notin \mathcal{E}$ as otherwise the instance is trivial. From the incidence bipartite graph $\mathcal{H}$ (see Theorem 4), we complete both $V$ and $V_{\mathcal{E}}$ into cliques. Then, we add a new vertex $r$, which is adjacent to all vertices in $V$. Let $G$ be the graph obtained in this way. It is easy to see that $G$ is co-bipartite and has two cliques $V \cup \{r\}$ and $V_\mathcal{E}$.

Observe that every transversal $S \subseteq V$ of $\mathcal{H}$ is a connected dominating set of $G$. This follows from the facts that every vertex in $V_\mathcal{E}$ is adjacent to at least one vertex in $S$ and every vertex in $S$ is adjacent to other vertices in $V \cup \{r\}$. (Our assumption $\mathcal{E} \neq \emptyset$ implies that $S \neq \emptyset$.)

Let $S$ be a minimal connected dominating set of $G$. As $r$ has no neighbors in $V_\mathcal{E}$, $S$ contains at least one vertex from $V \cup \{r\}$. Moreover, as $V_\mathcal{E} \neq \emptyset$ and $G[S]$ is connected, $S$ contains at least one vertex from $V$, implying that $r \notin S$ by the minimality of $S$. Then, observe that either $S = \{v, w_e\}$ for some $v \in V$ and $w_e \in V_\mathcal{E}$ with $v \in e$ or $S \subseteq V$. To see this, suppose that $S$ contains at least one vertex $w_e$ of $V_\mathcal{E}$. Then, as $S$ contains at least one vertex of $V$, it contains a vertex $v \in V$ that is adjacent to $w_e$ in $G$. Since these two vertices $v$ and $w_e$ dominate all the vertices in $G$, by the minimality of $S$, $S$ contains no other vertices at all. By a similar argument used in Theorem 4, $S$ is a minimal transversal of $\mathcal{H}$ if $S \subseteq V$ holds.

Now, suppose that there is an output-polynomial time algorithm for MINIMAL CONNECTED DOMINATING SET ENUMERATION. By applying the algorithm to $G$, we can enumerate all minimal connected dominating sets $\mathcal{S}$ of $G$. Since $\mathcal{S}$ contains all minimal transversals $\mathcal{S}'$ of $\mathcal{H}$ and $|\mathcal{S} \backslash \mathcal{S}'| = O(n^2)$, the algorithm enumerates all minimal transversals of $H$ in output-polynomial time as well. □

## 6  Capacitated Vertex Cover and Dominating Set

This section is devoted to showing polynomial-delay algorithms for the capacitated variants on bounded-degree graphs.

The basic idea to prove Theorems 2 and 5 is that for distinct minimal solutions $X$ and $Y$ and for $v \in X \backslash Y$, there always exists a constant size (depending on $\Delta$) vertex set $W \subseteq Y$ such that $(X \backslash \{v\}) \cup W$ is a (not necessarily minimal) solution. For MINIMAL CAPACITATED VERTEX COVER ENUMERATION and MINIMAL CAPACITATED DOMINATING SET ENUMERATION, we can find such a set $W$ in bounded degree graphs as well. Let $G$ be a graph with maximum degree $\Delta$ and let $c\colon V(G) \to \mathbb{N}$. Let $q = \max_{v \in V(G)} c(v)$. Without loss of generality, we can assume that $q \leq \Delta$.

**Lemma 4.** *Let $X, Y$ be distinct minimal capacitated vertex covers of $(G, c)$ and let $v \in X \backslash Y$. Then, there is a vertex set $W \subseteq Y$ of size at most $q$ such that $(X \backslash \{v\}) \cup W$ is a capacitated vertex cover of $(G, c)$.*

*Proof.* Let $\alpha_X\colon E(G) \to X$ (resp. $\alpha_Y\colon E(G) \to Y$) be a function such that $(X, \alpha_X)$ (resp. $(Y, \alpha_Y)$) is a capacitated vertex cover of $(G, c)$. As $c(v) \leq q$, $\alpha_X^{-1}(v)$ contains at most $q$ edges. For each $e = \{v, w\} \in \alpha_X^{-1}(v)$, we define a vertex $v_e \in Y$, and show that $(X \backslash \{v\}) \cup W$ with $W = \{v_e : e \in \alpha_X^{-1}(v)\}$ is a capacitated vertex cover of $(G, c)$.

To define such a vertex $v_e$ for $e \in \alpha_X^{-1}(v)$, we use a similar strategy as used in Proposition 2. Let $H$ be a bipartite graph with $V(H) = V_E \cup V$ such that $V_E = \{w_e : e \in E(G)\}$ and $V = \{w_u^i : u \in V(G), 1 \leq i \leq c(u)\}$. As shown in Proposition 2, the capacitated vertex cover $(X, \alpha_X)$ (resp. $(Y, \alpha_Y)$) forms a matching $M_X$ (resp. $M_Y$) saturating $V_E$ in $H$ and vice versa. As $v \in X \backslash Y$, there is a vertex $w_v^i$ that is matched in $M_X$ but not matched in $M_Y$. Since both $M_X$ and $M_Y$ are maximum matchings in $H$, there is an alternating path $P$ between $w_v^i$ and $w_u^j$ for some $w_u^j$ that is not matched in $M_X$ but matched in $M_Y$. This implies that $u \in Y$, and we define $v_e := u$. Then we obtain a new matching $(M_X \backslash E(P)) \cup (E(P) \cap M_Y)$ saturating $V_E$, which induces a capacitated vertex cover $(X \cup \{u\}, \alpha_{X \cup \{u\}})$ such that $|\alpha_{X \cup \{u\}}^{-1}(v)| < |\alpha^{-1}(v)|$. We repeat this for all other edges in $\alpha_{X \cup \{u\}}^{-1}(v)$, yielding a desired set $W$.    $\square$

An analogous lemma also holds for minimal capacitated dominating sets with a sightly more involved argument. The proof is deferred to the full version due to the space limitation.

**Lemma 5 (♣).** *Let $X, Y$ be distinct minimal capacitated dominating sets of $(G, c)$ and let $v \in X \backslash Y$. Then, there is a vertex set $W \subseteq Y$ of size at most $q + 1$ such that $(X \backslash \{v\}) \cup W$ is a capacitated dominating set of $(G, c)$.*

The above two lemmas together with Propositions 2 and Theorem 1 yield polynomial-delay algorithms on bounded-degree graphs.

**Theorem 7.** *There are $n^{\min\{q, \Delta\} + O(1)}$-delay enumeration algorithms for MIN-IMAL CAPACITATED VERTEX COVER ENUMERATION and MINIMAL CAPACITATED DOMINATING SET ENUMERATION, where $q = \max_{v \in V(G)} c(v)$ and $\Delta$ is the maximum degree of $G$.*

Similarly to Theorem 4, MINIMAL CAPACITATED VERTEX COVER ENUMER-ATION is "not so easy".

**Theorem 8.** *If there is an output-polynomial time (or a polynomial-delay) algorithm for MINIMAL CAPACITATED VERTEX COVER ENUMERATION on bipartite graphs, then there is an output-polynomial time (or a polynomial-delay) algorithm for enumerating minimal transversals in hypergraphs.*

*Proof.* The underlying idea of the proof is analogous to Theorem 4. From a hypergraph $\mathcal{H} = (V, \mathcal{E})$, we construct a bipartite graph $G$ and function $c: V(G) \to \mathbb{N}$ as follows.

We start with the incident bipartite graph of $\mathcal{H}$ (see Theorem 4 for details). For each $w_e \in V_{\mathcal{E}}$, we add a pendant vertex $w_e'$, and the obtained graph is denoted by $G$. See Fig. 2 for an illustration. We define a capacity function $c: V(G) \to \mathbb{N}$ as follows. For a vertex $v \in V$, $c(v) = d(v)$, for a vertex $v \in V_{\mathcal{E}}$, $c(v) = d(v) - 1$, and otherwise, $c(v) = 0$. Here, $d(v)$ is the degree of $v$ in $G$.

Let $C$ be an arbitrary minimal capacitated vertex cover of $(G, c)$ and let $\alpha: E(G) \to C$ be a function such that $(C, \alpha)$ is a capacitated vertex cover of $(G, c)$. As $c(w_e') = 0$ for $e \in \mathcal{E}$, we have $V_{\mathcal{E}} \subseteq C$. Moreover, $\alpha$ maps the incident

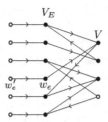

**Fig. 2.** The figure illustrates the constructed graph $G$. Filled circles indicate a minimal capacitated vertex cover of $(G, c)$ and arrows indicate the function $\alpha$.

edge of $w'_e$ to $w_e$ for each $e \in \mathcal{E}$. Since $c(w_e) = d(v) - 1$, $\alpha$ maps at least one edge incident to $w_e$ to a vertex $v$ in $C \cap V$. By the construction of $G$, such a vertex $v$ is contained in the hyperedge $e$. This implies that $S = C \cap V$ is a transversal of $\mathcal{H}$. The remaining part of this proof is analogous to that in Theorem 4.     □

## 7  Concluding Remarks

In this paper, we have developed polynomial-delay algorithms for enumerating minimal connected/capacitated vertex covers/dominating sets on bounded-degree graphs. Moreover, the algorithm for enumerating minimal connected vertex covers can be extended into one working on $d$-claw free graphs. In contrast to these positive results, we show that the problems of enumerating minimal connected/capacitated vertex covers in bipartite graphs are at least as hard as that of enumerating minimal transversals in hypergraphs, which has no known output-polynomial time algorithm until now.

We would like to mention that we can enumerate minimal vertex covers satisfying both connectivity and capacity constraints in polynomial delay on bounded-degree graphs. This can be done by just observing that for each minimal connected and capacitated vertex covers $X, Y$ of $(G, c)$ and for $v \in X$, there is a vertex set $W \subseteq Y$ of size at most $\Delta$ such that $(X \backslash \{v\}) \cup W$ is a capacitated vertex cover of $(G, c)$ (by Lemma 4). As $(X \backslash \{v\}) \cup W$ is a vertex cover of $G$, there is a vertex set $W' \subseteq Y$ of size at most $q - 1$ such that $(X \backslash \{v\}) \cup W \cup W'$ is a connected vertex cover of $G$, where $q$ is the number of connected components in $G[(X \backslash \{v\}) \cup W]$ (by Lemma 1). Since $G[(X \backslash \{v\}) \cup W]$ has at most $2\Delta$ components, the claim follows. A similar argument also holds for the case of minimal connected and capacitated dominating sets in bounded-degree graphs.

A natural extension of our results is to consider the problems on bounded-degeneracy graphs. It seems difficult to apply a similar technique to enumerate minimal dominating sets on such a class of graphs [15].

**Acknowledgements.** We are grateful to the anonymous reviewers for their careful reading of our manuscript and helpful comments.

# References

1. Abu-Khzam, F.N., Fernau, H., Gras, B., Liedloff, M., Mann, K.: Enumerating minimal connected dominating sets. In: Chechik, S., Navarro, G., Rotenberg, E., Herman, G. (eds.) 30th Annual European Symposium on Algorithms, ESA 2022. LIPIcs, Berlin/Potsdam, Germany, 5–9 September 2022, vol. 244, pp. 1:1–1:15. Schloss Dagstuhl - Leibniz-Zentrum für Informatik (2022)
2. Balakrishnan, H., Rajaraman, A., Rangan, C.P.: Connected domination and steiner set on asteroidal triple-free graphs. In: Dehne, F., Sack, J.-R., Santoro, N., Whitesides, S. (eds.) WADS 1993. LNCS, vol. 709, pp. 131–141. Springer, Heidelberg (1993). https://doi.org/10.1007/3-540-57155-8_242
3. Bioch, J.C., Ibaraki, T.: Complexity of identification and dualization of positive Boolean functions. Inf. Comput. 123(1), 50–63 (1995)
4. Bonamy, M., Defrain, O., Heinrich, M., Pilipczuk, M., Raymond, J.-F.: Enumerating minimal dominating sets in $k_t$-free graphs and variants. ACM Trans. Algorithms 16(3):39:1–39:23 (2020)
5. Boros, E., Borys, K., Elbassioni, K., Gurvich, V., Makino, K., Rudolf, G.: Generating minimal k-vertex connected spanning subgraphs. In: Lin, G. (ed.) COCOON 2007. LNCS, vol. 4598, pp. 222–231. Springer, Heidelberg (2007). https://doi.org/10.1007/978-3-540-73545-8_23
6. Bron, C., Kerbosch, J.: Finding all cliques of an undirected graph (algorithm 457). Commun. ACM 16(9), 575–576 (1973)
7. Cohen, S., Kimelfeld, B., Sagiv, Y.: Generating all maximal induced subgraphs for hereditary and connected-hereditary graph properties. J. Comput. Syst. Sci. 74(7), 1147–1159 (2008)
8. Conte, A., Grossi, R., Marino, A., Uno, T., Versari, L.: Proximity search for maximal subgraph enumeration. SIAM J. Comput. 51(5), 1580–1625 (2022)
9. Conte, A., Grossi, R., Marino, A., Versari, L.: Sublinear-space and bounded-delay algorithms for maximal clique enumeration in graphs. Algorithmica 82(6), 1547–1573 (2020)
10. Conte, A., De Virgilio, R., Maccioni, A., Patrignani, M., Torlone, R.: Finding all maximal cliques in very large social networks. In: Pitoura, E., et al. (eds.) Proceedings of the 19th International Conference on Extending Database Technology, EDBT 2016, Bordeaux, France, 15–16 March 2016, pp. 173–184. OpenProceedings.org (2016)
11. Courcelle, B.: Linear delay enumeration and monadic second-order logic. Discrete Appl. Math. 157(12), 2675–2700 (2009)
12. Cygan, M.: Deterministic parameterized connected vertex cover. In: Fomin, F.V., Kaski, P. (eds.) SWAT 2012. LNCS, vol. 7357, pp. 95–106. Springer, Heidelberg (2012). https://doi.org/10.1007/978-3-642-31155-0_9
13. Eiter, T., Gottlob, G.: Identifying the minimal transversals of a hypergraph and related problems. SIAM J. Comput. 24(6), 1278–1304 (1995)
14. Eiter, T., Gottlob, G.: Hypergraph transversal computation and related problems in logic and AI. In: Flesca, S., Greco, S., Ianni, G., Leone, N. (eds.) JELIA 2002. LNCS (LNAI), vol. 2424, pp. 549–564. Springer, Heidelberg (2002). https://doi.org/10.1007/3-540-45757-7_53
15. Eiter, T., Gottlob, G., Makino, K.: New results on monotone dualization and generating hypergraph transversals. SIAM J. Comput. 32(2), 514–537 (2003)
16. Eiter, T., Makino, K., Gottlob, G.: Computational aspects of monotone dualization: a brief survey. Discrete Appl. Math. 156(11), 2035–2049 (2008)

17. Escoffier, B., Gourvès, L., Monnot, J.: Complexity and approximation results for the connected vertex cover problem in graphs and hypergraphs. J. Discrete Algorithms **8**(1), 36–49 (2010)
18. Fernau, H., Golovach, P.A., Sagot, M.-F.: Algorithmic enumeration: output-sensitive, input-sensitive, parameterized, approximative (Dagstuhl Seminar 18421). Dagstuhl Rep. **8**(10), 63–86 (2018)
19. Fomin, F.V., Grandoni, F., Kratsch, D.: Solving connected dominating set faster than $2^n$. Algorithmica **52**(2), 153–166 (2008)
20. Fox, J., Roughgarden, T., Seshadhri, C., Wei, F., Wein, N.: Finding cliques in social networks: a new distribution-free model. SIAM J. Comput. **49**(2), 448–464 (2020)
21. Fredman, M.L., Khachiyan, L.: On the complexity of dualization of monotone disjunctive normal forms. J. Algorithms **21**(3), 618–628 (1996)
22. Golovach, P.A., Heggernes, P., Kratsch, D.: Enumeration and maximum number of minimal connected vertex covers in graphs. Eur. J. Comb. **68**, 132–147 (2018)
23. Golovach, P.A., Heggernes, P., Kratsch, D., Villanger, Y.: An incremental polynomial time algorithm to enumerate all minimal edge dominating sets. In: Fomin, F.V., Freivalds, R., Kwiatkowska, M., Peleg, D. (eds.) ICALP 2013. LNCS, vol. 7965, pp. 485–496. Springer, Heidelberg (2013). https://doi.org/10.1007/978-3-642-39206-1_41
24. Guha, S., Khuller, S.: Approximation algorithms for connected dominating sets. Algorithmica **20**(4), 374–387 (1998)
25. Kanté, M.M., Limouzy, V., Mary, A., Nourine, L.: On the enumeration of minimal dominating sets and related notions. SIAM J. Discrete Math. **28**(4), 1916–1929 (2014)
26. Kanté, M.M., Limouzy, V., Mary, A., Nourine, L., Uno, T.: A polynomial delay algorithm for enumerating minimal dominating sets in chordal graphs. In: Mayr, E.W. (ed.) WG 2015. LNCS, vol. 9224, pp. 138–153. Springer, Heidelberg (2016). https://doi.org/10.1007/978-3-662-53174-7_11
27. Kanté, M.M., Limouzy, V., Mary, A., Nourine, L., Uno, T.: Polynomial delay algorithm for listing minimal edge dominating sets in graphs. In: Dehne, F., Sack, J.-R., Stege, U. (eds.) WADS 2015. LNCS, vol. 9214, pp. 446–457. Springer, Cham (2015). https://doi.org/10.1007/978-3-319-21840-3_37
28. Khachiyan, L., Boros, E., Borys, K., Elbassioni, K., Gurvich, V., Makino, K.: Enumerating spanning and connected subsets in graphs and matroids. In: Azar, Y., Erlebach, T. (eds.) ESA 2006. LNCS, vol. 4168, pp. 444–455. Springer, Heidelberg (2006). https://doi.org/10.1007/11841036_41
29. Kobayashi, Y., Kurita, K., Wasa, K.: Efficient constant-factor approximate enumeration of minimal subsets for monotone properties with weight constraints. CoRR, abs/2009.08830 (2020)
30. Kobayashi, Y., Kurita, K., Wasa, K.: Polynomial-delay enumeration of large maximal common independent sets in two matroids and beyond. CoRR, abs/2307.08948 (2023)
31. Makino, K., Uno, T.: New algorithms for enumerating all maximal cliques. In: Hagerup, T., Katajainen, J. (eds.) SWAT 2004. LNCS, vol. 3111, pp. 260–272. Springer, Heidelberg (2004). https://doi.org/10.1007/978-3-540-27810-8_23
32. Schwikowski, B., Speckenmeyer, E.: On enumerating all minimal solutions of feedback problems. Discrete Appl. Math. **117**(1–3), 253–265 (2002)
33. Tomita, E., Tanaka, A., Takahashi, H.: The worst-case time complexity for generating all maximal cliques and computational experiments. Theor. Comput. Sci. **363**(1), 28–42 (2006)

34. Tsukiyama, S., Ide, M., Ariyoshi, H., Shirakawa, I.: A new algorithm for generating all the maximal independent sets. SIAM J. Comput. **6**(3), 505–517 (1977)

35. Ueno, S., Kajitani, Y., Gotoh, S.: On the nonseparating independent set problem and feedback set problem for graphs with no vertex degree exceeding three. Discrete Math. **72**(1–3), 355–360 (1988)

# Making the Interval Membership Width of Temporal Graphs Connected and Bidirectional

Filippos Christodoulou[1], Pierluigi Crescenzi[1], Andrea Marino[2], Ana Silva[3(✉)], and Dimitrios M. Thilikos[4]

[1] Gran Sasso Science Institute, L'Aquila, Italy
`{filippos.christodoulou,pierluigi.crescenzi}@gssi.it`
[2] Dipartimento di Statistica, Informatica, Applicazioni, Università degli Studi di Firenze, Firenze, Italy
`andrea.marino@unifi.it`
[3] Departamento de Matemática, Universidade Federal do Ceará, Fortaleza, Brazil
`anasilva@mat.ufc.br`
[4] LIRMM, Université Montpellier, CNRS, Montpellier, France
`sedthilk@thilikos.info`

**Abstract.** Temporal graphs are graphs that evolve over time. Many problems which are polynomial-time solvable in standard graphs become NP-hard when appropriately defined in the realm of temporal graphs. This suggested the definition of several parameters for temporal graphs and to prove the fixed-parameter tractability of several problems with respect to these parameters. In this paper, we introduce a hierarchy of parameters based on the previously defined interval membership width and on the temporal evolution of the connected components of the underlying static graph. We then show that the Eulerian trail problem and the temporal 2-coloring problem are both fixed-parameter tractable (in short, FPT) with respect to any of the parameters in the hierarchy. We also introduce a vertex-variant of the parameters and we show that the firefighter problem (which was known to be FPT with respect to the vertex-variant of the interval membership width) is also FPT with respect to one of the parameters in the second level of the hierarchy.

**Keywords:** Temporal graphs · Eulerian trails · Temporal coloring · Firefighter · Parameterized complexity · Width measures

## 1 Introduction

A temporal graph is a graph whose underlying topology is subject to discrete changes over time. Several real-world networks, such as social networks, transportation networks, and information and communication networks, can be modeled as temporal graphs. This is usually done by associating time labels to the edges of a graph, in order to indicate the moments of existence of the edges themselves (with the vertex set of the graph remaining unchanged).

A. A. Rescigno and U. Vaccaro (Eds.): IWOCA 2024, LNCS 14764, pp. 247–258, 2024.
https://doi.org/10.1007/978-3-031-63021-7_19

A temporal graph $\mathcal{G}$ with lifetime $\tau$ is a pair $(G = (V, E), \lambda)$, where $\lambda : E \to 2^{[\tau]}$ is a *time-labeling* function that assigns a set of integer time labels to each edge of the graph $G$.[1] An edge $e \in E$ is available only at the times specified by $\lambda(e)$. Due to their relevance and applicability in many areas, temporal graphs have attracted a lot of attention in the past decade (we refer the reader to the book of Holme and Saramäki [13], to the survey of Michail [18], and to the seminal paper of Kempe, Kleinberg, and Kumar [14]). Temporal graphs have also appeared under different names in the literature, such as time-varying graphs [10], dynamic graphs [5], evolving networks [2], and link streams [15].

Paths and walks in a temporal graph have to traverse a sequence of adjacent edges $e_1, \ldots, e_k$ at increasing times $t_1 < \ldots < t_k$, respectively, with $t_i \in \lambda(e_i)$ for every $i \in [k]$.[2] By referring to this kind of paths and walks, several polynomial-time problems on standard graphs become intractable when transferred to the temporal realm, such as, for example, the computation of connected components [1] and the identification of Eulerian walks [17]. Moreover, the introduction of the time dimension suggests new problems connected to paths and walks, such as, for example, determining the existence of a restless (that is, respecting specific waiting constraints) connection between two nodes, which is intractable for paths and polynomial-time solvable for walks [6], and the analysis of simple spreading processes [9].

The temporal version of other classical graph problems have been considered in the last ten years, such as, for example, the well-known coloring problem. The temporal coloring problem consists in deciding whether there exists a coloring of the temporal nodes (that is, the nodes at different time steps) such that each edge of the graph is properly colored in at least one time step. It is known that deciding whether a temporal graph admits a temporal coloring using 2 colors is NP-complete (even under very strict constraints of the temporal graph) [16].

When dealing with temporal graphs, the intractability of several problems holds also when the underlying static graph has small well-known parameters, such as the tree-width or the feedback vertex/edge number. This prompts the need to develop parameters that not only consider the underlying graph structure but also account the temporal structure of the input graph. Some such measures, like temporal variations of the above two parameters, have already been proposed [6,11]. In this paper, we focus on the parameter introduced by Bumpus and Meeks [3], called interval membership width, which intuitively quantifies the extent to which the set of intervals defined by the first and last appearance of each edge can overlap. The value of this parameter, hence, does not depend on the structure of the underlying static graph (other than its number of edges), but it is instead influenced only by the temporal structure of the input graph. In [3] it is shown that TEMPORAL EULERIAN TRAIL is FPT when parameterized by the interval membership width, while in [12] it is shown that TEMPORAL FIREFIGHTER RESERVE is FPT when parameterized by its 'vertex variant' (see below for the definitions of all considered problems).

---

[1] For every positive integer $k$, $[k]$ denotes the set $\{1, 2, \ldots, k\}$.

[2] Similarly, one can consider non-decreasing sequences, i.e. with $t_1 \leq \ldots \leq t_k$. In this paper, however, we focus on the strictly increasing case.

Motivated by these latter results, we here modify the definition of the interval membership width by taking into account the evolution (both forward and backward) of the connected components of the temporal graph and by introducing a new parameter called connected interval membership width (together with its vertex-variant). We show that both TEMPORAL EULERIAN TRAIL and TEMPORAL 2-COLORING are FPT with respect to this new parameter. We also show that TEMPORAL FIREFIGHTER RESERVE is FPT with respect to its vertex-variant. We then introduce another parameter which is based on the search for the best combination of the forward and the backward connected interval membership width and we prove that TEMPORAL EULERIAN TRAIL and TEMPORAL 2-COLORING are FPT with respect to this latter parameter.

**Preliminaries.** We use standard definitions and notation of graph theory (we refer the unfamiliar reader to [20]). We will also make use of the following notations. Given an undirected graph $G = (V, E)$ and a vertex $v \in V$, the *neighborhood* of $v$ is defined as $N_G(v) = \{u \mid \{u, v\} \in E\}$. For a set $X \subseteq V(G)$, we also use $N_G(X)$ to denote the set $\bigcup_{x \in X} N_G(x) \setminus X$. We omit $G$ when it is clear from the context. For any edge set $A \subseteq E$, $V(A)$ denotes the set of vertices with an incident edge in $A$, that is, $V(A) = \bigcup_{e = \{u, v\} \in A} \{u, v\}$. For any connected component $C$ of $G$, $V(C)$ (respectively, $E(C)$) denotes the set of nodes (respectively, edges) in $C$.

Concerning temporal graphs, we use and extend the notation in [18]. A *temporal graph* $\mathcal{G}$ is a pair $(G, \lambda)$, where $G = (V, E)$ is the *underlying (undirected) graph* of $\mathcal{G}$, and $\lambda : E \to 2^{\mathbb{N}}$ is a *time-labeling* function which assigns to every edge of $G$ a finite set of integer time labels. Without loss of generality, we can assume that $\min \bigcup_{e \in E} \lambda(e) = 1$. The *lifetime* $\tau$ of $\mathcal{G}$ is defined as $\tau = \max \bigcup_{e \in E} \lambda(e)$. For every edge $e = \{u, v\} \in E$ and every $t \in \lambda(e)$, the triple $(u, v, t)$ (or, equivalently, the pair $(e, t)$) is said to be a *temporal edge* of $\mathcal{G}$.

Given a temporal graph $\mathcal{G} = (G = (V, E), \lambda)$ with lifetime $\tau$, the *snapshot at time $t$*, with $t \in [\tau]$, is the graph $G_t = (V, E_t)$ where $E_t = \{e \in E : t \in \lambda(e)\}$. The temporal graph $\mathcal{G}$ can then also be defined as the sequence $G_1, \ldots, G_\tau$ of its snapshots. In the following, $\mathcal{G}_{\leq}(t)$ (respectively, $\mathcal{G}_{\geq}(t)$) will denote the temporal graph formed by the first $t$ (respectively, last $\tau - t + 1$) snapshots of $\mathcal{G}$. Moreover, $G_{\leq}(t)$ (respectively, $G_{\geq}(t)$) will denote the underlying graph of $\mathcal{G}_{\leq}(t)$ (respectively, $\mathcal{G}_{\geq}(t)$). For $S \subseteq V$ and $t \in [\tau]$, we define $N_t(S)$ to be the set of all vertices that are temporally adjacent at time $t$ to the vertices in $S$, excluding $S$, that is, $N_t(S) = N_{G_t}(S)$. A *(strict) temporal walk* from $u$ to $v$ in $\mathcal{G}$ is a sequence of temporal edges $(u_1, v_1, t_1), \ldots, (u_k, v_k, t_k)$ such that $u_1 = u$, $v_k = v$, and, for any $i \in [k-1]$, $v_i = u_{i+1}$ and $t_i < t_{i+1}$. A temporal walk is said to be a *temporal trail* if no edge in $E$ is traversed twice. A temporal trail is said to be a *temporal path* if no node in $V$ is visited twice.

The following lemma introduces a transformation of a temporal graph that will allow us to easily design a backward version of our algorithms (for the sake of brevity, all proofs have been omitted in this extended abstract).

**Lemma 1 ([4]).** *Given a temporal graph* $\mathcal{G} = (G = (V,E), \lambda)$, *let* $\mathcal{G}^R = (G = (V,E), \rho)$ *be the* reverse temporal graph *of* $\mathcal{G}$ *obtained by setting, for each* $e \in E$, $\rho(e) = \cup_{t \in \lambda(e)} \{-t - 1\}$. *Then, there exists a temporal walk in* $\mathcal{G}$ *starting from* $u$ *and arriving at* $v$ *at time at most* $t$ *if and only if there exists a temporal walk in* $\mathcal{G}^R$ *starting from* $v$ *at time* $-t$ *and arriving at* $u$.

*Edge Exploration.* An *Eulerian walk* (respectively, *trail*) in a graph is a walk which traverses every edge at least (respectively, exactly) once. A *temporal Eulerian walk* (respectively, *trail*) in a temporal graph is a temporal walk (respectively, trail) $(e_1, t_1), \ldots, (e_m, t_m)$ such that $e_1, \ldots, e_m$ is an Eulerian walk (respectively, trail) in the underlying graph. The TEMPORAL EULERIAN WALK (respectively, TEMPORAL EULERIAN TRAIL) problem consists of deciding whether a temporal graph admits a temporal Eulerian walk (respectively, trail). It is known that these two problems are both NP-complete [17]. The following lemma, instead, immediately follows from the proof of Lemma 1.

**Lemma 2.** *Given a temporal graph* $\mathcal{G} = (G = (V,E), \lambda)$ *and its reverse temporal graph* $\mathcal{G}^R$, *there exists an Eulerian walk in* $\mathcal{G}$ *if and only if there exists an Eulerian walk in* $\mathcal{G}^R$.

*Temporal Coloring.* Given a temporal graph $\mathcal{G} = (G = (V,E), \lambda)$ with lifetime $\tau$, an integer $k \geq 2$, and a function $f : V \times [\tau] \to [k]$, we say that $e = \{u, v\} \in E$ is *properly colored* by $f$ (or that $f$ *properly colors* $e$) if there exists $t \in \lambda(e)$ such that $f(u, t) \neq f(v, t)$. We say that $f$ is a *temporal $k$-coloring* of $\mathcal{G}$ if $f$ properly colors every edge in $E$. TEMPORAL 2-COLORING consists of deciding whether a temporal graph $\mathcal{G} = (G = (V,E), \lambda)$ admits a temporal 2-coloring (similarly, for any $k > 2$ we define TEMPORAL $k$-COLORING). It is known that TEMPORAL 2-COLORING is NP-complete [16] even if $G$ has bounded treewidth.

*Temporal Firefighter Reserve.* Given a temporal graph $\mathcal{G} = (G = (V,E), \lambda)$ with lifetime $\tau$, a *defence strategy* (in $\mathcal{G}$) is a sequence $\mathcal{S} = (S_1, \ldots, S_\tau)$ of pairwise disjoint subsets of $V$ such that $|S_t| \leq t - \sum_{i=1}^{t-1} |S_i|$, for each $t \in [\tau]$. Given a vertex $s \in V$, the set $B_s(\mathcal{S}) = B_s^\tau(\mathcal{S})$ of *burnt nodes* by a *fire starting in* $s$ is recursively defined as follows: $B_s^1(\mathcal{S}) = \{s\} \cup (N_1(s) \setminus S_1)$, and, for any $t \in [\tau]$ with $t > 1$, $B_s^t(\mathcal{S}) = B_s^{t-1}(\mathcal{S}) \cup (N_t(B_s^{t-1}(\mathcal{S})) \setminus \bigcup_{i=1}^t S_i)$. In words, starting in $s$, at each time step the fire spreads to the temporal neighbors of the current fire, except that defended vertices can never catch on fire (note that the condition on the cardinality of $S_t$ is due to the fact that, at each time step $t$, it is possible to decide not to defend a new node and to use this saving for future time steps). Given a temporal graph $\mathcal{G} = (G = (V,E), \lambda)$, a vertex $s \in V$, and an integer $k \leq |V|$, the TEMPORAL FIREFIGHTER RESERVE problem consists of deciding whether there exists a defence strategy $\mathcal{S}$ such that $|B_s(\mathcal{S})| \leq |V| - k$ (that is, at least $k$ vertices have not been burnt). It is known that TEMPORAL FIREFIGHTER RESERVE is NP-complete [12].

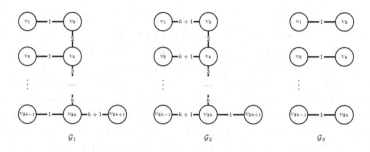

**Fig. 1.** For each edge $e$, the label of $e$ denotes the unique time label in $\lambda(e)$. For any $k \geq 2$, we have that $\mathrm{imw}_\leq(\mathcal{G}_1) = 1$ since, for any $t \in [k+1]$, each connected component of $G_\leq(t)$ contains only one edge of $\Psi_t^e$. On the other hand, $\mathrm{imw}_\geq(\mathcal{G}_1) = k$ since $G_\geq(1) = \mathcal{G}_1$ is connected and $\Psi_1^e$ contains all the $k$ edges $\{v_i, v_{i+1}\}$ for $i = \{1, 3, \ldots, 2k - 1\}$. Similarly, we have that $\mathrm{imw}_\leq(\mathcal{G}_2) = k$ while $\mathrm{imw}_\geq(\mathcal{G}_2) = 1$. Finally, $\mathrm{imw}(\mathcal{G}_3) = k$ (since $\Psi_1^e$ contains all edges), while, for any $d \in \{\leq, \geq\}$, $\mathrm{imw}_d(\mathcal{G}_3) = 1$ (since each connected component of $G_d(1)$ contains only one edge).

*Parameterized Complexity.* We use standard notation and terminology from parameterized complexity theory [7] and we say that a problem is *fixed-parameter tractable* (FPT) with respect to a parameter $k$ if it can be solved in time $f(k) \cdot n^{O(1)}$, where $n$ is the size of the input.

## 2 Connected (Vertex) Interval Membership Width

Given a temporal graph $\mathcal{G} = (G = (V, E), \lambda)$ with lifetime $\tau$, let the *active window* of an edge $e \in E$ be the interval $\mathrm{aw}(e) = [\min \lambda(e), \max \lambda(e)]$. Then, for each $t \in [\tau]$, let $\Psi_t^e = \{f \in E : t \in \mathrm{aw}(f)\}$ be the *activity bag at time $t$* of $\mathcal{G}$. The interval membership width [3] of $\mathcal{G}$ is defined as $\mathrm{imw}(\mathcal{G}) = \max_{t \in [\tau]} |\Psi_t^e|$. For each direction $d \in \{\leq, \geq\}$ and for each connected component $C$ of $G_d(t)$, let $\Psi_d^e(\mathcal{G}, t, C) = E(C) \cap \Psi_t^e$ be the *$d$-connected bag at time $t$* of $\mathcal{G}$ and $C$. Moreover, let $\mathcal{F}_d^e(t) = \{\Psi_d^e(\mathcal{G}, t, C) : C$ is a connected component of $G_d(t)\}$ be the family of $d$-connected bags at time $t$ of $\mathcal{G}$. The $d$-connected interval membership width of $\mathcal{G}$ is defined as $\mathrm{imw}_d(\mathcal{G}) = \max_{t \in [\tau], \Psi^e \in \mathcal{F}_d^e(t)} |\Psi^e|$. Note that these parameters can be computed in (almost) linear time by using disjoint-set data structures [19] for implementing incremental connectivity algorithms.

We first observe that the $\leq$-connected interval membership width and the $\geq$-connected interval membership width are two incomparable measures. That is, there exists an infinite family of temporal graphs $\mathcal{G}_1$ for which $\mathrm{imw}_\leq(\mathcal{G}_1)$ is arbitrarily smaller than $\mathrm{imw}_\geq(\mathcal{G}_1)$, and there exists an infinite family of temporal graphs $\mathcal{G}_2$ for which $\mathrm{imw}_\geq(\mathcal{G}_2)$ is arbitrarily smaller than $\mathrm{imw}_\leq(\mathcal{G}_2)$ (see Fig. 1).

For any temporal graph $\mathcal{G}$ and for each $d \in \{\leq, \geq\}$, we have that $\mathrm{imw}_d(\mathcal{G})$ is at most $\mathrm{imw}(\mathcal{G})$, since each bag in $\mathcal{F}_d^e(t)$ is a subset of $\Psi_t^e$. Moreover, the $d$-connected interval membership width can be arbitrarily smaller than the interval membership width (see the family of temporal graphs $\mathcal{G}_3$ shown in Fig. 1).

For any direction $d \in \{\leq, \geq\}$, we will also make use of the following vertex version of the $d$-connected interval membership width. Given a temporal graph $\mathcal{G} = (G = (V, E), \lambda)$ with lifetime $\tau$, let the *active window* of a vertex $v \in V(G)$ be the interval $\mathrm{aw}(v) = [\min \bigcup_{u \in N_G(v)} \lambda(\{u, v\}), \max \bigcup_{u \in N_G(v)} \lambda(\{u, v\})]$. Then, for each $t \in [\tau]$, let $\Psi_t^{\mathrm{v}} = \{u \in V : t \in \mathrm{aw}(u)\}$ be the *activity vertex bag* at time $t$ of $\mathcal{G}$. The vertex interval membership width [3] is defined as $\mathrm{vimw}(\mathcal{G}) = \max_{t \in [\tau]} |\Psi_t^{\mathrm{v}}|$. For each connected component $C$ of $G_d(t)$, let $\Psi_d^{\mathrm{v}}(\mathcal{G}, t, C) = V(C) \cap \Psi_t^{\mathrm{v}}$ be the vertex $d$-connected bag at time $t$ of $\mathcal{G}$ and $C$. Moreover, let $\mathcal{F}_d^{\mathrm{v}}(t) = \{\Psi_d^{\mathrm{v}}(\mathcal{G}, t, C) : C$ is a connected component of $G_d(t)\}$ be the family of vertex $d$-connected bags at time $t$ of $\mathcal{G}$. The $d$-connected vertex interval membership width of $\mathcal{G}$ is defined as $\mathrm{vimw}_d(\mathcal{G}) = \max_{t \in [\tau], \Psi^{\mathrm{v}} \in \mathcal{F}_d^{\mathrm{v}}(t)} |\Psi^{\mathrm{v}}|$. Once again, for any direction $d$, the $d$-connected vertex interval membership width is not greater than the vertex interval membership width (since each bag in $\mathcal{F}_d^{\mathrm{v}}(t)$ is a subset of $\Psi_t^{\mathrm{v}}$), and the $\leq$-connected vertex interval membership width and the $\geq$-connected vertex interval membership width are two incomparable measures, since the same examples given for the connected interval membership width work also for the connected vertex interval membership width.

## 3    Eulerian Trails Parameterized by $\mathrm{imw}_\leq$ and by $\mathrm{imw}_\geq$

In this section we consider the TEMPORAL EULERIAN TRAIL problem. This problem is FPT when parameterized by the interval membership width [3]. In this section, we show that this result can be improved by considering as a parameter the $d$-connected interval membership width, for any direction $d \in \{\leq, \geq\}$. Let us first consider the case $d = \leq$ (in the following, without loss of generality, we can assume that the underlying graph of the input temporal graph is connected). A $(u, v, t)$-trailset is a set of edges of $\mathcal{G}$ such that there exists a temporal trail from $u$ to $v$ in $\mathcal{G}$ which arrives in $v$ at time at most $t$ and which uses all and only the edges in the set. Given a temporal graph $\mathcal{G} = (G = (V, E), \lambda)$, let us define the following (dynamic programming) table $\overrightarrow{M}$. For each $t \in [\tau]$, each connected component $C$ of $G_\leq(t)$, each $u, v \in V(C)$, and each subset $F$ of $\Psi_\leq^{\mathrm{e}}(\mathcal{G}, t, C)$, we let $\overrightarrow{M}[t, C, u, v, F] = 1$ if and only if $E(C) \setminus F$ is a $(u, v, t)$-trailset.

**Fact 1.** $\mathcal{G} = (G = (V, E), \lambda)$ *has an Eulerian trail if and only if there exists* $u, v$ *such that* $\overrightarrow{M}[\tau, G, u, v, \emptyset] = 1$.

Now, given $t \in [\tau]$, a connected component $C$ of $G_\leq(t)$, a pair $u, v \in V(C)$, and $F \subseteq \Psi_\leq^{\mathrm{e}}(\mathcal{G}, t, C)$, we show how to recursively compute $\overrightarrow{M}[t, C, u, v, F]$.

**Base case.** If $t = 1$, then $\Psi_1^{\mathrm{e}} = E(G_1)$ and $G_\leq(1) = G_1$: hence, $\Psi_\leq^{\mathrm{e}}(\mathcal{G}, 1, C) = E(C) \cap \Psi_1^{\mathrm{e}} = E(C) \cap E(G_1) = E(C)$. We set $\overrightarrow{M}[1, C, u, v, F]$ to 1 if and only if $|F| = |E(C)| - 1$, $\{u, v\} \notin F$, and $1 \in \lambda(\{u, v\})$ (that is, $\{u, v\} \in E(C) \setminus F$).

**Recursive step.** Let $t \in [\tau]$ with $t > 1$. We set $\overrightarrow{M}[t, C, u, v, F]$ to 1 if and only if one of the following two cases occurs.

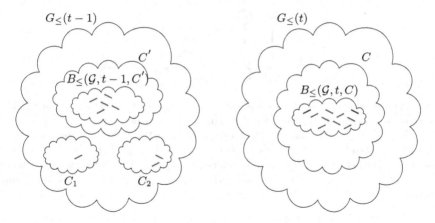

**Fig. 2.** Case 1 of the recursive step to compute table $\overrightarrow{M}$ for the TEMPORAL EULERIAN TRAIL problem. The red edges in $\Psi_{\leq}^{e}(\mathcal{G}, t, C)$ are the edges in $F$ which distribute among the three connected components of $G_{\leq}(t-1)$ in which the connected component $C$ of $G_{\leq}(t)$ is split. The blue edges in $\Psi_{\leq}^{e}(\mathcal{G}, t, C)$ are the edges in $F$ which are not contained in $G_{\leq}(t-1)$. If the $(u, v, t)$-trailset in $G_{\leq}(t)$ is equal to $E(C')$ minus the red edges in $\Psi_{\leq}^{e}(\mathcal{G}, t-1, C')$ (which form the set $F'$), then $\overrightarrow{M}[t, C, u, v, F]$ is set to 1 (the red edges of $F$ in the other two connected components $C_1$ and $C_2$ are not included in $F'$). (Color figure online)

**Case 1** There exists a connected component $C'$ of $G_{\leq}(t-1)$ and a subset $F'$ of $\Psi_{\leq}^{e}(\mathcal{G}, t-1, C')$ such that $\overrightarrow{M}[t-1, C', u, v, F'] = 1$ and $E(C) \setminus F = E(C') \setminus F'$ (see Fig. 2). In words, there is a desired trail finishing at time at most $t-1$.

**Case 2** There exists $e = \{w, v\} \in E(C) \setminus F$ such that $t \in \lambda(e)$, a connected component $C'$ of $G_{\leq}(t-1)$, and a subset $F'$ of $\Psi_{\leq}^{e}(\mathcal{G}, t-1, C')$ such that $\overrightarrow{M}[t-1, C', u, w, F'] = 1$ and $E(C) \setminus (F \cup \{\{w, v\}\}) = E(C') \setminus F'$. In words, the desired trail can be obtained through a trail from $u$ to some $w$ finishing at time at most $t-1$, and then using the edge $\{w, v\}$ active at time $t$.

By executing a "backward" version of the previously described dynamic programming algorithm, we can also solve TEMPORAL EULERIAN TRAIL parameterized by $\mathtt{imw}_{\geq}(\mathcal{G})$.

**Theorem 2.** *Given a temporal graph $\mathcal{G} = (G = (V, E), \lambda)$ with lifetime $\tau$, the* TEMPORAL EULERIAN TRAIL *problem with input $\mathcal{G}$ can be solved in time $O(2^w \tau n^4 m)$ where $n = |V|$, $m = |E|$, and $w \in \{\mathtt{imw}_{\leq}(\mathcal{G}), \mathtt{imw}_{\geq}(\mathcal{G})\}$.*

## 4    Vertex Coloring Parameterized by $\mathtt{imw}_{\leq}$ and by $\mathtt{imw}_{\geq}$

In this section, we consider TEMPORAL 2-COLORING. Deciding whether a temporal graph $\mathcal{G}$ admits a temporal 2-coloring is FPT when parameterized by the

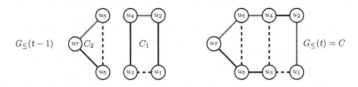

**Fig. 3.** The recursive step to compute table $\overrightarrow{M}$ for the vertex coloring problem. The dashed edges on the left (respectively, right) are the ones which are not in $\Psi^e_{t-1}$ (respectively, $\Psi^e_t$). The red edges on the left (respectively, on the right) are the ones in $F_1$ and $F_2$ (respectively, $F$). In this case, $(E(C) \setminus F) \setminus ((E(C_1) \setminus F_1) \cup (E(C_2) \setminus F_2)) = \{\{u_3, u_5\}, \{u_2, u_4\}\}$. Since the graph induced by these two edges is bipartite, if $\overrightarrow{M}[t-1, C_1, F_1] = \overrightarrow{M}[t-1, C_2, F_2] = 1$, then we set $\overrightarrow{M}[t, C, F] = 1$. (Color figure online)

treewidth of the underlying graph and the lifetime $\tau$ [16]. In the following, we show that this problem is FPT when parameterized by the $d$-connected interval membership width, for any direction $d \in \{\leq, \geq\}$. Let us first consider the case $d = \leq$. Given a temporal graph $\mathcal{G} = (G = (V, E), \lambda)$, let us define the following (dynamic programming) table $\overrightarrow{M}$. For each $t \in [\tau]$, each connected component $C$ of $G_\leq(t)$ with edge set $E(C)$, and each subset $F$ of $\Psi^e_\leq(\mathcal{G}, t, C)$, we let $\overrightarrow{M}[t, C, F] = 1$ if and only if there exists a function $f : V(C) \times [t] \to [2]$ which properly colors every edge in $E(C) \setminus F$.

**Fact 3.** Let $\mathcal{G} = (G = (V, E), \lambda)$ be a temporal graph with lifetime $\tau$. There exists a temporal 2-coloring of $\mathcal{G}$ if and only if $\overrightarrow{M}[\tau, G, \emptyset] = 1$ for every connected component $C$ of $G$.

Now, given $t \in [\tau]$, a connected component $C$ of $G_\leq(t)$ and a subset $F$ of $\Psi^e_\leq(\mathcal{G}, t, C)$, we show how to recursively compute $\overrightarrow{M}$.

**Base case.** If $t = 1$, then we set $\overrightarrow{M}[1, C, F]$ to 1 if and only if the graph $(V(C), E(C) \setminus F)$ is bipartite.

**Recursive step.** Let $t \in [\tau]$ with $t > 1$. We set $\overrightarrow{M}[t, C, F]$ to 1 if and only if there exist $q$ connected components $C_1, \ldots, C_q$ of $G_\leq(t-1)$ and $q$ sets $F_1, \ldots, F_q$ with $F_i \subseteq \Psi^e_\leq(\mathcal{G}, t-1, C_i)$, for $i \in [q]$, such that the following two properties are satisfied (see Fig. 3 for an example).

1. For every $i \in [q]$, $\overrightarrow{M}[t-1, C_i, F_i] = 1$.
2. The graph $(V(C), (E(C) \setminus F) \setminus \bigcup_{i=1}^q (E(C_i) \setminus F_i))$ is bipartite. Intuitively, this graph contains all the edges in $E(C) \setminus F$ which have not been properly colored before time $t$, and, hence, do not belong to any set $E(C_i) \setminus F_i$, for $i \in [q]$.

Again, by executing a "backward" version of the above algorithm, we can solve TEMPORAL 2-COLORING problem parameterized by $\mathrm{imw}_\geq(\mathcal{G})$.

**Theorem 4.** *The* TEMPORAL 2-COLORING *problem with input a temporal graph* $\mathcal{G} = (G = (V,E), \lambda)$ *with lifetime* $\tau$ *can be solved in time* $O(4^w \tau n^2 m)$ *where* $n = |V|$, $m = |E|$, *and* $w \in \{\mathtt{imw}_{\leq}(\mathcal{G}), \mathtt{imw}_{\geq}(\mathcal{G})\}$.

Observe, finally, that, in order to generalize our approach for higher values of $k$, it suffices to test whether $\Psi_d^e(\mathcal{G}, 1, C) \setminus F = E(C) \setminus F$ is $k$-colorable in the base case, and whether $f$ is a proper $k$-coloring of $(\Psi_d^e(\mathcal{G}, t, C) \setminus F) \setminus \bigcup_{i=1}^q (E(C_i) \setminus F_i)$ in item 2 of the recursive step. This clearly increases the time complexity of the algorithm by factor which is exponential in the $d$-connected interval membership width, giving an FPT algorithm for the TEMPORAL $k$-COLORING problem when parameterized by the $d$-connected interval membership width, for any direction $d \in \{\leq, \geq\}$, and $k$.

## 5  Firefighter Parameterized by $\mathtt{vimw}_{\leq}$

TEMPORAL FIREFIGHTER RESERVE is FPT when parameterized by the vertex interval membership width [12]. In the following, we show that this result can be improved by considering as a parameter the $\leq$-connected vertex interval membership width. Consider a temporal graph $\mathcal{G} = (G = (V,E), \lambda)$ with lifetime $\tau$ and a source node $s \in V(G)$. For each $t \in [\tau]$, we denote the connected component of $G_{\leq}(t)$ containing $s$ by $C_s^t$ and we denote by $\mathcal{G}_{\leq}^s(t)$ the related temporal graph (i.e., the temporal subgraph of $\mathcal{G}_{\leq}(t)$ constrained to $C_s^t$). Also, let $n$ denote $|V(G)|$. Observe that if $A(s) = [t_1, t_2]$ and $t_1 > 1$, then $C_s^t$ is the trivial component containing only $s$ for every $t \in [t_1 - 1]$. In such case, we can apply the following procedure starting at $t_1$ instead. To make presentation simpler, in what follows we consider that $t_1 = 1$. Let us now define the following (dynamic programming) table $\overrightarrow{M}$. For each pair $d, t \in [\tau]$, each two subsets $D$ and $B$ of $\Psi_{\leq}^v(\mathcal{G}, t, C_s^t)$, and each $b \in [n]$, we let $\overrightarrow{M}[t, D, B, d, b] = 1$ if and only if there exists a defence strategy $\mathcal{S} = (S_1, \ldots, S_t)$ in $\mathcal{G}_{\leq}^s(t)$ such that:

1.  $D$ is the set of vertices defended in bag $t$. Formally $D = \Psi_{\leq}^v(\mathcal{G}, t, C_s^t) \cap \bigcup_{i=1}^t S_i$;
2.  $B$ is the set of burnt vertices in bag $t$. Formally $B = \Psi_{\leq}^v(\mathcal{G}, t, C_s^t) \cap B_s^t(\mathcal{S})$;
3.  $d$ is the total number of defended vertices up to time $t$. Formally, $d = \sum_{i=1}^t |S_i|$. Observe that $d \leq t$ by definition; and
4.  $b$ is the total number of burnt vertices at time $t$. Formally, $b = |B_s^t(\mathcal{S})|$.

**Fact 5.** $(\mathcal{G}, s, k)$ *is a yes-instance of the* TEMPORAL FIREFIGHTER RESERVE *problem if and only if there exists* $D, B, d, b$ *such that* $\overrightarrow{M}[\tau, D, B, d, b] = 1$ *and* $n - b \geq k$.

Now, given $d, t \in [\tau]$, two subsets $D$ and $B$ of $\Psi_{\leq}^v(\mathcal{G}, t, C_s^t)$, and $b \in [n]$, we show how to recursively compute $\overrightarrow{M}[t, D, B, d, b]$.

**Base case.** Let $t = 1$. If $D = \emptyset$ (i.e., no vertex is defended at time 1), then we set $\overrightarrow{M}[1, D, B, d, b]$ to 1 if $B = \{s\} \cup N_1(s)$, $d = 0$ and $b = |B|$. If $D = \{u\}$ for some $u \in V(C_s^1)$, then we set $\overrightarrow{M}[1, D, B, d, b]$ to 1 if $B = \{s\} \cup (N_1(s) \setminus \{u\})$, $d = 1$ and $b = |B|$. In all other cases, we set $\overrightarrow{M}[1, D, B, d, b]$ to 0.

**Recursive step.** Let $t \in [\tau]$ with $t > 1$. We set $\overrightarrow{M}[t, D, F, d, b]$ to 1 if and only if there exist $D', B', d', b'$ such that $\overrightarrow{M}[t-1, D', B', d', b'] = 1$ and the following properties are satisfied.

(I)  $D = (D' \cap \Psi^{\vee}_{\leq}(\mathcal{G}, t, C^t_s)) \cup A$, where $A \subseteq \Psi^{\vee}_{\leq}(\mathcal{G}, t, C^t_s)) \setminus (B' \cup D')$.

(II)  $B = (B' \cap \Psi^{\vee}_{\leq}(\mathcal{G}, t, C^t_s)) \cup (N_t(B') \setminus D)$.

(III)  $d = d' + |A|$.

(V)  $b = b' + |N_t(B') \setminus D|$.

**Theorem 6.** *The* TEMPORAL FIREFIGHTER RESERVE *problem, with input a temporal graph* $\mathcal{G} = (G = (V, E), \lambda)$ *with lifetime* $\tau$, $s \in V$, *and* $k \in [n]$, *can be solved in time in time* $O(16^w \tau^4 n^2)$, *where* $n = |V|$ *and* $w = \mathtt{vimw}_{\leq}(\mathcal{G})$.

## 6    Bidirectional Connected Interval Membership Width

Given a temporal graph $\mathcal{G} = (G = (V, E), \lambda)$ with lifetime $\tau$, for each $t \in [\tau]$, we define the bidirectional connected interval membership width at time $t$ as

$$\mathtt{imw}_{\sim}(t) = \begin{cases} \max\{\mathtt{imw}_{\leq}(\mathcal{G}_{\leq}(t-1)), \mathtt{imw}_{\geq}(\mathcal{G}_{\geq}(t+1)), |\Psi^e_t|\} \,, & \text{if } 1 < t < \tau \\ \mathtt{imw}_{\geq}(\mathcal{G}) & \text{, if } t = 1 \\ \mathtt{imw}_{\leq}(\mathcal{G}) & \text{, if } t = \tau. \end{cases}$$

The bidirectional connected interval membership width of $\mathcal{G}$ is then defined as $\mathtt{imw}_{\sim}(\mathcal{G}) = \min_{t \in [\tau]} \mathtt{imw}_{\sim}(t)$. Note that also this parameter can be computed in (almost) linear time and that there exists an infinite family of temporal graphs $\mathcal{G}$ for which $\mathtt{imw}_{\sim}(\mathcal{G})$ is arbitrarily smaller than $\mathtt{imw}_d(\mathcal{G})$, for any direction $d$ (see Fig. 4). In order to solve a problem by referring to the bidirectional connected interval membership width, we can first solve it with input $\mathcal{G}_{\leq}(t-1)$ and with input $\mathcal{G}_{\geq}(t+1)$, by referring to the connected interval membership width of $\mathcal{G}_{\leq}(t-1)$ with direction $\leq$ and the connected interval membership width of $\mathcal{G}_{\geq}(t+1)$ with direction $\geq$, and then combine the two solutions. This approach can be followed in the case of the TEMPORAL EULERIAN TRAIL problem and of the TEMPORAL 2-COLORING problem, as stated by the following two results.

**Theorem 7.** *Given a temporal graph* $\mathcal{G} = (G, \lambda)$ *with* $n$ *nodes,* $m$ *edges, and lifetime* $\tau$, *the* TEMPORAL EULERIAN TRAIL *problem can be solved by an algorithm running in time* $O(2^{\mathtt{imw}_{\sim}(\mathcal{G})} \tau n^4 m)$.

**Theorem 8.** *Given a temporal graph* $\mathcal{G} = (G, \lambda)$ *with* $n$ *nodes,* $m$ *edges, and lifetime* $\tau$, *the* TEMPORAL 2-COLORING *problem can be solved by an algorithm running in time* $O(4^{\mathtt{imw}_{\sim}(\mathcal{G})} \tau n^2 m)$.

## 7    Conclusion

We have introduced a three-level hierarchy of polynomial-time computable parameters starting from the interval membership width up to the bi-directional version of the connected interval membership width (see Fig. 5). We proved that

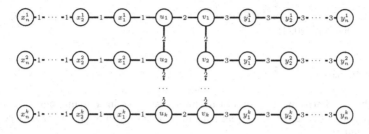

**Fig. 4.** For each edge $e$, the label of $e$ denotes the unique time label in $\lambda(e)$. For each $k, n \geq 1$ and for any direction $d$, we have that $\mathtt{imw}_d(\mathcal{G}) = (k+1)n$ (since $|\Psi_1^e| = |\Psi_3^e| = (k+1)n$, $G_{\leq}(3) = G_{\geq}(1) = G$, and $G$ is connected), while $\mathtt{imw}_{\sim}(\mathcal{G}) = n$ (since $|\Psi_2^e| = 2(k+1)$ and $\mathtt{imw}_{\leq}(\mathcal{G}_{\leq}(1)) = \mathtt{imw}_{\geq}(\mathcal{G}_{\geq}(3)) = n$).

the hierarchy is strict (that is, each parameter at one level can be arbitrarily smaller than the ones in the levels below), and that the two parameters at the second level are not comparable (that is, each of them can be arbitrarily smaller than the other). We also proved that TEMPORAL EULERIAN TRAIL and TEMPORAL 2-COLORING are FPT with respect to any of the parameters in the hierarchy, and that TEMPORAL FIREFIGHTER RESERVE is FPT with respect to the vertex-variant of the parameter at the first level and of one of the parameters of the second level. A natural research direction will be to design FPT algorithms with respect to our new parameters for other temporal graphs problems (such as counting temporal paths [8]). We also leave as an open question the classification of TEMPORAL FIREFIGHTER RESERVE when parameterized by $\mathtt{vimw}_{\geq}(\mathcal{G})$.

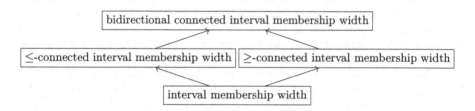

**Fig. 5.** The hierarchy of interval membership width parameters.

**Acknowledgement.** P.C. was supported by PNRR MIUR project GAMING "Graph Algorithms and MinINg for Green agents" (PE0000013, CUP D13C24000430001), A.M. by PNRR CN4 Centro Nazionale per la Mobilità Sostenibile, NextGeneration EU-CUP, B13C22001000001, by MUR of Italy, under PRIN Project n. 2022ME9Z78-NextGRAAL: Next-generation algorithms for constrained GRAph visuALization, and by PRIN PNRR Project n. P2022NZPJA-DLT-FRUIT: A user centered framework for facilitating DLTs FRUITion, A.S. by CNPq 303803/2020-7, 404479/2023-5, FUNCAP MLC-0191-00056.01.00/22, COFECUB 88887.712023/2022-00, and D.M.T. by the French-German Collaboration ANR/DFG Project UTMA (ANR-20-CE92-0027) and by the MEAE and the MESR, via the Franco-Norwegian project PHC Aurora projet n. 51260WL (2024).

# References

1. Bhadra, S., Ferreira, A.: Computing multicast trees in dynamic networks and the complexity of connected components in evolving graphs. J. Internet Serv. Appl. **3**(3), 269–275 (2012)
2. Bui-Xuan, B.M., Ferreira, A., Jarry, A.: Computing shortest, fastest, and foremost journeys in dynamic networks. Int. J. Found. Comput. Sci. **14**(02), 267–285 (2003)
3. Bumpus, B.M., Meeks, K.: Edge exploration of temporal graphs. Algorithmica **85**(3), 688–716 (2023)
4. Calamai, M., Crescenzi, P., Marino, A.: On computing the diameter of (weighted) link streams. ACM J. Exp. Algorithmics **27**, 1–28 (2022)
5. Casteigts, A., Flocchini, P., Quattrociocchi, W., Santoro, N.: Time-varying graphs and dynamic networks. Int. J. Parallel Emergent Distrib. Syst. **27**(5), 387–408 (2012)
6. Casteigts, A., Himmel, A.S., Molter, H., Zschoche, P.: Finding temporal paths under waiting time constraints. Algorithmica **83**, 2754–2802 (2021)
7. Cygan, M., et al.: Parameterized Algorithms. Springer, Cham (2015). https://doi.org/10.1007/978-3-319-21275-3
8. Enright, J.A., Meeks, K., Molter, H.: Counting temporal paths. In: 40th STACS. LIPIcs, vol. 254, pp. 30:1–30:19 (2023)
9. Finbow, S., MacGillivray, G.: The firefighter problem: a survey of results, directions and questions. Australas. J. Comb. **43**, 57–78 (2009)
10. Flocchini, P., Mans, B., Santoro, N.: Exploration of periodically varying graphs. In: International Symposium on Algorithms and Computation, pp. 534–543 (2009)
11. Fluschnik, T., Molter, H., Niedermeier, R., Renken, M., Zschoche, P.: As time goes by: reflections on treewidth for temporal graphs. In: Treewidth, Kernels, and Algorithms, pp. 49–77 (2020)
12. Hand, S.D., Enright, J.A., Meeks, K.: Making life more confusing for firefighters. In: International Conference on Fun with Algorithms, pp. 15:1–15:15 (2022)
13. Holme, P., Saramäki, J.: Temporal networks. Phys. Rep. **519**(3), 97–125 (2012)
14. Kempe, D., Kleinberg, J., Kumar, A.: Connectivity and inference problems for temporal networks. J. Comput. Syst. Sci. **64**(4), 820–842 (2002)
15. Latapy, M., Viard, T., Magnien, C.: Stream graphs and link streams for the modeling of interactions over time. Soc. Netw. Anal. Min. **8**(1), 61 (2018)
16. Marino, A., Silva, A.: Coloring temporal graphs. J. Comput. Syst. Sci. **123**, 171–185 (2022)
17. Marino, A., Silva, A.: Eulerian walks in temporal graphs. Algorithmica **85**(3), 805–830 (2022)
18. Michail, O.: An introduction to temporal graphs: an algorithmic perspective. Internet Math. **12**, 308–343 (2015)
19. Tarjan, R.E.: Efficiency of a good but not linear set union algorithm. J. ACM **22**(2), 215–225 (1975)
20. West, D.B., et al.: Introduction to graph theory, vol. 2. Prentice Hall, Upper Saddle River (2001)

# Efficient Algorithms for Decomposing Integers as Sums of Few Tetrahedral Numbers

Tong-Nong Lin[1], Yu-Cheng Lin[2], Cheng-Chen Tsai[1], Meng-Tsung Tsai[1(✉)], and Shih-Yu Tsai[1]

[1] Institute of Information Science, Academia Sinica, Taipei, Taiwan
{wilsonlin,saiss2022,mttsai,shihyutsai}@iis.sinica.edu.tw
[2] The Affiliated Senior High School of National Taiwan Normal University, Taipei, Taiwan
lance991144@gmail.com

**Abstract.** Pollock conjectures that every natural number can be expressed as a sum of at most five tetrahedral numbers. It remains unknown whether this conjecture holds, and Watson proved that sums of at most eight tetrahedral numbers suffice to express all natural numbers.

We devise two algorithms to decompose integers as sums of few tetrahedral numbers. Our first algorithm can decompose any given integer $n$ into a sum of at most eight tetrahedral numbers in $O(\log^3 n / \log \log n)$ time with probability $1 - 1/n^{\Omega(1)}$, assuming the extended Riemann hypothesis. Our second algorithm can deterministically decompose integers in $[1, \ell]$ into sums of the fewest possible tetrahedral numbers in $O(\ell)$ time using $O(\ell^{2/3})$ space, assuming a conjecture over the integers in $[1, \ell]$ and the Pollock's conjecture on tetrahedral numbers over the integers in $[1, \ell]$. While the conjectures for all integers are unproven, their validity for all integers in $[1, \ell]$ can be verified in $O(\ell)$ time.

As a result of our second algorithm, we can show that the Pollock's conjecture holds for all natural numbers up to $4.71 \times 10^{20}$. This significantly improves upon the previous known bound of $3.77 \times 10^{15}$.

**Keywords:** Pollock's conjecture · image set of cubic congruences · empirical verification

## 1 Introduction

Pollock conjectured in 1851 that every natural number can be expressed as a sum of at most five *tetrahedral numbers* [8], where for each $i \geq 1$ the $i$th tetrahedral number $T(i)$ is defined as

$$\binom{i+2}{3}.$$

This research was supported in part by the National Science and Technology Council under contract NSTC grants 112-2221-E-001-007.

In 1928, Yang proved that sums of at most nine tetrahedral numbers suffice to express all natural numbers [15], using Legendre's three-square theorem [6]. Subsequently, James proved in 1934 that sums of at most eight tetrahedral numbers suffice to express all sufficiently large natural numbers [4], using a method due to Landau [5]. Then, in 1935, Hua generalized James' result by showing that, for each integer $D$, sums of at most eight values in the form of $(i+1)D + T(i)$ suffice to express all sufficiently large natural numbers. Watson further strengthened James' and Hua's results in 1952, removing the requirement for the input integers to be sufficiently large [14]. In Watson's proof, the key observation is that the intersection of the image sets of some cubic congruences is non-empty (implicitly) using Hensel's lemma [11].

In this paper, we strengthen Watson's result by showing that this intersection is not only non-empty but also comprises a constant fraction of all possible incongruent values, as detailed in Lemma 4. We then use this finding to devise an algorithm to decompose any given integer $n$ into at most eight tetrahedral numbers, as stated in Theorem 1. The time complexity analyses in this paper are conducted on the standard model, WordRAM, where basic arithmetic operations over $O(\log n)$-bit operands take $O(1)$ time.

**Theorem 1.** *There exists an algorithm capable of representing an integer $n$ as a sum of at most eight tetrahedral numbers in $O(\log^3 n/\log\log n)$ time with probability $1 - 1/n^{\Omega(1)}$, assuming the extended Riemann hypothesis (ERH). This assumption can be removed if the decomposition of integers into three squares, a building block of this algorithm, can be done efficiently without assuming ERH.*

In addition to the mathematical attempts to prove the Pollock's conjecture, there have been computational efforts. In what follows, k-**numbers** refer to integers that cannot be expressed as sums of fewer than $k$ tetrahedral numbers, but can be expressed as sums of exactly $k$ tetrahedral numbers, following the notations in [1,2].

Salzer and Levine in 1958 [12] verified that the Pollock's conjecture holds for all integers in the closed interval $[1, 10^6]$. They reported that each integer in $[343868, 10^6]$ is a k-number for some $k \le 4$ and provided a list of 241 5-numbers. Deng and Yang in 1994 [2] improved the upper bound of the computation from $10^6$ to $10^9$ and reported that no new 5-number was found in their computation. It may be worth noting that, if there exists a k-number $n$ for some $k \ge 6$ and $n - n/k > 343,867$, then there exists a new 5-number $n' \in [n - n/k, n]$. Hence, reporting no new 5-number in a range implies no new k-number for $k \ge 5$ in the range was found. Chou and Deng in 1997 [1] further improved the upper bound of the computation to $4 \times 10^{10}$. Again, no new 5-number was found. Given these empirical results, it is conjectured in [1,2] that 343,867 is the largest 5-number. Furthermore, combining the above empirical result and a technique introduced in [1,2], they show that every positive integer up to $3,771,207,667,368,141$ is a k-number for some $k \le 5$. This gives the known best upper bound of integers for which the Pollock's conjecture holds.

As our second contribution in this paper, we devise an algorithm that can express all integers in $[1, \ell]$ as sums of the fewest possible tetrahedral numbers in

$O(\ell)$ time and $O(\ell^{2/3})$ space, assuming a conjecture over the integers in $[1, \ell]$ and the Pollock's conjecture on tetrahedral numbers over the integers in $[1, \ell]$. This result is formally stated in Theorem 2. Though the conjectures are unproven, their validity for all integers in $[1, \ell]$ can be verified in $O(\ell)$ time. A similar algorithm to express all integers in $[1, \ell]$ as sums of the fewest possible tetrahedral numbers with the same time complexity was claimed in [1], but the analysis is based on the following heuristic. For each 4-number $n$, let $r(n)$ be the rank of the smallest tetrahedral number $T(r(n))$ such that $n - T(r(n))$ is a 3-number. Their $O(\ell)$-time algorithm requires the assumption that $\lim_{n \to \infty} r(n) = O(1)$. They reported that $r(n) \leq 68$ for all 4-numbers less than or equal to $4 \times 10^{10}$, which we found to be incorrect. Indeed, there are 68 4-numbers $n \leq 4 \times 10^{10}$ with $r(n) > 68$, as listed in Table 2, and the largest $r(n)$ among them is 98.

**Theorem 2.** *There exists a deterministic algorithm capable of representing all integers $n \in [1, \ell]$ as sums of the fewest possible tetrahedral numbers in $O(\ell)$ time and $O(\ell^{2/3})$ space, assuming that the number of the 4-numbers $n$ in $[1, \ell]$ whose $r(n) = O(\log n)$ is at least $\ell - \ell^{1/3}$ and the Pollock's conjecture on tetrahedral numbers over the integers in $[1, \ell]$. This assumption over the integers in $[1, \ell]$ is unproven, but it can be empirically verified in $O(\ell)$ time with probability $1 - 1/\ell^{\Omega(1)}$.*

As a result of our second algorithm, we empirically verify that for all integers up to $10^{14}$ there exists no 5-number other than the known 241 ones. Together with an auxiliary computation, each integer up to $4.71 \times 10^{20}$ is a $k$-number for some $k \leq 5$. This gives a new upper bound on the integers for which the Pollock's conjecture holds and significantly improves upon the previous known bound. As a remark,

**Theorem 3.** *The Pollock's conjecture on tetrahedral numbers holds for all integers up to*

$$4.71 \times 10^{20}.$$

**Paper Organization.** In Sect. 2, we strengthen Watson's result by showing that the intersection of the image sets of some cubic congruences comprises a constant fraction of all possible incongruent values. Based on this findings we devise an efficient randomized algorithm to represent any given integer $n$ as a sum of at most eight tetrahedral numbers. Then, in Sect. 3, we devise an $O(\ell)$-time algorithm that can express all integers in $[1, \ell]$ as sums of the fewest possible tetrahedral numbers in $O(\ell)$ time and $O(\ell^{2/3})$ space, assuming two conjectures over integers in $[1, \ell]$. We also present how to empirically verify the unproven conjectures over the integers in $[1, \ell]$ in $O(\ell)$ time. Finally, in Sect. 4, we draw conclusions from our work.

## 2  Expressing Integers as Sums of at Most Eight Tetrahedral Numbers

We will prove Theorem 1 in this section. Our techniques are mainly based on Watson's techniques [14], Hensel's Lemma [11], and Chinese Remainder Theorem [11]; however, the key lemma (Lemma 4) that we obtain is very different from those lemmas in Watson's paper [14]. Lemma 4 and some simple arguments together yield a proof for Theorem 1.

**Lemma 1 (A restatement of [14, Lemma 3]).** *Let $n, m$ be positive integers so that $(m, 6) = 1$ and $m$ is a multiple of some square number. If $m$ and $n$ satisfy*

$$\left(\frac{1}{8} + \frac{9}{125}\right) m^3 < n < \frac{1}{4} m^3, \tag{1}$$

*then there exist integers $x, y, k$ so that $0 \leq x, y < 3m/5$, $0 \leq k < m^2/8$, and*

$$6n = x^3 - x + y^3 - y + \frac{1}{8}\left(6m^3 - 24\,m + 6m(8k + 3)\right). \tag{2}$$

By Legendre's three-square theorem [6], $8k + 3$ can be expressed as a sum of three squares for any integer $k \geq 1$. Hence, we can rewrite Eq. (2) as, for some $u, v, w \geq 0$,

$$6n = x^3 - x + y^3 - y + \frac{1}{8}\left(6m^3 - 24\,m + 6m(u^2 + v^2 + w^2)\right)$$

$$= 6T(x - 1) + 6T(y - 1) + \frac{1}{8}\sum_{i \in \{u,v,w\}} 2m^3 - 8m + 6i^2\, m$$

$$= 6T(x - 1) + 6T(y - 1) + \sum_{i \in \{u,v,w\}} \left(\frac{m+i}{2}\right)^3 - \frac{m+i}{2} + \left(\frac{m-i}{2}\right)^3 - \frac{m-i}{2}$$

$$= 6T(x - 1) + 6T(y - 1) + \sum_{i \in \{u,v,w\}} 6T\left(\frac{m+i}{2} - 1\right) + 6T\left(\frac{m-i}{2} - 1\right).$$

Because $u^2 + v^2 + w^2 \equiv 3 \pmod 8$, it follows that $u, v, w$ are odd integers. Additionally, since $0 \leq k < m^2/8$, we have $0 \leq u, v, w < m$. Combining that $m$ is an odd integer, both $(m + i)/2 - 1$ and $(m - i)/2 - 1$ for $i \in \{u, v, w\}$ are non-negative integers. Hence, $n$ can be expressed as a sum of at most eight tetrahedral numbers.

To implement the above decomposition, we need the following building blocks:

1. Computing a proper $m$.
2. Computing $x$ and $y$ such that $0 \leq x, y < 3m/5$ and satisfying the congruence relation $6n \equiv x^3 - x + y^3 - y \pmod m$, noting that $(6m^3 - 24m + 6m(8k + 3))/8 = sm$ has an integral solution for $k$ for any integer $s$,
3. Representing $8k + 3$ as a sum of three squares.

In Sects. 2.1 to 2.3, we show how to realize these building blocks in $O(\log^3 n/\log \log n)$ time, thereby proving Theorem 1.

## 2.1   Computing a Proper $m$

In this section, we present how to compute a proper $m$ in $O(\log n)$ time.

**Lemma 2.** *For every sufficiently large $n$, there exists an integer*

$$m = 5^a \cdot 7^b \cdot 11^c \text{ for some integers } a \geq 2, b \geq 0, c \geq 0$$

*so that Eq. (1) holds.*

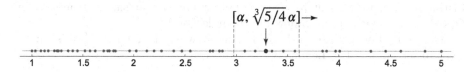

**Fig. 1.** An illustration of $I$, in which the dots represent the reals in $I$.

*Proof.* Let $I := \left\{ 5^a \cdot 7^b \cdot 11^c : a, b, c \text{ are integers in } [-4, 4] \right\} \cap [1, 5]$, as depicted in Fig. 1. It can be verified that $1, 5 \in I$ and every two consecutive reals in $I$ have a ratio within the range $(1, \sqrt[3]{5/4})$ where $\sqrt[3]{5/4}$ is selected to make (3) hold. Hence, for any real $\alpha \in [1, 5]$, the interval $[\alpha, \sqrt[3]{5/4}\alpha]$ contains at least one real number in $I$.

To find an $m$ satisfying Eq. (1), it suffices to have an $m$ such that

$$\sqrt[3]{4n} < m < \sqrt[3]{5n}. \tag{3}$$

As $n$ is sufficiently large, we can find some integer $a \geq 6$ such that $\ell = 5^a \cdot 7^4 \cdot 11^4$ and

$$\sqrt[3]{4n}/\ell \in [1, 5].$$

Let $\alpha = \sqrt[3]{4n}/\ell$. Then $[\sqrt[3]{4n}/\ell, \sqrt[3]{5n}/\ell] = [\alpha, \sqrt[3]{5/4}\alpha]$. By the property of $I$, $[\alpha, \sqrt[3]{5/4}\alpha]$ contains a real of the form $5^a \cdot 7^b \cdot 11^c$ for some integer $a, b, c \geq -4$. We are done.     □

The $m$ mentioned in Lemma 2 can be obtained by finding the $\ell$ mentioned in the proof of Lemma 2 followed by checking whether $\ell x$ is a proper $m$ by scanning all reals $x \in I$. $\ell$ can be found in $O(\log n)$ time by starting $\ell'$ with $5^6 \cdot 7^4 \cdot 11^4$ and iteratively multiplying $\ell'$ with 5 until $\ell'$ satisfies the need of $\ell$. Since $I$ contains only 47 real numbers, checking whether $\ell x$ is a proper $m$ can be done in $O(1)$ time.

**Remark.** One may wonder why not pick an $m = 5^2 \cdot p$ for some prime $p$. The reason is that the running time of the above is faster than using $m$ in the form $m = 5^2 \cdot p$.

Here are details. It is shown in [3, p. 494] that, for every real $\varepsilon > 0$,

$$\lim_{n \to \infty} \frac{\pi((1 + \varepsilon)n) - \pi(n)}{n/\log n} = \varepsilon.$$

Then, we sample an integer uniformly at random from the interval $[\sqrt[3]{4n}/25, \sqrt[3]{5n}/25]$ and, using Miller-Rabin's algorithm [9], test whether the sampled integer is a prime in $O(\log^2 n)$ time. To have the failure probability bounded by the claimed $1/n^{\Omega(1)}$, we require to sample $O(\log^2 n)$ integers in the interval. Hence, the total running time is $O(\log^4 n)$, exceeding our budget for the time complexity.

## 2.2    Computing $x$ and $y$

In this section, we present how to compute $x$ and $y$ that satisfy Lemma 1 in $O(\log^2 n)$ time with failure probability bounded by $1/n^{\Omega(1)}$.

We extend Lemma 1 from [14] to Lemma 3, where Lemma 1 of [14] is equivalent to setting $r = 1$ in Lemma 3.

**Lemma 3.** *Let $p$ be a prime number at least 5 and $r$ be a positive integer. If $a \not\equiv 0 \pmod{p}$, the image set $\mathcal{I}_r$ of the congruence*

$$x^3 + ax \pmod{p^r}$$

*contains exactly $p^{r-1}|\mathcal{I}_1|$ incongruent values where $|\mathcal{I}_1| = \lfloor (2p+1)/3 \rfloor$; that is,*

$$|\mathcal{I}_r := \{x^3 + ax \pmod{p^r} : x \in [p^r]\}| = p^{r-1}|\mathcal{I}_1|.$$

*Proof.* We will use Hensel's Lemma [11, Theorem 4.15] to have a many-to-one mapping from the image set $\mathcal{I}_r$ to $\mathcal{I}_1$. Let $f_n(x) = x^3 + ax - n$. We study in the following claim whether there exists a solution $v$ of $f_n(x) \equiv 0 \pmod{p}$ such that $f_n'(v) \not\equiv 0 \pmod{p}$.

*Claim 1.* For every $n \in \mathcal{I}_1$, there exists a value $v \in [p]$ such that

$$f_n'(v) \not\equiv 0 \pmod{p}.$$

*Proof.* We classify $n \in \mathcal{I}_1$ into the following three categories, where $\mathcal{I}_{1,t}$ for each $t \in [3]$ is the collection of all the values $n \in [p]$ such that the number of incongruent solutions to $f_n(x)$ is exactly $t$. Because $p$ is a prime, $f_n(x) \equiv 0 \pmod{p}$ has at most three incongruent solutions. Hence, $\mathcal{I}_1 = \mathcal{I}_{1,1} \cup \mathcal{I}_{1,2} \cup \mathcal{I}_{1,3}$.

- If $n \in \mathcal{I}_{1,1}$, then for some $v_0 \in [p]$ we have

$$f_n(x) \equiv (x - v_0)g(x) \pmod{p}$$

There are two subcases to discuss:
  - Case I: $g(v_0) \equiv 0 \pmod{p}$. In this case, $f_n(x) \equiv (x - v_0)^2(x - t) \pmod{p}$. Since $f_n(x)$ does not have the quadratic term, $-t - 2v_0 \equiv 0 \pmod{p}$. Since $a \not\equiv 0 \pmod{p}$, $v_0 \not\equiv 0 \pmod{p}$. Consequently, $t \not\equiv v_0 \pmod{p}$. Thus, we have $n \in \mathcal{I}_2$, a contradiction. Such a subcase cannot happen.
  - Case II: $g(v_0) \not\equiv 0 \pmod{p}$. Hence, $f_n'(v_0) \equiv g(v_0) + (v_0 - v_0)g'(v_0) \not\equiv 0 \pmod{p}$, as desired.

- If $n \in \mathcal{I}_{1,2}$, then for some $v_0 \not\equiv v_1$ (mod $p$) we have

$$f_n(x) \equiv (x - v_0)^2 (x - v_1).$$

Since $f_n(x)$ does not have the quadratic term, $-2v_0 \equiv v_1$ (mod $p$). Since $a \not\equiv 0$ (mod $p$), $v_0 \not\equiv 0$ (mod $p$). Thus, $v_0 \not\equiv -v_1$ (mod $p$). The incongruent solutions to $f'_n(x) \equiv 0$, if they exist, must be $v'$ and $-v'$ for some $v' \in [p]$. Since $v_0 \not\equiv -v_1$, at least one of them is not a solution to $f'_n(x)$, as desired.
- If $n \in \mathcal{I}_{1,3}$, then for some incongruent values $v_0, v_1, v_2 \in [p]$ we have

$$f_n(x) \equiv (x - v_0)(x - v_1)(x - v_2).$$

Since $f'_n(x) \equiv 0$ (mod $p$) has at most two incongruent solutions, at least one value $v' \in \{v_0, v_1, v_2\}$ satisfies $f'_n(v') \not\equiv 0$ (mod $p$), as desired.    ∎

The above claim ensures that, for every $n \in \mathcal{I}_1$, there exists $v \in [p]$ such that

$$\begin{cases} f_n(v) \equiv 0 \pmod{p} \\ f'_n(v) \not\equiv 0 \pmod{p} \end{cases}$$

By Hensel's Lemma [11, Theorem 4.15], there exists an unique $t \in [p]$ such that $v + tp$ is a solution to $f_n(x) \equiv 0$ (mod $p^2$). Since $f'_n(v + tp) \equiv f'_n(v)$ (mod $p$). Inductively, we get $n \in \mathcal{I}_r$ as well. Conversely, for every $n' \in \mathcal{I}_r$, $(n' \bmod p) \in \mathcal{I}_1$.

For each $n' \in [p^r]$, its analysis is the same as $n' \bmod p$. Thus, $|\mathcal{I}_r| = |\mathcal{I}_1| p^{r-1}$. In [14, Lemma 1], $|\mathcal{I}_1| = \lfloor (2p+1)/3 \rfloor$ is shown. This completes the proof.    □

By applying the Pigeonhole Principle and Chinese Remainder Theorem, we derive the following lemma:

**Lemma 4.** *Let $m = p_1^{r_1} p_2^{r_2} \cdots p_t^{r_t}$ where $p_i$ are distinct prime numbers at least 5 and $r_i$ are positive integers for all $i \in [t]$. If $(m, a) = 1$, for every $n \in [m]$ the congruence*

$$x^3 + ax + y^3 + ay \equiv n \pmod{m}$$

*is soluble by setting $x^3 + ax \equiv n'$ (mod $m$) for every $n' \in \mathcal{I}$ where $\mathcal{I} \subset [m]$ and*

$$|\mathcal{I}| \geq \prod_{i \in [t]} (2 \lfloor (2p_i + 1)/3 \rfloor - p_i) \, p_i^{r_i - 1}.$$

*Proof.* By Lemma 3 (where we require $a \not\equiv 0$ (mod $p_i^{r_i}$)), for each $i \in [t]$, the image sets of $x^3 + ax$ (mod $p_i^{r_i}$) and $y^3 + ay$ (mod $p_i^{r_i}$) both contain $\lfloor (2p_i + 1)/3 \rfloor p_i^{r_i - 1}$ incongruent values. By the Pigeonhole Principle, for each $i \in [t]$, there exist at least

$$(2 \lfloor (2p_i + 1)/3 \rfloor - p_i) \, p_i^{r_i - 1} \tag{4}$$

incongruent values $s_i$ in the image set of $x^3 + ax$ (mod $p_i^{r_i}$) such that $y^3 + ay \equiv n - s_i$ (mod $p_i^{r_i}$) is soluble.

For each $i \in [t]$, let $\alpha_i$ denote a solution to $x^3 + ax \equiv s_i$ (mod $p_i^{r_i}$), and let $\beta_i$ denote a solution to $y^3 + ay \equiv n - s_i$ (mod $p_i^{r_i}$). Since $p_i$ for all $i \in$

$[t]$ are distinct primes, by the Chinese Remainder Theorem for each $2t$-tuple $(\alpha_1, \alpha_2, \ldots, \alpha_t, \beta_1, \beta_2, \ldots, \beta_t))$, we can find a unique $(\alpha, \beta)$ such that for all $i \in [t]$

$$\alpha \equiv \alpha_i \text{ and } \beta \equiv \beta_i \pmod{p_i^{r_i}}.$$

Thus, for any choices of $s_i$ for $i \in [t]$, we can find a 2-tuple $(\alpha, \beta)$ as a solution to $x^3 + ax + y^3 + ay \equiv n \pmod{m}$.

Note that, for each $i \in [t]$, the number of choices of $s_i$ is lower-bounded by Eq. (4). Each possible $t$-tuple $(s_1, s_2, \ldots, s_t)$ determines a unique $s^\dagger \pmod{m}$ such that

$$s^\dagger \equiv s_i \pmod{p_i^{r_i}} \text{ for } i \in [t].$$

Let $\mathcal{I}$ be the collection of all such $s^\dagger$s. Thus $\mathcal{I}$ has size at least the product of Eq. (4) for all $i \in [t]$. We are done. □

Thus, by Lemma 4, we can sample an $n' \in \mathcal{I}$ with success probability $> (1/3)^3 - \varepsilon$ for some small constant $\varepsilon > 0$ because the $m$ we picked in Lemma 2 has at most three prime factors. Given $n'$, to check whether it is valid it suffices to find a solution to $x^3 + ax \equiv n' \pmod{p}$ and $y^3 + ay \equiv n - n' \pmod{p}$ for each $p \in \{5, 7, 11\}$. The reason is stated in the proof of Lemma 4. As each $p \in \{5, 7, 11\}$ is a constant, checking the validity of $n'$ takes $O(1)$ time for each random guess and $O(\log n)$ time to succeed with probability $1 - 1/n^{\Omega(1)}$.

Given a valid $n'$, we need to find a solution to $x^3 + ax \equiv n' \pmod{m}$ and another to $y^3 + ay \equiv n - n' \pmod{m}$. To accomplish this, we solve $x^3 + ax \equiv n' \pmod{p}$ and $y^3 + ay \equiv n - n' \pmod{p}$ for each $p \in \{5, 7, 11\}$, the set of all prime factors of the picked $m$. This can be done in $O(1)$ time, as $p = O(1)$. Then, we use Hensel's Lemma [11, Theorem 4.15] to lift the roots to the right powers of $p$. The total number of times to lift the powers is $O(\log n)$ and each takes $O(1)$ time, as $p = O(1)$. Finally, we use an efficient quadratic-time algorithm for Chinese Remainder Theorem [13] to combine the found solutions to $x^3 + ax \equiv n' \pmod{p^r}$ and $y^3 + ay \equiv n - n' \pmod{p^r}$ for each $p \in \{5, 7, 11\}$, for some integer $r \geq 1$. The final step takes $O(\log^2 n)$ time, dominating the total running time. To get $0 \leq x, y < 3m/5$, we require a fix by the following lemma Lemma 5.

**Lemma 5.** *Let* $m = p_1^{r_1} p_2^{r_2} \cdots p_t^{r_t}$ *where* $p_i$ *are distinct prime numbers at least 5 and* $r_i$ *are positive integers for all* $i \in [t]$. *If* $(m, a) = 1$ *and* $r_1 \geq 2$, *given a solution* $(x_0, y_0)$ *to the congruence*

$$x^3 + ax + y^3 + ay \equiv n \pmod{m/p_1},$$

*then in* $O(p_1)$ *time one can obtain a solution* $(x_1, y_1)$ *to the congruence*

$$x^3 + ax + y^3 + ay \equiv n \pmod{m}$$

*such that* $0 \leq x_1, y_1 < 3m/5$.

*Proof.* Let $x_1 = x_0 + u(m/p_1)$ and $y_1 = y_0 + v(m/p_1)$ for some $u, v$ to be determined. To let $(x_1, y_1)$ satisfy the congruence

$$x^3 + ax + y^3 + ay \equiv n \pmod{m},$$

it suffices to require that

$$(3x_0^2 + a)u + (3y_0^2 + a)v \equiv 0 \pmod{p_1}. \tag{5}$$

As stated in the proof of Lemma 3, $x_0$ and $y_0$ are picked so that $3x_0^2 + a \not\equiv 0$ (mod $p_1$) and $3y_0^2 + a \not\equiv 0$ (mod $p_1$). Thus, we can rewrite Eq. (5) as

$$u + \lambda v \equiv 0 \text{ for some } \lambda \not\equiv 0 \pmod{p_1}. \tag{6}$$

Thus, by setting $u$ as a value $u_0$ in $[p]$, there is a unique $v_0$ such that $(u_0, v_0)$ is a solution to Eq. (6). Hence, by the Pigeonhole Principle, there exists a value $u_0$ in $[0, (p_1 - 1)/2]$, by setting $u$ as $u_0$ the corresponding $v_0$ also in $[0, (p_1 - 1)/2]$. Consequently, $x_1 = x_0 + u_0(m/p_1) < (u_0 + 1)(m/p_1) \leq m(p_1 + 1)/(2p_1) \leq 3m/5$. The last inequality holds because $p_1 \geq 5$.

To find $(u_0, v_0)$ that satisfies Eq. (6) and $u_0, v_0 \in [0, (p_1 - 1)/2]$, one can enumerate all $u_0 \in [p]$, which takes $O(p_1)$ time.     □

### 2.3  Representing $8k + 3$ as a Sum of Three Squares

In this section, we present how to express $8k + 3$ for any $k \geq 1$ as a sum of three squares in $O(\log^3 n/\log\log n)$ time with failure probability bounded by $1/n^{\Omega(1)}$.

In the literature, there are two algorithms [7,10] for this task in $O(\text{polylog}\, n)$ time. We opt to use the algorithm from [7] for this decomposition task, as it relies only on the Extended Riemann Hypothesis. Although its expected running time is $O(\log^2 n/\log\log n)$, to ensure a successful outcome with probability $1 - 1/n^{\Omega(1)}$, the running time increases to $O(\log^3 n/\log\log n)$, as desired.

## 3  Expressing Each Integer as a Sum of the Fewest Possible Tetrahedral Numbers

We will prove Theorem 2 in this section by devising an $O(\ell)$-time algorithm that, for all $n \in [1, \ell]$, determine the integer $k$ such that $n$ is a $k$-number. Then we report some empirical results obtained from this algorithm, which yields Theorem 3.

Our $O(\ell)$-time algorithm relies on the validity of the Pollock's conjecture on tetrahedral numbers and Conjecture 1 over the integers in $[1, \ell]$. It is still unknown whether these conjectures hold for all integers. However, its validity over integers in $[1, \ell]$ can be empirically verified in $O(\ell)$ time with probability $1 - 1/\ell^{\Omega(1)}$. See Sect. 3.3 for details. Conjecture 1 is motivated by an observation in [1] that almost all 4-numbers $n$ up to $4 \times 10^{10}$ have value $r(n) \leq 30$, and by our observation that the distribution of the values $r(n)$ of all 4-numbers $n$ up to $10^{12}$, as depicted in Fig. 2. Besides, Conjecture 1 holds for all the ranges in our conducted experiments, as shown in Table 1.

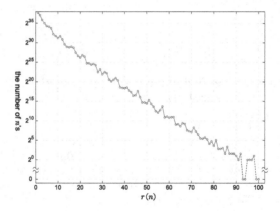

**Fig. 2.** The distributions of $r$ values for all 4-numbers up to $10^{12}$.

*Conjecture 1.* At least $\ell - \ell^{1/3}$ 4-numbers $n$ in $[1, \ell]$ have $r(n) = O(\log n)$.

### 3.1    A Simple Algorithm Based on Conjecture 1 and the Pollock's Conjecture

Our simple $O(\ell)$-time algorithm works as follows. To determine the integer $k$ for all $n \in [1, \ell]$ such that $n$ is a $k$-number, we need only the first $O(\ell^{1/3})$ 1-numbers because any 1-number appears later is greater than $\ell$. We use the formula of tetrahedral numbers to obtain the first $O(\ell^{1/3})$ numbers in $O(\ell^{1/3})$ time. Then, the computation of all 2-numbers and 3-numbers in $[1, \ell]$ can be achieved by enumerating the sums of any two and any three of the first $O(\ell^{1/3})$ 1-numbers, which takes $O(\ell)$ time.

To compute all the 4-numbers in $[1, \ell]$, we proceed as follows:

- Step 1. Let $A$ be the indicator bit-array of all the $k$-numbers in $[1, \ell]$ for all $k \leq 3$. Compute the logical-OR of $A$ with shifts by $\delta$ for all $\delta \in \{0\} \cup \{T(i) : i \in O(\log \ell)\}$. The logical OR of these $O(\log \ell)$-many $O(\ell)$-bit arrays takes $O(\ell)$ time because the model of computation is WordRAM where basic arithmetic operations (e.g. logical OR and shift) over $O(\log \ell)$-bit operands take $O(1)$ time. Let $B$ be the result of the above logical OR operations. By definition, $B$ is an indicator bit-array of all the 3-numbers in $[1, \ell]$ and all the 4-numbers $n$ in $[1, \ell]$ with $r(n) = O(\log \ell)$.
- Step 2. By Conjecture 1, $B$ has $O(\ell^{1/3})$ 0-bits. Each 0-bit corresponds to either a 4-number $x$ with $r(x) > \Delta$ for some $\Delta = O(\log \ell)$ or a $k$-number for some $k \geq 5$. Let $n$ be the index of one of the 0-bits in $B$. To verify whether $n$ is a 4-number, it suffices to check whether $n$ can be expressed as a sum of two 2-numbers. This verification can be done in $O(\ell^{2/3})$ time, given the sorted list of 2-numbers. To verify whether $n$ is a 5-number, it suffices to check if $n$ subtracted by any 2-numbers is a 3-number. Given $A$ and the sorted list of 2-numbers, this verification also can be done in $O(\ell^{2/3})$. For all 0-bits,

**Table 1.** The distribution of $k$-numbers for all positive integers up to $10^{14}$. The unreported numbers are 4-numbers.

|       | # 1-numbers | # 2-numbers | # 3-numbers   | # 5-numbers | # $n$ with $r(n) > 68$ |
|-------|-------------|-------------|---------------|-------------|------------------------|
| 2T    | 22893       | 231360135   | 892533269700  | 241         | 983                    |
| 4T    | 5950        | 135938272   | 892566316867  | 0           | 742                    |
| 6T    | 4175        | 114024249   | 892579370745  | 0           | 733                    |
| 8T    | 3323        | 101777580   | 892586662215  | 0           | 725                    |
| 10T   | 2806        | 93543895    | 892594264156  | 0           | 768                    |
| 12T   | 2453        | 87465970    | 892597268218  | 0           | 743                    |
| 14T   | 2194        | 82715101    | 892600240260  | 0           | 763                    |
| 16T   | 1993        | 78854043    | 892605925192  | 0           | 678                    |
| 18T   | 1834        | 75627012    | 892607810496  | 0           | 694                    |
| 20T   | 1702        | 72871243    | 892611249359  | 0           | 684                    |
| 22T   | 1592        | 70477495    | 892611703794  | 0           | 699                    |
| 24T   | 1498        | 68370546    | 892613608688  | 0           | 718                    |
| 26T   | 1418        | 66496077    | 892615100836  | 0           | 685                    |
| 28T   | 1346        | 64810643    | 892616483208  | 0           | 704                    |
| 30T   | 1284        | 63284787    | 892619064152  | 0           | 711                    |
| 32T   | 1227        | 61891868    | 892620842289  | 0           | 705                    |
| 34T   | 1178        | 60616223    | 892621019193  | 0           | 670                    |
| 36T   | 1133        | 59437585    | 892621494377  | 0           | 671                    |
| 38T   | 1091        | 58348054    | 892623074380  | 0           | 660                    |
| 40T   | 1053        | 57331847    | 892625811511  | 0           | 733                    |
| 42T   | 1019        | 56383995    | 892624729147  | 0           | 647                    |
| 44T   | 987         | 55495820    | 892626082051  | 0           | 645                    |
| 46T   | 958         | 54661584    | 892627506471  | 0           | 668                    |
| 48T   | 930         | 53874411    | 892629429638  | 0           | 696                    |
| 50T   | 905         | 53131427    | 892628890795  | 0           | 729                    |
| 52T   | 881         | 52427429    | 892629999069  | 0           | 704                    |
| 54T   | 858         | 51758363    | 892629240944  | 0           | 656                    |
| 56T   | 838         | 51124557    | 892630608375  | 0           | 673                    |
| 58T   | 818         | 50519244    | 892631146215  | 0           | 669                    |
| 60T   | 799         | 49941156    | 892631761094  | 0           | 666                    |
| 62T   | 782         | 49390221    | 892633870857  | 0           | 656                    |
| 64T   | 765         | 48861174    | 892633446685  | 0           | 702                    |
| 66T   | 750         | 48355792    | 892634312041  | 0           | 650                    |
| 68T   | 734         | 47868742    | 892634383912  | 0           | 691                    |
| 70T   | 720         | 47402034    | 892635181693  | 0           | 681                    |
| 72T   | 707         | 46953191    | 892636150470  | 0           | 648                    |
| 74T   | 693         | 46519325    | 892636792060  | 0           | 732                    |
| 76T   | 682         | 46103203    | 892636618322  | 0           | 777                    |
| 78T   | 669         | 45700380    | 892637949388  | 0           | 703                    |
| 80T   | 658         | 45311135    | 892636248600  | 0           | 682                    |
| 82T   | 647         | 44934977    | 892637894497  | 0           | 704                    |
| 84T   | 637         | 44571423    | 892638390967  | 0           | 663                    |
| 86T   | 626         | 44218320    | 892638667205  | 0           | 641                    |
| 88T   | 617         | 43877585    | 892640243280  | 0           | 673                    |
| 90T   | 608         | 43546706    | 892639526727  | 0           | 704                    |
| 92T   | 599         | 43225186    | 892639595445  | 0           | 707                    |
| 94T   | 590         | 42912932    | 892640997740  | 0           | 680                    |
| 96T   | 582         | 42609741    | 892641631626  | 0           | 699                    |
| 98T   | 574         | 42314723    | 892641152283  | 0           | 683                    |
| 100T  | 566         | 42027949    | 892642642346  | 0           | 715                    |

the above takes $O(\ell)$ time. If any number passes both tests, then it is a $k$-number for some $k \geq 6$. This refutes the Pollock's conjecture. Otherwise, the decomposition of all the integers in $[1, \ell]$ into the fewest possible tetrahedral numbers is completed.

To sum up, we have an $O(\ell)$-time algorithm as claimed in Theorem 2, assuming Conjecture 1 and the Pollock's conjecture over the integers in $[1, \ell]$.

## 3.2  Reducing the Space from $O(\ell)$ to $O(\ell^{2/3})$

The number of all 2-numbers in $[1, \ell]$ is $O(\ell^{2/3})$ and the length of the indicator bit-array $A$ is $O(\ell)$. Our approach is to divide $A$ into subintervals, each of length $O(\ell^{2/3})$. Then the algorithm in Sect. 3 operates on $A$ subinterval by subinterval. Constructing $A$ subinterval by subinterval does not incur more running time than constructing $A$ entirely at once. Note that $A$ can be constructed by enumerating all $x$ in the list of 1-numbers in $[1, \ell]$ and all $y$ in the list of 2-numbers in $[1, \ell]$ and summing each pair of $x$ and $y$. While constructing $A$ subinterval by subinterval, for each $x$ we use binary search to locate the lower- and upper-bounds of $y$ in the list of 2-numbers in $[1, \ell]$ such that $x + y$ falls within the current subinterval of $A$. Therefore, each pair of $x$ and $y$ joins at most one subinterval. The total running time is thus $O(\ell^{2/3} \log \ell + \ell) = O(\ell)$.

## 3.3  Empirical Verification of Conjecture 1

Below, we discuss how to verify Conjecture 1 over the integers in $[1, \ell]$ in $O(\ell)$ time. Let $\Delta = O(\log \ell)$ be an integer. Let $X$ be a random variable indicating that

$$X = \begin{cases} 1 & \text{if a uniform random sample } x \text{ from } [1, \ell] \text{ is a 4-number with } r(x) > \Delta \\ 0 & \text{otherwise} \end{cases}$$

Thus, it suffices to show that $E[X] \leq \ell^{-2/3}$. Let $S$ be the sum variable of $\Omega(\ell^{2/3} \log \ell)$ independent copies of $X$. By Chernoff bound, we have $\Pr\{|S - E[S]| \geq \varepsilon E[S]\} = e^{\Omega(\varepsilon^2 E[S]/(2+\varepsilon))}$. Thus, if $E[X] = \Omega(\ell^{-2/3})$, Chernoff bound yields a constant approximation for $E[X]$ with probability $1 - 1/\ell^{\Omega(1)}$. Otherwise $E[X] = \ell^{-2/3}/k$ for some $k = \omega(1)$, setting $\varepsilon E[S] = \log \ell$ also suffices to tell whether $E[X] < \ell^{-2/3}$ with probability $1 - 1/\ell^{\Omega(1)}$.

As a result, $O(\ell^{2/3} \log \ell)$ independent copies of $X$ suffice to verify Conjecture 1. Evaluating a copy of $X$ takes $O(\Delta)$ time if $A$ is given, which takes $O(\ell)$ time to prepare.

## 3.4  Verification of the Pollock's Conjecture up to $10^{20}$

Table 1 shows that there are only 241 5-numbers in $[1, 10^{14}]$. Thus, for every two consecutive tetrahedral numbers $T(i)$ and $T(i+1)$ whose difference is at most $10^{14}$, each integer between $T(i)$ and $T(i+1)$ is a $k$-number for some $k \leq 5$ except

**Table 2.** Errata.

| n | $r(n)$ | n | $r(n)$ | n | $r(n)$ | n | $r(n)$ |
|---|---|---|---|---|---|---|---|
| 837293 | 96 | 11332666483 | 80 | 23547651727 | 71 | 31820242053 | 72 |
| 1751787 | 81 | 12115559363 | 71 | 24041633103 | 69 | 32783549982 | 76 |
| 468164933 | 69 | 12636741707 | 72 | 24218940783 | 81 | 33723523927 | 78 |
| 725334878 | 70 | 12709829722 | 71 | 25902476693 | 69 | 33792979677 | 73 |
| 726409283 | 69 | 13646925297 | 73 | 26867883863 | 69 | 34001374367 | 70 |
| 1872385653 | 79 | 14533010753 | 75 | 27429950322 | 70 | 34085489982 | 72 |
| 1999255043 | 74 | 14780388803 | 69 | 28488454738 | 70 | 34118155573 | 71 |
| 2390056433 | 72 | 16212264602 | 81 | 28954173022 | 71 | 34492104467 | 73 |
| 3281447262 | 70 | 16490211457 | 71 | 29020094637 | 70 | 35040177258 | 71 |
| 4269476262 | 70 | 18131378533 | 71 | 29375497387 | 71 | 35248597663 | 69 |
| 5631140023 | 72 | 18685629958 | 75 | 30106439547 | 81 | 35780935678 | 72 |
| 6240760377 | 70 | 21042275158 | 69 | 30224854987 | 69 | 37306165122 | 71 |
| 6349723882 | 70 | 21282661867 | 71 | 30404017737 | 81 | 38970420417 | 71 |
| 8798752737 | 71 | 21400673177 | 71 | 31249750702 | 71 | 38978879802 | 98 |
| 9074616777 | 76 | 21909670998 | 76 | 31391571118 | 84 | 39282652547 | 74 |
| 9518151603 | 71 | 22491659283 | 69 | 31413964447 | 71 | 39389678018 | 75 |
| 10873443037 | 69 | 23084057267 | 69 | 31720955863 | 69 | 39981386443 | 69 |

for $T(i) + x$ for all $x$ in the set of 241 5-numbers. However, these exceptional numbers all fall within $[T(i), T(i-1) + 10^{14}]$, which are not exceptional numbers w.r.t. $T(i-1)$ for sufficiently large $i$. This yields a proof of Theorem 3. This technique is introduced in [1].

## 4 Conclusion

In this paper, we devise two algorithms to decompose integers as sums of few tetrahedral numbers. Our results leave some interesting directions to further explore.

The algorithm in Sect. 2 is the first one for decomposing integers into sums of at most eight tetrahedral numbers in time polynomial in the input size, assuming the extended Riemann hypothesis. This assumption can be removed if the decomposition of integers into three squares, a building block of this algorithm, can be done efficiently without assuming ERH. It is natural to ask whether ERH is necessary for this task.

Our second algorithm relies on an assumption on the distribution of $r$ values. We wonder whether this assumption for all integers can be rigorously proved.

**Acknowledgment.** We want to thank Dr. Pangfeng Liu for introducing us to the Pollock's conjecture on tetrahedral numbers.

# References

1. Chou, C., Deng, Y.: Decomposing 40 billion integers by four tetrahedral numbers. Math. Comput. **66**(218), 893–901 (1997)
2. Deng, Y.F., Yang, C.N.: Waring's problem for pyramidal numbers. Sci. China (Scientia Sinica) Ser. A **37**(3), 277–283 (1994)
3. Hardy, G.H., Wright, E.M.: An Introduction to the Theory of Numbers, 6th edn. Oxford University Press, Oxford (2008)
4. James, R.D.: The representation of integers as sums of pyramidal numbers. Math. Ann. **109**, 196–199 (1934)
5. Landau, E.: Über eine anwendung der primzahltheorie auf das waringsche problem in der elementaren zahlentheorie. Math. Ann. **66**, 102–105 (1908)
6. Legendre, A.M.: Essai sur la Théorie des Nombres. Cambridge Library Collection - Mathematics, 2nd edn. Cambridge University Press, Cambridge (2009)
7. Pollack, P., Schorn, P.: Dirichlet's proof of the three-square theorem: an algorithmic perspective. Math. Comput. **88**(316), 1007–1019 (2019)
8. Pollock, J.F.: On the extension of the principle of Fermat's theorem of the polygonal numbers to the higher orders of series whose ultimate differences are constant. with a new theorem proposed, applicable to all the orders. Proc. Roy. Soc. London **5**, 922–924 (1851)
9. Rabin, M.O.: Probabilistic algorithm for testing primality. J. Number Theory **12**(1), 128–138 (1980)
10. Rabin, M.O., Shallit, J.O.: Randomized algorithms in number theory. Comm. Pure Appl. Math. **39**, 239–256 (1986)
11. Rosen, K.: Elementary Number Theory and Its Applications. Pearson, London (2011)
12. Salzer, H.E., Levine, N.: Table of integers not exceeding 1,000,000 that are not expressible as the sum of four tetrahedral numbers. Math. Comput. **12**, 141–144 (1958)
13. Selianinau, M.: Computationally efficient approach to implementation of the Chinese remainder theorem algorithm in minimally redundant residue number system. Theory Comput. Syst. **65**, 1117–1140 (2021)
14. Watson, G.L.: Sums of eight values of a cubic polynomial. J. London Math. Soc. **s1-27**(2), 217–224 (1952)
15. Yang, K.C.: Various generalization of Waring's problem. Thesis, Chicago University (1928)

# Approximation Algorithms for Node-Weighted Directed Steiner Problems

Gianlorenzo D'Angelo[1]([✉])[ID] and Esmaeil Delfaraz[2]([✉])[ID]

[1] Gran Sasso Science Institute (GSSI), L'Aquila, Italy
gianlorenzo.dangelo@gssi.it
[2] Centre of EXcellence EX-Emerge, University of L'Aquila, L'Aquila, Italy
esmaeil.delfarazpahlevanloo@univaq.it

**Abstract.** Guha et al. [STOC, 1999] and Moss and Rabani [SIAM J. Comput., 2007] introduced two variants of the Steiner problem in undirected graphs in which the nodes are associated with two values, called costs and prizes. In the *budgeted rooted node-weighted Steiner tree* problem, we are given an undirected graph $G$ with $n$ nodes, a predefined node $r$, costs and prizes associated to the nodes of $G$, and a budget $B$. The aim is to find a tree in $G$ rooted at $r$ such that the total cost of its nodes is at most $B$ and the total prize is maximized. In the *quota rooted node-weighted Steiner tree* problem, we are given a quota $Q$, instead of the budget, and we aim at minimizing the cost of a tree rooted at $r$ whose overall prize is at least $Q$.

If the graph is undirected both problems can be approximated within polylogarithmic factors, possibly with a constant-factor budget violation, [Bateni et al. SIAM J. Comput., 2018][Moss and Rabani SIAM J. Comput., 2007]. If the graph is directed, the budgeted problem can be approximated within a $O\left(\frac{\log n'}{\log \log n'}\right)$ factor in quasi-polynomial time, where $n'$ is the number of vertices in an optimal solution [Ghuge and Nagarajan SODA 2020], and within a factor $O(\frac{1}{\epsilon^3}\sqrt{B})$ with a budget violation of $1 + \epsilon$, for any $\epsilon \in (0, 1]$, in polynomial time [D'Angelo, Delfaraz, and Gilbert ISAAC 2022].

In this paper, we provide two algorithms for the budgeted and quota problems on directed graphs that achieve, respectively, an $O(\frac{1}{\epsilon^2}n^{2/3}\ln n)$-approximation at the cost of a budget violation of a factor of at most

This work has been partially funded by the European Union - NextGenerationEU under the Italian Ministry of University and Research (MUR) National Innovation Ecosystem grant ECS00000041 - VITALITY - CUP: D13C21000430001; by PNRR MIUR project GAMING "Graph Algorithms and MinIng for Green agents" (PE0000013, CUP D13C24000430001); by ARS01_00540 - RASTA project, funded by the Italian Ministry of Research PNR 2015-2020; by Project EMERGE: innovation agreement between Ministry of Economical Development, Abruzzo Region, Radiolabs, Elital, Leonardo, Telespazio, University of L'Aquila and Centre of EXcellence EX-Emerge, funded by Italian Government under CIPE n. 70/2017 (Aug. 7, 2017); The first author acknowledges the support of the MUR (Italy) Department of Excellence 2023–2027.

© The Author(s), under exclusive license to Springer Nature Switzerland AG 2024
A. A. Rescigno and U. Vaccaro (Eds.): IWOCA 2024, LNCS 14764, pp. 273–286, 2024.
https://doi.org/10.1007/978-3-031-63021-7_21

$1 + \epsilon$, for any $\epsilon \in (0, 1]$, and an approximation factor of $O(n^{2/3} \ln n)$ at the cost of a violation of the quota constraint by a factor of at most 2. We develop a technique resorting on a standard flow-based linear programming relaxation to compute a tree with good trade-off between prize and cost, which allows us to provide polynomial time approximation algorithms for both problems. We provide the first approximation algorithms for these problems that run in polynomial time and guarantee an approximation factor depending only on the number of vertices $n$.

# 1   Introduction

*Prize Collecting Steiner Tree* (**PCST**) refers to a wide class of combinatorial optimization problems, involving variants of the *Steiner tree* problem and *traveling salesperson* problem, with many practical applications in computer and telecommunication networks, VLSI design, computational geometry, wireless mesh networks, and cancer genome studies [5,10,16,20,25].

In **PCST**, we are given a (directed) graph $G$, two functions modelling costs and prizes (or penalties) associated to the edges and/or to the nodes of the graph, and we want to find a connected subgraph $T$ of $G$ (usually a tree or an out-tree) which optimizes an objective function defined as a combination of its cost and prize and/or is subject to some constraints on its cost and prize. By considering different constraints and objective functions, one obtains distinct optimization problems. In *budgeted* problems, we are given a budget $B$ and we require that the cost of $T$ is at most $B$ and its prize is maximum. In *quota* problems, we require the prize of $T$ to be at least some quota $Q$ and its cost to be minimum. Additional constraints can be required, for example in *rooted* variants we are given a specific node, called *root*, which has to be part of $T$ and reach all the nodes in $T$.

While there is a vast literature providing approximation algorithms for many variants of **PCST** on undirected graphs, e.g. [1,2,11,13,15,17,18,23], the case of directed graphs received less attention [4,7,12,20,26]. Furthermore, prize collecting Steiner tree problems are usually much harder on directed graphs than on undirected graphs. A well-known example is the *Steiner tree* problem, for which there is a simple polynomial time 2-approximation algorithm for its undirected version, but for its directed version, unless $NP \subseteq \bigcap_{0<\epsilon<1} \text{ZPTIME}(2^{n^\epsilon})$ or the Projection Game Conjecture is false, there is no quasi-polynomial time algorithm that achieves an approximation ratio of $o(\frac{\log^2 k}{\log \log k})$ [14], where $k$ is the number of terminal nodes.

In this paper, we focus on *budgeted* and *quota* problems on *directed* graphs and, motivated by applications in the deployment of wireless relay networks [10, 20] and in the detection of mutated pathways in cancer [16,25], we study *node-weighted* problems, that is both costs and prizes are associated to the nodes of the graph. We consider the more general *rooted* variant of this problems.

In both budgeted and quota variants, we are given a directed graph $D = (V, A)$ with $|V| = n$, two nonnegative real-valued functions, namely, cost $c(v)$

and prize $p(v)$ that are associated to each vertex $v \in V$, and a root vertex $r$. In the *Budgeted Directed Rooted Tree problem* (**B-DRT**), we are given a budget $B$, and we aim at finding an out-tree (a.k.a. out-arborescence) $T$ of $D$ rooted at $r$ such that the sum of the costs of its vertices is at most $B$ and the sum of the prizes of its vertices is maximum. In the *Quota Directed Rooted Tree problem* (**Q-DRT**), we are given a quota $Q$, and aim at finding an out-tree $T$ of $D$ rooted at $r$ whose total prize is at least $Q$ and total cost is minimum.

**Related Work.** Guha et al. [15] introduced the undirected version of **B-DRT**, called **B-URT**. They gave a polynomial time $O(\log^2 n)$-approximation algorithm that violates the budget constraint by a factor of at most 2. Moss and Rabani [22] improved the approximation factor to $O(\log n)$, with the same budget violation. Their algorithm is based on a Lagrangian multiplier preserving $O(\ln n)$-approximation algorithm, proposed in the same paper, for the problem of minimizing the cost of the nodes in the resulting tree plus the prizes of vertices not spanned by the tree. We call this problem **PC-URT**. However, Könemann et al. [18] pointed out a flaw in their algorithm for **PC-URT** and proposed an alternative Lagrangian multiplier preserving algorithm with the same guarantee. Later, Bateni et al. [2] proposed an $O(\frac{1}{\epsilon^2} \log n)$-approximation algorithm for **B-URT** which requires a budget violation of only $1 + \epsilon$, for any $\epsilon \in (0, 1]$. Kortsarz and Nutov [19] showed that the unrooted version of **B-URT**, so does **B-URT**, admits no $o(\log \log n)$-approximation algorithm, unless $NP \subseteq DTIME(n^{\text{polylog}(n)})$, even if the algorithm is allowed to violate the budget constraint by a factor equal to a universal constant. Ghuge and Nagarajan [12] provided a tight quasi-polynomial time $O(\frac{\log n'}{\log \log n'})$-approximation algorithm for the edge-cost version of **B-DRT**, where $n'$ is the number of vertices in an optimal solution, the prize function is a monotone submodular function on subsets of nodes, and the edge costs are positive integers. D'Angelo et al. [7] provided a polynomial time $O(\frac{1}{\epsilon^2} \sqrt{B})$-approximation algorithm for the same problem that violates the budget constraint by a factor of $1 + \epsilon$, where $\epsilon \in (0, 1]$. By using a simple reduction, it is easy to see that these results for the directed graphs with edge-costs also hold for their node-cost version. Bateni et al. [2] showed that the integrally gap of the standard flow-based LP for **B-URT**, so is for **B-DRT**, is unbounded.

To the best of our knowledge, the quota problem on directed graphs has not been studied explicitly before. For the node-weighted quota problem on undirected graphs, the algorithm by Moss and Rabani [22] provides an $O(\log n)$-approximation algorithm by using as a black-box the algorithm of Könemann et al. [18] (or the one by Bateni et al. [2]) for **PC-URT**.

It is worth pointing out that by binary searching the budget (resp. quota) space, one can also show that any $\alpha$-approximation algorithm for a budgeted (resp. quota) problem results in a $(1 + \epsilon)$-approximation ($(1 - \epsilon)$-approximation) for its quota (budgeted) version, for any $\epsilon > 0$, that violates the quota (budget) constraint by a factor of at most $\alpha$.

**Our Results.** We introduce a new technique, resorting on flow-based linear programming relaxations, which allows us to find out-trees with a good trade-off between cost and prize. By computing suitable values of quota and budget and applying known tree trimming procedures, we achieve good bicriteria approximations for both **B-DRT** and **Q-DRT**. For **B-DRT**, we introduce a polynomial-time approximation algorithm that violates the budget constraint by a factor of at most $1 + \epsilon$, for $\epsilon \in (0, 1]$ and achieve an approximation factor of $O(\frac{1}{\epsilon^2} n^{2/3} \ln n)$. For **Q-DRT** we present an $O(n^{2/3} \ln n)$-approximation polynomial-time algorithm that violates the quota constraint by a factor of at most 2.

To the best of our knowledge, our techniques yield the first approximation algorithms for these problems whose approximation ratio depends only on the number of vertices $n$ and whose running time is polynomial in the input size. Indeed, the algorithms in [7] and [12] consider a more general version of our budgeted problem, in which the prize function is monotone and submodular. However, the approximation guarantee of the algorithm in [7] for the budgeted problem depends on the budget $B$, which can be much higher than $n$, while the algorithm in [12] requires quasi-polynomial running time. To the best of our knowledge, there are no known approximation algorithms for quota problems in directed graphs or submodular prize functions. Moreover, our algorithms allow the costs to be any non-negative real numbers, whereas the algorithms in [7] and [12] require the costs to be strictly positive integers.

## 2   Notation and Problem Statement

For an integer $s$, let $[s] := \{1, \ldots, s\}$. Let $D = (V, A)$ a directed graph with a distinguished vertex $r \in V$ and $c : V \to \mathbb{R}^{\geq 0}$ be a nonnegative cost function on nodes.

A *directed path* is a directed graph made of a sequence of distinct vertices $(v_1, \ldots, v_s)$ and a sequence of directed edges $(v_i, v_{i+1})$, $i \in [s-1]$. An *out-tree* (a.k.a. out-arborescence) is a directed graph in which there is exactly one directed path from a specific vertex $r$, called *root*, to each other vertex. If a subgraph $T$ of a directed graph $D$ is an out-tree, then we say that $T$ is an out-tree of $D$. For simplicity of reading, we may will refer to out-trees simply as trees.

Given two nodes $u, v \in V$, the cost of a path from $u$ to $v$ in $D$ is the sum of the cost of its nodes. A directed path from $u$ to $v$ with the minimum cost is called a *shortest path* and its cost, denoted by $dist(u, v)$, is called the *distance* from $u$ to $v$ in $D$. Let $F$ be the maximum distance from $r$ to a node in $V$, $F := \max_{v \in V} \{dist(r, v)\}$. Let $B \in \mathbb{R}^{>0}$, a graph $D$ is called $B$-*proper* for the vertex $r$ if $dist(r, v) \leq B$ for any $v$ in $D$.

For any subgraph $D'$ of $D$, we denote by $V(D')$ and $A(D')$ the set of nodes and edges in $D'$, respectively. Given a subset $S \subseteq V$ of nodes, $D[S]$ denotes the graph induced by set $S$, i.e., $A(D[S]) = \{(u, v) \in A | u, v \in S\}$.

In what follows, we state the problems we consider in this paper. Let $D = (V, A)$ be a directed graph with $n$ nodes, $c : V \to \mathbb{R}^{\geq 0}$ be a cost function on nodes, $p : V \to \mathbb{R}^{\geq 0}$ be a prize function on nodes, $r \in V$ be a root vertex. We consider two variants of node-weighted **PCST**.

1. The Quota Directed Rooted Tree problem (**Q-DRT**): Given a quota $Q \in \mathbb{R}^{>0}$, find an out-tree $T$ of $D$ rooted at $r$ that minimizes $c(T) = \Sigma_{v \in V(T)} c(v)$ subject to $p(T) = \Sigma_{v \in V(T)} p(v) \geq Q$.
2. The Budgeted Directed Rooted Tree problem (**B-DRT**): Given a budget $B \in \mathbb{R}^{>0}$, find an out-tree $T$ of $D$ rooted at $r$ that maximizes $p(T) = \Sigma_{v \in V(T)} p(v)$ subject to $c(T) = \Sigma_{v \in V(T)} c(v) \leq B$.

*Bicriteria Approximation.* For $\alpha, \beta \geq 1$, a bicriteria $(\beta, \alpha)$-approximation algorithm for **B-DRT** (resp. **Q-DRT**) is one that, for any instance $I_B$ (resp. $I_Q$) of the problem, returns a solution $Sol_{I_B}$ (resp. $Sol_{I_Q}$) such that $p(Sol_{I_B}) \geq \frac{OPT_{I_B}}{\alpha}$ (resp. $c(Sol_{I_Q}) \leq \alpha OPT_{I_Q}$) and $c(Sol_{I_B}) \leq \beta B$ (resp. $p(Sol_{I_Q}) \geq \frac{Q}{\beta}$), where $OPT_{I_B}$ (resp. $OPT_{I_Q}$) is the optimum for $I_B$ (resp. $I_Q$).

# 3 Directed Rooted Tree Problems

In this section we present a polynomial time bicriteria $\left(2, O(n^{2/3} \ln n)\right)$-approximation algorithm for **Q-DRT** and a polynomial time bicriteria $\left(1 + \epsilon, O(\frac{n^{2/3} \ln n}{\epsilon^2})\right)$-approximation algorithm for **B-DRT**, where $\epsilon$ is an arbitrary number in $(0, 1]$. Through the section we let $I_Q = < D = (V, A), c, p, r, Q >$ and $I_B = < D = (V, A), c, p, r, B >$ be two instances of **Q-DRT** and **B-DRT**, respectively, and we let $T_Q^*$ and $T_B^*$ be two optimal solutions for $I_Q$ and $I_B$, respectively. Both algorithms for **Q-DRT** and **B-DRT** use the same technique and can be summarized in the following three steps:

1. We define a set of linear constraints, denoted as (DRT), over fractional variables, that takes a given quota $\Pi$ and budget $\Lambda$ as parameters and admits a feasible solution if there exists a subtree $T$ of $D$ rooted at $r$ such that $p(T) \geq \Pi$ and $c(T) \leq \Lambda$. In particular, an optimal tree $T_Q^*$ for $I_Q$ has prize $p(T_Q^*) \geq Q$ and cost $c(T_Q^*)$, therefore the linear constraints (DRT) where parameters $\Pi$ and $\Lambda$ are set to $\Pi = Q$ and $\Lambda = c(T_Q^*)$ admits a feasible solution. It follows that the minimum value $OPT_Q$ of $\Lambda$ for which the set of linear constraints (DRT) admits a feasible solution when $\Pi = Q$ is a lower bound to $c(T_Q^*)$, i.e. $OPT_Q \leq c(T_Q^*)$. Similarly, the maximum value $OPT_B$ of $\Pi$ for which the set of linear constraints (DRT) admits a feasible solution when $\Lambda = B$ is an upper bound to $p(T_B^*)$, i.e. $OPT_B \geq p(T_B^*)$.
2. We give a polynomial time algorithm that takes as input a feasible solution to (DRT) and computes a subtree $T$ of $D$ rooted at $r$ such that $c(T) = O((F + \Lambda)n^{2/3} \ln n)$ and $p(T) \geq \Pi/2$, where $F := \max_{v \in V}\{dist(r, v)\}$.
3. **Q-DRT**: We first compute a solution for (DRT) for which $\Pi = Q$ and $\Lambda$ is minimum, i.e. $\Lambda = OPT_Q$. Then we use such a solution for (DRT) as input to the algorithm in the previous step and obtain a tree $T$ such that $c(T) = O((F + OPT_Q)n^{2/3} \ln n)$ and $p(T) \geq Q/2$. We can show that we can assume

w.l.o.g. that $F \leq (1+\epsilon)c(T_Q^*)$, for any $\epsilon > 0$. Since $OPT_Q \leq c(T_Q^*)$, then the computed tree $T$ satisfies $c(T) = O((1+\epsilon)c(T_Q^*)n^{2/3}\ln n)$ and $p(T) \geq Q/2$.

**B-DRT:** We compute a solution for (DRT) for which $\Lambda = B$ and $\Pi$ is maximum, i.e. $\Pi = OPT_B \geq p(T_Q^*)$ and use such a solution as input to the algorithm in the previous step. Since we can assume w.l.o.g. that $F \leq B$, we obtain a tree $T$ such that $c(T) = O(Bn^{2/3}\ln n)$ and $p(T) \geq OPT_B/2 \geq p(T_B^*)/2$.

The tree $T$ may violate the budget constraint by a large factor, however the ratio $\gamma$ between its prize and its cost is $\Omega\left(p(T_B^*)/(Bn^{2/3}\ln n)\right)$. The bound on the value of $\gamma$ allows us to apply the trimming process given in [2] to $T$ and obtain another tree $\hat{T}$ with cost $\frac{\epsilon}{2}B \leq c(\hat{T}) \leq (1+\epsilon)B$, for any $\epsilon \in (0,1]$, and prize-to-cost ratio $\frac{p(\hat{T})}{c(\hat{T})} = \epsilon\frac{\gamma}{4}$. Tree $\hat{T}$ achieves an approximation ratio of $O(\frac{n^{2/3}\ln n}{\epsilon^2})$ at the cost of a budget violation of $1+\epsilon$, for any $\epsilon \in (0,1]$.

In the following, we will detail each step of our algorithms.

### Step 1: Bounding the optimal cost and prize

Here we define a set of linear constraints that admits a feasible solution if there exists a tree $T$ rooted in $r$ in $D$ such that $c(T) \leq \Lambda$ and $p(T) \geq \Pi$, for given parameters $\Lambda, \Pi \in \mathbb{R}^{>0}$. For each $v \in V$, let $p_v = p(v)$, $c_v = c(v)$, and $\mathcal{P}_v$ be the set of simple paths in $D$ from $r$ to $v$. Our set of constraints (DRT) is defined as follows.

$$\sum_{v \in V} x_v p_v \geq \Pi \tag{D.1}$$

$$\sum_{v \in V} x_v c_v \leq \Lambda \tag{D.2}$$

$$\sum_{P \in \mathcal{P}_v} f_P^v = x_v, \qquad \forall v \in V \setminus \{r\} \tag{D.3}$$

$$\sum_{P \in \mathcal{P}_v : w \in P} f_P^v \leq x_w, \qquad \forall v \in V \setminus \{r\} \text{ and } \forall w \in V \setminus \{v\} \tag{D.4}$$

$$0 \leq x_v \leq 1, \qquad \forall v \in V$$

$$0 \leq f_P^v \leq 1, \qquad \forall v \in V \setminus \{r\}, P \in \mathcal{P}_v$$

We use variables $f_P^v$ and $x_v$, for each $v \in V$ and $P \in \mathcal{P}_v$, where $f_P^v$ represents the amount of flow sent from $r$ to $v$ using path $P$ and $x_v$ represents both the capacity of node $v$ and the overall amount of flow sent from $r$ to $v$.

The constraints in (DRT) are as follows. Constraints (D.1) and (D.2) ensure that any feasible (fractional) solution to (DRT) has a prize at least $\Pi$ and a cost at most $\Lambda$. Constraints (D.3) and (D.4) formulate a connectivity constraint through standard flow encoding, that is they ensure that the nodes $v$ with $x_v > 0$ induce subgraph in which all nodes are reachable from $r$. In particular, constraint (D.3) ensures that the amount of flow that is sent from $r$ to any vertex $v$ must be equal to $x_v$ and constraint (D.4) ensures that the total flow from $r$ to $v$ passing through a vertex $w$ cannot exceed $x_w$.

Note that (DRT) has an exponential number of variables. However, it can be solved efficiently as we only need to find, independently for any $v \in V \setminus \{r\}$, a flow from $r$ to $v$ of value $x_v$ that does not exceed the capacity $x_w$, for each vertex $w \in V \setminus \{r, v\}$.

We can show that a feasible solution to (DRT) can be used to find a lower bound to an optimal solution for **Q-DRT** and an upper bound to an optimal solution for **B-DRT**. The next lemma shows that if there exists a tree $T$ rooted in $r$ in $D$ such that $c(T) \leq \Lambda$ and $p(T) \geq \Pi$, then there exists a feasible solution $x$ for (DRT).

**Lemma 1.** *Given a directed graph $D = (V, A)$, $r \in V$, $c : V \to \mathbb{R}^{\geq 0}$, $p : V \to \mathbb{R}^{\geq 0}$, $\Lambda \in \mathbb{R}^{>0}$ and $\Pi \in \mathbb{R}^{>0}$, if there exists a tree $T$ rooted in $r$ in $D$ such that $c(T) \leq \Lambda$ and $p(T) \geq \Pi$, then there exists a feasible solution $x$ for (DRT).*

*Proof.* Let us consider a solution to (DRT) in which $x_v = 1$ for all $v \in V(T)$, while $x_v$ is set to 0 for all $v \notin V(T)$. As $p(T) \geq \Pi$ and $c(T) \leq \Lambda$, then the quota and budget constraints (D.1)–(D.2) are satisfied. Since $T$ is connected and for any $v \in V(T)$, there exists only one path $P$ from $r$ to $v$ in $T$, constraints (D.3) and (D.4) are satisfied by setting $f_P^v = 1$ and any other flow variable to 0. $\square$

Since $T_Q^*$ is a feasible solution for $I_Q$, we have that $p(T_Q^*) \geq Q$, which implies that there exists a feasible solution to (DRT) when parameters $\Lambda$ and $\Pi$ are set to $\Lambda = c(T_Q^*)$ and $\Pi = Q$. Hence, the optimum $OPT_Q$ to the linear program of minimizing $\Lambda$ subject to constraints (DRT) in which parameter $\Pi$ is set to $\Pi = Q$, gives a lower bound to $c(T_Q^*)$, i.e. $OPT_Q \leq c(T_Q^*)$. Similarly, the optimum $OPT_B$ to the linear program of maximizing $\Pi$ subject to constraints (DRT) with $\Lambda = B$, gives an upper bound to $p(T_B^*)$, $OPT_B \geq p(T_B^*)$. We denote these two linear programs by **Q-DRT-LP** and **B-DRT-LP**, respectively.

## Step 2: Finding a good tree from a feasible fractional solution to (DRT)

Here we elaborate the second step and show how to compute a tree with a good trade-off between prize and cost starting from a feasible fractional solution to the set of constraints (DRT).

**Theorem 1.** *Given a feasible solution $x$ to (DRT), then there exists a polynomial time algorithm that computes a tree $T$ rooted at $r$ such that $c(T) = O((F + \Lambda)n^{2/3} \ln n)$ and $p(T) \geq \Pi/2$.*

We now prove Theorem 1. Let $x$ be a feasible solution for (DRT) and let $S \subseteq V$ be the set of vertices $v$ with $x_v > 0$, i.e., $S = \{v \in V : x_v > 0\}$. We partition $S$ into two subsets $S_1, S_2 \subseteq S$, where $S_1 = \{v \in S | x_v \geq n^{-1/3}\}$ and $S_2 = \{v \in S | x_v < n^{-1/3}\}$. We first focus on nodes in $S_1$ and, in the following lemma, we show how to compute a tree $T$ rooted at $r$ spanning all vertices in $S_1$ with cost $c(T) = O((F + \Lambda)n^{2/3} \ln n)$.

**Lemma 2.** *There exists a polynomial time algorithm that finds a tree $T$ rooted at $r$ spanning all vertices in $S_1$ with cost $c(T) = O((F + \Lambda)n^{2/3} \ln n)$.*

*Proof.* Let $U := \{v \in S : x_v \geq n^{-2/3}\}$. We call a vertex $v \in S_1$ a *cheap vertex* if there exists a path from $r$ to $v$ in $D[U]$. We call a vertex $v \in S_1$ an *expensive vertex* otherwise. Note that $r$ belongs to $U$ if $U \neq \emptyset$ since we need to send $n^{-2/3}$ amount of flow from $r$ to any vertex in $U$ by constraint (D.3). Let $CH$ and $EX$ be the sets of cheap and expensive vertices, respectively.

In the following, we first show that we can compute in polynomial time two trees $T^{CH}$ and $T^{EX}$ spanning all nodes in $CH$ and $EX$, respectively, and having cost $c(T^{CH}) = O(\Lambda n^{2/3})$ and $c(T^{EX}) = O((F + \Lambda)n^{2/3} \ln n)$, then we show how to merge the two trees into a single tree with cost $O((F + \Lambda)n^{2/3} \ln n)$.

We first focus on $T^{CH}$. By definition, all vertices in $CH$ are reachable from $r$ through paths that contain only vertices in $U$. Thus, for each $v \in CH$, we compute a shortest path from $r$ to $v$ in $D[U]$ and find a tree $T^{CH}$ rooted at $r$ spanning all and only the vertices in the union of these shortest paths. Since $\Sigma_{u \in U} x_u c_u \leq \Lambda$ (by constraint (D.2)) and $x_u \geq n^{-2/3}$ for any $u \in U$ (by definition of $U$), then $c(U) = \Sigma_{u \in U} c_u \leq \Lambda n^{2/3}$. Hence $c(T^{CH}) \leq c(U) = O(\Lambda n^{2/3})$.

Now we show that there exists a tree $T^{EX}$ rooted at $r$ spanning all the expensive vertices $EX$ with cost $c(T^{EX}) = O((F + \Lambda)n^{2/3} \ln n)$ and that we can compute $T^{EX}$ in polynomial time. The algorithm to build $T^{EX}$ can be summarized as follows. We first compute, for each $v \in EX$, the set $X_v$ of vertices $w \in S \setminus U$ for which there exists a path from $w$ to $v$ that uses only vertices in $U \cup \{w\}$. Then we compute a small-size hitting set $X'$ of all $X_v$, i.e. $X' = \arg\min\{|Y| : Y \subseteq \bigcup_{v \in EX} X_v \text{ and } \forall v \in EX, Y \cap X_v \neq \emptyset\}$. Finally, we connect $r$ to the vertices of $X'$ and the vertices of $X'$ to those in $EX$ in such a way that each node $v$ in $EX$ is reached from one of the vertices $x \in X'$ that hits $X_v$, i.e., $x \in X_v$. The bound on the cost of $T^{EX}$ follows from the size of $X'$ and from the cost of nodes in $U$. We now detail on the construction of $T^{EX}$ and its cost analysis.

Let $U' \subseteq S$ be the set of all vertices $w$ with $x_w < n^{-2/3}$, i.e., $U' = S \setminus U$. For any expensive vertex $v \in EX$, we define $X_v$ as the set of vertices $w$ in $U'$ such that there exists a path from $w$ to $v$ in $D[U \cup \{w\}]$. Note that for any $w \in X_v$, $v$ is reachable from $w$ through a path $P$ such that $V(P) \setminus \{w\} \subseteq U$, i.e., $V(P) \setminus \{w\}$ only contains vertices from $U$. The following claim gives a lower bound on the size of $X_v$, for each $v \in EX$.

**Claim 1.** $|X_v| \geq n^{1/3}$, for each $v \in EX$.

*Proof.* We know that (i) the amount of flow that each vertex $v \in S_1$ should receive is at least $n^{-1/3}$ (by definition of $S_1$ and constraint (D.3) of (DRT)), (ii) any path $P$ from $r$ to any $v \in EX$ in the graph $D[S]$ contains at least one vertex $w \in U'$ (by definition of expensive vertices), and (iii) in any path $P$ from $r$ to any $v \in EX$, the node $w \in U'$ in $P$ that is closest to $v$ is a member of $X_v$, i.e. $w \in X_v$ (by definition of $X_v$), therefore any flow from $r$ to $v$ should pass through a vertex $w \in X_v$. This implies that the vertices in $X_v$ must send at least $n^{-1/3}$ amount of flow to $v$ in total. Formally, we have

$$n^{-1/3} \leq x_v = \Sigma_{P \in \mathcal{P}_v} f_P^v \leq \Sigma_{w \in X_v} \Sigma_{P \in \mathcal{P}_v : w \in P} f_P^v$$
$$\leq \Sigma_{w \in X_v} x_w \leq \Sigma_{w \in X_v} n^{-2/3} = |X_v| n^{-2/3},$$

which implies that $|X_v| \geq n^{1/3}$. Note that the first inequality follows from the definition of $S_1$, the first equality follows from constraint (D.3) of (DRT), the second inequality is due to the fact that, by definition of $X_v$, any path $P \in \mathcal{P}_v$ contains a vertex $w \in X_v$, the third inequality is due to constraint (D.4) of (DRT), the last inequality is due to $x_w < n^{-2/3}$, for each $w \in U'$, and $X_v \subseteq U'$. This concludes the proof of the claim. $\qquad\square$

We use the following well-known result (see e.g. Lemma 3.3 in [3]) to find a small set of vertices that hits all the sets $X_t$, for all $t \in EX$.

*Claim 2.* Let $V'$ be a ground set of $M$ elements and $\Sigma = (X_1', \ldots, X_N')$ be a collection of subsets of $V'$ such that $|X_i'| \geq R$, for each $i \in [N]$. There is a deterministic algorithm which runs in polynomial time in $N$ and $M$ and finds a subset $X' \subseteq V'$ with $|X'| \leq (M/R)\ln N$ and $X' \cap X_i' \neq \emptyset$ for all $i \in [N]$.

Using the bound of Claim 1 on the size of sets $X_v$, we can exploit the algorithm of Claim 2, using $\bigcup_{v \in EX} X_v$ as ground set and $\{X_v\}_{v \in EX}$ as collection of subsets, to find a set $X' \subseteq \bigcup_{v \in EX} X_v$ such that $X' \cap X_v \neq \emptyset$, for all $v \in EX$, whose size is at most $|X'| \leq \frac{n \ln n}{n^{1/3}} = n^{2/3} \ln n$. In this case the parameters of Claim 2 are $R = n^{1/3}$, $M = |\bigcup_{v \in EX} X_v| \leq n$, and $N = |EX| \leq n$. Since for any $v \in EX$ there exists a path from any $w \in X_v$ to $v$ in $D[U \cup \{w\}]$ and $X'$ contains at least one vertex in $X_v$, then there exists a path from one of the vertices $w \in X'$ to $v$ in $D[U \cup \{w\}]$.

Now we find a shortest path from $r$ to any $w \in X'$ in $D$. Let $\mathcal{P}_1$ be the set of all these shortest paths. We also find, for each $v \in EX$, a shortest path from an arbitrary vertex $w \in X' \cap X_v$ to $v$ in $D[U \cup \{w\}]$. Let $\mathcal{P}_2$ be the set of all these shortest paths. Let $V(\mathcal{P}_1)$ and $V(\mathcal{P}_2)$ be the union of all the nodes of the paths in $\mathcal{P}_1$ and $\mathcal{P}_2$, respectively. Then, we find a tree $T^{EX}$ rooted at $r$ spanning graph $D^{EX} := D[V(\mathcal{P}_1) \cup V(\mathcal{P}_2)]$. Note that such a tree exists as in $D^{EX}$ there exists a path from $r$ to any $w \in X'$ and, for each vertex $v \in EX$, at least a path from one of the vertices in $X'$ to $v$.

We next move to bounding the cost of $T^{EX}$, i.e. we bound the total cost of nodes in $V(\mathcal{P}_1) \cup V(\mathcal{P}_2)$. Since $|X'| \leq n^{2/3} \ln n$ and the maximum distance from $r$ to any other node is $F$, then $c(V(\mathcal{P}_1)) \leq Fn^{2/3} \ln n$. As $\Sigma_{u \in U} c(v) \leq \Lambda n^{2/3}$ (by constraint (D.2) of (DRT) and $x_u \geq n^{-2/3}$ for any $u \in U$) and $V(\mathcal{P}_2) \setminus X' \subseteq U$, then $c(V(\mathcal{P}_2) \setminus X') \leq \Lambda n^{2/3}$. Overall, $T^{EX}$ costs at most $O((F + \Lambda)n^{2/3} \ln n)$.

Since both $T^{EX}$ and $T^{CH}$ are rooted at $r$, we can find a tree $T$ rooted at $r$ that spans all vertices $V(T^{EX}) \cup V(T^{CH})$. Since $c(T^{CH}) + c(T^{EX}) = O((F + \Lambda)n^{2/3} \ln n)$, we have $c(T) = O((F + \Lambda)n^{2/3} \ln n)$, which concludes the proof. $\square$

We are now ready to prove Theorem 1.

*Proof (Proof of Theorem 1).* As $\Sigma_{v \in S} x_v p_v \geq \Pi$, either $\Sigma_{v \in S_1} x_v p_v \geq \Pi/2$ or $\Sigma_{v \in S_2} x_v p_v \geq \Pi/2$.

If $\Sigma_{v \in S_1} x_v p_v \geq \Pi/2$, then by Lemma 2, we can find in polynomial time a tree $T$ that spans all vertices in $S_1$ such that $c(T) = O((F + \Lambda)n^{2/3} \ln n)$. Since $T$ spans all vertices of $S_1$, then $p(T) \geq p(S_1) \geq \Pi/2$.

If $\Sigma_{v \in S_2} x_v p_v \geq \Pi/2$, we have that $p(S_2) = \Sigma_{v \in S_2} p_v \geq n^{1/3} \cdot \Sigma_{v \in S_2} x_v p_v \geq n^{1/3}\Pi/2$, where the first inequality holds since $0 < x_v < n^{-1/3}$ for any $v \in S_2$

and the second inequality holds by the case assumption. We partition $S_2$ into $M$ groups $U_1, \ldots, U_M$ in such a way that for each $i \in [M-1]$, $|U_i| = 2|S_2|^{2/3}$, and $|U_M| \leq 2|S_2|^{2/3}$. Hence the number of selected groups is at most $M \leq \left\lceil \frac{|S_2|}{2|S_2|^{2/3}} \right\rceil = \left\lceil \frac{|S_2|^{1/3}}{2} \right\rceil \leq \left\lfloor \frac{|S_2|^{1/3}}{2} \right\rfloor + 1 \leq |S_2|^{1/3} \leq n^{1/3}$. Now among $U_1, \ldots, U_M$, we select the group $U_z$ that maximizes the prize, i.e., $z = \arg\max_{i \in [M]} p(U_i)$. We know that $p(U_z) \geq \frac{1}{M} \Sigma_{i=1}^{M} p(U_i) = \frac{p(S_2)}{M} \geq \frac{n^{1/3}\Pi}{2n^{1/3}} = \Pi/2$, where the first inequality is due to an averaging argument, the first equality is due to the additivity of $p$, and the second inequality is due to $M \leq n^{1/3}$ and $p(S_2) \geq n^{1/3}\Pi/2$.

We now find for each vertex $v$ in $U_z$ a shortest path from $r$ to $v$ and compute a tree $T$ that spans all the vertices in the union of these shortest paths. Clearly, $c(T) \leq 2Fn^{2/3}$ as $|U_z| \leq 2n^{2/3}$ and the cost of a shortest path from $r$ to any $v \in V$ in $D$ is at most $F$. Furthermore, $p(T) \geq p(U_z) \geq \Pi/2$, by additivity of $p$. This concludes the proof.    □

## Step 3: Approximating Q-DRT and B-DRT

We next show how to use Theorem 1 to devise bicriteria approximation algorithms for **Q-DRT** and **B-DRT**. First we prove our result for **Q-DRT**.

**Theorem 2.** *Q-DRT admits a polinomial time bicriteria* $\left(2, O(n^{2/3}\ln n)\right)$*-approximation algorithm.*

*Proof.* We can assume w.l.o.g. that for all nodes $v$ we have $dist(r, v) \leq (1 + \epsilon)c(T_Q^*)$, i.e. that $F \leq (1+\epsilon)c(T_Q^*)$, for any $\epsilon > 0$. Indeed, we can guess an $(1+\epsilon)$-approximation of $c(T_Q^*)$ using the following procedure. Let $c_{\min}$ be the minimum positive cost of a vertex and $c_M = \sum_{v \in V} c(v)$, we know that $c(T_Q^*) \leq c_M$. We estimate the value of $c(T_Q^*)$ by guessing $N$ possible values, where $N$ is the smallest integer for which $c_{\min}(1 + \epsilon)^{N-1} \geq c_M$. For each guess $i \in [N]$, we remove the nodes $v$ with $dist(r, v) > c_{\min}(1 + \epsilon)^{i-1}$, and compute a tree $T$ rooted in $r$ with an algorithm that will be explained later in the proof. Eventually, we output the computed tree $T$ with the smallest cost. Since $c_{\min}(1 + \epsilon)^{N-2} < c_M$, the number $N$ of guesses is smaller than $\log_{1+\epsilon}(c_M/c_{\min}) + 2$, which is polynomial in the input size and in $1/\epsilon$.

Let $i \in [N]$ be the smallest value for which $c_{\min}(1 + \epsilon)^{i-1} \geq c(T_Q^*)$. Then, $c(T_Q^*) > c_{\min}(1 + \epsilon)^{i-2}$ and for all the nodes $v$ in the graph used in guess $i$, we have $dist(r, v) \leq c_{\min}(1 + \epsilon)^{i-1} < (1 + \epsilon)c(T_Q^*)$. Since we output a solution with the minimum cost among those computed in the guesses, then the final solution will not be worse than the one computed at guess $i$. Therefore, from now on we focus on guess $i$ and assume that $F \leq (1 + \epsilon)c(T_Q^*)$.

We first find an optimal solution $x$ to **Q-DRT-LP**, let $OPT_Q$ be the optimal value of **Q-DRT-LP**. Observe that $x$ is a feasible solution for the set of constraints (DRT) in which $\Lambda = OPT_Q$ and $\Pi = Q$ and, by Lemma 1, $OPT_Q \leq c(T_Q^*)$. Therefore, we can use $x$ and apply the algorithm in Theorem 1 to obtain a tree $T$ such that $c(T) = O((F + OPT_Q)n^{2/3}\ln n) = O((1 + \epsilon)c(T_Q^*)n^{2/3}\ln n)$ and $p(T) \geq Q/2$, which concludes the proof.    □

Next we prove our result for **B-DRT**. Let us assume that for every $v \in V$, $dist(r,v) \le B$, i.e. $F \le B$, since otherwise we can remove from $D$ all the nodes $v$ such that $dist(r,v) > B$.

Let $x$ be an optimal solution for **B-DRT-LP** and let $T$ be a tree computed from $x$ by the algorithm in Theorem 1. Since $x$ is a feasible solution to (DRT) when $\Pi = OPT_B \ge p(T_B^*)$ and $\Lambda = B$, then the prize of $T$ is at least $\frac{p(T_B^*)}{2}$ but its cost can exceed the budget $B$. In this case, however, the cost of $T$ is bounded by $c(T) = O(Bn^{2/3} \ln n)$ and its prize-to-cost ratio is $\gamma = \frac{p(T)}{c(T)} = \Omega\left(\frac{p(T_B^*)}{Bn^{2/3} \ln n}\right)$. Therefore, we can use $T$ and a variant of the trimming process introduced by Bateni et al. [2] for undirected graphs, to compute another tree $\hat{T}$ with cost between $\frac{\epsilon B}{2}$ and $(1 + \epsilon)B$ and prize-to-cost ratio $\frac{\epsilon\gamma}{4}$, for any $\epsilon \in (0,1]$. The following lemma gives extends the trimming process by Bateni et al. to directed graphs. The proof can be found in the full version of the paper [6].

**Lemma 3.** *Let $D = (V, A)$ be a $B$-proper graph for a node $r$. Let $T$ be an out-tree of $D$ rooted at $r$ with the prize-to-cost ratio $\gamma = p(T)/c(T)$. Suppose that, for $\epsilon \in (0,1]$, $c(T) \ge \epsilon B/2$. One can find an out-tree $\hat{T}$ rooted at $r$ with the prize-to-cost ratio at least $\epsilon\gamma/4$ such that $\epsilon B/2 \le c'(\hat{T}) \le (1 + \epsilon)B$.*

**Theorem 3.** *B-DRT admits a polynomial time bicriteria $\left(1 + \epsilon, O(\frac{n^{2/3} \ln n}{\epsilon^2})\right)$-approximation algorithm, for any $\epsilon \in (0,1]$.*

*Proof.* We first find an optimal solution $x$ to **B-DRT-LP**. and then we apply the algorithm in Theorem 1 and use $x$ as input to obtain a tree $T$. Since $x$ is a feasible solution to (DRT) when $\Pi = OPT_B \ge p(T_B^*)$ and $\Lambda = B$, then $c(T) = O(Bn^{2/3} \ln n)$ and $p(T) \ge \frac{OPT_B}{2} \ge \frac{p(T_B^*)}{2}$ as discussed above. The prize-to-cost ratio of $T$ is $\gamma = \frac{p(T)}{c(T)} = \Omega\left(\frac{p(T_B^*)}{Bn^{2/3} \ln n}\right)$. Then, if $c(T) > B$, we can apply the algorithm of Lemma 3 to $T$ and compute another tree $\hat{T}$ with cost $c(\hat{T}) \le (1 + \epsilon)B$ and prize-to-cost ratio at least $\frac{\epsilon\gamma}{4}$. Moreover, $c(\hat{T}) \ge \epsilon B/2$, and therefore we have $p(\hat{T}) = \Omega\left(\frac{\epsilon^2 p(T_B^*)}{n^{2/3} \ln n}\right)$, concluding the proof. $\square$

## 4   Discussion and Future Work

*On Budget Violation.* There is some evidence showing that budget violation is needed to approximate budgeted problems in polynomial time. Kortsarz and Nutov [19] showed that the unrooted version of **B-URT**, admits no $o(\log \log n)$-approximation algorithm, unless $NP \subseteq DTIME(n^{\text{polylog}(n)})$, even if the algorithm is allowed to violate the budget constraint by a factor equal to a universal constant. This lower bound holds for all the budgeted problems in this paper. The only approximation algorithm for the budgeted problems that does not violate the budget constraint is the combinatorial quasi-polynomial time algorithm given by Ghuge and Nagarajan [12]. Most of the literature on these problems, focuses on approximation algorithms that exploit the standard low-based LP relaxation. Bateni et al. [2] showed that the integrality gap of this LP relaxation is unbounded if no budget violation is allowed, even for **B-URT**. We show that, by violating the budget constraint, one can use this LP and provide an approximation algorithm for the more general **B-DRT**.

*On Lower Bounds.* As **Q-DRT** is a more general version of the node-weighted variant of the directed Steiner tree problem, called **DSteinerT**, any lower bound for **DSteinerT** also holds for **Q-DRT**. Recently, Li and Laekhanukit [21] ruled out poly-logarithmic approximation algorithms for the directed Steiner tree problem using the standard flow-based LP. This result holds for **DSteinerT** (and **Q-DRT**) as, in the instance in [21], the incoming edges of each vertex have the same cost. We also know that any $\alpha$-approximation algorithm for the edge cost version of **B-DRT** results in an $O(\alpha \log n)$-approximation algorithm for the directed Steiner tree problem, which was also pointed out by [12]. This implies that the quasi-polynomial time approximation algorithm of [12] is tight for the edge cost version of **B-DRT**. Furthermore, by performing a binary search on the budget (resp. quota) space, one can use an $\alpha$-approximation algorithm for a budgeted (resp. quota) problem to obtain a $1 + \epsilon$ (resp. $1 - \epsilon$)-approximation algorithm for its quota (resp. budgeted) version, for any $\epsilon > 0$, that violates the quota (resp. budget) constraint by a factor of at most $\alpha$.

*Future Works.* We believe that finding a polynomial-time lower bound for the problems considered in this paper is an interesting future work as this question for the directed Steiner tree is open for a long-time. The integrality gap of the standard flow-based LP relaxation for **B-URT** is unbounded if no budget violation is allowed [2], and has a polynomial lower bound for the directed Steiner tree problem [21]. Finding an integrality gap for **B-URT** in case of budget violation would be an interesting future work. Extending our results to the edge-cost variants of **B-DRT** and **Q-DRT** would be a very interesting future work. The use of LP-hierarchies for approximation algorithms in directed Steiner trees [9,24] would be a potential direction to provide approximation algorithms for **B-DRT** and **Q-DRT** and their edge-cost variants. For Directed Steiner Network, it is known that the integrality gap of the Lasserre Hierarchy has a polynomial lower bound [8].

# References

1. Archer, A., Bateni, M.H., Hajiaghayi, M.T., Karloff, H.J.: Improved approximation algorithms for prize-collecting steiner tree and TSP. SIAM J. Comput. **40**(2), 309–332 (2011)
2. Bateni, M.H., Hajiaghayi, M.T., Liaghat, V.: Improved approximation algorithms for (budgeted) node-weighted steiner problems. SIAM J. Comput. **47**(4), 1275–1293 (2018)
3. Chan, T.M.: More algorithms for all-pairs shortest paths in weighted graphs. In: Proceedings of the Thirty-Ninth Annual ACM Symposium on Theory of Computing, STOC 2007, pp. 590–598. Association for Computing Machinery (2007)
4. Charikar, M., et al.: Approximation algorithms for directed steiner problems. J. Algorithms **33**(1), 73–91 (1999)

5. Cheng, X., Li, Y., Du, D.-Z., Ngo, H.Q.: Steiner trees in industry. In: Du, D.Z., Pardalos, P.M. (eds.) Handbook of Combinatorial Optimization, pp. 193–216. Springer, Boston (2004). https://doi.org/10.1007/0-387-23830-1_4

6. D'Angelo, G., Delfaraz, E.: Approximation algorithms for node-weighted steiner problems: digraphs with additive prizes and graphs with submodular prizes. arXiv preprint arXiv:2211.03653 (2022)

7. D'Angelo, G., Delfaraz, E., Gilbert, H.: Budgeted out-tree maximization with submodular prizes. In: Bae, S.W., Park, H. (eds.) 33rd International Symposium on Algorithms and Computation, ISAAC 2022, 19–21 December 2022, Seoul, Korea. LIPIcs, vol. 248, pp. 9:1–9:19. Schloss Dagstuhl - Leibniz-Zentrum für Informatik (2022)

8. Dinitz, M., Nazari, Y., Zhang, Z.: Lasserre integrality gaps for graph spanners and related problems. In: Kaklamanis, C., Levin, A. (eds.) WAOA 2020. LNCS, vol. 12806, pp. 97–112. Springer, Cham (2021). https://doi.org/10.1007/978-3-030-80879-2_7

9. Friggstad, Z., Könemann, J., Kun-Ko, Y., Louis, A., Shadravan, M., Tulsiani, M.: Linear programming hierarchies suffice for directed steiner tree. In: Lee, J., Vygen, J. (eds.) IPCO 2014. LNCS, vol. 8494, pp. 285–296. Springer, Cham (2014). https://doi.org/10.1007/978-3-319-07557-0_24

10. Gao, X., Junwei, L., Wang, H., Fan, W., Chen, G.: Algorithm design and analysis for wireless relay network deployment problem. IEEE Trans. Mob. Comput. $18(10)$, 2257–2269 (2019)

11. Garg, N.: Saving an epsilon: a 2-approximation for the k-MST problem in graphs. In: Gabow, H.N., Fagin, R. (eds.) Proceedings of the 37th Annual ACM Symposium on Theory of Computing, pp. 396–402. ACM (2005)

12. Ghuge, R., Nagarajan, V.: Quasi-polynomial algorithms for submodular tree orienteering and other directed network design problems. In: Chawla, S. (ed.) Proceedings of the 2020 ACM-SIAM Symposium on Discrete Algorithms, SODA, pp. 1039–1048. SIAM (2020)

13. Goemans, M.X., Williamson, D.P.: A general approximation technique for constrained forest problems. SIAM J. Comput. $24(2)$, 296–317 (1995)

14. Grandoni, F., Laekhanukit, B., Li, S.: $O(\log^2 k \ / \ \log \log k)$-approximation algorithm for directed steiner tree: a tight quasi-polynomial-time algorithm. In: Charikar, M., Cohen, E. (eds.) Proceedings of the 51st Annual ACM SIGACT Symposium on Theory of Computing, STOC 2019, Phoenix, AZ, USA, 23–26 June 2019, pp. 253–264. ACM (2019)

15. Guha, S., Moss, A., Naor, J., Schieber, B.: Efficient recovery from power outage (extended abstract). In: Proceedings of the Thirty-First Annual ACM Symposium on Theory of Computing, pp. 574–582. ACM (1999)

16. Hochbaum, D.S., Rao, X.: Approximation algorithms for connected maximum coverage problem for the discovery of mutated driver pathways in cancer. Inf. Process. Lett. $158$, 105940 (2020)

17. Johnson, D. S., Minkoff, M., Phillips, S.: The prize collecting steiner tree problem: theory and practice. In: Shmoys, D.B. (ed.) Proceedings of the Eleventh Annual ACM-SIAM Symposium on Discrete Algorithms, pp. 760–769. ACM/SIAM (2000)

18. Könemann, J., Sadeghabad, S.S., Sanità, L.: An LMP o(log n)-approximation algorithm for node weighted prize collecting steiner tree. In: 54th Annual IEEE Symposium on Foundations of Computer Science, FOCS 2013, 26–29 October 2013, Berkeley, CA, USA, pp. 568–577. IEEE Computer Society (2013)

19. Kortsarz, G., Nutov, Z.: Approximating some network design problems with node costs. Theor. Comput. Sci. $412(35)$, 4482–4492 (2011)

20. Kuo, T.-W., Lin, K.C.-J., Tsai, M.-J.: Maximizing submodular set function with connectivity constraint: theory and application to networks. IEEE/ACM Trans. Netw. **23**(2), 533–546 (2015)
21. Li, S., Laekhanukit, B.: Polynomial integrality gap of flow LP for directed steiner tree. In: Naor, J.S., Buchbinder, N. (eds.) Proceedings of the 2022 ACM-SIAM Symposium on Discrete Algorithms, SODA 2022, pp. 3230–3236. SIAM (2022)
22. Moss, A., Rabani, Y.: Approximation algorithms for constrained node weighted steiner tree problems. SIAM J. Comput. **37**(2), 460–481 (2007)
23. Paul, A., Freund, D., Ferber, A.M., Shmoys, D.B., Williamson, D.P.: Budgeted prize-collecting traveling salesman and minimum spanning tree problems. Math. Oper. Res. **45**(2), 576–590 (2020)
24. Rothvoß, T.: Directed steiner tree and the lasserre hierarchy. CoRR, abs/1111.5473 (2011)
25. Vandin, F., Upfal, E., Raphael, B.J.: Algorithms for detecting significantly mutated pathways in cancer. J. Comput. Biol. **18**(3), 507–522 (2011)
26. Zelikovsky, A.: A series of approximation algorithms for the acyclic directed steiner tree problem. Algorithmica **18**(1), 99–110 (1997)

# Resolving Sets in Temporal Graphs

Jan Bok[1,2], Antoine Dailly[1], and Tuomo Lehtilä[1,3,4(✉)]

[1] Université Clermont-Auvergne, CNRS, Mines de Saint-Étienne,
Clermont-Auvergne-INP, LIMOS, 63000 Clermont-Ferrand, France
[2] Department of Algebra, Faculty of Mathematics and Physics, Charles University,
Prague, Czechia
[3] Department of Computer Science, University of Helsinki, Helsinki, Finland
[4] Helsinki Institute for Information Technology (HIIT), Espoo, Finland
tualeh@utu.fi

**Abstract.** A *resolving set* $R$ in a graph $G$ is a set of vertices such that every vertex of $G$ is uniquely identified by its distances to the vertices of $R$. Introduced in the 1970's, this concept has been since then extensively studied from both combinatorial and algorithmic point of view. We propose a generalization of the concept of resolving sets to temporal graphs, *i.e.*, graphs with edge sets that change over discrete time-steps. In this setting, the *temporal distance from $u$ to $v$* is the earliest possible time-step at which a journey with strictly increasing time-steps on edges leaving $u$ reaches $v$, *i.e.*, the first time-step at which $v$ could receive a message broadcast from $u$. A *temporal resolving set* of a temporal graph $\mathcal{G}$ is a subset $R$ of its vertices such that every vertex of $\mathcal{G}$ is uniquely identified by its temporal distances from vertices of $R$. We study the problem of finding a minimum-size temporal resolving set, and show that it is NP-complete even on very restricted graph classes and with strong constraints on the time-steps: temporal complete graphs where every edge appears in either time-step 1 or 2, temporal trees where every edge appears in at most two consecutive time-steps, and even temporal subdivided stars where every edge appears in at most two (not necessarily consecutive) time-steps. On the other hand, we give polynomial-time algorithms for temporal paths and temporal stars where every edge appears in exactly one time-step, and give a combinatorial analysis and algorithms for several temporal graph classes where the edges appear in periodic time-steps.

**Keywords:** Temporal graphs · Resolving sets · Metric dimension · Trees · Graph algorithms · Complexity

This work was supported by the International Research Center "Innovation Transportation and Production Systems" of the I-SITE CAP 20-25 and by the ANR project GRALMECO (ANR-21-CE48-0004). Jan Bok was also funded by the European Union (ERC, POCOCOP, 101071674). Tuomo Lehtilä was also supported by Business Finland Project 6GNTF, funding decision 10769/31/2022.

A. A. Rescigno and U. Vaccaro (Eds.): IWOCA 2024, LNCS 14764, pp. 287–300, 2024.
https://doi.org/10.1007/978-3-031-63021-7_22

# 1 Introduction

For a set $R$ of vertices of a graph $G$, every vertex of $G$ can compute a vector of its distances from the vertices of $R$ (the distance, or number of edges, in a shortest path from $u$ to $v$ will be denoted by $\text{dist}(u, v)$). If all such computed vectors are unique, then we call $R$ a *resolving set* of $G$. This notion was introduced in the 1970s and gave birth to the notion of *metric dimension of* $G$, that is, the smallest size of a resolving set of $G$. Metric dimension is a well-studied topic, with both combinatorial and algorithmic results, see for example surveys [31,35].

A *temporal graph* can be defined as a graph on a given vertex set, and with an edge set that changes over discrete time-steps. Their study gained traction as a natural representation of dynamic, evolving networks [5,23,24,33]. However, in the temporal setting, the notion of distance differs from the static setting: two vertices can be topologically close, but the *journey* (*i.e.*, a path in the underlying graph with strictly increasing time-steps[1]) between them can be long, or even impossible. The shortest time-step at which a journey from $u$ reaches $v$ is called the *temporal distance from* $u$ *to* $v$ and is denoted by $\text{dist}_t(u, v)$. Note that in a temporal graph, there might be vertex pairs $(u, v)$ such that there is no temporal journey from $u$ to $v$, in which case we set the distance as infinite. Furthermore, the temporal distance of non-adjacent vertices is not necessarily symmetric.

The notion of temporal distance allows us to define a *temporal resolving set* as a set $R$ of vertices of a temporal graph such that every vertex has a unique vector of temporal distances from the vertices of $R$. We are interested in the problem of finding a minimum-size temporal resolving set.

Our motivation is both introducing a temporal variant of the well-studied problem of resolving sets and studying its combinatorial and algorithmic properties, as well as the problem of locating in dynamic networks. Indeed, resolving sets are an analogue of geolocation in discrete structures [35], and thus temporal resolving sets are similar: if we consider that transmitters placed on vertices of the temporal resolving set emit continuously, we can locate ourselves by waiting long enough to receive the signals and constructing the temporal distance vector.

In the remainder of this section, we give an overview of the static (*i.e.*, non-temporal) version and some variants of resolving set as well as an overview of temporal graphs, before giving a formal definition of the TEMPORAL RESOLVING SET problem and an outline of our results.

**Separating Vertices.** Standard resolving sets were introduced independently by Harary and Melter [21] and by Slater [34], and have been well-studied due to their various applications (robot navigation [29], detection in sensor networks [34], and more [35]). Their non-local nature makes them difficult to study from an algorithmic point of view: finding a minimum-size resolving set is NP-hard even on very restricted graph classes (planar graphs of bounded degree [8], bipartite graphs [9], and interval graphs of diameter 2 [15], to name a few) and

---

[1] Those are sometimes called *strict* journeys in the literature, but as argued in [30], strict journeys are naturally suited to applications where one cannot traverse multiple edges at the same time.

W[2]- and W[1]-hard when parameterized by solution size [22] and feedback vertex set [17], respectively. On the positive side, the problem is polynomial-time solvable on trees [6,21,29,34], outerplanar graphs [8] and cographs [9], to name a few. Note that there are two conditions for a resolving set: *reaching* (every vertex must be reached from some vertex of the resolving set) and *separating* (no two vertices may have the same distance vector).

A natural variant of resolving sets consists in limiting the distance at which transmitters can emit. A first constraint is that of a robot which can only perceive its direct neighborhood: an *adjacency resolving set* [26] is a resolving set using the following distance: $\text{dist}_a(u, v) = \min(\text{dist}(u, v), 2)$. More generally, a *k-truncated resolving set* [11] is a resolving set using the following distance: $\text{dist}_k(u, v) = \min(\text{dist}(u, v), k + 1)$ (we can see that an adjacency resolving set is a 1-truncated resolving set). Note that both variants were mostly studied in their weak version. The $k$-truncated resolving sets have quickly attracted attention on the combinatorial side [2,16,18,19]. On the algorithmic side, the corresponding decision problem is known to be NP-hard [11], even on trees (although, in this case, it becomes polynomial-time solvable when $k$ is fixed) [20].

**Temporal Graphs.** A temporal graph $\mathcal{G} = (V, E_1, \ldots, E_{t_{\max}})$ is described by a sequence of edge sets representing the graph at discrete *time-steps*, which are positive integers in $\{1, \ldots, t_{\max}\}$ [5] (note that the sequence might be infinite, in which case the number of time-steps is not bounded by $t_{\max}$, and we can adapt all definitions in consequence). An alternate, equivalent description of $\mathcal{G}$ is $\mathcal{G} = (V, E, \lambda)$ (or $(G, \lambda)$ where $G = (V, E)$) with $E = \bigcup_{i=1}^{t_{\max}} E_i$ is called the *underlying graph* and $\lambda : E \to 2^{\{1, \ldots, t_{\max}\}}$ is an edge-labeling function called *time labeling* such that $\lambda(e)$ is the set of time-steps at which the edge $e$ exists, i.e., $i \in \lambda(e)$ if and only if $e \in E_i$ [28]. We call a time labeling a *k-labeling* if $|\lambda(e)| \leq k$ for every edge $e$. Furthermore, we say that a temporal graph is a *temporal tree* (resp. *temporal star*, etc.) if its underlying graph, understood as the graph induced by the union of all its edge sets, is a tree (resp. star, etc.). For a subgraph $G'$ of graph $G$ and a time labeling $\lambda$ of $G$, we denote by $\lambda_{|G'}$ the restriction of $\lambda$ to subgraph $G'$.

A specific case of temporal graphs are those with a repeating sequence of edge sets, which have been studied in contexts such as routing [13,14,25,32], graph exploration [3], cops and robbers games [7,10] and others [1,36] due to their natural applications in *e.g.* transportation networks. Formally, a *p-periodic k-labeling* is a time labeling such that both $E_{i+p} = E_i$ for every $i \geq 1$ and $|\lambda(e) \cap \{1, \ldots, p\}| \leq k$ for every edge $e$. A temporal graph with a periodic time labeling has an infinite sequence of edge sets, but can be represented with its $p$ first time-steps, understanding that the sequences will repeat after this.

A vertex $v$ is said to be *reachable from* another vertex $u$ if there exists a journey from $u$ to $v$. For a given vertex $v$, the set of vertices which can be reached from vertex $v$ is denoted by $\mathbf{R}(v)$. For a vertex set $S$ such that $v \in S$, we denote by $\mathbf{R}^S(v) \subseteq \mathbf{R}(v)$ the set of vertices which can be reached from $v$ but not from any other vertex in $S$.

**Temporal Resolving Sets.** We extend the definition of resolving sets to the temporal setting: a resolving set in a temporal graph is a reaching and separating set. More formally, a set $R$ of vertices of a temporal graph $\mathcal{G} = (V, E, \lambda)$ is a *temporal resolving set* if $(i)$ for every vertex $v \in V$, there is a vertex $s \in R$ such that $v \in \mathbf{R}(s)$; $(ii)$ for every two different vertices $u, v \in V$, there is a vertex $s \in R$ such that $\text{dist}_t(s, u) \neq \text{dist}_t(s, v)$. Note that every vertex in a temporal resolving set is trivially separated from every other vertex. The problem we are studying is the following:

---
TEMPORAL RESOLVING SET
Instance: A temporal graph $\mathcal{G} = (V, E, \lambda)$; and an integer $k$.
Question: Is there a *temporal resolving set* of size at most $k$?

---

Due to the fact that temporal distance is not a metric in the usual sense (symmetry and the standard definition of the triangle inequality might not hold), we call the minimum size of a temporal resolving set of a graph $G$ the *temporal resolving number of* $G$ instead of temporal metric dimension.

Note that we can assume in the following that there is an edge $e$ such that $1 \in \lambda(e)$ (otherwise, let $m$ be the smallest time-step and decrease every time-step by $m - 1$). Temporal resolving set can be seen as a generalization of standard and $k$-truncated resolving sets: if $\lambda(e) = \{1, \dots, \text{diam}(G)\}$ for every edge $e$ (where $\text{diam}(G) = \max\{\text{dist}(u, v) \mid u, v \in V\}$, then a temporal resolving set is a standard resolving set; and if $\lambda(e) = \{1, \dots, k\}$ for every edge $e$, then a temporal resolving set is a $k$-truncated resolving set[2].

**Our Results and Outline.** We focus on time labelings with few labels per edge, mostly limiting ourselves to one or two labels per edge. Although the setting is more restricted than the general case, we shall prove that these scenarios already yield NP-complete problems or non-trivial polynomial algorithms.

We present three sets of results. First, we focus in Sect. 2 on the computational complexity of finding a minimum-size temporal resolving set in temporal graphs with 2-labelings. In particular, we prove that the problem is NP-complete on temporal complete graphs, which contrasts heavily with other resolving set problems. The problem is also NP-complete on temporal subdivided stars, and on temporal trees even when the two time-steps are consecutive.

In Sect. 3, we give polynomial-time algorithms for some classes with 1-labelings. However, even for temporal paths, while the algorithm is quite natural, proving optimality is non-trivial. We also give algorithms for temporal stars, and temporal subdivided stars when $t_{\max} = 2$.

Finally, in Sect. 4, we take a more combinatorial approach to periodic time labelings. We find the optimal bounds for the temporal resolving number of several graph classes under this setting, namely in temporal paths, cycles, complete graphs, complete binary trees, and subdivided stars. We also prove that TEMPORAL RESOLVING SET is FPT on trees with respect to number of leaves and XP on subdivided stars with $p$-periodic 1-labelings with respect to the period $p$.

Proofs marked with (**\***) are omitted due to space constraints, (see [4]).

---

[2] Also called a $k$-*truncated dominating resolving set* in [20].

# 2    NP-Hardness of Temporal Resolving Set

In this section, we give hardness results for Temporal Resolving Set on very restricted graph classes, and with strong constraints on the time labeling.

In the static setting, complete graphs tend to be easy to work with: since all the vertices are twins, they are indistinguishable from one another, and hence we need to take all of them but one in order to separate them. Indeed, the metric dimension and location-domination number of $K_n$ is $n - 1$. However, this is not the case with temporal complete graphs, since now the vertices are not necessarily twins anymore. We even prove that it is NP-hard:

**Theorem 1 (*).** Temporal Resolving Set *is NP-complete on temporal complete graphs with a 1-labeling, even when there are only two time-steps.*

*Proof (Sketch).* We reduce from the adjacency resolving set problem [12]. From a connected planar graph $G(V, E)$, we construct the temporal complete graph $\mathcal{G} = (V, E', \lambda)$, with $\lambda(e) = \{1\}$ if $e \in E$ and $\lambda(e) = \{2\}$ otherwise.    □

The next two results are both proved by reducing from 3-Dimensional Matching, one of the seminal NP-complete problems [27], and are inspired by the NP-completeness proof for $k$-truncated resolving sets on trees in [20], with nontrivial adaptations to constrain the setting as much as possible.

---

3-Dimensional Matching (3DM)

Instance: A set $S \subseteq X \times Y \times Z$, where $X$, $Y$, and $Z$ are disjoint subsets of $\{1, \ldots, n\}$ of size $p$; and an integer $\ell < |S|$.

Question: Does $S$ contain a *matching* of size at least $\ell$, *i.e.*, a subset $M \subseteq S$ such that $|M| \geq \ell$ and no two elements of $M$ agree in any coordinate?

---

**Theorem 2.** Temporal Resolving Set *is NP-complete on temporal stars in which every edge is subdivided twice, and when each edge appears in at most two time-steps.*

*Proof.* We note that the problem is clearly in NP: a certificate is a set of vertices, and for each vertex, we can compute the time vectors and check that they are all different and that every vertex is reached by at least one vertex from the set in polynomial time. To prove completeness, we reduce from 3DM.

Starting from an instance $(S, \ell)$ of 3DM, denoting $s = |S|$ and the $i$-th triple in $S$ by $(x_i, y_i, z_i)$ with $x_i \in X, y_i \in Y, z_i \in Z$, we construct an instance $(\mathcal{G}, s+2-\ell)$ of Temporal Resolving Set, where $\mathcal{G} = (T, \lambda)$. This construction is detailed below, see Fig. 1.

Let $V(T) = \{u, t_1, t_2, t_3\} \cup \bigcup_{i=1}^{s} \{a_i, b_i, c_i\}$. We will arrange the vertices in the following way as a twice subdivided star. The center vertex is $u$ and it is attached to vertices $a_i$ which are adjacent to vertices $b_i$ which are adjacent to vertices $c_i$ for each $1 \leq i \leq s$. Moreover, $u$ is also adjacent to $t_1$ which is adjacent to $t_2$ and which is adjacent to $t_3$. Vertices $\{a_i, b_i, c_i\}$ correspond to elements $\{x_i, y_i, z_i\}$ so that $a_i < b_i < c_i$. Edges are labeled as follows:

- For every $i \in \{1, \ldots, s\}$, $\lambda(ua_i) = \{2, a_i + 4\}$;
- For every $i \in \{1, \ldots, s\}$, $\lambda(a_i b_i) = \{3, b_i + 4\}$;
- For every $i \in \{1, \ldots, s\}$, $\lambda(b_i c_i) = \{4, c_i + 4\}$;
- We have $\lambda(ut_1) = \{2\}$, $\lambda(t_1 t_2) = \{1\}$ and $\lambda(t_2 t_3) = \{3\}$;

Observe that the constructed graph is a star whose every edge is subdivided exactly twice. Moreover, every edge has at most two labels.

We shall now prove that we decide YES for 3DM on $(S, \ell)$ if and only if we decide YES for TEMPORAL RESOLVING SET on $((T, \lambda), s + 1 - \ell)$.

($\Rightarrow$) Assume that $S$ contains a matching $M$ of size at least $\ell$. We construct the following set: $R = \bigcup_{i \notin M} \{a_i\} \cup \{t_1\}$. Note that $R$ contains $t_1$ and each $a_i$ such that the corresponding element of $S$ is not in $M$. Furthermore, every vertex of $T$ is reached from $t_1$, so we need only consider the separation part.

First observe that $t_1$ separates $u$ and each $t_i$. Furthermore, it reaches each vertex of type $a_i$ at moment $a_i + 4$, vertices $b_i$ at moment $b_i + 4$ and vertices $c_i$ at moment $c_i + 4$. Consider then some $a_h \in R$. It reaches vertex $b_h$ at moment 3 and $c_h$ at moment 4. Hence, together with vertex $t_1$, vertex $a_h$ separates vertices $a_h, b_h$ and $c_h$ from all other vertices. Furthermore, vertex $a_h$ reaches other vertices of types $a_i, b_i$ and $c_i$ at the same moment as vertex $t_1$. Hence, we have uniquely separated vertices $u, t_1, t_2, t_3$ and every $a_h, b_h, c_h$ such that $a_h \in R$. Recall that sets $X, Y$ and $Z$ are disjoint. Thus, each vertex $a_i$ is separated from vertices of type $b_j$ for any $i$ and $j$ and the same is true for $a_i$ and $c_j$ as well as $b_i$ and $c_j$. Let us next consider when we might not separate $a_i$ from $a_j$ (the same argument holds for pairs $b_i, b_j$ and $c_i, c_j$). We may assume that $\{a_i, a_j\} \cap R = \emptyset$. Thus, corresponding elements belong to $M$. Therefore $a_i \neq a_j$ and hence, they are reached at different time moments from vertex $t_1$, a contradiction. Therefore, $R$ is a temporal resolving set of the claimed cardinality.

($\Leftarrow$) Assume that there is a temporal resolving set $R$ of size at most $s - \ell + 1$. Since every vertex must be reached from a vertex of $R$, we must have one of vertices $t_1, t_2$ or $t_3$ in $R$. Now, as in the previous case, only the $a$'s $b$'s and $c$'s must be reached and resolved. If each $a_i$, $b_i$ and $c_i$ is unique, then $t_1$ is a resolving set of size $1 \le s - \ell + 1$. Moreover, if for example $a_i = a_j$ (similar argument holds for $b_i = b_j$ and $c_i = c_j$), then at most one of corresponding tuples can belong to the matching. Moreover, to separate $a_i$ and $a_j$, we need a vertex in resolving set to belong to one of the branches. Hence, we may choose as our matching $M$ the sets corresponding to branches which contain no members of the temporal resolving set. There are at least $\ell$ such branches and the claim follows.    □

**Theorem 3 (*).** TEMPORAL RESOLVING SET *is NP-complete on temporal trees, even with only one vertex of degree at least 5, and when each edge appears in at most two, consecutive, time-steps.*

## 3    Polynomial-Time Algorithms for Subclasses of Trees

In this section, we give polynomial-time algorithms for TEMPORAL RESOLVING SET. We study temporal paths and stars with one time label per edge, and

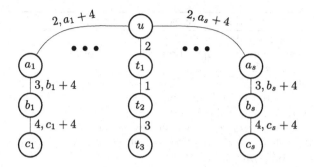

**Fig. 1.** The construction of the proof of Theorem 2. Only the branches 1 and $s$ are detailed together with the control branch. We have $\{a_i, b_i, c_i\} = \{x_i, y_i, z_i\}$ where $a_i < b_i < c_i$.

temporal subdivided stars with one time label per edge and where every label is in $\{1, 2\}$. Recall that TEMPORAL RESOLVING SET is already NP-complete on temporal subdivided stars with 2-labeling (Theorem 2), so these results are a first step for bridging the gap between polynomial-time and NP-hard.

### 3.1 Temporal Paths

Throughout this subsection, we denote by $P_n$ a path on $n$ vertices $v_1, \ldots, v_n$, with edges $v_i v_{i+1}$ for $1 \leq i \leq n - 1$. Furthermore, we assume that $\lambda$ is a 1-labeling. Algorithm 1 constructs a minimum-size temporal resolving set $R$ of $\mathcal{P} = (P_n, \lambda)$. The core of the algorithm consists in adding to $R$ the last vertex that can reach a leaf, then check if it separates everything in the two directions. If so, we can iterate on the vertices it cannot reach, and otherwise we have to add a vertex that separates the conflicting vertices before iterating. We denote $\lambda(v_i v_{i+1})$ by $t_i$ and by $\{r_1, \ldots, r_{|r|}\}$ the elements of $R$, and we assume that if $r_i = v_j$ and $r_h = v_k$ for $h > i$, then $k > j$. Consider vertex $v_i$, we say that $v_j$ is on its right (resp. left) side if $j > i$ (resp. $j < i$). The set of vertices on the left side of vertex $v_j$ is denoted by $\ell(v_j)$.

**Lemma 1 (\*).** Let $\mathcal{P} = (P_n, \lambda)$ be a temporal path, $P_m$ be a subpath of $P_n$ containing one leaf of $P_n$, and $\mathcal{P}' = (P_m, \lambda_{|P_m})$. The temporal resolving number of $\mathcal{P}$ is at least as large as the temporal resolving number of $\mathcal{P}'$.

**Theorem 4.** There is an $\mathcal{O}(n)$ algorithm solving TEMPORAL RESOLVING SET on temporal paths on $n$ vertices where each edge appears only once.

*Proof (Sketch).* In the following, we first show that Algorithm 1 returns a temporal resolving set $R$ of $\mathcal{P} = (P_n, \lambda)$. After that, we prove that $R$ is minimum-sized and finally, that the algorithm has linear-time complexity.

First of all, consider vertices in $\mathbf{R}(r_1)$. Note that if there are vertices $u, v_b \in \mathbf{R}(r_1)$ such that $\mathrm{dist}_t(r_1, u) = \mathrm{dist}_t(r_1, v_b)$, then one of them is on the left side of $r_1$ and other one is on its right side. Let us assume, without loss of generality,

---

**Algorithm 1:** Temporal resolving set for temporal paths with 1-labeling

**Input**   : A temporal path $\mathcal{P} = (P_n, \lambda)$.
**Output:** A minimum-size temporal resolving set $R$ of $\mathcal{P}$.

1  Set $v = v_1$ and $R = \emptyset$.
2  **while** *true* **do**
3  | Let $s = v_i$ where $i$ is the largest integer such that $v_i$ reaches $v$. Add $s$ to $R$ and set $a = i$.
4  | **if** *within* $\mathbf{R}^R(s)$ *every vertex has unique distance to $s$* **then**
5  | | $w = v_j$ where $j$ is the smallest integer with $j > i$ for each $v_i \in \mathbf{R}(s)$.
6  | | **if** $v_n \in \mathbf{R}^R(s)$ *or* $v_n \in R$ **then**
7  | | | **return** $R$.
8  | **else**
9  | | Let $w = v_b$ where $v_b \in \mathbf{R}^R(s)$ is the vertex which does not have unique distance among vertices in $\mathbf{R}^R(s)$ to $s$ and among those vertices $b$ is minimal such that $b > a$.
10 | Let $v = w$.

---

that $u$ is on the left side of $r_1$. Moreover, no third vertex $w$ can have $\text{dist}_t(r_1, u) = \text{dist}_t(r_1, v_b) = \text{dist}_t(r_1, w)$. Let us assume that $v_b$ is the vertex with the smallest index on the right side of $r_1$ such that it is not separated by $r_1$ from some other vertex (in this case, from $u$).

In this case, Algorithm 1 has chosen $w = v_b$ on Step 8 and set $v = w$ after that on Step 10. Hence, in the following while-loop, we choose $r_2$ as the rightmost vertex which reaches $v(= v_b)$. Furthermore, $r_2$ cannot reach $u$. Indeed, since $r_1$ cannot separate $u$ and $v_b$, there are two edges with the same time label on the path from $u$ to $v_b$. Therefore, on the path from $r_2$ to $u$, there are two edges with the same time label. Hence, $r_2$ separates $v_b$ from $u$. Consequently, if there were any other vertices in $\mathbf{R}^{\{r_1\}}(r_1)$ which were not separated by $r_1$, then they would be separated by $r_2$. A similar argument works for all pairs $r_i, r_{i+1}$. Note that in the end, either we choose $r_{|R|} = v_n$, or $r_{|R|}$ separates every vertex in $\mathbf{R}^R(r_{|R|})$. Hence, we eventually enter the if-clause on Step 6 and return $R$.

We now show that there does not exist any resolving set of smaller size than $R$ in $\mathcal{P}$. We do this by induction on the number $n$ of vertices. First of all, Algorithm 1 outputs a temporal resolving set of size 1 when $n \in \{1, 2\}$, which is optimal. Thus, we assume from now on that it outputs a minimum-size temporal resolving set for $n \leq n'$.

Let $n = n' + 1$, and $R$ be the temporal resolving set constructed using Algorithm 1 on $\mathcal{P} = (P_n, \lambda)$. Assume first that $r_1$ separates every vertex in $\mathbf{R}(r_1)$. Observe that if $|R| = 1$, then it is minimum-size. Hence, we may assume that $\mathcal{P}' = \mathcal{P} \backslash \mathbf{R}(r_1)$ (where $\mathcal{G} \backslash V'$ for a temporal graph $\mathcal{G} = (V, E, \lambda)$ denotes the temporal subgraph $\mathcal{G}' = (V \backslash V', E \backslash \{uv : u \in V' \text{ or } v \in V'\}, \lambda)$) is nonempty. By induction, Algorithm 1 outputs a minimum-size temporal resolving set $R \backslash \{r_1\}$ of $\mathcal{P}'$. Observe that $\ell(r_1) \cap \mathbf{R}(w) = \emptyset$ for any $w$ on the rightside of $r_1$. Moreover, we require at least one vertex in set $\ell(r_1) \cup \{r_1\}$ in any resolving set of $\mathcal{P}$ to reach vertex $v_1$. Observe that Algorithm 1 returns the temporal resolving set $R\setminus\{r_1\}$ for $\mathcal{P}\setminus\mathbf{R}(r_1)$. By induction, set $R\setminus\{r_1\}$ has minimum size. By Lemma 1, we require in any temporal path containing $\mathcal{P}\setminus\mathbf{R}(r_1)$ (and having $v_n$ as a leaf) at least $|R| - 1$ vertices. Furthermore, by our observations, we require in a set $\ell(r_1) \cup \{r_1\}$ at least one vertex to reach $v_1$. Furthermore, since these vertices do not reach any vertex in $V(\mathcal{P})\setminus\mathbf{R}(r_1)$, we require at least $|R|$ vertices in a resolving set of $\mathcal{P}$, as claimed.

Assume next that there are vertices $v_\ell$ and $v_r$ in $\mathbf{R}(r_1)$ which are not separated by $r_1$. Further assume that $v_\ell$ (resp. $v_r$) is on the left (resp. right) side of $r_1$. Consequently, $r_2$ is the rightmost vertex which reaches $v_r$. We show that if $v_\ell \neq v_1$, then the claim follows. Assume that $\ell \geq 2$. Consider the temporal path $\mathcal{P}' = \mathcal{P}\setminus\{v_1\}$. Note that $r_1$ is the rightmost vertex which reaches $v_2$. Moreover, $r_1$ does not separate vertices $v_\ell$ and $v_r$. Thus, Algorithm 1 outputs $R$ as a resolving set for $\mathcal{P}'$. By our induction hypothesis, $R$ is minimum-size. By Lemma 1, we know that $R$ is at least as large as a minimum-size temporal resolving set of $\mathcal{P}'$. Thus, $R$ is also a minimum-size temporal resolving set for $\mathcal{P}$. Note that this implies that $r_1$ separates every pair of vertices in $\mathbf{R}(r_1)$ except for $(v_1, v_r)$.

We now need to analyze several cases depending on whether $t_{r-1} \leq t_r$ and whether $t_r \leq t_{r+1}$. We distinguish four cases and for all of them, we conclude that at least $|R|$ vertices are necessary in a temporal resolving set of $\mathcal{P}$. The analysis is quite technical and is omitted due to space constraints.

Finally, we show that Algorithm 1 has linear time complexity. First of all, the while-loop ends at some point since every temporal path has a temporal resolving set by taking every vertex in the underlying path. Secondly, Step 3 uses at most $i + 1 - a$ comparisons and in total at most $2n$ comparisons. In Step 4, observe that sets $\mathbf{R}^R(s)$ do not overlap. Thus, each vertex is considered only once. Moreover, the time labels on the left and the right side of $s$ are ordered from small to large. Thus, checking if each time label has a unique value can be done in linear time on $|\mathbf{R}^R(s)|$. Again, in Step 5, the sets $\mathbf{R}(s)\setminus\ell(s)$ do not overlap. Thus, this step takes at most linear time on $n$ in total. Finally, all the other steps take at most constant time. Hence, the algorithm has linear-time complexity. $\qquad\qquad\square$

The following lemma gives some structure on the minimum-size temporal resolving sets of temporal paths, with respect to the one output by Algorithm 1. In particular, it states that the constructed temporal resolving set $R$ places each vertex in $R$ as far away from the leaf $v_1$ as possible. It will be used in the case of subdivided stars, allowing us to reuse Algorithm 1 to find a partial solution.

**Lemma 2 (*).** *Let $\mathcal{P}$ be a temporal path on $n \geq 2$ vertices with 1-labeling $\lambda$ on vertices $v_1, \ldots, v_n$ and edges $v_i v_{i+1}$ for $1 \leq i \leq n - 1$ where $v_i$ is on the left side of $v_{i+1}$ for each $i$. Let $R = \{r_1, \ldots, r_{|R|}\}$ be the temporal resolving set output by Algorithm 1, where $r_i$ is on the left side of $r_{i+1}$ for each $i$. Let $R' = \{r'_1, \ldots, r'_{|R|}\}$ be another temporal resolving set of $\mathcal{P}$. We have $r'_i \in \ell(r_i) \cup \{r_i\}$ for each $i$.*

## 3.2 Temporal Stars

In this subsection, we give polynomial-time algorithms for finding minimum-size temporal resolving sets for temporal stars with 1-labeling $\lambda$ and temporal subdivided stars with 1-labeling using only values 1 and 2. For both these classes, the central vertex is denoted by $c$.

**Theorem 5 (*).** *Let $S$ be a star, $\lambda$ be a 1-labeling, and $X$ be a maximum-size set of leaves of $S$ such that, for distinct $u, v \in X$, $\lambda(cu) \neq \lambda(cv)$. The set $V(S) \backslash X$ is a minimum-size temporal resolving set of $(S, \lambda)$.*

We now consider subdivided stars together with 1-labeling $\lambda$ using only values 1 and 2. In particular, we present a polynomial-time algorithm for this case. By a branch of a subdivided star, we mean a path starting from vertex $c$ without the vertex $c$ itself. Branches are denoted by $B_i, \ldots, B_\Delta$ and the leaves by $\ell_1, \ldots, \ell_\Delta$. Furthermore, the vertex in $B_i$ adjacent to $c$ is denoted by $v_i$, the vertex adjacent to $v_i$ by $u_i$ and the third vertex of branch by $w_i$ (if they exist, note that some of these vertices might also be denoted $\ell_i$). We further assume that branches are ordered so that $\lambda(cv_i) \leq \lambda(cv_{i+1})$ for each $i \leq \Delta - 1$. We assume that $\lambda(cv_i) = 1$ for each $i \leq I_1$ where $0 \leq I_1 \leq \Delta$ and denote $B^1 = \bigcup_{i=1}^{I_1} B_i$ and $B^2 = \bigcup_{i=I_1+1}^{\Delta} B_i$.

The following theorem shows that Algorithm 2 returns a minimum-size temporal resolving set in polynomial time for a given temporal subdivided star using only values 1 and 2 in its 1-labeling.

**Theorem 6 (*).** *Given a subdivided star $SS_\Delta$ of maximum degree $\Delta \geq 3$ and 1-labeling of edges restricted to values 1 and 2, Algorithm 2 returns a minimum-size temporal resolving set of $SS_\Delta$ in polynomial-time on the number of vertices.*

# 4    Combinatorial Results for $p$-Periodic 1-Labelings

In this section, we focus on $p$-periodic 1-labelings, and bound the minimum size of temporal resolving sets for several graph classes. When $p = 1$ or $\lambda(e)$ is the same for every edge $e$, those are exactly the usual resolving sets.

Given a temporal graph $\mathcal{G} = (G, \lambda)$ where $\lambda$ is a $p$-periodic 1-labeling, we denote by $M_p(\mathcal{G})$ the minimum size of a temporal resolving set of $\mathcal{G}$. Furthermore, if $\lambda(e) = \{i, i+p, i+2p, \ldots\}$, then, by abuse of notation, we denote $\lambda(e) = i$.

Note that, in this section, reachability is trivially assured (since the time-steps repeat indefinitely and considered graphs are connected), so to prove that a given set is a temporal resolving set, we only need to prove that it is separating.

**Theorem 7 (*).** *Let $P_n$ be a path on $n$ vertices, $\lambda$ a $p$-periodic 1-labeling, and $\mathcal{P} = (P_n, \lambda)$. We have $M_p(\mathcal{P}) = 1$.*

In particular, the proof of Theorem 7 implies that in a temporal tree $\mathcal{T}$ with $p$-periodic 1-labeling if we have a path from $r$ to $u$ to $v$, then $r$ separates vertices $u$ and $v$. In this case, we say that vertices $u$ and $v$ are *path-separated* (by $r$). In the following two theorems, we introduce combinatorial results for some simple graph classes.

---

**Algorithm 2:** Temporal metric dimension for subdivided star $SS_\Delta$ with 1-labels using values 1 and 2

---

**Input**  : Subdivided star $SS_\Delta$ of degree $\Delta \geq 3$ together with time labeling $\lambda$ for each edge $e \in E(SS_\Delta)$ such that $\lambda(e) \in \{1,2\}$.

**Output:** An optimal temporal resolving set $R$ on $SS_\Delta$.

1   For each path from $\ell_i$ to $c$, we create temporal resolving set $R_i$ using Algorithm 1.

2   Let $R' = (\bigcup_{i=1}^{\Delta} R_i) \setminus \{c\}$.

3   Let $B_c = \bigcup_{c \in R_i} B_i$.

4   **if** $B_c = \emptyset$ **then**

5     $\lfloor$   **return** $R'$.

6   **for** $j \in \{1,2\}$ **do**

7     **if** $v \in N(c) \cap V(B_c)$ *and* $v \in V(B_i)$ *and* $B_i \in B^j$, *for some* $i \leq \Delta$, *and* $R_i \setminus \{c\}$ *does not separate* $v$ *from some other vertex in* $B_i$ *but* $v \in \mathbf{R}(R_i \setminus \{c\})$ **then**

8       $\lfloor$   add $v$ to $Q_j$.

9   **for** $1 \leq j \leq \Delta$ **do**

10   $\lfloor$   Remove branch $B_j$ from $B_c$ if $Q_2 \cap V(B_j) \neq \emptyset$.

11   Let $r = |\{i \mid Q_1 \cap V(B_i) \neq \emptyset\}|$.

12   **for** $0 \leq j \leq 2$ **do**

13     **for** *each vertex set* $R''$ *such that* $R'' \subseteq \mathbf{R}(c) \cap (V(B_c) \cup \{c\})$, $|R'' \cap V(B_i)| \leq 1$ *for every* $B_i \in B_c$, $|R'' \cap Q_1| \leq 1$, $|R'' \cap \{r \mid \mathrm{dist}(c,r) = 2\}| \leq 1$ *and* $|R''| = |B_c| - 1 - r + j$ **do**

14       **if** $R' \cup R''$ *is a temporal resolving set of* $SS_\Delta$ **then**

15         $\lfloor$   **return** $R' \cup R''$

---

**Theorem 8 (\*).** *Let* $C_n$ *be a cycle on* $n$ *vertices,* $\lambda$ *a* $p$-*periodic 1-labeling, and* $\mathcal{C} = (C_n, \lambda)$. *We have* $1 \leq M_p(\mathcal{C}) \leq 2$.

**Theorem 9 (\*).** *Let* $K_n$ *be a complete graph on* $n = b + p^b$ *vertices with* $b \geq 1$, $\lambda$ *a* $p$-*periodic 1-labeling, and* $\mathcal{K} = (K_n, \lambda)$. *We have* $b \leq M_p(\mathcal{K}) \leq n - 1$ *and both bounds are tight.*

The following lemma will help us to simplify the remaining results on trees, as we will be able to consider only temporal resolving sets composed of leaves.

**Lemma 3 (\*).** *Let* $T$ *be a tree,* $\lambda$ *a* $p$-*periodic 1-labeling,* $\mathcal{T} = (T, \lambda)$ *with* $M_p(\mathcal{T}) \geq 2$. *There is a temporal resolving set of* $\mathcal{T}$ *of size* $M_p(\mathcal{T})$ *containing only leaves of* $T$.

Lemma 3 gives us an FPT algorithm for TEMPORAL RESOLVING SET in trees with respect to the number of leaves. It also allows us to prove a bruteforce-like polynomial-time algorithm for temporal subdivided stars with a $p$-periodic 1-labeling for fixed $p$. This shows that TEMPORAL RESOLVING SET is in XP with respect to the period of the time labeling in this context.

**Theorem 10.** TEMPORAL RESOLVING SET *is polynomial-time solvable in temporal subdivided stars with a p-periodic labeling for a fixed constant p.*

*Proof.* Let $S$ be a subdivided star with central vertex $c$ and $\ell$ leaves, $\lambda$ be a $p$-periodic 1-labeling, and $\mathcal{S} = (S, \lambda)$. By Lemma 3, there exists a minimum-size temporal resolving set in $\mathcal{S}$ which contains only leaves of $S$, if $M_p(\mathcal{S}) > 1$.

Note that if some branches, say $b$ of them, share the same time label on their edges incident with $c$, then, to separate vertices in these branches, we need to select at least $b - 1$ leaves in a temporal resolving set. Since there are at most $p$ distinct labels, we have, for any temporal resolving set $R$ of $\mathcal{S}$, $|R| \geq \ell - p$. Furthermore, checking whether a given vertex set is a temporal resolving set of $\mathcal{S}$ can be done in polynomial time.

If $n$ is the order of $S$, then there are $\binom{\ell}{\ell-p} = \binom{\ell}{p} \in O(n^p)$ ways to select a vertex set containing exactly $\ell - p$ leaves. Since $p$ is a fixed constant, we can check all these sets in polynomial time. If any of them is a temporal resolving set, then we found a minimum-size temporal resolving set of $\mathcal{S}$. Otherwise, we iterate the process by examining all vertex sets of size $\ell - p + 1$, $\ell - p + 2$, ..., until we find a temporal resolving set. The process will eventually stop as the set containing all leaves of $S$ is a temporal resolving set of $\mathcal{S}$. Thus, we need to check at most $\sum_{i=0}^{p} \binom{\ell}{p-i}$ vertex sets, that is, we need to do $O(n^p)$ polynomial-time operations. □

We end this section with two combinatorial results for other two subclasses of temporal trees: subdivided stars and complete binary trees.

**Theorem 11 (\*).** *Let $S$ be a subdivided star on $\ell \geq 2$ leaves, $\lambda$ a p-periodic 1-labeling, and $\mathcal{S} = (S, \lambda)$. We have $\max(1, \ell - p) \leq M_p(\mathcal{S}) \leq \ell - 1$ and both bounds are tight.*

**Theorem 12 (\*).** *Let $T$ be a complete binary tree on $2^n - 1$ vertices, $\lambda$ a 2-periodic 1-labeling, and $\mathcal{T} = (T, \lambda)$. We have $2^{n-3} \leq M_2(\mathcal{T}) \leq 2^{n-2}$. Both bounds are tight.*

## 5 Conclusion

We extended the definition of resolving sets to temporal graphs. We have proved that TEMPORAL RESOLVING SET is hard even in very restricted graph classes for some labelings. On the other hand, it seems that the problem might be easier with $p$-periodic labelings. In particular, the fact that the standard resolving sets are somewhat easier (they can be solved in polynomial-time in some classes for which TEMPORAL RESOLVING SET is NP-hard, such as complete graphs or trees) suggests that having time labelings containing a large number of labels or where the highest possible time-step is low might make the problem more tractable. In particular, since minimum $k$-truncated resolving sets can be found in polynomial-time in trees when $k$ is fixed, a potential direction would be to study the parameterized complexity of temporal resolving sets, with the total number of available time-steps as a natural parameter. This parameter is unbounded in our NP-hardness reductions for trees.

# References

1. Arrighi, E., Grüttemeier, N., Morawietz, N., Sommer, F., Wolf, P.: Multi-parameter analysis of finding minors and subgraphs in edge-periodic temporal graphs. In: Gąsieniec, L. (ed.) SOFSEM 2023. LNCS, vol. 13878, pp. 283–297. Springer, Cham (2023). https://doi.org/10.1007/978-3-031-23101-8_19
2. Bartha, Z., Komjáthy, J., Raes, J.: Sharp bound on the truncated metric dimension of trees. Discrete Math. **346**(8), 113410 (2023)
3. Bellitto, T., Conchon-Kerjan, C., Escoffier, B.: Restless exploration of periodic temporal graphs. In: 2nd Symposium on Algorithmic Foundations of Dynamic Networks (SAND 2023). Schloss Dagstuhl-Leibniz-Zentrum für Informatik (2023)
4. Bok, J., Dailly, A., Lehtilä, T.: Resolving sets in temporal graphs. arXiv preprint arXiv:2403.13183 (2024)
5. Casteigts, A., Flocchini, P., Quattrociocchi, W., Santoro, N.: Time-varying graphs and dynamic networks. Int. J. Parallel Emergent Distrib. Syst. **27**(5), 387–408 (2012)
6. Chartrand, G., Eroh, L., Johnson, M.A., Oellermann, O.R.: Resolvability in graphs and the metric dimension of a graph. Discrete Appl. Math. **105**(1–3), 99–113 (2000)
7. De Carufel, J.L., Flocchini, P., Santoro, N., Simard, F.: Cops & robber on periodic temporal graphs: characterization and improved bounds. In: Rajsbaum, S., Balliu, A., Daymude, J.J., Olivetti, D. (eds.) SIROCCO 2023. LNCS, vol. 13892, pp. 386–405. Springer, Cham (2023). https://doi.org/10.1007/978-3-031-32733-9_17
8. Díaz, J., Pottonen, O., Serna, M., van Leeuwen, E.J.: Complexity of metric dimension on planar graphs. J. Comput. Syst. Sci. **83**(1), 132–158 (2017)
9. Epstein, L., Levin, A., Woeginger, G.J.: The (weighted) metric dimension of graphs: hard and easy cases. Algorithmica **72**(4), 1130–1171 (2015)
10. Erlebach, T., Spooner, J.T.: A game of cops and robbers on graphs with periodic edge-connectivity. In: Chatzigeorgiou, A., et al. (eds.) SOFSEM 2020. LNCS, vol. 12011, pp. 64–75. Springer, Cham (2020). https://doi.org/10.1007/978-3-030-38919-2_6
11. Estrada-Moreno, A., Yero, I.G., Rodríguez-Velázquez, J.A.: On the $(k, t)$-metric dimension of graphs. Comput. J. **64**(5), 707–720 (2021)
12. Fernau, H., Rodríguez-Velázquez, J.A.: On the (adjacency) metric dimension of corona and strong product graphs and their local variants: combinatorial and computational results. Discrete Appl. Math. **236**, 183–202 (2018)
13. Flocchini, P., Kellett, M., Mason, P.C., Santoro, N.: Searching for black holes in subways. Theory Comput. Syst. **50**, 158–184 (2012)
14. Flocchini, P., Mans, B., Santoro, N.: On the exploration of time-varying networks. Theor. Comput. Sci. **469**, 53–68 (2013)
15. Foucaud, F., Mertzios, G.B., Naserasr, R., Parreau, A., Valicov, P.: Identification, location-domination and metric dimension on interval and permutation graphs. II. Algorithms and complexity. Algorithmica **78**, 914–944 (2017)
16. Frongillo, R.M., Geneson, J., Lladser, M.E., Tillquist, R.C., Yi, E.: Truncated metric dimension for finite graphs. Discrete Appl. Math. **320**, 150–169 (2022)
17. Galby, E., Khazaliya, L., Mc Inerney, F., Sharma, R., Tale, P.: Metric dimension parameterized by feedback vertex set and other structural parameters. SIAM J. Discrete Math. **37**(4), 2241–2264 (2023)
18. Geneson, J., Yi, E.: The distance-$k$ dimension of graphs. arXiv preprint arXiv:2106.08303 (2021)

19. Geneson, J., Yi, E.: Broadcast dimension of graphs. Australas. J. Comb. **83**, 243 (2022)
20. Gutkovich, P., Yeoh, Z.S.: Computing truncated metric dimension of trees. arXiv preprint arXiv:2302.05960 (2023)
21. Harary, F., Melter, R.A.: On the metric dimension of a graph. Ars Comb. **2**(191–195), 1 (1976)
22. Hartung, S., Nichterlein, A.: On the parameterized and approximation hardness of metric dimension. In: 2013 IEEE Conference on Computational Complexity, pp. 266–276. IEEE (2013)
23. Holme, P.: Modern temporal network theory: a colloquium. Eur. Phys. J. B **88**, 1–30 (2015)
24. Holme, P., Saramäki, J.: Temporal Network Theory. Computational Social Sciences, Springer, Cham (2019)
25. Ilcinkas, D., Wade, A.M.: On the power of waiting when exploring public transportation systems. In: Fernàndez Anta, A., Lipari, G., Roy, M. (eds.) OPODIS 2011. LNCS, vol. 7109, pp. 451–464. Springer, Heidelberg (2011). https://doi.org/10.1007/978-3-642-25873-2_31
26. Jannesari, M., Omoomi, B.: The metric dimension of the lexicographic product of graphs. Discrete Math. **312**(22), 3349–3356 (2012)
27. Karp, R.M.: Reducibility among combinatorial problems. In: Jünger, M., Liebling, T.M., Naddef, D., Nemhauser, G.L., Pulleyblank, W.R., Reinelt, G., Rinaldi, G., Wolsey, L.A. (eds.) 50 Years of Integer Programming 1958-2008, pp. 219–241. Springer, Heidelberg (2010). https://doi.org/10.1007/978-3-540-68279-0_8
28. Kempe, D., Kleinberg, J., Kumar, A.: Connectivity and inference problems for temporal networks. In: Proceedings of the Thirty-Second Annual ACM Symposium on Theory of Computing (STOC 2000), pp. 504–513 (2000)
29. Khuller, S., Raghavachari, B., Rosenfeld, A.: Landmarks in graphs. Discrete Appl. Math. **70**(3), 217–229 (1996)
30. Kunz, P., Molter, H., Zehavi, M.: In which graph structures can we efficiently find temporally disjoint paths and walks? In: Elkind, E. (ed.) Proceedings of the Thirty-Second International Joint Conference on Artificial Intelligence, IJCAI-2023, pp. 180–188. International Joint Conferences on Artificial Intelligence Organization (2023)
31. Kuziak, D., Yero, I.G.: Metric dimension related parameters in graphs: a survey on combinatorial, computational and applied results. arXiv preprint arXiv:2107.04877 (2021)
32. Liu, C., Wu, J.: Scalable routing in cyclic mobile networks. IEEE Trans. Parallel Distrib. Syst. **20**(9), 1325–1338 (2008)
33. Michail, O.: An introduction to temporal graphs: an algorithmic perspective. Internet Math. **12**(4), 239–280 (2016)
34. Slater, P.: Leaves of trees. Congr. Numer. **14**, 549–559 (1975)
35. Tillquist, R.C., Frongillo, R.M., Lladser, M.E.: Getting the lay of the land in discrete space: a survey of metric dimension and its applications. SIAM Rev. **65**(4), 919–962 (2023)
36. Zschoche, P., Fluschnik, T., Molter, H., Niedermeier, R.: The complexity of finding small separators in temporal graphs. J. Comput. Syst. Sci. **107**, 72–92 (2020)

# On the Finiteness of $k$-Vertex-Critical $2P_2$-Free Graphs with Forbidden Induced Squids or Bulls

Melvin Adekanye[1], Christopher Bury[1], Ben Cameron[2]([✉]) [iD],
and Thaler Knodel[1]

[1] The King's University, Edmonton, AB, Canada
{melvin.adekanye.stu,christopher.bury.stu,thaler.knodel.stu}@kingsu.ca
[2] University of Prince Edward Island, Charlottetown, PE, Canada
brcameron@upei.ca

**Abstract.** A graph $G$ is $k$-vertex-critical if $\chi(G) = k$ but $\chi(G - v) < k$ for all $v \in V(G)$, where $\chi(G)$ is the chromatic number of $G$. A graph is $(H_1, H_2)$-free if it contains no induced subgraph isomorphic to $H_1$ or $H_2$. We show that there are only finitely many $k$-vertex-critical $(2P_2, H)$-free graphs for all $k$ when $H$ is isomorphic to any of the following graphs:

- $(m, \ell)$-*squid* for $m = 3, 4$ and any $\ell \geq 1$ (where an $(m, \ell)$-*squid* is the graph obtained from an $m$-cycle by attaching $\ell$ leaves to a single vertex of the cycle),
- *bull*,
- *chair*,
- $claw + P_1$, or
- $\overline{diamond + P_1}$.

The latter three are corollaries of the $(m, \ell)$-*squid* results, while the *bull* is handled on its own. For each of the graphs $H$ as above and any fixed $k$, our results imply the existence of polynomial-time certifying algorithms for deciding the $k$-colourability problem for $(2P_2, H)$-free graphs. Further, our structural classifications allow us to exhaustively generate, with the aid of a computer search, all $k$-vertex-critical $(2P_2, H)$-free graphs for $k \leq 7$ when $H = bull$ or $H = (4, 1)$-*squid* (also known as *banner*).

**Keywords:** critical graph · graph colouring · certifying algorithm · forbidden induced subgraph

## 1 Introduction

The $k$-COLOURING decision problem is to determine, for fixed $k$, if a given graph admits a proper $k$-colouring. This problem is of considerable interest in computational complexity theory since for any $k \geq 3$, it is one of the most intuitive of Karp's [32] original 21 NP-complete problems. Research has focused on many computational aspects of $k$-COLOURING including approximation [22] and

© The Author(s), under exclusive license to Springer Nature Switzerland AG 2024
A. A. Rescigno and U. Vaccaro (Eds.): IWOCA 2024, LNCS 14764, pp. 301–313, 2024.
https://doi.org/10.1007/978-3-031-63021-7_23

heuristic algorithms [4], but we are most interested in the substructures that can be forbidden to produce polynomial-time algorithms to solve $k$-COLOURING for all $k$. The substructures we are interested in forbidding are induced subgraphs. We say a graph is $H$-free if it contains no induced copy of $H$. One of the most impressive results to this end is Hoàng et al.'s 2010 result that $k$-COLOURING can be solved in polynomial-time on $P_5$-free graphs for all $k$ [24]. The full strength of this result was later demonstrated by Huang's [27] result that $k$-COLOURING remains NP-complete for $P_6$-free graphs when $k \geq 5$, and for $P_7$-free graphs when $k \geq 4$. A polynomial-time algorithm to solve 4-COLOURING for $P_6$-free graphs was later developed by Chudnovsky et al. [16–18]. It has long been known that $k$-COLOURING remains NP-complete on $H$-free graphs when $H$ contains an induced cycle [30,34] or claw [26,33]. Thus, $P_5$ is the largest connected subgraph that can be forbidden for which $k$-COLOURING can be solved in polynomial-time for all $k$ (assuming P$\neq$NP). As such, deeper study of $P_5$-free graphs in relation to $k$-COLOURING has attracted significant attention. One of these deeper areas is determining which subfamilies of $P_5$-free graphs admit polynomial-time $k$-COLOURING algorithms that are fully *certifying*.

An algorithm is certifying if it returns with each output, an easily verifiable witness (called a *certificate*) that the output is correct. For $k$-COLOURING, a certificate for "yes" output is a $k$-colouring of the graph, and indeed the $k$-COLOURING algorithms for $P_5$-free graphs from [24] return $k$-colourings if they exist. To discuss certificates for "no" output for $k$-COLOURING, we must first define that a graph is $k$-*vertex-critical* if it is not $(k-1)$-colourable, but every proper induced subgraph is. Since every graph that is not $k$-colourable must contain an induced $(k+1)$-vertex-critical graph, such an induced subgraph is a certificate for "no" output of $k$-COLOURING. Indeed, if there are only finitely many $(k+1)$-vertex-critical graphs in a family $\mathcal{F}$, then a polynomial-time algorithm to decide $k$-COLOURING that certifies "no" output for all graphs in $\mathcal{F}$ can be implemented by searching for each $(k+1)$-vertex-critical graph in $\mathcal{F}$ as an induced subgraph of the input graph (see [6], for example, for the details). Thus, proving there are only finitely many $(k+1)$-vertex critical graphs in a subfamily of $P_5$-free graphs implies the existence of polynomial-time certifying $k$-COLOURING algorithms when paired with the algorithms that certify "yes" output from [24].

It should also be noted that classifying vertex-critical graphs is of interest in resolving the Borodin-Kostochka Conjecture [2] that states if $G$ is a graph with $\Delta(G) \geq 9$ and $\omega(G) \leq \Delta(G) - 1$, then $\chi(G) \leq \Delta(G) - 1$. This connection comes in the proof technique often employed (see for example [19,23,35]) to consider a minimal counterexample to the conjecture which must be vertex-critical [21].

A natural question to ask then is for which $k$ are there only finitely many $k$-vertex-critical $P_5$-free graphs. Bruce et al. [5] showed that there are only finitely many 4-vertex-critical $P_5$-free graphs, and this result was later extended by Maffray and Morel [34] to develop a linear-time certifying algorithm for 3-COLOURING $P_5$-free graphs. Unfortunately, for each $k \geq 5$, Hoàng et al. [25] constructed infinitely many $k$-vertex-critical $P_5$-free graphs. This result caused attention to shift to subfamilies of $P_5$-free graphs, usually defined by forbidding additional induced subgraphs. A graph is $(H_1, H_2, \ldots, H_\ell)$-free if it does not contain an induced copy of $H_i$ for all $i \in \{1, 2, \ldots, \ell\}$. It is known that there are only finitely many 5-vertex-critical $(P_5, H)$-free graphs if $H$ is isomorphic to $C_5$ [25], *bull* [28], or *chair* [29] (where *bull* is the graph in Fig. 3 and *chair* is the graph in Fig. 1c). When moving beyond $k = 5$ to larger values of $k$, the most comprehensive result is K. Cameron et al.'s [13] dichotomy theorem that there are infinitely many $k$-vertex-critical $(P_5, H)$-free graphs for $H$ of order 4 if and only if $H$ is $2P_2$ or $K_3 + P_1$. An open question was also posed in [13] to prove a dichotomy theorem for $H$ of order 5, and there has been substantial work toward this end. On the positive side of resolving this open question, it is known that there are only finitely many $k$-vertex-critical $(P_5, H)$-free graphs for all $k$ if $H$ is isomorphic to *banner* [3], *dart* [36], $K_{2,3}$ [31], $\overline{P_5}$ [20], $P_2 + 3P_1$ [10], $P_3 + 2P_1$ [1], *gem* [7,11], or $\overline{P_3 + P_2}$ [7] (where we refer to the table in [12] for the adjacencies of each of these graphs). On the negative side of resolving the open question is that there are infinitely many $k$-vertex-critical $(P_5, H)$-free graphs for every graph $H$ of order 5 containing an induced $2P_2$ or $K_3 + P_1$ (a corollary from the dichotomy theorem in [13]) for $k \geq 5$, and for $H = C_5$ for all $k \geq 6$ [12]. It therefore remains unknown for which graphs $H$ of order 5 there are only finitely many $(P_5, H)$-free graphs for all $k$ when $H$ is one of the following[1]:

- *claw* + $P_1$
- $P_4 + P_1$
- *chair*
- $\overline{diamond + P_1}$

- $C_4 + P_1$
- *bull*
- $\overline{K_3 + 2P_1}$
- $\overline{P_3 + 2P_1}$

- $W_4$

- $K_5 - e$

- $K_5$

In this paper, we prove that there are only finitely many $k$-vertex-critical $(2P_2, H)$-free graphs for all $k$ for four of the graphs above. Namely, *claw* + $P_1$, $\overline{diamond + P_1}$, *chair*, and *bull* (see Fig. 1 and Fig. 3). Three of these are corollaries of more general results that there are only finitely many $k$-vertex-critical $(2P_2, (4, \ell)$-*squid*)-free graphs and $(2P_2, (3, \ell)$-*squid*)-free graphs for all $k, \ell$ (see Fig. 2). We also show that there are only finitely many $(2P_2, K_3 + P_1, P_4 + P_1)$-free graphs for all $k$. These results, while not as strong as they would be if they were for $P_5$-free graphs instead of $2P_2$-free graphs, are nonetheless important as every infinite family of $k$-vertex-critical $P_5$-free graphs known (i.e., the families from [15,25] and the generalized families from [12]) are actually $(2P_2, K_3 + P_1)$-free as well. Thus, there is no known $H$ for which there are infinitely many $k$-vertex-critical $(P_5, H)$-free graphs and only finitely many that are $(2P_2, H)$-free for some $k$. Therefore, our results provide strong evidence for the finiteness of $k$-vertex-critical $(P_5, H)$-free graphs for each $H$ we consider, but will also still be of interest if it turns out that there are infinitely many that are $(P_5, H)$-free but only finitely many that are $(2P_2, H)$-free.

In addition, while most proof techniques for showing finiteness of vertex-critical graphs involve bounding structure around odd holes or anti-holes using

---

[1] Since this paper was submitted, the *claw* + $P_1$ and $\overline{K_3 + 2P_1}$ cases were both proved in [37].

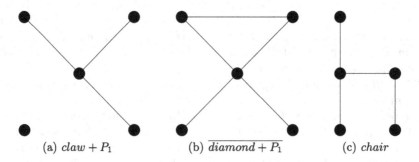

Fig. 1. Three of the new graphs $H$ of order 5 for which it is shown there are only finitely many $k$-vertex-critical $(2P_2, H)$-free graphs for all $k$.

Strong Perfect Graph Theorem [14], or using a clever application of Ramsey's Theorem, our techniques are novel and simple. We show that $k$-vertex-critical graphs in each of the families in question are necessarily $(P_3 + mP_1)$-free for some $m$ depending only on $k$ and the orders of the forbidden induced subgraphs. We then apply the result from [1] that there are only finitely many $k$-vertex-critical $(P_3 + mP_1)$-free graphs for all $m$ and $k$ to get the immediate corollary of finiteness. This approach results in short and easy-to-follow proofs that only require basic facts about vertex-critical graphs to prove finiteness.

## 1.1 Outline

The rest of this paper is structured as follows. We include the background and preliminary results that are used in the proofs of our main results in Sect. 2. We prove that there are only finitely many $k$-vertex-critical $(2P_2, bull)$-free graphs for all $k$ in Sect. 3. We then prove two more general results which have as corollaries that there are only finitely many $k$-vertex-critical $(2P_2, H)$-free graphs for all $k$ when $H$ isomorphic to *chair*, *claw* $+ P_1$ (Sect. 4), and $\overline{diamond + P_1}$ (Sect. 5). In Sect. 6, we discuss methods for using a computer-aided search to exhaustively generate all $k$-vertex-critical graphs in certain families and include our enumerations for the complete sets of all such graphs that are $(2P_2, H)$-free graphs for $H = bull$ or $H = banner$ for all $k \leq 7$. Finally, we conclude the paper with a discussion on future research directions and we conclude this section with a brief subsection outlining definitions and notation.

## 1.2 Notation

If vertices $u$ and $v$ are adjacent in a graph we write $u \sim v$ and if they are nonadjacent we write $u \nsim v$. For a vertex $v$ in a graph, $N(v)$, $N[v]$ and $\overline{N[v]}$ denote the open neighbourhood, closed neighbourhood, and set of nonneighbours of $v$, respectively. More precisely, $N(v) = \{u \in V(G) : u \sim v\}$, $N[v] = N(v) \cup \{v\}$, and $\overline{N[v]} = V(G) \setminus N[v]$. We call a vertex $v$ of a graph $G$ *universal* if $N[v] = V(G)$, and *nonuniversal* otherwise. We let $\delta(G)$ and $\Delta(G)$ denote the minimum and maximum degrees of $G$, respectively. We let $\alpha(G)$ denote the independence number of $G$. For subsets $A$ and $B$ of $V(G)$, we say $A$ is *(anti)complete* to $B$ if $a$ is (non)adjacent to $b$ for all $a \in A$ and $b \in B$. If $A = \{a\}$ then we simplify notation and say $a$ is (anti)complete to $B$. We use $\chi(G)$ and $\omega(G)$ to denote the *chromatic number* and *clique number* of $G$, respectively.

## 2 Preliminaries

We will make extensive use of the following lemma and theorem throughout the paper.

**Lemma 1** ([25]). *Let $G$ be a graph with chromatic number $k$. If $G$ contains two disjoint $m$-cliques $A = \{a_1, a_2, \ldots, a_m\}$ and $B = \{b_1, b_2, \ldots, b_m\}$ such that $N(a_i) \setminus A \subseteq N(b_i) \setminus B$ for all $1 \le i \le m$, then $G$ is not $k$-vertex-critical.*

We stated Lemma 1 here in its full generality for interested readers, but we will only use it for the case where $m = 1$. For easier reference in this case, we call vertices $a$ and $b$ *comparable* if $N(a) \subseteq N(b)$. The contrapositive of Lemma 1 for $m = 1$, then can be restated as there are no comparable vertices in a vertex-critical graph.

**Theorem 1** ([1]). *There are only finitely many $k$-vertex-critical $(P_3 + \ell P_1)$-free graphs for all $k \ge 1$ and $\ell \ge 0$.*

We also require two new technical lemmas that will be used multiple times throughout the paper as we apply Theorem 1 to prove the finiteness of vertex-critical graphs in many families.

**Lemma 2.** *If $G$ is a $k$-vertex-critical $2P_2$-free graph, then for every nonuniversal vertex $v \in V(G)$, $\overline{N[v]}$ induces a connected graph with at least two vertices.*

*Proof.* Let $G$ be a $k$-vertex-critical $2P_2$-free graph, $v \in V(G)$ be nonuniversal, and $H$ be the graph induced by $\overline{N[v]}$. If $u \in \overline{N[v]}$ such that $u$ is an isolated vertex in the graph induced by $H$, then $N(u) \subseteq N(v)$ contradicting $G$ being $k$-vertex-critical by Lemma 1. Therefore, if $H$ has at least two components, then each component has at least one edge and therefore taking an edge from each component induces a $2P_2$. This contradicts $G$ being $2P_2$-free. □

**Lemma 3.** *Let $G$ be a $2P_2$-free graph that contains an induced $P_3 + \ell P_1$ for some $\ell \ge 1$. Let $S \cup \{v_1, v_2, v_3\} \subseteq V(G)$ induce a $P_3 + \ell P_1$ where $v_1v_2v_3$ is the induced $P_3$ and $S$ contains the vertices in the $\ell P_1$. Then for ever vertex $u \in \overline{N[v_2]}$ such that $u$ has a neighbour in $S$, $u$ is complete to $\{v_1, v_3\}$.*

*Proof.* Let $s \in S$ and $u \in \overline{N[v_2]}$ with $u \sim s$. Let $i \in \{1, 3\}$. If $u \nsim v_i$, then $\{s, u, v_i, v_2\}$ induces a $2P_2$ in $G$, a contradiction. Thus, $u$ is complete to $\{v_1, v_3\}$. □

Finally, there are special families of graphs that will be used in our results. For $\ell \ge 1$ and $m \ge 3$, let the $(m, \ell)$-*squid* be the graph obtained from $C_m$ by attaching $\ell$ leaves to one of its vertices. Figure 2 shows these graphs for $m = 3, 4$, which are the ones of interest to us, since for $m > 4$ $(m, \ell)$-*squid* is not $2P_2$-free.

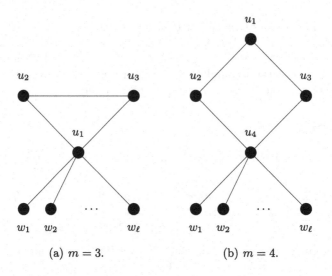

(a) $m = 3$.          (b) $m = 4$.

**Fig. 2.** The general form of the $(m, \ell)$-*squid* graphs for $m = 3$ and $m = 4$.

## 3   $(2P_2, \text{bull})$-Free

Let the *bull* be the graph obtained from a $K_3$ by attaching two leaves, each to a different vertex of the $K_3$ and shown in Fig. 3.

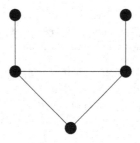

**Fig. 3.** The *bull* graph.

**Lemma 4.** *Let* $k \geq 1$. *If* $G$ *is* $k$-*vertex-critical* $(2P_2, \text{bull})$-*free, then* $G$ *is* $(P_3 + P_1)$-*free.*

*Proof.* Let $G$ be a $k$-vertex-critical $(2P_2, \text{bull})$-free graph and by way of contradiction let $\{v_1, v_2, v_3, s_1\}$ induce a $P_3 + P_1$ in $G$ where $\{v_1, v_2, v_3\}$ induces the $P_3$, in that order, of the $P_3 + P_1$. We will show that $v_1$ and $v_3$ are comparable which will contradict Lemma 1.

We first show that $s_1$ has no neighbours in $(N(v_1) \cap N(v_2)) - N(v_3)$ (and by symmetry no neighbours in $(N(v_3) \cap N(v_2)) - N(v_1)$). Suppose $n \in (N(v_1) \cap$

$N(v_2)) - N(v_3)$ such that $s_1 \sim n$. Now, $\{s_1, v_1, v_2, v_3, n\}$ induces a *bull* in $G$, a contradiction.

Since $s_1$ is not comparable with $\underline{v_1}$, then it must have a neighbour $u_1$ such that $u_1 \not\sim v_1$. By Lemma 3, $u_1 \notin \overline{N(v_2)}$ (or else $u_1 \sim v_1$) and therefore $u_1 \sim v_2$. Now, since $s_1$ has no neighbours in $(N(v_3) \cap N(v_2)) - N(v_1)$ as argued above, it must be that $u_1 \not\sim v_3$.

If $v_1$ and $v_3$ are not comparable, then they must have distinct neighbours. Let $v_1'$ be such that $v_1 \sim v_1'$ and $v_1' \not\sim v_3$ and $v_3'$ be such that $v_3 \sim v_3'$ and $v_3' \not\sim v_1$. Note that we must have $v_1' \sim v_3'$, otherwise $\{v_1, v_1', v_3, v_3'\}$ induces a $2P_2$ in $G$. Further $u_1$ must be complete to $\{v_1', v_3'\}$, else $\{s_1, u_1, v_1, v_1'\}$ or $\{s_1, u_1, v_3, v_3'\}$ will induce a $2P_2$ in $G$. But now, $\{u_1, v_1', v_3', v_1, v_3\}$ induces a *bull* in $G$, a contradiction. Figure 4 illustrates the induced *bull* in $G$ q, although the reader should note that while all adjacencies and nonadjacencies are complete within the set $\{u_1, v_1', v_3', v_1, v_3\}$, there could be other missing adjacencies in the figure (for example, we might have $v_1' \sim v_2$). Thus, $v_1$ and $v_3$ are comparable, contradicting $G$ being *k*-vertex-critical.

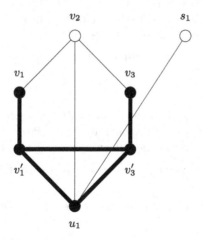

**Fig. 4.** An illustration of part of the proof of Lemma 4 with the induced *bull* in bold.

The following theorem follows directly from Lemma 4 and Theorem 1.

**Theorem 2.** *There are only finitely many k-vertex-critical $(2P_2, bull)$-free graphs for all $k \geq 1$.*

## 4   $(2P_2, (4, \ell)\text{-}squid)$-Free

Recall that the $(4, \ell)$-*squid* is the graph obtained from a $C_4$ by adding $\ell$ leaves to one vertex, see Fig. 2b.

**Lemma 5.** *Let $\ell, k \geq 1$ and $c = (\ell - 1)(k - 1) + 1$. If $G$ is a $k$-vertex-critical $(2P_2, (4, \ell)$-squid)-free graph, then $G$ is $(P_3 + cP_1)$-free.*

*Proof.* Let $G$ be a $k$-vertex-critical $(2P_2, (4, \ell)$-squid)-free graph for some $k, \ell \geq 1$ and let $c = (\ell - 1)(k - 1) + 1$. Suppose by way of contradiction that $G$ contains an induced $P_3 + cP_1$ with $\{v_1, v_2, v_3\}$ inducing the $P_3$ in that order and $S = \{s_1, s_2 \ldots, s_c\}$ the $cP_1$ of the induced $P_3 + cP_1$.

By Lemma 2, each $s_i$ must have a neighbour in $\overline{N[v_2]}$. Further, by Lemma 3, for every $u \in \overline{N[v_2]} - S$, such that $u \sim s_i$ for some $s_i \in S$, we must have that $u$ is complete to $\{v_1, v_3\}$. Therefore, each $u \in \overline{N[v_2]} - S$ has at most $\ell - 1$ neighbours in $S$, else $u, v_1, v_2, v_3$ together with any $\ell$ of $u$'s neighbours in $S$ would induce an $(4, \ell)$-squid. Let $U = \{u_1, u_2, \ldots u_m\}$ be a subset of $N(S) \cap \overline{N[v_2]}$ such that $(\bigcup_{i=1}^{m} N(u_i)) \cap S = S$ and such that $N(u_i) \cap S \nsubseteq N(u_j) \cap S$ for all $i \neq j$. Such a set $U$ exists since each vertex in $S$ has at least one neighbour in $\overline{N[v_2]}$. Since each $u_i$ can have at most $\ell - 1$ neighbours in $S$ we must have that $|U| \geq k$ by the Pigeonhole Principle. Let $u_i, u_j \in U$ for $i \neq j$ and without loss of generality let $s_i \in N(u_i) - N(u_j)$ and $s_j \in N(u_j) - N(u_i)$. If $u_i \nsim u_j$, then $\{s_i, u_i, s_j, u_j\}$ induces a $2P_2$ in $G$, a contradiction. Therefore, $u_i \sim u_j$ for all $i \neq j$ and therefore $U$ induces a clique with at least $k$ vertices in $G$. But now $U$ induces a proper subgraph of $G$ that requires at least $k$ colours, which contradicts $G$ being $k$-vertex-critical. Therefore, $G$ must be $(P_3 + cP_1)$-free. □

The following theorem follows directly from Lemma 5 and Theorem 1.

**Theorem 3.** *There are only finitely many $k$-vertex-critical $(2P_2, (4, \ell)$-squid)-free graphs for all $k, \ell \geq 1$.*

Since *chair* is an induced subgraph of $(4, 2)$-*squid*, *claw* $+ P_1$ is an induced subgraph of $(4, 3)$-*squid*, and more generally $K_{1,\ell} + P_1$ is an induced subgraph of $(4, \ell)$-*squid* for all $\ell \geq 1$, we get the following immediate corollaries of Theorem 3

**Corollary 1.** *There are only finitely many $k$-vertex-critical $(2P_2, chair)$-free graphs for all $k \geq 1$.*

**Corollary 2.** *There are only finitely many $k$-vertex-critical $(2P_2, claw + P_1)$-free graphs for all $k \geq 1$.*

**Corollary 3.** *There are only finitely many $k$-vertex-critical $(2P_2, K_{1,\ell} + P_1)$-free graphs for all $k \geq 1$.*

We note as well that *banner* is another name for $(4, 1)$-*squid*, so our results also imply that there are only finitely many $k$-vertex-critical $(2P_2, banner)$-free graphs for all $k$. This result is not new though; it was recently shown in [3] that every $k$-vertex-critical $(P_5, banner)$-free graph has independence number less than 3 and therefore there only finitely many such graphs by Ramsey's Theorem. However, a special case of Lemma 5 implies that every $k$-vertex-critical $(2P_2, banner)$-free graph is $(P_3 + P_1)$-free and therefore by Theorem 3.1 in [10], every such graph has independence number less than 3. Thus our results give a new and extremely short proof (in fact the proof of Lemma 5 only requires the first four sentences if restricted to *banner*-free graphs) of a slightly weaker version of the result in [3].

# 5    $(2P_2, (3, \ell)\text{-}squid)$-Free

Recall that $(3, \ell)$-*squid*, shown in Fig. 2a, is the graph obtained from a triangle by attaching $\ell$ leaves to one of its vertices.

**Lemma 6.** *Let $\ell, k \geq 1$. If $G$ is $k$-vertex-critical $(2P_2, (3, \ell)$-squid$)$-free, then $G$ is $(4, 2\ell - 1)$-squid-free.*

*Proof.* Let $G$ be a $k$-vertex-critical $(2P_2, (3, \ell)$-*squid*)-free graph and suppose by way of contradiction that $G$ contains an induced $(4, 2\ell - 1)$-*squid* with the same labelling as Fig. 2b. It must be the case that there are $u_2', u_3' \in V(G)$ such that $u_2' \sim u_2$, $u_2' \nsim u_3$, $u_3' \sim u_3$, and $u_3' \nsim u_2$ otherwise $u_2$ and $u_3$ are comparable, contradicting Lemma 1. Now, $\{u_2, u_3, u_2', u_3'\}$ induces a $2P_2$ unless $u_2' \sim u_3'$. There are now two cases to consider.

*Case 1: $u_2' \sim u_4$.*

In this case, $u_2'$ must be adjacent to at least $\ell$ of the $w_i$'s, or else $\{u_2', u_2, u_4\}$ together with any $\ell$ of the $w_i$'s that are nonadjacent to $u_2'$ induce a $(3, \ell)$-*squid*, a contradiction. Let $W$ be the subset of the $w_i$'s that are adjacent to $u_2'$. We now have $\{u_2', w, u_1, u_3\}$ inducing a $2P_2$ for all $w \in W$ unless $u_2' \sim u_1$. Since $G$ is $2P_2$-free, we must have $u_2' \sim u_1$. But now $\{u_2, u_2', u_1\}$ together with any subset of $W$ with at least $\ell$ vertices induces a $(3, \ell)$-*squid*, a contradiction.

*Case 2: $u_2' \nsim u_4$.*

In this case, $u_2' \sim w_i$ for all $i \in \{1, \ldots, \ell\}$, else $\{u_2, u_2', u_4, w_i\}$ would induce a $2P_2$. We now have $u_2' \sim u_1$, else $\{u_2', w_i, u_1, u_3\}$ induces a $2P_2$ for any $i \in \{1, \ldots, \ell\}$. But now $\{u_2, u_2', u_1\} \cup \{w_1, w_2, \ldots, w_\ell\}$ induces $(3, \ell)$-*squid*, a contradiction.

Since we reach contradictions in either case, we must contradict that fact that $G$ is $(4, 2\ell - 1)$-*squid*-free.

Thus, we obtain the following theorem directly from Lemma 6 and Theorem 3.

**Theorem 4.** *There are only finitely many $k$-vertex-critical $((3, \ell)$-squid$)$-free graphs for all $k, \ell \geq 1$.*

Since the graph $\overline{diamond + P_1}$ is another name for to $(3, 2)$-*squid*, we get the following immediate corollary.

**Corollary 4.** *There are only finitely many $k$-vertex-critical $(\overline{diamond + P_1})$-free graphs for all $k \geq 1$.*

# 6    Exhaustive Generation for Small $k$

Our results imply that there are several new families of graphs for which $k$-COLOURING can be solved in polynomial-time and certified with a $k$-colouring or a $(k + 1)$-vertex-critical induced subgraph graph for any fixed $k$. The implementation of any of these algorithms, however, requires the complete set of all $(k + 1)$-vertex-critical graphs in that family. For two of the families, we can give these complete sets for all $k \leq 6$, and therefore the corresponding polynomial-time certifying $k$-COLOURING algorithms can be readily implemented. These two families are $(2P_2, bull)$-free and $(2P_2, banner)$-free since $k$-vertex-critical graphs in both families were shown to be $(P_3 + P_1)$-free. In [10] it was shown that every $k$-vertex-critical $(P_3 + P_1)$-free graph has independence number at most 2 and order at most $2k - 1$, which allowed for the exhaustive generation of all such graphs for $k \leq 7$. These graphs, available at [8], were searched for those that are also $2P_2$-free to provide the complete sets of all $k$-vertex-critical $(2P_2, H)$-free graphs for $H = banner$ or $H = bull$ for all $k \leq 7$. Note that since $\alpha(banner) = \alpha(bull) = 3$, banner and bull are already forbidden induced subgraphs from each of the $k$-vertex-critical $(P_3 + P_1)$-free graphs since they necessarily have independence number at most 2. The graphs in graph6 format are available at [9] and the number of each order is summarized in Table 1.

**Table 1.** Number of $k$-vertex-critical $(2P_2, H)$-free graphs of order $n$ for $k \leq 7$ where $H$ is banner or bull.

| $n$ | 4-vertex-critical | 5-vertex-critical | 6-vertex-critical | 7-vertex-critical |
|---|---|---|---|---|
| 4 | 1 | 0 | 0 | 0 |
| 5 | 0 | 1 | 0 | 0 |
| 6 | 1 | 0 | 1 | 0 |
| 7 | 2 | 1 | 0 | 1 |
| 8 | 0 | 2 | 1 | 0 |
| 9 | 0 | 11 | 2 | 1 |
| 10 | 0 | 0 | 12 | 2 |
| 11 | 0 | 0 | 126 | 12 |
| 12 | 0 | 0 | 0 | 128 |
| 13 | 0 | 0 | 0 | 3806 |
| total | 4 | 15 | 142 | 3947 |

# 7    Conclusion

Of the 11 graphs $H$ of order 5 where it remains unknown if there are only finitely many $k$-vertex-critical $(P_5, H)$-free graphs for all $k$, we have shown finiteness for four of them in the more restricted $2P_2$-free case. A clear open question that remains is if these still hold when forbidding $P_5$ instead of $2P_2$, or (perhaps even more interestingly) if for some they do not. The question also remains completely open for the other 7 graphs of order 5; the one we find most interesting is $H = P_4 + P_1$, as it is still unknown whether there are only finitely many $k$-vertex-critical $(P_4 + P_1)$-free for any given $k \geq 5$ even without additional restrictions. We fully expect that our new proof technique of reducing to $(P_3 + \ell P_1)$-free for some $\ell$ will be useful for proving the finiteness of $k$-vertex-critical graphs in these open cases and indeed for other families of graphs. One place to start for interested readers might be by forbidding additional induced subgraphs and, in particular, considering $k$-vertex-critical $(2P_2, K_3 + P_1, H)$-free or even $(2P_2, K_3 + P_1, C_5, H)$-free graphs. These additional forbidden induced subgraphs may seem to be far too restrictive, but we feel they are justified entry points as all known infinite families of $k$-vertex-critical graphs that are $2P_2$-free are also $(K_3 + P_1)$-free and, for $k \geq 6$, there are families of $k$-vertex-critical $(2P_2, K_3 + P_1, C_5)$-free graphs (see the aforementioned construction in [12]).

**Acknowledgments.** The first two authors were supported by the Natural Sciences and Engineering Research Council of Canada (NSERC) USRA program. The third author also gratefully acknowledges research support from NSERC (grants RGPIN-2022-03697 and DGECR-2022-00446) and Alberta Innovates. The research support from Alberta Innovates was used by the third author to hire the fourth author.

**Disclosure of Interests.** The authors have no competing interests to declare that are relevant to the content of this article.

# References

1. Abuadas, T., Cameron, B., Hoàng, C.T., Sawada, J.: Vertex-critical $(P_3 + \ell P_1)$-free and vertex-critical (gem, co-gem)-free graphs. Discrete Appl. Math. **344**, 179–187 (2024). https://doi.org/10.1016/j.dam.2023.11.042
2. Borodin, O., Kostochka, A.: On an upper bound of a graph's chromatic number, depending on the graph's degree and density. J. Combin. Ser. B **23**(2), 247–250 (1977). https://doi.org/10.1016/0095-8956(77)90037-5
3. Brause, C., Geißer, M., Schiermeyer, I.: Homogeneous sets, clique-separators, critical graphs, and optimal $\chi$-binding functions. Discrete Appl. Math. **320**, 211–222 (2022). https://doi.org/10.1016/j.dam.2022.05.014
4. Brélaz, D.: New methods to color the vertices of a graph. Commun. ACM **22**(4), 251–256 (1979). https://doi.org/10.1145/359094.359101
5. Bruce, D., Hoàng, C.T., Sawada, J.: A certifying algorithm for 3-colorability of $P_5$-free graphs. In: Dong, Y., Du, D.Z., Ibarra, O. (eds.) ISAAC 2009. LNCS, vol. 5878, pp. 594–604. Springer, Heidelberg (2009). https://doi.org/10.1007/978-3-642-10631-6_61
6. Cai, Q., Huang, S., Li, T., Shi, Y.: Vertex-critical $(P_5$, banner)-free graphs. In: Chen, Y., Deng, X., Lu, M. (eds.) FAW 2019. LNCS, vol. 11458, pp. 111–120. Springer, Cham (2019). https://doi.org/10.1007/978-3-030-18126-0_10

7. Cai, Q., Goedgebeur, J., Huang, S.: Some results on $k$-critical $P_5$-free graphs. Discrete Appl. Math. **334**, 91–100 (2023). https://doi.org/10.1016/j.dam.2023.03.008

8. Cameron, B.: P3P1free_critical (2021). https://github.com/benrkcameron/P3P1free_critical

9. Cameron, B.: 2P2bullfree (2023). https://github.com/benrkcameron/2P2bull

10. Cameron, B., Hoàng, C.T., Sawada, J.: Dichotomizing $k$-vertex-critical $H$-free graphs for $H$ of order four. Discrete Appl. Math. **312**, 106–115 (2022). https://doi.org/10.1016/j.dam.2021.11.001. Ninth Workshop on Graph Classes, Optimization, and Width Parameters

11. Cameron, B., Hoàng, C.T.: A refinement on the structure of vertex-critical $(P_5,$ gem)-free graphs. Theoret. Comput. Sci. **961**, 113936 (2023). https://doi.org/10.1016/j.tcs.2023.113936

12. Cameron, B., Hoàng, C.T.: Infinite families of $k$-vertex-critical $(P_5, C_5)$-free graphs. Graphs Combin. **40**(30) (2024). https://doi.org/10.1007/s00373-024-02756-x

13. Cameron, K., Goedgebeur, J., Huang, S., Shi, Y.: $k$-Critical graphs in $P_5$-free graphs. Theoret. Comput. Sci. **864**, 80–91 (2021). https://doi.org/10.1016/j.tcs.2021.02.029

14. Chudnovsky, M., Robertson, N., Seymour, P., Thomas, R.: The strong perfect graph theorem. Ann. Math. **164**(1), 51–229 (2006). https://doi.org/10.4007/annals.2006.164.51

15. Chudnovsky, M., Goedgebeur, J., Schaudt, O., Zhong, M.: Obstructions for three-coloring graphs without induced paths on six vertices. J. Combin. Ser. B **140**, 45–83 (2020). https://doi.org/10.1016/j.jctb.2019.04.006

16. Chudnovsky, M., Spirkl, S., Zhong, M.: Four-coloring $P_6$-free graphs. In: Proceedings of the Thirtieth Annual ACM-SIAM Symposium on Discrete Algorithms, SODA 2019, pp. 1239–1256. Society for Industrial and Applied Mathematics, USA (2019)

17. Chudnovsky, M., Spirkl, S., Zhong, M.: Four-coloring $P_6$-free graphs. I. Extending an excellent precoloring. SIAM J. Comput. **53**(1), 111–145 (2024). https://doi.org/10.1137/18M1234837

18. Chudnovsky, M., Spirkl, S., Zhong, M.: Four-coloring $P_6$-free graphs. II. Finding an excellent precoloring. SIAM J. Comput. **53**(1), 146–187 (2024). https://doi.org/10.1137/18M1234849

19. Cranston, D.W., Lafayette, H., Rabern, L.: Coloring $(P_5,$gem)-free graphs with $\Delta - 1$ colors. J. Graph Theory **101**(4), 633–642 (2022). https://doi.org/10.1002/jgt.22845

20. Dhaliwal, H.S., Hamel, A.M., Hoàng, C.T., Maffray, F., McConnell, T.J.D., Panait, S.A.: On color-critical $(P_5,$co-$P_5)$-free graphs. Discrete Appl. Math. **216**, 142–148 (2017). https://doi.org/10.1016/j.dam.2016.05.018

21. Gupta, U.K., Pradhan, D.: Borodin-Kostochka's conjecture on $(P_5, C_4)$-free graphs. J. Appl. Math. Comput. **65**(1), 877–884 (2021). https://doi.org/10.1007/s12190-020-01419-3

22. Halldórsson, M.M.: A still better performance guarantee for approximate graph coloring. Inform. Process. Lett. **45**(1), 19–23 (1993). https://doi.org/10.1016/0020-0190(93)90246-6

23. Haxell, P., Naserasr, R.: A note on $\Delta$-critical graphs. Graphs Combin. **39**, 101 (2023). https://doi.org/10.1007/s00373-023-02696-y

24. Hoàng, C.T., Kamiński, M., Lozin, V., Sawada, J., Shu, X.: Deciding $k$-colorability of $P_5$-free graphs in polynomial time. Algorithmica **57**, 74–81 (2010). https://doi.org/10.1007/s00453-008-9197-8

25. Hoàng, C.T., Moore, B., Recoskie, D., Sawada, J., Vatshelle, M.: Constructions of *k*-critical $P_5$-free graphs. Discrete Appl. Math. **182**, 91–98 (2015). https://doi.org/10.1016/j.dam.2014.06.007

26. Holyer, I.: The NP-completeness of edge-coloring. SIAM J. Comput. **10**(4), 718–720 (1981). https://doi.org/10.1137/0210055

27. Huang, S.: Improved complexity results on *k*-coloring $P_t$-free graphs. European J. Combin. **51**, 336–346 (2016). https://doi.org/10.1016/j.ejc.2015.06.005

28. Huang, S., Li, J., Xia, W.: Critical ($P_5$, bull)-free graphs. Discrete Appl. Math. **334**, 15–25 (2023). https://doi.org/10.1016/j.dam.2023.02.019

29. Huang, S., Li, Z.: Vertex-critical ($P_5$, *chair*)-free graphs. Discrete Appl. Math. **341**, 9–15 (2023). https://doi.org/10.1016/j.dam.2023.07.014

30. Kamiński, M., Lozin, V.: Coloring edges and vertices of graphs without short or long cycles. Contrib. Discrete Math. **2**(1), 61–66 (2007). https://doi.org/10.11575/cdm.v2i1.61890

31. Kamiński, M., Pstrucha, A.: Certifying coloring algorithms for graphs without long induced paths. Discrete Appl. Math. **261**, 258–267 (2019). https://doi.org/10.1016/j.dam.2018.09.031

32. Karp, R.M.: Reducibility among combinatorial problems. In: Complexity of Computer Computations (Proc. Sympos., IBM Thomas J. Watson Res. Center, Yorktown Heights, N.Y., 1972), pp. 85–103 (1972)

33. Leven, D., Gail, Z.: NP completeness of finding the chromatic index of regular graphs. J. Algorithms **4**, 35–44 (1983)

34. Maffray, F., Morel, G.: On 3-colorable $P_5$-free graphs. SIAM J. Discrete Math. **26**(4), 1682–1708 (2012). https://doi.org/10.1137/110829222

35. Wu, D., Wu, R.: Borodin-Kostochka conjecture for a class of $P_6$-free graphs. preprint, arXiv: arXiv:2306.12062 (2023)

36. Xia, W., Jooken, J., Goedgebeur, J., Huang, S.: Critical ($P_5$, *dart*)-free graphs. In: Wu, W., Guo, J. (eds.) COCOA 2023, Part II. LNCS, vol. 14462, pp. 390–402. Springer, Heidelberg (2023). https://doi.org/10.1007/978-3-031-49614-1_29

37. Xia, W., Jooken, J., Goedgebeur, J., Huang, S.: Some Results on Critical ($P_5$, $H$)-free Graphs. preprint, arXiv: arXiv:2403.05611 (2024)

# Directed Path Partition Problem on Directed Acyclic Graphs

Hiroshi Eto[1], Shunsuke Kawaharada[1], Guohui Lin[2], Eiji Miyano[1(✉)], and Tugce Ozdemir[3]

[1] Kyushu Institute of Technology, Iizuka, Japan
{eto,miyano}@ai.kyutech.ac.jp, kawaharada.shunsuke319@mail.kyutech.jp
[2] University of Alberta, Edmonton, Canada
guohui@ualberta.ca
[3] The CUNY Graduate Center, New York, USA
tozdemir@ccny.cuny.edu

**Abstract.** We study the problem of partitioning the vertex set of a given directed graph $G = (V, E)$ into a small number of vertex-disjoint directed paths each of order at most a prescribed $k$, called the DIRECTED $k$-PATH PARTITION problem ($k$-PP, for short). The $k$-PP problem with $k = |V|$ is equivalent to the DIRECTED PATH PARTITION problem (PP), where the goal of PP is to find a minimum collection of vertex-disjoint directed paths of any order to cover all the vertices of $V$. The decision version of $k$-PP includes the DIRECTED HAMILTONIAN PATH problem as a special case. Therefore, when $k$ is part of the input, the $k$-PP is APX-hard, and it is not approximable within two unless P = NP. Although $k$-PP on general graphs are intractable, several tractable cases are known. The $k$-PP problem with $k = 2$ (i.e., 2-PP) is essentially equivalent to the MAXIMUM MATCHING problem on directed graphs, which is solvable in polynomial time. Also, it is claimed that if the input is restricted to directed acyclic graphs (DAGs), then PP can be solved in polynomial time. This implies that if the input graph is a DAG of height at most $k$, then $k$-PP can be solved in polynomial time since the order of every path in the directed path partition is at most $k$. In this paper, we prove that $k$-PP is NP-hard even if the input graph is restricted to planar, bipartite DAGs of height $k + 1$ and degree at most three. Also, we present the 1.0075-inapproximability of $k$-PP for any fixed $k \geq 3$. In contrast, we show that $k$-PP on directed graphs of degree at most two can be solved in polynomial time.

## 1 Introduction

Graph partitioning problems constitute fundamental classes of well-studied problems in the field of graph theory and combinatorial algorithms. One of the most important classes must be graph partitioning problems by paths [8,12], which have received considerable attention in the literature. The problems are motivated by applications in diverse areas such as broadcasting problems in computer

© The Author(s), under exclusive license to Springer Nature Switzerland AG 2024
A. A. Rescigno and U. Vaccaro (Eds.): IWOCA 2024, LNCS 14764, pp. 314–326, 2024.
https://doi.org/10.1007/978-3-031-63021-7_24

or communication networks [15], and postal delivery [10]. Several types of paths have been considered so far [13], for example, unrestricted paths, paths containing at most $k$ vertices, induced paths, directed paths in directed graphs, and so on.

The order of a path is the number of vertices on the path. In this paper we mainly study the directed $k$-path partition problem ($k$-PP, for short) for a prescribed positive integer $k$: The input of $k$-PP is a directed graph $G = (V, E)$. The goal of $k$-PP is to find a minimum collection of vertex-disjoint directed paths each of order at most $k$ to cover all the vertices in $G$. The $k$-PP problem with $k = |V|$ is equivalent to the DIRECTED PATH PARTITION problem (PP, for short), i.e., the goal of PP is to find a minimum collection of vertex-disjoint directed paths of any order to cover all the vertices of $V$. One sees that the decision version of $k$-PP includes the DIRECTED HAMILTONIAN PATH problem as a special case [6]. Therefore, when $k$ is part of the input, the $k$-PP is APX-hard and is not approximable within 2 unless P = NP. As for the approximability of the undirected variants of $k$-PP (i.e., ignoring the edge directions in the input graph and the path partition), Monnot and Toulouse [14] provided a 3/2-approximation algorithm for 3-PP. Subsequently, the approximation ratio has been improved to 13/9 [1], 4/3 [2], and the current best approximation ratio is 21/16 [3]. For the $k$-PP problem on directed graphs, Chen, Chen, Kennedy, Lin, Xu and Zhang [1] presented a $k/2$-approximation algorithm for any $k \geq 3$, and an improved $(k + 2)/3$-approximation algorithm for $k \geq 7$. Furthermore, they proposed an improved 13/9-approximation algorithm for $k = 3$ [1].

Although the general PP and also $k$-PP problems are intractable, several tractable cases are known. The $k$-PP problem with $k = 2$ (i.e., 2-PP) is essentially equivalent to the MAXIMUM MATCHING problem on directed graphs, which is solvable in polynomial time. Also, if the input graph is restricted to directed acyclic graphs (DAGs), then PP can be solved in polynomial time by using a flow-based algorithm [4,5].

In this paper, we investigate the intractability and the inapproximability of the $k$-PP on DAGs for a prescribed integer $k \geq 3$: Here is a summary of our main results:

- As shown in the above, the PP problem on DAGs can be solved in polynomial time. This implies that if the height of the input DAG is at most $k$, then $k$-PP can be solved in polynomial time since the order of every path in the directed path partition is at most $k$. In a sharp contrast to the tractability, we show that $k$-PP is NP-hard even if the input graph is restricted to planar, bipartite DAGs of height $k+1$ and degree at most three, by giving the polynomial-time reduction from the PLANAR 3-SAT problem.
- As mentioned before, if $k$ is part of the input, then $k$-PP can not be approximated within ratio 2 unless P = NP. Chen et al. [1] conjectured that $k$-PP is also APX-hard for any fixed $k \geq 3$, and they pointed out that giving a non-trivial lower bound on the approximation ratio of $k$-PP is interesting. We show the 1.0075-inapproximability of $k$-PP on bipartite DAGs of height $k+1$ and degree at most three by using a similar polynomial-time reduction from the MAXIMUM 2-SAT problem.

- As another contrast, we show that $k$-PP on directed graphs of degree at most two can be solved in polynomial time.

## 2 Preliminaries

Let $G = (V, E)$ be a *directed* unweighted graph, where $V$ and $E$ are sets of vertices and directed edges, respectively. We assume without loss of generality that there are no self-loops or multiple directed edges in the graph. We sometimes denote by $V(G)$ and $E(G)$ the vertex and the directed edge sets of $G$, respectively. Unless otherwise described, $n$ and $m$ denote the cardinality of $V$ and the cardinality of $E$, respectively, for $G = (V, E)$. A *directed edge* from a vertex $u$ to another vertex $v$ is denoted by $(u, v)$. An undirected graph obtained by replacing all directed edges of the directed graph $G$ with undirected edges is called the *underlying graph* of $G$. A *directed path* $P$ (*directed cycle* $C$, resp.) of $k$ vertices from a vertex $v_1$ to a vertex $v_k$ is represented by a sequence $P = \langle v_1, v_2, \ldots, v_k \rangle$ ($C = \langle v_1, v_2, \ldots, v_k, v_1 \rangle$, resp.) of distinct $k$ vertices. That is, $P$ ($C$, resp.) contains $k - 1$ directed edges $(v_1, v_2), \ldots, (v_{k-1}, v_k)$ ($k$ directed edges $(v_1, v_2), \ldots,$ $(v_{k-1}, v_k), (v_k, v_1)$, resp.). Let $P^{rev} = \langle v_1 v_2 \ldots v_k \rangle^{rev}$ be the reverse of $P$, i.e., $P^{rev}$ contains $k - 1$ directed edges, $(v_2, v_1), \ldots, (v_k, v_{k-1})$. Similarly, let $C^{rev}$ be the reverse of $C$. An *order* (a *length*, respectively) of the directed path $P$ is defined by the number of vertices (directed edges, respectively) contained in $P$. If the order of a directed path $P$ is $k$, then we say that $P$ is a $k$-path. A directed path is said to be *maximal* if it is not contained in any path with more vertices. If the underlying graph of a directed graph $G$ is a path, then $G$ is called the *alternate directed path*. Note that the alternate directed path $AP$ can be seen as the concatenation of maximal directed subpaths, i.e., $AP = P_1 \circ P_2 \circ \cdots \circ P_p$ where the $i$th subpath $P_i$ for every $1 \leq i \leq p$ is the maximal directed path, and the direction of $P_i$ is reverse of the direction of the next directed path $P_{i+1}$ for $1 \leq i \leq p - 1$. The *alternate directed cycle* is similarly defined by the concatenation of maximal directed subpaths.

For a vertex $v$ in $G$, the number of edges entering (leaving, resp.) $v$ is denoted by $d_G^-(v)$ ($d_G^+(v)$, resp.), which is referred to as the *in-degree* (*out-degree*, resp.) of $v$. When the graph $G$ is clear from the context, $d_G^-(v)$ and $d_G^+(v)$ are simplified as $d^-(v)$ and $d^+(v)$, respectively. The *open out-neighborhood* of a vertex $v \in V$ in $G = (V, E)$ is the set $N^+(v) = \{u \in V : (u, v) \in E\}$. That is, $d^+(v) = |N^+(v)|$. Similarly, the *open in-neighborhood* of $v \in V$ is the set $N^-(v) = \{u \in V : (v, u) \in E\}$, and $d^-(v) = |N^-(v)|$.

A directed graph $G$ is *subcubic* if its underlying graph is subcubic, i.e., the total number $d_G^+(v) + d_G^-(v)$ of the in-degree and the out-degree of every $v$ in $G$ is at most three. A directed graph $G$ is *bipartite* if vertices in $G$ can be partitioned into two sets $V_1$ and $V_2$ such that every directed edge connects a vertex in $V_1$ to a vertex in $V_2$ or vice versa. A directed graph $G$ is *planar* if $G$ can be embedded in the plane, i.e., it can be drawn on the plane in such a way that no directed edges cross each other. A directed graph $G$ is *acyclic* if $G$ does not contain any directed cycle as an induced subgraph. An acyclic directed graph is often called

*DAG* for short. A *height* of a DAG is the maximum length of directed paths in the graph. A DAG $G = (V, E)$ is *layered* if its vertex set $V$ can be partitioned into $\{V_1, V_2, \ldots, V_t\}$ such that $N^-(V_1) = \emptyset$, $N^+(V_t) = \emptyset$, and $(u, v) \in E$ implies that $u \in V_i$ and $v \in V_{i+1}$ for $1 \leq i \leq t - 1$. Note that a layered DAG is bipartite.

Let $k \geq 1$ be a fixed integer. By the $k$-*path partition* we mean a collection of vertex-disjoint directed paths each of order at most $k$ such that every vertex is on exactly one path in the collection. Our problem, called the DIRECTED $k$-PATH PARTITION problem ($k$-PP for short), is defined as follows:

DIRECTED $k$-PATH PARTITION ($k$-PP)
**Instance:** A directed graph $G = (V, E)$.
**Goal:** Find a minimum $k$-path partition $\mathcal{P}^*$ of vertex-disjoint directed paths each of order at most $k$ such that every vertex is on exactly one path in $\mathcal{P}^*$.

The $k$-PP problem with $k = 1$ is trivial. The 2-PP problem is essentially equivalent to the MAXIMUM MATCHING problem on undirected graphs, and therefore, 2-PP can be solved in polynomial time [7]. From now on, we focus on $k$-PP for a fixed $k \geq 3$.

## 3    Degree 3 and Height $k + 1$

In this section, we first prove that 3-PP is NP-hard even if the input graph is subcubic, planar, layered (therefore, bipartite) DAGs of height four. Then, we show the NP-hardness of $k$-PP on subcubic, planar, layered DAGs of height $k+1$. Furthermore, we show the approximation hardness of $k$-PP.

Here, let us consider the following decision version $k$-PP($L$) of $k$-PP:

$k$-PP($L$)
**Instance:** A directed graph $G = (V, E)$ and a positive integer $L$.
**Question:** Is there a $k$-path partition $\mathcal{P}$ of vertex-disjoint directed paths each of order at most $k$ such that every vertex is on exactly one path in $\mathcal{P}$, and $|\mathcal{P}| \leq L$?

To prove the NP-hardness, we give a polynomial-time reduction to 3-PP($L$) from PLANAR 3-SAT [11]. PLANAR 3-SAT is a restriction of 3-SAT in which a particular graph that is associated with an input 3CNF-formula is planar. More precisely, for any 3CNF-formula $\phi$ consisting of a collection $\mathcal{C}$ of clauses over the variables $U = \{u_1, u_2, \ldots, u_n\}$, the *associated graph* $G_\phi = (V_\phi, E_\phi)$ is defined by $V_\phi = U \cup \mathcal{C}$ and $E_\phi = \{(u, C) \mid u \in C \text{ or } \overline{u} \in C \text{ for } u \in U \text{ and } C \in \mathcal{C}\}$. PLANAR 3-SAT takes as input a 3CNF-formula $\phi$ consisting of a collection of $m$ clauses $\mathcal{C} = \{C_1, C_2, \ldots, C_m\}$ over $n$ variables $U = \{u_1, u_2, \ldots, u_n\}$ such that each clause contains exactly three variables and such that the associated graph $G_\phi$ admits a planar embedding, and asks whether there exists a truth assignment for $U$ that makes all clauses in $\mathcal{C}$ true.

In the following, given a 3CNF-formula $\phi$, we construct a subcubic, planar, layered DAG $G$ such that there is a truth assignment that satisfies all the clauses

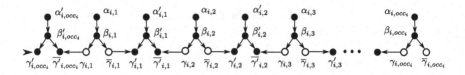

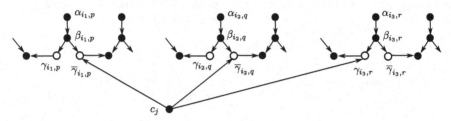

**Fig. 1.** $i$th variable gadget $G_{u_i}$

**Fig. 2.** $j$th subgraph $G_{c_j}$ in the clause gadget has three outgoing edges

in $\phi$ if and only if there is a collection $\mathcal{P}$ of vertex-disjoint directed paths each of order at most three such that $\mathcal{P}$ covers all the vertices in $V(G)$ and $|\mathcal{P}| \leq L$.

Suppose that the 3CNF-formula $\phi$ has $n$ variables, $U = \{u_1, u_2, \ldots, u_n\}$, and contains $m$ clauses, $\mathcal{C} = \{C_1, C_2, \ldots, C_m\}$, where the $j$th clause includes exactly three variables $u_{j_1}$, $u_{j_2}$, and $u_{j_3}$ in $U$. Let $occ_i$ be the number of occurrences of the $i$th variable in $\phi$, i.e., the total number of positive literals $u_i$'s and negative literals $\overline{u_i}$'s. The $p$th occurrence of $u_i$ is denoted by $u_i^p$ if it is positive, and $\overline{u_i}^p$ if it is negative.

The reduced directed graph $G$ consists of two subgraphs, (1) $G_u$ and (2) $G_\mathcal{C}$, and (3) directed edges from $G_\mathcal{C}$ to $G_u$. $G_u$ is the *variable gadget* associated with variable set $U$, and $G_\mathcal{C}$ is the *clause gadget* associated with the clause set $\mathcal{C}$.

(1) The variable gadget $G_u$ is further divided into $n$ subgraphs, $G_{u_1}$ through $G_{u_n}$, corresponding to $n$ variables in $\phi$, $u_1$ through $u_n$, respectively. For each $i$, the $i$th subgraph $G_{u_i}$ is illustrated in Fig. 1, which consists of $8 \times occ_i$ vertices and $8 \times occ_i$ directed edges as follows:

$$V(G_{u_i}) = \{\alpha_{i,1}, \alpha'_{i,1}, \beta_{i,1}, \beta'_{i,1}, \gamma_{i,1}, \overline{\gamma}_{i,1}, \gamma'_{i,1}, \overline{\gamma'}_{i,1}\}$$
$$\cup \{\alpha_{i,2}, \alpha'_{i,2}, \beta_{i,2}, \beta'_{i,2}, \gamma_{i,2}, \overline{\gamma}_{i,2}, \gamma'_{i,2}, \overline{\gamma'}_{i,2}\} \cup \quad \ldots$$
$$\cup \{\alpha_{i,occ_i}, \alpha'_{i,occ_i}, \beta_{i,occ_i}, \beta'_{i,occ_i}, \gamma_{i,occ_i}, \overline{\gamma}_{i,occ_i}, \gamma'_{i,occ_i}, \overline{\gamma'}_{i,occ_i}\}$$

$$E(G_{u_i}) = \Big\{ (\alpha_{i,1}, \beta_{i,1}), (\beta_{i,1}, \gamma_{i,1}), (\beta_{i,1}, \overline{\gamma}_{i,1}), (\gamma_{i,1}, \overline{\gamma'}_{i,occ_i}),$$
$$(\alpha'_{i,1}, \beta'_{i,1}), (\beta'_{i,1}, \gamma'_{i,1}), (\beta'_{i,1}, \overline{\gamma'}_{i,1}), (\overline{\gamma'}_{i,1}, \gamma'_{i,1}) \Big\}$$
$$\cup \Big\{ (\alpha_{i,2}, \beta_{i,2}), (\beta_{i,2}, \gamma_{i,2}), (\beta_{i,2}, \overline{\gamma}_{i,2}), (\gamma_{i,2}, \overline{\gamma'}_{i,1}),$$
$$(\alpha'_{i,2}, \beta'_{i,2}), (\beta'_{i,2}, \gamma'_{i,2}), (\beta'_{i,2}, \overline{\gamma'}_{i,2}), (\overline{\gamma'}_{i,2}, \gamma'_{i,2}) \Big\} \cup \ldots$$
$$\cup \Big\{ (\alpha_{i,occ_i}, \beta_{i,occ_i}), (\beta_{i,occ_i}, \gamma_{i,occ_i}), (\beta_{i,occ_i}, \overline{\gamma}_{i,occ_i}), (\gamma_{i,occ_i}, \overline{\gamma'}_{i,occ_i-1}),$$
$$(\alpha'_{i,occ_i}, \beta'_{i,occ_i}), (\beta'_{i,occ_i}, \gamma'_{i,occ_i}), (\beta'_{i,occ_i}, \overline{\gamma'}_{i,occ_i}), (\overline{\gamma'}_{i,occ_i}, \gamma'_{i,occ_i}) \Big\}$$

(2) The clause gadget $G_C$ consists of $m$ isolated vertices $\{c_1, c_2, \ldots, c_m\}$, which are associated with $m$ clauses in $\phi$, $C_1$ through $C_m$. Let $G_{c_j} = (\{c_j\}, \emptyset)$ be the $j$th subgraph in the clause gadget for $1 \leq j \leq m$.

(3) We add $3m$ directed edges from the clause gadget to the variable gadget as follows: Consider the $j$th clause $C_j$ for some $1 \leq j \leq m$. Suppose that three variables $u_{j_1}$, $u_{j_2}$, and $u_{j_3}$ are the $p$th positive literal $u_{i_1}$ (negative literal $\overline{u_1}$, resp.), the $q$th positive literal $u_{i_2}$ (negative literal $\overline{u_2}$, resp.), and the $r$th positive literal $u_{i_3}$ (negative literal $\overline{u_3}$, resp.), where $1 \leq p \leq occ_{i_1}$, $1 \leq q \leq occ_{i_2}$, and $1 \leq r \leq occ_{i_3}$. Then, we add the following three directed edges from the clause gadget $G_C$ to the variable gadget $G_u$: (i) $(c_j, \gamma_{i_1,p})$ if $u_{i_1}$ is positive $((c_j, \overline{\gamma_{i_1,p}})$ if $u_{i_1}$ is negative), (ii) $(c_j, \gamma_{i_2,q})$ if $u_{i_2}$ is positive $((c_j, \overline{\gamma_{i_2,q}})$ if $u_{i_2}$ is negative), and (iii) $(c_j, \gamma_{i_3,r})$ if $u_{i_3}$ is positive $((c_j, \overline{\gamma_{i_3,r}}, c_j)$ if $u_{i_3}$ is negative). For example, suppose that $C_j = \overline{u_{i_1}} \vee \overline{u_{i_2}} \vee u_{i_3}$, where $\overline{u_{i_1}}$, $\overline{u_{i_2}}$, and $u_{i_1}$ are the $p$th, the $q$th, and the $r$th occurrences, respectively. Then, the $j$th gadget $G_{c_j} = (\{c_j\}, \emptyset)$ (i.e., vertex $c_j$) has three outgoing edges as illustrated in Fig. 2.

(4) Finally, we set the integer $L$ as follows:

$$L = 3 \times \sum_{i=1}^{n} occ_i.$$

This completes the construction, which is done in polynomial time.

One sees that the directed graph $G$ constructed above is subcubic. Furthermore, it is important to note that if the associated graph $G(\phi)$ of the instance $\phi$ of PLANAR 3-SAT is planar, then $G$ can be mapped in the plane without edge-crosses. Also, $G$ is acyclic and layered. Let $V_1$, $V_2$, $V_3$, and $V_4$ be the top, the second, the third, and the bottom layers, respectively. One can verify that the layers have the following vertices and there are no directed edges between $V_1$ and $V_3$, $V_1$ and $V_4$, or $V_2$ and $V_4$:

$$V_1 = \bigcup_{i=1,\ldots,n} \{\alpha_{i,1}, \ldots, \alpha_{i,occ_i}\}$$

$$V_2 = \bigcup_{i=1,\ldots,n} \{\alpha'_{i,1}, \ldots, \alpha'_{i,occ_i}, \beta_{i,1}, \ldots, \beta_{i,occ_i}\} \cup \bigcup_{j=1,\ldots,m} \{c_j\}$$

$$V_3 = \bigcup_{i=1,\ldots,n} \{\beta'_{i,1}, \ldots, \beta'_{i,occ_i}, \gamma_{i,1}, \ldots, \gamma_{i,occ_i}, \overline{\gamma_{i,1}}, \ldots, \overline{\gamma_{occ_i}}\}$$

$$V_4 = \bigcup_{i=1,\ldots,n} \{\gamma'_{i,1}, \ldots, \gamma'_{i,occ_i}, \overline{\gamma'_{i,1}}, \ldots, \overline{\gamma'_{occ_i}}\}.$$

Therefore, the maximum order of directed paths in the variable gadget $G_{u_i}$ is four for every $i$. See Fig. 1 again. For example, a directed path $\langle \alpha_{i,1}, \beta_{i,1}, \gamma_{i,1}, \overline{\gamma'}_{i,occ_i} \rangle$ is one of the maximum order directed paths. Also, $\langle \alpha_{i,1}, \beta_{i,1}, \overline{\gamma}_{i,1}, \gamma'_{i,i} \rangle$ is another maximum order directed path.

Now, consider two types of $k$-path partitions on the $i$th variable gadget $G_{u_i}$, illustrated in Figs. 3-(A) and (B). In Fig. 3-(A), for each $1 \leq j \leq occ_i$, two 3-paths $\langle \alpha_{i,j}, \beta_{i,j}, \overline{\gamma'}_{i,j} \rangle$ and $\langle \alpha'_{i,j}, \beta'_{i,j}, \gamma'_{i,j} \rangle$ form the $\vee$-shape and thus we call the $k$-path partition the "$\vee$-partition". The 2-paths starting from $\gamma_{i,j}$, $j = 1, \ldots, occ_i$, are called *true-2-paths*. Similarly, the $k$-path partition in Fig. 3-(B) is called the "$\wedge$-partition". The 2-paths starting from $\overline{\gamma}_{i,j}$, $j = 1, \ldots, occ_i$, are called *false-2-paths*. Note that the number of directed paths in the $\vee$-partition or the $\wedge$-partition on $G_{u_i}$ is $3 \times occ_i$.

The following lemma plays a key role to prove the correctness of our reduction:

**Lemma 1.** *(i) The number of vertex-disjoint directed paths each of order at most three is at least $3 \times occ_i$ for every $G_{u_i}$. (ii) The minimum $k$-path partition is one of the two partitions, the $\vee$-partition in Fig. 3-(A) or the $\wedge$-partition in Fig. 3-(B).*

*Proof.* (i) Intuitively, to minimize the number of vertex-disjoint directed paths, we want to find as many 3-paths as possible. One sees that all the 3-paths includes exactly one vertex in the set $\bigcup_{j=1,\ldots,occ_i} \{\beta_{i,j}, \beta'_{i,j}\}$ of $2 \times occ_i$ vertices for every $i$. Hence, the number of 3-paths in $G_{u_i}$ is at most $2 \times occ_i$. The total number of vertices contained in those $2 \times occ_i$ 3-paths is $6 \times occ_i$. If all the remaining $2 \times occ_i$ vertices can be partitioned into 2-paths as the best choice, then the number of 2-paths is $occ_i$. As a result, the number of vertex-disjoint directed paths is at least $3 \times occ_i$.

(ii) If we select a 3-path starting from $\beta_{i,j}$ (resp. $\beta'_{i,j}$) for some $j$ ($1 \leq j \leq occ_i$), then a vertex $\alpha_{i,j}$ (resp. $\alpha'_{i,j}$) is isolated, which implies that the number of vertex-disjoint directed paths is at least $3 \times occ_i + 1$ in $G_{u_i}$. Hence, all the 3-paths must start from $\alpha_{i,j}$ (resp. $\alpha'_{i,j}$). Furthermore, for example, if we select two 3-paths $\langle \alpha_{i,1}, \beta_{i,1}, \overline{\gamma}_{i,1} \rangle$ and $\langle \alpha'_{i,1}, \beta_{i,1}, \overline{\gamma'}_{i,1} \rangle$, then the vertex $\gamma'_{i,1}$ is isolated and thus the number of directed paths is at least $3 \times occ_i + 1$ in $G_{u_i}$. As a result, the minimum $k$-path partition must be one of the two partitions, the $\vee$-partition or the $\wedge$-partition. $\square$

*Remark 1.* See Fig. 4. One can verify that if we select the $\vee$-partition of the top in Fig. 3, then the true-2-path $\langle \gamma_{i,j}, \overline{\gamma'}_{i-1} \rangle$ for every $2 \leq j \leq occ_i$ or $\langle \gamma_{i,1}, \gamma'_{i,occ_i} \rangle$ can be added one inward edge from the clause gadget without increasing the total number of paths; on the other hand, if we select the $\wedge$-partition of the bottom in Fig. 3, then the false-2-path $\langle \overline{\gamma}_{i,j}, \gamma'_{i,j} \rangle$ for every $1 \leq j \leq occ_i$ can be added one inward edge from the clause gadget without increasing the total number of paths.

We are now ready to prove the correctness of the reduction:

**Lemma 2.** *For the 3CNF-formula $\phi$ and the constructed DAG $G$ by the above reduction, the following is satisfied: There is a collection $\mathcal{P}$ of vertex-disjoint directed paths each of order at most three such that $\mathcal{P}$ covers all the vertices in $V(G)$ and $|\mathcal{P}| \leq L = 3 \times \sum_{i=1}^{n} occ_i$ if and only if there is a truth assignment that satisfies all the clauses in $\phi$.*

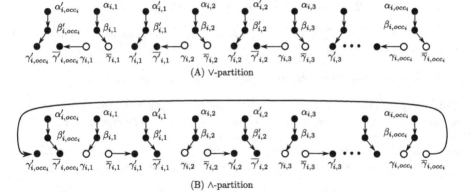

Fig. 3. Most efficient directed path partition on $G_{u_i}$: For $1 \leq j \leq occ_i$, (A) in the V-partition, every pair of two 3-paths $\langle \alpha_{i,j}, \beta_{i,j}, \overline{\gamma}'_{i,j} \rangle$ and $\langle \alpha'_{i,j}, \beta'_{i,j}, \gamma'_{i,j} \rangle$ forms the V-shape, and (B) in the $\wedge$-partition, every pair of $\langle \alpha_{i,j}, \beta_{i,j}, \overline{\gamma}'_{i,j} \rangle$ and $\langle \alpha'_{i,j}, \beta'_{i,j}, \gamma'_{i,j} \rangle$ forms the $\wedge$-shape.

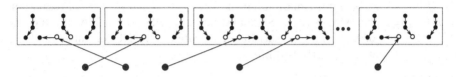

Fig. 4. Directed edges from the clause gadget can be added to true-2-paths or false-2-paths in the variable gadget without increasing the total number of paths.

*Proof.*(1) Suppose that $\phi$ is satisfiable. For each $1 \leq i \leq n$, if the variable $u_i$ is true, we chose the V-partition; otherwise we take the $\wedge$-partition. Recall that the number of directed paths in the V-partition or the $\wedge$-partition is $3 \times \sum_{i=1}^{n} occ_i$. For each $1 \leq j \leq m$, we add the $j$th clause gadget $G_{c_j}$ (i.e., $j$th isolated vertex $c_j$) to one of the 2-paths, true-2-paths or false-2-paths. The additions of $m$ vertices in the clause gadget to the 2-paths do not increase the total number of directed paths as mentioned in Remark 1. Hence, the total number of directed paths in this $k$-path partition must be $3 \times \sum_{i=1}^{n} occ_i$.

(2) Suppose that there is a collection $\mathcal{P}$ of vertex-disjoint directed paths each of order at most three such that $\mathcal{P}$ covers all the vertices in $V(G)$ and $|\mathcal{P}| \leq L = 3 \times \sum_{i=1}^{n} occ_i$. For each $1 \leq i \leq n$, the $k$-path partition of the $i$th variable gadget $G_{u_i}$ must be either the V-partition or the $\wedge$-partition from Lemma 1. Furthermore, as observed in Remark 1, for each $1 \leq j \leq m$, the $j$th clause gadget $c_j$ must be connected to a true-2-path or a false-2-path. If the $k$-path partition for the $i$th variable gadget $G_{u_i}$ is the V-partition, then we set the $i$th variable $u_i$ in $\phi$ to true; if the partition is the $\wedge$-partition, then we set $u_i$ to false. After doing so, since at least one literal in each clause has been set to true, the truth assignment satisfies $\phi$. □

From Lemma 2, we obtain the main result of this section:

**Theorem 1.** *The 3-PP problem is NP-hard even if the input graph is restricted to subcubic, planar, layered (therefore, bipartite) DAGs of height four.*

With a small modification to the above reduction, we obtain:

**Corollary 1.** *The k-PP problem is NP-hard even if the input graph is restricted to subcubic, planar, layered (therefore, bipartite) DAGs of height $k + 1$.*

*Proof.* By adding directed paths of $k - 3$ vertices to every $\alpha_{i,j}$ and every $\alpha'_{i,j}$ for each $1 \leq i \leq n$ and each $1 \leq j \leq occ_i$, we can show this corollary.    $\Box$

The reduction used to prove Theorem 1 also yields:

**Corollary 2.** *There is no approximation algorithm for k-PP with an approximation guarantee of* 1.0075, *unless $P = NP$, even if the input graph is restricted subcubic, layered (therefore, bipartite) DAGs of height $k + 1$.*

*Proof.* To show the approximation hardness, we give a *gap-preserving* reduction to $k$-PP from the MAXIMUM 2-SATISFIABILITY problem (Max2SAT), which is the restriction of the maximum satisfiability problem to CNF-formulae in which each clause contains exactly two literals:

Max2SAT
   **Instance:** A CNF-formula $\phi_2$ in conjunctive normal form such that each clause has exactly two literals (called 2CNF-formula).
   **Goal:** Find a truth assignment that maximizes the total number of satisfied clauses in $\phi_2$.

Assume that $\phi_2$ consists of $n$ variables and $m$ clauses. Then, $\sum_{i=1}^{n} occ_i = 2m$ since each clause has exactly two literals. It is known that Max2SAT is not approximable within 1.0476 $\geq 22/21$ [9]. Similarly to the previous reduction, given a 2CNF-formula $\phi_2$, we construct a subcubic, layered DAG $G$, which consists of the variable gadget, the clause gadget, and the directed edges from the clause to the variable gadgets. The variable and the clause gadgets are the same as the previous gadgets, but, the number of directed edges from each clause is at most two since each clause contains at most two literals. Note that the associated graph of $\phi_2$ might not be planar and thus the reduced directed graph $G$ is not always planar.

   Let $OPT_{2SAT}(\phi_2)$ and $OPT_{PP}(G)$ be the number of satisfied clauses in $\phi_2$ obtained by an optimal algorithm for Max2SAT and the size of the $k$-path partition in $G$ obtained by an optimal algorithm for $k$-PP, respectively.

   We show that the reduction satisfies:

– If $OPT_{2SAT}(\phi_2) \geq \ell$, then $OPT_{PP}(G) \leq 3\sum_{i=1}^{n} occ_i + m - \ell = 7m - \ell$; and
– If $OPT_{2SAT}(\phi_2) < \frac{1}{c}\ell$, then $OPT_{PP}(G) > 3\sum_{i=1}^{n} occ_i + m - \frac{1}{c}\ell = 7m - \frac{1}{c}\ell$,

where $c$ is the hardness parameter of the approximation of Max2SAT, i.e., $c = 22/21$.

As shown before, the minimum $k$-path partition on the variable gadget is either the $\lor$-partition or the $\land$-partition, and the minimum number of vertex-disjoint directed paths is $3\sum_{i=1}^{n} occ_i = 6m$. If $OPT_{2SAT}(\phi_2) \geq \ell$, then at least $\ell$ vertices in the clause gadget can be added to $\ell$ 2-paths in the variable gadget. That is, at most $m - \ell$ vertices remain as 1-paths. As a result, $OPT_{PP}(G) \leq 7m - \ell$ holds. On the other hand, if $OPT_{2SAT}(\phi_2) < \frac{1}{c}\ell$, then $OPT_{PP}(G) > 7m - \frac{1}{c}\ell$ can be shown by using the similar arguments.

Since $\ell \leq m$ and $1/c = 21/22 < 1.0$, we get the following approximation gap:

$$\frac{7m - \frac{1}{c}\ell}{7m - \ell} \geq \frac{7m - \frac{1}{c}m}{6m} = \frac{7 - \frac{21}{22}}{6} = \frac{133}{132} \geq 1.0075$$

This completes the proof of the theorem. □

# 4   Tractable Cases for $k$-PP

## 4.1   DAGs of Height at Most $k$

Recall that the PP problem is to find a minimum collection of vertex-disjoint paths of *any order* to cover all the vertices in the given graph $G$. It is known [4,5] that PP can be solved in polynomial time for DAGs. If the input graph is a DAG and its height is at most $k$, then $k$-PP is equivalent to PP since the order of every path is at most $k$. Therefore, $k$-PP is tractable when restricted to DAGs of height at most $k$, while $k$-PP on DAGs of height $k + 1$ is intractable as shown in the previous section:

**Proposition 1 ([4]).** *The $k$-PP problem can be solved in polynomial time for DAGs of height at most $k$.*

## 4.2   Directed Graphs of Degree at Most Two

In Sect. 3, we have proved that $k$-PP on directed graphs of degree at most three is NP-hard. In this section, we show that $k$-PP on directed graphs of degree at most two, i.e., $k$-PP on alternate directed paths or cycles can be solved in polynomial time.

Let $AP = P_1 \circ P_2 \circ \cdots \circ P_p$ be an alternate directed path, where the $i$th directed subpath $P_i$ for every $1 \leq i \leq p$ is the maximal directed path, and the direction of $P_i$ is reverse of the direction of the next directed path $P_{i+1}$ for $1 \leq i \leq p-1$. Note that $P_i$ and $P_{i+1}$ share one vertex of in-degree or out-degree two. We call the vertex shared by the two consecutive maximal directed paths the *connecting-vertex*.

**Algorithm ALG_PATH$_k$.** Our path-partition algorithm ALG_PATH$_k$ for $k$-PP on alternate directed paths is based on the greedy approach, i.e., from the left to the right, we basically select disjoint directed paths of order $k$ into the $k$-path partition $\mathcal{P}$, but, ALG_PATH$_k$ follows the next exceptional rules:

$v_1$   $v_2$   $v_3$   $v_4$   $v_5$   $v_6$   $v_7$   $v_8$   $v_9$   $v_{10}$   $v_{11}$   $v_{12}$   $v_{13}$   $v_{14}$   $v_{15}$

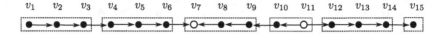

**Fig. 5.** Alternate directed path of 15 vertices and its 3-path partition

– If the order of the rightmost directed path in $P_i$ is shorter than $k$, then we select such the shorter directed path into the $k$-path partition $\mathcal{P}$.
– If there isolatedly remains the rightmost connecting vertex in the greedy approach, then it is regarded as the leftmost connecting vertex in the next subpath $P_{i+1}$, and it is merged into the next disjoint directed $k$-path.

As an example, see Fig. 5. The alternate directed path $AP$ consists of three directed paths, $P_1 = \langle v_1, v_2, v_3, v_4, v_5, v_6, v_7 \rangle$, $P_2^{rev} = \langle v_7, v_8, v_9 v_{10}, v_{11} \rangle^{rev}$, and $P_3 = \langle v_{11}, v_{12}, v_{13}, v_{14}, v_{15} \rangle$. Two vertices $v_7$ and $v_{11}$ are connecting vertices, the former is shared with $P_1$ and $P_2$, and the latter is shared with $P_2$ and $P_3$. Now consider the 3-PP problem for $AP$ and our algorithm ALG_PATH$_3$. The algorithm ALG_PATH$_3$ first finds two vertex-disjoint directed 3-paths $\langle v_1, v_2, v_3 \rangle$ and $\langle v_4, v_5, v_6 \rangle$ by scanning from the left to the right. Then, the connecting vertex $v_7$ is regarded as the head vertex of the next 3-path $\langle v_7, v_8, v_9 \rangle^{rev}$. Next, the 2-path $\langle v_{10}, v_{11} \rangle^{rev}$ is selected into the 3-path partition $\mathcal{P}$ since the direction of the directed edge $(v_{11}, v_{12})$ is reverse of the 2-path $\langle v_{10}, v_{11} \rangle^{rev}$. Subsequently, $\langle v_{12}, v_{13}, v_{14} \rangle$ and $\langle v_{15} \rangle$ are selected into $\mathcal{P}$.

**Theorem 2.** *The algorithm* ALG_PATH$_k$ *can find the minimum $k$-path partition for $k$-PP on alternate directed paths in linear time.*

*Proof.* The algorithm ALG_PATH$_k$ runs in linear time since it just scans an input alternate directed path from the left to the right.

Let a directed path of order $k$ and a directed path of order less than $k$ be denoted by $P^k$ and $P^{<k}$, respectively. The $k$-path partition obtained by ALG_PATH$_k$ for the input $AP$ is divided by ALG_PATH$_k$ as follows (by ignoring the directions of the directed paths):

$$AP = \overbrace{P^k \circ P^k \cdots \circ P^k \circ P^{<k}}^{AP_1} \circ \overbrace{P^k \circ P^k \circ \cdots \circ P^k \circ P^{<k}}^{AP_2} \circ \cdots \circ \overbrace{P^k \circ \cdots \circ P^k \circ P^{<k}}^{AP_p}.$$

That is, $AP$ is divided into subgroups of paths starting $P^k$ and ending the first $P^{<k}$. One sees that the rightmost vertex of $P^{<k}$ in each subset must be a connecting vertex. If it is not the connecting vertex, then $P^{<k}$ must include at least one more vertex since ALG_PATH$_k$ greedily works. Consider such the $i$th subgroup of paths, denoted by $AP_i = P^k \circ P^k \circ \cdots \circ P^k \circ P^{<k}$, assuming that $1 \leq i \leq p$. One can see that the number of paths in the $k$-path partition of every $AP_i$ is clearly the minimum. As a result, this theorem is satisfied. $\square$

Next, consider an alternate directed cycle. Let $AC = P_1 \circ P_2 \circ \cdots \circ P_p$ be an alternate directed cycle of $n$ vertices $v_1$ through $v_n$. Suppose that $P_1 =$

**Fig. 6.** Alternate directed cycle $AC$ and its connecting vertex $v_1$

$\langle v_1, v_2, \ldots, v_i \rangle$ and $P_p = \langle v_j, v_{j+1}, \ldots, v_n, v_1 \rangle^{rev}$ for $1 \leq i < j \leq n$, i.e., the leftmost vertex $v_1$ is shared by $P_1$ and $P_p$. The directions of the two directed edges of $v_1$ are opposite to each other.

**Algorithm** ALG_CYCLE$_k$. See Fig. 6. The $n$ vertices in an input directed cycle $AC$ are now arranged in clockwise order. First of all, our path-partition algorithm ALG_CYCLE$_k$ for $k$-PP on alternate directed cycles makes the alternate directed path from $AC$ by removing one of the two directed edges of $v_1$. Then, it computes a $k$-path partition by using ALG_PATH$_k$, but it must investigate the following two possible alternate directed paths regarding the connecting vertex $v_1$: (1) $v_1$ is the tail (or the head) of the first $k$-path in $P_1$, and (2) $v_1$ is the tail (or the head) of the last $k$-path $P_p$, as shown in Figs. 6-(1), and (2), respectively.

By Theorem 2, we immediately obtain:

**Theorem 3.** *The algorithm* ALG_CYCLE$_k$ *can find the minimum $k$-path partition for $k$-PP on alternate directed cycles in linear time.*

**Acknowledgments.** We are grateful to anonymous referees for their helpful suggestions. This work is partially supported by NSERC Canada, and JSPS KAKENHI Grant Numbers JP21K11755 and JP24K02902.

# References

1. Chen, Y., Chen, Z.Z., Kennedy, C., Lin, G., Xu, Y., Zhang, A.: Approximating the directed path partition problem. Inf. Comput. **297**, 105150 (2024). https://doi.org/10.1016/j.ic.2024.105150
2. Chen, Y., et al.: A local search 4/3-approximation algorithm for the minimum 3-path partition problem. J. Comb. Optim. **44**(5), 3595–3610 (2022). https://doi.org/10.1007/S10878-022-00915-5
3. Chen, Y., Goebel, R., Su, B., Tong, W., Xu, Y., Zhang, A.: A 21/16-approximation for the minimum 3-path partition problem. In: Lu, P., Zhang, G. (eds.) 30th International Symposium on Algorithms and Computation (ISAAC 2019). Leibniz International Proceedings in Informatics (LIPIcs), Dagstuhl, Germany, vol. 149, pp. 46:1–46:20. Schloss Dagstuhl – Leibniz-Zentrum für Informatik (2019). https://doi.org/10.4230/LIPIcs.ISAAC.2019.46
4. Cormen, T.H., Leiserson, C.E., Rivest, R.L., Stein, C.: Introduction to Algorithms, 3rd edn. MIT Press, Cambridge (2009). http://mitpress.mit.edu/books/introduction-algorithms
5. Erickson, J.: Algorithms (2019). http://jeffe.cs.illinois.edu/teaching/algorithms/
6. Garey, M.R., Johnson, D.S.: Computers and Intractability: A Guide to the Theory of NP-Completeness. W. H. Freeman, New York (1979)

7. Goldberg, A.V., Karzanov, A.V.: Maximum skew-symmetric flows and matchings. Math. Program. **100**(3), 537–568 (2004). https://doi.org/10.1007/S10107-004-0505-Z

8. Hartman, I.B.: Berge's conjecture on directed path partitions - a survey. Discret. Math. **306**(19–20), 2498–2514 (2006). https://doi.org/10.1016/J.DISC.2005.12.039

9. Håstad, J.: Some optimal inapproximability results. J. ACM **48**(4), 798–859 (2001). https://doi.org/10.1145/502090.502098

10. Korpelainen, N.: A boundary class for the k-path partition problem. Electron. Notes Discret. Math. **67**, 49–56 (2018). https://doi.org/10.1016/J.ENDM.2018.05.009

11. Lichtenstein, D.: Planar formulae and their uses. SIAM J. Comput. **11**(2), 329–343 (1982). https://doi.org/10.1137/0211025

12. Linial, N.: Covering digraphs by paths. Discret. Math. **23**(3), 257–272 (1978). https://doi.org/10.1016/0012-365X(78)90007-9

13. Manuel, P.: Revisiting path-type covering and partitioning problems. arXiv:1807.10613 (2018). https://doi.org/10.48550/arXiv.1807.10613

14. Monnot, J., Toulouse, S.: The path partition problem and related problems in bipartite graphs. Oper. Res. Lett. **35**(5), 677–684 (2007). https://doi.org/10.1016/J.ORL.2006.12.004

15. Yan, J., Chang, G.J., Hedetniemi, S.M., Hedetniemi, S.T.: k-path partitions in trees. Discret. Appl. Math. **78**(1–3), 227–233 (1997). https://doi.org/10.1016/S0166-218X(97)00012-7

# Computing Minimal Absent Words
# and Extended Bispecial Factors
# with CDAWG Space

Shunsuke Inenaga[1]([envelope])[ID], Takuya Mieno[2][ID], Hiroki Arimura[3][ID],
Mitsuru Funakoshi[1][ID], and Yuta Fujishige[1]

[1] Department of Informatics, Kyushu University, Fukuoka, Japan
inenaga.shunsuke.380@m.kyushu-u.ac.jp
[2] Department of Computer and Network Engineering, University of
Electro-Communications, Chofu, Japan
tmieno@uec.ac.jp
[3] Graduate School of IST, Hokkaido University, Sapporo, Japan
arim@ist.hokudai.ac.jp

**Abstract.** A string $w$ is said to be a *minimal absent word* (*MAW*)
for a string $S$ if $w$ does not occur in $S$ and any proper substring of
$w$ occurs in $S$. We focus on non-trivial MAWs which are of length at
least 2. Finding such non-trivial MAWs for a given string is motivated
by applications in bioinformatics and data compression. Fujishige et al.
[TCS 2023] proposed a data structure of size $\Theta(n)$ that can output the set
MAW($S$) of all MAWs for a given string $S$ of length $n$ in $O(n+|\text{MAW}(S)|)$
time, based on the *directed acyclic word graph* (*DAWG*). In this paper,
we present a more space efficient data structure based on the *compact
DAWG* (*CDAWG*), which can output MAW($S$) in $O(|\text{MAW}(S)|)$ time
with $O(\mathsf{e}_{\min})$ space, where $|\text{MAW}(S)|$ denotes the cardinality of MAW($S$)
and $\mathsf{e}_{\min}$ denotes the minimum of the sizes of the CDAWGs for $S$ and for
its reversal $S^R$. For any strings of length $n$, it holds that $\mathsf{e}_{\min} < 2n$, and
for highly repetitive strings $\mathsf{e}_{\min}$ can be sublinear (up to logarithmic) in
$n$. We also show that MAWs and their generalization *minimal rare words*
have close relationships with *extended bispecial factors*, via the CDAWG.

## 1 Introduction

A string $w$ is said to be a *minimal absent word* (*MAW*) for a string $S$ if $w$
does not occur in $S$ and any proper substring of $w$ occurs in $S$. Throughout this
paper, we focus on MAWs which are of form $aub$, where $a, b$ are characters and $u$
is a (possibly empty) string. Finding such MAWs for a given string is motivated
by applications including bioinformatics [1,12,19,23] and data compression [3,
14,15].

Fujishige et al. [16] proposed a data structure of size $\Theta(n)$ that can output the
set MAW($S$) of all MAWs for a given string $S$ of length $n$ in $O(n + |\text{MAW}(S)|)$
time, based on the *directed acyclic word graph* (*DAWG*) [9]. The DAWG for
string $S$, denoted DAWG($S$), is the smallest DFA recognizing all suffixes of $S$.

© The Author(s), under exclusive license to Springer Nature Switzerland AG 2024
A. A. Rescigno and U. Vaccaro (Eds.): IWOCA 2024, LNCS 14764, pp. 327–340, 2024.
https://doi.org/10.1007/978-3-031-63021-7_25

In this paper, we present a more space efficient data structure based on the *compact DAWG* (*CDAWG*) [10], which takes $O(\mathsf{e}_{min})$ space and can output $\mathsf{MAW}(S)$ in $O(|\mathsf{MAW}(S)|)$ time, where $\mathsf{e}_{min} = \min\{\mathsf{el}(S), \mathsf{er}(S)\}$, with $\mathsf{el}(S)$ and $\mathsf{er}(S)$ being the numbers of edges of the CDAWGs for $S$ and for its reversal $S^R$, respectively. For any string $S$ of length $n$ it holds that the number of edges of the CDAWG for $S$ is less than $2n$ and it can be sublinear (up to logarithmic) in $n$ for highly repetitive strings [5,24,25].

Our new data structure with the CDAWG is built on our deeper analysis on how the additive $O(n)$ term is required in the output time of Fujishige et al.'s DAWG-based algorithm [16]. Let $x$ be a node of $\mathsf{DAWG}(S)$ and let $u$ be the longest string represented by $x$. Suppose that there is a node $y$ of which the suffix link points to $x$, and let $a$ be the character such that $au$ is the shortest string represented by $y$. It follows that $aub$ is a MAW for $S$ if and only if node $x$ has an out-edge labeled $b$ but node $y$ does not. Fujishige et al.'s algorithm finds such characters $b$ by comparing the out-edges of $y$ and $x$ in sorted order. However, Fujishige et al.'s algorithm performs redundant comparisons even for every pair of *unary nodes* $x, y$ connected by a suffix link. This observation has led us to the use of the CDAWG for $S$, where unary nodes are compacted. Still, we need to manage the case where $y$ is unary but $x$ is not unary. We settle this issue by a non-trivial use of the *extended longest path tree* of the CDAWG [18].

In the previous work by Belazzougui and Cunial [5], they proposed another CDAWG-based data structure of size $O(\max\{\mathsf{el}(S), \mathsf{er}(S)\})$ that can output all MAWs for $S$ in $O(\max\{\mathsf{el}(S), \mathsf{er}(S)\} + |\mathsf{MAW}(S)|)$ time. The $\mathsf{el}(S)$ term comes from the *Weiner links* for all nodes of the CDAWG, which can be as large as $\Omega(\mathsf{er}(S)\sqrt{n})$ for some strings [18]. Note that our data structures uses only $O(\mathsf{e}_{min}) = O(\min\{\mathsf{el}(S), \mathsf{er}(S)\})$ space and takes only $O(|\mathsf{MAW}(S)|)$ time to report the output. We also present a simpler algorithm that restores the string $u$ in linear time in its length using the CDAWG-grammar [5], without using level-ancestor data structures [7,8], which can be of independent interest.

We also show how our CDAWG-based data structure of $O(\mathsf{e}_{min})$ space can output all *extended bispecial factors* (*EBFs*) [13] for the input string $S$ in output optimal time. This proposed method can be seen as a more space-efficient variant of Almirantis et al.'s method [2] that is built on suffix trees [26]. We also show that *minimal rare words* (*MRWs*) [4], which are generalizations of MAWs and *minimal unique substrings MUSs* [17], have close relationships with EBFs through the CDAWG.

## 2    Preliminaries

### 2.1    Basic Notations

Let $\Sigma$ be an alphabet. An element of $\Sigma$ is called a character. An element of $\Sigma^*$ is called a string. The length of a string $S$ is denoted by $|S|$. The empty string $\varepsilon$ is the string of length 0. Let $\Sigma^+ = \Sigma^* \setminus \{\varepsilon\}$. If $S = xyz$, then $x$, $y$, and $z$ are called a *prefix*, *substring* (or *factor*), and *suffix* of $S$, respectively. Let $\mathsf{Prefix}(S)$, $\mathsf{Substr}(S)$, and $\mathsf{Suffix}(S)$ denote the sets of the prefixes, substrings, and suffixes of

$S$, respectively. An element in $\mathsf{Suffix}(S) \setminus \{S\}$ is called a *proper suffix* of $S$. For any $1 \leq i \leq |S|$, $S[i]$ denotes the $i$-th character of string $S$. For any $1 \leq i \leq j \leq |S|$, $S[i..j]$ denotes the substring of $S$ that begins at position $i$ and ends at position $j$. For convenience, let $T[i..j] = \varepsilon$ for any $i > j$. We say that string $w$ *occurs* in string $S$ iff $w$ is a substring of $S$. Let $\mathsf{occ}_S(w) = |\{i \mid w = S[i..i+|w|-1]\}|$ denote the number of occurrences of $w$ in $S$. For convenience, let $\mathsf{occ}_S(\varepsilon) = |S| + 1$. For a string $S$, let $S^R = S[|S|] \cdots S[1]$ denote the reversed string of $S$.

For a set $\mathsf{S}$ of strings, let $|\mathsf{S}|$ denote the number of strings in $\mathsf{S}$.

## 2.2   MAW, MUS, MRW, and EBF

A string $w$ is said to be an *absent word* for another string $S$ if $\mathsf{occ}_S(w) = 0$. An absent word $w$ for $S$ is called a *minimal absent word* (*MAW*) for $S$ if $\mathsf{occ}_S(w[1..|w|-1]) \geq 1$ and $\mathsf{occ}_S(w[2..|w|]) \geq 1$. By definition, any character $c$ not occurring in $S$ is a MAW for $S$. In the rest of this paper, we exclude such trivial MAWs and we focus on non-trivial MAWs of form $aub$, where $a, b \in \Sigma$ and $u \in \Sigma^*$. We denote by $\mathsf{MAW}(S)$ the set of all non-trivial MAWs for $S$.

A string $w$ is called *unique* in another string $S$ if $\mathsf{occ}_S(w) = 1$. A unique substring $w$ for $S$ is called a *minimal unique substring* (*MUS*) for $S$ if $\mathsf{occ}_S(w[1..|w|-1]) \geq 2$ and $\mathsf{occ}_S(w[2..|w|]) \geq 2$. Let $\mathsf{MUS}(S)$ denote the set of non-trivial MUSs of form $aub$ for string $S$. Since a unique substring $w$ of $S$ has exactly one occurrence in $S$, each $w$ element in $\mathsf{MUS}(S)$ can be identified with a unique interval $[i..j]$ such that $1 \leq i \leq j \leq n$ and $w = S[i..j]$.

A string $w$ is called a *minimal rare word MRW* for another string $S$ if

(1) $w = c \in \Sigma$, or
(2) $w = aub$ with $a, b \in \Sigma$ and $u \in \Sigma^*$, such that $\mathsf{occ}_S(au) > \mathsf{occ}_S(aub)$ and $\mathsf{occ}_S(ub) > \mathsf{occ}_S(aub)$.

In particular, an MRW $w$ for $S$ is a MAW for $S$ when $\mathsf{occ}_S(w) = 0$. Also, an MRW $w$ is a MUS for $S$ when $\mathsf{occ}_S(w) = 1$. Let $\mathsf{MRW}(S)$ denote the set of non-trivial MRWs of type (2) for string $S$. Then, $\mathsf{MAW}(S) \subseteq \mathsf{MRW}(S)$ and $\mathsf{MUS}(S) \subseteq \mathsf{MRW}(S)$ hold.

A string $w = aub \in \mathsf{Substr}(S)$ with $a, b \in \Sigma$ and $u \in \Sigma^*$ is called an *extended bispecial factor* (*EBF*) for string $S$ if $\mathsf{occ}_S(a'u) \geq 1$ and $\mathsf{occ}_S(ub') \geq 1$ for some characters $a' \neq a$ and $b' \neq b$. Let $\mathsf{EBF}(S)$ denote the set of EBFs for string $S$.

## 2.3   Maximal Substrings and Maximal Repeats

A substring $u$ of a string $S$ is said to be *left-maximal* in $S$ if (1) there are distinct characters $a, b$ such that $au, bu \in \mathsf{Substr}(S)$, or (2) $u \in \mathsf{Prefix}(S)$. Symmetrically, a substring $u$ of $S$ is said to be *right-maximal* in $S$ iff (1) there are distinct characters $a, b$ such that $ua, ub \in \mathsf{Substr}(S)$, or (2) $u \in \mathsf{Suffix}(S)$. A substring $u$ of $S$ is said to be *maximal* in $S$ iff $u$ is both left-maximal and right-maximal in $S$. A maximal substring $u$ is said to be a *maximal repeat* if $\mathsf{occ}_S(u) \geq 2$. Let $\mathsf{MRep}(S)$ denote the set of maximal repeats in $S$. For a maximal repeat $u$

of a string $S$, the characters $a \in \Sigma$ such that $\mathrm{occ}_S(au) \geq 1$ are called the *left-extensions* of $u$, and the characters $b \in \Sigma$ such that $\mathrm{occ}_S(ub) \geq 1$ are called the *right-extensions* of $u$.

The following relationship between MAWs and maximal repeats is known:

**Lemma 1 ([4] and Theorem 1 of [22]).** *For any MRW aub for string $S$ with $a, b \in \Sigma$ and $u \in \Sigma^*$, $u$ is a maximal repeat in $S$.*

## 2.4    CDAWG

The *compact directed acyclic word graph* (*CDAWG*) of string $S$, that is denoted $\mathrm{CDAWG}(S) = (\mathsf{V}, \mathsf{E})$, is a path-compressed smallest partial DFA that represents $\mathrm{Suffix}(S)$, such that there is a one-to-one correspondence between the nodes of $\mathrm{CDAWG}(S)$ and the maximal substrings in $S$. Intuitively, $\mathrm{CDAWG}(S)$ can be obtained by merging isomorphic subtrees of the suffix tree [26] for $S$. Therefore, each string represented by a node $u$ of $\mathrm{CDAWG}(S)$ is a suffix of all the other longer strings represented by the same node $u$. For convenience, for any CDAWG node $u \in \mathsf{V}$, let $\mathrm{str}(u)$ denote the *longest* string represented by the node $u$. The *suffix link* of a CDAWG node $u$ points to another CDAWG node $v$ iff $\mathrm{str}(v)$ is the longest proper suffix of $\mathrm{str}(u)$ that is not represented by $u$. We denote it by $\mathrm{slink}(u) = v$. See Fig. 1 for an example of $\mathrm{CDAWG}(S)$.

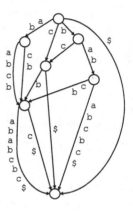

**Fig. 1.** CDAWG$(S)$ for string $S = $ ababcbababcbc$.

We remark that every internal node in $\mathrm{CDAWG}(S)$ corresponds to a maximal repeat in $S$, that is, $\mathrm{MRep}(S) = \{\mathrm{str}(v) \mid v \text{ is an internal node of } \mathrm{CDAWG}(S)\}$. The first characters of the labels of the out-going edges of an internal node of $\mathrm{CDAWG}(S)$ are the right-extensions of the corresponding maximal substring for that node. Hence the number of edges in $\mathrm{CDAWG}(S)$ is equal to the total number of right-extensions of maximal repeats in $\mathrm{MRep}(S)$, which is denoted by $\mathrm{er}(S)$. Similarly, we denote by $\mathrm{el}(S)$ the number of left-extensions of the maximal repeats in $\mathrm{MRep}(S)$. Blumer et al. [10] showed that $\min\{\mathrm{er}(S), \mathrm{el}(S)\} < 2n$ for any string $S$ of length $n$. On the other hand, it is known that there is an $\Omega(\sqrt{n})$ gap between $\mathrm{er}(S)$ and $\mathrm{el}(S)$:

**Lemma 2 (Lemma 9 of [18]).** *There exists a family of strings $S$ of length $n$ such that $\mathrm{er}(S) = \Theta(\sqrt{n})$ and $\mathrm{el}(S) = \Theta(n)$.*

The *length* of an edge $(u, y, v) \in \mathsf{V} \times \Sigma^+ \times \mathsf{V}$ in $\mathrm{CDAWG}(S)$ is its string label length $|y|$. An edge $(u, y, v)$ is called a *primary edge* if $|\mathrm{str}(u)| + |y| = |\mathrm{str}(v)|$, and $(u, y, v)$ is called a *secondary edge* otherwise ($|\mathrm{str}(u)| + |y| < |\mathrm{str}(v)|$). Namely, the primary edges are the edges in the *longest* paths from the source to all the nodes of CDAWG($S$). Thus, there are exactly $|V| - 1$ primary edges. Let LPT($S$) denote the spanning tree of CDAWG($S$) that consists of the primary edges, i.e., LPT($S$) is the tree of the longest paths from the source to all nodes of CDAWG($S$).

Since MRep($S^R$) = $\{w^R \mid w \in$ MRep($S$)$\}$ holds for any string $S$, the CDAWGs for $S$ and $S^R$ can share the same nodes. Let $E^R$ denote the set of edges of CDAWG($S^R$). The primary (resp. secondary) edges in $E^R$ are also called *hard* (resp. *soft*) *Weiner links* of CDAWG($S$). Each Weiner link is labeled by the first character of the label of the corresponding edge in $E^R$. Note that $|E^R| = $ el($S$).

## 3   CDAWG Grammars

A *grammar-compression* is a method that represents an input string $S$ with a context-free grammar $\mathcal{G}$ that only generates $S$. The size of $\mathcal{G}$ is evaluated by the total length of the right-hand sides of the productions, which is equal to the size of the minimal DAG obtained by merging isomorphic subtrees of the derivation tree for $\mathcal{G}$. Hence the more repeats $S$ has, the smaller grammar $S$ tends to have.

Assume that $S$ terminates with a unique character \$ not occurring elsewhere in $S$. Belazzougui and Cunial [5] proposed a grammar-compression based on the CDAWGs. We use a slightly modified version of their grammar: The *CDAWG-grammar* for string $S$ of length $n$, denoted $\mathcal{G}_{\mathsf{CDAWG}}$, is such that:

- The non-terminals are the nodes in CDAWG($S$);
- The root of the derivation tree for $\mathcal{G}_{\mathsf{CDAWG}}$ is the sink of CDAWG($S$);
- The derivation tree path from its root to its leaf representing the $i$-th character $S[i]$ is equal to the reversed path from the sink to the source of CDAWG($S$) that spells out the $i$-th (text positioned) suffix $S[i..n]$.

By definition, the DAG for the CDAWG-grammar $\mathcal{G}_{\mathsf{CDAWG}}$ is isomorphic to the reversed DAG for CDAWG($S$), and hence, the size for $\mathcal{G}_{\mathsf{CDAWG}}$ is er($S$). See Fig. 2 for an example.

Belazzougui and Cunial [5] showed how to extract the string label $x$ of a given CDAWG edge in $O(|x|)$ time in a sequential manner (i.e. from $x[1]$ to $x[|x|]$), by augmenting the CDAWG-grammar with a level-ancestor data structure [7,8] that works on the word RAM. This allows one to represent CDAWG($S$) in $O(\mathrm{er}(S))$ space and perform pattern matching queries in optimal time, without explicitly storing string $S$. They also used this technique to output a MAW $aub$ as a string: Since $u$ is a maximal repeat, it suffices to decompress the path labels in LPT($S$) from the source to the node $v$ that corresponds to $u$ (i.e. str($v$) = $u$).

We show that, when our goal is simply to obtain the longest string str($v$) for a given CDAWG node $v$, then level-ancestor data structures are not needed.

**Lemma 3.** *Given* CDAWG($S$) = $(V, E)$ *with suffix links and grammar* $\mathcal{G}_{\mathsf{CDAWG}}$ *of total size* $O(\mathrm{er}(T))$, *one can decompress the longest string* str($v$) *represented by a given node* $v$ *in* $O(|\mathsf{str}(v)|)$ *time.*

*Proof.* For each CDAWG node $v \in V$, let $X(v)$ denote the non-terminal in the grammar $\mathcal{G}_{\mathsf{CDAWG}}$ that corresponds to $v$. We first decompress $X(v)$ and let $p_v$

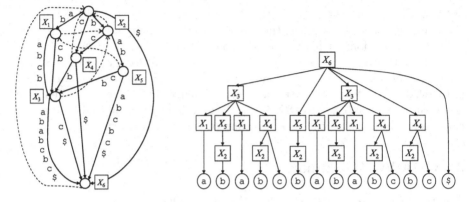

**Fig. 2.** CDAWG($S$) (left) and the CDAWG-grammar $\mathcal{G}_{\text{CDAWG}}$ (right) for the running example from Fig. 1. The dashed arcs represent the suffix links of the nodes of CDAWG($S$). To obtain str($X_3$) = ababcb for the CDAWG node $X_3$, we first decompress the non-terminal $X_3$ and obtain ababc. We move to the node $X_2$ by following the suffix link of $X_3$. We then decompress the non-terminal $X_2$ and obtain the remaining b.

be the decompressed string. Then, the length of $p_v$ is equal to the number of (reversed) paths between the source and $v$ in the CDAWG. Thus, if the suffix link slink($v$) points to the source, then $p_v$ is the whole string str($v$). Otherwise, then let $v_0 = v$. We take the suffix link slink($v_0$) = $v_1$, obtain string $p_{v_1}$ by decompressing $X(v_1)$. This gives us a longer prefix $p_{v_0}p_{v_1}$ of str($v$), since in any interval $[i..i + |\text{str}(v)| - 1]$ of leaves of the derivation tree for $\mathcal{G}_{\text{CDAWG}}$ that represents str($v$), $p_{v_1}$ immediately follows $p_{v_0}$. We continue this process until we encounter the last node $v_k$ in the suffix link chain from $v$, of which the suffix link points to the source. Then, the concatenated string $p_{v_0}p_{v_1} \cdots p_{v_k}$ is str($v$).

When there are no unary productions of form $X \to Y$, then the total number of nodes visited for obtaining $p_{v_0}p_{v_1} \cdots p_{v_k} = \text{str}(v)$ is linear in its length. Otherwise, then for each maximal sequence of unary productions $X_1 \to X_2 \to \cdots \to X_\ell \to Y$, we create a pointer from $X_i$ to $Y$ for every $1 \leq i \leq \ell$ so that we can extract str($v$) in time linear in its length[1].    □

# 4   Computing MAWs and EBFs with CDAWG

## 4.1   Computing MAWs

In this section, we first show the following:

**Theorem 1.** *There exists a data structure of $O(\mathsf{e}_{\min})$ space that can output all MAWs for string $S$ in $O(|\mathsf{MAW}(S)|)$ time with $O(1)$ working space, where* $\mathsf{e}_{\min} = \min\{\mathsf{er}(S), \mathsf{el}(S)\}.$

---

[1] A similar technique was used in the original CDAWG-grammar by Belazzougui and Cunial [5], where they explicitly removed such unary productions.

For ease of description of our algorithm, let us assume that our input string $S$ begins and ends with unique terminal symbols $\sharp$ and $\$$ not occurring inside $S$, respectively. We also assume $\sharp, \$ \in \Sigma$ so that they can be the first or the last characters for MAWs and EBFs for $S$, respectively. The case without $\sharp$ and $\$$ can be treated similarly.

We remark that MAWs have symmetric structures such that $aub \in \mathsf{MAW}(S)$ iff $bu^R a \in \mathsf{MAW}(S^R)$. This implies that one can work with the smaller one of the two: $\mathsf{CDAWG}(S)$ of $O(\mathsf{er}(S))$ space and $\mathsf{CDAWG}(S^R)$ of $O(\mathsf{el}(S))$ space. In what follows, we assume w.l.o.g. that $\mathsf{e}_{\min} = \mathsf{er}(S)$, unless otherwise stated.

On top of the edge-sorted $\mathsf{CDAWG}(S)$ augmented with suffix links and the CDAWG-grammar $\mathcal{G}_{\mathsf{CDAWG}}$ for $S$, we use the extended version of $\mathsf{LPT}(S)$ proposed in [18], that is denoted $\mathsf{LPT}^+(S)$. Let $v$ be any node in the longest path tree $\mathsf{LPT}(S)$, and suppose that its corresponding node in $\mathsf{CDAWG}(S)$ has a secondary out-going edge $(v, u)$. Then, we create a copy $\ell$ of $u$ as a leaf, and add a new edge $(v, \ell)$ to the tree. $\mathsf{LPT}^+(S)$ is the tree obtained by adding all secondary edges of $\mathsf{CDAWG}(S)$ to $\mathsf{LPT}(S)$ in this way, and thus, $\mathsf{LPT}^+(S)$ has exactly $\mathsf{er}(S)$ edges. For any node $v$ in $\mathsf{LPT}^+(S)$, let $\mathsf{str}(v)$ denote the string label in the path from the root to $v$.

Let $u$ be any node in $\mathsf{CDAWG}(S)$ and suppose it has $k$ in-coming edges. The above process of adding secondary edges to $\mathsf{LPT}(S)$ can be regarded as splitting $u$ into $k + 1$ nodes $u', \ell_1, \ldots, \ell_k$, such that $|\mathsf{str}(u')| > |\mathsf{str}(\ell_1)| > \cdots > |\mathsf{str}(\ell_k)|$. We define the suffix links of these nodes such that $\mathsf{slink}(u') = \ell_1$, $\mathsf{slink}(\ell_i) = \ell_{i+1}$ for $1 \leq i < k$, and $\mathsf{slink}(\ell_k) = v'$ such that $v'$ is the node in $\mathsf{LPT}^+(S)$ that corresponds to the destination node $v = \mathsf{slink}(u)$ of the suffix link of $u$ in $\mathsf{CDAWG}(S)$. We call these copied nodes $\ell_1, \ldots, \ell_k$ as *gray nodes*, and the other node $u'$ as a *white node*. For convenience, we assume that the source is a white node. See Fig. 3 for illustration.

In what follows, we show how to compute MAWs for a given edge $(\hat{v}, \hat{u})$ in $\mathsf{LPT}^+(S)$. Note that $\hat{v}$ is always a white node since every gray node has no children. Let $X$ be the string label of $(\hat{v}, \hat{u})$. Let $v$ be the *first* white node in the chain of suffix links from $\hat{v}$ and let $u$ be the descendant of $v$ in $\mathsf{LPT}^+(S)$ such that the path from $v$ to $u$ spells out $X$. We maintain a *fast link* from the edge $(\hat{v}, \hat{u})$ to the path $\langle v, u \rangle$ so we can access from $(\hat{v}, \hat{u})$ to the terminal nodes $v$ and $u$ in the path in $O(1)$ time. We have the four following cases (See also Fig. 4):

(A) $\hat{u}$ and $u$ are both white nodes;
(B) $\hat{u}$ is a white node and $u$ is a gray node;
(C) $\hat{u}$ is a gray node and $u$ is a white node;
(D) $\hat{u}$ and $u$ are both gray nodes.

**How to deal with Case (A):** In Case (A), $u = \mathsf{slink}(\hat{u})$.

Let $m$ be the number of nodes in the path from $v$ to $u$, with $v$ exclusive. In the sequel, we consider the case where $m > 1$, as the case with $m = 1$ is simpler to show. Let $u_1 (= u), u_2, \ldots, u_m$ denote these $m$ nodes in the path arranged in decreasing order of their depths. We process each $u_i$ in increasing order of $i$. For simplicity, we will identify each node $u_i$ as the string $\mathsf{str}(u_i)$ represented by $u_i$.

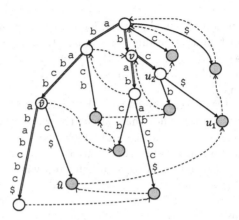

**Fig. 3.** LPT$^+(S)$ for the running example from Fig. 1 and 2. The double-lined arcs represent primary edges, and the single-lined arcs represent secondary edges. For edge $(\hat{v}, \hat{u})$ with string label c$, consider the path $\langle v, u_1 \rangle$ that spells out c$ and is obtained by the fast link. The CDAWG node that corresponds to $u_2 = $ bc has a virtual soft Weiner link with label c pointing to the CDAWG node that corresponds to $\hat{u}$. Node $u_2$ has an out-edge with b. Therefore, $cu_2b = $ cbcb is a MAW for the string $S = $ ababcbababcbc$.

For $i = 1$, node $u_1 (= u)$ has a hard Weiner link to $\hat{u}$. Let $a$ be the character label of this hard Weiner link. Let child$(\hat{u})$ and child$(\text{slink}(\hat{u})) = $ child$(u_1)$ be the sorted lists of the first characters of the out-edge labels from $\hat{u}$ and from $u_1$. Note that child$(\hat{u}) \subseteq $ child$(u_1)$. Then, it follows from the definition of MAWs and Lemma 1 that $au_1b \in \text{MAW}(S)$ iff $b \in \text{child}(u_1) \setminus \text{child}(\hat{u})$.

For each $i = 2, \ldots, m$, observe that node $u_i$ has a (virtual) *soft* Weiner link with the same character label $a$, and leading to the same node $\hat{u}$. We emphasize that this soft Weiner link is only virtual and is *not* stored in our $O(\text{er}(S))$-space data structure. We have the following lemma:

**Lemma 4.** *For each internal node $u_i$ with $2 \leq i \leq m$ in the path between $v$ and $u$, there exists at least one MAW $au_ib$ for $S$, where $a, b \in \Sigma$.*

*Proof.* Let $c$ be the first character of the label of the edge $(u_{i+1}, u_i)$. Let $\hat{u}_i$ denote the (virtual) implicit node on the edge $(\hat{v}, \hat{u})$, such that slink$(\hat{u}_i) = u_i$. Note that $c$ is the unique character that immediately follows the locus of $\hat{u}_i$ in the string label of edge $(\hat{v}, \hat{u})$. Since $u_i$ is branching but $\hat{u}_i$ is non-branching, the set child$(u_i) \setminus \text{child}(\hat{u}_i) = \text{child}(u_i) \setminus \{c\}$ is not empty. Thus, for any character $b \in \text{child}(u_i) \setminus \{c\}$, we have $au_ib \in \text{MAW}(S)$. □

Due to Lemma 4, we output MAWs $au_ib$ for every character $b \in \text{child}(u_i) \setminus \{c\}$.

**How to deal with Case (B):** In Case (B), $u$ is a gray node, and there might be other node(s) in the suffix link chain from $\hat{u}$ to $u$. No MAW is reported for the gray node $u_1 (= u)$. The rest is the same as Case (A).

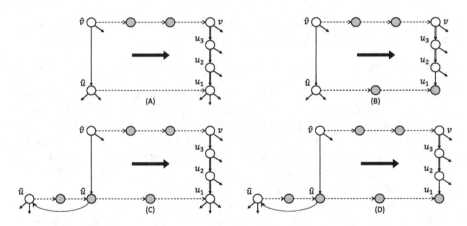

**Fig. 4.** Cases (A), (B), (C), and (D) for computing MAWs from a given edge $(\hat{v}, \hat{u})$ on $\mathsf{LPT}^+(S)$. The bold arc represents the fast link from edge $(\hat{v}, \hat{u})$ to path $\langle v, u \rangle$, where $u = u_1$. The dashed arcs represent suffix links. The dotted arcs in Cases (C) and (D) are additional pointers for $O(1)$-time access from $\hat{u}$ to $\tilde{u}$.

**How to deal with Cases (C) and (D):** In Cases (C) and (D) where node $\hat{u}$ is a gray node, we use the white node $\tilde{u}$, which corresponds to the CDAWG node from which $\hat{u}$ is copied. We output MAW(s) by comparing out-edges of $\tilde{u}$ and $u_1$. We store a pointer from $\hat{u}$ to $\tilde{u}$, so that we can access $\tilde{u}$ from $\hat{u}$ in $O(1)$ time (see also Fig. 4). The rest is the same as Cases (B) and (A).

**Putting All Together.** By performing the above procedure for all edges $(\hat{v}, \hat{u})$ in $\mathsf{LPT}^+(S)$, we can output all elements in $\mathsf{MAW}(S)$.

Let us analyze the time complexity and the data structure space usage. In the case with $i \geq 2$, by Lemma 4, accessing each node $u_i$ can be charged to any one of the MAWs reported with $u_i$. The time required for $u_i$ is thus linear in the number of reported MAWs $au_ib$. In the case with $i = 1$, we compare the sorted lists $\mathsf{child}(\hat{u})$ and $\mathsf{child}(u_1)$. Each successful character comparison with a character $b \in \mathsf{child}(u_1) \setminus \mathsf{child}(\hat{u})$ reports a MAW $au_1b$. On the other hand, each unsuccessful character comparison with a character $b \in \mathsf{child}(u_1) \cap \mathsf{child}(\hat{u})$ can be charged to the corresponding out-edge of $\hat{u}$ whose string label begins with $b$. Thus, we can output all MAWs in $O(\mathsf{er}(S) + |\mathsf{MAW}(S)|)$ time. Further, when $|\mathsf{MAW}(S)| \leq \mathsf{er}(S)$, then we can store the set $\mathsf{MAW}(S)$ with $O(|\mathsf{MAW}(S)|) \subseteq O(\mathsf{er}(S))$ space and output each MAW in $O(1)$ time, using Lemma 3. This leads to our $O(|\mathsf{MAW}(S)|)$-time $O(\mathsf{er}(S))$-space data structure.

To analyze the working memory usage, let us consider the number of pointers. For each given edge $(\hat{v}, \hat{u})$, we climb up the path from $\mathsf{slink}(\hat{u}) = u_1$ to $u_m$ one by one, until reaching $\mathsf{slink}(\hat{v}) = v$, which requires $O(1)$ pointers. We also use $O(1)$ additional pointers for the character comparisons on the sorted lists of each pair of nodes. Thus the total working space is $O(1)$. We have proven Theorem 1.

**Combinatorics on MAWs with CDAWG.** Theorem 1 also leads to the following combinatorial property of MAWs:

**Theorem 2.** *For any string* $S$, $|\mathsf{MAW}(S)| = O(\sigma\mathsf{e}_{\min})$. *This bound is tight.*

*Proof.* For each node $u_i$ of $\mathsf{LPT}^+(S)$, the number of distinct characters $a$ such that $au_ib \in \mathsf{MAW}(S)$ is at most the number of (hard or soft) Weiner links from $u_i$, which is bounded by $\sigma$. The number of distinct characters $b$ such that $au_ib \in \mathsf{MAW}(S)$ does not exceed the number of out-edges from $u_i$. By summing them up for all nodes $u_i$ of $\mathsf{LPT}^+(S)$, we obtain $|\mathsf{MAW}(S)| = O(\sigma \cdot \mathsf{er}(S))$.

Since the same argument holds for the reversed string $S^R$ and $\mathsf{LPT}^+(S^R)$, we have $|\mathsf{MAW}(S)| = O(\sigma \cdot \min\{\mathsf{er}(S), \mathsf{el}(S)\}) = O(\sigma\mathsf{e}_{\min})$.

It is known [20] that for the de Bruijn sequence $B_{k,\sigma}$ of order $k$ over an alphabet of size $\sigma$, $\mathsf{MAW}(B_{k,\sigma}) = \Omega(\sigma n)$, where $n = |B_{k,\sigma}|$. Since $\mathsf{e}_{\min} \le \mathsf{er}(S) < 2n$ for any string $S$ of length $n$, we have $\Omega(\sigma\mathsf{e}_{\min})$ for $B_{k,\sigma}$. □

## 4.2 Computing EBFs

We first prove the following combinatorial property for EBFs:

**Lemma 5.** *For any string* $S$, $|\mathsf{EBF}(S)| \le \mathsf{er}(S) + \mathsf{el}(S) - |V| + 1$.

*Proof.* Recall our algorithm for computing $\mathsf{MAW}(S)$ in Sect. 4.1. Let $u_i$ be a node that has a (soft or hard) Weiner link with label $a \in \Sigma$ and has an out-edge starting with $b \in \Sigma$. Assuming $S[1] = \sharp$ and $S[|S|] = \$$, there are characters $a', b' \in \Sigma$ such that $a' \ne a$, $b' \ne b$, and $a'u_i, u_ib' \in \mathsf{Substr}(S)$. Note also that $au_i, u_ib \in \mathsf{Substr}(S)$. Thus, if $au_ib \in \mathsf{Substr}(S)$, then $au_ib \in \mathsf{EBF}(S)$. If the locus for $au_i$ is the node $\hat{u}$, in which case the Weiner link from $u_i$ $(= u_1)$ to $\hat{u}$ is hard, then the EBF $au_ib$ is charged to the out-edge of $\hat{u}$ that starts with $b$. If the locus for $au_i$ is on the edge $(\hat{v}, \hat{u})$, then the EBF $au_ib$ is charged to the soft-Weiner link from $u_i$ that labeled $a$. Since the number of hard Weiner links is $|V| - 1$, the number of soft-Weiner links is $\mathsf{el}(S) - |V| + 1$. Thus we have $|\mathsf{EBF}(S)| = \mathsf{er}(S) + \mathsf{el}(S) - |V| + 1$ for any string $S$ with $S[1] = \sharp$ and $S[|S|] = \$$.

If $S[1] \ne \sharp$ and/or $S[|S|] \ne \$$, then some node(s) $u_i$ in $\mathsf{CDAWG}(S)$ may only have unique characters $a, b$ such that $au_i, u_ib \in \mathsf{Substr}(S)$. Thus the bound is $|\mathsf{EBF}(S)| \le \mathsf{er}(S) + \mathsf{el}(S) - |V| + 1$ for arbitrary strings $S$. □

By modifying our algorithm for computing MAWs for $S$ as is described in the proof for Lemma 5, we can obtain the following:

**Theorem 3.** *There exists a data structure of* $O(\mathsf{e}_{\min})$ *space that can output all EBFs for string* $S$ *in* $O(|\mathsf{EBF}(S)|)$ *time with* $O(1)$ *working space.*

## 4.3 Computing Length-Bounded MAWs and EBFs

For any set $A \subseteq \Sigma^*$ of strings and an integer $\ell > 1$, let $A^{\ge \ell} = \{w \in A \mid |w| \ge \ell\}$ and $A^{\le \ell} = \{w \in A \mid |w| \le \ell\}$.

**Theorem 4.** *There exists a data structure of* $O(\mathsf{e}_{min})$ *space which, given a query length* $\ell$, *can output*

- $\mathsf{MAW}(S)^{\geq \ell}$ *in* $O(|\mathsf{MAW}(S)^{\geq \ell}| + 1)$ *time,*
- $\mathsf{MAW}(S)^{\leq \ell}$ *in* $O(|\mathsf{MAW}(S)^{\leq \ell}| + 1)$ *time,*
- $\mathsf{EBF}(S)^{\geq \ell}$ *in* $O(|\mathsf{EBF}(S)^{\geq \ell}| + 1)$ *time,*
- $\mathsf{EBF}(S)^{\leq \ell}$ *in* $O(|\mathsf{EBF}(S)^{\leq \ell}| + 1)$ *time,*

*with* $O(1)$ *working space.*

*Proof.* We only describe how to compute length-bounded MAWs. We can compute length-bounded EBFs similarly.

In our data structure for reporting $\mathsf{MAW}(S)^{\leq \ell}$, we precompute the largest integer $m$ with $|\mathsf{MAW}(S)^{\leq m}| \leq \mathsf{e}_{min}$, and store the elements in $\mathsf{MAW}(S)^{\leq m}$ using Lemma 3, sorted by their lengths. Given a query length $\ell$, if $\ell \leq m$, then we can output the elements of $\mathsf{MAW}(S)^{\leq \ell}$ by reporting the stored MAWs in increasing order, in $O(|\mathsf{MAW}(S)^{\leq \ell}| + 1)$ time. Otherwise (if $\ell > m$), then we process each edge $(\hat{v}, \hat{u})$ in $\mathsf{LPT}^{+}(S)$ in increasing order of $|\mathsf{str}(\hat{u})|$. We need to access the internal nodes in the path from $\mathsf{slink}(\hat{v})$ to $\mathsf{slink}(\hat{u})$ topdown, in $O(1)$ time each. This can be done in $O(1)$ time per node by augmenting $\mathsf{LPT}^{+}(S)$ with the level-ancestor (LA) data structure [6,8]. We stop processing the edge $(\hat{v}, \hat{u})$ as soon as we find the shallowest ancestor $u_i$ of $u = \mathsf{slink}(\hat{u})$ such that $|\mathsf{str}(u_i)| > \ell - 2$. Since $|\mathsf{MAW}(S)^{\leq \ell}| > \mathsf{e}_{min}$ for any $\ell < m$, this algorithm outputs $\mathsf{MAW}(S)^{\leq \ell}$ in $O(|\mathsf{MAW}(S)^{\leq \ell}| + 1)$ time.

Computing $\mathsf{MAW}(S)^{\geq \ell}$ is symmetric, except that the LA data structure is not needed: We can check the nodes between $\mathsf{slink}(\hat{u})$ and $\mathsf{slink}(\hat{v})$ in decreasing order of their lengths, in $O(1)$ time each, by simply climbing up the path.    □

Theorem 4 is an improvement and a generalization over the previous work [11] that gave an $O(\ell n)$-space data structure that reports $\mathsf{MAW}(S)^{\leq \ell}$ in $O(\sigma n)$ time.

## 5   Computing MRWs

Let $\mathsf{MRW}_k(S) = \{w \in \mathsf{MRW}(S) \mid \mathsf{occ}_S(w) = k\}$. Then $\mathsf{MRW}_0(S) = \mathsf{MAW}(S)$ and $\bigcup_{k \geq 0} \mathsf{MRW}_k(S) = \mathsf{MRW}(S)$ hold. We show the following lemma. Recall that any string in $\mathsf{MAW}(S)$ or $\mathsf{EBF}(S)$ is of length at least 2.

**Lemma 6.** *For any string* $S$ *with* $S[1] = \sharp$ *and* $S[|S|] = \$$, $\mathsf{MRW}(S) \setminus \mathsf{EBF}(S) = \mathsf{MAW}(S)$. *In other words,* $\mathsf{MRW}(S) \cap \mathsf{EBF}(S) = \bigcup_{k \geq 1} \mathsf{MRW}_k(S)$.

*Proof.* Clearly, $\mathsf{MAW}(S) \cap \mathsf{EBF}(S) = \emptyset$. It suffices to show that $\bigcup_{k \geq 1} \mathsf{MRW}_k(S) \subseteq \mathsf{EBF}(S)$. Let $aub$ be an element in $\mathsf{MRW}_k(S)$ for some $k \geq 1$ where $a, b \in \Sigma$ and $u \in \Sigma^{\star}$. Since $aub$ is a minimal rare word occurring $k$ times in $S$, $\mathsf{occ}_S(au) \geq k+1$ and $\mathsf{occ}_S(ub) \geq k+1$. Also, since $k + 1 \geq 2$, the last character of $au$ is not equal to $\$$, which is a unique character. Similarly, the first character of $ub$ is not equal to $\sharp$. Thus, any occurrence of $au$ (resp., $ub$) in $S$ is not a suffix (resp., prefix) of $S$. Hence, there must be characters $b' \neq b$ and $a' \neq a$ such that $aub', a'ub \in \mathsf{Substr}(S)$. Therefore, $aub$ is an extended bispecial factor.    □

Next, we estimate the number of minimal rare words occurring in $S$.

**Lemma 7.** $|\bigcup_{k \geq 1} \mathsf{MRW}_k(S)| \leq \mathsf{e}_{\min}$.

*Proof.* For any $w \in \bigcup_{k \geq 1} \mathsf{MRW}_k(S)$, the locus of $w$ is the position on some edge $(v_1, v_2)$ exactly one character down from $v_1$ since $\mathrm{occ}_S(w[1..|w| - 1]) > \mathrm{occ}_S(w)$. For the sake of contradiction, we assume that the locus of $w_1 = a_1 u_1 b \in \bigcup_{k \geq 1} \mathsf{MRW}_k(S)$ is the same as that of another $w_2 = a_2 u_2 b \in \bigcup_{k \geq 1} \mathsf{MRW}_k(S)$ where $a_1, a_2, b \in \Sigma$ and $u_1, u_2 \in \Sigma^*$. We further assume $|w_1| \leq |w_2|$ w.l.o.g. Since $a_1 u_1$ and $a_2 u_2$ reach the same CDAWG node, $a_1 u_1$ is a suffix of $a_2 u_2$. Since $w_1 \neq w_2$, $|w_1| < |w_2|$ holds, and thus, $w_1$ is a proper suffix of $w_2$. This contradicts that $\mathrm{occ}_S(w_2) = \mathrm{occ}_S(w_1)$ and $w_2$ is an MRW. Hence, the number of MRWs in $\bigcup_{k \geq 1} \mathsf{MRW}_k(S)$ is never larger than $|\mathsf{E}|$. The above argument holds even if the reversal of $S$ is considered. Therefore, the upper bound is $\min\{|\mathsf{E}|, |\mathsf{E}^R|\} = \mathsf{e}_{\min}$. $\square$

The proof of Lemma 7 implies that if the locus of an MRW $aub$ is on an edge $(v_1, v_2)$, then $au$ is the *shortest* string represented by the node $v_1$. Namely, $\mathsf{slink}(v_1) = u$. From this fact, we can represent each MRW $aub$ as its corresponding edge $(v_1, v_2)$ where the character label of the hard Weiner link from $\mathsf{slink}(v_1)$ to $v_1$ is $a$, the longest string represented by $\mathsf{slink}(v_1)$ is $u$, and the first character of the edge label of $(v_1, v_2)$ is $b$. Also, by Lemma 3, we can restore the MRW $aub$ from $(v_1, v_2)$ in time linear in its length. Thus, the following lemma holds.

**Lemma 8.** *There exists a data structure of $O(\mathsf{e}_{\min})$ space that can output all MRWs occurring in $S$ in $O(|\bigcup_{k \geq 1} \mathsf{MRW}_k(S)|)$ time with $O(1)$ working space.*

If $k$ is fixed in advance, we can store $\mathsf{MRW}_k(S)$ in $O(\mathsf{e}_{\min})$ space and can output them in $O(|\mathsf{MRW}_k(S)|)$ time. We obtain the next corollary by fixing $k = 1$.

**Corollary 1.** *There exists a data structure of $O(\mathsf{e}_{\min})$ space that can output all MUSs for string $S$ in $O(|\mathsf{MUS}(S)|)$ time with $O(1)$ working space.*

# 6  Conclusions and Future Work

In this paper, we proposed space-efficient $O(\mathsf{e}_{\min})$-size data structures that can output MAWs, EBFs, and MRWs for an input string $S$ of length $n$ in time linear in the output size, where $\mathsf{e}_{\min}$ denotes the minimum of the left-extensions and the right-extensions of the maximal repeats in $S$. The key tools of our method are the $O(\mathsf{e}_{\min})$-size grammar-representation of the CDAWG, and the extended longest path tree of size $O(\mathsf{e}_{\min})$. In the full version of this article, we will consider how to build our data structures. We believe that our data structures can be built in $O(n \log \sigma)$ time with $O(\mathsf{e}_{\min})$ working space.

Our future work includes the following: Is it possible to extend our method for reporting length-bounded MAWs/EBFs of Sect. 4.3 to the problem of reporting MAWs/EBFs of length *exactly* $\ell$ for a query length $\ell > 1$? This would be a space-improvement of the previous work [13]. Okabe et al. [21] proposed a (non-compacted) DAWG-based data structure that can report the symmetric difference of the sets of MAWs for multiple strings, using $O(n)$ space and time linear in the output size, where $n$ is the total length of the strings. Can our CDAWG-based data structure be extended to this problem?

# References

1. Almirantis, Y., et al.: On avoided words, absent words, and their application to biological sequence analysis. Algorithms Mol. Biol. **12**(1), 5 (2017)
2. Almirantis, Y., et al.: On overabundant words and their application to biological sequence analysis. Theor. Comput. Sci. **792**, 85–95 (2019)
3. Ayad, L.A.K., Badkobeh, G., Fici, G., Héliou, A., Pissis, S.P.: Constructing anti-dictionaries of long texts in output-sensitive space. Theory Comput. Syst. **65**(5), 777–797 (2021)
4. Belazzougui, D., Cunial, F.: Space-efficient detection of unusual words. In: Iliopoulos, C., Puglisi, S., Yilmaz, E. (eds.) SPIRE 2015. LNCS, vol. 9309, pp. 222–233. Springer, Cham (2015). https://doi.org/10.1007/978-3-319-23826-5_22
5. Belazzougui, D., Cunial, F.: Fast label extraction in the CDAWG. In: Fici, G., Sciortino, M., Venturini, R. (eds.) SPIRE 2017. LNCS, vol. 10508, pp. 161–175. Springer, Cham (2017). https://doi.org/10.1007/978-3-319-67428-5_14
6. Bender, M.A., Farach-Colton, M.: The LCA problem revisited. In: Gonnet, G.H., Viola, A. (eds.) LATIN 2000. LNCS, vol. 1776, pp. 88–94. Springer, Heidelberg (2000). https://doi.org/10.1007/10719839_9
7. Bender, M.A., Farach-Colton, M.: The level ancestor problem simplified. Theor. Comput. Sci. **321**(1), 5–12 (2004)
8. Berkman, O., Vishkin, U.: Finding level-ancestors in trees. J. Comput. Syst. Sci. **48**(2), 214–230 (1994)
9. Blumer, A., Blumer, J., Haussler, D., Ehrenfeucht, A., Chen, M., Seiferas, J.: The smallest automation recognizing the subwords of a text. Theoret. Comput. Sci. **40**, 31–55 (1985)
10. Blumer, A., Blumer, J., Haussler, D., McConnell, R., Ehrenfeucht, A.: Complete inverted files for efficient text retrieval and analysis. J. ACM **34**(3), 578–595 (1987)
11. Chairungsee, S., Crochemore, M.: Using minimal absent words to build phylogeny. Theoret. Comput. Sci. **450**, 109–116 (2012)
12. Charalampopoulos, P., Crochemore, M., Fici, G., Mercas, R., Pissis, S.P.: Alignment-free sequence comparison using absent words. Inf. Comput. **262**, 57–68 (2018)
13. Charalampopoulos, P., Crochemore, M., Pissis, S.P.: On extended special factors of a word. In: Gagie, T., Moffat, A., Navarro, G., Cuadros-Vargas, E. (eds.) SPIRE 2018. LNCS, vol. 11147, pp. 131–138. Springer, Cham (2018). https://doi.org/10.1007/978-3-030-00479-8_11
14. Crochemore, M., Mignosi, F., Restivo, A., Salemi, S.: Data compression using antidictionaries. Proc. IEEE **88**(11), 1756–1768 (2000)
15. Crochemore, M., Navarro, G.: Improved antidictionary based compression. In: Proceedings of 12th International Conference of the Chilean Computer Science Society, pp. 7–13. IEEE (2002)
16. Fujishige, Y., Tsujimaru, Y., Inenaga, S., Bannai, H., Takeda, M.: Linear-time computation of DAWGs, symmetric indexing structures, and MAWs for integer alphabets. Theoret. Comput. Sci. **973**, 114093 (2023). https://doi.org/10.1016/J.TCS.2023.114093
17. Ilie, L., Smyth, W.F.: Minimum unique substrings and maximum repeats. Fund. Inform. **110**(1–4), 183–195 (2011)
18. Inenaga, S.: Linear-size suffix tries and linear-size CDAWGs simplified and improved. CoRR abs/2401.04509 (2024)

19. Koulouras, G., Frith, M.C.: Significant non-existence of sequences in genomes and proteomes. Nucleic Acids Res. **49**(6), 3139–3155 (2021)
20. Mignosi, F., Restivo, A., Sciortino, M.: Words and forbidden factors. Theor. Comput. Sci. **273**(1–2), 99–117 (2002)
21. Okabe, K., Mieno, T., Nakashima, Y., Inenaga, S., Bannai, H.: Linear-time computation of generalized minimal absent words for multiple strings. In: Nardini, F.M., Pisanti, N., Venturini, R. (eds.) SPIRE 2023. LNCS, vol. 14240, pp. 331–344. Springer, Cham (2023). https://doi.org/10.1007/978-3-031-43980-3_27
22. Pinho, A.J., Ferreira, P.J.S.G., Garcia, S.P., Rodrigues, J.M.O.S.: On finding minimal absent words. BMC Bioinform. **10**, 1–11 (2009)
23. Pratas, D., Silva, J.M.: Persistent minimal sequences of SARS-CoV-2. Bioinformatics **36**(21), 5129–5132 (2020)
24. Radoszewski, J., Rytter, W.: On the structure of compacted subword graphs of Thue-Morse words and their applications. J. Discrete Algorithms **11**, 15–24 (2012)
25. Rytter, W.: The structure of subword graphs and suffix trees of Fibonacci words. Theor. Comput. Sci. **363**(2), 211–223 (2006)
26. Weiner, P.: Linear pattern matching algorithms. In: Proceedings of the 14th Annual Symposium on Switching and Automata Theory, pp. 1–11. IEEE (1973)

# Lower Bounds for Leaf Rank of Leaf Powers

Svein Høgemo[✉]

University of Bergen, Bergen, Norway
svein.hogemo@uib.no

**Abstract.** Leaf powers and $k$-leaf powers have been studied for over 20 years, but there are still several aspects of this graph class that are poorly understood. One such aspect is the *leaf rank* of leaf powers, i.e. the smallest number $k$ such that a graph $G$ is a $k$-leaf power. Computing the leaf rank of leaf powers has proved a hard task, and furthermore, results about the asymptotic growth of the leaf rank as a function of the number of vertices in the graph have been few and far between. We present an infinite family of rooted directed path graphs that are leaf powers, and prove that they have leaf rank exponential in the number of vertices (utilizing a type of subtree model first presented by Rautenbach [Some remarks about leaf roots. Discrete mathematics, 2006]). This answers an open question by Brandstädt et al. [Rooted directed path graphs are leaf powers. Discrete mathematics, 2010].

## 1 Introduction

A graph $G$ is a $k$-*leaf power* if there is a tree $T$ such that $G$ is isomorphic to the subgraph of $T^k$ induced by its leaves. $T$ is referred to as the $k$-*leaf root* of $G$. The original motivation for studying leaf powers comes from computational biology, particularly the problem of reconstructing phylogenetic trees – if $T$ is interpreted as the "tree of life", then $G$ constitutes a simplified model of relationships between known species, where those species within distance $k$ in $T$ are deemed "closely related" and become neighbors in $G$, and those that have larger distance are deemed "not closely related". Checking if a graph is a $k$-leaf power for some $k$ is thereby analogous to the task of fitting an evolutionary tree to these simplified relationships. $k$-leaf powers were first introduced by Nishimura, Ragde and Thilikos in 2000 [23], although the connection between powers of trees and the task of (re)constructing phylogenetic trees was explored by several authors at the time [19]. Since then, the class of leaf powers – all graphs that are $k$-leaf powers for some $k$ – has become a well-studied graph class in its own right. A survey on leaf powers published in a recent anthology on algorithmic graph theory [25] gives a more or less up-to-date introduction to the most important results on this graph class.

---

The author is a student at the time of submission.

© The Author(s), under exclusive license to Springer Nature Switzerland AG 2024
A. A. Rescigno and U. Vaccaro (Eds.): IWOCA 2024, LNCS 14764, pp. 341–353, 2024.
https://doi.org/10.1007/978-3-031-63021-7_26

Two closely related questions concern the characterization of $k$-leaf powers for some constant $k$; and the problem of computing the leaf rank (the smallest integer $k$ such that $G$ is a $k$-leaf power) of a leaf power $G$. The first problem has been addressed by several authors, most notably by Lafond [17], who announced an algorithm for recognizing $k$-leaf powers that runs in polynomial time for each fixed $k$ (though, admittedly, with a runtime that depends highly on $k$). Complete characterizations in terms of forbidden subgraphs are, on the other hand, known only for 2- and 3-leaf powers [9] and partially for 4-leaf powers [24] (see also [5]). The second problem seems even harder. A few graph classes have bounded leaf rank, for example block graphs or squares of trees. The only subclass of leaf powers with unbounded leaf rank, for which leaf rank is shown to be easy to compute, is the chordal cographs (also known as the trivially perfect graphs); this was shown very recently by Le and Rosenke [18].

Though deciding the exact leaf rank of a leaf power seems hard, the asymptotic growth of the leaf rank as a function of the number of vertices has shown to be at most linear for most subclasses of leaf powers [4] (also implicit in [2], see further below). This could lead one to conjecture at most linear – or at least polynomial – growth on the leaf rank of any leaf power. In this paper, we show that this is not the case. In particular, we show that there exists an infinite family of leaf powers $\{R_m \mid m \geq 3\}$ that have leaf rank proportional to $2^{\frac{n}{4}}$, where $n$ is the number of vertices.

The broader problem of recognizing leaf powers has been addressed more recently: Leaf powers, being induced subgraphs of powers of trees, are strongly chordal, as first noted in [6]. Nevries and Rosenke [22] find a forbidden structure in the clique arrangements of leaf powers, and find the seven forbidden strongly chordal graphs exhibiting this structure. Lafond [16] furthermore finds an infinite family of strongly chordal graphs that are not leaf powers, and shows that deciding if a chordal graph contains one of these graphs as an induced subgraph is NP-complete. Jaffke et al. [15] point out that leaf powers have mim-width 1, a trait shared with several other classes of intersection graphs. Mengel's [20] observation that strongly chordal graphs can have unbounded (linear in the number of vertices) mim-width suggests that the gap between leaf powers and strongly chordal graphs is quite big [14].

It has also been observed [7] that leaf powers are exactly the induced subgraphs of powers of trees (so-called *Steiner powers* [19]). This, and the observation that $k$-leaf powers without true twins are induced subgraphs of $(k - 2)$-powers of trees, forms the basis for the algorithms to recognize 5- and 6-leaf powers [8,10], which until Lafond's breakthrough result [17] were the state of the art in recognizing $k$-leaf powers. One peculiar interpretation of our result is therefore that there exist induced subgraphs of powers of trees whose smallest tree powers that contain them are exponentially bigger than themselves.

In another direction, Bergougnoux et al. [2] look at subclasses of leaf powers admitting leaf roots with simple structure: In particular, they show that the leaf powers admitting leaf roots that are subdivided caterpillars are exactly the *co-threshold tolerance* graphs, a graph class lying between interval graphs and tolerance graphs ([21], see Fig. 1). The leaf roots constructed in [2] had rational weights; however, it is not hard to see that they can be modified into $k$-leaf roots for some $k \leq 2n$. Interestingly, this shows that there is a big difference between caterpillar-shaped leaf roots and caterpillar-shaped RS models (defined in Sect. 2): As we will see, the graphs considered in this paper have RS models that are caterpillars, but exponential leaf rank.

The rest of the paper is organized as follows: In Sect. 2 we develop basic terminology regarding leaf powers, chordal graphs and subtree models. In Sect. 3 we show how each graph $R_n$ is built and show that these graphs are leaf powers, in particular rooted directed path graphs. In Sect. 4 we prove the main result, that $R_n$ has exponential leaf rank for every $n$. In the end, we provide a brief discussion on possible upper bounds on the leaf rank of leaf powers.

## 2   Basic Notions

We use standard graph theory notation. All trees are unrooted unless stated otherwise.

In this paper, we will assume that all trees we work with have at least three leaves, and therefore there exists at least one node of degree at least 3. For two nodes $x, y$ in a tree $T$, $p_T(x, y)$ denotes the unique path between $x$ and $y$ in $T$.

Some more specialized notions follow here:

**Definition 1 (Caterpillar).** *A caterpillar is a tree in which every internal node lies on a single path. This path is called the* spine *of the caterpillar.*

**Definition 2 (Connector).** *For a leaf $v$ in a tree $T$, there is one unique node with degree at least 3 that has minimum distance to $v$. We call this node the* connector *of $v$, or* conn$(v)$.

**Definition 3 ($k$-leaf power, $k$-leaf root, leaf rank).** *For some positive integer $k$, a graph $G$ is a $k$-leaf power if there exists a tree $T$ and a bijection $\tau$ from $V(G)$ to $L(T)$, the set of leaves of $T$, such that any two vertices $u, v$ are neighbors in $G$ if and only if $\tau(u)$ and $\tau(v)$ have distance at most $k$ in $T$. $T$ is called a $k$-leaf root of $G$. The leaf rank of $G$, $lrank(G)$, is the smallest value $k$ such that $G$ is a $k$-leaf power, or $\infty$ if $G$ is not a leaf power.*

**Definition 4 (Leaf power).** *A graph $G$ is a leaf power if there exists a positive integer $k$ for which $G$ is a $k$-leaf power.*

**Definition 5 (Leaf span).** *Given a graph class $\mathcal{F}$, the leaf span of $\mathcal{F}$, $ls_{\mathcal{F}}$, is a function on the positive integers that, for each $n$, outputs the smallest $k$ such that every graph in $\mathcal{F}$ on $n$ vertices has a $k$-leaf root. Clearly, this definition only makes sense if $\mathcal{F}$ is the class of leaf powers, or a subclass thereof. Alternatively, one can define $ls_{\mathcal{F}}(n) = \infty$ if $\mathcal{F}$ contains a graph on $n$ vertices which is not a leaf power.*

In our case, we will only look at leaf powers, so in our case, the leaf span is well defined regardless.

Leaf powers are known to be *chordal graphs* [24], graphs with no induced cycles of four or more vertices. A famous theorem by Gavril [11] says that the chordal graphs are the intersection graphs of subtrees of a tree, i.e. the graphs admitting a *subtree model*:

**Definition 6 (Subtree model).** *Given a graph $G$, a subtree model of $G$ is a pair $(T, \mathcal{S})$, where $T$ is a tree and $\mathcal{S} = \{S_v \mid v \in V(G)\}$ is a collection of connected subtrees of $T$ with the property that for any two vertices $u, v \in V(G)$, the subtrees $S_u$ and $S_v$ have non-empty intersection if and only if $uv \in E(G)$.*

**Definition 7 (Cover).** *Let $G$ be a chordal graph and $(T, \mathcal{S})$ a subtree model of $G$. For a node $x \in V(T)$, the cover of $x$, $V_{G,T}(x)$ (subscripts may be omitted), is defined as the set of vertices in $G$ whose subtrees in $T$ include $x$: $V_{G,T}(x) = \{v \in V(G) \mid x \in S_v\}$.*

The cover of any node must be a clique in $G$.

It is a well-known fact that subtrees of a tree have the Helly property (see e.g. [13]). Therefore, given a chordal graph $G$ and a subtree model $(T, \mathcal{S})$, we define the following subtrees:

**Definition 8 (Clique subtree).** *Given $G$ and $(T, \mathcal{S})$ as above, for every maximal clique $C \subseteq V(G)$, the clique subtree $S_T(C) := \bigcap_{v \in C} S_v$ is non-empty (we can omit the subscript $T$ if the tree is obvious from context).*

It is therefore clear that every maximal clique of $G$ is the cover of some node in $T$. Furthermore, for any two maximal cliques $C \neq C'$, $S_T(C) \cap S_T(C') = \emptyset$.

We give an alternative characterization of leaf powers here, that we will use to prove our result:

**Definition 9 (Radial Subtree model).** *Given a graph $G$, a radial subtree model (henceforth called RS model) of $G$ is a subtree model $(T, \mathcal{S})$, where for each $v \in V(G)$ there exists a node $c_v \in V(T)$ (the center) and integer $r_v \geq 0$ (the radius) such that $S_v = T[\{u \in V(T) \mid dist(u, c_v) \leq r_v\}]$. In other words, $S_v$ is spanned by exactly the nodes in $T$ having distance at most $r_v$ from $c_v$. Each $S_v$ is called a radial subtree.*

RS models are a special case of the much more general *NeST* (Neighborhood Subtree Tolerance) models, introduced by Bibelnieks and Dearing in [3]. NeST models are more complicated, involving trees embedded in the plane with rational distances, as well as tolerances on each vertex. We will therefore not define them here. In any case, if one removes the tolerances, the graphs admitting the resulting "NeS models" [2] are, again, exactly the leaf powers [4]. NeST graphs thus generalize leaf powers in much the same way that tolerance graphs generalize interval graphs (see Fig. 1).

The following lemma is implicit in Rautenbach ([24], Lemma 1) as a proof that leaf powers are chordal – though RS models were not explicitly defined in that paper. We will repeat the proof here, since its contrapositive (stated below as Corollary 1) is crucial for our proof that $R_n$ has high leaf rank.

**Lemma 1.** *If a graph $G$ admits a $k$-leaf root, then it admits a RS model where* $\max_{v \in V(G)} r_v \leq k$.

*Proof.* Given a $k$-leaf root $(T, \tau)$ of $G$, we make a RS model of $G$ by:

- Subdividing every edge in $T$ once.
- Setting $c_v := \tau(v)$ and $r_v := k$ for every $v \in V(G)$.

Two subtrees $S_v, S_u$ intersect iff $dist(v, u) \leq 2k$; in other words, iff $u$ and $v$ had distance at most $k$ before subdivision of the edges.  □

**Corollary 1.** *Let $G$ be a leaf power. If there is some integer $r$ such that every RS model of $G$ contains a subtree with radius at least $r$, then $G$ is not a $k$-leaf power for any $k < r$.*

The radial subtrees constructed in this proof are all centered on leaves and have the same radius. However, the definition of RS models is more general, so we must prove that the implication holds in the other direction as well:

**Lemma 2.** *Let $G$ be a graph. If $G$ admits an RS model $(T, \mathcal{S})$, then $G$ is a leaf power.*

*Proof.* Let $k$ be the maximum radius among the subtrees in $\mathcal{S}$. For each $v \in V(G)$, we add a new leaf to $T$ that is fastened to $c_v$ with a path of length $k+1-r_v$, and let $\tau(v)$ point to this new leaf. Afterwards, as long as $T$ contains a leaf $x$ that is not one of these new leaves, delete the path from $x$ to $conn(x)$. Now, it is evident that two subtrees $S_u, S_v$ overlap if and only if $dist(\tau(u), \tau(v)) \leq 2k + 2$. In other words, $(T, \tau)$ is a $(2k + 2)$-leaf root of $G$.  □

## 3   Construction of $R_n$

The graphs with high leaf rank that we construct are *rooted directed path graphs*:

**Definition 10 (Rooted directed path graph).** *A graph $G$ is a rooted directed path graph (RDP graph) if it admits an intersection model consisting of paths in an arborescence (a DAG in the form of a rooted tree where every edge points away from the root).*

**Theorem 1 ([4], Theorem 5).** *RDP graphs are leaf powers.*

The leaf roots shown to exist by Brandstädt et al. in [4] had, in the worst case, $k$ exponential in $n$. They left it as an open question whether the leaf span of RDP graphs actually is significantly smaller.

Here we show that it is not: specifically, for every $n \geq 3$, there is a RDP graph $R_n$ with $4n$ vertices that has leaf rank proportional to $2^n$. In other words, if $RDP$ is the class of RDP graphs and $LP$ the class of leaf powers, then $ls_{LP} \geq$

346     S. Høgemo

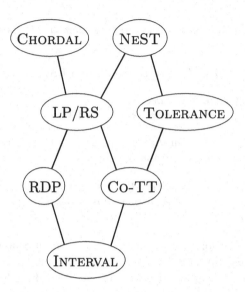

**Fig. 1.** A Hasse diagram of inclusions between leaf powers and some related graph classes. (Abbreviations: LP = Leaf Powers; RDP = Rooted Directed Path graphs; Co-TT = Co-Threshold Tolerance graphs; RS = Graphs with RS models; NeST = Neighborhood Subtree Tolerance graphs.) All inclusions are strict and all non-inclusions are between incomparable graph classes. For more information, see [2–4,12,21].

$ls_{RDP} = 2^{\Omega(n)}$. This is the first result that shows that the leaf span of leaf powers is non-polynomial.

For some $n \geq 3$, the graph $R_n$ has $4n$ vertices: $V(R_n) = \bigcup_{i=1}^{n}\{a_i, b_i, c_i, d_i\}$. We define $E(R_n)$ through its maximal cliques: The family of maximal cliques is

$$\mathcal{C}(R_n) = \{C_i \mid 1 \leq i \leq n\} \cup \{C_i' \mid 1 \leq i \leq n-1\}$$

where

$$C_i = \{a_i, b_i, c_i, d_i\}$$

and

$$C_i' = \{a_j \mid i \leq j \leq n\} \cup \{b_i, b_{i+1}, c_i\}$$

*Remark 1.* For any $n$, $R_n$ is a rooted directed path graph.

*Proof.* Let our arborescence be a rooted caterpillar $T$ with spine $x_1, x_2, \ldots, x_n$ and one leaf $y_i$ fastened to each $x_i$. The root is $x_1$. Each $a_i$ corresponds to the path from $x_1$ to $y_i$; each $b_i$ corresponds to the path from $x_{i-1}$ to $y_i$ (except $b_1$, whose path starts at $x_1$); each $c_i$ corresponds to the path from $x_i$ to $y_i$; and each $d_i$ corresponds to the path consisting only of $y_i$. We can easily check that, for each $1 \leq i \leq n$, $V(y_i) = C_i$; and for each $1 \leq i \leq n-1$, $V(x_i) = C_i'$; and there

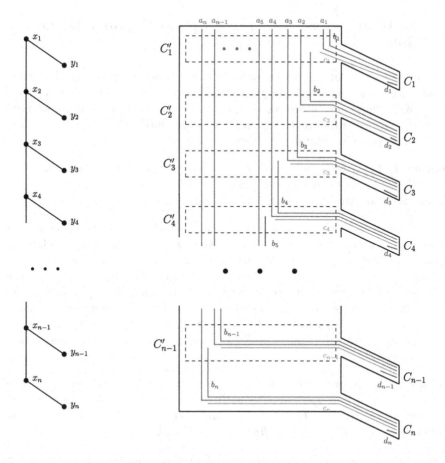

**Fig. 2.** A rooted directed path model of $R_n$. The arborescence $T$ is a caterpillar; on the right, the tree has been fattened into a box diagram so we can see all the paths.

can be no other maximal cliques since every maximal clique must be the cover of some node in $T$. (see also Fig. 2 for a visual representation of the paths.)    □

The construction of $R_n$ shows the ingredients we need in order to prove that some graphs have exponential leaf rank. The $b_i$'s form an induced path in $R_n$, forcing a linear topology on any subtree model of $R_n$. In other words, any subtree model, and specifically any RS model of $R_n$ must have the overall shape of a caterpillar, where the spine contains $C_1', \ldots, C_{n-1}'$ and the hairs contain $C_1, \ldots, C_n$. The $a_i$'s have a large neighborhood, which in turn give their subtrees in any RS model a large diameter. The $d_i$'s give each $a_i$ a private neighbor, while the $c_i$'s and $b_i$'s together force every hair to branch off the spine at different points.

# 4  The Graph Class $\{R_n \mid n \geq 3\}$ has Exponential Leaf Rank

In order to prove that the aforementioned construction leads to high leaf rank, we must formalize the intuitions given earlier, and explicitly show that every RS model of $R_n$ must have a subtree with big radius. To be able to do so, we need quite a bit of new infrastructure regarding subtree models.

The first piece is a simple, but very useful lemma:

**Lemma 3.** *Let $G$ be a chordal graph and $(T, \mathcal{S})$ a subtree model of $G$. Let $P$ be any path in $G$ with endpoints $u, v$, and let $x_u$ and $x_v$ be two arbitrary nodes in $S_u$ and $S_v$, respectively. For any node $x \in p_T(x_u, x_v)$, $V(x) \cap P \neq \emptyset$.*

*Proof.* Assume towards a contradiction that there is a node $x_0 \in p_T(x_u, x_v)$ whose cover does not intersect with $P$. Now, $x_u, x_v \neq x_0$, and therefore $x_0$ separates $x_u$ and $x_v$. We enumerate the vertices in $P$ $(p_1, p_2, \ldots, p_k)$ where $p_1 = u$ and $p_k = v$. Since every subtree in a subtree model must be connected, every $S_{p_i}$ must be contained in one component of $T \setminus x_0$. Also, since $p_i$ and $p_{i+1}$ are neighbors, their respective subtrees intersect and must therefore be contained in the same component. But this leads to a contradiction, since $S_u$ and $S_v$ are located in different components of $T \setminus x_0$ (namely, the ones containing $x_u$ and $x_v$ respectively). □

**Definition 11 (Connecting Path).** *Given two disjoint subtrees of a tree $S, S' \subseteq T$, we define a connecting path from $S$ to $S'$, denoted $cp(S, S')$, as the minimal subtree $P$ of $T$ (i.e. a path) such that $S \cup S' \cup P$ is connected. Note that $cp(S, S')$ contains one node from each of $S$ and $S'$.*

**Lemma 4.** *Let $G$, $(T, \mathcal{S})$ and recall the definition of clique subtrees. Given two maximal cliques $C, C' \in \mathcal{C}(G)$ and any node $x$ on $cp(S_T(C), S_T(C')) \setminus (S_T(C) \cup S_T(C'))$, then $V(x)$ is a separator in $G$.*

*Proof.* Since $C$ and $C'$ are maximal cliques, there exist two non-adjacent vertices $v \in C \setminus C'$ and $v' \in C' \setminus C$. By Lemma 3, every path from $v$ to $v'$ must contain a vertex in $V(x)$. Therefore, $V(x)$ separates $v$ and $v'$ in $G$. □

The above lemma is useful for us because of its contrapositive. Specifically, one can verify that in $R_n$, none of the cliques $C_1, \ldots, C_n$ are separators. This means that in any subtree model $(T, \mathcal{S})$ of $R_n$, the subtree $S_T(C_i)$ does not intersect the connecting path between any two other cliques. In other words, for each $1 \leq i \leq n$, the clique subtree $S_T(C_i)$ is situated at a leaf of $T$.

**Definition 12 (Median).** *Given a tree $T$ and three nodes $u, v, w \in V(T)$, the median of the nodes $\mathsf{med}(u, v, w)$ is the unique node $m$ that lies on all three paths $p_T(u, v)$, $p_T(u, w)$ and $p_T(v, w)$. The median is equal to one of the nodes (say, $v$) iff $v$ is on $p_T(u, w)$; otherwise, it separates $u, v, w$ in $T$ (and consequently, has degree at least 3) (Fig. 3).*

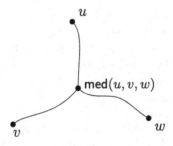

**Fig. 3.** The median of three nodes in a tree.

Now we get to the meat of the proof. We will need the following definitions:
Given $R_n$ and any RS model $(T, \mathcal{S})$, we note the following *branch points* in
$T$: Let $m_1$ and $m_n$ be the endpoints of $cp(S(C_1), S(C_n))$, and for every $1 <$
$i < n$, let $m_i$ be the common node between $cp(S(C_1), S(C_i))$, $cp(S(C_i), S(C_n))$
and $cp(S(C_1), S(C_n))$. Also, $m_i$ is the median $\mathsf{med}(m_1, m_n, s_i)$ where $s_i$ is the
endpoint of $cp(S(C_1), S(C_i))$ (or $cp(S(C_n), S(C_i))$) in $S(C_i)$. From Lemma 4
we know that none of $m_1$, $m_n$ and $s_i$ lie on the path between the two others;
therefore $m_i$ separates $S(C_1)$, $S(C_i)$ and $S(C_n)$. By definition, every $m_i$ is on
$p_T(m_1, m_n)$ (Fig. 4).

Take note of the nodes $m_1, \ldots, m_n$ and $s_2, \ldots, s_{n-1}$; these will all be used
later on. It is worth to note that since $s_i \in S(C_i)$, the cover of $s_i$ is equal to $C_i$.

We now prove a series of lemmas, concluding with Theorem 2, showing that
the leaf rank of $R_n$ is exponential in $n$. We will assume that $(T, \mathcal{S})$ is an RS
model of $R_n$ containing the branching points mentioned above.

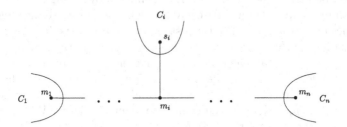

**Fig. 4.** The branching point $m_i$ in a subtree model of $R_n$, and the three cliques it
separates.

**Lemma 5.** *For every* $1 < i < n$, $V(m_i)$ *is equal to the union of* $\{a_j \mid i \leq j \leq n\}$,
$b_i$ *and at least one of* $c_i$ *and* $b_{i+1}$.

*Proof.* From the definition, $m_i$ separates the three subtrees $S(C_1)$, $S(C_i)$ and
$S(C_n)$, represented by the three nodes $m_1$, $s_i$ and $m_n$ respectively. This means
that for each of the three cliques, at least one of their vertices are not in $V(m_i)$.

We start by showing that $a_j, b_j \notin V(m_i)$ for any $j < i$: Consider the path $P = (c_i, b_{i+1}, b_{i+2}, \ldots, b_n)$ in $R_n$. Since $c_i \in C_i$ and $b_n \in C_n$, and $m_i$ is on the path in $T$ between those two cliques, by Lemma 3, $V(m_i)$ contains one of the vertices in $P$. But none of these are adjacent to $a_j$ or $b_j$, therefore none of these can be in $V(m_i)$. Furthermore, as $\bigcup_{j=1}^{i-1}\{a_j, b_j\}$ induce a connected subgraph of $R_n$, all their respective subtrees must lie in the same component of $T \setminus m_i$; namely, the one containing $m_1$.

Next, we show that $a_i \in V(m_i)$. This is done by applying Lemma 3 to the path $(a_1, a_i)$ in $R_n$ and noting that $m_i$ is on the path $p_T(m_1, s_i)$ in $T$. Since we have established that $a_1 \notin V(m_i)$, $a_i$ must be in $V(m_i)$.

Now we show $b_j, c_j, d_j \notin V(m_i)$ for any $i + 2 \leq j \leq n$, but at least one of $c_i$ and $b_{i+1}$ is. Consider the path $P$ from before. We know at least one vertex in $P$ is in $V(m_i)$, but since $c_i$ and $b_{i+1}$ are the only ones adjacent to $a_i$, they are the only ones that can be in $V(m_i)$. Furthermore, since $\bigcup_{j=i+1}^{n}\{b_j, c_j, d_j\}$ induce a connected subgraph of $R_n$, all of their respective subtrees must be located in the same component of $T \setminus m_i$; namely, the one containing $m_n$.

Next, we show that $a_j \in V(m_i)$ for every $i < j \leq n$. We have established that the subtrees $S_{a_1}$ and $S_{d_j}$ do not contain $m_i$, and furthermore, they are located in different components of $T \setminus m_i$. Taking the node $s_j \in S(C_j)$, we see that $m_i \in p_T(m_1, s_j)$. Therefore, we can apply Lemma 3 to the path $(a_1, a_j, d_j)$ and conclude that $a_j \in V(m_i)$.

Finally, we show $d_i \notin V(m_i)$. This is deduced by noting that $d_i$ and $a_n$ are not adjacent, and $a_n \in V(m_i)$. □

**Lemma 6.** *None of the nodes $m_1, m_2, \ldots, m_n$ are equal. Furthermore, the path $p_T(m_1, m_n)$ visits all of these nodes in that order.*

*Proof.* The first claim follows straight from Lemma 5 by noting that the cover of each branching point is unique. Also, by the definition of the branching points, every $m_i$ lies on $p_T(m_1, m_n)$. For the last claim, we prove the following, equivalent formulation: For any $1 \leq r < s < t \leq n$, $m_s$ lies on $p_T(m_r, m_t)$.

From the previous statement, it is clear that these three nodes lie on a single path, and therefore one of them lies in the middle. However, we see that $V(m_r)$ is not a separator (in $G$) of $V(m_s) \setminus V(m_r)$ and $V(m_t) \setminus V(m_r)$. By Lemma 4, $m_r$ cannot lie on $p_T(m_s, m_t)$. The same argument applies to $V(m_t)$; thus the only remaining choice is that $V(m_s)$ lies on $p_T(m_r, m_t)$. □

**Lemma 7.** *For any $2 < i < n$, $dist(m_i, m_{i+1}) > dist(m_2, m_i)$.*

*Proof.* Recall that $(T, \mathcal{S})$ is an RS model; therefore, for any $v \in V(R_n)$, $S_v$ is characterized by a center $c_v$ and radius $r_v$.

Look then at the vertex $a_i$ for some $2 < i < n$. From Lemma 5, we know that $S_{a_i}$ contains $m_2$ and (by definition) $s_i$, but not $m_{i+1}$. Also, by Lemma 6, $m_i$ separates those three nodes. Given the node $c_{a_i}$, we therefore know that $dist(c_{a_i}, m_{i+1}) > \max(dist(c_{a_i}, m_2), dist(c_{a_i}, s_i))$.

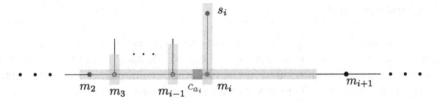

**Fig. 5.** The purple shadow represents the radial subtree $S_{a_i}$ inside $T$, with center $c_{a_i}$ (the solid purple block). It reaches $m_2$ and $s_i$, but not $m_{i+1}$. (Color figure online)

Now, since $(T, \mathcal{S})$ is an arbitrary RS model, we do not know where in $T$ the node $c_{a_i}$ is situated, but we will employ two cases, based on which component of $T \setminus m_i$ we find $c_{a_i}$ in.

*Case 1:* $c_{a_i}$ is not in the component of $T \setminus m_i$ containing $m_2$.

This includes the case $c_{a_i} = m_i$. In this case, we see that

$$dist(m_i, m_{i+1}) \geq (dist(c_{a_i}, m_{i+1}) - dist(c_{a_i}, m_i)) >$$
$$(dist(c_{a_i}, m_2) - dist(c_{a_i}, m_i)) = dist(m_i, m_2)$$

The first inequality is a strict inequality iff $c_{a_i}$ is in the component of $T \setminus m_i$ containing $m_{i+1}$; otherwise it is an equality.

*Case 2:* $c_{a_i}$ is in the component of $T \setminus m_i$ containing $m_2$.

Now we see that

$$dist(m_i, m_{i+1}) = (dist(c_{a_i}, m_{i+1}) - dist(c_{a_i}, m_i)) >$$
$$(dist(c_{a_i}, s_i) - dist(c_{a_i}, m_i)) = dist(m_i, s_i)$$

(This corresponds to the scenario in Fig. 5.)

To complete the proof, we look at the center of another vertex, namely $c_{a_{i+1}}$. From Lemma 5, we know that $S_{a_{i+1}}$ contains $m_2$ and $m_{i+1}$, but not $s_i$. Since $dist(m_i, m_{i+1}) > dist(m_i, s_i)$, $c_{a_{i+1}}$ must be placed in the component of $T \setminus m_i$ containing $m_{i+1}$. But now

$$dist(m_i, s_i) = (dist(c_{a_{i+1}}, s_i) - dist(c_{a_{i+1}}, m_i)) >$$
$$(dist(c_{a_{i+1}}, m_2) - dist(c_{a_{i+1}}, m_i)) = dist(m_i, m_2)$$

Now we have $dist(m_i, m_{i+1}) > dist(m_i, s_i) > dist(m_i, m_2)$ and the proof is complete.    □

**Theorem 2.** *The leaf rank of $R_n$ is at least $2^{n-2}$.*

*Proof.* By Lemma 5, the subtree $S_{a_n}$ contains both $m_2$ and $m_n$, and therefore has diameter at least $dist(m_2, m_n)$. From Lemma 6 we see that $dist(m_2, m_n) = dist(m_2, m_3) + dist(m_3, m_4) + \ldots + dist(m_{n-1}, m_n)$. From Lemma 7 and the fact that $dist(m_2, m_3) \geq 1$, we see that $dist(m_2, m_n) \geq 2^{n-1} - 1$. This implies that $r_{a_n} \geq 2^{n-2}$, which by Corollary 1 and the fact that $(T, \mathcal{S})$ is an arbitrary RS model, implies that $R_n$ has leaf rank at least $2^{n-2}$.    □

**Corollary 2.** *Let $\mathcal{R} = \{R_m \mid m \geq 3\}$. Then, $ls_{RDP} \geq ls_{\mathcal{R}} = \Omega(2^{\frac{n}{4}})$.*

# 5  Conclusion

We have shown that the leaf rank of leaf powers is not upper bounded by a polynomial function in the number of vertices. While such an upper bound has never been explicitly conjectured in the literature, we nevertheless believe that this result is surprising. The only previously established lower bounds for leaf rank are linear in the number of vertices [4], and, as previously noted, most graph classes that have been shown to be leaf powers have linear upper bounds on their leaf rank as well. Though the $k$-leaf roots of RDP graphs found by Brandstädt et al. in [4] had $k$ exponential in the number of vertices, the authors left it as an open question to "determine better upper bounds on their leaf rank".

Single exponential upper bounds on leaf rank of leaf powers generally have not been found, and we leave it as an open question whether the leaf span of leaf powers is $2^{\Theta(n)}$. However, we will finish with the following nice observation noted by B. Bergougnoux [1], that shows that recognizing leaf powers is in NP. This implies a not much worse upper bound on the leaf span of leaf powers:

Given a graph $G$, a positive certificate for $G$ being a leaf power consists of a candidate leaf root $(T, \tau)$, where every internal node of $T$ has degree at least 3; and a linear program that (say) maximizes the sum of weights on each edge in $T$, while fulfilling constraints that every pair of adjacent vertices in $G$ has distance at most 1 in $T$, and every pair of non-adjacent vertices in $G$ has distance higher than 1 in $T$. If the linear program is feasible, then $(T, \tau)$ is a weighted leaf root of $G$.

The above linear program can be solved in polynomial time, outputting a feasible solution (if one exists) with rational weights with a polynomial number of bits. Therefore, if $G$ admits a leaf root, it admits a $k$-leaf root where $k \leq 2^{n^c}$ for some (fairly small) constant $c$. This observation also implies that if recognizing $k$-leaf powers is strongly in P for arbitrary $k$, then computing leaf rank is also in P, since given a polynomial-time algorithm for recognizing $k$-leaf powers, one could compute leaf rank by way of binary search on the value of $k$. Recognizing leaf powers would also be in P. Conversely, if either of these questions turn out to be NP-hard, then computing leaf rank must be at least weakly NP-hard.

# References

1. Bergougnoux, B.: Personal communication (2023)
2. Bergougnoux, B., Høgemo, S., Telle, J.A., Vatshelle, M.: Recognition of linear and star variants of leaf powers is in p. In: Bekos, M.A., Kaufmann, M. (eds.) WG 2022. LNCS, vol. 13453, pp. 70–83. Springer, Cham (2022). https://doi.org/10.1007/978-3-031-15914-5_6
3. Bibelnieks, E., Dearing, P.M.: Neighborhood subtree tolerance graphs. Discret. Appl. Math. **43**(1), 13–26 (1993). https://doi.org/10.1016/0166-218X(93)90165-K
4. Brandstädt, A., Hundt, C., Mancini, F., Wagner, P.: Rooted directed path graphs are leaf powers. Discret. Math. **310**(4), 897–910 (2010)
5. Brandstädt, A., Le, V.B., Sritharan, R.: Structure and linear-time recognition of 4-leaf powers. ACM Trans. Algorithms (TALG) **5**(1), 1–22 (2008)

6. Brandstädt, A., Le, V.B.: Structure and linear time recognition of 3-leaf powers. Inf. Process. Lett. **98**(4), 133–138 (2006). https://doi.org/10.1016/j.ipl.2006.01.004

7. Brandstädt, A., Rautenbach, D.: Exact leaf powers. Theoret. Comput. Sci. **411**, 2968–2977 (2010). https://doi.org/10.1016/j.tcs.2010.04.027

8. Chang, M.-S., Ko, M.-T.: The 3-steiner root problem. In: Brandstädt, A., Kratsch, D., Müller, H. (eds.) WG 2007. LNCS, vol. 4769, pp. 109–120. Springer, Heidelberg (2007). https://doi.org/10.1007/978-3-540-74839-7_11

9. Dom, M., Guo, J., Huffner, F., Niedermeier, R.: Error compensation in leaf power problems. Algorithmica **44**, 363–381 (2006)

10. Ducoffe, G.: The 4-steiner root problem. In: Sau, I., Thilikos, D.M. (eds.) WG 2019. LNCS, vol. 11789, pp. 14–26. Springer, Cham (2019). https://doi.org/10.1007/978-3-030-30786-8_2

11. Gavril, F.: The intersection graphs of subtrees in trees are exactly the chordal graphs. J. Comb. Theory Ser. B **16**(1), 47–56 (1974)

12. Golumbic, M.C., Monma, C.L., Trotter, W.T., Jr.: Tolerance graphs. Discret. Appl. Math. **9**(2), 157–170 (1984)

13. Golumbic, M.: Algorithmic Graph Theory and Perfect Graphs, 3rd edn. Elsevier Science (2004)

14. Jaffke, L.: Bounded width graph classes in parameterized algorithms. Ph.D. thesis, University of Bergen (2020)

15. Jaffke, L., Kwon, O.J., Strømme, T.J., Telle, J.A.: Mim-width III. Graph powers and generalized distance domination problems. Theor. Comput. Sci. **796**, 216–236 (2019)

16. Lafond, M.: On strongly chordal graphs that are not leaf powers. In: Bodlaender, H.L., Woeginger, G.J. (eds.) WG 2017. LNCS, vol. 10520, pp. 386–398. Springer, Cham (2017). https://doi.org/10.1007/978-3-319-68705-6_29

17. Lafond, M.: Recognizing k-leaf powers in polynomial time, for constant k. ACM Trans. Algorithms **19**(4), 1–35 (2023). https://doi.org/10.1145/3614094

18. Le, V.B., Rosenke, C.: Computing optimal leaf roots of chordal cographs in linear time. In: Bodlaender, H., Woeginger, G. (eds.) FCT 2023. LNCS, pp. 348–362. Springer, Cham (2023). https://doi.org/10.1007/978-3-319-68705-6_29

19. Lin, G.-H., Kearney, P.E., Jiang, T.: Phylogenetic k-root and steiner k-root. In: Goos, G., Hartmanis, J., van Leeuwen, J., Lee, D.T., Teng, S.-H. (eds.) ISAAC 2000. LNCS, vol. 1969, pp. 539–551. Springer, Heidelberg (2000). https://doi.org/10.1007/3-540-40996-3_46

20. Mengel, S.: Lower bounds on the mim-width of some graph classes. Discret. Appl. Math. **248**, 28–32 (2018)

21. Monma, C.L., Reed, B., Trotter, W.T., Jr.: Threshold tolerance graphs. J. Graph Theory **12**(3), 343–362 (1988)

22. Nevries, R., Rosenke, C.: Towards a characterization of leaf powers by clique arrangements. Graphs Comb. **32**, 2053–2077 (2016)

23. Nishimura, N., Ragde, P., Thilikos, D.M.: On graph powers for leaf-labeled trees. J. Algorithms **42**(1), 69–108 (2002)

24. Rautenbach, D.: Some remarks about leaf roots. Discret. Math. **306**(13), 1456–1461 (2006). https://doi.org/10.1016/j.disc.2006.03.030

25. Rosenke, C., Le, V.B., Brandstädt, A.: Leaf powers. In: Beineke, L.W., Golumbic, M.C., Wilson, R.J. (eds.) Topics in Algorithmic Graph Theory, pp. 168–188. Encyclopedia of Mathematics and its Applications, Cambridge University Press (2021)

# Perfect Roman Domination: Aspects of Enumeration and Parameterization

Kevin Mann$^{(\boxtimes)}$⊙ and Henning Fernau⊙

Universität Trier, Fachbereich 4 – Abteilung Informatikwissenschaften, 54286 Trier, Germany
{mann,fernau}@uni-trier.de

**Abstract.** Perfect Roman dominating functions and unique response Roman dominating functions are two ways to translate *perfect code* into the framework of Roman dominating functions. We also consider the enumeration of minimal perfect Roman dominating functions and show a tight relation to minimal Roman dominating functions. Furthermore, we consider the complexity of the underlying decision problems PERFECT ROMAN DOMINATION and UNIQUE RESPONSE ROMAN DOMINATION on special graph classes. Split graphs are the first graph class for which UNIQUE RESPONSE ROMAN DOMINATION is polynomial-time solvable while PERFECT ROMAN DOMINATION is NP-complete. Beyond this, we give polynomial-time algorithms for PERFECT ROMAN DOMINATION on interval graphs and for both decision problems on cobipartite graphs. However both problems are NP-complete on chordal bipartite graphs. We show that both problems are W[1]-complete if parameterized by solution size and FPT if parameterized by the dual parameter or by clique-width.

## 1 Introduction

There are many ways to get along with NP-complete problems. For example, you could consider exponential-time algorithms or approximation algorithms. Here, we want to deal with PERFECT ROMAN DOMINATION and UNIQUE RESPONSE ROMAN DOMINATION by looking into enumeration, into the complexity of related decision problems in special graph classes and into parameterized complexity.

We only consider simple undirected graphs $G = (V, E)$ throughout this paper. We recall some notions next. $N_G(v) \coloneqq \{u \in V \mid uv \in E\}$ describes the *open neighborhood* of $v \in V$ with respect to $G$. The *closed neighborhood* of $v \in V$ with respect to $G$ is defined by $N_G[v] \coloneqq N_G(v) \cup \{v\}$. $D \subseteq V$ is a *dominating set* (or a *perfect code*, respectively) of $G$ if each vertex $v$ finds a (unique) vertex $u$ in $D \cap N_G[v]$. A function $f : V \to \{0, 1, 2\}$ is called a *Roman dominating function* (Rdf for short) if each $v \in V$ with $f(v) = 0$ is adjacent to a $u \in V$ with $f(u) = 2$. The weight of a function $f : V \to \{0, 1, 2\}$ is defined by $\omega(f) = \sum_{v \in V} f(v)$. The decision problem ROMAN DOMINATION has as input a graph $G$ and a positive number $k$, and the question is if there exists a Rdf such that the weight of the function is at most $k$. ROMAN DOMINATION is a well-studied variation of DOMINATING SET. It is motivated by a defense strategy of the Roman Empire.

© The Author(s), under exclusive license to Springer Nature Switzerland AG 2024
A. A. Rescigno and U. Vaccaro (Eds.): IWOCA 2024, LNCS 14764, pp. 354–368, 2024.
https://doi.org/10.1007/978-3-031-63021-7_27

The idea was to position the armies on regions in such a way that either (1) one army is in this region or (2) two armies are on one neighbored region. We model the distribution of the armies on the region (viewed as vertices of the graph) by a function that maps each region to the number of armies placed on this region.

In recent years, notable attention [15, 19, 20, 34, 35, 38, 40, 41] was paid to this problem. As for dominating set, there are also many variants of Roman dominating functions considered in the literature. Examples are Roman-{2}-domination (also known as Italian domination) [14] or double Roman domination [1, 6, 8]. Clearly, as testified by books written on this topic [28, 29, 31], there are even more variations of the basic concept of domination in graphs.

Here, we will consider two further variations of Roman domination, perfect and unique response Roman domination. A *perfect Roman dominating function* (pRdf for short) (introduced by Henning *et al.* [30]) is a Roman dominating function where each vertex with value 0 has *exactly* one neighbor with value 2. Moreover, if all vertices with at least 1 as a value have no neighbor with value 2, then we have a *unique response Roman dominating function* (urRdf for short) (as introduced in [39]). Both variations can be seen as a way to translate the idea of perfect code into the realm of Rdf. A further motivation can also be the idea of positioning armies on regions: If the armies are placed according to a perfect Rdf and a region without any army is attacked, then it is clear from which region an army moves to secure the attacked region, so no time is wasted to first agree on who is the one to take action and move to the endangered region.

The decision problems are given as follows:

**Problem name:** UNIQUE RESPONSE ROMAN DOMINATION (URRD)
**Given:** A graph $G = (V, E)$ and $k \in \mathbb{N}$
**Question:** Is there a urRdf $f$ on $G$ with $\omega(f) \leq k$?

**Problem name:** PERFECT ROMAN DOMINATION (PRD)
**Given:** A graph $G = (V, E)$ and $k \in \mathbb{N}$
**Question:** Is there a pRdf $f$ on $G$ with $\omega(f) \leq k$?

Let $u_R(G)$ denote the smallest weight of any urRdf $f$ on $G$. Furthermore, $\gamma_R^p(G)$ denotes the smallest weight of any pRdf on $G$. The following is known about the complexity of these decision problems. URRD is NP-complete on regular bipartite graphs [9]. Banerjee *et al.* [5] showed that it is NP-complete on chordal graphs, and polynomial-time solvable on distance-hereditary and interval graphs. They also prove that there is no URRD polynomial-time approximation algorithm within a factor of $n^{1-\epsilon}$ for any constant $\epsilon > 0$ and any $n$-vertex input graph, unless NP = P. PRD is NP-complete on chordal graphs, planar graphs, and bipartite graphs and polynomial-time solvable on block graphs, cographs, series-parallel graphs, and proper interval graphs, as shown in [7].

Enumeration is a wide area of research, as testified by specialized workshops like [22]. For some examples, we refer to the survey [44]. From a practical point of view, enumeration can be interesting if not all aspects of the problem have been satisfactorily modeled. This makes also sense for the motivation of ROMAN DOMINATION (variants). It could be the case that between neighbored regions there is a conflict unknown to the strategic planner of the army deployment. Such a conflict could be a reason why the regions do not want to defend each other. Enumerating minimal Rdfs was considered by Abu-Khzam *et al.* [3]. There, it is shown that all pointwise (*i.e.*, for $f, g : V \to \{0, 1, 2\}$, $f \leq g$ iff $f(v) \leq g(v)$ for all $v \in V$) minimal Rdfs of a graph of order $n$ can be enumerated in time $\mathcal{O}(1.9332^n)$ with *polynomial delay*. An enumeration algorithm has *polynomial delay* if the time between two subsequent outputs of the algorithm can be bounded by some polynomial, as well as the time from the start to the first output and the time between the last output and the termination of the algorithm. This property is interesting if a user (or another program) processes the results in a pipeline fashion. Also, it is not known if there is such an algorithm for enumerating (inclusion-wise) minimal dominating sets. In this context, we remark:

**Observation 1.** *Unless* $\mathsf{P} = \mathsf{NP}$, *there exists no polynomial-delay enumeration algorithm for (inclusion-wise minimal) perfect codes.*

Namely, such a polynomial-delay enumeration algorithm could also be used to decide (in polynomial time) if a perfect code exists, which is NP-hard, see [33].

We now collect more notation that we will use in the paper. Let $\mathbb{N}$ denote the set of all nonnegative integers (including 0). For $n \in \mathbb{N}$, we will use the notation $[n] := \{1, \dots, n\}$. Let $A, B$ be two sets. $B^A$ is the set of all function $f : A \to B$. For $A \subseteq B$, $\chi_A : B \to \mathbb{N}$ denotes the characteristic function ($\chi_A(a) = 1$ iff $a \in A$; $\chi_A(a) = 0$, otherwise). Let $G = (V, E)$ be a graph. For a set $A \subseteq V$, the open neighborhood is defined as $N_G(A) := \left( \bigcup_{v \in A} N_G(v) \right)$. The closed neighborhood of $A$ is given by $N_G[A] := N_G(A) \cup A$. The *private neighborhood* of a vertex $v \in A$ with respect to $G$ and $A$ is denoted by $P_{G,A}(v) := N_G[v] \setminus N_G[A \setminus \{v\}]$. Let $G = (V, E)$ be a graph and $f \in \{0, 1, 2\}^V$. We define $V_i(f) := \{v \in V \mid f(v) = i\}$ for $i \in \{0, 1, 2\}$. If $v \in V$ obeys $f(v) = 2$, then $u \in N_G(v)$ with $f(u) = 0$ is called a *private neighbor* of $v$ (with respect to $f$) if $|N_G(u) \cap V_2(f)| = 1$.

**Observation 2.** *Let* $G = (V, E)$ *be a graph and* $f : V \to \{0, 1, 2\}$. *(1) If* $f$ *is a pRdf, then for all* $v \in V_2(f)$ *and* $u \in N(v) \cap V_0(f)$, $u$ *is a private neighbor of* $v$. *(2) If* $f$ *is a Rdf such that for all* $v \in V_2(f)$ *and* $u \in N(v) \cap V_0(f)$, $u$ *is a private neighbor of* $v$, *then* $f$ *is a pRdf.*

**Observation 3.** *Each urRdf on a graph without isolates is a minimal Rdf.*

A final note on parameterized complexity: If P is some problem and $\kappa$ is some parameter, $\kappa$-P denotes the problem P parameterized by $\kappa$, see [18,24].

## 2  Enumerating Minimal Perfect Roman Domination

The main goal of this section is to find a polynomial-delay algorithm to enumerate all minimal pRdfs even when solving the extension problem is hard, which blocks the standard approach to getting such enumeration algorithms by a branching algorithm that checks extendibility and prunes unnecessary branches.

---

**Problem name:** EXTENSION PERFECT ROMAN DOMINATION (EXT PRD)
**Given:** A graph $G = (V, E)$ and $f \in \{0, 1, 2\}^V$.
**Question:** Is there a (pointwise) minimal pRdf $g$ on $G$ with $f \leq g$?

**Theorem 1.** *[23]* EXT PRD *is NP-complete.* $\omega(f)$-EXT PRD *is* $W[1]$-*complete and* $|V_0(f)|$-EXT PRD *is* $W[2]$-*complete (even on bipartite graphs).*

We often use the following characterization result that is of independent interest.

**Theorem 2.** *Let* $G = (V, E)$ *be a graph. A function* $f : V \to \{0, 1, 2\}$ *is a minimal pRdf iff the following conditions are true:*

1. $\forall v \in V_0(f) : |N_G(v) \cap V_2(f)| = 1,$
2. $\forall v \in V_1(f) : |N_G(v) \cap V_2(f)| \neq 1,$
3. $\forall v \in V_2(f) : |N_G(v) \cap V_0(f)| \neq 0.$

*Proof.* Let $f$ be a minimal pRdf. This implies the first condition. If there exists a $v \in V_1(f)$ with $|N_G(v) \cap V_2(f)| = 1$, then $f - \chi_{\{v\}}$ is also a pRdf, cf. Observation 2. For each $v \in V_2(f)$, $|N_G(v) \cap V_0(f)| \neq 0$, as otherwise $f - \chi_{\{v\}}$ is also a pRdf.

Now assume $f$ is a function that fulfills these conditions. By Condition 1, we know $f$ is a pRdf. Let $g$ be a minimal pRdf with $g \leq f$. Thus, $V_0(f) \subseteq V_0(g)$ and $V_2(g) \subseteq V_2(f)$ hold. Assume there exists a $v \in V$ with $g(v) < f(v) = 2$. By Condition 3, there exists some $u \in N_G(v) \cap V_0(f)$. Because of condition 1, $v$ is the only neighbor of $u$ with value 2. Hence, $N_G(u) \cap V_2(g) \subseteq N_G(u) \cap (V_2(f) \setminus \{v\}) = \emptyset$. This contradicts the construction of $g$ as a pRdf. Therefore, $V_2(g) = V_2(f)$. If there were a $v \in V$ with $g(v) = 0 < 1 = f(v)$, then this contradicts that $g$ is a pRdf, since $|N_G(v) \cap V_2(g)| = |N_G(v) \cap V_2(f)| \neq 1$. □

*Remark 1.* This result implies that the constant function $\chi_V$ is the maximum minimal pRdf with respect to $\omega$ for each graph $G = (V, E)$, as for Rdfs, see [13].

We use Theorem 2 also to obtain the announced polynomial-delay enumeration algorithm. This algorithm uses the idea that there is a bijection between all minimal pRdfs and all minimal Rdfs of a graph. To show this bijection, we need the following characterization theorem from [2].

**Theorem 3.** *[2] Let* $G = (V, E)$ *be a graph,* $f : V \to \{0, 1, 2\}$ *be a function and let* $G' := G[V_0(f) \cup V_2(f)]$. *Then,* $f$ *is a minimal Rdf iff the following holds:*
    *(1)* $N_G[V_2(f)] \cap V_1(f) = \emptyset$, *(2)* $\forall v \in V_2(f) : P_{G', V_2(f)}(v) \not\subseteq \{v\}$, *and (3)* $V_2(f)$ *is a minimal dominating set of* $G'$.

Let $G = (V, E)$ be graph. Define the sets of functions
$\mu - \mathcal{RDF}(G) := \{ f : V \to \{0, 1, 2\} \,|\, f$ is a minimal RDF $\}$ and
$\mu - \mathcal{PRDF}(G) := \{ f : V \to \{0, 1, 2\} \,|\, f$ is a minimal pRdf $\}$.

**Theorem 4.** *Let* $G = (V, E)$ *be graph. There is a bijection* $B : \mu - \mathcal{RDF}(G) \to \mu - \mathcal{PRDF}(G)$. *Furthermore,* $B(f)$ *and* $B^{-1}(g)$ *can be computed in polynomial time (with respect to* $G$*) for each* $f \in \mu - \mathcal{RDF}(G)$ *and* $g \in \mu - \mathcal{PRDF}(G)$.

*Proof.* Let $f \in \mu - \mathcal{RDF}(G)$ and $g \in \mu - \mathcal{PRDF}(G)$. Define $B(f)$ by the sets

$V_0(B(f)) = \{v \in V_0(f) \mid |N_G(v) \cap V_2(f)| = 1\},$
$V_1(B(f)) = \{v \in V_0(f) \mid |N_G(v) \cap V_2(f)| \geq 2\} \cup V_1(f), \quad V_2(B(f)) = V_2(f).$

By definition of $B$, $|N_G(v) \cap V_2(f)| = 1$ holds for all $v \in V_0(B(f))$. Since each $v \in V$ with $f(v) = 2$ needs a private neighbor except itself by Theorem 3, $N_G(v) \cap V_0(B(f)) \neq \emptyset$ holds. As any $v \in V_1(f)$ has no neighbor in $V_2(f)$ and any $v \in V_1(B(f)) \setminus V_1(f)$ has at least two neighbors in $V_2(f)$, all conditions of Theorem 2 are met. Therefore, $B(f) \in \mu - \mathcal{PRDF}(G)$. Define $B^{-1}(g)$ by

$$V_0(B^{-1}(g)) = V_0(g) \cup \{v \in V_1(g) \mid |N_G(v) \cap V_2(g)| \geq 2\},$$

$$V_1(B^{-1}(g)) = \{v \in V_1(g) \mid |N_G(v) \cap V_2(g)| = 0\}, \quad V_2(B^{-1}(g)) = V_2(f).$$

By definition of $B^{-1}$, $N_G(v) \cap V_2(B^{-1}(g)) = \emptyset$ for each $v \in V_1(B^{-1}(g))$. By Theorem 2, each $v \in V_0(g)$ has exactly one neighbor in $V_2(g)$. Hence, each vertex in $V_0(B^{-1}(g))$ has a neighbor in $V_2(B^{-1}(g))$ and $V_2(B^{-1}(g))$ is a dominating set of $G[V_0(B^{-1}(g)) \cup V_2(B^{-1}(g))]$. By Theorem 2, each $v \in V_2(g)$ has a $u \in N_G(v) \cap V_0(g)$ with $\{v\} = N_G(u) \cap V_2(g)$. So, $u$ is a private neighbor of $v$ and $B^{-1}(g) \in \mu - \mathcal{RDF}(G)$. We leave to the reader to prove: $B$ is bijective. $\quad\square$

This bijection is a so-called *parsimonious reduction*. This is a class of reductions designed for enumeration problems. For more information, we refer to [42]. Even without this paper, Theorem 4 implies some further results thanks to [2].

**Corollary 1.** *There are graphs of order $n$ that have at least $\sqrt[5]{16}^n \in \Omega(1.7441^n)$ many minimal pRdfs.*

**Theorem 5.** *There is a polynomial-space, polynomial-delay algorithm that enumerates all minimal pRdfs of a given graph of order $n$ in time $\mathcal{O}^*(1.9332^n)$.*

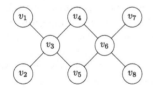

$f = 2\chi_{\{v_3,v_6\}}$ is a minimum Rdf $(\omega(f) = 4)$ and $g = 2\chi_{\{v_3\}} + \chi_{\{v_6,v_7,v_8\}}$ is a minimum pRdf $(\omega(g) = 5)$, but $\omega(B(f)) = 6$, $\omega(B^{-1}(g)) = 5$.

**Fig. 1.** Counter-example of Remark 2

*Remark 2.* The function $B$ defined in the proof of Theorem 4 inherits inclusion-wise minimality, but this does not mean that, if $f$ is a minimum Roman domi-nating function, then $B(f)$ is a minimum pRdf. Consider Fig. 1.

According to [4], Theorem 4 implies some further results for special graph classes. Let $G = (V, E)$, with $n := |V|$ being the order of $G$. If $G$ is a forest or interval graph, then there is a pRdf enumeration algorithm that runs in time $\mathcal{O}^*(\sqrt{3}^n)$ with polynomial delay. For both graph classes, a graph with many isolated edges is example of a graph with $\sqrt{3}^n$ many minimal pRdfs. This is also the worst-case example for chordal graphs which is known so far. We can enumerate all pRdfs of these graphs in time $\mathcal{O}(1.8940^n)$. If $G$ is a split or cobipartite graph, then we can enumerate all pRdfs of $G$ in $\mathcal{O}^*(\sqrt[3]{3}^n)$ with polynomial delay. Further, both classes include graphs of order $n$ with $\Omega(\sqrt[3]{3}^n)$ many pRdfs. There is also a polynomial-time recursive algorithm to count all pRdfs of paths. Enumeration of urRdfs in split graphs is considered in Subsect. 3.4.

# 3    Complexity and Enumeration in Special Graph Classes

In this section, we will consider the complexity of our problems PRD and URRD on different graph classes: interval, split, cobipartite and chordal bipartite.

## 3.1    Interval Graphs

We first start with a polynomial-time algorithm for PRD for interval graphs. A graph $G = (V, E)$ is called an *interval graph* (see [25]) if, for $v \in V$, there is an interval $I_v = [l_v, r_v]$ such that $\{v, u\} \in E$ iff $I_v \cap I_u \neq \emptyset$. We can assume that the interval representation obeys $\forall u, v \in V : r_u = r_v \implies u = v$.

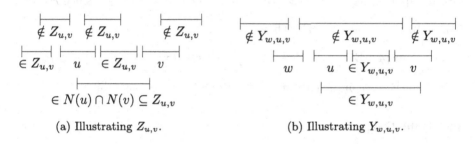

(a) Illustrating $Z_{u,v}$.                    (b) Illustrating $Y_{w,u,v}$.

Fig. 2. Examples illustrating our notions.

**Theorem 6.** PRD *is polynomial-time solvable on interval graphs.*

*Proof.* Let $G = (V, E)$ be an interval graph such that $I_v = [l_v, r_v]$ is the interval representation of $v \in V$. The dynamic programming works from "left to right" according to some linear extension of the partial order defined as $w \leq v$ iff $l_w \leq l_v \wedge r_w \leq r_v$. For $v \in V$, $L_v := \{w \in V \mid l_w \leq r_v\}$ denotes the set of vertices which are dominated by $v$ or "to the left" of $I_v$. To simplify our notation, we define $P_v := \{w \in V \setminus \{v\} \mid w \leq v\}$ to easily refer to the already considered vertices in the dynamic program. $W_v := \{w \in V \mid r_v < r_w\}$ is the set of not yet considered vertices. For $v \in V$, $u \in P_v$ and $w \in P_u$, define $Z_{u,v} := (L_v \setminus (N[u] \cup N[v])) \cup (N(u) \cap N(v))$ and $Y_{w,u,v} := ((N(u) \cap N(v)) \setminus N[w]) \cup (L_v \setminus (L_u \cup N[v]))$, see Fig. 2. All these sets and their cardinalities can be computed before the actual dynamic programming starts. Also, we can compute the pRdfs that set at most one vertex to 2 and store their minimum weight in $\min_{\leq 1}$.

We need some more notation to formulate the dynamic programming algorithm and prove its correctness. For $v \in V$ and $u \in P_v$, define $\Gamma_{u,v} := \{f \in \{0,1,2\}^V \mid f|_{L_v} \in \mathcal{PRDF}(G[L_v]) \wedge f(u) = f(v) = 2 \wedge \forall w \in W_u \setminus \{v\} : f(w) \neq 2\}$ and $\gamma_{u,v} = \min_{f \in \Gamma_{u,v}} \omega(f)$. In the following, $f'_{u,v}$ will denote one function from $\Gamma_{u,v}$ that attains this minimum, i.e., $\omega\left(f'_{u,v}\right) = \gamma_{u,v}$. For the inductive computation, we will reuse the values $\gamma_{w,u}$ with $w \in P_u$. For the correctness of this approach, the following observations are helpful.

1. $Z_{u,v}$ is tied to the function $f_{u,v} \in \Gamma_{u,v}$ by $V_2(f_{u,v}) = \{u,v\}$, $V_1(f_{u,v}) = Z_{u,v}$.
2. $Y_{w,u,v} = V_1(f_{w,u,v}) \setminus V_1(f'_{w,u})$, where $f_{w,u,v}$ is the auxiliary pRdf which we will construct from $f'_{w,u}$ in the induction step.
3. If $f \in \Gamma_{u,v}$, then $\tilde{f} := f + \chi_{V \setminus L_v}$ is a pRdf on $G$. If $f$ is minimum, then $\tilde{f}$ is a minimum pRdf on $G$.

By the third property, after the inductive computations, $\min\{\min_{\leq 1}, \min\{\gamma_{u,v} + |V \setminus L_v| \mid v \in V, u \in P_v\}\}$ gives the minimum weight of any pRdf for $G$.

Assume now that $v$ is a vertex for which we want to compute all relevant $\gamma_{u,v}$ next, $u \in P_v$. The following claim contains a formula that describes how to do this. It implicitly also covers the initial step: then $\gamma_{u,v} = 4 + |Z_{u,v}|$ since $P_u = \emptyset$.

*Claim.* Let $v \in V$ and $u \in P_v$. Then

$$\gamma_{u,v} = \min\{4 + |Z_{u,v}|, 2 - |\{v\} \cap N(u) \cap N(w)| + \gamma_{w,u} + |Y_{w,u,v}| \mid w \in P_u\}.$$

Formal correctness arguments are omitted due to space constraints.    □

## 3.2    Split Graphs

A *split graph* $G = (V, E)$ is a graph in which the vertex set $V$ can be partitioned into $C, I$, where $C$ is a clique and $I$ is an independent set of $G$. It seems to be open to which complexity class PRD and URRD belong on split graphs, as explicitly asked in [5]. We will answer this question in a somewhat surprising manner: while URRD can be solved in linear time, PRD turns out to be NP-complete. First, we will consider urRdfs on split graphs and prove a combinatorial result.

**Lemma 1.** *Let $G = (V, E)$ be a connected split graph with a clique $C$ and an independent set $I$ such that $V = C \cup I$ and $C \cap I = \emptyset$. For each urRdf $f : V \to \{0,1,2\}$, one of the following conditions holds: (1) $V_2(f) \cap I = \emptyset$ and $|V_2(f) \cap C| \leq 1$; or (2) $V_2(f) \subseteq I$.*

*Proof.* Suppose that $V_2(f) \nsubseteq I$. By the definition of urRdf, two vertices of $C$ cannot have the value 2, since $C$ is a clique. Hence, $|V_2(f) \cap C| \leq 1$. For the sake of contradiction, assume there are $v \in I$ and $u \in C$ with $f(u) = f(v) = 2$. As $f$ is a urRdf, $u$ and $v$ cannot be neighbors. Since $G$ is connected, $\emptyset \subsetneq N_G(v) \subseteq C \setminus \{u\} \subseteq N_G(u)$. This contradicts the assumption that $f$ is a urRdf.    □

This lemma implies that we can solve URRD by looking for a vertex $v$ of maximum degree (which must be in an inclusion-wise maximal clique), set the value of $v$ to two and the values of the other vertices accordingly. This implies:

**Theorem 7.** (∗) URRD *can be solved in linear time on split graphs.*

Interestingly, the seemingly very similar problem PRD on split graphs is harder to solve. To see this, we reduce from a problem called PERFECT CODE.

**Theorem 8.** (∗) PRD *on split graphs is NP-complete.*

We contrast this by giving a branching algorithm for PRD. Our algorithm relies on combinatorial insights from Theorem 4 and [4], where Rdfs have been discussed on split graphs, so that $V_2(f') \cap I = \emptyset$ or $V_2(f') \cap C = \emptyset$ for minimum $f'$.

**Theorem 9.** (∗) $\omega(f) - $ PRD *can be solved in time* $\mathcal{O}^*(\varphi^{\omega(f)})$ *on split graphs, where* $\varphi$ *is the golden ratio.*

### 3.3   Cobipartite Graphs

Now, we consider *cobipartite graphs*, the complementary graphs of bipartite graphs, i.e., the vertex set can be partitioned into two cliques. On this graph class, PRD as well as URRD are solvable in polynomial time.

**Theorem 10.** PRD *is polynomial-time solvable on cobipartite graphs.*

*Proof.* Let $G = (V, E)$ be a cobipartite graph, with $V$ partitioned into the two cliques $C_1, C_2 \subseteq V$. For $v \in V$, define the pRdf $g_v := 2 \cdot \chi_{\{v\}} + \chi_{V \setminus N[v]}$.

Let $f \in \{0, 1, 2\}^V$ be a pRdf with $|V_2(f) \cap C_1| \geq 2$ and $v \in C_2$. Next, we prove that $\omega(g_v) \leq \omega(f)$. For each $u \in C_1 \setminus V_2(f)$, $|V_2(f) \cap N(u)| \geq 2$. Hence $C_1 \subseteq V_1(f) \cup V_2(f)$. Since $C_2 \subseteq N[v] = V_0(g_v) \cup \{v\}$, $\omega(g_v) \leq |C_1| + 2 \leq \omega(f)$. Symmetrically, $\omega(g_v) \leq |C_1| + 2 \leq \omega(f)$ for $v \in C_1$ and a pRdf $f$ with $|V_2(f) \cap C_2| \geq 2$. Let $f$ be a minimum pRdf on $G$. By the previous thoughts, we can assume $|V_2(f) \cap C_i| \leq 1$ for each $i \in \{1, 2\}$. This leaves only $|V|^2 + |V|$ many possibilities to check.    □

Cabrera *et al.* [9] proved that URRD is NP-complete on bipartite graphs. For cobipartite graphs, this problem is polynomial-time solvable. For this result, we use the idea of 2-packings. Let $G = (V, E)$ be a graph. A set $S \subseteq V$ is called a *2-packing* if the distance between any two vertices in $S$ is at least 3. Targhi *et al.* [43] presented a proof for $u_r(G) = \min\{2|S| + |V(G) \setminus N_G[S]| \mid S$ is a 2-packing$\}$. Analogously to this, we can prove that, for each graph $G = (V, E)$, there exists a bijection $\psi_G$ between all 2-packings of the graph and all urRdfs. Here, for a 2-packing $S$ and for $x \in V$, $\psi_G(S) := 2 \cdot \chi_S + \chi_{V \setminus N(S)}$. This also yields that, for each urRdf $f$, $V_2(f)$ is a 2-packing. This is the main idea of our algorithm.

**Lemma 2.** (∗) URRD *is polynomial-time solvable on cobipartite graphs. Furthermore, there are at most* $\frac{|V|^2}{4}$ *many 2-packings, or urRdfs, in* $G = (V, E)$.

The upper bound $\frac{|V|^2}{4}$ is tight: consider the complement of the complete bipartite graph $K_{t,t}$, where both classes have the same size $t$. This is interesting as there could be exponentially many urRdfs even on connected split graphs, which is quite a related class of graphs that we consider next.

### 3.4   Remarks on Enumeration of urRdfs in Split Graphs

Enumerating urRdfs in general graphs in time $\mathcal{O}(1.5399^n)$ was shown in [32].

**Corollary 2.** *All urRdfs of a connected split graph of order $n$ can be enumerated in time $\mathcal{O}(\sqrt[3]{3}^n) \subseteq \mathcal{O}(1.4423^n)$.*

*Proof.* Non-trivial connected split graphs have no isolates. By Observation 3, we can hence use the enumeration algorithm for minimal Rdfs on split graphs of order $n$ from [4] which runs in time $\mathcal{O}(\sqrt[3]{3}^n)$.                    $\square$

$\mathcal{O}(\sqrt[3]{3}^{|V|})$ is even a tight upper bound for connected split graphs as shown next.

**Theorem 11.** *There is an infinite family of connected split graphs $G_t = (V_t, E_t)$ with $\Omega(\sqrt[3]{3}^{|V_t|})$ many urRdfs.*

*Proof.* We only need to consider the split graph family $G_t = (V_t, E_t)$ with $V_t :=$ $C_t \cup I_t$, $C_t := \{c_1, \ldots, c_t\}$, $I_t := \{v_1, \ldots, v_{2t}\}$, $E_t := \binom{C_t}{2} \cup \{\{c_i, v_{2i-1}\}, \{c_i, v_{2i}\} \mid i \in \{1, \ldots, t\}\}$. Clearly, $|V_t| = 3t$. By arguments from Theorem 7, for each urRdf $f$ on $G_t$ with $|V_2(f) \cap C_t| = 1$, in fact $|V_2(f)| = 1$. There are $t$ many such urRdfs. Let $f$ be a urRdf with $V_2(f) \cap C_t = \emptyset$. In this situation, for each $i \in \{1, \ldots, t\}$, there are three ways to dominate $c_i$: $f(c_i) = f(v_{2i-1}) = f(v_{2i}) = 1$, $f(c_i) = f(v_{2i}) = 0$ and $f(v_{2i-1}) = 2$, or $f(c_i) = f(v_{2i-1}) = 0$ and $f(v_{2i}) = 2$. Hence, there are $t + 3^t = \frac{|V|}{3} + \sqrt[3]{3}^{|V|}$ many urRdfs.                    $\square$

For general graphs, the best lower bound [32] is $\mathcal{O}(1.4970^n)$.

The polynomial-delay property of the Rdf-enumeration algorithm from [4] is not inherited for enumerating urRdf, as not each minimal Rdf is a urRdf. Nonetheless, we will present a sketch of a direct polynomial-delay branching enumeration algorithm for enumerating urRdfs that is optimal due to Theorem 11. For this purpose, we consider each 2-packing on a connected split graph and use the bijection between urRdfs and 2-packings. Our algorithm is a modification of the recursive backtracking algorithm of [37] and has polynomial delay with the same arguments. The algorithm of [37] is a general approach to enumerate all sets of $\mathcal{F} \subseteq 2^U$ for a universe $U$, where $\mathcal{F}$ fulfills the following downward closure property: if $X \in \mathcal{F}$ and $Y \subseteq X$, then $Y \in \mathcal{F}$.

**Theorem 12.** (∗) *All urRdfs of a connected split graph of order $n$ can be enumerated in time $\mathcal{O}(\sqrt[3]{3}^n)$ with polynomial delay and polynomial space.*

One of the rather surprising aspects of the omitted proof is that we do not need any sophisticated measure-and-conquer methodology to analyze the runtime of the branching algorithm by a simple case analysis concerning the vertex degrees. In the core part of the algorithm, we decide if a vertex in the independent set is put into the related 2-packing or not, also cf. [37].

### 3.5  Chordal Bipartite Graphs

These are bipartite graphs where each induced cycle has length 4. To prove NP-hardness of PRD and URRD on chordal bipartite graphs, we use the problem ONE-IN-THREE 3SAT, with a reduction inspired by [36, Theorem 2.2] for the NP-hardness of PC. However, our proof is much more sophisticated in order to deal with the weights of the pRdfs.

**Theorem 13.** (∗) PRD& URRD *are NP-complete on chordal bipartite graphs.*

## 4  Parameterized Complexity

In this section we, consider parameterized complexity aspects of PRD and URRD. More precisely, we consider the parameters *clique width* (cwd for short, defined in [16]), solution size and its dual, i.e., order minus solution size.

First, we give a LinEMSOL($\tau_1$) formula for the optimization problems corresponding to PRD and to URRD in order to show that these are linear-time FPT if parameterized by cwd, using now classical results from [17,26]. cwd is an interesting parameter, since an FPT algorithm with cwd as parameter implies FPT algorithms with respect to tw, nd or vc as parameters. However, the algorithmic results of the previous section do not follow from these considerations, as it is known that even unit interval graphs have unbounded cliquewidth, see [26].

**Lemma 3.** MINIMUM PRD *on the graph* $G = (V, E)$ *can be expressed as*

$$\operatorname*{argmin}_{X_1 \subseteq V, X_2 \subseteq V} \{|X_1| + 2|X_2| \mid \langle G, X_1, X_2 \rangle \vDash \theta(X_1, X_2)\}, \ \textit{with } \theta(X_1, X_2) \textit{ equal to}$$

$$\forall v \, (X_1(v) \lor X_2(v) \lor \exists u \, [X_2(u) \land E(v, u) \land \forall w \, ([X_2(w) \land E(w, v)] \Rightarrow u = w)]) \, .$$

*Proof.* Let $G = (V, E)$ be a graph for which we want to determine the pRdf of smallest weight. First, if $f$ is some pRdf of $G$, then $\theta(V_1(f), V_2(f))$ is obviously true. Now we want to show why $X_1, X_2 \subseteq V$ as a minimum solution would describe a pRdf. Define $f := \chi_{X_1} + 2\chi_{X_2}$. We can then see that for all $v \in V$, $v \in X_1$, $v \in X_2$, or there exists a $u \in X_2 \cap N(v)$ such that there is no other neighbor of $v$ in $X_2$. Therefore, we only need to show $X_1 \cap X_2 = \emptyset$. So, assume there is a $v \in X_1 \cap X_2$. Then for $X_1' := X_1 \setminus \{v\}$, $(X_1', X_2)$ is a smaller solution. Hence, $f$ is a pRdf.  □

**Theorem 14.** cwd − PRD ∈ *FPT.*

By [26], distance-hereditary graphs have cliquewidth bounded by three, so that we get the following supplement to the previous section.

**Corollary 3.** PRD *can be solved in linear time on distance-hereditary graphs.*

With similar techniques, we can also prove an analogous result for URRD.

**Theorem 15.** cwd − URRD ∈ *FPT.*

Beside structural parameters, also the solution size is an interesting often-used parameter. We show next that $\omega(f)$-URRD and $\omega(f)$-PRD are W[1]-complete. We do this by using Steps-SHORT NON-DETERMINISTIC TURING MACHINE COMPUTATION (SNTMC) which is known to be W[1]-complete (see [10–12,18]), as well as PC, parameterized by solution size. So, we cannot expect FPT-results. We start our presentation with a W[1]-membership proof, which is in a mathematical sense the most interesting proof. As a side-product, we also get membership in XP. To this end, we need the following interesting structural characterization lemma that is proven by employing some double-counting arguments.

**Lemma 4.** (∗) *Let $G = (V, E)$ be a graph and $f \in \{0, 1, 2\}^V$ be a function such that $N(u) \cap N(v) \cap V_0(f) = \emptyset$ for all $u, v \in V_2(f)$, $u \neq v$. Then $f$ is a pRdf iff*

$$|V| = \left( \sum_{u \in V_1(f)} (1 - |N(u) \cap V_2(f)|) \right) + \left( \sum_{v \in V_2(f)} (|N[v]| - |N(v) \cap V_2(f)|) \right).$$

**Theorem 16.** $\omega(f)$-PRD *belongs to* W[1].

*Proof.* (Sketch) We will use Steps-SNTMC. So let $G = (V, E)$ be a graph of order $n$ and $k \in \mathbb{N}$. The reduction will compute a Turing machine $M$ with tape alphabet containing $V \cup \{\zeta\}$, with $\zeta \notin V$. In its internal memory, $M$ stores a table $T$ that contains, for $u, v \in V$, $u \neq v$, either $N(u) \cap N(v)$ (if $|N(u) \cap N(v)| \leq k$) or $\infty$. Moreover, the reduction machine also computes addition and subtraction tables for numbers up to $n$ and the alphabet contains also these numbers, so that $M$ can do basic arithmetics, like reading two number symbols from the tape and printing the result, in constant time. Finally, $M$ has access to a table of the degrees of all vertices in its internal memory. The Turing machine $M$ will work in 5 phases as follows:

1. Guess the vertices in $V_2(f)$ and then in $V_1(f)$ and write them (as symbols) on the tape. We separate the vertices via the symbol $\zeta$. This takes at most $k + 1$ many steps.
2. Also, in time $\mathcal{O}(k)$ $M$ can check if twice the number of $V_2$-symbols (left of the separator $\zeta$) plus the number of $V_1$-symbols (right to $\zeta$) is at most $k$. If not, $M$ stops in a non-final state.
3. Check if we enumerated any vertex in $V_1(f) \cup V_2(f)$ twice (in $\mathcal{O}(k^2)$ steps). If this is the case, then the machine stops in a non-final state.
   Otherwise, we know now that the $V_1$-symbols together with the $V_2$-symbols that we guess do define a function $f$ of weight $\omega(f) \leq k$.
4. Check for each pair of vertices $u, v \in V_2(f)$ ($u \neq v$) if $N(u) \cap N(v) \subseteq V_1(f) \cup V_2(f)$. This condition is clearly necessary if $f$ should be a pRdf. For this check, we need table $T$. If $T[u, v] = \infty$, $N(u) \cap N(v) \subseteq V_1(f) \cup V_2(f)$ is impossible and we can stop in a non-final state. Otherwise, $M$ checks element by element of $N(u) \cap N(v)$ if it is in $V_1(f) \cup V_2(f)$. Altogether, this needs at most $\mathcal{O}(k^3)$ many steps.

5. For each $v \in V_1(f) \cup V_2(f)$, $M$ computes $|N(v) \cap V_2(f)|$ (at most $\mathcal{O}(k^2)$ steps). Use these numbers to compute (with basic arithmetics)

$$n' := \left( \sum_{u \in V_1(f)} (1 - |N(u) \cap V_2(f)|) \right) + \left( \sum_{v \in V_2(f)} (|N[v]| - |N(v) \cap V_2(f)|) \right)$$

Finally, $M$ uses Lemma 4 to check if $f$ is a pRdf, i.e., if $n = n'$.    □

Using a construction simpler than the previous one for W[1]-membership together with a relatively straightforward reduction from PC for the W[1]-hardness that basically introduces $2k$ twins per vertex, we can prove:

**Theorem 17.** $(*)$ $\omega(f)$-URRD is W[1]-complete.

The previous W[1]-hardness result could also be used to show that $\omega(f)$-PRD is W[1]-complete. We can show this type of result even on bipartite graphs.

**Theorem 18.** $(*)$ $\omega(f)$-PRD is W[1]-complete even on bipartite graphs.

Different from PC, it makes sense to consider the dual parameterization $(|V| - k)$, as it is even NP-hard to find out if the graph has any perfect code (see [33]). So, the problem is para-NP-hard for the dual parameterization. This is not the case for PRD and URRD, since the constant 1 function is always a pRdf/urRdf. Therefore, these problems are trivially solvable if $n - k = 0$. We could even see this parameter as a *distance from triviality* parameter as in [27].

**Theorem 19.** $(n - k) - \mathrm{PRD}$, $(n - k) - \mathrm{URRD} \in FPT$.

*Proof.* Let $G = (V, E)$ be graph and $f \in \{0, 1, 2\}^V$. Then

$$n - \omega(f) = \sum_{v \in V} (1 - f(v))$$

$$= \left( \sum_{v \in V_0(f)} 1 - 0 \right) + \left( \sum_{v \in V_1(f)} 1 - 1 \right) + \left( \sum_{v \in V_2(f)} 1 - 2 \right)$$

$$= |V_0(f)| - |V_2(f)|.$$

We will use this equality to bound the diameter and degree by $n - k$.

Assume there is a vertex $v$ with $\deg(v) \geq n - k + 1$. Then $f_v := \chi_V + \chi_{\{v\}} - \chi_{N(v)}$ is clearly a urRdf and therefore a pRdf with

$$n - \omega(f_v) = |V_0(f_v)| - |V_2(f_v)| = \deg(v) - 1 \geq n - k.$$

Hence, $\omega(f_v) \leq k$ and we observe a yes-instance and can hence use the consideration of a vertex of maximum degree as a reduction rule.

Assume there is a shortest path $p = v_1, \ldots v_{3 \cdot (n-k)}$ of length $3 \cdot (n - k)$. Define $f_p := \chi_V + \sum_{i=0}^{n-k-1} (\chi_{\{v_{3i+2}\}} - \chi_{N(v_{3i+2})})$. Clearly, each vertex in $V_0(f)$

has at least one neighbor in $V_2(f)$. Assume there exists a vertex $u \in V_0(f)$ with $v_{3i+2}, v_{3j+2} \in V_2(f) \cap N(u)$ with $i \neq j$ (w.l.o.g., $i < j$). This would imply that $v_1, \ldots, v_{3i+2}, u, v_{3j+2}, \ldots, v_{3\cdot(n-k)}$ is a shorter path. Hence, $f_p$ is a pRdf. Analogously, $V_2(f_p)$ is an independent set. Therefore, $f_p$ is even a urRdf. Further,

$$n - \omega(f_p) = |V_0(f_p)| - |V_2(f_p)| \geq 2 \cdot (n - k) - n - k \geq n - k.$$

So, $\omega(f) \leq k$ and we observe a yes-instance and can hence use the consideration of the graph diameter as a reduction rule. As maximum degree and diameter are bounded in $n - k$, the number of vertices is bounded by the *Moore bound*, giving $\mathcal{O}\left((n - k)^{3 \cdot (n-k)}\right)$ as the kernel size, see [21]. $\qquad \square$

## 5    Final Remarks

We have enhanced the understanding of the combinatorial structure and the complexity of PRD and URRD in many ways. But many aspects are still open to explore. For instance, we did not touch the question of kernel sizes in any of our FPT results. Also, the fact that the 'bijection approach' that we took for enumeration necessitates parsimonious reductions could be used towards investigating counting complexities, again an untouched ground in this area.

We employed ($*$) when proofs have been omitted due to space constraints.

## References

1. Abdollahzadeh Ahangar, H., Chellali, M., Sheikholeslami, S.M.: On the double Roman domination in graphs. Discret. Appl. Math. **232**, 1–7 (2017)
2. Abu-Khzam, F.N., Fernau, H., Mann, K.: Minimal Roman dominating functions: extensions and enumeration. In: Bekos, M.A., Kaufmann, M. (eds.) WG 2022. LNCS, vol. 13453, pp. 1–15. Springer, Cham (2022). https://doi.org/10.1007/978-3-031-15914-5_1
3. Abu-Khzam, F.N., Fernau, H., Mann, K.: Roman census: enumerating and counting Roman dominating functions on graph classes. Technical report, 2208.05261, Cornell University, arXiv/CoRR (2022). https://arxiv.org/abs/2208.05261
4. Abu-Khzam, F.N., Fernau, H., Mann, K.: Roman census: enumerating and counting Roman dominating functions on graph classes. In: Leroux, J., Lombardy, S., Peleg, D. (eds.) 48th International Symposium on Mathematical Foundations of Computer Science, MFCS. Leibniz International Proceedings in Informatics (LIPIcs), vol. 272, pp. 6:1–6:15. Schloss Dagstuhl – Leibniz-Zentrum für Informatik (2023)
5. Banerjee, S., Chaudhary, J., Pradhan, D.: Unique response Roman domination: complexity and algorithms. Algorithmica **85**, 3889–3927 (2023). https://doi.org/10.1007/s00453-023-01171-7
6. Banerjee, S., Henning, M.A., Pradhan, D.: Algorithmic results on double Roman domination in graphs. J. Comb. Optim. **39**(1), 90–114 (2020)
7. Banerjee, S., Keil, J.M., Pradhan, D.: Perfect Roman domination in graphs. Theoret. Comput. Sci. **796**, 1–21 (2019)

8. Beeler, R.A., Haynes, T.W., Hedetniemi, S.T.: Double Roman domination. Discret. Appl. Math. **211**, 23–29 (2016)
9. Cabrera Martínez, A., Puertas, M., Rodríguez-Velázquez, J.: On the 2-packing differential of a graph. Results Math. **76**, 175:1–175:24 (2021)
10. Cai, L., Chen, J., Downey, R., Fellows, M.: On the parameterized complexity of short computation and factorization. Arch. Math. Logic **36**, 321–337 (1997)
11. Cattanéo, D., Perdrix, S.: The parameterized complexity of domination-type problems and application to linear codes. In: Gopal, T.V., Agrawal, M., Li, A., Cooper, S.B. (eds.) TAMC 2014. LNCS, vol. 8402, pp. 86–103. Springer, Cham (2014). https://doi.org/10.1007/978-3-319-06089-7_7
12. Cesati, M.: The Turing way to parameterized complexity. J. Comput. Syst. Sci. **67**, 654–685 (2003)
13. Chellali, M., Haynes, T.W., Hedetniemi, S.M., Hedetniemi, S.T., McRae, A.A.: A Roman domination chain. Graphs Combin. **32**(1), 79–92 (2016)
14. Chellali, M., Haynes, T.W., Hedetniemi, S.T., McRae, A.A.: Roman 2-domination. Discret. Appl. Math. **204**, 22–28 (2016)
15. Cockayne, E.J., Dreyer, P., Jr., Hedetniemi, S.M., Hedetniemi, S.T.: Roman domination in graphs. Discret. Math. **278**, 11–22 (2004)
16. Courcelle, B., Engelfriet, J., Rozenberg, G.: Handle-rewriting hypergraph grammars. J. Comput. Syst. Sci. **46**(2), 218–270 (1993)
17. Courcelle, B., Makowsky, J., Rotics, U.: Linear time solvable optimization problems on graphs of bounded clique-width. Theory Comput. Syst. **33**, 125–150 (2000)
18. Downey, R.G., Fellows, M.R.: Fundamentals of Parameterized Complexity. Texts in Computer Science, Springer, London (2013). https://doi.org/10.1007/978-1-4471-5559-1
19. Dreyer, P.A.: Applications and variations of domination in graphs. Ph.D. thesis, Rutgers University, New Jersey, USA (2000)
20. Fernau, H.: Roman domination: a parameterized perspective. Int. J. Comput. Math. **85**, 25–38 (2008)
21. Fernau, H.: Extremal kernelization: a commemorative paper. In: Brankovic, L., Ryan, J., Smyth, W.F. (eds.) IWOCA 2017. LNCS, vol. 10765, pp. 24–36. Springer, Cham (2018). https://doi.org/10.1007/978-3-319-78825-8_3
22. Fernau, H., Golovach, P.A., Sagot, M.: Algorithmic enumeration: output-sensitive, input-sensitive, parameterized, approximative (Dagstuhl Seminar 18421). Dagstuhl Rep. **8**(10), 63–86 (2018)
23. Fernau, H., Mann, K.: Parameterized complexity aspects of extension problems (2024). Unpublished manuscript; parts are contained in [?]
24. Flum, J., Grohe, M.: Parameterized Complexity Theory. Springer, Heidelberg (2006). https://doi.org/10.1007/3-540-29953-X
25. Golumbic, M.C.: Interval graphs. In: Annals of Discrete Mathematics, vol. 57, chap. 8, pp. 171–202. Elsevier (2004)
26. Golumbic, M.C., Rotics, U.: On the clique-width of some perfect graph classes. Int. J. Found. Comput. Sci. **11**(3), 423–443 (2000)
27. Guo, J., Hüffner, F., Niedermeier, R.: A structural view on parameterizing problems: distance from triviality. In: Downey, R., Fellows, M., Dehne, F. (eds.) IWPEC 2004. LNCS, vol. 3162, pp. 162–173. Springer, Heidelberg (2004). https://doi.org/10.1007/978-3-540-28639-4_15
28. Haynes, T.W., Hedetniemi, S.T., Henning, M.A.: Structures of Domination in Graphs. Developments in Mathematics, vol. 66. Springer, Cham (2021). https://doi.org/10.1007/978-3-030-58892-2

29. Haynes, T.W., Hedetniemi, S., Henning, M.A. (eds.): Topics in Domination in Graphs. Developments in Mathematics, vol. 64. Springer, Cham (2020). https://doi.org/10.1007/978-3-030-51117-3

30. Henning, M.A., Klostermeyer, W.F., MacGillivray, G.: Perfect Roman domination in trees. Discret. Appl. Math. **236**, 235–245 (2018)

31. Hennings, M., Yeo, A.: Total Domination in Graphs. Springer, New York (2013). https://doi.org/10.1007/978-1-4614-6525-6

32. Junosza-Szaniawski, K., Rzążewski, P.: On the number of 2-packings in a connected graph. Discret. Math. **312**, 3444–3450 (2012)

33. Kratochvíl, J., Křivánek, M.: On the computational complexity of codes in graphs. In: Chytil, M., Janiga, L., Koubek, V. (eds.) MFCS 1988. LNCS, vol. 324, pp. 396–404. Springer, Heidelberg (1988). https://doi.org/10.1007/BFb0017162

34. Liedloff, M., Kloks, T., Liu, J., Peng, S.L.: Efficient algorithms for Roman domination on some classes of graphs. Discret. Appl. Math. **156**(18), 3400–3415 (2008)

35. Liu, C.H., Chang, G.J.: Roman domination on strongly chordal graphs. J. Comb. Optim. **26**(3), 608–619 (2013)

36. Lu, C.L., Tang, C.Y.: Weighted efficient domination problem on some perfect graphs. Discret. Appl. Math. **117**(1–3), 163–182 (2002)

37. Marino, A.: Analysis and enumeration. Algorithms for biological graphs, Atlantis Studies in Computing, vol. 6. Atlantis Press, Paris (2015)

38. Peng, S.L., Tsai, Y.H.: Roman domination on graphs of bounded treewidth. In: The 24th Workshop on Combinatorial Mathematics and Computation Theory, pp. 128–131 (2007)

39. Rubalcaba, R.R., Slater, P.J.: Roman dominating influence parameters. Discret. Math. **307**(24), 3194–3200 (2007)

40. Shang, W., Wang, X., Hu, X.: Roman domination and its variants in unit disk graphs. Discrete Math. Algorithms Appl. **2**(1), 99–106 (2010)

41. Stewart, I.: Defend the Roman Empire. Scientific American, pp. 136, 137, 139 (1999)

42. Strozecki, Y.: Enumeration complexity. EATCS Bull. **129** (2019)

43. Targhi, E.E., Rad, N.J., Volkmann, L.: Unique response Roman domination in graphs. Discret. Appl. Math. **159**(11), 1110–1117 (2011)

44. Wasa, K.: Enumeration of enumeration algorithms. Technical report, 1605.05102, Archiv, Cornell University (2016)

# Computing Longest Common Subsequence Under Cartesian-Tree Matching Model

Taketo Tsujimoto[1], Hiroki Shibata[2]◉, Takuya Mieno[3]◉, Yuto Nakashima[4]◉, and Shunsuke Inenaga[4(✉)]◉

[1] Department of Information Science and Technology, Kyushu University, Fukuoka, Japan
tsujimoto.taketo.852@s.kyushu-u.ac.jp
[2] College of Information Science, University of Tsukuba, Tsukuba, Japan
s2213586@s.tsukuba.ac.jp
[3] Department of Computer and Network Engineering, University of Electro-Communications, Chofu, Japan
tmieno@uec.ac.jp
[4] Department of Informatics, Kyushu University, Fukuoka, Japan
{nakashima.yuto.003,inenaga.shunsuke.380}@m.kyushu-u.ac.jp

**Abstract.** Two strings of the same length are said to *Cartesian-tree match* (*CT-match*) if their Cartesian-trees are isomorphic [Park et al., TCS 2020]. Cartesian-tree matching is a natural model that allows for capturing similarities of numerical sequences. Oizumi et al. [CPM 2022] showed that subsequence pattern matching under CT-matching model can be solved in polynomial time. This current article follows and extends this line of research: We present the first polynomial-time algorithm that finds the *longest common subsequence under CT-matching* of two given strings $S$ and $T$ of length $n$, in $O(n^6)$ time and $O(n^4)$ space for general ordered alphabets. We then show that the problem has a faster solution in the binary case, by presenting an $O(n^2/\log n)$-time and space algorithm.

# 1 Introduction

The *longest common subsequence* (*LCS*) is one of the most fundamental models for measuring string similarities. It is well known that (the length of) an LCS of two given strings $S$ and $T$ of length $n$ can be computed in $O(n^2)$ time and space by standard dynamic programming, or in $O(n^2/\log n)$ time and space [13] by the so-called "Four-Russians" method in the word RAM [3]. These quadratic and weakly sub-quadratic bounds are believed to be essentially optimal, since a strongly sub-quadratic $O(n^{2-\epsilon})$-time solution to LCS with any constant $\epsilon > 0$ refutes the famous SETH [1,5]. Indeed, while there are a number of algorithms for computing LCS whose running times are dependent on other parameters (e.g. [2,10,14,17]), their worst-case time complexities remain $\Omega(n^2)$.

The existing string alignments including LCS can be useful for natural language text and biological sequences, however, these methods have limitations in dealing with sequences for which non-exact matching models are more suitable.

© The Author(s), under exclusive license to Springer Nature Switzerland AG 2024
A. A. Rescigno and U. Vaccaro (Eds.): IWOCA 2024, LNCS 14764, pp. 369–381, 2024.
https://doi.org/10.1007/978-3-031-63021-7_28

For instance, in the analysis of numerical sequences such as time series, capturing "structural" similarities is more important than simply comparing them with standard alignments under the exact matching model.

*Order-preserving matching* (*OP-matching*) is a natural model for dealing with numerical sequences: Two strings $A$ and $B$ of length $n$ are said to be OP-match iff the lexicographical rank of $A[i]$ in $A$ and that of $B[i]$ in $B$ are equal for all $1 \leq i \leq n$. While *substring* matching is polynomially solvable under OP-matching [6,11,12], it is known that subsequence matching is NP-hard under OP-matching [4]. It is immediate from the latter that *order-preserving longest common subsequence* (*OP-LCS*) is also NP-hard.

*Cartesian-tree matching* (*CT-matching*), first proposed by Park et al. [16], is another model for dealing with numerical sequences: Two strings $A$ and $B$ of length $n$ are said to be CT-match iff the (unlabeled) Cartesian trees [7] of $A$ and $B$ are isomorphic. The CT-matching model is a relaxation of the OP-matching model, i.e., any OP-matching strings also CT-match, but the opposite is not true (for instance, $A = \langle 1, 1, 2 \rangle$ and $B = \langle 1, 1, 1 \rangle$ CT-match, but they do not OP-match). CT-matching has attracted attention in terms of pattern matching [16,18], string periodicity [16], and indeterminate strings [8]. The recent work by Oizumi et al. [15] has revealed that this relaxation enables us to perform *subsequence* matching under CT-matching in polynomial time: Given a text $T$ of length $n$ and a pattern $P$ of length $m$, one can find all minimal intervals $[i, j]$ in $T$ such that $T[i..j]$ contains a subsequence $Q$ that CT-matches $P$ in $O(nm \log \log n)$ time and $O(n \log m)$ space.

The aforementioned result poses the natural question - Is the *CT-LCS problem* also polynomial-time solvable? Here, the CT-LCS problem is, given two strings $S$ and $T$ of length $n$, to compute (the length) of a longest string $Q$ such that both $S$ and $T$ have subsequences that CT-match $Q$. We answer this question affirmatively, by presenting an algorithm for computing CT-LCS in $O(n^6)$ time and $O(n^4)$ space for general ordered alphabets. We then present an $O(n^2 / \log n)$-time and space algorithm for computing CT-LCS in the case of binary alphabets. While the $O(n^6)$-time solution in the general case is based on the idea of *pivoted Cartesian-trees* from Oizumi et al. [15], the $O(n^2 / \log n)$-time solution for the binary case is built on a completely different approach that exploits interesting properties of CT-matching of binary strings.

## 2   Preliminaries

### 2.1   Strings

For any positive integer $i$, we define a set $[i] = \{1, \ldots, i\}$ of $i$ integers. Let $\Sigma$ be an *ordered alphabet* of size $\sigma$. For simplicity, let $\Sigma = \{0, \ldots, \sigma - 1\}$. An element of $\Sigma$ is called a *character*. A sequence of characters is called a *string*. The *length* of string $S$ is denoted by $|S|$. The empty string $\varepsilon$ is the string of length 0. If $S = XYZ$, then $X$, $Y$, and $Z$ are respectively called a *prefix*, *substring*, and *suffix* of $S$. For a string $S$, $S[i]$ denotes the $i$th character of $S$ for each $i$ with $1 \leq i \leq |S|$. For each $i, j$ with $1 \leq i \leq j \leq |S|$, $S[i..j]$ denotes the substring of $S$ that begins at position $i$ and ends at position $j$. For convenience, let $S[i..j] = \varepsilon$ for $i > j$. We write $\min(S) = \min\{S[i] \mid i \in [n]\}$ for the minimum value contained in the string $S$. For any $0 \leq m \leq n$, let $\mathcal{I}_m^n$ be the set consisting of all *subscript sequence* $I = (i_1, \ldots, i_m) \in [n]^m$ in ascending order satisfying $1 \leq i_1 < \cdots < i_m \leq n$. For subscript sequence $I = (i_1, \ldots, i_m) \in \mathcal{I}_m^n$, we denote by $S_I = \langle S[i_1], \ldots, S[i_m] \rangle$ the *subsequence* of $S$ corresponding to $I$. For a subscript sequence $I$ and its elements $i_s, i_t \in I$ with $i_s \leq i_t$, $I[i_s : i_t]$ denotes the substring of $I$ that starts with $i_s$ and ends with $i_t$.

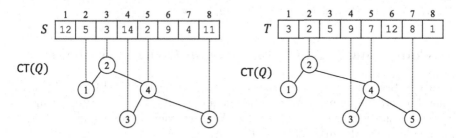

**Fig. 1.** $Q = \langle 5, 3, 14, 2, 11 \rangle$ is a CT longest common subsequence of $S = \langle 12, 5, 3, 14, 2, 9, 4, 11 \rangle$ and $T = \langle 3, 2, 5, 9, 7, 12, 8, 1 \rangle$. Each node in $\mathsf{CT}(Q)$ is labeled by the corresponding position in $Q$.

## 2.2  Cartesian Tree Matching and CT-LCS

For a string $S$, let $\mathrm{min\_id}(S)$ denote the least index $i$ such that $S[i]$ is the smallest element in $S$. The *Cartesian tree* of a string $S$, denoted by $\mathsf{CT}(S)$, is the ordered binary tree recursively defined as follows: If $S = \varepsilon$, then $\mathsf{CT}(S)$ is an empty tree, and otherwise, $\mathsf{CT}(S)$ is the tree rooted at $i = \mathrm{min\_id}(S)$ such that the left subtree of $i$ is $\mathsf{CT}(S[1..i-1])$ and the right subtree of $i$ is $\mathsf{CT}(S[i+1..|S|])$. For two strings $S$ and $T$ of equal length, the two Cartesian trees $\mathsf{CT}(S)$ and $\mathsf{CT}(T)$ are *isomorphic* if they have the same tree topology as ordered trees [9]. We denote it by $\mathsf{CT}(S) = \mathsf{CT}(T)$. We say that two strings $S$ and $T$ *CT-match* if $\mathsf{CT}(S) = \mathsf{CT}(T)$.

A string $Q$ is said to be a *CT-subsequence* of a string $S$ if there is a subsequence $P$ of $S$ such that $\mathsf{CT}(Q) = \mathsf{CT}(P)$. A string $Q$ is said to be a *common CT-subsequence* of two strings $S$ and $T$ if $Q$ is a CT subsequence of both $S$ and $T$. A string $Q$ is said to be a *longest common CT-subsequence (CT-LCS)* of $S$ and $T$ if there are no common CT-subsequences of $S$ and $T$ longer than $Q$. We show an example of CT-LCS in Fig. 1. The length of CT-LCS of strings $S$ and $T$ is denoted by $\mathsf{ct\_lcs}(S, T)$. We solve the following problem.

*Problem 1.* Given two strings $S$ and $T$ of length $n$, compute $\mathsf{ct\_lcs}(S, T)$.

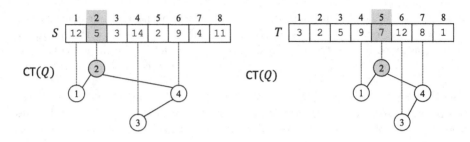

**Fig. 2.** $Q = \langle 12, 5, 14, 9 \rangle$ is a fixed CT longest common subsequence of $S = \langle 12, 5, 3, 14, 2, 9, 4, 11 \rangle$ and $T = \langle 3, 2, 5, 9, 7, 12, 8, 1 \rangle$ with pivot $(2, 5)$.

# 3   Computing CT-LCS for General Ordered Alphabets

In this section, we propose an algorithm for solving the CT-LCS problem for general ordered alphabets. An $O(n^6)$-time and $O(n^4)$-space algorithm, which is our main result will be given in Sect. 3.2. We start from explaining an $O(n^8)$-time and $O(n^6)$-space algorithm for simplicity (Sect. 3.1).

For each $c \in \Sigma$, let $P_c(S) = \{i \mid S[i] = c\}$ and let $S'$ be the string of length $|S|$ such that $S'[i] = (S[i], r_i)$ for $1 \le i \le |S|$, where $r_i$ is the rank of $i$ in $P_{S[i]}(S)$. For ordered pairs $(c, r)$ and $(c', r')$ of characters and integers, let $(c, r) < (c', r')$ iff (1) $c < c'$ or (2) $c = c'$ and $r < r'$. Then, it holds that $\mathsf{CT}(S)$ and $\mathsf{CT}(S')$ are isomorphic. Thus, without loss of generality, we can assume that all characters in the string $S$ are distinct. The same assumption applies to the other string $T$, but $S$ and $T$ may share the same characters.

## 3.1   $O(n^8)$-Time and $O(n^6)$-Space Algorithm

We refer to a pair $(i, j) \in [n]^2$ of positions in $S$ and $T$ as a *pivot*. Our algorithm in the general case is based on the idea of *pivoted Cartesian-trees* from Oizumi et al. [15] defined as follows.

**Definition 1 (Fixed CT (longest) common subsequence).** *Let $(i, j) \in [n]^2$ be a pivot of strings $S$ and $T$. A string $Q$ is said to be a* fixed CT common subsequence *(f-CT-CS) of $S$ and $T$ with pivot $(i, j)$ if there exist subscript sequences $I, J \in \mathcal{I}_{|Q|}^n$ such that*

- $\mathsf{CT}(Q) = \mathsf{CT}(S_I) = \mathsf{CT}(T_J)$,
- $S[i] = \min(S_I)$, *and*
- $T[j] = \min(T_J)$.

*Moreover, a string $Q$ is said to be the* fixed CT longest common subsequence *(f-CT-LCS) of $S$ and $T$ with pivot $(i, j)$ if there are no f-CT-CS with pivot $(i, j)$ longer than $Q$ (see also Fig. 2).*

Our solution is a dynamic programming based on the f-CT-LCS. We also consider the f-CT-LCS for substrings of $S$ and $T$. We will use positions $i, j \in [n]$

of the input strings to indicate a pivot for substrings, namely, we say pivot $(i, j) \in [n]^2$ of substrings $S' = S[\ell_1..r_1]$ and $T' = T[\ell_2..r_2]$ instead of pivot $(i - \ell_1 + 1, j - \ell_2 + 1) \in [r_1 - \ell_1 + 1] \times [r_2 - \ell_2 + 1]$ of $S'$ and $T'$, where $\ell_1 \leq i \leq r_1$, $\ell_2 \leq j \leq r_2$. Let $C(i, j, \ell_1, r_1, \ell_2, r_2)$ be the length of the f-CT-LCS of substrings $S[\ell_1..r_1]$ and $T[\ell_2..r_2]$ with pivot $(i, j)$. It is clear from the definition that

$$\mathsf{ct\_lcs}(S, T) = \max\{C(i, j, 1, n, 1, n) \mid 1 \leq i \leq n, 1 \leq j \leq n\}$$

holds. The following lemma shows the main idea of computing $C(i, j, 1, n, 1, n)$ by a dynamic programming (see also Fig. 3 for an illustration).

**Lemma 1.** *For any* $(i, j, \ell_1, r_1, \ell_2, r_2) \in [n]^6$ *that satisfies* $\ell_1 \leq i \leq r_1$, $\ell_2 \leq j \leq r_2$, *define* $\mathcal{M}_L$ *and* $\mathcal{M}_R$ *as follows:*

$$\mathcal{M}_L = \{C(i', j', \ell_1, i - 1, \ell_2, j - 1)$$
$$\mid S[i'] > S[i], T[j'] > T[j], \ell_1 \leq i' \leq i - 1, \ell_2 \leq j' \leq j - 1\} \cup \{0\},$$
$$\mathcal{M}_R = \{C(i', j', i + 1, r_1, j + 1, r_2)$$
$$\mid S[i'] > S[i], T[j'] > T[j], i + 1 \leq i' \leq r_1, j + 1 \leq j' \leq r_2\} \cup \{0\}.$$

*Then the recurrence* $C(i, j, \ell_1, r_1, \ell_2, r_2) = \max \mathcal{M}_L + \max \mathcal{M}_R + 1$ *holds.*

*Proof.* Let $Q$ be the f-CT-LCS of $S[\ell_1..r_1]$ and $T[\ell_2..r_2]$ with $(i, j)$. By Definition 1, there exist subscript sequences $I = (i_1, \ldots, i_{|Q|})$ and $J = (j_1, \ldots, j_{|Q|})$ that satisfy $\mathsf{CT}(Q) = \mathsf{CT}(S_I) = \mathsf{CT}(T_J)$. It is also clear from the definition that $S[i] = \min(S_I)$ and $T[j] = \min(T_J)$ hold. Let $q = \mathsf{min\_id}(Q)$. We show that $q - 1 = \max \mathcal{M}_L$ holds.

- Assume that $q = 1$. In this case, we need to show $\mathcal{M}_L = \{0\}$. Suppose on the contrary that there exists $(i', j')$ such that $S[i'] > S[i], T[j'] > T[j], \ell_1 \leq i' \leq i - 1$, and $\ell_2 \leq j' \leq j - 1$. Let $I^* = (i', i_1, \ldots, i_{|Q|}), J^* = (j', j_1, \ldots, j_{|Q|})$, and $Q^* = \alpha \cdot Q$ where $\alpha$ is a character in $\Sigma$ that satisfies $\alpha > Q[q]$. Also, $\mathsf{CT}(Q^*) = \mathsf{CT}(S_{I^*}) = \mathsf{CT}(T_{J^*})$ holds since $\mathsf{CT}(Q) = \mathsf{CT}(S_I) = \mathsf{CT}(T_J)$, $\alpha > Q[q], S[i'] > S[i]$, and $T[j'] > T[j]$ hold. Moreover, $Q^*$ is an f-CT-CS of $S[\ell_1..r_1]$ and $T[\ell_2..r_2]$ with $(i, j)$, since $S[i] = \min(S_{I^*})$ and $T[j] = \min(T_{J^*})$. This contradicts the fact that $Q$ is the f-CT-LCS of $S[\ell_1..r_1]$ and $T[\ell_2..r_2]$ with $(i, j)$ (from $|Q^*| > |Q|$). Thus, $\mathcal{M}_L = \{0\}$ and $\max \mathcal{M}_L = 0$.
- Assume that $q > 1$. Let $p_1$ (resp., $p_2$) be the predecessor of $i$ in $I$ (resp., the predecessor of $j$ in $J$). By the definition of the Cartesian tree, $\mathsf{CT}(Q[1..q - 1]) = \mathsf{CT}(S_{I[i_1:p_1]}) = \mathsf{CT}(T_{J[j_1:p_2]})$ holds, and there exist $i^*$ and $j^*$ such that $S[i^*] = \min(S_{I[i_1:p_1]})(> S[i])$ and $T[j^*] = \min(T_{J[j_1:p_2]})(> T[j])$. This implies that $Q[1..q - 1]$ is an f-CT-CS of $S[\ell_1..i - 1]$ and $T[\ell_2..j - 1]$ with $(i^*, j^*)$. Thus, $|Q[1..q - 1]| = q - 1 \leq \max \mathcal{M}_L$ holds. In the rest of this case, we show $q - 1 \geq \max \mathcal{M}_L$ to prove the equality. Suppose on the contrary that $q - 1 < \max \mathcal{M}_L$. Since $0 < q - 1 < \max \mathcal{M}_L$ (from assumptions), there exists $(i'', j'') \in [n]^2$ such that $C(i'', j'', \ell_1, i - 1, \ell_2, j - 1) = \max \mathcal{M}_L$. Let $Q''$ be an f-CT-LCS string of $S[\ell_1..i - 1]$ and $T[\ell_2..j - 1]$ with $(i'', j'')$. Then there exist subscript sequences $I''$ and $J''$ over $\{\ell_1, \ldots, i - 1\}$ and $\{\ell_2, \ldots, j - 1\}$,

respectively, such that $\mathsf{CT}(Q'') = \mathsf{CT}(S_{I''}) = \mathsf{CT}(T_{J''})$. Let $\hat{I}$ denote the subscript sequence that is the concatenation of $I''$ and $I[i : i_{|Q|}]$, and $\hat{J}$ denote the subscript sequence that is the concatenation of $J''$ and $J[j : j_{|Q|}]$. Then $\mathsf{CT}(Q'' \cdot Q[q..|Q|]) = \mathsf{CT}(S_{\hat{I}}) = \mathsf{CT}(T_{\hat{J}})$, $S[i] = \min(S_{\hat{I}})$, and $T[j] = \min(T_{\hat{J}})$ hold. This implies that $Q'' \cdot Q[q..|Q|]$ is an f-CT-LCS of $S[\ell_1..i-1]$ and $T[\ell_2..j-1]$ with $(i,j)$. However, $|Q'' \cdot Q[q..|Q|]| = |Q''| + |Q[q..|Q|]| = \max\mathcal{M}_L + |Q| - q + 1 > q - 1 + |Q| - q + 1 = |Q|$ holds. This contradicts to the fact that $Q$ is the f-CT-LCS of $S[\ell_1..i-1]$ and $T[\ell_2..j-1]$ with $(i,j)$. Thus $q - 1 = \max\mathcal{M}_L$ also holds for $q > 1$.

We can also prove $|Q| - q = \max\mathcal{M}_R$ by a symmetric manner. Therefore, $|Q| = q - 1 + |Q| - q + 1 = \max\mathcal{M}_L + \max\mathcal{M}_R + 1$ holds.     □

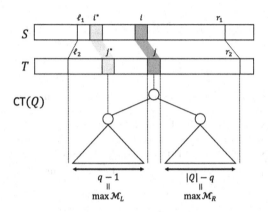

**Fig. 3.** Sketch of our recurrence by Lemma 1.

Then we can obtain an $O(n^8)$-time and $O(n^6)$-space algorithm for solving the CT-LCS problem based on Lemma 1 (see Algorithm 2). Our algorithm computes a six-dimensional table $C$ for any $(i, j, \ell_1, r_1, \ell_2, r_2) \in [n]^6$ that satisfies $\ell_1 \le i \le r_1$, $\ell_2 \le j \le r_2$. Notice that the processing order $i_1, i_2, \ldots, i_n$ (resp., $j_1, j_2, \ldots, j_n$) w.r.t. index $i$ (resp., $j$) has to satisfy $S[i_1] > S[i_2] > \cdots > S[i_n]$ (resp., $T[j_1] > T[j_2] > \cdots > T[j_n]$). The algorithm finally returns $\mathsf{ct\_lcs}(S, T) = \max\{C(i, j, 1, n, 1, n) \mid 1 \le i \le n, 1 \le j \le n\}$. For each fixed $(i, j, \ell_1, r_1, \ell_2, r_2) \in [n]^6$ (i.e., $O(n^6)$ iterations), we can compute $\mathcal{M}_L$ and $\mathcal{M}_R$ in $O(n^2)$ time. Therefore, we can compute table $C$ in $O(n^8)$ time and $O(n^6)$ space.

**Theorem 1.** *The CT-LCS problem can be solved in $O(n^8)$ time and $O(n^6)$ space*

### 3.2  $O(n^6)$-Time $O(n^4)$-Space Algorithm

In the sequel, we propose an improved algorithm that is based on the previous algorithm and runs in $O(n^6)$ time and $O(n^4)$ space. The key observation is that

$\mathcal{M}_L$ and $\mathcal{M}_R$ actually depend on only four variables. Namely, $\mathcal{M}_L$ depends on $(i, j, \ell_1, \ell_2)$, and $\mathcal{M}_R$ depends on $(i, j, r_1, r_2)$. Let $L(i, j, \ell_1, \ell_2) = \max \mathcal{M}_L$ and $R(i, j, r_1, r_2) = \max \mathcal{M}_R$. Then we can represent $C(i, j, \ell_1, r_1, \ell_2, r_2)$ as

$$C(i, j, \ell_1, r_1, \ell_2, r_2) = L(i, j, \ell_1, \ell_2) + R(i, j, r_1, r_2) + 1.$$

Based on this recurrence, we can obtain the following alternative lemma.

**Lemma 2.** *For any* $(i, j, \ell_1, r_1, \ell_2, r_2) \in [n]^6$ *that satisfies* $\ell_1 \le i \le r_1$, $\ell_2 \le j \le r_2$, *the following recurrences hold:*

$$L(i, j, \ell_1, \ell_2) = \max\{L(i', j', \ell_1, \ell_2) + R(i', j', i - 1, j - 1) + 1 \mid S[i'] > S[i],$$
$$T[j'] > T[j], \ell_1 \le i' \le i - 1, \ell_2 \le j' \le j - 1\} \cup \{0\},$$
$$R(i, j, r_1, r_2) = \max\{L(i', j', i + 1, j + 1) + R(i', j', r_1, r_2) + 1 \mid S[i'] > S[i],$$
$$T[j'] > T[j], i + 1 \le i' \le r_1, j + 1 \le j' \le r_2\} \cup \{0\}.$$

It follows from the definitions that

$$\mathsf{ct\_lcs}(S, T) = \max\{L(i, j, 1, 1) + R(i, j, n, n) + 1 \mid 1 \le i \le n, 1 \le j \le n\}.$$

Then we can obtain an $O(n^6)$-time and $O(n^4)$-space algorithm for solving the CT-LCS problem based on Lemma 2. Our algorithm computes two dimensional tables $L$ and $R$. For each fixed $(i, j) \in [n]^2$ (i.e., $O(n^2)$ iterations), we can compute $L(i, j, \cdot, \cdot)$ and $R(i, j, \cdot, \cdot)$ in $O(n^4)$ time. The algorithm finally returns $\mathsf{ct\_lcs}(S, T) = \max\{L[i][j][1][1] + R[i][j][n][n] + 1 \mid 1 \le i \le n, 1 \le j \le n\}$. Therefore, we can compute table $C$ in $O(n^6)$ time and $O(n^4)$ space.

**Theorem 2.** *The CT-LCS problem can be solved in* $O(n^6)$ *time and* $O(n^4)$ *space.*

We can compute a CT-LCS string by storing the following additional information: We store pivot $(i', j')$ with $L(i, j, \ell_1, \ell_2)$ that satisfies $L[i][j][\ell_1][\ell_2] = L[i'][j'][\ell_1][\ell_2] + R[i'][j'][i - 1][j - 1] + 1$ (also for $R$). If we do so, we can compute a CT-LCS by tracking back the tables from the pivot $(i, j)$ that gives $\mathsf{ct\_lcs}(S, T)$ in $O(|\mathsf{ct\_lcs}(S, T)|) = O(n)$ time.

**Corollary 1.** *A CT-LCS string can be computed in* $O(n^6)$ *time and* $O(n^4)$ *space.*

# 4   Computing CT-LCS for Binary Alphabets

In this section, we propose an algorithm for solving CT-LCS problem for the binary alphabet $\{0, 1\}$. Throughout this section, we assume that the strings $S$ and $T$ are binary strings and discard the assumption that all characters are distinct in $S$ and in $T$.

We first recall the *parent-distance representation* presented by Park et al. [16]: Given a string $S[1..n]$, the parent-distance representation of $S$ is an integer string $PD(S)[1..n]$, which is defined as follows:

$$PD(S)[i] = \begin{cases} i - \max_{1 \le j < i}\{j \mid S[j] \le S[i]\} & \text{if such } j \text{ exists,} \\ 0 & \text{otherwise.} \end{cases}$$

For example, the parent-distance representation of string $S = \langle 1, 0, 1, 1, 0, 0, 1 \rangle$ is $PD(S) = \langle 0, 0, 1, 1, 3, 1, 1 \rangle$.

**Lemma 3 ([16]).** *Two strings $S_1$ and $S_2$ CT-match if and only if $S_1$ and $S_2$ have the same parent-distance representations.*

Lemma 3 allows for determining whether two strings CT-match or not. We will only use this representation to guarantee the correctness of our algorithm for the binary case, and do not explicitly compute it.

We start from a simple observation of CT-matching for binary strings. The following lemmas will support our algorithm for the binary alphabet.

**Lemma 4.** *Let $S_1 = 1^n$ and $S_2$ be a binary string of length $n$. Then, $S_1$ and $S_2$ CT-match if and only if $S_2 = 0^i 1^{n-i}$ for some integer $i \ge 0$.*

It is easy to see that $PD(1^n) = 01^{n-1}$. This implies that $S_2$ should be a non-decreasing binary string. From now on, we discuss the case where 0 appears in both $S_1$ and $S_2$.

**Lemma 5.** *For two binary strings $S_1$ and $S_2$ of length $n$ both containing 0, $S_1$ and $S_2$ CT-match if and only if there exist a string $w$ and two integers $i \ge 1$, $j \ge 1$ such that $S_1 = w0^i 1^{n-|w|-i}$ and $S_2 = w0^j 1^{n-|w|-j}$.*

*Proof* ($\Longrightarrow$). Since $S_1$ and $S_2$ CT-match, $PD(S_1) = PD(S_2)$ (by Theorem 3). Let $p$ (resp., $q$) be the smallest integer such that $S_1[p..n] = 0^i 1^j$ (resp., $S_2[q..n] = 0^i 1^j$) for some $i \ge 1$, $j \ge 0$. Assume that $1 \le q < p$. Since $S_1[p-1] = 1$ and $S_1[p] = 0$, either $PD(S_1)[p] = 0$ or $PD(S_1)[p] > 1$ holds (i.e., $PD(S_1)[p] \ne 1$). On the other hand, $PD(S_2)[p] = 1$ holds since $S_2[p-1] \le S_2[p]$. Then $PD(S_1)[p] \ne PD(S_2)[p]$, which is a contradiction. By a similar discussion, $1 \le p < q$ also leads a contradiction. Next, we assume that $p = q = 1$. This assumption implies that the statement holds since $w = \varepsilon$. Assume that $1 < p = q$. Suppose on the contrary that $S_1[1..p-1] \ne S_2[1..p-1]$. There exists an integer $j^*$ such that $j^* = \max_{1 \le j \le p-1}\{j \mid S_1[j] \ne S_2[j]\}$. This means that $S_1[j^*+1..p-1] = S_2[j^*+1..p-1]$. We assume w.l.o.g. that $S_1[j^*] = 0$ and $S_2[j^*] = 1$ (the other case is symmetric).

- If 0 does not appear in $S_1[j^*+1..p-1]$, $PD(S_1)[p] = p - j^* > 0$ holds. On the other hand, either $PD(S_2)[p] = 0$ or $PD(S_2)[p] > p - j^*$ holds. Thus $PD(S_1) \ne PD(S_2)$, which is a contradiction.
- If 0 appears in $S_1[j^*+1..p-1]$, there exists an integer $i^*$ such that $i^* = \min_{j^*+1 \le j \le p-1}\{j \mid S_1[j] = S_2[j] = 0\}$ holds. This implies that $PD(S_1)[i^*] = i^* - j^* > 0$ holds. On the other hand, either $PD(S_2)[i^*] = 0$ or $PD(S_2)[i^*] > i^* - j^*$ holds. Thus $PD(S_1) \ne PD(S_2)$, which is a contradiction.

Therefore, $S_1 = w0^i1^{n-p+1-i}$ and $S_2 = w0^j1^{n-p+1-j}$ hold for some integers $i, j$ where $w = S_1[1..p-1] = S_2[1..p-1]$.

($\Longleftarrow$) If $w = \varepsilon$, it is clear that $S_1$ and $S_2$ CT-match. We consider the case of $w \neq \varepsilon$. We show that $PD(S_1) = PD(S_2)$ holds, which is suffice due to Lemma 3. It is easy to see that $PD(S_1)[i] = PD(S_2)[i]$ for all $i$ that satisfies $1 \leq i \leq |w|$. Moreover, $PD(S_1)[i] = PD(S_2)[i] = 1$ also holds for all $i$ with $|w| + 1 < i \leq n$. If 0 does not appear in $w$, $PD(S_1)[|w| + 1] = PD(S_2)[|w| + 1] = 0$. We assume that 0 appears in $w$ for the remaining case. Let $j^* = \max_{1 \leq j \leq |w|}\{j \mid w[j] = 0\}$. Since $S_1[|w| + 1] = S_2[|w| + 1] = 0$ and $S_1[1..|w|] = S_2[1..|w|]$, $PD(S_1)[|w| + 1] = PD(S_2)[|w| + 1]$ holds. Therefore $PD(S_1) = PD(S_2)$.  $\square$

Based on the above two lemmas, we propose an algorithm for the binary alphabet case. Let $N_1(S)$ be the number of occurrences of 1 in string $S$, and $L_{01}(S)$ the length of the longest non-decreasing subsequence of $S$ that contains 0. If there is no such subsequence, let $L_{01}(S) = 0$. We also define $\mathsf{cand}(S, T)$ as the maximum integer $k = |w| + i + j = |w| + i' + j'$ such that $w0^i1^j$ is a subsequence of $S$ and $w0^{i'}1^{j'}$ is a subsequence of $T$ for some string $w$, and integers $i, i' \geq 1$, $j, j' \geq 0$. Then the following properties hold for subsequences $S'$ and $T'$ of $S$ and $T$ that give $\mathsf{ct\_lcs}(S, T)$.

- If 0 appears in both $S'$ and $T'$, $\mathsf{ct\_lcs}(S, T) = \mathsf{cand}(S, T)$ (by Lemma 5).
- If either $S'$ or $T'$ does not contain 0, $\mathsf{ct\_lcs}(S, T)$ equals $\min(N_1(S), L_{01}(T))$ or $\min(N_1(T), L_{01}(S))$, respectively (from Lemma 4).
- If 0 does not appears in both $S'$ and $T'$, $\mathsf{ct\_lcs}(S, T) = \min(N_1(S), N_1(T))$.

Due to the above properties, $\mathsf{ct\_lcs}(S, T) = \max(\mathsf{cand}(S, T), m_1, m_2, m_3)$ holds for $m_1 = \min(N_1(S), L_{01}(T))$, $m_2 = \min(N_1(T), L_{01}(S))$, and $m_3 = \min(N_1(S), N_1(T))$.

Now we are ready to describe our algorithm. Let $LND_S(i) = L_{01}(S[i..n])$ for any integer $i$ with $1 \leq i \leq n$. For convenience, let $LND_S(n + 1) = 0$. Firstly, we compute $N_1(S[i..n])$, $N_1(T[i..n])$, $LND_S(i)$, and $LND_T(i)$ for all $i$ that satisfies $1 \leq i \leq n$. We can easily compute $N_1(S[i..n])$ and $N_1(T[i..n])$ for all $i$ in $O(n)$ time and space. We can also compute $LND_S(i)$ (and $LND_T(i)$ in a similar way) by using the following recurrence:

$$LND_S(i) = \begin{cases} \max(LND_S(i+1) + 1, N_1(S[i+1..n]) + 1) & \text{if } S[i] = 0, \\ LND_S(i+1) & \text{if } S[i] = 1. \end{cases}$$

These values can also be computed in $O(n)$ time and space.

Since $\mathsf{cand}(S, T)$ requires the length of $w$ which is described in the above discussion, we use a data structure for computing the longest common subsequence $\mathsf{LCS}(i, j)$ of $S[1..i]$ and $T[1..j]$. For convenience, we set $\mathsf{LCS}(i, 0) = \mathsf{LCS}(0, i) = 0$ for all $i \in [n] \cup \{0\}$. By using the Four-Russians method [13], we can compute an $O(n^2/\log n)$-space data structure in $O(n^2/\log n)$ time that can answer $\mathsf{LCS}(i, j)$ in $O(\log^2 n)$ time for any $i \in [n] \cup \{0\}$ and $j \in [n] \cup \{0\}$. By the definition of $\mathsf{cand}$, the following equation can be obtained:

$$\mathsf{cand}(S, T) = \max_{1 \leq \ell \leq n} \{\mathsf{LCS}(p_\ell - 1, q_\ell - 1) + \ell\}$$

where $p_\ell = \max\{p \mid LND_S(p) = \ell\}$ and $q_\ell = \max\{q \mid LND_T(q) = \ell\}$ (see also Fig. 4 for an illustration). Since we have already computed $LND_S(i)$, $LND_S(i)$, and the data structure for LCS, we can compute $\mathsf{cand}(S,T)$ in $O(n \log^2 n)$ time based on the above equation. Finally, we can obtain $\mathsf{ct\_lcs}(S,T)$ by computing $\max(\mathsf{cand}(S,T), m_1, m_2, m_3)$ in constant time (see also Algorithm 1).

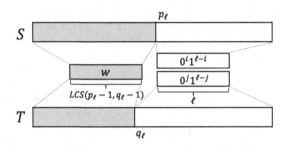

**Fig. 4.** Illustration for our idea for computing $\mathsf{cand}(S,T)$.

**Theorem 3.** *The CT-LCS problem on binary can be solved in $O(n^2/\log n)$ time and $O(n^2/\log n)$ space.*

We can reconstruct a CT-LCS of $S$ and $T$ in $O(n \log n)$ time as follows: If one of $m_1$, $m_2$, and $m_3$ gives $\mathsf{ct\_lcs}(S,T)$, we can easily obtain a CT-LCS string in $O(n)$ time by using $LND_S(i)$ and $LND_T(i)$. Otherwise, two subsequences $S'$ and $T'$ which give $\mathsf{ct\_lcs}(S,T)$ can be represented as $S' = w0^i1^{\ell-i}$ and $T' = w0^j1^{\ell-j}$ for some $i$ and $j$. Integers $i$ and $j$ can be obtained by $LND_S(i)$ and $LND_T(i)$. In the Four-Russians method, $(n \times n)$-table LCS is factorized into $(n/\log n \times n/\log n)$-blocks. The data structure actually stores LCS values on boundaries of blocks. Thus we can obtain string $w$ by tracing back in $O((n/\log n) \cdot \log^2 n) = O(n \log n)$ time (see also Fig. 5).

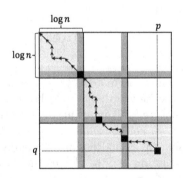

**Fig. 5.** Illustration for an idea for reconstructing the LCS part.

---

**Algorithm 1:** Algorithm for solving CT-LCS problem on binary

---

**Input:** Binary strings $S[1..n], T[1..n] \in \{0,1\}^*$

**Output:** ct_lcs$(S, T)$

1 Precompute data structure $LCS[i][j]$ that can answer $\mathsf{LCS}(i, j)$ in $O(\log^2 n)$ time for any $i \in [n] \cup \{0\}$ and $j \in [n] \cup \{0\}$;

2 $N_S[n + 1] \leftarrow 0$;

3 **for** $i \leftarrow n$ **to** 1 **do**

4     **if** $S[i] = 0$ **then**

5         $N_S[i] \leftarrow N_S[i + 1]$;

6     **else**

7         $N_S[i] \leftarrow N_S[i + 1] + 1$;

8 $LND_S[n + 1] \leftarrow 0$;

9 **for** $i \leftarrow n$ **to** 1 **do**

10     **if** $S[i] = 0$ **then**

11         $LND_S[i] \leftarrow \max(LND_S[i + 1] + 1, N_S[i + 1] + 1)$;

12     **else**

13         $LND_S[i] \leftarrow N_S[i + 1]$;

14 $p[\ell] \leftarrow 0$ for all $\ell \in [n]$;

15 **for** $i \leftarrow 1$ **to** $n$ **do**

16     $p[LND_S[i]] \leftarrow i$;

17 Compute $N_T[i] = N_1(T[i..n])$, $LND_T[i] = LND_T(i)$, $q[\ell] = \max\{q \mid LND_T[q] = \ell\}$ for all $i \in [n + 1]$ and $\ell \in [n]$ in the same way;

18 $cand \leftarrow 0$;

19 **for** $\ell \leftarrow 1$ **to** $n$ **do**

20     **if** $p[\ell] \neq 0$ **and** $q[\ell] \neq 0$ **then**

21         $cand \leftarrow \max(cand, LCS[p[\ell] - 1][q[\ell] - 1] + \ell)$;

22 $m_1 \leftarrow \min(N_S[1], LND_T[1])$;

23 $m_2 \leftarrow \min(LND_S[1], N_T[1])$;

24 $m_3 \leftarrow \min(N_S[1], N_T[1])$;

25 **return** $\max(cand, m_1, m_2, m_3)$

---

**Corollary 2.** *The CT-LCS string of two binary strings can be computed in* $O(n^2/\log n)$ *time and* $O(n^2/\log n)$ *space.*

---

**Algorithm 2:** Algorithm for solving CT-LCS problem

**Input:** Strings $S[1..n], T[1..n] \in \Sigma^*$
**Output:** ct_lcs$(S, T)$

1 Find $i_1, i_2, \ldots, i_n$ that satisfy $S[i_1] > S[i_2] > \cdots > S[i_n]$;
2 Find $j_1, j_2, \ldots, j_n$ that satisfy $T[j_1] > T[j_2] > \cdots > T[j_n]$;
3 $ctlcs \leftarrow 0$;
4 **for** $i \leftarrow i_1$ **to** $i_n$ **do**
5    **for** $j \leftarrow j_1$ **to** $j_n$ **do**
6       **for** $\ell_1 \leftarrow 1$ **to** $i$ **do**
7          **for** $r_1 \leftarrow i$ **to** $n$ **do**
8             **for** $\ell_2 \leftarrow 1$ **to** $j$ **do**
9                **for** $r_2 \leftarrow j$ **to** $n$ **do**
10                   $\mathcal{M}_L \leftarrow 0$;
11                   **if** $\ell_1 \neq i$ **and** $\ell_2 \neq j$ **then**
12                      **for** $i' \leftarrow \ell_1$ **to** $i - 1$ **do**
13                         **for** $j' \leftarrow \ell_2$ **to** $j - 1$ **do**
14                            **if** $S[i'] > S[i]$ **and** $T[j'] > T[j]$ **then**
15                               $\mathcal{M}_L \leftarrow$
                                $\max(\mathcal{M}_L, C[i'][j'][\ell_1][i-1][\ell_2][j-1])$;

16                   $\mathcal{M}_R \leftarrow 0$;
17                   **if** $r_1 \neq i$ **and** $r_2 \neq j$ **then**
18                      **for** $i' \leftarrow i + 1$ **to** $r_1$ **do**
19                         **for** $j' \leftarrow j + 1$ **to** $r_2$ **do**
20                            **if** $S[i'] > S[i]$ **and** $T[j'] > T[j]$ **then**
21                               $\mathcal{M}_R \leftarrow$
                                  $\max(\mathcal{M}_R, C[i'][j'][i+1][r_1][j+1][r_2])$;

22                   $C[i][j][\ell_1][r_1][\ell_2][r_2] \leftarrow \mathcal{M}_L + \mathcal{M}_R + 1$;

23    $ctlcs \leftarrow \max(ctlcs, C[i][j][1][n][1][n])$;

24 **return** $ctlcs$

---

**Acknowledgments.** The authors thank Shay Mozes, Oren Weimann, and Tsubasa Oizumi for discussions on subsequence CT-matching in the binary case.

This work was supported by JSPS KAKENHI Grant Numbers JP23H04381 (TM), JP21K17705, JP23H04386 (YN), JP20H05964, JP22H03551 (SI).

# References

1. Abboud, A., Backurs, A., Williams, V.V.: Tight hardness results for LCS and other sequence similarity measures. In: FOCS 2015, pp. 59–78 (2015)
2. Apostolico, A., Browne, S., Guerra, C.: Fast linear-space computations of longest common subsequences. Theor. Comput. Sci. **92**(1), 3–17 (1992)

3. Arlazarov, V., Dinic, E., Kronrod, M., Faradzev, I.: On economical construction of the transitive closure of a directed graph. Sov. Math. Dokl. **11**(5), 1209–1210 (1970)

4. Bose, P., Buss, J.F., Lubiw, A.: Pattern matching for permutations. Inf. Process. Lett. **65**(5), 277–283 (1998)

5. Bringmann, K., Künnemann, M.: Quadratic conditional lower bounds for string problems and dynamic time warping. In: FOCS 2015, pp. 79–97 (2015)

6. Crochemore, M., et al.: Order-preserving indexing. Theor. Comput. Sci. **638**, 122–135 (2016)

7. Gabow, H.N., Bentley, J.L., Tarjan, R.E.: Scaling and related techniques for geometry problems. In: STOC 1984, pp. 135–143 (1984)

8. Gawrychowski, P., Ghazawi, S., Landau, G.M.: On indeterminate strings matching. In: CPM 2020. LIPIcs, vol. 161, pp. 14:1–14:14 (2020)

9. Hoffmann, C.M., O'Donnell, M.J.: Pattern matching in trees. J. ACM **29**(1), 68–95 (1982). https://doi.org/10.1145/322290.322295

10. Hunt, J.W., Szymanski, T.G.: A fast algorithm for computing longest common subsequences. Commun. ACM **20**(5), 350–353 (1977)

11. Kim, J., et al.: Order-preserving matching. Theor. Comput. Sci. **525**, 68–79 (2014)

12. Kubica, M., Kulczynski, T., Radoszewski, J., Rytter, W., Walen, T.: A linear time algorithm for consecutive permutation pattern matching. Inf. Process. Lett. **113**(12), 430–433 (2013)

13. Masek, W.J., Paterson, M.S.: A faster algorithm computing string edit distances. J. Comput. Syst. Sci. **20**(1), 18–31 (1980). https://doi.org/10.1016/0022-0000(80)90002-1

14. Nakatsu, N., Kambayashi, Y., Yajima, S.: A longest common subsequence algorithm suitable for similar text strings. Acta Inf. **18**, 171–179 (1982)

15. Oizumi, T., Kai, T., Mieno, T., Inenaga, S., Arimura, H.: Cartesian tree subsequence matching. In: CPM 2022. LIPIcs, vol. 223, pp. 14:1–14:18 (2022)

16. Park, S.G., Bataa, M., Amir, A., Landau, G.M., Park, K.: Finding patterns and periods in Cartesian tree matching. Theor. Comput. Sci. **845**, 181–197 (2020)

17. Sakai, Y.: Computing the longest common subsequence of two run-length encoded strings. In: Chao, K.-M., Hsu, T., Lee, D.-T. (eds.) ISAAC 2012. LNCS, vol. 7676, pp. 197–206. Springer, Heidelberg (2012). https://doi.org/10.1007/978-3-642-35261-4_23

18. Song, S., Gu, G., Ryu, C., Faro, S., Lecroq, T., Park, K.: Fast algorithms for single and multiple pattern Cartesian tree matching. Theor. Comput. Sci. **849**, 47–63 (2021)

# Output-Sensitive Enumeration of Potential Maximal Cliques in Polynomial Space

Caroline Brosse[1,3], Alessio Conte[2(✉)] (ID), Vincent Limouzy[3] (ID),
Giulia Punzi[2,4] (ID), and Davide Rucci[2] (ID)

[1] CNRS, Université Côte d'Azur, Inria, I3S, Sophia-Antipolis, Nice, France
caroline.brosse@inria.fr
[2] University of Pisa, Pisa, Italy
{alessio.conte,giulia.punzi}@unipi.it, davide.rucci@phd.unipi.it
[3] Université Clermont Auvergne, Clermont Auvergne INP, CNRS, Mines
Saint-Etienne, LIMOS, 63000 Clermont-Ferrand, France
vincent.limouzy@uca.fr
[4] National Institute of Informatics, Chiyoda City, Japan

**Abstract.** A set of vertices in a graph forms a *potential maximal clique* if there exists a minimal chordal completion in which it is a maximal clique. Potential maximal cliques were first introduced as a key tool to obtain an efficient, though exponential-time algorithm to compute the treewidth of a graph. As a byproduct, this allowed to compute the treewidth of various graph classes in polynomial time. In recent years, the concept of potential maximal cliques regained interest as it proved to be useful for a handful of graph algorithmic problems. In particular, it turned out to be a key tool to obtain a polynomial time algorithm for computing maximum weight independent sets in $P_5$-free and $P_6$-free graphs (Lokshtanov et al., SODA '14 and Grzeskik et al., SODA '19). In most of their applications, obtaining all the potential maximal cliques constitutes an algorithmic bottleneck, thus motivating the question of how to efficiently enumerate all the potential maximal cliques in a graph $G$.

The state-of-the-art algorithm by Bouchitté & Todinca can enumerate potential maximal cliques in output-polynomial time by using exponential space, a significant limitation for the size of feasible instances. In this paper, we revisit this algorithm and design an enumeration algorithm that preserves an output-polynomial time complexity while only requiring polynomial space.

**Keywords:** Potential Maximal Cliques · Graph Enumeration

## 1 Introduction

Potential maximal cliques are fascinating objects in graph theory. A *potential maximal clique*, or *PMC* for short, of a graph is a set of vertices inducing a maximal clique in some minimal triangulation of a graph (see Fig. 1 for an example). These objects were originally introduced by Bouchitté and Todinca in the late

A. A. Rescigno and U. Vaccaro (Eds.): IWOCA 2024, LNCS 14764, pp. 382–395, 2024.
https://doi.org/10.1007/978-3-031-63021-7_29

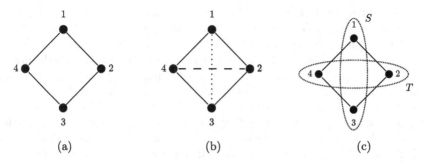

**Fig. 1.** (a) a simple graph on 4 vertices. (b) Two possible minimal triangulations of the graph (dashed line or dotted line). (c) Two minimal separators of the graph: $\{1,3\}$ and $\{2,4\}$. The potential maximal cliques of (a) include all the maximal cliques of all the different minimal triangulations, that is to say, $\{1,2,3\}$ and $\{1,3,4\}$ from the dotted triangulation and $\{1,2,4\}$, $\{2,3,4\}$ from the dashed triangulation.

1990s [3] as a key tool to handle the computation of treewidth and minimum fill-in of a graph, which are both NP-complete problems. To deal with these problems, the authors adopted an enumerative approach: their idea is to compute the list of all PMCs of the input graph before processing them as quickly as possible. Just like for maximal cliques, the number of potential maximal cliques of a graph can be large, typically exponential in the number of vertices. Therefore, an algorithm enumerating the potential maximal cliques of a graph cannot in general run in polynomial time. However, the algorithm proposed by Bouchitté and Todinca can be used to build efficient exact exponential algorithms for the NP-complete problems they considered. The guarantee of efficiency is given by the *output-polynomial* complexity measure. An enumeration algorithm is said to have output-polynomial time complexity if its running time is polynomial both in the input size and in the number of solutions that have to be returned. This allows a finer-grained analysis than simply bounding the total execution time by an exponential function. The output-polynomial complexity also guarantees a running time that is polynomial in the input size only when the number of objects to be generated is known to be polynomial, as it is the case for PMCs, or equivalently for *minimal separators*, in some classes of graphs [3].

Due to their links with tree-decompositions and the computation of graph parameters such as treewidth, being able to list the potential maximal cliques of a graph efficiently is still at the heart of several algorithms, including recent ones [8,9,11,15,16]. Being able to guarantee the fastest possible complexity for the enumeration of PMCs is thus of great interest. The currently best known algorithm is the one proposed by Bouchitté and Todinca [5], which is quadratic in the number of PMCs. In their series of papers, they asked whether a linear dependency can be achieved, but this problem is still open. On the other hand, space usage is also an issue: their algorithm needs to remember all the (exponentially many) solutions that have already been found. As memory is in practice limited this can make the computation unfeasible, thus seeking a faster algorithm without improving the space usage may not lead to practical benefits.

Despite growing interest on PMC enumeration (see, e.g., [10,12]), few advances have been made on improving the complexity of the problem [13,14]. It is worth mentioning the enumeration of minimal triangulations, linked with tree-decomposition, which received at least two algorithmic improvements during the past decade [6,7]. However, it is still unknown if improving the enumeration of minimal triangulations can help to list the PMCs more efficiently.

*Our Contribution.* The main result of the current paper is an algorithm that generates all the potential maximal cliques of a graph in polynomial space, while keeping the output-polynomial time complexity status. First, in Sect. 3 we show how to modify the original algorithm [5] to avoid duplicates: we refine the tests inside the algorithm to be able not to explore the same solution twice. Then in Sect. 4, we reduce the space usage of our version of Bouchitté and Todinca's algorithm by modifying the exploration strategy. The main effect of this modification is that remembering all the solutions already found is no longer needed. Of course, the execution time is affected, but in the end our algorithm still has output-polynomial complexity, and uses only polynomial space.

# 2   Definitions and Theoretical Results

## 2.1   Notations and Basic Concepts

We denote a graph by $G = (V, E)$, using $n$ for the number of vertices ($|V|$) and $m$ for the number of edges ($|E|$). The neighborhood of a vertex $u$ is the set $N(u) = \{v \in V \mid uv \in E\}$, and the neighborhood of a set of vertices $S$ is the set of vertices that have a neighbor in $S$: $N(S) = \{v \in V \setminus S \mid \exists u \in S, \ uv \in E\}$.

A *minimal triangulation* of the graph $G = (V, E)$ is a chordal graph $H = (V, E \cup F)$ – that is, a graph without induced cycles of length 4 or more – such that for any proper subset $F'$ of $F$, the graph $H' = (V, E \cup F')$ is not chordal. The *potential maximal cliques*, or *PMCs*, of the graph $G$ are the sets of vertices inducing an inclusion-wise maximal clique in some minimal triangulation of $G$ (see Fig. 1b). The set of all PMCs of the graph $G$ is denoted by $\Pi_G$.

As highlighted by Bouchitté and Todinca [3], potential maximal cliques are closely related to *minimal separators*. In the graph $G$, for two vertices $a$ and $b$, a *minimal $(a, b)$ separator* is a set $S$ of vertices of $G$ such that $a$ and $b$ are in distinct connected components of $G \setminus S$, and is minimal for this property. The set of *minimal separators* of $G$, denoted by $\Delta_G$, is made of all the minimal $(a, b)$ separators for all pairs $(a, b)$ of vertices (see Fig. 1c). This will be a crucial ingredient to build new PMCs.

Finally, for algorithmic purposes, we consider the vertices of a graph $G$ to be arbitrarily ordered, that is to say, $V = \{v_1, \ldots, v_n\}$. Then, for any $1 \leq i \leq n$, we define the graph $G_i$ as the subgraph of $G$ induced by the vertex set $\{v_1, \ldots, v_i\}$. Therefore, $G_1 = (\{v_1\}, \varnothing)$ and for any $i < n$, $G_{i+1}$ contains exactly one more vertex than $G_i$, together with all the edges of $G$ between it and vertices of $G_i$. For simplicity of notation, we will use $\Pi_i$ and $\Delta_i$ instead of $\Pi_{G_i}$ and $\Delta_{G_i}$ to denote the sets of PMCs and minimal separators of $G_i$.

## 2.2    Background

To enumerate the potential maximal cliques of a graph, we base our algorithm on the first one proposed by Bouchitté and Todinca. It uses an incremental approach: the principle is to add the vertices one by one and generate at each step $i$ the PMCs of the graph $G_i$ by extending the PMCs of $G_{i-1}$ and generating new ones from minimal separators of $G_i$.

As our algorithm builds up on the original one given by Bouchitté and Todinca [5], we rely on their proofs and terminology. Namely, to decide efficiently if a set of vertices is a PMC, we will need the notion of *full component*. Initially introduced for minimal separators, this definition can in fact be given for any subset of the vertex set, even if its removal leaves the graph connected.

**Definition 1 (Full Component).** *Given a set $S$ of vertices of the graph $G$, a connected component $C$ of $G \setminus S$ is full for $S$ if every $v \in S$ has a neighbor in $C$. That is to say, for any $v \in S$ there exists $u \in C$ such that $uv \in E(G)$.*

The potential maximal cliques have been characterized by Bouchitté and Todinca using the full components [4, Theorem 3.15]. This characterization, that we state thereafter, is very useful from an algorithmic point of view since it provides a polynomial test for a subset of the vertices of a graph to be a PMC.

**Theorem 1 (Characterization of PMCs [4]).** *Given a graph $G = (V, E)$, a subset $K \subseteq V$ is a PMC of $G$ if and only if*

*(a) there is no full component for $K$ in $G$, and*
*(b) for any two vertices $x$ and $y$ of $K$ such that $xy \notin E$, there exists a connected component $C$ of $G \setminus K$ such that $x, y \in N(C)$.*

Theorem 1 provides an efficient test to decide if a given subset of the vertices of $G$ is a PMC. The test can be run in time $O(mn)$.

In our algorithm, as in the original one by Bouchitté and Todinca, the potential maximal cliques will be computed incrementally. Bouchitté and Todinca proved that for any vertex $a$ of a graph $G$, the PMCs of $G$ can be obtained either from PMCs in $G \setminus \{a\}$, or from minimal separators of $G$ [5, Theorem 20]. However, they did not prove that this condition was sufficient. From Theorem 1, we are able to deduce the following property of potential maximal cliques: any PMC of $G \setminus \{a\}$ can be uniquely extended to a PMC of $G$.

**Proposition 1.** *Let $G$ be a graph and $a$ be any vertex of $G$. For any PMC $K$ of $G \setminus \{a\}$, exactly one between $K$ and $K \cup \{a\}$ is a PMC of $G$.*

*Proof.* We define $G' := G \setminus \{a\}$, so that $G = G' + a$. Let $K$ be a PMC of $G'$ and suppose that $K$ is not a PMC of $G$.

We start by proving that condition (b) of Theorem 1 remains true for $K$ in $G$. Let $x$ and $y \in K$ such that $xy \notin E$ (if two such vertices exist, otherwise (b) is true by emptiness). Since $K$ is a PMC of $G'$, there exists a connected component $C'$ of $G' \setminus K$ such that $x$ and $y$ both have neighbors in $C'$. We know that any connected component of $G' \setminus K$ is contained in some connected component of $G \setminus K$; in particular there exists a connected component $C$ of $G \setminus K$ such that $C' \subseteq C$. Consequently, $x$ and $y$ both have neighbors in $C$ and item (b) is true for $K$ in $G = G' + a$.

Therefore, since $K$ is not a PMC of $G$, it means that there exists a full component for $K$ in $G = G' + a$. Necessarily this component contains $a$, otherwise it would also be a connected component of $G' \setminus K$, and by hypothesis it is not full. In this case, we prove using Theorem 1 that $K \cup \{a\}$ is a PMC of $G$.

(a) The connected components of $G \setminus (K \cup \{a\})$ are the same as those of $G' \setminus K$, so for any connected component $C$ there exist elements of $K \subseteq K \cup \{a\}$ that do not have neighbors in $C$. Consequently, there are no full components for $K \cup \{a\}$ in $G$.

(b) Let $x$ and $y$ be two vertices of $K \cup \{a\}$ such that $xy \notin E$. If $x \neq a$ and $y \neq a$, then condition (b) is satisfied by $x$ and $y$, since the connected components of $G \setminus (K \cup \{a\})$ are the same as those of $G' \setminus K$. Otherwise, we can assume $x = a$ and $ay \notin E(G)$. Since there exists a full component for $K$ in $G = G' + a$, in particular there exists a connected component of $G \setminus (K \cup \{a\})$ such that both $a$ and $y$ have neighbors in this component. Thus, condition (b) is satisfied.

Both conditions are satisfied, so by Theorem 1, $K \cup \{a\}$ is a PMC of $G$ and the proposition is proved. Moreover, since the PMCs of a graph are incomparable sets, $K$ and $K \cup \{a\}$ cannot be both PMCs of the same graph.    □

*Potential Maximal Cliques Generation and Minimal Separators.* In the current best known algorithm for enumerating the PMCs, it is crucial to be able to pass quickly through the list of all minimal separators, in what we called subroutine GEN. The complexity status of the minimal separators enumeration problem has evolved since the introduction of PMCs. In 2000, Berry et al. [2] provided a *polynomial delay* algorithm for the minimal separators enumeration, meaning that the running time needed between two consecutive outputs is polynomial in the input size only. However, it needs exponential space. In 2010, Takata [17] managed to enumerate the minimal $(a, b)$-separators for one pair $(a, b)$ with polynomial delay in polynomial space. He also gave an output-polynomial algorithm in polynomial space for minimal $(a, b)$-separators for all pairs $(a, b)$. More recently (WEPA 2019), Bergougnoux, Kanté and Wasa [1] presented an algorithm enumerating the minimal $(a, b)$-separators for all pairs $(a, b)$ in polynomial space, with (amortized) polynomial delay. From a theoretical point of view, it is the currently best known algorithm for the enumeration of all minimal separators.

---

**Algorithm 1:** Potential maximal cliques enumeration by [5].

---

**Input:** A graph $G = (V, E)$
**Output:** The set of all potential maximal cliques $\Pi_n$

1  $\Pi_1 \leftarrow \{v_1\}$
2  **for** $i = 2, \ldots, |V|$ **do** $\Pi_i \leftarrow \varnothing$; ONE_MORE_VERTEX$(G, i)$
3  **return** $\Pi_n$

4  **Function** ONE_MORE_VERTEX$(G, i)$
5      **forall** $\pi \in \Pi_{i-1}$ **do**
6          **if** IsPMC$(\pi \cup \{v_i\}, G_i)$ **then add** $\pi \cup \{v_i\}$ to $\Pi_i$
7          **if** IsPMC$(\pi, G_i)$ **then add** $\pi$ to $\Pi_i$
8      **forall** $S \in \Delta_i$ **do**
9          **if** IsPMC$(S \cup \{v_i\}, G_i)$ **then add** $S \cup \{v_i\}$ to $\Pi_i$
10         **if** $v_i \notin S$ **and** $S \notin \Delta_{i-1}$ **then**
11             **forall** $T \in \Delta_i$ **do**
12                 **forall** $C$ *full component associated to $S$ in $G_i$* **do**
13                     **if** IsPMC$(S \cup (T \cap C), G_i)$ **then add** $S \cup (T \cap C)$ to $\Pi_i$

---

*Original Algorithm.* Algorithm 1 shows the original strategy proposed in [5] for PMC enumeration, based on an iterative argument: generate the set $\Pi_i$ of PMCs of $G_i$ and keep it in memory to use it to compute $\Pi_{i+1}$ at the next step. New potential maximal cliques can be found in two ways, either by expanding an existing PMC from the previous step, or by extending a minimal separator. The algorithm follows both roads, first try to expand existing cliques, then generate new ones from minimal separators. This idea can be implemented by storing the family of sets $\mathcal{P} = \{\Pi_i \mid i \in [1, n]\}$. The algorithm stops at the end of step $n$, when the set $\Pi_n$ containing all the PMCs of $G_n = G$ has been generated, and returns the whole set of solutions at the end of its execution. The strategy is summarized in [5, Theorem 23], by the function One_More_Vertex, which is called for all $i = 1, \ldots, n$. The total time complexity is $O(n^2 m |\Delta_G|^2)$. However, the additional space required by the algorithm is $O(|\mathcal{P}|) = O(\sum_{i=1}^{n} |\Pi_i|) = O(n|\Pi_n|)$ as all solutions for all $G_i$ are stored, and this is exponential in $n$.

# 3  Duplication Avoidance in the B&T Algorithm

Since our goal is to have a polynomial-space algorithm, our first task is to rethink the One_More_Vertex strategy so that it does not output the same solution twice.

Algorithm 2 still stores the family of sets $\mathcal{P}$, but contains additional checks for duplication avoidance, so that we never try to add the same potential maximal clique twice to the same set. In particular, for a set $D^*$ generated at line 15 from the sets $S^*$, $T^*$ and $C^*$, the Not_Yet_Seen$(D^*)$ check works as follows: the nested loops on $S$, $T$ and $C$ are run again to generate all the possible sets $D$ until $D^*$ is found for the first time. If $D^*$ is found from $S^*$, $T^*$ and $C^*$ during this procedure, then Not_Yet_Seen$(D^*)$ returns *true*, otherwise it returns *false*.

---

**Algorithm 2:** PMC enumeration without duplication based on [5].

---

**Input:** A graph $G = (V, E)$
**Output:** The set of all potential maximal cliques $\Pi_n$

1  $\Pi_1 \leftarrow \{v_1\}$
2  **for** $i = 2, \ldots, |V|$ **do** $\Pi_i \leftarrow \varnothing$; NONDUP_ONE_MORE_VERTEX$(G, i)$
3  **return** $\Pi_n$

4  **Function** NONDUP_ONE_MORE_VERTEX$(G, i)$
5      **forall** $\pi \in \Pi_{i-1}$ **do** // EXT(i)
6          **if** IsPMC$(\pi \cup \{v_i\}, G_i)$ **then add** $\pi \cup \{v_i\}$ to $\Pi_i$
7          **if** IsPMC$(\pi, G_i)$ **then add** $\pi$ to $\Pi_i$

8      **forall** $S \in \Delta_i$ **do** // GEN(i)
9          **if** IsPMC$(S \cup \{v_i\}, G_i)$ **then**
10             **if** !IsPMC$(S, G_{i-1})$ **then add** $S \cup \{v_i\}$ to $\Pi_i$

11         **if** $v_i \notin S$ **and** $S \notin \Delta_{i-1}$ **then**
12             **forall** $T \in \Delta_i$ **do**
13                 **forall** $C$ *full component associated to $S$ in $G_i$* **do**
14                     **if** IsPMC$(S \cup (T \cap C), G_i)$ **then**
15                         $D \leftarrow S \cup (T \cap C)$
16                         **if** Not_Yet_Seen$(D)$ **then**
17                             **if** !IsPMC$(D, G_{i-1})$ **then**
18                               **if** $T \cap C \neq \{v_i\}$ **then**
19                                 **if** $v_i \in T \cap C$ **then**
20                                   **if** !IsPMC$(D \setminus \{v_i\}, G_{i-1})$ **and**
                                        $D \setminus \{v_i\} \notin \Delta_i$ **then**
21                                     **add** $D$ to $\Pi_i$
22                             **else**
23                               **add** $D$ to $\Pi_i$

---

**Lemma 1.** *Algorithm 2 is correct and produces the same potential maximal cliques as the original Bouchitté and Todinca's algorithm [5], without duplicates.*

*Proof.* Algorithm 2 differs from the initial algorithm by [5] only in the highlighted parts. The changes consist in introducing additional checks before adding some given PMC to $\Pi_i$. Specifically, we add the following checks:

(i) Not_Yet_Seen$(D)$ at line 16;
(ii) !IsPMC$(S, G_{i-1})$ at line 10;
(iii) !IsPMC$(D, G_{i-1})$ at line 17;
(iv) $T \cap C \neq \{v_i\}$ at line 18;
(v) !IsPMC$(D \setminus \{v_i\}, G_{i-1})$ **and** $D \setminus \{v_i\} \notin \Delta_i$ at line 20.

The *PMC corresponding to a check* is the PMC that is not added to $\Pi_i$ when such check fails. Specifically, the PMC corresponding to check (ii) is $S \cup \{v_i\}$ (line 10), while for all other checks the corresponding PMC is set $D$ defined at line 15 (see lines 21 and 23). As the underlying enumeration strategy is unchanged, the correctness of our algorithm follow from these two statements:

1. check (i) fails if and only if $D$ was previously processed by another choice of $S, T, C$ over the three for loops. The validity of this statement directly follows from the definition of Function Not_Yet_Seen().
2. a check among (ii)–(v) fails in the call of Nondup_One_More_Vertex$(G, i)$ if and only if the corresponding potential maximal clique already belongs to $\Pi_i$ at that moment of computation.

These statements also guarantee that each solution is inserted only once into $\Pi_n$ during the execution of Algorithm 2, that is to say, no duplicate solution is processed. In what follows we prove item 2 by analyzing checks (ii)-(v) separately.

First, assume that check (ii) is reached and fails: this happens if and only if $S \cup \{v_i\}$ is a potential maximal clique of $G_i$, and $S \in \Pi_{i-1}$, which is true if and only if the for loop at line 5 considered $\pi = S$ at some point, and the check at line 6 was successful, meaning that $S \cup \{v_i\}$ was added to $\Pi_i$ at that time.

Let us now consider checks (iii)–(v), all corresponding to the same potential maximal clique $D$. For these checks we have a shared setting, i.e., checks (iii)–(v) happen when $S \in \Delta_i \setminus \Delta_{i-1}$ such that $v_i \notin S$; $T \in \Delta_i$; $C$ is a full component associated to $S$ in $G_i$, and finally $D = S \cup (T \cap C)$ is a PMC of $G_i$. We refer to this specific setting as the *common checks setting*, and it will serve as a set of hypotheses for the rest of the proof. Notice that this setting is the same set of checks performed by Algorithm 1 to determine if $D$ is a PMC in $G_i$.

In the common checks setting, condition (iii) is reached and fails if and only if IsPMC$(D, G_{i-1})$, i.e., $D \in \Pi_{i-1}$. This happens if and only if $D$ was considered as $\pi$ during the forall loop at line 5 and the check at line 7 was successful (so $D$ is a PMC of $G_i$), thus if and only if $D$ was added to $\Pi_i$ at that time.

Consider now check (iv): assuming the common checks setting, we reach and fail this check if and only if (iii) succeeds and $T \cap C = \{v_i\}$, which is equivalent to $D = S \cup \{v_i\}$. Thus, this check fails if and only if $D$ was already added at line 6 or at line 10 (according to whether $S \in \Pi_{i-1}$ or not).

Finally, check (v) is reached, in the common checks setting, whenever (iii) and (iv) succeed, and $v_i \in T \cap C$. Thus for (v) to fail it is either that $D \setminus \{v_i\}$ is a PMC of $G_{i-1}$ or $D \setminus \{v_i\}$ is a minimal separator of $G_i$. We have that $D \setminus \{v_i\}$ is a PMC of $G_{i-1}$ if and only if the algorithm already processed the same set $D$ on line 7, thus we do not add it again to $\Pi_i$. As for the second part, we only need to consider what happens when $D \setminus \{v_i\}$ is not a PMC of $G_{i-1}$, but $D \setminus \{v_i\}$ is a minimal separator of $G_i$. This happens if and only if $S = D \setminus \{v_i\}$ was already processed during the forall loop of line 8, and was added to $\Pi_i$ at line 10. □

We observe that deduplication could be achieved relying on the principle of the Not_Yet_Seen procedure for all cases considered (i.e., even for $(ii)$-$(v)$), preserving output-polynomial complexity. However, limiting the number of deduplication re-runs can lead to practical improvements (as the alternative checks we perform take only polynomial time), and may support future efforts for incremental-polynomial time guarantees.

**Lemma 2.** *Algorithm 2 has output polynomial time complexity of $O(n^8|\Pi_G|^4)$.*

*Proof.* Before describing the complexity of the algorithm, note that the IsPMC check function can be implemented to run in $O(nm)$ time, using Theorem 1 and [5, Corollary 12]. Additionally, we assume that we can generate the set $\Delta_i$ with $O(n^3m)$ delay and polynomial space, using the algorithm from [1]. Finally, in what follows we assume that the sets $\Pi_i$, $\Delta_i$ and each $S \in \Delta_i$ are implemented as linked lists, so that we can append a new potential maximal clique to them in $O(1)$ time, while membership checks require linear time.

We start by analyzing the Nondup_One_More_Vertex function: this can be calculated by summing the complexity of the first forall loop on line 5 and the second forall loop on line 8, plus the $O(n^3m|\Delta_i|)$ time required to compute $|\Delta_i|$. The first loop, performs two IsPMC checks per element of $\Pi_{i-1}$. Thus, its total running time is $O(|\Pi_{i-1}|nm) = O(nm|\Pi_G|)$.

The costly part is the second forall loop. It executes $|\Delta_i|$ iterations, during each of which we (a) perform two IsPMC checks ($O(nm)$ time) at lines 9–10; (b) check if $v_i$ belongs to $S$ and if $S$ belongs to $\Delta_{i-1}$, by scanning respectively $S$ and $|\Delta_{i-1}|$ in $O(|S|) = O(n)$ and $O(|\Delta_{i-1}|)$; (c) perform the forall loop at line 12. The complexity of the second loop is therefore $O(|\Delta_i|(nm + |\Delta_{i-1}| + n + C_{12}))$, where $C_{12}$ is the complexity of the loop at line 12, which we now analyze. The loop over $T \in \Delta_i$ counts $|\Delta_i|$ iterations, each of which costs $O(n \cdot (n + nm + n^2|\Delta_i|^2))$. Indeed, we iterate over all full components $C$ associated to $S$ in $G_i$, which can be $n$ in the worst case (when each vertex is a separate full connected component); for each of these we need to compute $S \cup (T \cap C)$ and later check if $v_i \in T \cap C$, which can both be done in $O(n)$ time, then perform up to three IsPMC calls in $O(nm)$ time. Finally, we need further $O(n^2|\Delta_i|^2)$ time for computing Not_Yet_Seen, as implemented with the loop restart mentioned above, and another $O(|\Delta_i|)$ to check whether $D \setminus \{v_i\}$ belongs to $\Delta_i$ at line 20. Putting everything together we have:

$$O(\underbrace{|\Delta_i|}_{8}(\underbrace{nm}_{9\text{-}10} + \underbrace{|\Delta_i| + n}_{11} + \underbrace{|\Delta_i|}_{12}(\underbrace{n}_{13}(\underbrace{nm + n + n^2|\Delta_i^2|}_{14\text{-}15} + \underbrace{nm}_{16} + \underbrace{n}_{17} + \underbrace{nm + |\Delta_i|}_{18\text{-}19})))\underbrace{\phantom{)}}_{20})$$

$$= O(|\Delta_i|(n^3|\Delta_i|^3 + n^2m|\Delta_i|)) = O(n^2|\Delta_i|^2(n|\Delta_i|^2 + m)).$$

This is the cost of a call to the second loop of Function Nondup_One_More_Vertex with fixed $i$. Summing this cost with the cost of the first forall loop and of the computation of $\Delta_i$ for all the $n$ calls to the function from line 2 we obtain the overall time complexity for Algorithm 2: $O(n^2m|\Pi_G| + n^4m|\Delta_G| + n^3|\Delta_G|^2 \cdot (n|\Delta_G|^2 + m)) = O(n^2m|\Pi_G| + n^4m|\Delta_G| + n^3|\Delta_G|^2 \cdot (n|\Delta_G|^2 + m))$.

The total complexity of Algorithm 2 is polynomial in the number of PMCs of the graph. To highlight this, we use the known inequality $|\Pi_G| \geq |\Delta_G|/n$ [5]. Thus we obtain the final output-sensitive complexity of

$$O(n^2m|\Pi_G| + n^4m|\Delta_G| + n^3|\Delta_G|^2(n|\Delta_G|^2 + m))$$
$$= O(n^2m|\Pi_G| + n^5m|\Pi_G| + n^3(n|\Pi_G|)^2 \cdot ((n|\Pi_G|)^2n + m))$$
$$= O(n^5m|\Pi_G| + n^5m|\Pi_G| + n^5|\Pi_G|^2 \cdot (n^3|\Pi_G|^2 + m)) = O(n^8|\Pi_G|^4).$$

$\square$

# 4  Polynomial Space Algorithm

Algorithm 2 has output-sensitive time complexity and lists all the potential maximal cliques without duplicates, but it still uses more than polynomial space because it has to store the sets of solutions $\Pi_i$ before returning $\Pi_n$. Therefore, in this section we show how to adapt the algorithm to output solutions as soon as they are found, without having to store $\Pi_i$ for duplicate detection. The key idea is to change the way in which we traverse the solution space from breadth-first to depth-first. Said otherwise, once we get a PMC $K$ in $G_i$, we immediately extend it to a PMC of $G_n$, which is always possible according to Proposition 1, before looking for a new one. This idea is summarized in Algorithm 3.

---

**Algorithm 3:** Depth-first algorithm for output-polynomial time, polynomial space potential maximal clique enumeration.

---

    **Input:** An integer $i$ corresponding to $v_i \in V$, a PMC $\pi$ in $G_{i-1}$ ($G$ is implicit)

1  **for** $i = 1, \ldots, |V|$ **do**
2      **forall** $\pi' \in \texttt{GEN}(G, i)$ **do** $\texttt{EXT}(i+1, \pi')$

3  **Function** $\texttt{EXT}(i, \pi)$
4      **if** $i = n$ **then** output $\pi$
5      **if** $\texttt{IsPMC}(\pi \cup \{v_i\}, G_i)$ **then** $\texttt{EXT}(i+1, \pi \cup \{v_i\})$
6      **else if** $\texttt{IsPMC}(\pi, G_i)$ **then** $\texttt{EXT}(i+1, \pi)$

7  **Function** $\texttt{GEN}(G, i)$
8      **if** $i = 1$ **then** yield $\{v_1\}$ **and return**
9      **forall** $S \in \Delta_i$ **do** // $S$ is a minimal separator in $G_i$
10        **if** $!\texttt{IsPMC}(S, G_{i-1})$ **and** $\texttt{IsPMC}(S \cup \{v_i\}, G_i)$ **then**
11          yield $S \cup \{v_i\}$
12        **if** $v_i \notin S$ **and** $S \notin \Delta_{i-1}$ **then**
13          **forall** $T \in \Delta_i$ **do**
14            **forall** $C$ *full component associated to $S$ in $G_i$* **do**
15              **if** $\texttt{IsPMC}(S \cup (T \cap C), G_i)$ **then**
16                $D \leftarrow S \cup (T \cap C)$
17                **if** $\texttt{Not\_Yet\_Seen}(D)$ **then**
18                  **if** $!\texttt{IsPMC}(D, G_{i-1})$ **then**
19                    **if** $T \cap C \neq \{v_i\}$ **then**
20                      **if** $v_i \in T \cap C$ **then**
21                        **if** $!\texttt{IsPMC}(D \setminus \{v_i\}, G_{i-1})$ **and**
                                $D \setminus \{v_i\} \notin \Delta_i$ **then**
22                        yield $D$
23                    **else**
24                      yield $D$

---

First, we split the One_More_Vertex function into two parts: the EXT routine extends a given potential maximal clique $\pi$ in $G_i$ to a PMC of $G$ by recurring on $i+1, i+2, \ldots, n$. The EXT procedure is presented here as a recursive function in order to highlight the key idea of a depth-first traversal of the search space, but it is easy to rewrite it iteratively, saving the additional space required for handling recursion. The second part of the original function is the GEN routine, which generates all the new PMCs of the graph $G_i$, yielding them one by one to the EXT procedure. In particular, we see the GEN function as a *generator* of PMCs, i.e., the solutions are yielded during the execution while keeping the internal state[1]. This way the total running time of the function does not change, and we process each PMC immediately after it has been generated.

**Proposition 2.** *Let $K$ be a potential maximal clique of $G_i$. If $K$ is not a PMC of $G_{i+1}$, then it cannot be a PMC of any $G_j$ with $j > i$.*

*Proof.* If $K$ is a PMC of $G_i$ but not of $G_{i+1}$, then by Proposition 1 $K \cup \{v_{i+1}\}$ must be a PMC of $G_{i+1}$. So, $K$ is strictly included in a PMC of $G_{i+1}$. By repeatedly applying this reasoning, $K$ is also strictly included in some PMC of $G_j$ for any $j > i$. Therefore, since PMCs are not included in each other, $K$ cannot be a PMC of $G_j$ for any $j > i$.    $\square$

**Proposition 3.** *For two indices $i < j$, let $K_i$ and $K_j$ be two PMCs of $G_i$ and $G_j$ respectively. If $K_i$ and $K_j$ are extended to the same $K$ in $G$, then $K_i \subseteq K_j$.*

*Proof.* We prove this result by contrapositive. Suppose that $K_i$ is not included in $K_j$. As we view the vertices of $G$ as an ordered set, there exists $k \leq i$ such that $v_k \in K_i$ and $v_k \notin K_j$. Since Algorithm 3 considers the vertices in order, we are sure that $K_j$ is a PMC of $G_j$ that does not contain $v_k$, with $k < j$. Therefore, by Proposition 1, $K_j$ will be extended into a PMC of $G$ that does not contain $v_k$. Since $v_k \in K_i$, and $K_i$ will be extended into a PMC of $G$ that contains $v_k$, the two sets $K_i$ and $K_j$ will be extended into different PMCs of $G$.    $\square$

**Lemma 3.** *Algorithm 3 outputs all and only the potential maximal cliques of $G$, without duplication.*

*Proof.* First, as Algorithm 3 is a rearrangement of Algorithm 2 in a depth-first fashion, it must also output every PMC at least once. It remains to prove that after splitting the algorithm in two distinct functions, every PMC of $G$ is produced by Algorithm 3 in exactly one way. Note that line 11 will never output a duplicated PMC, since vertex $v_i$ is different $\forall i \leq n$. Now suppose that Algorithm 3 outputs a duplicated PMC $K$, we have two cases.

(a) The same set $D$ is found twice by GEN at steps $i$ and $j$: it must be that $i \neq j$ (we assume $i < j$ w.l.o.g.), i.e., the duplicated $D$ comes from different calls

---

[1] Consider GEN($G, i$) as an iterator over the set $\Pi_i$. The first call, for each $i$, will compute everything needed to output $\pi_{i,1} \in \Pi_i$. Then, when called with the same $i$, it will produce $\pi_{i,2} \in \Pi_i$ without recomputing everything from scratch and so on.

GEN($i$) and GEN($j$). This is because of the check at line 17 of Algorithm 3 that explicitly prevents this. Thus, $D$ is a PMC in each step between $i$ and $j$ by Proposition 2, including step $j - 1$, so it will be filtered by line 18.

(b) $K$ is output twice by EXT: in this case $K$ comes from two different sets $K_i$ and $K_j$ produced by GEN at different steps $i$ and $j$ (with $i < j$). By Proposition 3, $K_i \subseteq K_j$. Since $K_i$ and $K_j$ have been produced by GEN, they are PMCs of $G_i$ and $G_j$ respectively. Then, as a consequence of Proposition 1, for any $i < i' < j$ there exists $K_i \subseteq K_{i'} \subseteq K_j$ such that $K_{i'}$ is a PMC of $G_{i'}$, where the second containment comes again from Proposition 3. In particular, consider $i' = j - 1$, then $K_j \setminus v_j$ (that is, $K_j$ if $v_j \notin K_j$) is a PMC of $G_{j-1}$ and is therefore filtered at line 18 or 21.

Thus we conclude that Algorithm 3 cannot output duplicated solutions.     □

**Lemma 4.** *Algorithm 3 has output-polynomial time complexity and uses polynomial space. Namely, it uses $O(n^9 m^2 |\Pi_G|^4)$ time and $O(n^3)$ space.*

*Proof.* We first recall that, at each step $i$, we can enumerate separators of $\Delta_i$ with $O(n^3 m)$ delay by [1]. We start by analyzing the time complexity of the body of the outermost for loop of line 1. We see the function GEN as an iterator(See footnote 1), so that it yields new PMCs during its execution. The time complexity of GEN($G, i$) for a fixed $i$ is $O(n^3 m |\Delta_i| + n^2 |\Delta_i|^2 \cdot (n |\Delta_i|^2 + m))$ (same as Algorithm 2). The corresponding EXT call has maximum depth of $O(n)$ and a cost of $O(nm)$ due to the IsPMC checks, yielding $O(n^2 m)$ time worst case and returning a new solution, by Proposition 1. Thus, for a fixed $i$, the total cost of step $i$ is $O(n^4 m^2 |\Delta_i|^4)$. Since $|\Delta_i| \leq |\Delta_G|$ for all $i$, the total cost of Algorithm 3 is $O(n^5 m^2 |\Delta_G|^4)$. As $|\Pi_G| \geq |\Delta_G|/n$ we obtain the final output polynomial complexity of $O(n^9 m^2 |\Pi_G|^4)$.

The space usage is $O(n^3 + m) = O(n^3)$ due to the space complexity of the listing algorithm for minimal separators [1]. Note that we do not explicitly store any solution, as we immediately output it at the end of the EXT computation. □

# 5   Conclusions

This paper shows that potential maximal cliques can be enumerated in output-sensitive time using only polynomial space, rather than exponential space as required by existing approaches. While the complexity of our algorithm is still significant, this approach opens the way for the development of practical enumeration algorithms for PMCs, which would lead in turn to advancements on related graph problems such as treewidth decomposition and maximum-weight independent sets.

**Acknowledgements.** This work was started during the WEPA 2022 workshop in Clermont-Ferrand, France. The work is partially supported by MUR of Italy, PRIN Project n. 2022TS4Y3N - EXPAND and the complementary actions to NRRP "Fit4MedRob - Fit for Medical Robotics" Grant n. #PNC0000007, the French Agence Nationale de la Recherche, contract Digraphs ANR-19-CE48-0013-01, and JSPS KAKENHI Grant n. JP20H05962.

# References

1. Bergougnoux, B., Kanté, M.M., Wasa, K.: Disjunctive minimal separators enumeration. WEPA (2019). https://www.mimuw.edu.pl/~bbergougnoux/pdf/bkw19.pdf
2. Berry, A., Bordat, J.P., Cogis, O.: Generating all the minimal separators of a graph. Int. J. Found. Comput. Sci. **11**(03), 397–403 (2000)
3. Bouchitté, V., Todinca, I.: Minimal triangulations for graphs with "few" minimal separators. In: Bilardi, G., Italiano, G.F., Pietracaprina, A., Pucci, G. (eds.) ESA 1998. LNCS, vol. 1461, pp. 344–355. Springer, Heidelberg (1998). https://doi.org/10.1007/3-540-68530-8_29
4. Bouchitté, V., Todinca, I.: Treewidth and minimum fill-in: grouping the minimal separators. SIAM J. Comput. **31**(1), 212–232 (2001)
5. Bouchitté, V., Todinca, I.: Listing all potential maximal cliques of a graph. Theoret. Comput. Sci. **276**(1–2), 17–32 (2002)
6. Brosse, C., Limouzy, V., Mary, A.: Polynomial delay algorithm for minimal chordal completions. In: 49th International Colloquium on Automata, Languages, and Programming, ICALP 2022, 4–8 July 2022, Paris, France. LIPIcs, vol. 229, pp. 33:1–33:16. Schloss Dagstuhl - Leibniz-Zentrum für Informatik (2022)
7. Carmeli, N., Kenig, B., Kimelfeld, B., Kröll, M.: Efficiently enumerating minimal triangulations. Discret. Appl. Math. **303**, 216–236 (2021)
8. Chudnovsky, M., Pilipczuk, M., Pilipczuk, M., Thomassé, S.: On the maximum weight independent set problem in graphs without induced cycles of length at least five. SIAM J. Discrete Math. **34**(2), 1472–1483 (2020). https://doi.org/10.1137/19M1249473
9. Fomin, F.V., Todinca, I., Villanger, Y.: Large induced subgraphs via triangulations and CMSO. SIAM J. Comput. **44**(1), 54–87 (2015). https://doi.org/10.1137/140964801
10. Fomin, F.V., Villanger, Y.: Treewidth computation and extremal combinatorics. In: Aceto, L., Damgård, I., Goldberg, L.A., Halldórsson, M.M., Ingólfsdóttir, A., Walukiewicz, I. (eds.) ICALP 2008. LNCS, vol. 5125, pp. 210–221. Springer, Heidelberg (2008). https://doi.org/10.1007/978-3-540-70575-8_18
11. Grzesik, A., Klimosová, T., Pilipczuk, M., Pilipczuk, M.: Polynomial-time algorithm for maximum weight independent set on $P_6$-free graphs. In: Chan, T.M. (ed.) Proceedings of the Thirtieth Annual ACM-SIAM Symposium on Discrete Algorithms, SODA 2019, San Diego, California, USA, 6–9 January 2019, pp. 1257–1271. SIAM (2019). https://doi.org/10.1137/1.9781611975482.77
12. Korhonen, T.: Finding optimal triangulations parameterized by edge clique cover. In: Cao, Y., Pilipczuk, M. (eds.) 15th International Symposium on Parameterized and Exact Computation (IPEC 2020). Leibniz International Proceedings in Informatics (LIPIcs), vol. 180, pp. 22:1–22:18. Schloss Dagstuhl – Leibniz-Zentrum für Informatik, Dagstuhl (2020). https://doi.org/10.4230/LIPIcs.IPEC.2020.22
13. Korhonen, T., Berg, J., Järvisalo, M.: Enumerating potential maximal cliques via sat and asp. In: Proceedings of the twenty-eigth International Joint Conference on Artificial Intelligence (IJCAI 2019). International Joint Conferences on Artifical Intelligence (2019)
14. Korhonen, T., Berg, J., Järvisalo, M.: Solving graph problems via potential maximal cliques: An experimental evaluation of the bouchitté-todinca algorithm. J. Exp. Algorithmics (JEA) **24**, 1–19 (2019)
15. Liedloff, M., Montealegre, P., Todinca, I.: Beyond classes of graphs with "few" minimal separators: FPT results through potential maximal cliques. Algorithmica **81**(3), 986–1005 (2019). https://doi.org/10.1007/s00453-018-0453-2

16. Lokshtanov, D., Vatshelle, M., Villanger, Y.: Independent set in $P_5$-free graphs in polynomial time. In: Chekuri, C. (ed.) Proceedings of the Twenty-Fifth Annual ACM-SIAM Symposium on Discrete Algorithms, SODA 2014, Portland, Oregon, USA, 5–7 January 2014, pp. 570–581. SIAM (2014). https://doi.org/10.1137/1. 9781611973402.43

17. Takata, K.: Space-optimal, backtracking algorithms to list the minimal vertex separators of a graph. Discret. Appl. Math. **158**(15), 1660–1667 (2010)

# Linear Search for an Escaping Target with Unknown Speed

Jared Coleman[1][(✉)], Dmitry Ivanov[2], Evangelos Kranakis[2], Danny Krizanc[3], and Oscar Morales-Ponce[4]

[1] University of Southern California, Los Angeles, CA, USA
jaredcol@usc.edu
[2] Carleton University, Ottawa, ON, Canada
dimaivanov@cmail.carleton.ca, kranakis@scs.carleton.ca
[3] Wesleyan University, Middletown, CT, USA
dkrizanc@wesleyan.edu
[4] California State University, Long Beach, USA
Oscar.MoralesPonce@csulb.edu

**Abstract.** We consider linear search for an escaping target whose speed and initial position are unknown to the searcher. A searcher (an autonomous mobile agent) is initially placed at the origin of the real line and can move with maximum speed 1 in either direction along the line. An oblivious mobile target that is moving away from the origin with an unknown constant speed $v < 1$ is initially placed by an adversary on the infinite line at distance $d$ from the origin in an unknown direction. We consider two cases, depending on whether $d$ is known or unknown. The main contribution of this paper is to prove a new lower bound and give algorithms leading to new upper bounds for search in these settings. This results in an optimal (up to lower order terms in the exponent) competitive ratio in the case where $d$ is known and improved upper and lower bounds for the case where $d$ is unknown. Our results solve an open problem proposed in [Coleman et al., Proc. OPODIS 2022].

**Keywords:** Autonomous agent · Competitive ratio · Linear Search · Oblivious Mobile Target · Searcher

## 1 Introduction

Linear search is concerned with a searcher that wants to find a target located on the infinite real line. The *searcher* is an autonomous mobile agent, henceforth also referred to as the *robot*, that can move on the line in any direction with max speed 1. A second agent, henceforth also referred to as the *target*, is oblivious (in particular, it cannot change direction), starts at an unknown (to the searcher) location on the line, and moves with constant speed $v < 1$ away from the origin. The goal is to design search algorithms that minimize the ratio of the target

---

E. Kranakis—Research supported in part by NSERC Discovery grant.

capture time (i.e., the first time when the searcher and target are co-located on the line) under the algorithm versus the minimum time required by an agent that knows the speed and starting position of the target. We refer to this ratio as the *competitive ratio*. For many problems, it is possible to design an algorithm with a finite supremum competitive ratio amongst all problem instances. In our case, however, the competitive ratio grows arbitrarily large for certain instances; so, our goal will be to minimize the growth of the competitive ratio with respect to the target's initial parameters.

One would expect that the resulting competitive ratio of a search algorithm should depend not only on the moving target's initial parameters (speed $v$ and starting distance $d$), but also on whether those parameters are known to the searcher prior to the execution of the algorithm. It would be natural to speculate that the less the searcher knows about the target, the worse the resulting competitive ratio. Our main focus in this paper is to understand how the knowledge available to the searcher about $d$, $v$ affects the competitive ratio of linear search. More specifically, we ask: what is the competitive ratio of search if the target is mobile and its speed is fixed but *unknown* to the searcher? This question was first considered in [11], where the authors study the resulting competitive ratios of search under the four possible knowledge scenarios of the pair $d, v$. The main contribution of the present paper is to give new algorithms and determine new upper and lower bounds for the competitive ratios when $v$ is unknown to the searcher, and thus answer one of the main open problems left in [11].

## 1.1 Notation and Terminology

A robot with maximum speed 1 starts at the origin 0 on the real number line. A target starts a distance $d \geq 1$ from the origin in an unknown direction and moves away from the origin with an unknown speed $0 \leq v < 1$. The goal is for the robot to catch the target in a manner that is as efficient as possible. With the knowledge that it has, the robot executes an algorithm until it becomes collocated with the target. For simplicity, we define the positive direction of the number line to be that in which the robot's position reaches magnitude $d$ first.

The only feedback that the robot receives during its search is either the instruction to keep searching or the instruction to terminate (once it is collocated with the target). As such, its motion is predetermined by the initial parameters that are available to it. We will refer to this predetermined pattern of motion as a "strategy" and define it as follows.

**Definition 1.** *A strategy $\mathcal{A}$ is a function $\mathcal{A}(t)$ for $t \geq 0$ that denotes the robot's position on the number line at time $t$. $\mathcal{A}(t)$ is continuous, does not violate the speed limit of 1, and $\mathcal{A}(0) = 0$. A strategy $\mathcal{A}$ is "successful" if for any possible target, there exists some $t$ at which the robot is collocated with that target.*

Since one strategy must account for all problem instances, it is useful to imagine the targets from all problem instances as existing simultaneously. In such a paradigm, the goal of a strategy is to "catch" all possible targets (whereas only one is caught in reality). The time taken to catch each potential target can be used to evaluate the efficiency of the strategy. We will use the word "reach" to mean becoming collocated with a target whereas the word "catch" will mean reaching a target for the first time.

The fraction $u = \frac{1}{1-v}$ represents the so-called *evasiveness* of a target of speed $v$. The evasiveness of a target is proportional to the minimum amount of time that it takes for the robot to close a gap between itself and the target. Note that $v = 1 - \frac{1}{u}$ and by the bounds of $v$, we have $1 \leq u < \infty$. From here on out, it is useful to frame the conversation in terms of evasiveness rather than speed. Every variable denoted by the letter $u$ will represent an evasiveness and will implicitly be associated with a speed variable denoted by the letter $v$ with the same subscript or lack thereof.

The variable $\sigma \in \{0, 1\}$ will be used to represent the positive side of the real line if $\sigma = 0$ or the negative side if $\sigma = 1$. We define

$$T_{\mathcal{A}}^{\sigma}(u, d) = \inf \left\{ t \geq 0 : A(t) = (-1)^{\sigma} \left( d + \left( 1 - \frac{1}{u} \right) t \right) \right\}$$

as the time taken by strategy $\mathcal{A}$ to catch $u$ on side $\sigma$ and $T_{opt}(u, d) = \frac{d}{1-v} = ud$ as the time taken by a robot running an offline algorithm to accomplish the same task. The competitive ratio is defined as

$$CR_{\mathcal{A}}^{\sigma}(u, d) = \frac{T_{\mathcal{A}}^{\sigma}(u, d)}{T_{opt}(u, d)} \text{ and } CR_{\mathcal{A}}(u, d) = \max \left( CR_{\mathcal{A}}^{0}(u, d), CR_{\mathcal{A}}^{1}(u, d) \right).$$

In all cases, we will be using $CR_{\mathcal{A}}(u, d)$ to evaluate the performance of search strategies. When the starting distance $d$ is known to the searcher (as is the case in Sect. 2), we will suppress $d$ in the definitions above and simply write $T_{\mathcal{A}}^{\sigma}(u)$, $T_{opt}(u)$, $CR_{\mathcal{A}}^{\sigma}(u)$, or $CR_{\mathcal{A}}(u)$, respectively to simplify the notation.

## 1.2   Related Work

The study of linear search for a static target was first proposed independently in [6,8] for stochastic and game theoretic models, while deterministic linear search by a single mobile agent was investigated in [3,4]. There is extensive literature on linear search for both stochastic and deterministic search models. For additional information, the interested reader could consult the seminal books by Ahlswede and Wegener [1] and Alpern and Gal [2].

The ultimate goal of linear search is to determine "precise bounds" on the competitive ratio of a specific search model. As such, the competitive ratio may play a primary role in determining the computational strength of a search model as well as its relation to other existing models. This is particularly evident when analyzing the impact of communication in various settings including multi-agent (group) search with or without faulty agents, different communication abilities, etc. For example, see the work of [14,16] for linear group search with crash faults, and [12,13,17] for the case of byzantine faults. Another example, is when two agents can communicate only in the F2F (Face-to-Face) model (see [9]) as well as [5] which analyzes the competitive ratio of linear search for two robots with different speeds. It should also be noted that there is extensive literature when different cost models are considered, e.g., [15] which takes into account the number of turns.

Most of the known research on analyzing and determining the competitive ratio of linear search is concerned with a static target. Alpern and Gal [2, p. 134, Eq. 8.25] were the first to analyze linear search for catching an escaping oblivious target and proved that the optimal competitive ratio of search for a moving target starting at an unknown starting distance and moving with known (to the searcher) speed $0 \leq v < 1$ (where 1 is the speed of the searcher) away from the origin is exactly $1 + 8\frac{1+v}{(1-v)^2}$. If both starting distance $d$ and speed $v$ of the mobile target are known to the searcher, the competitive ratio is shown in [11] to be $1 + \frac{2}{1-v}$. This last publication was also the first work to consider knowledge/competitive ratio tradeoffs in linear search for a moving target. In the present paper, we consider the general problem of the impact of knowing either $d$ or $v$ on the competitive ratio of search. In particular, we focus on determining the competitive ratio of search for a moving target when its speed is unknown to the searcher and solve one of the main open problems proposed in [11].

### 1.3   Contributions and Outline

Recall that, throughout the paper, we assume that the speed $v$ of the mobile target is unknown to the searcher and $u = \frac{1}{1-v}$ is the evasiveness of $v$. The main contributions are as follows.

In Sect. 2, we assume the starting distance $d$ of the mobile target is known to the searcher. In Subsect. 2.1, we prove a lower bound in Theorem 1: that no search strategy $\mathcal{A}$ can satisfy $CR_{\mathcal{A}}(u) \in O\left(u^{4-\varepsilon}\right)$, for any constant $\varepsilon > 0$ (improving the lower bound of $1 + 2u$ in [11]). In Subsect. 2.2, we define strategy $\mathcal{A}_1$ (see Algorithm 1) and prove, in Theorem 2, that it obeys $CR_{\mathcal{A}_1}(u) \leq 56.18u^{4-(\log_2 \log_2 u)^{-2}}$ for $u > 4$, while $CR_{\mathcal{A}_1}(u) = 9$ for $1 \leq u \leq 4$ (improving the previous upper bound of $1 + 16u^4(\log_2 u)^2$ from [11] for all $u > 4$). Our results for Sect. 2 are tight up to lower order terms in the exponent, thus answering one of the main open problems proposed in [11].

In Sect. 3, we analyze the no-knowledge case whereby the searcher knows neither the speed $v$ nor the starting distance $d$ of the target. We define a strategy $\mathcal{A}_2$ (see Algorithm 2) and prove, in Theorem 3, that it obeys $CR_{\mathcal{A}_2}(u,d) \leq 1 + \frac{1}{d}\left(56.18(ud)^{4-(\log_2 \log_2(ud))^{-2}} - 1\right)$ for $ud > 4$, while $CR_{\mathcal{A}_2}(u,d) \leq 1 + \frac{8}{d}$ for $ud \leq 4$ (improving the previous best upper bound: $O\left(\frac{1}{d}M^8 \log_2^2 M \log_2 \log_2 M\right)$, where $M = \max(u,d)$, in [11]). We also note that the lower bound in the case where $d$ is known extends trivially to the case where $d$ is unknown, yielding that no strategy $\mathcal{A}$ could satisfy $CR_{\mathcal{A}}(u,d) \in O\left(u^{4-\varepsilon}\right)$, for any constant $\varepsilon > 0$. We conclude in Sect. 4 by summarizing our contributions and proposing some open problems. All proofs omitted due to space constraints can be found in the full version of the paper [10].

## 2   Unknown Speed and Known Starting Distance

In this section, we consider search when the speed $0 \leq v < 1$ of the target is unknown but the starting distance $d$ is known to the searcher. First, we prove a lower bound in Subsect. 2.1, followed by an upper bound in Subsect. 2.2.

### 2.1   Lower Bound

In Definition 1, we defined the general concept of a search strategy. Next, we define a particular class of search strategies: so-called zigzag strategies.

**Definition 2.** *A zigzag strategy $\mathcal{A}$ is defined in terms of a sequence of evasiveness values: $(u_i)_{i=0}^{\infty}$. This sequence must satisfy the additional stipulations that $u_{i+2} > u_i$ for all $i \geq 0$ and that $\lim_{i \to \infty} u_i = \infty$. The movement of the robot is divided into rounds with indices $i \geq 0$. On round $i$, the robot travels from the origin to position $(-1)^i x_i$ and back to the origin where $x_i$ is the distance that is just large enough to catch the target with evasiveness $u_i$. For future convenience, let $s_i = \sum_{n=0}^{i} x_n$ for all $i \geq -1$. To be clear, $s_{-1} = 0$.*

For any successful strategy, there exists a zigzag strategy that performs just as well or better. This is because it is optimal for the robot to always be travelling to the next new target that it intends to catch at top speed (see [2]). For this reason, we will only be considering zigzag strategies as they are the best-performing class of strategies. This subsection is dedicated to proving the following theorem.

**Theorem 1.** *In the case where initial distance $d$ is known but speed $v$ is unknown, no strategy $\mathcal{A}$ can satisfy $CR_{\mathcal{A}}(u) \in O\left(u^{4-\varepsilon}\right)$, for any constant $\varepsilon > 0$.*

*Proof.* This proof will show that a contradiction arises from the assumption that $CR_{\mathcal{A}}(u) \in O\left(u^k\right)$ for an arbitrary zigzag strategy $\mathcal{A}$ and for an arbitrary constant $k < 4$. Part of the proof borrows work done by Beck and Newman [7] regarding the cow path problem. Here is an outline of the proof:

1. Lemmas 1 through 3 will give a lower bound for $CR_{\mathcal{A}}(u)$ in terms of $u_i$ for various $i$.
2. This bound will be distilled into a key condition, but only under the assumption that $CR_{\mathcal{A}}(u) \in O\left(u^k\right)$ for $k < 4$.
3. Lemma 4 will show that this condition is impossible to satisfy.

Note that Lemma 2 will also be useful in later proofs.

**Lemma 1.** *For some zigzag strategy $\mathcal{A}$, for any round number $i \geq 0$, and for any evasiveness value $u \in (u_i, u_{i+2}]$, $CR_{\mathcal{A}}(u) \geq 1 + \frac{2}{d}s_{i+1}$.*

*Proof.* Consider a round $i \geq 0$ with parity $\sigma$. During round $i+2$, all $u \in (u_i, u_{i+2}]$ on side $\sigma$ are caught (for the first time). At the start of round $i + 2$, the robot is at the origin but has spent $2s_{i+1}$ time on previous rounds, giving all targets an additional $2vs_{i+1}$ distance head start. The target with evasiveness $u$ is caught in time

$$T_{\mathcal{A}}^{\sigma}(u) = 2s_{i+1} + \frac{d + 2vs_{i+1}}{1 - v} = \frac{d + 2vs_{i+1} + 2s_{i+1}(1 - v)}{1 - v} = ud + 2us_{i+1}.$$

It is true that $T_{\mathcal{A}}^{\sigma}(u) \leq \max\left(T_{\mathcal{A}}^0(u), T_{\mathcal{A}}^1(u)\right)$. Hence, by definition, $CR_{\mathcal{A}}(u) \geq \frac{T_{\mathcal{A}}^{\sigma}(u)}{T_{opt}(u)} = \frac{ud + 2us_{i+1}}{ud} = 1 + \frac{2}{d}s_{i+1}$ . This completes the proof of Lemma 1. $\qquad \square$

**Lemma 2.** *For any zigzag strategy $\mathcal{A}$, for all $i \geq 0$, $s_i = u_i d + (2u_i - 1)s_{i-1}$.*

*Proof.* At the start of round $i$, the robot is at the origin and must catch evasiveness values up to $u_i$. The robot has spent $2s_{i-1}$ time on previous rounds. Therefore, the target with evasiveness $u_i$ is a distance $d + 2v_i s_{i-1}$ away from the robot at the start of round $i$. Catching this target will take an amount of time (and distance) equal to

$$x_i = u_i \left(d + 2v_i s_{i-1}\right) = u_i \left(d + 2\left(1 - \frac{1}{u_i}\right)s_{i-1}\right) = u_i d + (2u_i - 2)s_{i-1}.$$

By definition, $s_i = x_i + s_{i-1} = u_i d + (2u_i - 2)s_{i-1} + s_{i-1} = u_i d + (2u_i - 1)s_{i-1}$. This proves Lemma 2. $\qquad \square$

**Lemma 3.** *For any zigzag strategy $\mathcal{A}$, for all $i \geq 0$, $s_i \geq d \prod_{n=0}^{i} u_n$.*

*Proof.* Lemma 2 states that $s_i = u_i d + (2u_i - 1)s_{i-1}$. Since $u_i \geq 1$,

$$s_i = u_i d + (2u_i - 1)s_{i-1} \geq u_i d + (2u_i - u_i)s_{i-1} = u_i d + u_i s_{i-1} \geq u_i s_{i-1}.$$

In summary, for all $i \geq 0$, $s_i \geq u_i s_{i-1}$. Using $s_0 = u_0 d$ as the base case, we can unwrap the recurrence to yield $s_i \geq d \prod_{n=0}^{i} u_n$. This completes the proof of Lemma 3. $\qquad \square$

Now, for the sake of contradiction, assume that there exists a zigzag strategy $\mathcal{A}$ satisfying $CR_{\mathcal{A}}(u) \in O\left(u^k\right)$ for some $k < 4$. More specifically, assume that there exists some $a > 0$ such that for all $u \geq c_1$ for some constant $c_1$,

$$CR_{\mathcal{A}}(u) \leq au^k. \tag{2.1}$$

We now examine a subset of all $u$ values on which this condition must hold. For each round numbered $i \geq 0$, let us select an evasiveness $u = \alpha_i u_i$ where

$$\alpha_i = \min\left(\frac{u_{i+2}}{u_i}, \sqrt[k]{2}\right). \tag{2.2}$$

Note that $u_i < u_{i+2}$ by definition and that we need only consider $k > 0$. With this in mind, one can verify that $\alpha_i u_i$ satisfies the following properties:

$$u_i < \alpha_i u_i \leq u_{i+2}, \tag{2.3}$$

$$(\alpha_i u_i)^k \leq 2u_i^k. \tag{2.4}$$

Lemma 1 in conjunction with Eq. (2.3) yields: $CR_\mathcal{A}(\alpha_i u_i) \geq 1 + \frac{2}{d}s_{i+1}$. Lemma 3, Eq. (2.1), and Eq. (2.4) yield:

$$1 + 2\prod_{n=0}^{i+1} u_n \leq 1 + \frac{2}{d}s_{i+1} \leq CR_\mathcal{A}(\alpha_i u_i) \leq a(\alpha_i u_i)^k \leq 2au_i^k.$$

Now, let $u_i = 2^{w_i}$ where $w_i \geq 0$. Note that since $u_i$ goes to infinity by the definition of a zigzag strategy, so does $w_i$. Hence, $2a \cdot 2^{kw_i} \geq 1 + 2\prod_{n=0}^{i+1} 2^{w_n} \geq 2\prod_{n=0}^{i+1} 2^{w_n}$. Dividing by 2 and applying $\log_2$ to both sides produces $kw_i + \log_2(a) \geq \sum_{n=0}^{i+1} w_n$. We can remove $\log_2(a)$ by swapping $k$ with $k_2 = \frac{1}{2}(k+4)$, which is bigger than $k$ but still less than 4. Since $\log_2(a)$ is a constant and $w_i$ goes to infinity, there will necessarily be some round $j$ such that for all $i \geq j$, $kw_i + \log_2(a) \leq \frac{1}{2}(k+4)w_i$. Alternatively, for all $i \geq j$, $w_i \geq \frac{2\log_2(a)}{4-k}$. Let $c_2 = u_j$. We now have that for all $i$ satisfying $u_i \geq \max(c_2, c_1)$,

$$k_2 w_i \geq \sum_{n=0}^{i+1} w_n. \tag{2.5}$$

We will now show that Eq. (2.5) cannot hold using Lemma 4, which adapts part of Lemma 2 from [7]. Though [7] works with a doubly infinite sequence, all of the reasoning is analogous and requires little change.

**Lemma 4.** *Consider a sequence $(z_i)_{i=0}^{\infty}$ where $z_i > 0$ for all $i \geq 2$ and $z_i \geq 0$ for $i = 0, 1$. For $h < 4$, no such sequence is able to satisfy the following condition for all $i$ larger than a sufficiently large $m$:*

$$hz_i \geq \sum_{n=0}^{i+1} z_n. \tag{2.6}$$

Lemma 4 shows that Eq. (2.5) is contradictory: take $z_i = w_i$, $m = \max(c_1, c_2)$, and $h = k_2 < 4$. Note that $w_i > w_{i-2} \geq 0$ for all $i \geq 2$. The only unfounded assumption made in the derivation of Eq. (2.5) was that $CR_\mathcal{A}(u) \in O(u^k)$ for $k < 4$. Therefore, this assumption is false. This completes the proof of Theorem 1.  □

## 2.2  Upper Bound

In this subsection, we present an algorithm and prove an upper bound on its performance in Theorem 2. Algorithm 1 follows a zigzag strategy $\mathcal{A}$ given by $u_i = 2^{3 \cdot 2^i \sqrt{i+1} - 1}$.

---

**Algorithm 1** (Search with unknown speed and known initial distance)

---

1: **input:** initial target distance $d$
2: terminate at any point if collocated with target
3: $t \leftarrow 0$
4: **for** $i \leftarrow 0, 1, 2, ..., \infty$ **do**
5:      $u_i \leftarrow 2^{3 \cdot 2^i \sqrt{i+1} - 1}$
6:      $v_i \leftarrow 1 - u_i^{-1}$
7:      $x_i \leftarrow u_i \left( d + v_i t \right)$
8:      move to position $(-1)^i x_i$ and back to the origin
9:      $t \leftarrow t + 2x_i$
10: **endfor**

---

**Theorem 2.** *In the case where initial distance $d$ is known but speed $v$ is unknown, Algorithm 1 has the following bounds on its performance:*

$$\forall \, 1 \le u \le 4 \qquad CR_{\mathcal{A}}(u) = 9 \qquad\qquad (2.7)$$

$$\forall \, u > 4 \qquad CR_{\mathcal{A}}(u) \le 56.18u^{4 - (\log_2 \log_2 u)^{-2}} \qquad (2.8)$$

The entire proof of Theorem 2 can be found in the full version of the paper [10], though an outline is presented here. The proof relies on the six lemmas 5 through 10. Lemma 6 is of particular importance.

**Lemma 5.** *For any zigzag strategy $\mathcal{A}$, $CR_{\mathcal{A}}^\sigma(u) = 1 + \frac{2}{d}s_{k-1}$, where $k$ is the round on which the target with evasiveness $u$ on side $\sigma$ is caught.*

**Lemma 6.** *Consider a zigzag strategy $\mathcal{A}$ whose defining sequence of evasiveness values $(u_i)_{i=0}^\infty$ is strictly increasing. For a round number $k \ge 0$, if a function $F(u)$ is increasing for all $u \ge u_k$, then statement (2.9) implies statement (2.10).*

$$1 + \frac{2}{d}s_{i+1} \le F(u_i) \qquad\qquad \forall i \ge k, \qquad (2.9)$$

$$CR_{\mathcal{A}}(u) \le F(u) \qquad\qquad \forall u > u_k. \qquad (2.10)$$

*Proof.* Consider any round $i$ with parity $\sigma$. We know that all evasiveness values $u_i < u \le u_{i+2}$ on side $\sigma$ are caught on round $i + 2$. By Lemma 5, $CR_{\mathcal{A}}^\sigma(u) = 1 + \frac{2}{d}s_{i+1}$ for all $u_i < u \le u_{i+2}$. Consequently, if $F(u_i) \ge 1 + \frac{2}{d}s_{i+1}$ and $F(u_i)$ is increasing within the interval $[u_i, u_{i+2}]$, then $F(u) \ge CR_{\mathcal{A}}^\sigma(u)$ for all $u_i < u \le u_{i+2}$. Let $\sigma_k$ denote the parity of $k$. Statement (2.9) allows us to apply our prior reasoning to all rounds $i \ge k$, collectively yielding the following:

$$F(u) \ge CR_{\mathcal{A}}^{\sigma_k}(u) \qquad\qquad \forall u > u_k,$$

$$F(u) \ge CR_{\mathcal{A}}^{1-\sigma_k}(u) \qquad\qquad \forall u > u_{k+1}.$$

The latter statement can be extended to all $u > u_k$ as follows. Statement (2.9) yields the following in conjunction with the fact that $s_{k+1} \ge s_k$:

$$F(u) \ge 1 + \frac{2}{d}s_{k+1} \ge 1 + \frac{2}{d}s_k \qquad\qquad \forall u_k < u \le u_{k+2}.$$

Recall the stipulation that $(u_i)_{i=0}^{\infty}$ is strictly increasing. We can reduce the applicable range of the statement above to $u_k < u \leq u_{k+1}$, which is a subset of the range $u_{k-1} < u \leq u_{k+1}$ on which $1 + \frac{2}{d}s_k = CR_{\mathcal{A}}^{1-\sigma_k}(u)$. In the event that $k = 0$, this range is instead $1 \leq u \leq u_{k+1}$, which still contains all $u_k < u \leq u_{k+1}$. From this, we can conclude that $F(u) \geq CR_{\mathcal{A}}^{1-\sigma_k}(u)$ for all $u_k < u \leq u_{k+1}$.

$$F(u) \geq CR_{\mathcal{A}}^{\sigma_k}(u) \, , \; F(u) \geq CR_{\mathcal{A}}^{1-\sigma_k}(u) \qquad \qquad \forall u > u_k,$$

$$F(u) \geq \max\left(CR_{\mathcal{A}}^{\sigma_k}(u), CR_{\mathcal{A}}^{1-\sigma_k}(u)\right) = CR_{\mathcal{A}}(u) \qquad \forall u > u_k.$$

This proves Lemma 6. $\qquad \qquad \qquad \qquad \qquad \qquad \qquad \qquad \qquad \qquad \qquad \qquad \square$

**Lemma 7.** *Given a zigzag strategy, for any $i \geq 0$, $s_i \leq 2^{i+1} \cdot d \prod_{j=0}^{i} u_j$.*

The following notation will be used in the upcoming Lemmas. For $n \geq 0$, let $f^{<n>}(x)$ denote the "$n$-th difference" of a function $f$ at $x$ as $f^{<0>}(x) = f(x)$, $f^{<n>}(x) = f^{<n-1>}(x) - f^{<n-1>}(x - 1)$. There are obvious parallels between $f^{<n>}(x)$ and the $n$-th derivative, $f^{(n)}(x)$. Note that difference operations tend to alter the domain of $f$ more significantly than differentiation. The useful properties listed below apply only where the result is defined. Just like with regular derivatives, $(f^{<a>})^{<b>}(x) = f^{<a+b>}(x)$. Additionally, by the linearity of differentiation,

$$\left(f^{<1>}\right)^{(1)}(x) = \frac{\mathrm{d}}{\mathrm{d}x}\left(f(x) - f(x - 1)\right) = f^{(1)}(x) - f^{(1)}(x - 1) = \left(f^{(1)}\right)^{<1>}(x).$$

In conjunction with the properties of regular derivatives, the above two properties reveal that when a function $f(x)$ is differentiated $a$ times and has $b$ first difference operations applied, the order in which all $a + b$ operations are applied does not matter. Let us write the result of such a procedure as $f^{(a)<b>}(x)$ or $f^{<b>(a)}(x)$.

**Lemma 8.** *Consider two arbitrary functions $f(x)$ and $g(x)$ defined on integers $x \geq 0$. Let $G(x)$ be shorthand for $\sum_{j=0}^{x} g(j)$ where $G(0) = g(0)$ and $G(-1) = 0$. For any constants $n \geq 0$ and $m \geq 0$,*

$$\sum_{j=0}^{n} g(j)f^{<m>}(j) = G(n)f^{<m>}(n) - \sum_{j=0}^{n-1} G(j)f^{<m+1>}(j+1).$$

**Lemma 9.** *For the function $f(x) = \sqrt{x}$ and for any integers $k \geq 1$ and $0 \leq m \leq k$, $(-1)^{k+1}f^{(k-m)<m>}(x)$ is non-negative for all $x > m$.*

**Lemma 10.** *For the function $f(x) = \sqrt{x}$ specifically, $f^{<k>}(x)$ is subject to the following upper and lower bounds for any integer $k \geq 1$ for all real $x > k$:*

$$(-1)^{k+1}f^{(k)}\left(x - \frac{1}{2}k\right) \leq (-1)^{k+1}f^{<k>}(x) \leq (-1)^{k+1}f^{(k)}(x - k).$$

*Instances of $(-1)^{k+1}$ can be removed by appropriately alternating the direction of the inequality.*

*Proof* (Theorem 2 - sketch). We begin by imagining an upper bound of the form $CR_A(u) \leq F(u) = cu^{4-h(u)}$. Lemma 6 reveals that this condition holds for all $u > u_1$ if $F(u)$ is increasing on that range and $1 + \frac{2}{d}s_{i+1} \leq F(u_i)$ for all $i \geq 1$. From $F(u_i)$, we isolate $h(u_i)$ to obtain the following:

$$h(u_i) \leq \frac{4\log_2(u_i) - \log_2\left(1 + \frac{2}{d}s_{i+1}\right) + \log_2 c}{\log_2(u_i)} \qquad \forall i \geq 1.$$

We continue to transform this condition so that it is more strict but easier to work with. Lemma 7 converts $s_{i+1}$ into a product, which becomes a sum under the logarithm. We apply Lemma 8 to the resulting sum three times, which leaves us with many terms, a few of which cleanly cancel the $4\log_2(u_i)$ term. We then reduce all but one of the remaining terms to constants with the help of Lemma 10; $c = 56.18$ is large enough that the $\log_2 c$ term cancels those constants. The remaining term is simplified using Lemma 10. It can then cancel out the denominator, leaving $(i+1)^{-2}$ on the right-hand side, which is larger than $h(u_i) = (\log_2 \log_2 u_i)^{-2}$. We note that for such an $h(u)$, the resulting function $F(u)$ is increasing where needed. We conclude that $F(u) = 56.18u^{4-(\log_2 \log_2 u)^{-2}}$ is a valid upper bound on $CR_A(u)$ for all $u > u_1$.

In the case where $u \leq u_1$, we calculate $CR_A(u)$ precisely using Lemma 5. As a consequence of Lemma 5, $CR_A(u)$ only has two values in this range. We find that $F(u)$ remains a valid upper bound for all $4 < u \leq u_1$. Meanwhile, $CR_A(u) = 9$ for all $u \leq 4$. This concludes the proof sketch for Theorem 2. □

# 3   Unknown Speed and Starting Distance

In this section, we analyze the competitive ratio of search when both the speed and starting distance of the mobile target are unknown to the searcher. Our proof of the upper bound will make use of the following lemma.

**Lemma 11.** $T_A^\sigma(u,d) \leq T_A^\sigma(ud,1)$ *for* $\sigma \in \{0,1\}$ *and* $CR_A(u,d) \leq CR_A(ud,1)$, *for all strategies* $A$.

*Proof.* Consider two moving targets: the first with $u_1 = u, d_1 = d$ and the second with $u_2 = ud, d_2 = 1$, and respective speeds $v_1 = 1 - \frac{1}{u}$ and $v_2 = 1 - \frac{1}{ud}$. The first target is not further from the origin than the second target at all times $t \geq ud$, and this is also the earliest time at which either target can be caught by the searcher. If we note that

$$t \geq ud \Leftrightarrow t\left(u^{-1} - (ud)^{-1}\right) \geq d - 1 \Leftrightarrow 1 + (1 - (ud)^{-1})t \geq d + (1 - u^{-1})t$$

then we conclude that the first target cannot be caught after the second target. This implies that $T_A^\sigma(u,d) \leq T_A^\sigma(ud,1)$ for $\sigma \in \{0,1\}$. It follows that

$$\frac{\max\left\{T_A^0(u,d), T_A^1(u,d)\right\}}{ud} \leq \frac{\max\left\{T_A^0(ud,1), T_A^1(ud,1)\right\}}{ud}$$

implying that $CR_A(u,d) \leq CR_A(ud,1)$, which proves Lemma 11. □

Motivated by Lemma 11, one is led to consider running the previous search Algorithm 1 under the assumption that the target's initial distance is 1.

---

**Algorithm 2** (Search with unknown speed and initial distance)

---

1: Execute Algorithm 1 as though $d = 1$ (regardless of the true value of $d$)

---

**Theorem 3.** *In the case where both initial distance $d$ and speed $v$ are unknown, the competitive ratio of Algorithm 2 satisfies the following bounds:*

$$CR_{\mathcal{A}}(u,d) \leq \begin{cases} 1 + \frac{8}{d} & \text{if } ud \leq 4 \\ 1 + \frac{1}{d}\left(56.18(ud)^{4-(\log_2 \log_2(ud))^{-2}} - 1\right) & \text{if } ud > 4 \end{cases} \quad (3.1)$$

*Proof.* By the same reasoning as in Lemma 5, $CR_{\mathcal{A}}(u,d) = 1 + \frac{2}{d}s_{k-1}$ where $k$ is the round on which the target with evasiveness $u$ and initial distance $d$ on side $\sigma$ is caught. By extension, $s_{k-1} = \frac{d}{2}\left(CR_{\mathcal{A}}(u,d) - 1\right)$. Consider two targets: the first with $u_1 = u, d_1 = d$ and the second with $u_2 = ud, d_2 = 1$. By Lemma 11, $T_{\mathcal{A}}(u,d) \leq T_{\mathcal{A}}(ud, 1)$, meaning that target 2 is caught no sooner than target 1. If we define $k_1, k_2$ to be the rounds on which each respective target is caught, we can conclude that $k_1 \leq k_2$ and that $s_{k_1-1} \leq s_{k_2-1}$. It follows from our prior reasoning that $\frac{d}{2}\left(CR_{\mathcal{A}}(u,d) - 1\right) \leq \frac{1}{2}\left(CR_{\mathcal{A}}(ud,1) - 1\right)$. Finally, isolating $CR_{\mathcal{A}}(u,d)$ yields $CR_{\mathcal{A}}(u,d) \leq 1 + \frac{1}{d}\left(CR_{\mathcal{A}}(ud,1) - 1\right)$.

Algorithm 2 runs Algorithm 1 under the assumption that the starting distance of the mobile target is 1, meaning that Theorem 2 provides an upper bound on $CR_{\mathcal{A}}(u,1)$. Since we are interested in $CR_{\mathcal{A}}(ud,1)$, we replace occurrences of $u$ with $ud$. Doing so provides us with statement (3.1), concluding the proof of Theorem 3.    □

We note that since $ud \leq \max\{u,d\}^2 = M^2$, the upper bound presented in Theorem 3 is a strict improvement (asymptotically) over the previous best known bound, $O\left(\frac{1}{d}M^8 \log_2^2 M \log_2 \log_2 M\right)$, in [11]. Also the lower bound in the case where $d$ is known extends trivially to the case where $d$ is unknown, yielding that no strategy $\mathcal{A}$ could satisfy $CR_{\mathcal{A}}(u,d) \in O\left(u^{4-\varepsilon}\right)$, for any constant $\varepsilon > 0$.

## 4    Conclusion

In this paper, we considered linear search for an escaping oblivious mobile target by an autonomous mobile agent. Based on the competitive ratio, our algorithm and its analysis indicates optimality up to low order terms in the exponent for the case when the speed $0 \leq v < 1$ of the mobile target is unknown to the searcher, thus answering an open problem in [11]. We also analyzed and improved on previous results in [11] when both $d, v$ are unknown; however, tight bounds for this case remain elusive. A most interesting (and challenging) direction for future research is group search (in the context of linear search) by a multi-agent system of searchers with various communication behaviours and capabilities.

# References

1. Ahlswede, R., Wegener, I.: Search Problems. Wiley-Interscience (1987)
2. Alpern, S., Gal, S.: The Theory of Search Games and Rendezvous. International series in Operations Research and Management Science, vol. 55. Kluwer (2003)
3. Baeza-Yates, R., Culberson, J., Rawlins, G.: Searching in the plane. Inf. Comput. **106**(2), 234–252 (1993)
4. Baeza-Yates, R., Schott, R.: Parallel searching in the plane. Comput. Geom. **5**(3), 143–154 (1995)
5. Bampas, E., et al.: Linear search by a pair of distinct-speed robots. Algorithmica **81**(1), 317–342 (2019)
6. Beck, A.: On the linear search problem. Israel J. Math. **2**(4), 221–228 (1964)
7. Beck, A., Newman, D.J.: Yet more on the linear search problem. Israel J. Math. **8**(4), 419–429 (1970)
8. Bellman, R.: An optimal search. SIAM Rev. **5**(3), 274–274 (1963)
9. Chrobak, M., Gąsieniec, L., Gorry, T., Martin, R.: Group search on the line. In: Italiano, G.F., Margaria-Steffen, T., Pokorný, J., Quisquater, J.-J., Wattenhofer, R. (eds.) SOFSEM 2015. LNCS, vol. 8939, pp. 164–176. Springer, Heidelberg (2015). https://doi.org/10.1007/978-3-662-46078-8_14
10. Coleman, J.R., Ivanov, D., Kranakis, E., Krizanc, D., Morales-Ponce, O.: Linear search for an escaping target with unknown speed. CoRR, abs/2404.14300 (2024)
11. Coleman, J.R., Kranakis, E., Krizanc, D., Morales-Ponce, O.: Line search for an oblivious moving target. In: Hillel, E., Palmieri, R., Rivière, E. (edis.) 26th International Conference on Principles of Distributed Systems, OPODIS 2022, 13–15 December 2022, Brussels, Belgium. LIPIcs, vol. 253, pp. 12:1–12:19. Schloss Dagstuhl - Leibniz-Zentrum für Informatik (2022)
12. Czyzowicz, J., et al.: Search on a line by byzantine robots. Int. J. Found. Comput. Sci. **32**(4), 369–387 (2021)
13. Czyzowicz, J., Killick, R., Kranakis, E., Stachowiak, G.: Search and evacuation with a near majority of faulty agents. In: SIAM ACDA21 (Applied and Computational Discrete Algorithms), Proceedings, Seattle, USA, 19–21 July 2021, pp. 217–227. SIAM (2021)
14. Czyzowicz, J., Kranakis, E., Krizanc, D., Narayanan, L., Opatrny, J.: Search on a line with faulty robots. Distrib. Comput. **32**(6), 493–504 (2019)
15. Demaine, E.D., Fekete, S.P., Gal, S.: Online searching with turn cost. Theoret. Comput. Sci. **361**(2), 342–355 (2006)
16. Kupavskii, A., Welzl, E.: Lower bounds for searching robots, some faulty. In: PODC 2018, Egham, UK, pp. 447–453. ACM (2018)
17. Sun, X., Sun, Y., Zhang, J.: Better upper bounds for searching on a line with byzantine robots. In: Du, D.-Z., Wang, J. (eds.) Complexity and Approximation. LNCS, vol. 12000, pp. 151–171. Springer, Cham (2020). https://doi.org/10.1007/978-3-030-41672-0_9

# Dominance for Enclosure Problems

Waseem Akram[(✉)] and Sanjeev Saxena

Department of Computer Science and Engineering, Indian Institute of Technology,
Kanpur, Kanpur 208 016, India
{akram,ssax}@iitk.ac.in

**Abstract.** Given a set of homothetic triangles (i.e., closed under scaling and translating operations) in the plane, we study the problem of computing the triangles enclosing a query object. The query objects considered are points, line segments, trapezoids, and ellipses. We show that the problem can be solved using the 3-d dominance reporting problem. The result can also be extended to higher dimensions.

**Keywords:** Geometric Intersection · Dominance · Algorithms · Data Structures

## 1 Introduction

In $d$-dimensional Euclidean space, an object $o_1$ encloses (or contains) another object $o_2$ if every point of $o_2$ is also a point of $o_1$. In general, the enclosure searching problem is defined as follows.

Given a set $S$ of geometric objects and a query object $q$, find all the objects $p \in S$ enclosing the query object $q$.

The problem has been well-studied for orthogonal objects (e.g., intervals, rectangles, hyper-rectangles in three and higher dimensions) [5,8,16]. However, despite its theoretical and practical importance, the enclosure problem has received little attention for non-orthogonal objects. For instance, in VLSI design, chip designs deal with orthogonal objects and non-orthogonal objects like $c$-oriented objects [4].

In this paper, we study the enclosure searching problem for a class of non-orthogonal objects (homothetic triangles) in $\mathbb{R}^2$ in the pointer machine model. Two triangles in the plane are homothetic if one can be obtained from the other using scaling and translation operations [6]. Formally, we consider the following problem: Given a set $S$ of $n$ homothetic triangles in the plane, preprocess the set $S$ so that whenever a query object $q$ is supplied, we can efficiently report all the objects $p \in S$ enclosing $q$. The query objects considered are points, line segments, ellipses, and trapezoids.

We present a linear space solution that can answer an enclosure query with any of the objects (points, line segments, trapezoids, and ellipses) in optimal $O(\log n + k)$ time, where $k$ is the number of reported triangles. We solve the

© The Author(s), under exclusive license to Springer Nature Switzerland AG 2024
A. A. Rescigno and U. Vaccaro (Eds.): IWOCA 2024, LNCS 14764, pp. 408–420, 2024.
https://doi.org/10.1007/978-3-031-63021-7_31

problem by transforming it into another geometric intersection problem, namely, *the dominance reporting problem in* $\mathbb{R}^3$. A point $p$ dominates another point $q$, denoted by "$q \preceq p$", if each coordinate of $p$ is greater than or equal to the corresponding coordinate of $q$ [10]. In the dominance reporting problem, the goal is to preprocess a given set of points in $\mathbb{R}^d$ so that we can quickly report all the points dominated by a query point [2, 10–13]. We show that after preprocessing the set $S$, an enclosure query with any of the objects (points, line segments, ellipses, and trapezoids) can be answered using a 3-d dominance reporting query. Thus, the problem can be solved within the same time and space bounds as the 3-d dominance reporting problem. Markis and K. Tsakalidis [11] showed that the 3-d dominance reporting problem, in the pointer machine model, can be solved using linear space and $O(\log n + k)$ query time, where $n$ and $k$ are input and output sizes, respectively. The preprocessing algorithm takes $O(n \log n)$ worst-case time.

Tan, Hirata, and Inagaki [14] solve the problem of computing $c$-oriented polygons enclosing a query point using the 3-d dominance reporting problem. They gave a transformation using a non-standardized oblique coordinate system to solve the problem. Our transformation is simpler and uses a standard coordinate system. Moreover, our solution works for various query objects, including points, line segments, trapezoids, and ellipses.

Moreover, our results can easily be extended to higher dimensions. We can preprocess a set of homothetic simplices in $\mathbb{R}^d$ so that all the simplices enclosing a query object (points, line segments, hyper-rectangles, hyper-ellipsoids in $\mathbb{R}^d$) can be found using a $(d + 1)$-dimensional dominance query.

The problem of computing homothetic polygons enclosing a query point has earlier been studied [3] and an optimal $O(\log n + \#output)$-query time solution which can be built in $O(n \log n)$ time and space have been obtained. However, the solution works only for (query) points. The solution we are describing here can answer an enclosing query with various kinds of query objects (points, line segments, trapezoids, and ellipses) in the same time bounds. We, however, consider only homothetic triangles (a special case of homothetic polygons) as input objects.

Chazelle and Edelsbrunner [6] investigated the following range searching problem: "preprocess a given set of $n$ points in the plane so that all the points enclosed in a homothetic query triangle can be efficiently reported". They gave a linear space data structure that supports queries in optimal $O(\log n + \#output)$ time. The problem we are studying in this paper is its "dual" where the input objects are homothetic triangles, and query objects are points, line segments, rectangles, ellipses, and trapezoids.

An outline of the paper is as follows. In Sect. 2, we first show that the interval overlapping problem can be solved using the 2-d dominance query problem. Section 3 considers the special case when input triangles are isosceles and right-angled triangles. We consider general triangles in Sect. 4. Section 5 deals with homothetic simplices in higher dimensions. We conclude our work in Sect. 6.

## 1.1  Previous and Related Works

A polygon in the plane is said to be a $c$-oriented polygon if each edge is parallel to one of the previously defined $c$ directions [9]. Given a set of $n$ $c$-oriented polygons in the plane, *the c-oriented polygonal point enclosure searching problem* is to compute all the polygons enclosing a query point. Tan, Hirata, and Inagaki [14] solved the problem; after representing the input set into an $O(n \log n)$ space data structure, each subsequent query can be answered in $O(c^2 \log n + \#output)$ time. We present a high-level sketch of their solution in the following paragraph.

Each polygon in the input set is decomposed, using a plane-sweep algorithm, into disjoint axes-parallel rectangles and triangles with one vertical edge and the other two edges parallel to any of the $c$ (fixed) directions. Thus, the main problem is reduced into two subproblems: (1) *rectangle point enclosure searching problem*, and (2) *triangle point enclosure searching problem*. The rectangle point enclosure problem asks to preprocess a given of rectangles in the plane so that all rectangles enclosing a query point can be reported efficiently [5,16]. In the triangle point enclosure searching problem, the goal is to represent a given set of triangles into a data structure such that, given a query point, the triangles enclosing the query point can be found efficiently [3]. For the rectangle case, they employ an existing solution that uses $O(n)$ space and $O(\log n)$ query time [5]. The collection of all the created triangles is split into $(c-1)*(c-2)$ subsets of triangles. Each subset consists of homothetic triangles whose one edge is vertical and the other two edges are oriented along two particular directions (from the set of $c$ (fixed) directions). Then, for each subset of homothetic triangles, the point enclosure searching problem is solved by transforming it into an instance of the 3-d dominance reporting problem. In the transformation process, a non-standardized oblique coordinate system is used.

Let $S$ be a set of axes-parallel rectangles in $\mathbb{R}^d$ and $R$ be any other such rectangle. Edelsbrunner and Overmars [8] considered the following three geometric intersection problems. The *rectangle enclosure problem* is to report the rectangles of $S$ enclosing $R$. In the *rectangle containment problem*, we are to report the rectangles of $S$ contained in $R$. The *rectangle intersection problem* is to report the rectangles of $S$, which have a common point with $R$. In the counting variant of these problems, we are to count the rectangles satisfying the respective properties. Using appropriate geometric transformations, they showed that the rectangle enclosure and containment searching/counting problems in a $d$-dimensional space are both equivalent to the $2d$-dimensional dominance searching/counting problem (see Theorem 2.1 in [8]). They also showed that the rectangle intersection counting problem and the dominance counting problem are equivalent in $\mathbb{R}^d$ (see Theorem 3.2 in [8]).

Edelsbrunner, Maurer, and Kirkpatrick [7] introduced the polygonal intersection searching problem, which asks to preprocess a given set $n$ of simple polygons with a bounded number (at most $K$) of edges in the plane so that, given a polygon of the same type, all the polygons of the set intersecting the query polygon can be found efficiently. They gave a solution that takes $O(\log n + k)$ time to answer a query, but the space used is $O(n + sn)$, where $s$ is the number of intersections among the edges of the input polygons, and $s$ is bounded above by $K^2 n(n-1)$.

## 2   The Interval Overlapping Problem

The one-dimensional point enclosure is a well-studied problem [5] where the goal is to preprocess a set of possibly overlapping intervals on the real line so that we can report all intervals containing (or overlapping) a query point. This problem can easily be transformed into the 2-d dominance query problem. Let $I$ be a set of $n$ (possibly overlapping) intervals on the real line. An interval $[a_i, b_i] \in I$ will contain a point $q$ on the line if and only if $a_i \leq q \leq b_i$. In other words, the point $(q, -q)$ should dominate the point $(a_i, -b_i)$. We create a set $S = \{(a_i, -b_i) : [a_i, b_i]) \in I\}$ of points in the plane and associate each interval in $I$ with the corresponding created point in $S$. After preprocessing the set $S$ into a 2-d dominance query structure [13], denoted by $\mathcal{D}_2$, we can find all the intervals containing a query point $q$ by querying the structure $\mathcal{D}_2$ with point $(q, -q)$. The intervals associated with the returned points are precisely those that contain the query point $q$.

A more general enclosure problem where query objects are intervals can also be solved using 2-d dominance queries. Let $[x_1, x_2]$ be an interval on the line. An interval $[a_i, b_i] \in I$ contains the interval $[x_1, x_2]$ if and only if $a_i \leq x_1$ and $x_2 \leq b_i$. In other words, the point $(x_1, -x_2)$ should dominate the point $(a_i, -b_i)$. Note that we can use the same structure $\mathcal{D}_2$ described above to find all the points dominated by $(x_1, -x_2)$, and hence the intervals enclosing the query segment $[x_1, x_2]$.

This result is a particular instance of the result by Edelsbrunner and Overmars [8]: the rectangle enclosure problem in $\mathbb{R}^d$ can be solved using the dominance query problem in $\mathbb{R}^{2d}$.

Remark: For the problem where one is interested in finding the intervals enclosed in (contained by) a query interval, we preprocess the points $(-a_i, b_i)$, for each $[a_i, b_i] \in I$, for the 2-d dominance queries. For a query interval $[x_1, x_2]$, we make a 2-d dominance query with point $(-x_1, x_2)$ and report all the intervals associated with the returned points.

Let us now consider the interval overlapping problem where the goal is to compute all intervals $p \in I$ intersecting a query interval $q$, i.e., $p \cap q \neq \phi$. An interval $[a_i, b_i] \in I$ intersects an interval $[x_1, x_2]$ if interval $[a_i, b_i]$ contains an endpoint of $[x_1, x_2]$ or $[a_i, b_i]$ is contained in $[x_1, x_2]$. So, in order to compute all the intervals intersecting the interval $[x_1, x_2]$, the following 2-dominance queries (to the structure $\mathcal{D}_2$) are needed.

- for intervals containing endpoint $x_j$, query with point $(x_j, -x_j)$, $j = 1, 2$
- for intervals containing the whole interval $[x_1, x_2]$, query with $(x_1, -x_2)$

**Theorem 1.** *The one-dimensional point enclosure problem, the interval enclosure problem, and the interval overlapping problem can be solved using a constant number of the 2-d dominance queries.*

As the 2-d dominance query problem can be solved in $O(\log n + k)$ time, where $k$ is the number of reported points [11] after $O(n \log n)$ preprocessing time and $O(n)$ space. By Theorem 1, each considered problem can be solved with the same bounds.

# 3  Isosceles Right-Angled Triangles

In this section, we consider an enclosure problem where the input objects are homothetic isosceles right-angled triangles, and query objects are points, line segments, ellipses, and trapezoids. We will show that after preprocessing the input triangles, each enclosure query (with any of the objects mentioned earlier) can be answered by employing a 3-d dominance query.

Let $S = \{T_1, T_2, ..., T_n\}$ be a set of $n$ homothetic isosceles right-angled triangles. Without loss of generality, we assume that their equal sides are parallel to the axes and the right-angled vertices are at the bottom-left position (i.e., minimum $x$ and minimum $y$ coordinates), see Fig. 1. We identify triangle $T_i \in S$ using the coordinates $(a_i, b_i)$ of its right-angled vertex and the length $\alpha_i$ of its equal sides.

For each triangle $T_i \in S$, we create a 3-d point $(a_i, b_i, -a_i - b_i - \alpha_i)$ and associate the triangle $T_i$ with it. We then preprocess these points for the 3-d dominance queries using the method given by Markis and K. Tsakalidis [11]. We denote the 3-d dominance query structure by $\mathcal{D}_3$.

## 3.1  Points

A point $(x, y)$ in the plane lies inside a triangle $T_i \in S$ if $x \geq a_i, y \geq b_i$, and the point lies below the supporting line of $T_i$'s hypotenuse. The point $(x, y)$ would be below the line if the coordinate sum $x + y$ is at most $a_i + b_i + \alpha_i$. Thus, the triangles of $S$ containing a point can be computed using a 3-dimensional dominance reporting query to the data structure $\mathcal{D}_3$.

## 3.2  Line Segments

We partition the query space into orthogonal (horizontal or vertical) segments, segments with positive slope, and segments with negative slope. We deal with each of them separately. A triangle $T_i \in S$ will contain a line segment with endpoints $(x_1, y_1)$ and $(x_2, y_2)$ iff both the endpoints lie inside the triangle (as a triangle is a convex object). Let $PQ$ be a line segment in the plane with endpoints $P(x_1, y_1)$ and $Q(x_2, y_2)$. Without loss of generality, we assume that $x_1 \leq x_2$.

**Orthogonal Segments:** We first consider the case when the segment $PQ$ is horizontal, i.e., $y_1 = y_2 = y_0$ (say). Observe that a triangle $T_i \in S$ will contain the horizontal segment $[x_1, x_2] \times y_0$ iff the left endpoint of the segment lies inside $T_i$, and the right endpoint lies below the hypotenuse of triangle $T_i$. See Fig. 1.

A point inside the triangle $T_i$ will always dominate the right-angled vertex $(a_i, b_i)$. A point would lie below the line through the hypotenuse of $T_i$ if and only if the sum of the point coordinates is less than or equal to $(a_i + b_i + \alpha_i)$. Therefore, the triangle $T_i$ will contain the segment only if

- point $(x_1, y_0)$ dominates the point $(a_i, b_i)$, and
- the value $x_2 + y_0$ is not more than $(a_i + b_i + \alpha_i)$

Thus, the triangle $T_i$ will contain the segment $[x_1, x_2] \times y_0$ only if the point $(a_i, b_i, -a_i - b_i - \alpha_i)$ is dominated by the point $(x_1, y_0, -x_2 - y_0)$. For a horizontal query segment $[x_1, x_2] \times y_0$, we query the structure $\mathcal{D}_3$ with the point $(x_1, y_0, -x_2 - y_0)$. We report the triangles associated with the points returned by $\mathcal{D}_3$ in response to the query. The queries with vertical segments can be handled symmetrically.

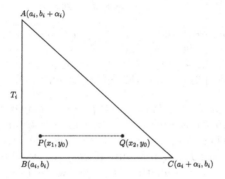

**Fig. 1.** Triangle $T_i$ containing horizontal segment $PQ$.

**Segments with Positive Slope.** Segment $PQ$ with a positive slope will always have $y_1 < y_2$ as $x_1 \leq x_2$ (by assumption), where $(x_1, y_1)$ and $(x_2, y_2)$ are coordinates of point $P$ and $Q$ respectively. See Fig. 2. Note that the endpoint $Q$ dominates the endpoint $P$. As a consequence, in order to check whether endpoints $P$ and $Q$ are inside a triangle $T_i \in S$, it is sufficient to check that $(i)$ $P$ is dominating the point $(a_i, b_i)$, and $(ii)$ $Q$ is below the line through the hypotenuse.

The triangle $T_i$ will contain segment $PQ$ if $x_1 \geq a_i$, $y_1 \geq b_i$, and $-x_2 - y_2 \geq -a_i - b_i - \alpha_i$. Interpreting it in 3-d space, the triangle $T_i$ will contain the segment $PQ$ if the point $(x_1, y_1, -x_2 - y_2)$ dominates the point $(a_i, b_i, -a_i - b_i - \alpha_i)$. Thus, we find all the triangles in $S$ containing the segment $PQ$ by querying the structure $\mathcal{D}_3$ with point $(x_1, y_1, -x_2 - y_2)$.

**Segments with Negative Slope.** Assume that the segment $PQ$ has a negative slope. Since $x_1 \leq x_2$ (by assumption), $y_1 > y_2$. We use the following lemma to categorize the queries into two categories.

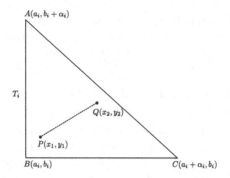

**Fig. 2.** Triangle $T_i$ containing segment $PQ$ with a positive slope.

**Lemma 1.** *The endpoints $P$ and $Q$ will lie inside a triangle $T_i \in S$ if $y_2 \geq b_i$, $x_1 \geq a_i$ and $\max\{x_1 + y_1, x_2 + y_2\}$ is less than or equal to $a_i + b_i + \alpha_i$.*

*Proof.* By assumption, we have $x_1 \leq x_2$. Since $PQ$ has a negative slope, $y_2 < y_1$. The endpoints $P$ and $Q$ will dominate $(a_i, b_i)$ if $x_1 \geq a_i$ and $y_2 \geq b_i$. The third condition $\max\{x_1 + y_1, x_2 + y_2\} \leq a_i + b_i + \alpha_i$ ensures that $P$ and $Q$ (hence the complete segment) are lying below the line through the hypotenuse.  □

If the slope of the segment $PQ$ is more (resp. less) than that of the hypotenuses of the triangles in $S$, then the value $x_2 + y_2$ will be greater (resp. smaller) than the value $x_1 + y_1$. Therefore, a triangle $T_i$ will contain the segment $PQ$ with slope larger than that of $T_i$'s hypotenuse if $x_1 \geq a_i$, $y_2 \geq b_i$, and $-x_2 - y_2 \geq -a_i - b_i - \alpha_i$. See Fig. 3(a). For the other case when $PQ$ has a slope less than that of the hypotenuse (see Fig. 3(b)), the triangle $T_i$ will contain the segment $PQ$ if $x_1 \geq a_i$, $y_2 \geq b_i$, and $-x_1 - y_1 \geq -a_i - b_i - \alpha_i$. Thus, we can find all the triangles containing a line segment with a negative slope using a 3-d dominance query to the data structure $\mathcal{D}_3$.

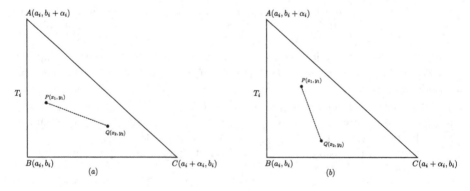

**Fig. 3.** Triangle $T_i$ containing segment $PQ$ with negative slopes. In $(a)$, the slope of $PQ$ is larger than that of $AC$, while in $(b)$ it is less than that of $AC$.

### 3.3 Ellipses

Let $E$ be an ellipse in the plane. Let $(x_1, y_1)$ (resp. $(x_2, y_2)$) be the coordinates of the point on its boundary with the minimum $y$-coordinate (resp. $x$-coordinate). See Fig. 4. The ellipse $E$ will be contained in a triangle $T_i \in S$ if all its boundary points lie inside the triangle. There will be exactly two tangents on the ellipse parallel to the hypotenuse of $T_i$. Given a slope, one can find a tangent to the ellipse using basic geometry [15]. Let $Q(x_0, y_0)$ be the tangent point with the larger coordinate sum. Observe that the triangle $T_i$ will contain the ellipse $E$ if $a_i \leq x_2, b_i \leq y_1$ and the line through its hypotenuse lies above the point $(x_0, y_0)$ i.e. $x_0 + y_0 \leq a_i + b_i + \alpha_i$. We arrive at a 3-d dominance query. Therefore, we can find all the triangles containing $E$ by querying the structure $\mathcal{D}_3$ with point $(x_2, y_1, -x_0 - y_0)$.

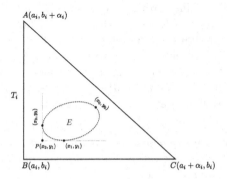

**Fig. 4.** Triangle $T_i$ containing ellipse $E$.

### 3.4 Trapezoids

Let $EFGH$ be a trapezoid with bases $EF$ and $GH$ parallel to the $x$-axis. Let $(x_1, y_1), (x_2, y_1), (x_3, y_2)$, and $(x_4, y_2)$ be the coordinates of vertices $E, F, G$, and $H$ respectively. Without loss of generality, we assume that the vertex with the minimum $x$-coordinate is on the lower base. See Fig. 5. In the other case, $E$ and $H$ do not dominate each other. Obviously, we can find the triangles containing the trapezoid using four individual 3-d dominance queries, one for each vertex. The intersection of the four computed sets will be the output set. However, this approach has an issue: the number of triangles enclosing a vertex can be much larger than the actual output size (i.e., the number of triangles enclosing the trapezoid).

Now we describe how to compute the output set of triangles using a single 3-d dominance query. As both triangle and trapezoid are convex geometric objects, an input triangle $T_i$ will contain the trapezoid if all four vertices of the trapezoid are inside the triangle. By the assumption, vertices $F, G$, and $H$ dominate vertex $E$. Since the dominance is a partial order relation, if $E$ dominates the point $(a_i, b_i)$, all other vertices will also dominate it.

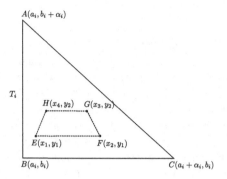

**Fig. 5.** Triangle $T_i$ containing trapezoid $EFGH$ with bases parallel to the $x$-axis.

A point lies below the line through the hypotenuse if the sum of coordinates is no more than the value $a_i + b_i + \alpha_i$. If the vertices $F(x_2, y_1)$ and $G(x_3, y_2)$ are lying below the line through the hypotenuse, then all other vertices will also lie below it as the sum of coordinates of each vertex is at most $\max\{x_2 + y_1, x_3 + y_2\}$. Therefore, the input triangles containing the trapezoid can be found using a 3-d dominance query with point $(x_1, y_1, \min\{-x_3 - y_2, -x_2 - y_1\})$.

We combine the results obtained in this section in the following lemma.

**Lemma 2.** *A set of homothetic isosceles right-angled triangles can be preprocessed such that, given a query object (point, line segment, ellipse, trapezoid with bases parallel to either axis), we can find all triangles containing the query object by employing a 3-d dominance query.*

In fact, the method covers a broader class of query objects. Consider a geometric object $O$ in the plane. Let $x_m$ and $y_m$ be the minimum $x$ and $y$ coordinates of the points of the object $O$. We refer to the point $(x_m, y_m)$ as the *canonical point* for object $O$. As a consequence, each point in $O$ will dominate the canonical point. We will call a point in $O$ with the maximum coordinate sum as a *maxpoint* of the object $O$.

A triangle $T_i \in S$ will enclose the object $O$ if all points in $O$ are also present in $T_i$. By definition, every point in $O$ dominates the canonical point $(x_m, y_m)$. If the canonical point $(x_m, y_m)$ dominates $(a_i, b_i)$, then all other points in $O$ will also dominate point $(a_i, b_i)$ (by transitive closure property of dominance). Let $(x', y')$ be a maxpoint of object $O$. If point $(x', y')$ lies below the line through the hypotenuse of $T_i$ (i.e. $x' + y' \leq a_i + b_i + \alpha_i$), other points in $O$ will also do. So, object $O$ will be enclosed by triangle $T_i$ if the canonical point $(x_m, y_m)$ dominates $(a_i, b_i)$ and a maxpoint $(x', y')$ lies below the hypotenuse. Therefore, triangles $T_i$ satisfying the following three inequalities will enclose the object $O$: $x_m \geq a_i, y_m \geq b_i$, and $x' + y' \leq a_i + b_i + \alpha_i$. Thus, we get the following result.

**Theorem 2.** *Let $O$ be a geometric object in the plane whose canonical point and a maxpoint of $O$ (point with the maximum coordinate sum) are known (or can be computed). We can compute the triangles of $S$ enclosing object $O$ by employing a 3-d dominance query.*

*Remark 1.* Let $Q$ be a trapezoid in the plane such that its bases are not parallel to either axis. Since a trapezoid is a convex object, one can compute the smallest $x$- and $y$-coordinates in $O(1)$-time by comparing the coordinates of the vertices of $Q$. A point in $Q$ with the maximum coordinate sum would be one of the vertices, hence can be computed in $O(1)$-time as well. Thus, we can compute both the canonical point and a maxpoint of $Q$ in constant time. Thus, a query with a general trapezoid can also be answered using a 3-d dominance query.

## 4   Generalizations

*Arbitrary Right-angled Triangles:* We first consider the case when the homothetic triangles in $S$ are right-angled (not necessarily isosceles). Let the input triangles have horizontal and vertical sides parallel to the axes, with their right-angled vertices at the bottom-left position. If not, we can achieve it using a rotation. A scaling operation as in [3] can transform a right-angled triangle into an isosceles right-angled triangle. Under rotate and scaling operations, each query object (line segment, trapezoid, ellipse) transforms into another object of the same type. Thus, we get the following lemma (see also Remark 1).

**Lemma 3.** *The problem of finding all homothetic right-angled triangles containing a query point, line segment, ellipse, or trapezoid can be solved using 3-d dominance query.*

*Arbitrary Homothetic Triangles:* We describe the case where homothetic triangles are acute-angled triangles. The other case can be handled analogously. A shear transformation can transform an acute-angled triangle into a right-angled triangle [3]. The objects considered (line segment, ellipse, and trapezoid) do not change their type on a shear transformation. Thus, we have the following result.

**Lemma 4.** *A set of homothetic triangles in the plane can be preprocessed so that all objects containing a query point, line segment, ellipse, or trapezoid can be obtained using 3-d dominance.*

As 3-d dominance queries can be answered in $O(\log n + k)$ time with $O(n \log n)$ preprocessing time [11]. Thus we obtain the following theorem.

**Theorem 3.** *We can preprocess a given set of $n$ homothetic triangles so that all the triangles enclosing a query object (line segment, ellipse, or trapezoid) can be reported in $O(\log n + k)$ time, where $k$ is the output size. The preprocessing algorithm uses $O(n)$ space and $O(n \log n)$ worst-case time.*

*Remark 2.* The dynamic (resp. counting) variant of the problem can also be solved using the dynamic (resp. counting) counterpart of the 3-dimensional dominance reporting query problem.

*Remark 3.* If the input set $S$ consists of 3-oriented triangles, the triangles in $S$ can be categorized into four types of homothetic triangles (the details are being omitted) and the result will also hold for general 3-oriented triangles.

*Remark 4.* If we are given a set of 3-oriented polygons and the query space consists of points only, the solution can be adapted without aggravating any bounds. This extension is possible because a point is a non-decomposable geometric object.

# 5    Extension to Higher Dimensions

In this section, we extend the result to higher dimensions; we show that the subset of a given set of simplices in $\mathbb{R}^d$ enclosing a query object (point, line segment, Parallelepiped, hyper ellipsoid) can be found using a $(d+1)$-dimensional dominance query. We describe the method for $d = 3$; it can analogously be adopted for higher dimensions.

In $\mathbb{R}^3$, simplices are tetrahedra. A tetrahedron is trirectangular if all three face angles at one vertex are right angles [1]. That vertex is called the right angle of the trirectangular tetrahedron, and the opposite face is called the base. The definition of homothetic tetrahedra is similar to the one for homothetic triangles. Let $\mathcal{T}$ be a trirectangular tetrahedron with equal perpendicular edges. We describe the result for the case where input objects are homothetic to $\mathcal{T}$. The solution can be easily modified for general homothetic tetrahedra, as in the case of homothetic triangles.

Let $\mathcal{S} = \{T_1, T_2, \ldots, T_n\}$ be a set of $n$ homothetic trirectangular tetrahedra in $\mathbb{R}^3$ with three faces parallel to the $x - y, y - z, z - x$ planes. We also assume that the three perpendicular edges of each $T_i$ have the same length, say $\alpha_i$. We identify a tetrahedron $T_i$ with right angle anchored corner $(f_i, g_i, h_i)$ and arm-length $\alpha_i$. Let $Q$ be a geometric object in $\mathbb{R}^3$, and $x_m, y_m$, and $z_m$ be the minimum $x, y$, and $z$-coordinates, respectively, of the points in $Q$. We will refer to $(x_m, y_m, z_m)$ as the canonical point of $Q$.

We can have the following theorem by using arguments analogous to those used in the 2-d case.

**Theorem 4.** *Let $Q \in \mathbb{R}^3$ be a geometric object whose canonical point and a point in $Q$ with the maximum coordinate sum are known (or can be computed). We can compute the tetrahedra in $\mathcal{S}$ enclosing object $Q$ by employing a 4-dominance query.*

Basically, for each tetrahedron $T_i$, we create a point $(f_i, g_i, h_i, -f_i - g_i - h_i - \alpha_i)$ in $\mathbb{R}^4$ and associate the tetrahedron $T_i$ with it. We preprocess these created $n$ points for 4-d dominance queries. We denote the structure by $\mathcal{D}_4$.

Depending on a query object, we answer a query as follows.

*Point*: A point $q \in \mathbb{R}^3$ will lie in (be enclosed by) $T_i$ if it dominates the corner point $(f_i, g_i, h_i)$ and lies below the base plane i.e. $q_x + q_y + q_z \leq f_i + g_i + h_i + \alpha_i$. It is a 4-d dominance query. In other words, the triangles $T_i$ in $\mathcal{S}$ with $f_i \leq q_x, g_i \leq q_y, h_i \leq q_z$, and $-f_i - g_i - h_i - \alpha_i \leq -q_x - q_y - q_z$ will enclose point $q$. We will query $\mathcal{D}_4$ with point $(q_x, q_y, q_z, -q_x - q_y - q_z)$.

*Line Segments*: Let $l$ be a line segment in the plane with endpoints $p$ and $q$. By comparing the coordinates of $p$ and $q$, we can determine the canonical point

$(x_m, y_m, z_m)$ for $l$. A maxpoint of $l$ will always be one of its endpoints, say $q$, as $l$ is a convex object. We make a 4-d dominance query to the structure $\mathcal{D}_4$ with point $(x_m, y_m, z_m, -q_x - q_y - q_z)$ and report the triangles associated with the points returned as query response.

*Parallelepiped*: Let $\mathcal{P}$ be a parallelepiped in $\mathbb{R}^3$ with edges parallel to the $x, y$, and $z$ axes. Note that its anchored point (at position '000') is also its canonical point. The point corresponding to position '111' is the maxpoint of the parallelepiped. Thus, we can find the required tetrahedra using a single 4-d dominance query to the structure $\mathcal{D}_4$.

*Ellipsoid*: We state the method for the sphere. For a sphere with center $(a, b, c)$ and radius $r$, the canonical point would be $(a - r, b - r, c - r)$. The point $(a + \frac{r}{\sqrt{3}}, b + \frac{r}{\sqrt{3}}, c + \frac{r}{\sqrt{3}})$ is the only maxpoint of the sphere. So we can compute the tetrahedra in $\mathcal{S}$ containing the sphere by querying the structure $\mathcal{D}_4$ with point $(a - r, b - r, c - r, -a - \frac{r}{\sqrt{3}}, -b - \frac{r}{\sqrt{3}}, -c - \frac{r}{\sqrt{3}})$.

## 6  Conclusion

In this paper, we considered the problem of finding homothetic triangles containing different query objects including point, line segment, ellipse, and trapezoid. We have shown that the problem can be solved using the 3-d dominance reporting problem, which admits optimal query time solutions. We also showed that the result can be extended to higher dimensions.

Like homothetic triangles, a circle in the plane can be uniquely defined using three parameters. It will be interesting to see whether the dominance problem can help in finding circles containing query objects like points, line segments, etc.

**Acknowledgement.** We would like to thank anonymous referees for their comments, which helped us to improve the clarity of the paper.

## References

1. Trirectangular tetrahedron. https://en.wikipedia.org/w/index.php?title=Trirecta ngular_tetrahedron&oldid=1101419105
2. Afshani, P.: On dominance reporting in 3D. In: Halperin, D., Mehlhorn, K. (eds.) ESA 2008. LNCS, vol. 5193, pp. 41–51. Springer, Heidelberg (2008). https://doi. org/10.1007/978-3-540-87744-8_4
3. Akram, W., Saxena, S.: Point enclosure problem for homothetic polygons. In: Hsieh, S.Y., Hung, L.J., Lee, C.W. (eds.) IWOCA 2023. LNCS, vol. 13889, pp. 13–24. Springer, Cham (2023). https://doi.org/10.1007/978-3-031-34347-6_2
4. Bozanis, P., Kitsios, N., Makris, C., Tsakalidis, A.: New results on intersection query problems. Comput. J. **40**(1), 22–29 (1997)
5. Chazelle, B.: Filtering search: a new approach to query-answering. SIAM J. Comput. **15**(3), 703–724 (1986)

6. Chazelle, B., Edelsbrunner, H.: Linear space data structures for two types of range search. Discrete Comput. Geom. **2**(2), 113–126 (1987)
7. Edelsbrunner, H., Maurer, H., Kirkpatrick, D.: Polygonal intersection searching. Inf. Process. Lett. **14**(2), 74–79 (1982)
8. Edelsbrunner, H., Overmars, M.H.: On the equivalence of some rectangle problems. Inf. Process. Lett. **14**(3), 124–127 (1982)
9. Güting, R.H.: Stabbing c-oriented polygons. Inf. Process. Lett. **16**(1), 35–40 (1983)
10. Makris, C., Tsakalidis, A.: Algorithms for three-dimensional dominance searching in linear space. Inf. Process. Lett. **66**(6), 277–283 (1998)
11. Makris, C., Tsakalidis, K.: An improved algorithm for static 3D dominance reporting in the pointer machine. In: Chao, K.-M., Hsu, T., Lee, D.-T. (eds.) ISAAC 2012. LNCS, vol. 7676, pp. 568–577. Springer, Heidelberg (2012). https://doi.org/10.1007/978-3-642-35261-4_59
12. Saxena, S.: Dominance made simple. Inf. Process. Lett. **109**(9), 419–421 (2009)
13. Shi, Q., JaJa, J.: Fast algorithms for 3-D dominance reporting and counting. Int. J. Found. Comput. Sci. **15**(04), 673–684 (2004)
14. Tan, X., Hirata, T., Inagaki, Y.: The intersection searching problem for C-oriented polygons. Inf. Process. Lett. **37**(4), 201–204 (1991)
15. Thomas, G., Finney, R.: Calculus and Analytic Geometry. World Student Series. Addison-Wesley Publishing Company (1984)
16. Vaishnavi, V.K.: Computing point enclosures. IEEE Trans. Comput. **31**(01), 22–29 (1982)

# Approximate Cycle Double Cover

Babak Ghanbari$^{(\boxtimes)}$ and Robert Šámal

Computer Science Institute, Charles University, Prague, Czech Republic
{babak,samal}@iuuk.mff.cuni.cz

**Abstract.** The Cycle double cover (CDC) conjecture states that for every bridgeless graph $G$, there exists a family $\mathcal{F}$ of cycles such that each edge of the graph is contained in exactly two members of $\mathcal{F}$. Given an embedding of a graph $G$, an edge $e$ is called a *bad edge* if it is visited twice by the boundary of one face. The CDC conjecture is equivalent to bridgeless cubic graphs having an embedding with no bad edge. In this work, we introduce nontrivial upper bounds on the minimum number of bad edges in an embedding of a cubic graph. Moreover, we present efficient algorithms to find embeddings satisfying these bounds.

**Keywords:** Cycle double cover · Embedding · Approximation

## 1 Introduction

The Cycle double cover conjecture (CDC for short) states that for every bridgeless graph $G$, there exists a family $\mathcal{F}$ of cycles such that each edge of the graph is contained in exactly two members of $\mathcal{F}$. It was made independently by Szekeres [Sze73] and Seymour [Sey79] in the 70 s and is now widely considered to be among the most important open problems in graph theory. A reason for this is its close relationship with topological graph theory, integer flow theory, graph coloring, and the structure of snarks. More detail about the conjecture and historical notes can be found in [Zha12].

Here, we mention just some of the partial results. It is known (and easy) that the CDC conjecture is true for Hamiltonian cubic graphs, more generally for 3-edge-colorable cubic graphs. Tarsi [Tar86] and Goddyn [God89] proved that graphs with a Hamilton path admit a 6-CDC, a CDC with six cycles (or, more precisely, cycles that can be combined into six Eulerian graphs). Goddyn [God89] and Häggkvist, McGuinness [HM05] construct a CDC, among else, in graphs that have a spanning subgraph that is a Kotzig graph. The complexity of testing whether a given graph has a cycle double cover is unknown. (But it is presumed that it can be done in a polynomial time, as we only need to check whether the graph has any bridges.)

Supported by the Czech Science Foundation project Flows and cycles in graphs on surfaces (grant number 22-17398S) and by the project GAUK182623 of the Charles University Grant Agency.
B.G. is a full-time Ph.D. student.

A. A. Rescigno and U. Vaccaro (Eds.): IWOCA 2024, LNCS 14764, pp. 421–432, 2024.
https://doi.org/10.1007/978-3-031-63021-7_32

It is known [Zha12, Section 1.2] that it is enough to prove the CDC conjecture for bridgeless cubic graphs. It is also known [Zha12, Section 1.4] that for a bridgeless cubic graph, each cycle double cover consists of facial boundaries of some embedding. However, not every embedding gives a cycle double cover – a face boundary may visit one edge twice. We will call such edges *bad* (also called *singular* [MT01]). Thus, the CDC conjecture is equivalent to every bridgeless cubic graph having an embedding with no bad edges. We approximate this statement by looking for an embedding with not too many bad edges.

We prove nontrivial upper bounds on the minimum number of bad edges in an embedding of a cubic graph in Theorem 4, Theorem 6, and Theorem 8. Our main technique is using a *partial CDC*, a collection of cycles that cover every edge once or twice (by two different cycles) such that no "angle" (pair of adjacent edges) is contained in two distinct cycles. In Lemma 1 we show that a partial CDC can always be extended to an embedding of the graph that gives an embedding with many good edges (all edges covered by the partial CDC will be good edges in the embedding). It would be interesting to know what the limit of this technique is- how many bad edges there must be in an extension of a given partial CDC. A known example showing that we cannot generate a CDC by extending an arbitrary partial CDC is the Petersen graph with a 2-factor consisting of two 5-circuits (see Observation 2).

The algorithmic point of view on the CDC problem is presumably trivial for the decision problem (does the given graph have a CDC?). However, finding a concrete CDC in a given graph may be nontrivial even if the CDC conjecture holds true. We start the exploration of this question by giving most of our results in an algorithmic version: we provide polynomial time algorithms for finding (in different contexts) an embedding of a given cubic graph where the number of bad edges is bounded by a linear function of the number of vertices.

## 2   Preliminaries

We follow the notation of the book Graphs on Surfaces [MT01]. Here, we list the definitions that are most important to us.

A graph in which all vertices have degree 3 is called *cubic*. A *circuit* is a connected 2-regular graph. A *cycle* is a graph in which the degree of each vertex is even. A *bridge* in a graph $G$ is an edge whose removal increases the number of components of $G$. Equivalently, a bridge is an edge that is not contained in any circuit of $G$. A graph is *bridgeless* if it contains no bridge. A connected graph is $k$-*edge-connected* if it remains connected whenever fewer than $k$ edges are removed. A graph is *cyclically* $k$-*edge-connected*, if at least $k$ edges must be removed to disconnect it into two components such that each component contains a cycle.

A graph $G$ is *embedded* in a topological space $X$ if the vertices of $G$ are distinct elements of $X$ and every edge of $G$ is a simple arc connecting in $X$ the two vertices which it joins in $G$, such that its interior is disjoint from other edges and vertices. An *embedding* of a graph $G$ in the topological space $X$ is an isomorphism of $G$ with a graph $G'$ embedded in $X$. If there is an embedding of $G$ into $X$, we say that $G$ can be embedded into $X$.

Let $G$ be a graph that is cellularly embedded in a surface $S$. Let $\pi = \{\pi_v | v \in V(G)\}$ where $\pi_v$ is the cyclic permutation of the edges incident with the vertex $v$ such that $\pi_v(e)$ is the successor of $e$ in the clockwise ordering around $v$. The cyclic permutation $\pi_v$ is called the *local rotation* at $v$, and the set $\pi$ is the *rotation system* of the given embedding of $G$ in $S$.

Let $G$ be a connected multigraph. A *combinatorial embedding* of $G$ is a pair $(\pi, \lambda)$ where $\pi = \{\pi_v | v \in V(G)\}$ is a rotation system, and $\lambda$ is a signature mapping which assigns to each edge $e \in E(G)$ a sign $\lambda(e) \in \{-1, 1\}$. If $e$ is an edge incident with $v \in V(G)$, then the cyclic sequence $e, \pi_v(e), \pi_v^2(e), \ldots$ is called the $\pi$-*clockwise ordering* around $v$ (or the *local rotation* at $v$). Given an embedding $(\pi, \lambda)$ of $G$ we say that $G$ is $(\pi, \lambda)$-*embedded*.

It is known that the combinatorial embedding uniquely determines an embedding to some surface, up to homeomorphism [MT01, Theorems 3.2.4 and 3.3.1].

A *closed walk* is a sequence $(v_0, e_0, v_1, e_1, \ldots, e_{n-1}, v_n)$ where $v_0 = v_n$ and for every $i$, $e_i = \{v_i, v_{i+1}\}$.

**Definition 1.** *Let $(\pi, \lambda)$ be an embedding of a graph $G$. A $(\pi, \lambda)$-facial walk (or* facial walk*) $F$ is a closed walk $(v_0, e_0, v_1, e_1, \ldots, e_{n-1}, v_n)$ where $v_0 = v_n$, $e_0 = e_n$, and for every $i \in \{1, \ldots, n\}$,*

$$
e_i = \begin{cases} \pi_{v_i}(e_{i-1}) & \text{if } \mu_i = 1 \\ \pi_{v_i}^{-1}(e_{i-1}) & \text{if } \mu_i = -1 \end{cases}
$$

*where $\mu_i = \varepsilon \lambda(e_0)\lambda(e_1)\ldots\lambda(e_i)$ and $\varepsilon \in \{-1, +1\}$ determine the side of the edge $e_0$ that is being used by $F$.*

**Observation 1** ([MT01]). *Let $(\pi, \lambda)$ be an embedding of a graph $G$. For every edge $e$ of $G$ and every $\varepsilon \in \{-1, +1\}$ there is a facial walk described in Definition 1. When we consider cyclic shifts and reversals of a facial-walk to be the same walk, then every edge appears once in two different walks or twice in the same facial walk.*

## 3  Partial CDC

Let $G = (V, E)$ be a cubic and bridgeless graph, $(\pi, \lambda)$ be an embedding of $G$, and $\mathcal{F} = \{F_1, \ldots, F_t\}$ be the set of all facial walks in this embedding (we take just one facial walk among cyclic shifts and reversals). As mentioned in Observation 1, each edge of the graph is covered twice by one or once by two members of $\mathcal{F}$. We say that an edge $e$ is *bad* if the former case occurs, that is, if there exists a facial walk $F_i \in \mathcal{F}$ that uses $e$ twice. Otherwise, it is called a *good* edge. We say that $\mathcal{F}$ is a *facial double cover* (or FDC) of $G$.

Edges $e$ and $f$ incident with vertex $v$ are $\pi$-*consecutive* if $\pi_v(e) = f$ or $\pi_v(f) = e$. Every such pair $\{e, f\}$ of edges forms a $\pi$-*angle*. We say that a collection of closed walks $C_1, \ldots, C_n$ forms a *partial circuit double cover* (or *partial CDC*) if **(C1)** each edge is covered at most once by one of $C_i$'s or exactly twice by two different $C_i$ and $C_j$, and **(C2)** for every vertex $v$ and edges $e$ and $f$ where $e \cap f = \{v\}$, there exists at most one closed walk $C_i$ such that $\{e, f\} \subseteq E(C_i)$. In other words, condition (C2) means that no angle is used twice.

The following lemma explains the term partial CDC.

**Lemma 1.** *Let $G$ be a bridgeless cubic graph, and $C_1, C_2, \ldots, C_t$ be a collection of closed walks in $G$. If $C_1, C_2, \ldots, C_t$ form a partial CDC, then there is an embedding $(\pi, \lambda)$ of $G$ where $C_1, \ldots, C_t$ are some of the facial walks of $(\pi, \lambda)$. Moreover, such an embedding can be found by a linear time algorithm.*

*Proof.* Let $C_1, C_2, \ldots, C_t$ be a partial CDC. To extend $C_1, C_2, \ldots, C_t$ to an embedding, we proceed as follows. For every $v \in V(G)$ we create a graph $D_v$ with $V(D_v) = \delta(v)$ i.e., vertices of $D_v$ are the three edges incident to the vertex $v$. For every $i$ and every $e, f \ni v$ if $C_i$ contains $e, v, f$ then we add the edge $\{e, f\}$ to $D_v$. Since $G$ is cubic (C1) implies that $\Delta(D_v) \leq 2$, and (C2) implies that it has no double edges. Now, for every $v \in V(G)$ we extend $D_v$ to a circuit and fix an orientation $\overrightarrow{D_v}$ of it. If $(e, f)$ is an arc of $\overrightarrow{D_v}$, we define $\pi_v(e) = f$. To create an embedding $(\pi, \lambda)$, we will define $\lambda$ as follows. For every $e \in E(G)$ if $e$ is not used by any $C_i$, define $\lambda(e) = 1$. Otherwise, let $C_i$ contain $f, u, e, v, g$ where $f, e, g \in E(G)$ and $u, v \in V(G)$.

$$
\lambda(e) = \begin{cases}
1 & \text{if } \pi_u(f) = e \And \pi_v(e) = g \\
-1 & \text{if } \pi_u(f) = e \And \pi_v^{-1}(e) = g \\
-1 & \text{if } \pi_u^{-1}(f) = e \And \pi_v(e) = g \\
1 & \text{if } \pi_u^{-1}(f) = e \And \pi_v^{-1}(e) = g
\end{cases}
$$

It remains to check that $\lambda$ is well defined. We need to show that if $e \in E(C_i) \cap E(C_j)$, then $C_i$ and $C_j$ define the same $\lambda(e)$. This is easy to check. For example, if in $C_i$ we have $\pi_u(f) = e$ and $\pi_v^{-1}(e) = g$ then $\lambda(e) = -1$. Now, let $\delta(u) = \{e, f, f'\}$, $\delta(v) = \{e, g, g'\}$. Since $C_1, C_2, \ldots, C_t$ is a partial CDC, by conditions (C1) and (C2), the walk $C_j$ must contain $f', u, e, v, g'$ with $\pi_u^{-1}(f') = e$ and $\pi_v(e) = g'$ (or alternatively, $C_j$ may be visited in the other direction as $g', v, e, u, f'$ i.e. $\pi_u(e) = f'$ and $\pi_v^{-1}(g') = e$) which gives $\lambda = -1$ (See Fig. 1). Similarly, the remaining three cases of the defined $\lambda(e)$ by $C_i$ give the same $\lambda(e)$ in $C_j$.

The algorithm proceeds as the proof above: First, we read the description of all of the cycles $C_1, \ldots, C_t$ and construct the graphs $D_v$ along the way. Then, we extend each $D_v$ to 3-circuit and orient it (constant time for each $v$), which defines $\pi$. Finally, we read again descriptions of the cycles and define $\lambda$ by the formula above. □

The following known observation shows that not every partial CDC can be extended to a CDC. We will prove it by using the fact that no angle is used twice in a CDC.

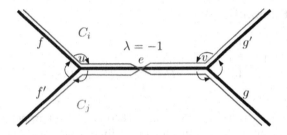

**Fig. 1.** $f, u, e, v, g \in C_i$ and $f', u, e, v, g' \in C_j$

**Observation 2.** *A 2-factor in Petersen graph consisting of two 5-circuits forms a partial CDC that can not be extended to a CDC.*

*Proof.* We have two 5-circuits $C_1$ and $C_2$ and a perfect matching $M$ between them. No angle in a CDC is used twice. Since $C_1$ and $C_2$ are disjoint, they satisfy conditions (C1) and (C2). Therefore, $C_1$ and $C_2$ forms a partial CDC. Now, we show that in any extension of $C_1$ and $C_2$ to an FDC $\mathcal{F}$, there exists at least one bad edge. If there is no bad edge, then every facial walk must be a circuit. Note that $C_1$ and $C_2$ cover 10 edges together. Since every edge is covered twice in a CDC, the remaining facial walks must cover 20 edges. Let $F$ be a facial walk in an extension $\mathcal{F}$ of $C_1$ and $C_2$ to an FDC. Every angle that is not used by $C_1$ and $C_2$, contains an edge from $M$. So, the edges of $F$ alternate between $C_1$, $M$, $C_2$, $M$, $C_1$, ... Therefore, the length of $F$ is divisible by 4. Since there is no 4-circuit and 12-circuit in Petersen, the length of $F$ must be 8. But it is not possible to cover exactly 20 edges with 8-circuits, a contradiction. $\quad\square$

**Lemma 2.** *For every FDC of an embedding of a graph $G$, each angle is part of some facial walk.*

*Proof.* Let $\{e, f\}$ be a $\pi$-angle in an FDC where $e \cap f = \{v\}$. By definition of $\pi$-angle, we have $\pi_v(e) = f$ or $\pi_v(f) = e$. Assume that $\pi_v(e) = f$ and let $e = (u, v)$ and $f = (v, w)$. Then $F = (u, e, v, f, w, \ldots, u)$ with $\epsilon = \lambda(e)$ is a facial walk in this FDC by Observation 1. $\quad\square$

We use the notation of the book Combinatorial Optimization [Sch03] to denote perfect matching polytopes. In an undirected graph $G = (V, E)$ and $U \subseteq V$, we denote $\delta(U)$ as the set of edges leaving $U$. For any set $Y$, we identify the function $x : Y \longrightarrow \mathbb{R}$ with the vectors $x \in \mathbb{R}^Y$. Equivalently, we denote $x(v)$ by $x_v$. For $U \subseteq Y$, we denote $x(U) := \sum_{v \in U} x_v$. For a matching $M$, we denote $\chi_M$ as the incidence vector of $M$ in $\mathbb{R}^E$. Then $\mathcal{P}_{PM}(G)$, the perfect matching polytope of a graph $G$, is defined as follows

$$\mathcal{P}_{PM}(G) = \mathrm{conv}\{\chi_M : M \text{ is a perfect matching in } G\}.$$

**Theorem 3 (Edmonds, [Edm65]).**

$$\mathcal{P}_{PM}(G) = \left\{ x \in \mathbb{R}^E : \begin{array}{l} x \geq 0, \\ \forall v \in V; \ x(\delta(v)) = 1, \\ \forall U \subseteq V; \ |U| \text{ is odd}, \ x(\delta(U)) \geq 1 \end{array} \right\}$$

In the next section, we use Edmonds' theorem to get upper bounds on the minimum number of bad edges in the embedding of bridgeless cubic graphs.

## 4   Bounds on the Minimum Number of Bad Edges Using Perfect Matchings

By Lemma 1, every partial CDC $C_1, C_2, \ldots, C_t$ can be extended to an FDC. Since in a partial CDC, each edge is either covered once by one facial walk or twice by two facial walks, the edges that are already covered can not be bad. Therefore, potential bad edges of the FDC are those that are not covered by $C_1, C_2, \ldots, C_t$.

**Theorem 4.** *Let $G$ be a bridgeless cubic graph. There exists an embedding of $G$ with at most $\frac{n}{2}$ bad edges. Moreover, such an embedding can be found in time $O(n \log^2 n)$.*

*Proof.* Let us start with any perfect matching $M$. (This can be found in time $O(n \log^2 n)$, see [DS10].) Then, $G - M$ is a cycle, that is, a collection of circuits. By Lemma 1, this collection can be extended to a double cover for $G$ in linear time. Therefore, the resulting embedding has at most $|M| = \frac{n}{2}$ bad edges.   □

The following theorem of Kaiser, Král, and Norine is proved by Theorem 3. This result is formulated only as existence of matchings $M_1$ and $M_2$. However, it is proved as formulated below. We will use it to improve upon Theorem 4.

**Theorem 5 ([KKN06]).** *Let $G$ be a bridgeless cubic graph, and let $M_1$ be a perfect matching in $G$ that contains no 3-edge-cuts in $G$. There exists perfect matching $M_2$ such that $|M_1 \cup M_2| \geq \frac{3}{5}|E(G)|$.*

We reformulate this theorem as below.

**Corollary 1.** *Let $G$ be a bridgeless cubic graph, and let $M_1$ be a perfect matching in $G$ that contains no 3-edge-cuts in $G$. There exists a perfect matching $M_2$ in $G$ with $|M_1 \cap M_2| \leq \frac{n}{10}$. Moreover, such $M_2$ can be found by a polynomial time algorithm.*

*Proof.* By Theorem 5, there exists a perfect matching $M_2$ such that

$$|M_1 \cup M_2| \geq \frac{3}{5}|E(G)| = \frac{3}{5} \cdot \frac{3n}{2} = \frac{9n}{10}.$$

On the other hand,

$$|M_1 \cup M_2| = |M_1| + |M_2| - |M_1 \cap M_2| = n - |M_1 \cap M_2|.$$

Therefore,

$$|M_1 \cap M_2| \le n - \frac{9n}{10} = \frac{n}{10}.$$

This proves the existence of $M_2$. To find it efficiently, we use the Edmonds' blossom algorithm to solve minimum weight perfect matching (the weights being $\chi_{M_1}$). □

**Theorem 6.** *Let $G$ be a bridgeless cubic graph. There exists an embedding of $G$ with at most $\frac{n}{10}$ bad edges.*

Unfortunately, we don't know how to find such an embedding by a polynomial time algorithm. (See Theorem 8 though.) The missing part is the following problem.

*Problem 1.* Given a bridgeless cubic graph. How fast can we find a perfect matching that contains no odd cut of size 3?

*Proof (of Theorem 6).* Let $M_1$ be a perfect matching in $G$ that contains no odd cut of size 3. (This is known to exist.) Let $M_1$ and $M_2$ be two perfect matchings that are provided by Theorem 5. Let $C_1 = G - M_1$ and $C_2 = (M_1 \cup M_2) - (M_1 \cap M_2)$. Next, we will show that $C_1$, $C_2$ is a partial CDC. Each edge of $M_2 \backslash M_1$ is used by both $C_1$ and $C_2$. Each edge of $M_1 \backslash M_2$ is used by $C_2$, and each edge of $E(G) \backslash (M_1 \cup M_2)$ is used by $C_1$. The remaining edges are not used by any of $C_1$ or $C_2$. Therefore, each edge is covered at most twice by $C_1$ and $C_2$. To prove that no angle is used twice by $C_1$ and $C_2$, let $e, f, g$ be three edges incident with a vertex $v$. Since $v$ is covered by an edge in both $M_1$ and $M_2$, we have two following cases

1. One of $e$, $f$, and $g$, let say $e$, is covered by both of $M_1$ and $M_2$. In this case, angle $\{f, g\}$ is used by $C_1$, and other angles are not used. (see Fig. 2)
2. Two of $e$, $f$, and $g$ are covered, one by $M_1$ and the other by $M_2$. Let $e \in M_1$ and $f \in M_2$. In this case, angle $\{e, g\}$ is not used by any of $C_1$ and $C_2$. Angle $\{e, f\}$ is used by $C_2$, and angle $\{f, g\}$ is used by $C_1$. Since $g \notin E(C_2)$, angle $\{f, g\}$ is not used by $C_2$, and since $e \notin E(C_1)$, angle $\{e, f\}$ is not used by $C_1$. (see Fig. 2)

Therefore, no angle is used twice, and the collection of circuits forming $C_1$ and $C_2$ is a partial CDC. By Lemma 1 and Corollary 1, the resulting embedding from extending $C_1$ and $C_2$ has at most $|E(G) \backslash (C_1 \cup C_2)| = |M_1 \cap M_2| \le \frac{n}{10}$ bad edges. □

Given a graph $G = (V, E)$, a *postman set* is a set $J \subseteq E$ such that the multi-set $J \cup E$ is Eulerian. The following known proposition is immediate.

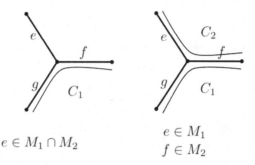

$$e \in M_1 \cap M_2$$

$$e \in M_1$$
$$f \in M_2$$

**Fig. 2.** Case 1 (left picture) and case 2 (right picture) in the proof of Theorem 6.

**Proposition 1.** *$J$ is a postman set of $G = (V, E)$ if and only if the following condition holds for every vertex $v \in V$:*

*$v$ is incident to an odd number of edges in $J$ if and only if $v$ has odd degree in $G$.*

Therefore, in a cubic graph $G = (V, E)$, a subset $J \subseteq E$ is a postman set if for every $v \in V(G)$, $v$ is incident with 1 or 3 edges in $J$. The following lemma is known. We include its proof for the reader's convenience.

**Lemma 3.** *For every spanning tree $T$ in $G$, there exists a postman set $J$ such that $J \subseteq E(T)$.*

*Proof.* Let $v_1$, $v_2$, ..., $v_{2k}$ be all of the odd vertices of $G$. Since $T$ is a spanning tree, there is a path between every two vertices of $T$. Consider the $k$ paths $v_i - v_{i+k}$, for $1 \le i \le k$. Let $J$ be the set of all edges that are used an odd number of times by these paths. Then, $J$ is a postman set, as adding one path changes the parity of exactly its end vertices.     □

**Lemma 4.** *Let $M$ be a perfect matching and $J$ be a postman set in a bridgeless cubic graph $G$. Then, $C_1 = G - J$ and $C_2 = M \triangle J$ form a partial CDC.*

*Proof.* It is easy to check that $G - J$ and $M \triangle J$ (the symmetric difference) are cycles. Next, we prove that these cycles form a partial CDC. Each edge of $M \backslash J$ is used by both $C_1$ and $C_2$. Each edge of $J \backslash M$ is used by $C_2$, and each edge of $E(G) \backslash (M \cup J)$ is used by $C_1$. The remaining edges ($J \cap M$) are not used by any of $C_1$ or $C_2$. Therefore, each edge is covered at most twice by $C_1$ and $C_2$. To prove that $C_1$ and $C_2$ form a partial CDC, it remains to show that no angle is used twice by $C_1$ and $C_2$. Let $e$, $f$, and $g$ be three edges incident with a vertex $v$. Every vertex $v \in V(G)$ is covered by one edge from $M$ and either one or three edges from $J$. Therefore, we have the following two cases.

1. There is only one edge from $J$ that is incident with $v$. The proof of this case is the same as the proof of case 1 and 2 in the proof of Theorem 6. Therefore, no angle is used twice.

2. Vertex $v$ is incident with three edges from $J$. Let $e$ be the edge that belongs to the perfect matching $M$. In this case, $e \in M \cap J$ and none of $C_1$ nor $C_2$ uses $e$. Therefore, angles $\{e, f\}$ and $\{e, g\}$ are used neither by $C_1$ nor by $C_2$. Both $f$ and $g$ are not in $M$ and are used by $C_2$. Therefore, the angle $\{f, g\}$ is used once by $C_2$. This completes the proof as no angle is used twice by $C_1$ and $C_2$ (Fig. 3). $\qquad\square$

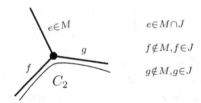

**Fig. 3.** Case 2 in the proof of Lemma 4.

**Theorem 7.** *Let $G$ be a cyclically $2k$-edge-connected cubic graph. There exists an embedding of $G$ with at most $\frac{n}{2k}$ bad edges.*

*Proof.* Construct a graph $H$ from $G$ as follows: select any perfect matching $M$ of $G$ and remove the edges of $M$ from $G$. Now, contract each component of $G - M$ to a vertex. Bring back the edges of $M$ and name the new graph $H$. Note that $E(H)$ is in natural $1-1$ correspondence with $M$. Clearly, $H$ is $2k$-edge-connected as the graph $G$ is cyclically $2k$-edge-connected. By Nash-Williams/Tutte's Theorem [NW61, Tut61], there exist $k$ edge-disjoint spanning trees in $G$. Therefore, a spanning tree in $H$ has at most

$$\frac{|E(H)|}{k} = \frac{|M|}{k} = \frac{n}{2k}$$

edges. Let $T_H$ be a spanning tree in $H$ and let $T_G$ be any extension of $T_H$ to a spanning tree of $G$. We use Lemma 3 to get a postman set $J \subseteq T_G$. By Lemma 4, $C_1 = G - J$ and $C_2 = M \triangle J$ forms a partial CDC. Then, the number of bad edges is at most $|M \cap J|$ as potential bad edges are those that are not covered by none of $C_1$ and $C_2$. We know that $J \subseteq T_G$ and $T_G$ is an extension of $T_H$ in $G$. On the other hand, $T_H$ correspond to a subset $M'$ of $M$ as $E(T_H) \subseteq E(H)$ and $E(H)$ is in a $1-1$ correspondence with $M$. Therefore, $M \cap J$ is at most $M'$ and

$$|M \cap J| \le |M'| = |T_H| \le \frac{n}{2k}.$$

$\qquad\square$

We improve our bound for all cyclically $k$-edge-connected cubic graphs in the following theorem. Note that for $k \le 5$, the bound from Theorem 6 is better. The next theorem has an algorithmic version, though.

**Theorem 8.** *Let G be a cyclically k-edge-connected cubic graph ($k \geq 3$). There exists an embedding of G with at most $\frac{n}{2k}$ bad edges. Moreover, such an embedding can be found in polynomial time.*

*Proof.* By the same technique as in the proof of Theorem 6, it is enough to find two perfect matchings $M$ and $M'$ with $|M \cap M'| \leq \frac{n}{2k}$. Let $M$ be any perfect matching in $G$. (We can find this in time $O(n \log^2 n)$ as above, see [DS10].) Define a function $f : E(G) \rightarrow \mathbb{R}$ as follows

$$f(e) = \begin{cases} \frac{1}{k} & \text{if} \quad e \in M \\ \frac{k-1}{2k} & \text{if} \quad e \notin M. \end{cases}$$

We check that $f$ is in $\mathcal{P}_{PM}(G)$. Since the graph is cubic, around each vertex $v$, one edge is in $M$ and has value $\frac{1}{k}$ and the other two edges connected to $v$ have value $\frac{k-1}{2k}$, and the sum of these three values is 1. It remains to show that for every set $U \subseteq V(G)$ with $|U|$ odd, $x(\delta(U)) = \sum_{e \in \delta(u)} f(e) \geq 1$. Then, by Theorem 3, $f \in \mathcal{P}_{PM}(G)$. So, let $U$ be an odd-size vertex set in $G$. Then, we have one of the following cases:

1. $\delta(U) \subseteq M$. By removing the matching edges, every vertex in $U$ is incident with exactly two edges in $G[U]$. Therefore, it consists of a collection of cycles. As the graph is cyclically k-edge-connected, $|\delta(U)| \geq k$, So

$$\sum_{e \in \delta(U)} f(e) = \sum_{e \in \delta(U)} \frac{1}{k} \geq k \cdot \frac{1}{k} = 1$$

2. $|\delta(U) \setminus M| = 1$. We have the following two sub-cases. There exists a vertex $u$ with either degree one or two in $G[U]$ (see the picture below). We show that each sub-case is impossible to happen. Note that in the following pictures, bold edges are matching edges.

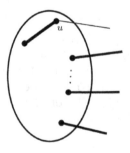

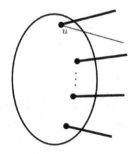

Sub-case 1                    Sub-case 2

In both sub-cases, there exists a vertex $u$ that has exactly one non-matching edge in $G[U]$. Let us remove the edges of $M$ from $G[U]$. Then, every vertex in $U \setminus \{u\}$ is incident with two edges in $G[U]$. So, $G[U] \setminus M$ has one vertex of degree 1 and the rest of degree 2, which is impossible.

3. $|\delta(U)\backslash M| \geq 2$. In this case, as $|U|$ is odd and $|\delta(U)\backslash M| \geq 2$, we have $|\delta(U)| \geq 3$. If $\delta(U)$ has three non-matching edges, then

$$\sum_{e \in \delta(U)} f(e) \geq 3 \cdot \frac{k-1}{2k} \geq 1.$$

Note that $k \geq 3$. If there is at least one matching edge, then

$$\sum_{e \in \delta(U)} f(e) \geq 2 \cdot \frac{k-1}{2k} + \frac{1}{k} = 1.$$

As $f$ is in the perfect matching polytope, $f$ is a convex combination of $\chi_{M_i}$, for some perfect matching $M_i$. Set $S = E(G)\backslash M$. By definition of $f$, we have $f(S) = \frac{k-1}{2k}|S|$, hence

$$\chi_{M_i}(S) \geq \frac{k-1}{2k}|S| = \frac{2}{3}|E(G)|\frac{k-1}{2k}$$

for some perfect matching $M_i$ involved in the convex combination for $f$. Now,

$$|M \cup M_i| = |M| + |M_i \backslash M| \geq |E(G)|\left(\frac{1}{3} + \frac{2}{3} \cdot \frac{k-1}{2k}\right) = \frac{2k-1}{3k}|E(G)|$$

Therefore, $|M \cap M_i| \leq \frac{n}{2k}$, and we can put $M' = M_i$. This proves the existence of $M'$. To find it efficiently, we use Edmonds' algorithm to find minimum weight perfect matching, as in the proof of Corollary 1.                                 $\square$

## 5  Conclusion

We studied a new approach to the cycle double cover conjecture. We first introduced the notion of partial CDC, and then we showed that by having a partial CDC, we can extend it to an embedding in which some of the edges might have the same face on both sides. We called these edges bad and then provided non-trivial upper bounds on the minimum number of bad edges in an embedding of a cubic graph. We showed that some of these embeddings can be found in polynomial time in Theorem 4 and Theorem 8. For the embeddings in Theorem 6 and Theorem 7, we don't know how to find such an embedding by a polynomial time algorithm.

In the future, it would be interesting to find an efficient algorithm to find an embedding with at most $\frac{n}{10}$ bad edges in general and at most $\frac{n}{2k}$ bad edges for the class of cyclically $2k$-edge-connected cubic graphs. We believe that polynomial-time algorithms exist. In another direction, it would be interesting to find better upper bounds on the number of bad edges. Our hope is that this will guide us to a discovery of new techniques that will eventually solve CDC conjecture, that is, to show that there is an embedding with no bad edge.

**Acknowledgment.** Special thanks should be given to Matt DeVos for helpful discussions that led to the extension of our results in Sect. 4.

We are grateful to the anonymous referees for their useful comments.

# References

[DS10] Diks, K., Stanczyk, P.: Perfect matching for biconnected cubic graphs in $O(n \log^2 n)$ time. In: van Leeuwen, J., Muscholl, A., Peleg, D., Pokorný, J., Rumpe, B. (eds.) SOFSEM 2010. LNCS, vol. 5901, pp. 321–333. Springer, Heidelberg (2010). https://doi.org/10.1007/978-3-642-11266-9_27

[Edm65] Edmonds, J.: Maximum matching and a polyhedron with 0,1-vertices. J. Res. Natl. Bureau Stand. Sect. B Math. Math. Phys. 125 (1965)

[God89] Goddyn, L.A.: Cycle double covers of graphs with Hamilton paths. J. Combin. Theory Ser. B **46**, 253–254 (1989)

[HM05] Häggkvist, R., McGuinness, S.: Double covers of cubic graphs with oddness 4. J. Combin. Theory Ser. B **93**(2), 251–277 (2005)

[KKN06] Kaiser, T., Král', D., Norine, S.: Unions of perfect matchings in cubic graphs. In: Klazar, M., Kratochvíl, J., Loebl, M., Matoušek, J., Valtr, P., Thomas, R. (eds.) Topics in Discrete Mathematics. Algorithms and Combinatorics, vol. 26, pp. 225–230. Springer, Heidelberg (2006). https://doi.org/10.1007/3-540-33700-8_14

[MT01] Mohar, B., Thomassen, C.: Graphs on Surfaces. Johns Hopkins Series in the Mathematical Sciences. Johns Hopkins University Press (2001)

[NW61] Nash-Williams, C.S.J.A.: Edge-disjoint spanning trees of finite graphs. J. Lond. Math. Soc. **1**(1), 445–450 (1961)

[Sch03] Schrijver, A.: Combinatorial Optimization. Polyhedra and Efficiency. Algorithms and Combinatorics. Springer, Berlin (2003)

[Sey79] Seymour, P.D.: Sums of circuits. Graph Theory Relat. Top. **1**, 341–355 (1979)

[Sze73] Szekeres, G.: Polyhedral decompositions of cubic graphs. Bull. Aust. Math. Soc. **8**(3), 367–387 (1973)

[Tar86] Tarsi, M.: Semiduality and the cycle double cover conjecture. J. Combin. Theory Ser. B **41**(3), 332–340 (1986)

[Tut61] Tutte, W.T.: On the problem of decomposing a graph into n connected factors. J. Lond. Math. Soc. **1**(1), 221–230 (1961)

[Zha12] Zhang, C.-Q.: Circuit Double Cover. London Mathematical Society Lecture Note Series. Cambridge University Press, Cambridge (2012)

# The Bottom-Left Algorithm for the Strip Packing Problem

Stefan Hougardy[1]([⊠])[iD] and Bart Zondervan[2]

[1] Research Institute for Discrete Mathematics and Hausdorff Center
for Mathematics, University of Bonn, Bonn, Germany
hougardy@dm.uni-bonn.de
[2] Faculty of Mathematics and Computer Science, University of Bremen,
Bremen, Germany
bart.zondervan@uni-bremen.de

**Abstract.** The bottom-left algorithm is a simple heuristic for the Strip
Packing Problem. Despite its simplicity, the exact approximation ratio of
the bottom-left algorithm remains unknown. We will improve the more-
than-40-year-old value for the lower bound from $5/4$ to $4/3 - \varepsilon$. Additionally, we will show that this lower bound holds even in the special case of
squares, where the previously known lower bound was $12/11 - \varepsilon$. When
squares are arranged in the worst possible order, we establish a constant
upper bound and a $10/3 - \varepsilon$ lower bound for the approximation ratio of
the BL algorithm. This bound also applies to some online setting and
yields an almost tight result there. Finally, we show that the approximation ratio of a local search algorithm based on permuting rectangles
in the ordering of the bottom-left algorithm is at least 2 and that such
an algorithm may need an exponential number of improvement steps to
reach a local optimum.

## 1 Introduction

In the Strip Packing Problem, a rectangular strip of fixed width and infinite
height is given. The task is to find an orthogonal packing of a given set of
rectangles into the strip such that no two rectangles overlap and the total height
of the packing is minimal. Rotation of rectangles by 90 degrees is not allowed.

A reduction from the Bin Packing Problem shows that Strip Packing is NP-
hard [9]. It is even strongly NP-hard [6]. Moreover, this reduction establishes
that unless P = NP, there cannot exist a $(3/2 - \varepsilon)$-approximation algorithm for
Strip Packing. Currently, the best-known approximation algorithm achieves an
approximation ratio of $5/3 + \varepsilon$ [5,7]. However, this algorithm is rather complicated and may not be of practical relevance.

In contrast, the bottom-left algorithm is extremely simple. It operates by
packing the rectangles in the given order, positioning them at the lowest available point within the strip. In situations where there are multiple lowest positions

Funded by the Deutsche Forschungsgemeinschaft (DFG, German Research Foundation)
under Germany's Excellence Strategy – EXC-2047/1 – 390685813.

possible, the BL algorithm selects the leftmost of these positions. An implementation of the BL algorithm with quadratic time complexity is presented in [3].

The approximation ratio of the BL algorithm heavily depends on the ordering of the rectangles. It is easy to construct instances and orderings of the rectangles such that the approximation ratio of the BL algorithm is arbitrarily bad. Baker, Coffman, and Rivest [1] have shown that this may even happen if the rectangles are ordered by increasing width. Contrary to this they proved [1] that when the rectangles are ordered by decreasing width then the BL algorithm has approximation ratio 3. In case that all rectangles are squares they proved an approximation ratio of 2 for the BL algorithm.

A natural question arising in this context is: *What is the approximation ratio of the bottom-left algorithm if we have a best possible ordering of the input rectangles?* It is tempting to expect that among the $n!$ possible orderings of the given $n$ rectangles there is always one such that the BL algorithm will find an optimum solution. However, this is not the case as was shown in 1980 by Brown [2]: There exist instances of the strip packing problem for which the BL algorithm cannot achieve an approximation ratio better than $5/4$ not even for the best ordering of the rectangles. In case of squares it was shown by Baker, Coffman, and Rivest [1] that for any fixed $\varepsilon > 0$ the BL algorithm cannot achieve an approximation ratio of $12/11 - \varepsilon$. Thus there remain large gaps between $5/4$ and 3 for the approximation ratio of the BL algorithm for the Strip Packing Problem in case of rectangles and between $12/11 - \varepsilon$ and 2 in case of squares. We will narrow these gaps by improving the lower bound in both cases to $4/3 - \varepsilon$. This is the first improvement on these bounds after more than 40 years.

**Theorem 1.** *For all $\varepsilon > 0$ the approximation ratio of the bottom-left algorithm for the Square Strip Packing Problem cannot be better than $4/3 - \varepsilon$ even if the squares are ordered in the best possible way.*

Additionally, instead of looking at the best ordering, we will also look at the worst ordering. As mentioned above the approximation ratio of the BL algorithm might be unbounded when the rectangles are badly ordered. On the contrary, we can show that for squares the approximation ratio of the BL algorithm is always bounded, regardless of the ordering. We also construct a $10/3 - \varepsilon$ lower bound for this case improving the so far best lower bound of $2 - \varepsilon$ [1].

**Theorem 2.** *The bottom-left algorithm has constant approximation ratio for the Square Strip Packing Problem, for all possible orderings of the squares. This approximation ratio cannot be better than $10/3 - \varepsilon$.*

Our lower bound also applies to an online version of the BL algorithm that was studied in [4] and was shown to have approximation ratio 3.5. We therefore get an almost tight result for this case:

**Corollary 1.** *The online BottomLeft algorithm from [4] has approximation ratio between $10/3$ and 3.5.*

Last of all, we study a local search variant of the BL algorithm. The bottom-left $k$-local search algorithm ($k$-BL) starts with an initial BL packing and in each iteration the algorithm tries to permute $k$ rectangles such that the BL algorithm on the new ordering returns a packing with strictly lower height. Firstly, we show a lower bound equal to 2 for the $k$-BL algorithm, implying that this algorithm cannot find an ordering such that the BL algorithm has approximation ratio better than the currently best-known $(5/3 + \varepsilon)$-approximation ratio from [7]. Secondly, we also show that the local search algorithm may need an exponential number of iterations before reaching a local optimum.

**Theorem 3.** *The approximation ratio of the $k$-local search bottom-left algorithm is bounded from below by 2, even in case of squares. Moreover, this algorithm may need an exponential number of iterations to find a local optimum.*

*Outline of the Paper.* After starting with some basic definitions in Sect. 2 we present in Sect. 3 our new lower bounds for the BL algorithm assuming a best possible ordering of the rectangles. We first consider the general Strip Packing case, and afterwards show that the lower bound $4/3 - \varepsilon$ also holds in the square case. Next, Sect. 4 shows that in case of squares the approximation ratio of the BL algorithm is bounded by a constant, even for the worst possible ordering. We also prove a lower bound of $10/3 - \varepsilon$ for this case. Last of all, Sect. 5 studies the $k$-BL algorithm. We prove that this novel local search algorithm has approximation ratio no better than 2. Moreover, we show that this algorithm might take an exponential number of improvement steps to reach a local optimum.

# 2  Preliminaries

A Strip Packing Problem (SPP) instance $\mathcal{I}$ consists of a vertical strip of fixed width $W$ and infinite height together with a set $R = \{r_1, \ldots, r_n\}$ of $n$ closed rectangles. Each rectangle $r_i$ has a given height $h_i := h(r_i)$ and width $w_i := w(r_i)$. Assume that $\max\{w_i : r_i \in R\} \leq W$. A packing of $R$ into the strip is defined by specifying the lower left coordinate $(x_i, y_i)$ for each $r_i \in R$. A packing of $R$ is *feasible* if all rectangles lie within the strip and no two rectangles overlap within their interior, i.e., the following two conditions have to be satisfied:

$$x_i \geq 0, \ x_i + w_i \leq W, \ y_i \geq 0 \quad \text{for all } 1 \leq i \leq n,$$
$$(x_i, x_i + w_i) \times (y_i, y_i + h_i) \cap (x_j, x_j + w_j) \times (y_j, y_j + h_j) = \emptyset \quad \text{for all } 1 \leq i < j \leq n.$$

The height of a feasible packing is the maximum of $\{y_i + h_i : 1 \leq i \leq n\}$. The goal of SPP is to compute a feasible packing of minimum height for a given instance $\mathcal{I}$. We denote this value by $h_{\text{OPT}}(\mathcal{I})$. Note that our definition of SPP does not allow to rotate the rectangles. The Square Strip Packing Problem (SSPP) is the special case of SPP where all rectangles are squares.

Given a SPP instance $\mathcal{I}$ with rectangles $R = \{r_1, \ldots, r_n\}$ the bottom-left algorithm places the rectangles in the given order at a lowest free position in the strip, using the leftmost position in case of ties. More formally, the BL algorithm will place rectangle $r_1$ at position $(0,0)$. This is a feasible packing of the first rectangle. Assume that the BL algorithm has obtained a feasible packing of the first $i-1$ rectangles into the strip. Then it chooses a position $(x_i, y_i)$ that results in a feasible packing for the first $i$ rectangles such that $(y_i, x_i)$ is lexicographically minimal among all possible choices for the position $(x_i, y_i)$. The height of the packing computed by the BL algorithm on an instance $\mathcal{I}$ is denoted by $h_{\mathrm{BL}}(\mathcal{I})$. This height may heavily depend on the ordering of the rectangles in the instance. We are therefore also interested in the best possible height that the BL algorithm can achieve for a given set of rectangles, i.e., the minimum height among all $n!$ orderings of the $n$ rectangles; denote this value by $h_{\mathrm{BL}}^{\mathrm{best}}$. Similarly, define the maximum height among all $n!$ orderings as $h_{\mathrm{BL}}^{\mathrm{worst}}$.

The approximation ratio achieved by the BL algorithm on an instance $\mathcal{I}$ is defined as the ratio $h_{\mathrm{BL}}(\mathcal{I})/h_{\mathrm{OPT}}(\mathcal{I})$. We are also interested in the best possible and worst possible approximation ratio of the BL algorithm which are defined as $h_{\mathrm{BL}}^{\mathrm{best}}(\mathcal{I})/h_{\mathrm{OPT}}(\mathcal{I})$ and as $h_{\mathrm{BL}}^{\mathrm{worst}}(\mathcal{I})/h_{\mathrm{OPT}}(\mathcal{I})$.

# 3    An Improved Lower Bound for the Best BL Packing

Even for the best possible ordering of rectangles, the BL algorithm might produce a non-optimal packing. Baker, Coffman, and Rivest [1] were the first to show that for all $\varepsilon > 0$ the BL algorithm cannot have an approximation ratio better than $12/11 - \varepsilon$. Later an improvement was given by Brown [2], showing that there exists a set of eight rectangles for which the BL algorithm cannot have an approximation ratio below $5/4$. Up to now, this was the best-known lower bound for the ratio between the height of a best BL packing and the height of an optimal packing. Furthermore, the value $12/11 - \varepsilon$ from [1] was the best-known lower bound for SSPP instances. In the following two subsections we improve both results.

## 3.1    Rectangular Case

The main idea in this section to show a better lower bound for the best BL algorithm is to construct an instance that has an optimum packing with bottom-left structure such that the optimum packing is unique up to symmetries. After that, the instance is slightly modified, preserving the uniqueness (up to symmetry) of the optimum packing, but loosing the BL structure for the optimum packing.

**Theorem 4.** *For all $\varepsilon > 0$ the approximation ratio of the bottom-left algorithm for the Strip Packing Problem cannot be better than $4/3 - \varepsilon$ even if the rectangles are ordered in the best possible way.*

*Proof.* Consider an instance with rectangles $(3,2)$, $(3,2)$, $(2,1)$, $(2,1)$, $(2,1)$, $(2,1)$, $(1,1)$ and strip width 7. The packings of this instance in Fig. 1 with height 3 are tight, i.e., all space is occupied. Hence these packings are optimal.

**Claim:** The packings from Fig. 1 are the only optimal packings. To prove this claim, consider the three disjoint $1 \times 7$ horizontal rows for an arbitrary optimal packing. Let the type of a row be a multiset of the sizes of rectangles that the row intersects. There are three possible types: (a) $\{3, 3, 1\}$, (b) $\{3, 2, 2\}$ and (c) $\{2, 2, 2, 1\}$.

Let $a$, $b$, and $c$ denote the number of rows of type $(a)$, resp. type $(b)$, resp. type $(c)$. There is a total of three rows, hence $a + b + c = 3$. Furthermore, there is exactly one rectangle of width 1 and this rectangle has height 1, hence $a + c = 1$. There are four rectangles of width 2, these rectangles all have height 1, so $2b + 3c = 4$. This implies that $a = 1$, $b = 2$, and $c = 0$.

Both rectangles of width 3 have height 2, hence the row of type $(a)$ must be in the middle, as otherwise there must be another row of type $(a)$, contradicting that $a = 1$. Furthermore, if the square of width 1 is placed at one of the sides of the strip, then either in the top or bottom row another rectangle of width 1 is needed. As there is no other rectangle of width 1, the square of width 1 must be placed in the middle. Thus the packings in Fig. 1 are the only possible packings with one row of type $(a)$ and two rows of type $(b)$. This proves the claim.

Next, the instance is modified slightly, making the two rectangles of width 3 thinner and the square of size 1 higher. More formally, for $\varepsilon > 0$ sufficiently small, let $(3 - \varepsilon, 2)$, $(3 - \varepsilon, 2)$, $(2, 1)$, $(2, 1)$, $(2, 1)$, $(2, 1)$, $(1, 1 + \varepsilon)$ be the rectangles of the modified instance $\mathcal{I}_\varepsilon$. The strip width remains 7. The preceding proof shows that (to within $\varepsilon$) the packings of Fig. 1 are still optimum and have height $3 + \varepsilon$. However, for each of the optimal solution, the packing cannot be a BL packing.

Consider the first packing of Fig. 1, the two top rectangles of size $(2, 1)$ have to be packed last in the BL algorithm. However, as the $(3, 2)$ rectangle shrinks to size $(3 - \varepsilon, 2)$, there is no space to fit two $(2, 1)$ rectangles, unless if the top right $(3 - \varepsilon, 2)$ is shifted a bit to the right, breaking the BL structure. The packing obtained by the BL algorithm is depicted in Fig. 2(a). Next, consider the second packing of Fig. 1. If the rectangle of size $(1, 1 + \varepsilon)$ is packed before the two top right rectangles of size $(2, 1)$, then the BL algorithm returns the packing shown in Fig. 2(b). Else if one of the rectangles of size $(2, 1)$ is placed before the rectangle of size $(1, 1 + \varepsilon)$, then the best packing that can be obtained by the BL algorithm is shown in Fig. 2(c).

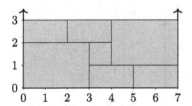 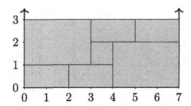

**Fig. 1.** The two optimum packings $(3, 2)$, $(3, 2)$, $(2, 1)$, $(2, 1)$, $(2, 1)$, $(2, 1)$, $(1, 1)$.

In conclusion, the best BL packing has height 4 while an optimum packing has height $3 + \varepsilon$. For $\varepsilon' := 4\varepsilon/(9 + 3\varepsilon)$ we get a lower bound of $\frac{4}{3} - \varepsilon'$ on the approximation ratio of the BL algorithm.                                  □

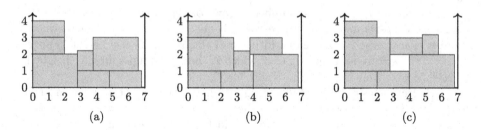

(a)                          (b)                          (c)

**Fig. 2.** The BL packing of $\mathcal{I}_{0.2}$ is not optimum.

## 3.2 Square Case

There is an easy modification of the instance from Theorem 4 such that all rectangles are squares. In Corollary 2 it is shown that this results in a $6/5 - \varepsilon$ lower bound for the BL algorithm for SSPP-instances. After that, Corollary 2 is generalized to get a $4/3 - \varepsilon$ lower bound for the BL algorithm for SSPP-instances.

**Corollary 2.** *For all $\varepsilon > 0$ the approximation ratio of the bottom-left algorithm for the Square Strip Packing Problem cannot be better than $6/5 - \varepsilon$ even if the squares are ordered in the best possible way.*

*Proof.* Consider the collection of squares of sizes $3, 3, 2, 2, 2, 2, 1$. Let the strip width be 7. These squares have the same widths as the rectangles from Theorem 4. Using similar reasoning, the two packings of Fig. 3(a) and Fig. 3(b) are the only optimal packings.

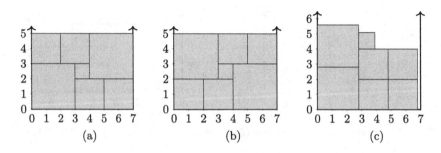

(a)                          (b)                          (c)

**Fig. 3.** (a),(b) are optimal packings of $\mathcal{I}_0$. (c) is the best BL packing of $\mathcal{I}_{0.1}$.

Let $\varepsilon > 0$, and consider the modified instance $\mathcal{I}_\varepsilon$ where the squares of size 3 get size $3 - 2\varepsilon$, and the unit square gets size $1 + \varepsilon$. Akin to Theorem 4, the packings from Fig. 3(a) and Fig. 3(b) are still optimum, but have no BL structure. Using similar arguments as in the proof of Theorem 4 one can show that a best BL-packing has height at least $6 - 4\varepsilon$, and as Fig. 3(c) shows there is a BL packing with height $6 - 4\varepsilon$.                                                                    □

Next Theorem 1 will improve the $6/5 - \varepsilon$ lower bound to $4/3 - \varepsilon$ by making the instance larger in width, height and number of squares. Besides the construction, the proof is similar to that of Theorem 4.

**Theorem 1.** *For all $\varepsilon > 0$ the approximation ratio of the bottom-left algorithm for the Square Strip Packing Problem cannot be better than $4/3 - \varepsilon$ even if the squares are ordered in the best possible way.*

*Proof.* Let $h \geq 2$ be an integer. Consider the instance with one square of size $h$, $4h$ squares of size $h + 1$ and $2h$ squares of size $2h + 1$. Let the strip width be $W = 4h^2 + 3h$. The packing in Fig. 4 uses $2h$ squares of size $h + 1$ on the bottom left, $h$ squares of size $2h + 1$ on the top left, and conversely $2h$ squares of size $h + 1$ on the top right, and $h$ squares of size $2h + 1$ on the bottom right. This leaves a gap in the middle for the square of size $h$. The packing is tight, hence an optimal packing has height $3h + 2$.

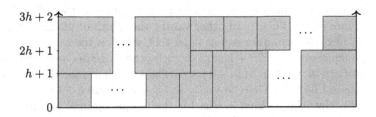

**Fig. 4.** The unique optimal square packing up to symmetry.

Consider the $4h^2 + 3h$ disjoint $1 \times (3h + 2)$ vertical columns. There are two possible types $(a)$ $\{h+1, h+1, h\}$ and $(b)$ $\{2h+1, h+1\}$. Let $a$ resp. $b$ denote their number. In total there are $4h^2 + 3h$ columns, hence $a + b = 4h^2 + 3h$. Furthermore, there is only one square of size $h$, thus $a = h$. This implies that $b = 4h^2 + 2h$.

From $a = h$ it follows that either the square of size $h$ is above (symmetrically below) two rows of squares of size $h + 1$ or it is between squares of size $h + 1$. The first case is impossible, because this creates a gap of height $h$ which cannot be filled by another square as demonstrated by the red area in Fig. 5(a). In case two, if the two squares of size $h + 1$ above and below the square of size $h$ extend over the same side of the square of size $h$ as in Fig. 5(b), then no other square can fill the red space. Thus the square above and below the square of size $h$ must go in different directions over the left resp. right boundary of the square of size $h$, as depicted in Fig. 5(c). As $b = 4h^2 + 2h$, it follows that each other column with a square of size $h + 1$ also contains a square of size $2h + 1$. Thus the structure in Fig. 5(d) must be part of the optimal solution.

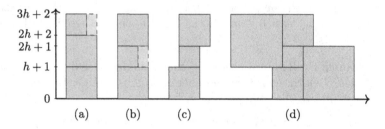

**Fig. 5.** Different cases that are possibly contained in an optimal packing. (Color figure online)

Building on the structure in Fig. 5(d), $b - 2$ more columns of type $(b)$ are required. An optimal packing is tight, therefore next to the square of size $h + 1$ in the bottom left there must be more squares of size $h + 1$, and also next to the square of size $2h + 1$ in the top left there must be more squares of size $2h + 1$. More precisely, there must be at least $2h - 1$ squares of size $h + 1$ next to the square of size $h + 1$ and at least $h - 1$ squares of size $2h + 1$ next to the square of size $2h + 1$, because those are the smallest numbers such that the left side of the leftmost square of size $h + 1$ is at the same place as the left side of the leftmost square of size $2h + 1$. In other words, the width of placing $2h$ squares of size $h + 1$ next to each other equals $2h(h + 1)$, which is the same as the width of placing $h$ squares of size $2h + 1$ next to each other together with the width of the square of size $h$, that is, $h(2h + 1) + h$. The same holds for the right side of the structure in Fig. 5(d). There are only $4h$ squares of size $h + 1$ and $2h$ squares of size $2h + 1$, thus the packing in Fig. 4 is the unique optimal packing.

Let $\varepsilon > 0$ be sufficiently small and consider the slight modification where the squares of size $2h + 1$ get size $2h + 1 - \varepsilon$ and the square of size $h$ gets size $h + \varepsilon$. The packing of Fig. 4 has height $3h + 2 - \varepsilon$ and still is the unique optimal packing (up to symmetry), but it does not have the BL structure. Similar to Corollary 2, the packing in Fig. 6 is an optimal BL packing and has height $4h + 2 - 2\varepsilon$. Because if the squares of size $2h + 1 - \varepsilon$ are not on top of each other, then the height of a BL packing is at least $4h + 2$, because there is a column containing the squares of size $2h + 1 - \varepsilon$, $h + 1$ and $h + \varepsilon$.

Thus for sufficiently small $\varepsilon > 0$, and every $h \geq 2$, there exists an instance $\mathcal{I}_{\varepsilon,h}$ such that $\frac{h_{\mathrm{BL}}^{\mathrm{best}}(\mathcal{I}_{\varepsilon,h})}{h_{\mathrm{OPT}}(\mathcal{I}_{\varepsilon,h})} = \frac{4h - 2\varepsilon + 2}{3h + 2 - \varepsilon}$. Setting $\epsilon' := (2 + 2\varepsilon)/(9h + 6 - 3\varepsilon)$ gives a lower bound of $4/3 - \varepsilon'$ on the approximation ratio of the BL algorithm for SSPP. □

## 4   Bounds for the Worst-Order BL Packing

For badly ordered rectangles, the approximation ratio of the BL algorithm can be arbitrarily large. This might even happen if they are ordered by increasing

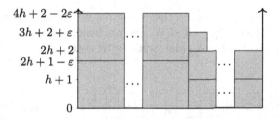

**Fig. 6.** An optimum BL packing of the modified instance.

width [1]. On the contrary, in the square case we can show that the approximation ratio of the BL algorithm remains bounded regardless of the ordering of the squares and also improve the best-known lower bound of $2 - \varepsilon$ [1]:

**Theorem 2.** *The bottom-left algorithm has constant approximation ratio for the Square Strip Packing Problem, for all possible orderings of the squares. This approximation ratio cannot be better than $10/3 - \varepsilon$.*

*Proof.* Our proof for the constant upper bound is very similar to a proof presented in [4] to show that the online *BottomLeft* algorithm has approximation ratio 3.5. The idea is to prove that holes in a BL packing can be covered by surrounding squares. Contrary to [4] we do not aim to get the best possible constant. This makes our proof shorter than the one presented in [4] at the cost of getting a worse constant (16 instead of 3.5). The details of the upper bound are presented in [8]. To prove the $10/3 - \varepsilon$ lower bound we start with the so-called checkerboard construction from [1].

**Construction 1 ($m$-checkerboard).** Let $m \geq 2$ be even and $\varepsilon = \frac{2}{m^3(m^2+1)}$. Consider the instance consisting of squares of size $2-i\varepsilon$ for $i = 1, \ldots, m^2$ together with $m^3 + \frac{(m-1)m}{2}$ unit squares. Let the strip width be

$$W = \sum_{i=1}^{m^2}(2 - i\varepsilon) = 2m^2 - \varepsilon\frac{m^2(m^2 + 1)}{2} = 2m^2 - \frac{1}{m}.$$

The BL packing of the squares ordered by decreasing size is shown in Fig. 7 (blue squares). By definition, all large squares fit precisely next to each other on the strip bottom. These squares are ordered by decreasing size, hence the unit squares on top of these large squares are placed from right to left. Since two unit squares are wider than any of the large squares, it follows that only one unit square is placed on top of each large square. This type of placement repeats in the second row because the gaps between the unit squares in the first row all have width less than 1. In general, the $i$-th row of unit squares alternates gaps and squares except for the initial and final squares of the row, these form almost solid triangles. More precisely, the $i$-th row contains $m^2 + i - 1$ unit squares, and there are precisely $m$ such rows, because $\sum_{i=1}^{m}(m^2 + i - 1) = m^3 + \frac{(m-1)m}{2}$. Last of all, each row fits into the strip. The $i$-th row consists of $m^2 + (i - 1)$ unit squares and $m^2 - i$ gaps between these squares of size less than 1, hence each row has width less than $2m^2 - 1 < W$. In conclusion, the BL packing of height $m + 2 - \varepsilon$ as depicted in Fig. 7 (blue squares) is correct.

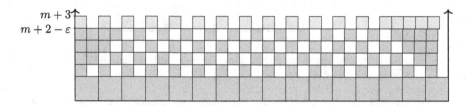

**Fig. 7.** In blue the 4-checkerboard, together with the reset row in green. (Color figure online)

There fit $2m^2 - 1$ unit squares in one row, thus we need $m/2 + 1$ rows to place all unit squares, because $(2m^2 - 1)(m/2 + 1) = m^3 + 2m^2 - m/2 - 1 > m^3 + (m - 1)m/2$. Place the squares of size $2 - i\varepsilon$ on top of this in one row, then $m/2 + 1 + 2 = m/2 + 3$ is an upper bound on $h_{\text{OPT}}$. Thus it holds that $\frac{h_{\text{BL}}^{\text{worst}}}{h_{\text{OPT}}} \geq \frac{m+2-\varepsilon}{m/2+3}$ for the $m$-checkerboard, which goes to 2 as $m$ becomes large.

Adding one large square of size $m/2$ at the end to the $m$-checkerboard already results in a lower bound approaching 3 for the worst-order BL algorithm. The lower bound can be improved further to $10/3 - \varepsilon$ by adding a construction with density $\frac{1}{3}$ to the checkerboard and ending with one large square. This $\frac{1}{3}$-dense construction requires a flat foundation on top of the checkerboard, therefore we add one extra row of squares to the checkerboard such that the height of the top face of these squares is the same and no large gaps exist in this row.

**Construction 2 (Reset row).** Let $m$ be even. There are $m^2 + m - 1$ unit squares in the $m$-th row of the $m$-checkerboard from Construction 1. Let $h_i$ be the height of the top face of the $i$-th square in the $m$-th row (from left to right). It holds that $h_i = m + 2 - \min\{i, m^2\}\varepsilon$ for $1 \leq i < m^2 + m$, because the $i$-th such unit square leans onto the $i$-th bottom row square of size $2 - i\varepsilon$. The reset row consists of $m + 1$ squares of size $1 + m^2\varepsilon$, and a square of size $1 + i\varepsilon$ for each $m < i < m^2$ and odd $1 \leq i \leq m$. Bottom-left place them in decreasing size order on top of the $m$-checkerboard, then the square of size $1 + i\varepsilon$ is going to be placed on top of $h_i$, hence the height of the top face of these almost unit squares equals $h_i + (1 + i\varepsilon) = (m + 2 - i\varepsilon) + (1 + i\varepsilon) = m + 3$. Thus the top face of all squares in the reset row have the same height. Also all gaps between squares in this construction have width less than 1. Figure 7 depicts the reset row in green.

The main idea to obtain the $10/3 - \varepsilon$ lower bound is to construct exponentially growing rows above the $m$-checkerboard with gaps of approximately twice the width of the squares in that row. To this end, consider an integer $n \geq 1$ and let $m$ be the largest even number such that $m \leq \frac{4}{3}2^n$. Consider the $m$-checkerboard together with the reset row from Construction 2. After that, BL pack the pattern consisting of four unit squares followed by a square of size $2 + (a_i + 1)\varepsilon$ for $1 \leq i \leq \lfloor \frac{W}{6} \rfloor$ where $a_i = 1$ if $i$ is odd and $a_i = 0$ otherwise. These squares fit into one row on top of the reset row and the gaps between these squares is 4. If the space at the end of this row is at least $4 + 2\varepsilon$, then add more squares of size $2 + \varepsilon$ at the end. For $2 \leq j \leq n - 1$, BL pack squares of size $2^j + (a_i + 2^{j-1})\varepsilon$ for $1 \leq i \leq \lfloor \frac{W}{2^j + 2^{j+1}} \rfloor$. Each $j$ constitutes a row because the gaps between the squares in the previous row have width less than

$$(2^{j-1} + (1 + 2^{j-1})\varepsilon) + (2^{j-1} + 2^{j-1}\varepsilon) + (2^j + j\varepsilon) - 2^j\varepsilon = 2^{j+1} + (j+1)\varepsilon$$

which can be shown by induction together with the observation that the squares in the $j$-th row are placed on the squares in the $(j-1)$-th row for which $a_i = 0$. If the space at the end of a row is larger than $2^{j+1} + 2^j\varepsilon$, then add extra squares of size $2^j + 2^{j-1}\varepsilon$ to this row. After placing the $n - 1$ rows, finish with bottom-left placing one square of size $2^n + 2^{n-1}\varepsilon$ on top of the packing. This BL packing is depicted in Fig. 8. It holds that $h_{\mathrm{BL}}^{\mathrm{worst}} \geq m + \sum_{i=1}^{n-1} 2^i + 2^n \geq \frac{10}{3} \cdot 2^n$, because the $m$-checkerboard has height at least $\frac{4}{3}2^n$, the squares in the middle of the structure of Fig. 8 have height at least $\sum_{i=1}^{n-1} 2^i$, and on top of that a square of size at least $2^n$ is placed.

As mentioned, the gap between the squares of size $2^j + (a_i + 2^{j-1})\varepsilon$ is at most $2^{j+1} + (j+1)\varepsilon$, hence each row is approximately for $\frac{1}{3}$ occupied by squares. The checkerboard is approximately half-occupied, and the top row with only a square of size $2^n + 2^{n-1}\varepsilon$ is almost entirely empty as $n$ becomes large, because the width of the strip is quadratic in $m$. Thus the total amount of occupied area is approximately $\frac{1}{2} \cdot \frac{4}{3}2^n + \frac{1}{3} \cdot \sum_{i=1}^{n-1} 2^i + 0 \cdot 2^n = \frac{2}{3}2^n + \frac{1}{3}2^n + \mathcal{O}(1) = 2^n + \mathcal{O}(1)$. As the squares get exponentially small, it is not difficult to construct an optimum packing of height $2^n + \mathcal{O}(1)$. Hence there exists $\varepsilon > 0$ such that the lower bound for the worst ordering in the BL algorithm is $\frac{h_{\mathrm{BL}}^{\mathrm{worst}}}{h_{\mathrm{OPT}}} \geq \frac{\frac{10}{3} \cdot 2^n}{2^n + \mathcal{O}(1)} = \frac{10}{3} - \varepsilon$. $\square$

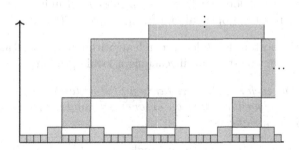

**Fig. 8.** Squares of size $2^j + (a_i + 2^{j-1})\varepsilon$ and gaps of size at most $2^{j+1} + (j+1)\varepsilon$.

The lower bound also holds in an online context, where the squares are placed in the order given, without having all squares available from the start. In this online setting, a 3.5-upper bound on the competitive ratio of the BL algorithm is known that additionally satisfies the Tetris and gravity constraints [4]. The best-known lower bound for the BL algorithm with Tetris and gravity constraints was $3/2$ [4]. Our $10/3 - \varepsilon$ lower bound also applies to the Tetris-gravity BL algorithm, closing the gap to the 3.5-upper bound significantly.

**Corollary 1.** *The online BottomLeft algorithm from* [4] *has approximation ratio between* $10/3$ *and* 3.5.

## 5    Bottom-left $k$-local Search

The bottom-left $k$-local search algorithm computes the height of the BL packing of an instance and tries to improve the height by permuting at most $k$ rectangles in the ordering and comparing if the new BL packing has a smaller height. The algorithm continues until a local optimum is reached. Denote by $h_{k\text{-BL}}(\mathcal{I})$ the height of a solution returned by this algorithm. It turns out that there are (S)SPP instances with approximation ratio at least 2 when starting with a bad ordering.

**Theorem 5.** *For every* $k \in \mathbb{N}$, *there exists a (S)SPP-instance* $\mathcal{I}$ *such that* $h_{k\text{-BL}}(\mathcal{I}) = 2 \cdot h_{OPT}(\mathcal{I})$.

*Proof.* Consider an instance of $2k+4$ unit squares and $2k+5$ squares of size $k+2$. Let the strip width be $W = (2k+4)(k+3)$. An optimum packing of height $k+2$ is obtained by placing all big squares in one row followed by two columns of unit squares. A $k$-local BL packing of height $2(k+2)$ is obtained by alternately packing unit squares and big squares, followed by one more big square. There does not exist a permutation of $k$ squares decreasing $h_{\text{BL}}$. First of all, even when $k$ of the unit squares are removed, the big top square does not fit in the resulting gap, because $(2k+5) \cdot (k+2) + (2k+4-k) \cdot 1 > W$. Secondly, permuting $k$ large squares with unit squares amounts to swapping at most $k/2$ such pairs, in which case at most $2 \cdot \frac{k}{2} + 1 = k+1$ unit squares are adjacent in the bottom row, thus the top square of size $k+2$ does not fit on top of those $k+1$ unit squares. Therefore, this BL packing is $k$-local optimal and $h_{k\text{-BL}}/h_{\text{OPT}} = 2(k+2)/(k+2) = 2$.    $\square$

Moreover, the bottom-left $k$-local search algorithm might need an exponential number of improvement steps until reaching a local optimum:

**Theorem 6.** *For each* $k \in \mathbb{N}$, *there exists a SPP-instance such that the bottom-left $k$-local search algorithm takes an exponential number of improvement steps with respect to the instance size.*

*Proof.* Consider an instance with rectangles $r_i = \left(\frac{1}{k}, 2^i\right)$ and $r'_i = \left(1, \frac{1}{k}\right)$ for $0 \leq i \leq k-1$. The rectangles $r_i$ are vertical and $r'_i$ are horizontal. Let the strip width be 1. Order the rectangles as $r_0, r'_0, r_1, r'_1, \ldots, r_{k-1}, r'_{k-1}$. The height of the BL packing with regards to this ordering equals $\sum_{i=0}^{k-1} (2^i + \frac{1}{k}) = 2^k - 1 + k \cdot \frac{1}{k} = 2^k$. There exists a sequence of $k$-permutations of the vertical rectangles such that the height of the BL packing counts down from $2^k$ to $2^{k-1} + 1$. Namely, after the $p$-th permutation the height of the BL packing must be $2^k - p$. Write the number $2^k - p - 1$ in its unique binary representation $\sum_{j=0}^{k-1} a_j 2^j$ with $a_j \in \{0, 1\}$. Now permute the vertical rectangles such that $2^j$ with $a_j = 1$ is placed before all $2^\ell$ with $\ell < j$ and $a_\ell = 0$ and after all $2^\ell$ with $\ell > j$ and $a_\ell = 1$. Obviously, the height of this BL packing is $2^k - p$, because the rectangles of size $2^\ell$ with $a_\ell = 0$ fit into holes and do not account to the height of the packing, while all rectangles with $a_\ell = 1$ do account to the height of the packing. As the height of the sequence counts down, it follows that the number of improvement steps is exponential in the input size. □

# References

1. Baker, B.S., Coffman, E.G., Jr., Rivest, R.L.: Orthogonal packings in two dimensions. SIAM J. Comput. **9**(4), 846–855 (1980). https://doi.org/10.1137/0209064
2. Brown, D.J.: An improved BL lower bound. Inf. Process. Lett. **11**(1), 37–39 (1980). https://doi.org/10.1016/0020-0190(80)90031-9
3. Chazelle, B.: The bottom-left bin-packing heuristic: an efficient implementation. IEEE Trans. Comput. **C-32**(8), 697–707 (1983). https://doi.org/10.1109/TC.1983.1676307
4. Fekete, S.P., Kamphans, T., Schweer, N.: Online square packing with gravity. Algorithmica **68**(4), 1019–1044 (2014). https://doi.org/10.1007/s00453-012-9713-8
5. Gálvez, W., Grandoni, F., Jabal Ameli, A., Jansen, K., Khan, A., Rau, M.: A tight $(3/2 + \varepsilon)$-approximation for skewed strip packing. Algorithmica **85**(10), 3088–3109 (2023). https://doi.org/10.1007/s00453-023-01130-2
6. Garey, M.R., Johnson, D.S.: "Strong" NP-completeness results: motivation, examples, and implications. J. Assoc. Comput. Mach. **25**(3), 499–508 (1978). https://doi.org/10.1145/322077.322090
7. Harren, R., Jansen, K., Prädel, L., van Stee, R.: A $(5/3 + \varepsilon)$-approximation for strip packing. Comput. Geom. **47**(2), 248–267 (2014). https://doi.org/10.1016/j.comgeo.2013.08.008
8. Hougardy, S., Zondervan, B.: The bottom-left algorithm for the strip packing problem (2024). https://doi.org/10.48550/arXiv.2402.16572
9. Karp, R.M.: Reducibility among combinatorial problems. In: Miller, R.E., Thatcher, J.W. (eds.) Complexity of Computer Computations (Proc. Sympos., IBM Thomas J. Watson Res. Center, Yorktown Heights, N.Y., 1972), pp. 85–103. Plenum Press, New York, London (1972)

# Parameterized Upper Bounds for Path-Consistent Hub Labeling

Stefan Funke[1] and Sabine Storandt[2(✉)]

[1] University of Stuttgart, Stuttgart, Germany
[2] University of Konstanz, Konstanz, Germany
sabine.storandt@uni-konstanz.de

**Abstract.** Hub labeling (HL) comprises a class of algorithms that construct fast distance oracles on weighted graphs. The goal of HL is to assign each node a label such that the shortest path distance between any node pair can be deduced solely based on their label information. HL has been extensively studied from a theoretical and practical perspective, including parameterized upper bounds. It was shown that average label sizes in $\mathcal{O}(\kappa \log n)$ are possible where $\kappa$ denotes the skeleton dimension of the graph. In this paper, we focus on a special type of HL, called path-consistent HL (PC-HL). This type of labeling is beneficial for fast shortest path extraction and compact storage. We prove novel parameterized upper bounds for path-consistent labelings (which also apply to HL). In particular, we show that label sizes can be bounded by $\mathcal{O}(gt \log n)$ where $gt$ denotes the geodesic transversal number. Furthermore, we propose a new variant $\kappa^+$ of the skeleton dimension, show that there are graphs where $\kappa^+$ is a factor of $\Theta(\sqrt{n})$ smaller than $\kappa$, and prove that $\mathcal{O}(\kappa^{(+)} \log n)$ constitutes a valid upper bound for the label size of a PC-HL. We devise polytime algorithms to construct labelings that adhere to the parameterized upper bounds. Those are the first non-hierarchical labelings that are path-consistent, which were not known to exist before. Furthermore, we prove that $gt$ and $\kappa^{(+)}$ are incomparable. We also compute their values on diverse benchmark graphs to assess which of them provides tighter upper bounds in practice.

**Keywords:** Hub Labeling · Geodetic Transversal Number · Skeleton Dimension

## 1 Introduction

Efficient shortest path computation on weighted graphs is an important building block for many algorithms and applications. When multiple shortest path queries are issued on the same input graph $G(V, E)$, preprocessing can help to reduce query times dramatically. One of the most successful preprocessing-based techniques is Hub Labeling (HL) [9]. It relies on the simple concept of precomputing and storing for each node $v \in V$ the shortest path cost to a selected set of other nodes $L$, called the label of $v$. More formally, the label function is defined as

$L : V \rightarrow 2^V$. The node labels have to satisfy the so called *cover property* which demands that for any node pair $s, t \in V$, there is at least one node $w \in L(s) \cap L(t)$ that lies on the shortest path from $s$ to $t$. The node $w$ is called a *hub* for $s, t$. If the cover property holds, shortest distance queries can be answered solely based on label information by identifying the node $w \in L(s) \cap L(t)$ with minimum summed shortest path cost from $s$ and $t$. This query procedure can be implemented to run in time linear in the maximum label size $L_{\max} := \max_{v \in V} |L(v)|$. The space consumption of HL is in $\mathcal{O}(n \cdot L_{avg})$ where $n = |V|$ and $L_{avg} = \frac{1}{n} \sum_{v \in V} |L(v)|$. Thus, for the benefit of both the query time and the space consumption, the optimization goal is to compute a HL with small label size. Computing an HL of minimum size poses an NP-hard problem [5]. Approximation algorithms that produce a HL within a factor of $\mathcal{O}(\log n)$ are known [4,5,9], but their running times are prohibitive for application to large real-world inputs.

Thus, HL is usually constructed in practice via heuristics [3,11,15]. Most of them produce a special type of HL called a *hierarchical* HL (HHL). In a HHL, nodes are ordered by a ranking function $r : V \rightarrow [n]$, and a label $L(v)$ is only allowed to contain nodes $w \neq v$ if $r(w) > r(v)$, that is, nodes further up in the hierarchy. Given a ranking function, the optimal HHL can be computed in polynomial time. It is called the canonical HHL with respect to $r$. As each HHL can be made canonical in polytime, we only consider canonical HHL in the remainder of the paper. Determining the optimal ranking function is still NP-hard, though. The currently best approximation algorithm computes a HHL within a factor of $\mathcal{O}(\sqrt{n} \log n)$ of the optimum. There are also graphs known in which the optimal HHL is larger than the optimal HL by a factor of $\sqrt{n}$ [5]. Nevertheless, heuristics usually produce decent label sizes on practical instances.

A novel class of HL, called *path-consistent* HL (PC-HL), was recently proposed in [17]. In a PC labeling, it holds that $w \in L(v) \Rightarrow w \in L(u)$ for all $u$ on the shortest path from $v$ to $w$. This property is beneficial for shortest path extraction. The query algorithm for HL as described above only returns the shortest path distance but not the path itself. Extracting the shortest path based on a general HL requires additional storage and substantial additional effort, either on query time or in the preprocessing phase. In a PC-HL, storing a pointer with each label $w \in L(s)$ that points at the label $w \in L(s')$ where $(s, s')$ is the first edge on the shortest path from $s$ to $w$ suffices. These pointers can be determined efficiently and enable a constant extraction time per edge on the shortest path [17]. Furthermore, the PC property implies that the set of nodes that contain a certain node $w$ in their label can always be described as a connected subtree of the shortest path tree emerging from $w$. This allows to apply label compression techniques more effectively [12] and thus reduces the space consumption.

While HHL currently only admits a $\mathcal{O}(\sqrt{n} \log n)$ approximation, PC-HL can be approximated within a factor of $\mathcal{O}(\log n)$ [17]. It was further proven in [17] that all canonical HHL are PC. Thus, canonical HHL poses a subclass of PC-HL. However, it was not known before whether HHL is a proper subclass. We will answer this question affirmatively by constructing labelings that are PC but not hierarchical. Furthermore, we present novel theoretical results for PC-HL

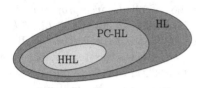

| HHL | PC-HL | HL |
|---|---|---|
| $\mathcal{O}(h \log D) \longrightarrow$ | | $\longrightarrow$ |
| $\mathcal{O}(tw \log n) \longrightarrow$ | | $\longrightarrow$ |
| | $\mathcal{O}(gt \log n)$ | $\longrightarrow$ |
| | $\mathcal{O}(\kappa^{(+)} \log D) \longrightarrow$ | |
| | | $\mathcal{O}(\kappa \log D)$ |

**Fig. 1.** Labeling classes: Hub Labeling (HL), path-consistent HL (PC-HL) and hierarchical HL (HHL). Parameterized upper bounds propagate from subsets to supersets. In the table on the right, $n$ refers to the number of nodes, $h$ denotes the highway dimension, $D$ the graph diameter, $tw$ the treewidth, $gt$ the geodesic transversal number, $\kappa$ the skeleton dimension and $\kappa^+$ our newly proposed skeleton dimension variant.

including several parameterized upper bounds for the label size. As PC-HL is a subclass of HL, these bounds transfer to label sizes of general HL.

## 1.1  Related Work

HL has been extensively studied through the lens of parameterized analysis. The goal is to characterize graphs which admit labelings of small size. It was proven in [9] that there exist graphs in which the average label size of any HL is in $\Omega(\sqrt{m})$ where $m$ is the number of edges in the input graph. Thus, smaller upper bounds can not be guaranteed on general graphs. However, on practical instances even simple (H)HL algorithms often construct small labels [10,15]. One way to provide a theoretical explanation for this presumed gap is to identify a parameter $p$ that captures the complexity of the graph structure, and to prove that label sizes are bound by some function $f(p)$. Graphs with small $p$ value are then guaranteed to admit concise labelings.

The first such bound for HHL was proven by Abraham et al. [2]. They designed a new parameter $h$ called *highway dimension*, specifically for the purpose of analysing preprocessing-based shortest path techniques. It was proven that $L_{\max} \in \mathcal{O}(h \log D)$ can be guaranteed where $D$ is the diameter of the input graph. Constructing a HHL that adheres to this bound relies on repeatedly solving the HittingSet problem, which is known to be NP-hard. But they also describe a polytime algorithm that constructs a HHL with label sizes in $\mathcal{O}(h \log h \log D)$.

In [7], it was shown that label sizes in a HHL can also be bound with respect to the *treewidth* $tw$, a classical and well-studied parameter. More precisely, it was proven that $L_{\max} \in \mathcal{O}(tw \log n)$ is possible and labels of size $\mathcal{O}(tw \log^2 n)$ can be computed in polytime. Furthermore, it was proven in [7] that there exist graphs with $h \geq (tw - 1)/(\log_{3/2} n + 2)$.

The *skeleton dimension* $\kappa$ was introduced as a new parameter in [14] to bound general HL. In contrast to the highway dimension and the treewidth, the skeleton dimension can be computed in polynomial time. Using a randomized construction algorithm, expected HL label sizes in $\mathcal{O}(\kappa \log D)$ were proven. The algorithm runs in polynomial time. It was further shown that $\kappa \in \mathcal{O}(h)$ for graphs of bounded maximum degree and that there exist graphs with $\kappa = \log n$ and $h \in \Omega(\sqrt{n})$. In [8], it was proven that there are graphs where $\kappa \geq (tw-1)/(\log_2 n+2)$.

Figure 1 provides an overview of the known parameterized upper bounds for the three labeling classes as well as the novel bounds we present in this paper.

### 1.2 Contribution

In this paper, we establish path-consistent HL (PC-HL) as its own labeling class by showing that it is a proper superclass of HHL.

Next, we prove novel parameterized upper bounds for label sizes. The first bound is based on the *geodesic transversal number* $gt$, a recently introduced graph parameter that characterizes the complexity of the shortest path structure [16]. We show that in an optimal PC-HL $L_{\max} \in \mathcal{O}(gt \log n)$ can be guaranteed and we discuss how to compute labels of size $\mathcal{O}(gt \log gt \log n)$ in polytime. This bound then also applies to general HL as PC-HL is a subclass of HL.

Furthermore, we show that the *skeleton dimension* based upper bound of $\mathcal{O}(\kappa \log D)$ for HL also holds for PC-HL. Note that this result is not by propagation as HL is the superclass. Instead, we need a dedicated proof that the labels produced by the randomized algorithm described in [14] indeed fulfill the PC property. We do not prove this directly. Instead, we first propose a new variant of the skeleton dimension, called *hop-skeleton dimension* $\kappa^+$, and prove that PC-HL label sizes are in $\mathcal{O}(\kappa^+ \log D)$. This novel parameter has several advantages compared to $\kappa$: It is more robust and simpler to compute, it allows for an easier analysis of the expected label size, and we also show that there are graphs where $\kappa^+$ is smaller than $\kappa$ by a factor of $\Theta(\sqrt{n})$. Moreover, it allows us to prove that the algorithm proposed in [14] indeed produces a PC labeling.

Finally, we investigate the question whether the geodesic transversal number based bound or the (hop-)skeleton dimension based bound for PC-HL is tighter. We prove that these parameters are incomparable. We also compute their (approximate) values on a diverse set of benchmark graphs and observe that our novel hop-skeleton dimension produces a better bound on average.

## 2    Preliminaries

Throughout the paper, the input graph $G(V, E)$ is assumed to be connected, weighted and (for ease of exposition) undirected. Edge costs $c : E \to \mathbb{R}^+$ are non-negative and metric. Furthermore, shortest paths are assumed to be unique, which is a common assumption (see e.g. [4,14]) that can be enforced via symbolic perturbation [13]. For a node pair $s, t \in V$, we use $\pi(s, t)$ to refer to the shortest path between them and $c(\pi(s, t)) = c(s, t) = c(t, s)$ to refer to the shortest path cost. As the main focus of the paper is on path-consistent HL, we provide the concise formal definition below.

**Definition 1 (PC-HL).** Given a graph $G(V, E)$, a path-consistent HL assigns each node $v \in V$ a label $L : V \to 2^V$ such that the following conditions are met:

- cover property: $\forall s, t \in V : \exists w \in L(s) \cap L(t) \cap \pi(s,t)$
- path-consistency (PC): $w \in L(v) \Rightarrow w \in L(u)\ \forall u \in \pi(v,w)$

For each node $w \in L(v)$, also the shortest path cost $c(v,w)$ is stored and a pointer to $w \in L(v')$ if $v'$ is the first node on the shortest path from $v$ to $w$ after $v'$. Note that $w \in L(v')$ is enforced by the PC property.

# 3     Geodesic Transversal Number

In this section, we prove that label sizes in a PC-HL can be upper bounded by $\mathcal{O}(gt \log n)$ where $gt$ denotes the *geodesic transversal number* of the graph.

**Definition 2** (*gt*). The geodesic transversal number $gt(G)$ denotes the size of a smallest node set $H \subset V$ that hits all maximal shortest paths in $G$.

This parameter was proposed in [16]. A maximal shortest path (aka geodesic) is defined as a shortest path for which there is no superpath that is itself a shortest path. The transversal number denotes the size of an optimal HittingSet for the collection of maximal geodesics. It was proven in [16] that $gt$ and the respective HittingSet $H$, called gt-set from now on, can be computed efficiently on trees and on spread cactus graphs. But on general graphs, the problem is NP-complete.

Interestingly, it was argued in [17] that label sizes in a HL are *not* bound by $gt$. The counterexample is a simple path graph of $n$ nodes. Here, $gt = 1$ as there is only a single maximal geodesic. However, it is well known that an optimal HL on paths requires $L_{max} \in \Theta(\log n)$ [6]. In the following, we prove that this logarithmic gap between $gt$ and the maximum label size in a HL is the largest possible. Indeed, we show the stronger result that this even applies to PC-HL. As the optimal HL on a path graph is a HHL and thus PC, this also implies that our upper bound of $\mathcal{O}(gt \log n)$ for PC-HL label sizes is asymptotically tight.

To prove this parameterized upper bound, we first need two helping lemmas.

**Lemma 1.** *Given two PC-labelings $L_1, L_2$, their union is also a PC-labeling.*

*Proof.* Let $L(v) := L_1(v) \cup L_2(v)$ be the union labeling. Now let $w \in L(v)$ be some node contained in the union label and let $u$ be a node on the shortest path from $v$ to $w$. W.l.o.g. we assume $w \in L_1(v)$. Then by the PC-property of $L_1$, we know that $w \in L_1(u)$ must hold and thus also $w \in L(u)$.

Thus, PC labeling is closed under unions. We remark that the same is not true for hierarchical labelings with different ranking functions.

**Lemma 2.** *Let $T_v$ denote the shortest path tree emerging from $v$. There is a PC-HL for $T_v$ with maximum label size in $\mathcal{O}(\log n)$.*

*Proof.* For HHL, a parameterized upper bound on the label size of $\mathcal{O}(tw \log n)$ is known [7]. In trees, we have $tw = 1$ and thus the maximum label size is in $\mathcal{O}(\log n)$. The respective canonical HHL can only have smaller labels and is always PC [17]. Thus, the lemma follows.

Note that the respective tree labeling can be computed in polynomial time and that the resulting average label size is within a factor of 2 of the optimum [4]. Now, we are ready to prove the main result.

**Theorem 1.** *The maximum label size in a PC-HL can be bounded by $\mathcal{O}(gt \log n)$.*

*Proof.* Let $H$ be an optimal gt-set for $G(V, E)$. Further, let $L(T_h)$ be a labeling for the shortest path tree emerging from $h \in H$. For each node $v \in V$, we define its label as the union of labels in $L(T_h), h \in H$ for that node. By virtue of Lemma 2, the maximum label size for each considered tree $T_h$ can be ensured to be in $\mathcal{O}(\log n)$. Thus, the total maximum label size is bounded by $\mathcal{O}(gt \log n)$. As shown in Lemma 1, the union of PC labelings is also a PC labeling. As the tree labelings are PC as discussed in Lemma 2, the PC-property for the overall labeling follows. It remains to show that the resulting labeling $L$ fulfills the cover property. For that, consider $s, t \in V$ and let $\pi^*$ be a maximal geodesic that contains both $s$ and $t$. By definition of a gt-set, there needs to be at least node $h \in H \cap p$. Thus, the shortest path $\pi(s, t)$ between $s$ and $t$ is contained in $T_h$ and the respective labeling $L(T_h)$ ensures that there is $w \in L(s) \cap L(t) \cap \pi(s, t)$. □

To construct the respective PC-HL, we need access to a gt-set. Computing an exact gt-set is not feasible in practice, though. But as there are $\mathcal{O}(n^2)$ maximal geodesics, the standard greedy algorithm provides a $\mathcal{O}(\log n)$ approximation factor. This allows to construct labels with a size in $\mathcal{O}(gt \log^2 n)$ in polytime. A more intricate approximation algorithm that relies on the set system having small VC-dimension can be used as well, as sets of shortest paths have constant VC-dimension [1]. This algorithm returns a gt-set of size $\mathcal{O}(gt \log gt)$. Accordingly, we can construct a PC-HL with label sizes in $\mathcal{O}(gt \log gt \log n)$ in polytime.

# 4 Skeleton Dimension

The *skeleton dimension* $\kappa$ was introduced in [14] to analyze label sizes in HL. The definition of the parameter relies on the concept of a tree skeleton.

**Definition 3.** Let $T_v$ be the shortest path tree from $v$. The trimmed shortest path from $v$ to $w$ is the prefix of the shortest path up to a distance of $\frac{2}{3}c(v, w)$. The tree skeleton $T_v*$ of $T_v$ is the union of all trimmed shortest paths in $T_v$.

Note that the skeleton relies on a geometric realization of the shortest path tree. That means, that the cut point at $\frac{2}{3}c(v, w)$ does not need to be a graph node.

**Definition 4.** The width of a tree $T$ for a given distance $d$ is the maximum number of points at distance $d$ from the root. The cutwidth of $T$ is the maximum width over all (continuous) distance values.

Based on these two concepts, the skeleton dimension is defined as follows:

**Definition 5** ($\kappa$). The skeleton dimension $\kappa$ of a graph $G(V, E)$ is defined as the maximum cutwidth of a tree skeleton $T_v^*$ with $v \in V$.

The skeleton dimension of a given graph can be computed in polytime.

For HL, it was proven that a randomized algorithm produces expected label sizes in $\mathcal{O}(\kappa \log D)$. We will show that the resulting labeling is PC but not necessarily hierarchical. This proves that PC-HL is a proper superclass of HHL.

To prove path-consistency, we first introduce a skeleton dimension variant which we call hop-skeleton dimension. This parameter is an interesting contribution on its own as it comes with many desirable properties that will be discussed in more detail below.

### 4.1 Hop-Skeleton Dimension

The hop-skeleton dimension $\kappa^+$ also measures the maximum cutwidth of any shortest path tree skeleton of the input graph. The only change is that for the definition of the tree skeleton, we do not use shortest path cost but hop distance, i.e. number of edges on the respective shortest path. To illustrate the difference of $\kappa$ and $\kappa^+$ and to show the usefulness of the novel parameter variant, we prove that there are graphs in which $\kappa^+$ is substantially smaller than $\kappa$.

**Lemma 3.** *There are graphs in which $\kappa^+$ is smaller than $\kappa$ by a factor of $\Theta(\sqrt{n})$.*

*Proof.* Consider a graph that consists of $\sqrt{n}$ stars of size $\sqrt{n}$ each, where one star is the central star. Each leaf of the central star is a shared leaf with exactly one other star. Figure 2 illustrates the graph structure. We divide the non-central stars into two groups of equal size, the 'cheap' and the 'expensive' stars. We set the edge costs in the cheap stars uniformly to 1, and in the expensive stars uniformly to 7. The edges in the central star also have a cost of 1.

Let us now consider the skeleton of a leaf node $v$ in a cheap star. To other leaves in the same star, the distance is 2 and thus the edge from $v$ to the star center is fully included in the skeleton, and the edges from the star center to the leaves are included up to a distance of $\frac{1}{3}$. For leaves in the other cheap stars, the distance is 6. Thus, the shortest path from $v$ to any such leaf is included up to a distance of 4. That means that the edges incident to the nodes in the

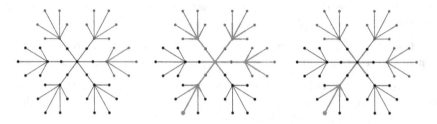

**Fig. 2.** Illustration of a graph structure that yields $\kappa^+ \ll \kappa$. Edges in the red stars have much higher cost than the other edges. In the middle image, the skeleton for the blue marked node is depicted when considering edge costs. The right image shows the skeleton of the same node when considering hop distance instead. (Color figure online)

central star are included, but none of the other edges in the other cheap stars. Now, let us consider the expensive stars. The path from $v$ to a leaf in such a star has a cost of 12. So the path prefix up to length 8 is included in the skeleton. Accordingly, all edges in the expensive stars are included up to distance 2. The cutwidth of this skeleton is maximized at any distance in $(6, 8)$, where all the edges in the expensive stars need to be cut. As there are $\sqrt{n}/2$ such stars with $\sqrt{n}$ edges each, the cutwidth – and therefore also $\kappa$ – is in $\Theta(n)$.

If we now ignore edge weights and consider uniform costs instead, all stars become cheap. The skeleton from any node of degree 1 contains only (partial) edges of its own star and of the ones incident to nodes in the central star. The argument is exactly the same as in the consideration of cheap stars above. Therefore, the cutwidth of such a leaf skeleton is $\sqrt{n}$. For centers of cheap stars, the skeleton includes the edges to its own leaves up to a distance of $\frac{2}{3}$. The distance to leaves from other stars is 5 and thus the path prefix up to distance $3\frac{1}{3}$ is included in the skeleton. This again does not include any edges not incident to nodes in the central star and thus the cutwidth does not exceed $\sqrt{n}$ as well. It remains to consider the nodes in the central star. For its center, the distance to all leaves is equal to 3 and thus the included path prefix ends at a distance of 2 which is precisely the distance to the other star centers. Again, no edges from other stars are included. For all nodes $v$ at distance 1 of the central star, we have a close cheap star with its center at distance 1 from $v$ and far cheap stars with their centers at distance 3 from $v$. For the close star, the distance to its leaves is 2 and thus the star edges are all included in the skeleton up to a distance of $\frac{1}{3}$. For the far stars, however, the distance from $v$ to their leaves is 4. The included prefix hence ends at distance $2\frac{2}{3}$ and does only include edges incident to nodes of the central star. At distances in $(1, 1\frac{1}{3})$, the cutwidth assumes its maximum of $2\sqrt{n}$ which then defines the hop-skeleton dimension $\kappa^+$ of the graph.

As $\kappa^+ \in \mathcal{O}(\sqrt{n})$ and $\kappa \in \Theta(n)$ in this example graph, the lemma follows.

Not only can $\kappa^+$ assume much smaller values than $\kappa$ but it is also more robust in the sense that edge cost changes that do not change the structure of shortest paths also do not change the parameter value. Moreover, it gives rise to an easier (PC-)HL analysis as will be discussed next.

## 4.2   Skeleton Labelings

The randomized algorithm to construct a valid HL proposed in [14] works as follows: First, all edge costs are scaled to integers. Subsequently, each edge $e$ is subdivided into $12 \cdot c(e)$ subedges of length $\frac{1}{12}$. As a consequence, all subedges have the same cost. This allows to treat the graph as unweighted for the remainder of the label construction. Then, a random priority $p(e) \in (0, 1)$ is assigned u.a.r. to each edge $e \in E$. For each node pair $s, t$, the central subpath of $\pi(s, t)$ consists of the edges $e \in \pi$ with at least one of their endpoints being at distance $d \in \left(\frac{k}{3}, \frac{2k}{3}\right)$ from $s$ where $k$ denotes the number of edges in $\pi(s, t)$ (aka the hop distance). Then, the edge $e^*$ with highest priority in the central subpath is selected and included in the label of $s$ and $t$ (ensuring the cover property).

Including an edge means that both of its endpoints are included in the label. But for easier discussion and to comply with [14], we stick to the notion of labels being edge sets instead of node sets for the remainder of this section.

As proven in [14], it is guaranteed by construction that all edge labels selected for a node $s$ are part of its tree skeleton $T_s^*$. And the probability of an edge to be included in $L(s)$ decreases proportional to its hop-distance to $s$. A thorough analysis of these observations yields expected label sizes in $\mathcal{O}(\kappa \log D)$.

The only aspect of construction that needs to change when we consider $\kappa^+$ instead of $\kappa$ is that we omit the cost scaling to integers and the subdivision of the edges. Instead, we directly consider hop distances in the shortest path tree to define the central subpath. The analysis then works exactly the same as it was cost-oblivious anyway. Thus, we obtain expected HL label sizes in $\mathcal{O}(\kappa^+ \log D)$. Next, we prove that the resulting hop-skeleton labeling is always path-consistent. This allows to transfer the label size bound to PC-HL.

**Theorem 2.** *Hop-skeleton labelings are path-consistent.*

*Proof.* Let $\pi(s,t)$ be the shortest path from $s$ to $t$ with hop distance $k$ (not to be confused with $\kappa$ or $\kappa^+$). Further, let $e^*$ be the edge of highest priority on the central subpath of $\pi$, which implies that $e^* \in L(s) \cap L(t)$. To prove that the labeling is path-consistent, we need to show that for each $v \in \pi(s,t)$, we also have $e^* \in L(v)$.

To show this, we interpret the path as interval $I = [0, k]$ where $s$ is located at 0, $t$ is located at $k$, and all intermediate nodes are located at the integers in this interval. Edges are open subintervals between consecutive integers. We will show that for $t'$ located at $k-1$, the edge $e^*$ is the most important edge on the subpath to a node $s'$ being located at either 0 or 1 or 2. Note that this suffices to prove the overall statement, as we can then apply the same argument to the shortest path from $s'$ and $t'$, and we can also inverse their roles.

Now, let us consider the central subpath interval of $I$. It can be expressed as $C_{0,k} = (\frac{k}{3}, \frac{2k}{3})$. The respective relevant subpath intervals for $t'$ are as follows:

$$C_{0,k-1} = (\tfrac{k}{3} - \tfrac{1}{3}, \tfrac{2k}{3} - \tfrac{2}{3}) \quad C_{1,k-1} = (\tfrac{k}{3} + \tfrac{1}{3}, \tfrac{2k}{3} - \tfrac{1}{3}) \quad C_{2,k-1} = (\tfrac{k}{3} + 1, \tfrac{2k}{3})$$

Let us distinguish three cases for the number $k$ of hops:

- $k \mod 3 = 0$: Let $k = 3f$, hence $C_{0,k} = (f, 2f)$. But we also have $C_{1,k-1} = (f + \tfrac{1}{3}, 2f - \tfrac{1}{3})$. This central subpath interval intersects the same edges as $(f, 2f) = C_{0,k}$, hence $e^*$ is also chosen.
- $k \mod 3 = 1$: Let $k = 3f + 1$, hence $C_{0,k} = (f + \tfrac{1}{3}, 2f + \tfrac{2}{3})$ and $C_{1,k-1} = (f + \tfrac{2}{3}, 2f + \tfrac{1}{3})$ which intersects the same edges as $(f + \tfrac{1}{3}, 2f + \tfrac{2}{3}) = C_{0,k}$. So again, $e^*$ must be chosen.
- $k \mod 3 = 2$: let $k = 3f + 2$, hence $C_{0,k} = (f + \tfrac{2}{3}, 2f + \tfrac{4}{3})$. But then we have $C_{0,k-1} = (f + \tfrac{1}{3}, 2f + \tfrac{2}{3})$ and $C_{2,k-1} = (f + \tfrac{5}{3}, 2f + \tfrac{4}{3})$. So clearly each edge that intersects $C_{0,k}$ also intersects $C_{0,k-1}$ or $C_{2,k-1}$ and all edges intersecting $C_{0,k-1}$ or $C_{2,k-1}$ also intersect $C_{0,k}$. That is, $e^*$ must be chosen in at least one of them.

As in all three cases the edge $e^*$ is included in the label of $t'$, the correctness of the theorem follows.

Accordingly, expected label sizes of a hop-skeleton PC-HL are in $\mathcal{O}(\kappa^+ \log D)$. Note that the respective construction algorithm runs in polytime.

To prove that the original skeleton labeling is PC as well, we need to show that reversing the edge subdivision process applied to get to a hop-skeleton labeling does not affect the PC property. Thus, we now argue that given a PC labeling, contracting a 2-degree node does not violate the PC property. In a contraction, we remove a node $v$ and its two incident edges $e' = \{u, v\}$ and $e'' = \{v, w\}$ and insert the edge $e = \{u, w\}$ with a cost value of $c(e) := c(e') + c(e'')$. For any node $v$ with $e' \in L(v)$ or $e'' \in L(v)$ we replace $e'$ and/or $e''$ by $e$.

**Lemma 4.** *Given a PC labeling and a degree-2 node $v$, the resulting labeling after contraction of $v$ is still a valid PC labeling.*

*Proof.* Let $s \in V$ be a node for which the shortest path $\pi(s, v)$ ends w.l.o.g. in $e' = \{u, v\}$ and $e' \in L(s)$. Then $\forall x \in \pi(s, v)$ it follows from the PC property that $e' \in L(x)$. Replacing $e'$ with $e$ in all these labels clearly maintains the PC property. Consider the set of targets $t$ for which $e'$ was the label in $L(s) \cap (t) \cap \pi(s, t)$. For any $t \neq v$, the edge $e''$ also needs to be part of the shortest path and thus the new edge $e$ is part of the shortest path after contraction of $v$.

Based on this lemma, we get the following corollary.

**Corollary 1.** *Skeleton labelings are path-consistent and thus expected label sizes in PC-HL are in $\mathcal{O}(\kappa \log D)$.*

Next, we prove by example that skeleton labelings do not need to be hierarchical.

**Lemma 5.** *(Hop-)skeleton labelings are not necessarily hierarchical.*

*Proof.* A labeling is clearly not hierarchical if there exists a pair of nodes $v, w$ with $w \in L(v)$ and $v \in L(w)$. Let now $a, b, c, d, e$ be a path graph with four edges and uniform costs. The central subpath of $\pi(a, d)$ is solely the edge $\{b, c\}$ and thus we have $\{b, c\} \subset L(d)$. Next, we consider the central subpath of $\pi(b, e)$ which is solely the edge $\{c, d\}$ and so we have $\{c, d\} \subset L(b)$. Accordingly, we have $d \in L(b)$ and $b \in L(d)$. Thus, there is no valid ranking function under which both of these label sets are feasible.

As we have proven that (hop-)skeleton labelings are always PC but not necessarily hierarchical, we have established the following corollary.

**Corollary 2.** *HHL is a proper subclass of PC-HL.*

## 5 Parameter Relationships

We have proven parameterized upper bounds for PC-HL label sizes based on the geodesic transversal number and the (hop-)skeleton dimension. The question now is whether one bound is always stronger than the other. We prove below that this is not the case as there are graphs in which $gt \ll \kappa^{(+)}$ and vice versa.

**Theorem 3.** *The geodesic transversal number $gt$ and the (hop-)skeleton dimension $\kappa^{(+)}$ are incomparable.*

*Proof.* To prove that the parameters are incomparable, we show that there are graphs with $gt \in \mathcal{O}(1)$ and $\kappa^{(+)} \in \Theta(n)$ as well as graphs with $gt \in \Theta(n)$ and $\kappa^{(+)} \in \mathcal{O}(1)$. See Fig. 3 for illustrations.

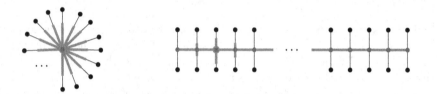

**Fig. 3.** Left: Star graph with $gt = 1$ (red node) and $\kappa^{(+)} = n - 1$ (skeleton from central node in blue). Right: Caterpillar with $gt = \frac{n}{3}$ (red nodes) and $\kappa^{(+)} = 6$. The skeleton of a selected path node is shown in blue. (Color figure online)

In a simple star graph with $n-1$ arms und uniform costs, all maximal shortest paths go through the center node. Therefore, we have $gt = 1$. However, if we consider the skeleton of the star center, it contains $\frac{2}{3}$ of each edge, see Fig. 3, left. Thus, the respective cutwidth equals $n - 1$, showing that $\kappa^{(+)} \in \Theta(n)$.

A caterpillar is a special type of tree graph in which we have a central path and all leaf nodes are incident to a node on the central path. We consider a caterpillar with uniform costs in which the central path contains $\frac{n}{3}$ nodes and where each path node is incident to exactly two leaf nodes, see Fig. 3, right. For any path node, there is a maximal shortest path of length 2 between its two incident leaves. All of these $\frac{n}{3}$ maximal shortest paths are disjoint from each other and thus require each another node to be hit. Accordingly, we have $gt \in \Theta(n)$. To bound $\kappa^{(+)}$, we need to consider the skeletons of leaf and path nodes. For leaf nodes, we observe that only edges on the central path are contained in the skeleton as well as a part of the edge to the leaf incident to the same central path node. All other leaf nodes have a distance of at least 3 to the node, and as $\frac{2}{3}d \leq d - 1$ for $d \geq 3$ their incident edges are not included. Thus, the cutwidth is 3. For central path nodes, the path edges are part of the skeleton as well as the leaf edges incident to the node itself and to its neighbors on the path. For all other leaf nodes, the distance is again at least 3 and the above argument applies. Here, the cutwidth is maximized at distance $d \in (1, 1\frac{1}{3})$ and equals 6.

# 6    Practical Parameter Values

As $gt$ and $\kappa^{(+)}$ are incomparable it is not clear which of them provides a tighter upper bound on a given instance. To see how these parameters behave on different types of graphs, we implemented the greedy approximation algorithm for $gt$ and the exact algorithm for $\kappa$ and $\kappa^+$. We also report the size of a largest independent set of maximal geodesics as a lower bound for $gt$ to narrow down the true parameter value. To conduct experiments on a diverse set of benchmark graphs, we chose the 2018 PACE challenge, Track 1 (T1) and Track 2 (T2). It feature graphs of different structure and with varying edge cost distributions[1].

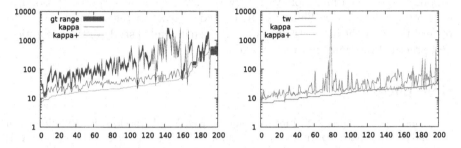

**Fig. 4.** Left: $gt, \kappa$ and $\kappa^+$ values on the T1 instances, sorted by $\kappa^+$. Right: Values for $tw, \kappa$ and $\kappa^+$ on the T2 instances, sorted by $tw$. Note the logscale of the y-axes.

The comparative results for $gt$, $\kappa$ and $\kappa^+$ are shown in Fig. 4, left, for the 200 instances of T1. We observe that our novel parameter $\kappa^+$ assumes the smallest value on almost all of them. There is a single instance for which $\kappa < \kappa^+$. Even the lower bound for $gt$ is in most instances significantly larger than the exact values of both $\kappa$ and $\kappa^+$. However, there is also one instance on which $gt$ is smaller than $\kappa^{(+)}$ by almost a factor of 10. Averaged over all instances, $\kappa$ is larger than $\kappa^+$ by a factor of 1.58, the lower bound of $gt$ is larger than $\kappa^+$ by a factor of 6.46 and the upper bound by factor of 9.78.

For T2, instances come with precomputed treewidth values. As discussed above, $\mathcal{O}(tw \log n)$ also bounds label sizes in PC-HL by propagation from the HHL subclass. The 200 instances in T2 were selected for the PACE challenge due to their small $tw$ values (on average 15.92). Thus, somewhat unsurprising, $tw$ values are on average smaller than those of $\kappa^+, \kappa$ and $gt$ on these instances. $\kappa^+$, as the second smallest, was on average a factor of 3.76 larger. However, there were also 48 instances with $\kappa^+ < tw$, see Fig. 4, right. Note that we omitted $gt$ values there for readability. In future work, a comparison of $\kappa^{(+)}$ and $tw$ on instances with larger treewidth would be interesting. But as $tw$ computation is NP-hard, one would need to resort to approximate values for bigger graphs.

Note that the only other parameter known to bound PC-HL label sizes which we did not consider here is the highway dimension. But there was already an

---

[1] https://github.com/PACE-challenge/SteinerTree-PACE-2018-instances.

extensive comparative study conducted on $h$ and $\kappa$ in [8] which revealed (in compliance with theoretical results) that $\kappa \ll h$.

# 7    Conclusions an Future Work

We have established that path-consistent HL is a proper superclass of the widely studied HHL and have shown novel parameterized upper bounds for label sizes in PC-HL and HL. One interesting open question is the maximum label size gap between an optimal HL and an optimal PC-HL. For HL and HHL the gap is known to be $\Omega(\sqrt{n})$ but it might be smaller for PC-HL. Furthermore, our newly proposed skeleton dimension variant was shown to have desirable properties in theory and practice. Thus, parameterized analysis of related problems with respect to $\kappa^+$ might be worthwhile as well as en extended study of the theoretical and practical relationship of $\kappa^+, \kappa$ and $tw$.

# References

1. Abraham, I., Delling, D., Fiat, A., Goldberg, A.V., Werneck, R.F.: VC-dimension and shortest path algorithms. In: Aceto, L., Henzinger, M., Sgall, J. (eds.) ICALP 2011. LNCS, vol. 6755, pp. 690–699. Springer, Heidelberg (2011). https://doi.org/10.1007/978-3-642-22006-7_58
2. Abraham, I., Delling, D., Fiat, A., Goldberg, A.V., Werneck, R.F.: Highway dimension and provably efficient shortest path algorithms. J. ACM (JACM) **63**(5), 1–26 (2016)
3. Abraham, I., Delling, D., Goldberg, A.V., Werneck, R.F.: Hierarchical hub labelings for shortest paths. In: Epstein, L., Ferragina, P. (eds.) ESA 2012. LNCS, vol. 7501, pp. 24–35. Springer, Heidelberg (2012). https://doi.org/10.1007/978-3-642-33090-2_4
4. Angelidakis, H., Makarychev, Y., Oparin, V.: Algorithmic and hardness results for the hub labeling problem. In: Proceedings of the Twenty-Eighth Annual ACM-SIAM Symposium on Discrete Algorithms, pp. 1442–1461. SIAM, Society for Industrial and Applied Mathematics (2017)
5. Babenko, M., Goldberg, A.V., Kaplan, H., Savchenko, R., Weller, M.: On the complexity of hub labeling (Extended Abstract). In: Italiano, G.F., Pighizzini, G., Sannella, D.T. (eds.) MFCS 2015. LNCS, vol. 9235, pp. 62–74. Springer, Heidelberg (2015). https://doi.org/10.1007/978-3-662-48054-0_6
6. Bauer, R., Columbus, T., Katz, B., Krug, M., Wagner, D.: Preprocessing speed-up techniques is hard. In: Calamoneri, T., Diaz, J. (eds.) CIAC 2010. LNCS, vol. 6078, pp. 359–370. Springer, Heidelberg (2010). https://doi.org/10.1007/978-3-642-13073-1_32
7. Bauer, R., Columbus, T., Rutter, I., Wagner, D.: Search-space size in contraction hierarchies. Theor. Comput. Sci. **645**, 112–127 (2016)
8. Blum, J., Storandt, S.: Computation and growth of road network dimensions. In: Wang, L., Zhu, D. (eds.) COCOON 2018. LNCS, vol. 10976, pp. 230–241. Springer, Cham (2018). https://doi.org/10.1007/978-3-319-94776-1_20
9. Cohen, E., Halperin, E., Kaplan, H., Zwick, U.: Reachability and distance queries via 2-hop labels. SIAM J. Comput. **32**(5), 1338–1355 (2003)

10. Delling, D., Goldberg, A.V., Pajor, T., Werneck, R.F.: Robust exact distance queries on massive networks. Microsoft Research, USA, Technical report, vol. 2 (2014)
11. Delling, D., Goldberg, A.V., Savchenko, R., Werneck, R.F.: Hub labels: theory and practice. In: Gudmundsson, J., Katajainen, J. (eds.) SEA 2014. LNCS, vol. 8504, pp. 259–270. Springer, Cham (2014). https://doi.org/10.1007/978-3-319-07959-2_22
12. Delling, D., Goldberg, A.V., Werneck, R.F.: Hub label compression. In: Bonifaci, V., Demetrescu, C., Marchetti-Spaccamela, A. (eds.) SEA 2013. LNCS, vol. 7933, pp. 18–29. Springer, Heidelberg (2013). https://doi.org/10.1007/978-3-642-38527-8_4
13. Hershberger, J., Suri, S.: Vickrey prices and shortest paths: what is an edge worth? In: Proceedings 42nd IEEE Symposium on Foundations of Computer Science, pp. 252–259. IEEE (2001)
14. Kosowski, A., Viennot, L.: Beyond highway dimension: small distance labels using tree skeletons. In: Proceedings of the Twenty-Eighth Annual ACM-SIAM Symposium on Discrete Algorithms, pp. 1462–1478. SIAM (2017)
15. Lakhotia, K., Dong, Q., Kannan, R., Prasanna, V.: Planting trees for scalable and efficient canonical hub labeling (2019). arXiv preprint arXiv:1907.00140
16. Manuel, P., Brešar, B., Klavžar, S.: The geodesic-transversal problem. Appl. Math. Comput. **413**, 126621 (2022)
17. Storandt, S.: Algorithms for landmark hub labeling. In: 33rd International Symposium on Algorithms and Computation (ISAAC 2022). Schloss Dagstuhl-Leibniz-Zentrum für Informatik (2022)

# The Hamiltonian Cycle Problem and Monotone Classes

Vadim Lozin$^{(\boxtimes)}$ [ID]

Mathematics Institute, University of Warwick, Coventry CV4 7AL, UK
V.Lozin@warwick.ac.uk

**Abstract.** We study the computational complexity of the Hamiltonian cycle problem on monotone classes of graphs, i.e. classes closed under taking subgraphs. We focus on classes defined by a single forbidden subgraph and present some necessary and some sufficient conditions for polynomial-time solvability of the problem in this case (assuming $P \neq NP$). The main result is a polynomial-time algorithm to solve the problem for graphs excluding a certain tree, called the long-$H$, as a subgraph.

**Keywords:** Hamiltonian cycle · Polynomial algorithm · Monotone class

## 1 Introduction

A Hamiltonian cycle in a graph is a cycle containing all vertices of the graph. Determining whether a graph has a Hamiltonian cycle is one of the first problems that has been shown to be NP-complete. Moreover, the problem remains NP-complete under substantial restrictions, for instance, for planar graphs, graphs of degree at most 3 or $K_{1,5}$-free split graphs [6]. On the other hand, in some classes it admits a polynomial-time solution, for instance, for graphs of bounded tree-[1] or clique-width [3]. From these examples we conclude that there is a simple criterion for polynomial-time solvability of the problem in the family of minor-closed classes of graphs: unless $P = NP$, the problem admits a polynomial-time solution in a minor-closed class $X$ if and only if $X$ excludes (does not contain) at least one planar graph. This conclusion follows from the above discussion and the result of Robertson and Seymour stating that graphs excluding a planar graph as a minor have bounded tree-width [7].

When we extend the family of minor-closed classes to the more general family of monotone classes, i.e. classes closed under taking (not necessarily induced) subgraphs, the situation becomes substantially more complex in the sense that within this family the problem admits polynomial-time solutions beyond classes of bounded tree-width. In particular, no criterion for polynomial-time solvability of the problem on monotone classes is available even in the case of a single excluded subgraph. Towards clarifying this situation, in the present paper we describe some necessary and some sufficient conditions for polynomial-time

© The Author(s), under exclusive license to Springer Nature Switzerland AG 2024
A. A. Rescigno and U. Vaccaro (Eds.): IWOCA 2024, LNCS 14764, pp. 460–471, 2024.
https://doi.org/10.1007/978-3-031-63021-7_35

solvability of the problem on monotone classes defined by a single forbidden subgraph. The main result is a polynomial-time algorithm to solve the problem for graphs excluding a certain tree, called the long-$H$, as a subgraph. We describe this solution in Sect. 3. All preliminaries and some preparatory results can be found in Sect. 2. Section 4 concludes the paper with a number of open problems. In the rest of the present section we introduce basic terminology and notation used in the paper.

We consider only simple graphs, i.e. undirected graphs without loops and multiple edges. The vertex set and the edge set of a graph $G$ are denoted $V(G)$ and $E(G)$, respectively. The *neighbourhood* of a vertex $v \in V(G)$ is the set of vertices adjacent to $v$, and the *degree* of $v$ is the number of its neighbours. A graph is *sub-cubic* if all its vertices have degree at most 3. Vertices of degree exactly 3 will be called *cubic*. We say that

- a graph $H$ is a *subgraph* of a graph $G$ if $H$ can be obtained from $G$ by deleting some (possibly none) vertices and edges,
- $H$ is an *induced subgraph* of $G$ if $H$ can be obtained from $G$ by deleting some (possibly none) vertices,
- $H$ is an *spanning subgraph* of $G$ if $H$ can be obtained from $G$ by deleting some (possibly none) edges.

A class of graphs is *monotone* if it is closed under taking subgraphs and is *hereditary* if it is closed under taking induced subgraphs. If a graph $G$ does not belong to a class $X$, then we say that $X$ *excludes* $G$, or that $G$ is forbidden for $X$.

Every hereditary class $X$ can be characterized by a set $M$ of minimal forbidden induced subgraphs, i.e. minimal (with respect to the induced subgraph relation) graphs that do not belong to $X$, in which case we say that graphs in $X$ are $M$-free. If $M$ is a finite set, then we call $X$ a *finitely-defined* class.

Similarly, every monotone class $X$ can be characterized by a set $M$ of minimal forbidden subgraphs, i.e. minimal (with respect to the subgraph relation) graphs that do not belong to $X$, in which case we say that graphs in $X$ are $M$-subgraph-free. Clearly, every monotone class is hereditary, and hence, a monotone class admits a characterization both in terms of minimal forbidden subgraphs and in terms of minimal forbidden induced subgraphs. The two sets of forbidden graphs for a monotone class may coincide or may be different, but what is important is that either both of them are finite or both are infinite, as the following observation shows.

**Observation 1.** *Let $X$ be a monotone class of graphs, let $M$ be the set of minimal forbidden subgraphs and $M^*$ be the set of minimal forbidden induced subgraphs for $X$. Then $M \subseteq M^*$, and every graph in $M^* - M$ contains at least one graph from $M$ as a spanning subgraph. In particular, $M$ is finite if and only if $M^*$ is finite.*

According to this observation, we can apply the term 'finitely-defined' to a monotone class without specifying which of the two forbidden sets we refer to.

## 2    Preliminaries and Preparatory Results

In this paper, we focus on the computational complexity of the Hamiltonian cycle problem on monotone classes of graphs defined by a single forbidden subgraph $F$. A helpful criterion for the NP-completeness of an algorithmic problem $\Pi$ in a finitely-defined hereditary class $X$ is the notion boundary classes: $\Pi$ is NP-complete in $X$ if and only if $X$ contains at least one boundary class for $\Pi$ [4].

One of the boundary classes for the Hamiltonian cycle problem was identified in [4]. It is the class $\Phi$ of graphs every connected component of which is a caterpillar with hairs of arbitrary length: these are sub-cubic trees in which all cubic vertices belong to the same path (see an example in Fig. 1) [4]. Therefore, the Hamiltonian cycle problem is polynomial-time solvable in a finitely-defined class $X$ *only if* $X$ excludes a graph from $\Phi$ (unless $P = NP$).

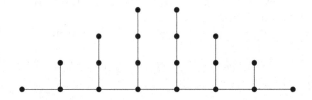

**Fig. 1.** A caterpillar with hairs of arbitrary length

Let $F$ be a graph from $\Phi$ in which every connected component contains at most one hair, i.e. at most one cubic vertex. If we forbid $F$ as a subgraph, then we obtain a class of graphs of bounded tree-width (see e.g. [5]) and hence the problem can be solved in polynomial time in such a class. Moreover, according to [5] this is the only condition allowing us to bound the tree-width by forbidding a graph from $\Phi$ as a subgraph. However, this is not the only condition allowing us to solve the problem in polynomial time.

To present more conditions, consider a connected graph $F$ in $\Phi$ with two *consecutive* hairs, i.e. with exactly two cubic vertices that are adjacent to each other. We will show that the problem can be solved for $F$-subgraph-free graphs in polynomial time. To this end, we introduce some terminology that will be used throughout the paper and prove a helpful result that will be needed at some point later.

Assume a graph $G$ has a Hamiltonian cycle $C$. We will call any edge of $G$ not in $C$ a *chord*. Any chord splits the edges of $C$ into two paths. The length of a shortest of these two paths will be called the *length of the chord*.

**Lemma 1.** *Let $G$ be a graph and $k$ a natural number. If $G$ contains a Hamiltonian cycle $C$ in which every chord has length at most $k$, then the tree-width of $G$ is at most $2k$.*

*Proof.* By deleting at most $k$ consecutive vertices of $C$ we transform $G$ into a graph $G'$ containing a Hamiltonian path, in which every chord (i.e. an edge not in the path) has length at most $k$. Clearly, the tree-width of $G'$ is at most $k$, and hence the tree-width of $G$ is at most $2k$.

**Corollary 1.** *Let $k$ be a natural number, and let $F$ be the graph obtained from two disjoint paths of length $2k$ by connecting their central vertices by an edge. The Hamiltonian cycle problem can be solved for $F$-subgraph-free graphs in polynomial time.*

*Proof.* Assume an $F$-subgraph-free graph $G$ has a Hamiltonian cycle $C$. Then every chord in $C$ has length at most $2k - 1$, since otherwise the forbidden subgraph arises. Therefore, by Lemma 1, the tree-width of $G$ is at most $4k - 2$. Deciding whether the tree-width of $G$ is at most $4k - 2$ (for a fixed $k$) can be done in linear time [2]. If the tree-width is bounded by $4k - 2$, then detecting a Hamiltonian cycle in $G$ can be done in polynomial time [1]. Otherwise, $G$ has no Hamiltonian cycle.

From this corollary we conclude that if $F$ is a connected graph from $\Phi$ with two consecutive hairs, then the problem can be solved in polynomial time for $F$-subgraph-free graphs. However, if $F$ contains three consecutive hairs, then the problem suddenly jumps to the NP-completeness. This is because $\Phi$ is not the *only* boundary class for the problem. One more boundary class was also identified in [4]: this is the class $\Phi^\Delta$ of graphs that can be obtained from graphs in $\Phi$ by replacing each cubic vertex with a triangle. More precisely, $\Phi^\Delta$ contains all graphs of this form and all their induced subgraphs, i.e. $\Phi^\Delta$ is hereditary. The operation of replacement of a cubic vertex with a triangle is illustrated in Fig. 2. Obviously, it transforms a graph containing a Hamiltonian cycle into a graph containing a Hamiltonian cycle, and vice versa.

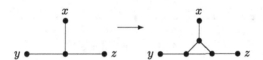

**Fig. 2.** Replacement of a cubic vertex with a triangle

When we forbid a graph $F$ as a subgraph, we exclude $F$ and all graphs containing $F$ as a spanning subgraph. If none of these excluded (forbidden) graphs belongs to $\Phi^\Delta$, then we conclude that the Hamiltonian cycle problem is NP-complete in the class of $F$-subgraph-free graphs. This observation allows us to derive the NP-completeness of the problem in $F$-subgraph-free graphs for infinitely many pairwise incomparable graphs $F$ in the class $\Phi$. In particular, this is the case for the two graphs in Fig. 3.

**Theorem 1.** *The Hamiltonian cycle problem is NP-complete in the class of $\Xi_1$-subgraph-free graphs and in the class of $\Xi_2$-subgraph-free graphs, where $\Xi_1$ and $\Xi_2$ are the graphs represented in Fig. 3.*

**Fig. 3.** The graphs $\Xi_1$ and $\Xi_2$ in Theorem 1

Theorem 1 substantially reduces the area of polynomial-time solvability of the problem in $F$-subgraph-free graphs. We believe (and conjecture) that with some more involved transformations this area can be reduced to connected graphs in $\Phi$ with finitely many hairs. However, complexity of the problem remains open even if we forbid a connected graph in $\Phi$ with only two hairs. As we have seen earlier, if the two hairs are consecutive, the problem admits a simple solution. However, if the two hairs are not consecutive, solving the problem becomes much more difficult. We solve one of these cases in the next section.

## 3    An Algorithm for Graphs with No Long-$H$-subgraphs

In graph theory literature, $H$ denotes the graph that looks like the capital letter H. In this section, we denote by long-$H$ the graph obtained from $H$ by subdividing each of its edges exactly once. This graph is represented in Fig. 4 as a caterpillar with two non-consecutive hairs of length 2. In what follows, we show that the Hamiltonian cycle problem can be solved for long-$H$-subgraph-free graphs in polynomial time.

**Fig. 4.** The graph long-$H$

Let $G$ be a long-$H$-subgraph-free graph. We call vertices of degree 2 in $G$ *white* and vertices of degree at least 3 *black*. The *black graph* is a subgraph of $G$ induced by black vertices and a *black component* is a connected component in the black graph.

To avoid trivialities, we assume without loss of generality that the input graph is sufficiently large, say $|V(G)| \geq 15$, since otherwise the problem can be solved by brute-force.

### 3.1    Rules and Reductions

We start by describing some helpful rules and a set of reductions simplifying the input graph, i.e. reductions transforming $G$ into a graph that has fewer edges and/or vertices and that has a Hamiltonian cycle if and only if $G$ has.

We emphasize that by deleting an edge or a vertex from a long-$H$-subgraph-free graph, we obtain a long-$H$-subgraph-free graph again. The first four rules and reductions are obvious:

(R1)  if the graph has vertices of degree 0 or 1, then stop: $G$ has no Hamiltonian cycle.

(R2)  if the graph is disconnected, then stop: $G$ has no Hamiltonian cycle.

(R3)  if the graph contains a vertex adjacent to more than two white vertices, then stop: $G$ has no Hamiltonian cycle.

(R4)  if the graph contains a vertex $v$ adjacent to exactly two white vertices, then delete the edges connecting $v$ to all other its neighbours (if there are any).

One more reduction is shown in Fig. 5. We will refer to this reduction as the *diamond reduction* and will denote it by (R5).

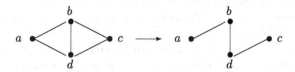

**Fig. 5.** The diamond reduction: it is applicable to a graph $G$ containing a subgraph shown on the left, in which vertices $b$ and $d$ have degree exactly 3 and vertices $a$ and $c$ have degree at least 3 in $G$.

**Lemma 2.** *Let $G'$ be a graph obtained from $G$ by the diamond reduction. Then $G'$ has a Hamiltonian cycle if and only if $G$ has.*

*Proof.* Clearly, if $G'$ has a Hamiltonian cycle, then the same set of edges form a Hamiltonian cycle in $G$.

Conversely, let $G$ have a Hamiltonian cycle $C$. Then the edge $bd$ necessarily belongs to $C$, which is easy to see. Therefore, either the pair of edges $ab$, $cd$ belongs to $C$ and the pair $ad$, $bc$ does not, or vice versa. Deleting either of these pairs transforms $G$ into a graph which also has a Hamiltonian cycle.

To define more helpful reductions, let us introduce some terminology. In a connected graph $G$, a *separating pair* is a pair of vertices $a, b$ such that $G - \{a, b\}$ is disconnected. If $a$ and $b$ are adjacent, then we call $ab$ a *separating edge*.

**Lemma 3.** *If $G$ has a separating edge $ab$, then $G$ has a Hamiltonian cycle if and only if $G - ab$ has.*

*Proof.* Clearly, if $G - ab$ has a Hamiltonian cycle, then the same edges form a Hamiltonian cycle in $G$.

Conversely, let $G$ have a Hamiltonian cycle $C$. Then the separating edge $ab$ does not belong to $C$, because the deletion of two consecutive vertices of $C$ cannot disconnect the graph. Therefore, $G - ab$ has a Hamiltonian cycle as well.

The deletion of a separating edge will be called the *separating edge reduction* and will be denoted (R6). Clearly, detecting separating edges and hence implementing (R6) can be done in polynomial time.

**Definition 1.** *A separating pair $a, b$ of non-adjacent vertices is called good if*

- *one of the components of $G - \{a, b\}$, say $X$, has tree-width at most 12 and is different from a path,*
- *the subgraph of $G$ induced by $X \cup \{a, b\}$ has a Hamiltonian path connecting $a$ to $b$.*

**Lemma 4.** *Let $G$ be a graph containing a good separating pair $a, b$, let $X$ be a connected component of $G - \{a, b\}$ of tree-width at most 12, and let $P_{ab}$ be a Hamiltonian path connecting $a$ to $b$ in the subgraph of $G$ induced by $X \cup \{a, b\}$. Also, let $G'$ be the graph obtained from $G$ by replacing the subgraph of $G$ induced by $X \cup \{a, b\}$ with $P_{ab}$. Then $G$ has a Hamiltonian cycle if and only if $G'$ has.*

*Proof.* Clearly, $G'$ is obtained from $G$ by deleting some edges. Therefore, if $G'$ has a Hamiltonian cycle, then the same edges form a Hamiltonian cycle in $G$.

Conversely, let $G$ have a Hamiltonian cycle $C$. Since $a$ and $b$ are not adjacent, they are not consecutive in $C$ and hence their removal splits the graph into two connected components, one of which is $X$. Denote the other component by $Y$. Then the edges of $P_{ab}$ together with the edges of $C$ connecting $a$ to $b$ through the vertices of $Y$ form a Hamiltonian cycle in $G'$.

It is known that detecting whether a graph has bounded tree-width and detecting whether a graph of bounded tree-width contains a Hamiltonian path can be done in polynomial time. Therefore, in polynomial time we can detect good pairs and implement the reduction of Lemma 4, which will be called the *good pair reduction* and denoted (R7).

In our algorithm to solve the problem we implement the above rules and reductions whenever they are applicable.

## 3.2    Long and Short Chords

To develop a solution for the problem, assume $G$ has a Hamiltonian cycle $C$. Any chord of length at most 6 will be called *short* and any chord of length at least 7 will be called *long*. We will reduce the problem to graphs without short chords. A key step in this direction is the fact that a long chord cannot cross a short chord. To define the notion of crossing chords, we represent the cycle $C$ as a circle on the plane and draw each chord *inside* the circle. If two chords have a non-empty intersection different from their endpoints, then we say that these chords *cross each other*.

In the proofs below, we distinguish a vertex $a_0$ on the cycle $C$ and denote the remaining vertices of the graph by $a_1, a_2, a_3, \ldots$ if we move from $a_0$ along the cycle clockwise, and by $a_{-1}, a_{-2}, a_{-3}, \ldots$ if we move from $a_0$ along the cycle anticlockwise. Also, if $a_k$ is a vertex on the cycle, then $a_{k+1}$ is the neighbour of $a_k$ in the clockwise direction and $a_{k-1}$ is the neighbour of $a_k$ in the anticlockwise direction.

**Lemma 5.** *A long chord cannot be crossed by a short chord.*

*Proof.* Let $a_0 a_k$ be a long chord and assume by contradiction that it crosses a short chord $a_i a_j$. Up to symmetry, we have to analyse the following cases.

*Case 0:* $i = -1$, $j = 1$. In this case, the edges $a_{-3} a_{-2}$, $a_{-2} a_{-1}$, $a_{-1} a_0$, $a_0 a_k$, $a_k a_{k+1}$, $a_{k+1} a_{k+2}$, $a_{-1} a_1$, $a_1 a_2$, $a_k a_{k-1}$, $a_{k-1} a_{k-2}$ form a forbidden subgraph. Observe that this contradiction arises for any $k \geq 5$, i.e. even if $a_0 a_k$ is not necessarily long. We distinguish this case separately as Case 0, because it will be used later in the proof.

*Case 1:* $i = -1$, $j = 2, 3, 4, 5$.

If $j = 2, 3, 4$ or $j = 5, k > 7$, then the edges $a_{-3} a_{-2}$, $a_{-2} a_{-1}$, $a_{-1} a_0$, $a_0 a_k$, $a_k a_{k+1}$, $a_{k+1} a_{k+2}$, $a_{-1} a_j$, $a_j a_{j-1}$, $a_k a_{k-1}$, $a_{k-1} a_{k-2}$ form a forbidden subgraph.

If $j = 5$ and $k = 7$, then the edges $a_3 a_4$, $a_4 a_5$, $a_5 a_6$, $a_6 a_7$, $a_7 a_8$, $a_8 a_9$, $a_5 a_{-1}$, $a_{-1} a_{-2}$, $a_7 a_0$, $a_0 a_1$ form a forbidden subgraph.

*Case 2:* $i = -2$, $j = 2, 3, 4$.

If $j = 3, 4$, then the edges $a_{-4} a_{-3}$, $a_{-3} a_{-2}$, $a_{-2} a_{-1}$, $a_{-1} a_0$, $a_0 a_1$, $a_1 a_2$, $a_{-2} a_j$, $a_j a_{j+1}$, $a_0 a_k$, $a_k a_{k+1}$ form a forbidden subgraph.

Now we analyse the case $j = 2$. According to (R4), at least one of $a_{-1}$ and $a_1$ is black, say $a_1$. Therefore, there must exist a chord $f = a_1 a_\ell$ incident to $a_1$. If $f$ is long, we are in Case 1 (with $f$ instead of $a_0 a_k$). Therefore, $f$ is short. If $f$ crosses $a_0 a_k$, then we are in Case 0 or Case 1. If $f$ does not cross $a_0 a_k$, then $\ell = 3, 4, 5, 6, 7$.

If $\ell = 3$, then the edges $a_4 a_3$, $a_3 a_1$, $a_1 a_0$, $a_0 a_k$, $a_k a_{k+1}$, $a_{k+1} a_{k+2}$, $a_1 a_2$, $a_2 a_{-2}$, $a_k a_{k-1}$, $a_{k-1} a_{k-2}$ form a forbidden subgraph.

If $\ell = 4, 5, 6$ or $\ell = 7, k > 7$, then the edges $a_{-4} a_{-3}$, $a_{-3} a_{-2}$, $a_{-2} a_{-1}$, $a_{-1} a_0$, $a_0 a_k$, $a_k a_{k+1}$, $a_{-2} a_2$, $a_2 a_3$, $a_0 a_1$, $a_1 a_\ell$ form a forbidden subgraph.

Finally, if $\ell = 7$ and $k = 7$, then the edges $a_5 a_6$, $a_6 a_7$, $a_7 a_1$, $a_1 a_2$, $a_2 a_3$, $a_3 a_4$, $a_7 a_8$, $a_8 a_9$, $a_2 a_{-2}$, $a_{-2} a_{-1}$ form a forbidden subgraph.

*Case 3:* $i = -3$, $j = 3$. According to (R4), at least one of $a_{-1}$ and $a_1$ is black, say $a_1$. Therefore, there must exist a chord $f = a_1 a_\ell$ incident to $a_1$, which is short and does not cross $a_0 a_k$ (as in case 2), i.e. $\ell = 3, 4, 5, 6, 7$.

If $\ell = 6, 7$, then the edges $a_{-1} a_0$, $a_0 a_1$, $a_1 a_2$, $a_2 a_3$, $a_3 a_4$, $a_4 a_5$, $a_1 a_\ell$, $a_\ell a_{\ell+1}$, $a_3 a_{-3}$, $a_{-3} a_{-2}$ form a forbidden subgraph.

If $\ell = 4$ or $\ell = 5, k > 7$, then the edges $a_{-2} a_{-1}$, $a_{-1} a_0$, $a_0 a_1$, $a_1 a_\ell$, $a_\ell a_{\ell+1}$, $a_{\ell+1} a_{\ell+2}$, $a_0 a_k$, $a_k a_{k+1}$, $a_\ell a_{\ell-1}$, $a_{\ell-1} a_{\ell-2}$ form a forbidden subgraph.

If $\ell = 5$ and $k = 7$, then the edges $a_3 a_4$, $a_4 a_5$, $a_5 a_6$, $a_6 a_7$, $a_7 a_8$, $a_8 a_9$, $a_5 a_1$, $a_1 a_2$, $a_7 a_0$, $a_0 a_{-1}$ form a forbidden subgraph.

Now we analyse the case $\ell = 3$. Since all other cases have been ruled out, vertex $a_1$ has degree exactly 3. If $a_2$ is white, then $a_1 a_3$ is a separating edge and hence it can be removed according to (R6). Otherwise, there exists a chord $a_2 a_m$ of length at most 4 (by Case 0) which does not cross $a_0 a_k$ (to avoid Case 2) with $m \neq 4$ (again by Case 0), i.e. $m = 0, 5, 6$.

If $m = 5, 6$, then the edges $a_{-2} a_{-1}$, $a_{-1} a_0$, $a_0 a_1$, $a_1 a_2$, $a_2 a_m$, $a_5 a_6$, $a_0 a_k$, $a_k a_{k+1}$, $a_2 a_3$, $a_3 a_{-3}$ form a forbidden subgraph.

We are left with the case when $m = 0$ and the degree of $v_2$ is exactly 3. But then the diamond reduction is applicable to the subgraph of $G$ with vertices $a_0, a_1, a_2, a_3$.

### 3.3  Reducing the Problem to Graphs Containing Hamiltonian Cycle without Short Chords

**Lemma 6.** *Assume the tree-width of $G$ is greater than 12 and (R1)–(R7) are not applicable to $G$. If $G$ has a Hamiltonian cycle $C$, then all chords of $C$ are long.*

*Proof.* Assume to the contrary that $C$ has a short chord $a_1 b_1$. It splits the edges of $C$ into a shorter path $L_{a_1 b_1}$ (the left part of $C$) and a longer path $R_{a_1 b_1}$ (the right part of $C$). Since the separating edge reduction is not applicable to $G$, there exists a chord $a_2 b_2$ crossing $a_1 b_1$. Then $a_2 b_2$ is short by Lemma 5. Without loss of generality, we assume that $a_2$ belongs to the left part of $C$ and that $b_2$ is closer to $b_1$ than to $a_1$. Now we redefine the two parts of $C$, this time with respect to the pair $a_1, b_2$, with $L_{a_1 b_2}$ being the left part and $R_{a_1 b_2}$ being the right part.

We observe that the left part has bounded tree-width, because all chords of $L_{a_1 b_2}$ are short. Indeed, if a chord $e$ in the left part $L_{a_1 b_2}$ is different from $a_1 b_2$, then either $e$ is shorter than $a_1 b_1$ or $a_2 b_2$, or $e$ crosses at least one of these short chords, in which case $e$ must be short itself by Lemma 5. Now assume $e = a_1 b_2$. Since $a_1 b_2$ is not a separating edge, it is crossed by a chord $f$, which must be short because it crosses one of the short chords of $L_{a_1 b_2}$. But then $e$ is also short and hence the left part of $C$ has bounded tree-width by Lemma 1.

The above discussion shows that the pair $a_1, b_2$ is not separating, since otherwise either (R6) (if $a_1$ is adjacent to $b_2$) or (R7) (if $a_1$ is not adjacent to $b_2$) is applicable to $G$. Therefore, there must exist a chord $a_3 b_3$ connecting a vertex $a_3$ in the left part to a vertex $b_3$ in the right part. The chord $a_3 b_3$ crosses at least one of the short chords $a_1 b_1$ and $a_2 b_2$ and hence is short. Again, without loss of generality, we assume that $b_3$ is closer to $b_2$ than to $a_1$, i.e. the two parts of $C$ are now defined by the pair $a_1, b_3$.

The same arguments as before show that all chord in the left part are short and hence it has bounded tree-width, and that the pair $a_1, b_3$ is not separating. Therefore, there exists a (short) chord connecting the left part and the right part of $C$, and this process continues until the right part becomes empty, in which case the left part contains the entire cycle $C$. At each step of the process all chords in the left part are short and hence the tree-width of $G$ is at most 12 by Lemma 1. This contradicts the assumption of the lemma and shows that all chords of $C$ are long.

### 3.4  Detecting Hamiltonian Cycles Without Short Chords

In this section, we assume that $G$ has a Hamiltonian cycle $C$ in which all chords are long. To develop a solution in this case, we will show that black components have simple structure. To this end, consider a black vertex $a_0$ and a chord $a_0 a_k$ incident to it.

(a) If vertex $a_2$ is black, then $a_2a_k$ is the only chord incident to $a_2$. To show this, assume by contradiction that $a_2$ is incident to a chord $a_2a_i$ with $i \neq k$. If $a_0a_k$ and $a_2a_i$ do not cross each other, then the edges $a_{-2}a_{-1}$, $a_{-1}a_0$, $a_0a_1$, $a_1a_2$, $a_2a_3$, $a_3a_4$, $a_0a_k$, $a_ka_{k+1}$, $a_2a_i$, $a_ia_{i-1}$ form a forbidden subgraph. The case of crossing chords $a_0a_k$ and $a_2a_i$ is similar.

By rule (R4), $a_0$ and $a_k$ have black neighbours on $C$. We analyse the case when $a_1$ and $a_{k+1}$ are black. The other cases can be analysed by analogy.

(b) $a_1$ cannot be incident to a chord $a_1a_i$ with $i \notin \{k-2, k-1, k, k+1, k+2\}$, since otherwise the edges $a_{k-2}a_{k-1}$, $a_{k-1}a_k$, $a_ka_0$, $a_0a_1$, $a_1a_2$, $a_2a_3$, $a_ka_{k+1}$, $a_{k+1}a_{k+2}$, $a_1a_i$, $a_ia_{i-1}$ (or $a_ia_{i+1}$) form a forbidden subgraph.
(c) $a_1$ cannot be incident to a chord $a_1a_i$ with $i = k-2$ or $i = k+2$ by (a).
(d) $a_1$ is incident to at most three chords $a_1a_{k-1}$, $a_1a_k$, $a_1a_{k+1}$, which follows from (b) and (c). By symmetry, $a_{k+1}$ is incident to at most three chords $a_{k+1}a_{-1}$, $a_{k+1}a_0$, $a_{k+1}a_1$.
(e) Assume $a_{k+2}$ is black. Then $a_{k+2}a_0$ is the only chord incident to $a_{k+2}$ (by (a)), $a_ka_0$ is the only chord incident to $a_k$ (for the same reason) and $a_1a_{k+1}$ is the only chord incident to $a_1$ (by (d) applied to chords $a_0a_k$, $a_0a_{k+2}$). This implies that $a_{k-1}$, $a_{k+3}$, $a_2$ are white (by (a) and the previous arguments). If $a_{-1}$ is black, then $a_{-1}a_{k+1}$ is the only chord incident to $a_{-1}$ (by (a)) and $a_{-2}$ is white (by symmetry with $a_2$). Therefore, if $a_{k+2}$ is black, then we have a black component which consists of at most 6 vertices. Observe that this component necessarily has a $C_4$ (a cycle of length 4) with two edges on the cycle $C$ and two chords.
(f) Assume $a_{k+2}$ is white, and by symmetry $a_2$ is white. We can also assume that $a_{k-2}$ is white, since otherwise $a_{k-2}a_0$ is the only chord incident to $a_{k-2}$ (by (a)) and $a_1a_{k-1}$ is the only chord incident to $a_1$ (by (d) applied to chords $a_0a_k$, $a_0a_{k-2}$), in which case $a_{k+1}$ cannot be black by (a). By symmetry $a_{-2}$ is white. Then we are in conditions similar to case (e), where a black component has at most 6 vertices and a cycle $C_4$ with two edges on the cycle $C$ and two chords. The cycle $C_4$ arises if $a_1$ is adjacent to $a_{k-1}$ or $a_{k+1}$, or if $a_1$ is adjacent to $a_k$ and (by symmetry) $a_{k+1}$ is adjacent to $a_0$.

From the above discussion, we conclude that in the presence of a Hamiltonian cycle every black component has at most 6 vertices and exactly four white neighbours. Moreover, every black component together with its white neighbours contains a spanning subgraph represented in Fig. 6.

**Fig. 6.** A spanning subgraph in a black component with 4 white neighbours. Dotted edges represent paths (possibly of length 1).

**Lemma 7.** *Assume the tree-width of $G$ is greater than 12 and (R1) – (R7) are not applicable to $G$. Then $G$ has a Hamiltonian cycle if and only if every black component of $G$ together with its white neighbours contains a spanning subgraph represented in Fig. 6.*

*Proof.* By Lemma 6, if $G$ has a Hamiltonian cycle, then all chords in this cycle are long, and therefore, the statement of the lemma follows from claims (a), (b), (c), (d), (e), (f).

Now assume that every black component of $G$ together with its white neighbours contains a spanning subgraph represented in Fig. 6. In each black component we choose (arbitrarily) such a spanning subgraph and delete all the other edges from the component. This transforms $G$ into a graph $G'$ which consists of a collection of cycles of length 4 (squares) connected to each other by white paths. Clearly, $G'$ is connected, since $G$ is connected.

In order to show that $G'$ contains a Hamiltonian cycle, we will show that in each square there is pair of edges (forming a matching) such that the deletion of this pair does not disconnect the graph. If there is only one such square, then the statement is easy to see.

If the number of squares is larger than one, then take one of them and assume by contradiction that the deletion of each of the two pairs of edges in this square disconnects the graph. Then the deletion of all four edges of the square splits the graph into four disjoint subgraphs with no edges between them. In each of these subgraphs the number of vertices of odd degree is odd. Indeed, in $G'$ each vertex of odd degree (of degree 3) appears in a square. If we delete all edges in one of the squares, then each vertex of this square becomes a leaf (a vertex of degree 1). The four leaves obtained in this way belong to four different subgraphs, according to our assumption. This is not possible by the Handshake Lemma, and therefore, we can always delete a pair of edges from each square without disconnecting the graph. In the resulting graph all vertices have degree 2 and hence it is Hamiltonian.

### 3.5    Solution

We summarize the above discussion in the following algorithm to solve the problem.

1. Apply rules and reductions (R1)–(R7) as long as they are applicable.
2. If the algorithm did not stop at Step 1 and the graph has tree-width at most 12, solve the problem by an algorithm for graphs of bounded tree-width and stop.
3. If the algorithm did not stop at Step 2, check the black components of the graph. If at least one of them has more than 6 vertices or does not contain a spanning subgraph represented in Fig. 6 (together with its white neighbours), then stop: the graph has no Hamiltonian cycle.
4. If the algorithm did not stop at Step 3, then find a Hamiltonian cycle according to Lemma 7.

It is not difficult to see that all steps of the algorithm can be implemented in polynomial time. The correctness of the algorithm follows from the proofs of the lemmas and claims in the preceding sections.

**Theorem 2.** *The Hamiltonian cycle problem can be solved in the class of long-H-subgraph-free graphs in polynomial time.*

## 4 Concluding Remarks and Open Problems

This paper leaves more open questions than it answers. One natural question to ask is whether the main result presented in the paper can be extended to a forbidden subgraph with longer hairs. More specifically, if we subdivide each of the pendant edges in the long-$H$ exactly once, can we apply similar arguments to solve the problem? Unfortunately, no. Indeed, a key step in the solution for the long-$H$-subgraph-free graphs is Lemma 5 stating that long and short chords do not cross each other. This argument does not work if we forbid a long-$H$ with subdivided hairs regardless of the threshold $k$ separating long and short chords. To see this, consider a long chord $a_0 a_k$ and a short chord $a_{-1} a_{k-3}$. They cross each other and do not create any long-$H$ with subdivided hairs as a subgraph.

One more open question is the complexity of the problem in case of a disconnected forbidden subgraph $F$. If each connected component of $F$ contains at most one hair, then the tree-width of any $F$-subgraph-free graph is bounded and hence the problem can be solved in polynomial time. But what if each component of $F$ contains two hairs? If the hairs are consecutive, then we claim that the problem can be solved in polynomial time by extending the solution for a connected graph $F$ presented in Corollary 1. However, the question remains open in case of non-consecutive hairs. In particular, the complexity of the problem remains open if $F$ consists of two disjoint copies of the long-$H$.

Finally, we believe that the NP-completeness results can be substantially strengthened. In particular, we conjecture that there is a finite value of $k$ such that the problem is NP-complete for $F$-subgraph-free graphs, where $F$ is a connected graph in $\Phi$ with at least two hairs of distance at least $k$ from each other.

## References

1. Arnborg, S., Proskurowski, A.: Linear time algorithms for NP-hard problems restricted to partial $k$-trees. Discret. Appl. Math. **23**, 11–24 (1989)
2. Bodlaender, H.L.: A linear time algorithm for finding tree-decompositions of small treewidth. SIAM J. Comput. **25**, 1305–1317 (1996)
3. Borie, R.B., Parker, R.G., Tovey, C.A.: Solving problems on recursively constructed graphs. ACM Comput. Surv. **41**, 4 (2008)
4. Korpelainen, N., Lozin, V., Malyshev, D., Tiskin, A.: Boundary properties of graphs for algorithmic graph problems. Theor. Comput. Sci. **412**(29), 3545–3554 (2011)
5. Lozin, V., Razgon, I.: Tree-width dichotomy. Eur. J. Comb. **103**, 103517 (2022)
6. Renjith, P., Sadagopan, N.: Hamiltonian cycle in $K_{1,r}$-free split graphs - a dichotomy. Int. J. Found. Comput. Sci. **33**(1), 1–32 (2022)
7. Robertson, N., Seymour, P.D.: Graph minors. V. Excluding a planar graph. J. Comb. Theory Ser. B **41** 92–114 (1986)

# The Minimum Algorithm Size of $k$-Grouping by Silent Oblivious Robots

Paola Flocchini[1] , Debasish Pattanayak[2(✉)] , Nicola Santoro[2] ,
and Masafumi Yamashita[3]

[1] University of Ottawa, Ottawa, Canada
[2] Carleton University, Ottawa, Canada
drdebmath@gmail.com
[3] Kyushu University, Fukuoka, Japan

**Abstract.** Consider a set of mobile computational elements, called robots, that are viewed as points, and operate in the Euclidean plane in synchronous rounds. The robots are oblivious (they forget all computations performed in previous rounds), silent (unable of direct communication), and anonymous (indistinguishable from the outside). Each robot is provided with a private coordinate system, and can determine the position of the other robots (performing a Look operation); it has an algorithm, which it executes (performing a Compute operation) to determine a destination point; and it can move towards the destination (performing a Move operation).

The $k$-Grouping problem requires the robots, starting from an arbitrary initial configuration in the plane, to gather at $k$ distinct locations, not chosen in advance, by performing Look-Compute-Move cycles, and no longer move. This simple problem is however unsolvable if all the robots execute the same algorithm. It has been recently shown that, were different subgroups of the robots to execute different algorithms, the problem remains still unsolvable if the number of the algorithms is less than $k$.

In this paper we prove that this number is minimum: we design $k$ distinct algorithms and prove that, if each is executed by an arbitrary non-empty subset of the robots, they will collectively be able to solve the problem; furthermore, they are able to do so under a weak assumption on the level of agreement among the local coordinate systems. We further prove that, without any agreement, the problem becomes unsolvable even with $k + 1$ different algorithms. However, if an unbounded number of algorithms are allowed such that each robot has a unique algorithm, then we can solve $k$-Grouping without any agreement.

**Keywords:** Autonomous mobile robot · Minimum algorithm size · Scattering · Gathering · Pattern formation

© The Author(s), under exclusive license to Springer Nature Switzerland AG 2024
A. A. Rescigno and U. Vaccaro (Eds.): IWOCA 2024, LNCS 14764, pp. 472–484, 2024.
https://doi.org/10.1007/978-3-031-63021-7_36

# 1   Introduction

## 1.1   Framework, Background, and Motivation

Consider a collection $\mathcal{R}$ of mobile computational entities, called *robots*, viewed as points, operating in synchronous rounds in the Euclidean plane. Time is divided into discrete intervals, called *rounds*; in each round a non-empty subset of robots is activated. All activated robots perform a *Look-Compute-Move* (LCM) cycle, consisting of three phases, in perfect synchrony: in the *Look* phase, each robot takes a snapshot of the environment, returning the positions of the other robots in its own local coordinate system; in the *Compute* phase, it executes its preprogrammed algorithm to compute a destination point; in the *Move* phase, it moves towards the destination. The decision of which robots are activated in a given round is under the control of an adversarial *scheduler*. Under the weakest adversarial scheduler, called *fully-synchronous* ($\mathcal{FSYNC}$), the entire set of robots is activated in every round. Under the strong adversary, called *semi-synchronous* ($\mathcal{SSYNC}$), the activated subset is arbitrary (provided that every robot is activated infinitely often).

Since its introduction in the distributed computing community [16], this setting has been extensively investigated, and the primary direction of research has been on determining the minimal assumptions on the robots' capabilities under which they are collectively capable to solve a given problem or to successfully perform a given task.

In the weakest (de-facto standard) model, $\mathcal{OBLOT}$, the robots are *oblivious* (i.e., have no memory of previous cycles), *silent* (i.e., have no means of direct communication), and *disoriented* (i.e., do not necessarily agree on a common coordinate system). They are also generally assumed to be *anonymous* (i.e. do not have distinct identifiers or visible markers) and *homogeneous* (i.e., execute the same algorithm – compute the same function).

Within this model, the research has been focused on the computability and complexity aspects of basic fundamental tasks, in particular the class of **Pattern Formation** problems, requiring the robots to move from the configuration of their initial (arbitrary) position in the plane to one congruent with a geometric shape (the pattern) given in input (e.g., see [1,4–6,9–11,16,17], as well as [13] and chapters therein). An extensive amount of research has been dedicated on identifying the impact that specific factors have on the solvability of these basic problems. These are, for instance, the studies on the impact of capabilities such as multiplicity detection (i.e. ability in the *Look* phase to determine the number of robots at the same point at the same time), some agreement on the coordinate system (e.g., compass, chirality, ...), rigidity (e.g., ability in the *Move* phase to always reach the computed destination), etc.

Because of the stringent limitations imposed by the model on the computational power of the robots, several problems remain unsolvable even in presence of some of these additional capabilities. For example, it is impossible to form an asymmetric pattern starting from a symmetric initial configuration, even if the robots are rigid, have chirality and multiplicity detection, and the scheduler is fully synchronous [16].

Consider the case where there is no single algorithm $A$ that would allow the robots to solve a given problem $\mathcal{P}$ under a given set of conditions. It might be however possible to solve the problem by designing two (or more) distinct algorithms, say $A_1$ and $A_2$, programming some robots $\mathcal{R}_1 \subset \mathcal{R}$ with $A_1$ while the others $\mathcal{R}_2 = \mathcal{R} \setminus \mathcal{R}_1$ with $A_2$, and having each robot execute its algorithm unaware of the composition of the two sets.

This observation, recently made in [2], leads to the following interesting research question:

*Given a problem $\mathcal{P}$ in a given setting, what is the minimum number of distinct algorithms that would allow a set $\mathcal{R}$ of robots to solve it?*

We shall denote such a measure, called the *minimum algorithm size* in [2], by $\Phi(\mathcal{P})$, and investigate it for a basic family of Pattern Formation problem.

## 1.2  Problem and Contributions

The class of problems we are investigating, called $k$-Grouping, requires the robots to gather within finite time, at $k$ distinct points, not chosen in advance, and no longer move; this has to be done regardless of the number of robots (provided $k \leq |\mathcal{R}|$) and of their initial configuration $C(0)$ (i.e., location of the robots at time $t = 0$).

In other words, for every $\mathcal{R}$ with $|\mathcal{R}| \geq k$, any solution algorithm must satisfy the following temporal geometric predicate in every possible execution:

$$k\text{-Grouping} \equiv \{\exists \hat{t} \geq 0 \ \{(|C(\hat{t})| = k) \textbf{ and } (\forall t' \geq \hat{t}, \ (C(t') = C(\hat{t})))\}\}$$

where $|C(t)|$ denotes the number of distinct points occupied by the robots at time $t$.

Observe that the special case 1-Grouping is precisely one of the most important and investigated problem in the field, called Gathering (or Rendezvous) (e.g., see [1,3,8,12,14,15]). Another special case of $k$-Grouping is the subclass of the Scattering family of problems, called $k$-Scattering, which assumes that, in the initial configuration $C(0)$, all robots are co-located in the same point (e.g., see [2,7]).

An interesting aspect of $k$-Grouping is that, in spite of its apparent simplicity, it is generally *unsolvable* in $\mathcal{OBLOT}$ (i.e., with a single algorithm). This impossibility is very easy to verify; e.g., consider forming two groups starting from an initial configuration $C(0)$ where all robots are co-located in the same point. Furthermore, this impossibility holds true even if the robots are very powerful and the adversary very weak; in particular, it holds also when the robots share a common orientation and direction of the axes (i.e., they have a compass), have strong multiplicity detection, movements are rigid, and the scheduler is fully synchronous $\mathcal{FSYNC}$.

Hence the need to determine $\Phi(k\text{-Grouping})$, i.e., the *minimum algorithm size* of $k$-Grouping. A recent result of [2] implies that

$$\Phi(k\text{-Grouping}) \geq k$$

even if the robots are very powerful and the adversary is very weak (see above).

The following questions thus naturally arise: is this bound tight? can it be achieved under less stringent conditions (i.e., with weaker robots and under a stronger adversary)?

In this paper, we first of all provide an affirmative answer to both questions. We prove that the lower-bound is tight, and indeed $k$ distinct algorithms are sufficient; the proof is constructive. This result holds even if the robots agree only on the direction of a single axis, have no multiplicity detection, movements are non-rigid (i.e., they can be stopped in their movement) and the adversarial scheduler is the stronger $\mathcal{SSYNC}$.

We then investigate whether such a result can be achieved in an even weaker setting: specifically, without any agreement among the local coordinate systems. We prove that the answer is negative: in this case, $k$-Grouping is unsolvable even if the robots had $k + 1$ algorithms, and the other conditions were the best possible (strong multiplicity detection, rigid movements, and fully synchronous scheduler).

Finally we show that the problem is always solvable, even without any agreement among the local coordinate systems, if the number of algorithms is unbounded in number; indeed it suffices with a distinct algorithm for each robot.

See Table 1 for a summary of the results.

**Table 1.** Summary of results

| $k$-Grouping | Algorithms | Axis | Rigidity | Scheduler | Reference |
|---|---|---|---|---|---|
| unsolvable | $< k$ | 2 | Yes | $\mathcal{FSYNC}$ | [2] |
| solvable | $k$ | 1 | No | $\mathcal{SSYNC}$ | this paper |
| unsolvable | $k + 1$ | No | Yes | $\mathcal{FSYNC}$ | this paper |
| solvable | $|\mathcal{R}|$ | No | No | $\mathcal{SSYNC}$ | this paper |

## 1.3 Organization

The paper is organized as follows. In Sect. 2, we describe the model under consideration. In Sect. 3.1, we show that the bound on the number of algorithms is tight with partial agreement on axes, and in Sect. 3.2 show impossibility without agreement even with a collective algorithm of size $k + 1$. Further, in Sect. 3.2, we show a possibility if the algorithm size is unbounded before concluding in Sect. 4. Due to space limitations, some proofs and details are omitted.

# 2    Model and Terminology

The system consists of a set $\mathcal{R} = \{r_1, r_2, \ldots, r_n\}$ of mobile computational entities, called robots, modeled as geometric points, that operate in $\mathbb{R}^2$. The robots are autonomous without a central control. Each robot has its own local coordinate system, and is equipped with devices that allow it to observe the positions of the other robots in its local coordinate system.

The robots operate in *Look-Compute-Move* (LCM) cycles. When activated, a robot executes a cycle by performing the following three operations:

1. *Look:* The robot obtains a snapshot of the positions occupied by robots expressed with respect to its own coordinate system; this operation is assumed to be instantaneous.
2. *Compute:* The robot executes the algorithm using the snapshot as input; the result of the computation is a destination point.
3. *Move:* The robot moves towards the computed destination. If the destination is the current location, the robot stays still.

The robots are *silent*: they have no explicit means of communication; furthermore they are *oblivious*: at the start of a cycle, a robot has no memory of observations and computations performed in previous cycles.

The system is *synchronous*; that is, time is divided into discrete intervals, called *rounds*. In each round a robot is either active or inactive. The robots active in a round perform their LCM cycle in perfect synchronization; if not active, the robot is idle in that round. All robots are initially idle.

The decision of which robots are activated in a given round is under the control of an adversarial *scheduler*. The weakest adversarial scheduler, called *fully-synchronous* ($\mathcal{FSYNC}$), activates the entire set of robots in every round. Under the strong adversary, called *semi-synchronous* ($\mathcal{SSYNC}$), the activated subset is arbitrary; however, every robot is activated infinitely often. In the following, we use round and time interchangeably.

Let $r_i(t)$ denote the location of robot $r_i$ at time $t$ in some global coordinate system (possibly unknown to the robots); let $C(t) = \{p_1, p_2, \ldots, p_n\}$, called *configuration* at time $t$, be the multi-set of robot positions $p_i = r_i(t)$, and let $Q = \{q_1, q_2, \ldots, q_m\}$ be the corresponding set of points occupied by the robots. Observe that $|Q| = m \leq n$ since some robots might be at the same locations, called multiplicity points. The robots are said to have strong/weak/no *multiplicity detection* capability if they are able to detect in a given point the exact number of robots, only whether or not there is more than one robot, or have no such a capability, respectively

Movements are said to be *rigid* if the robots always reach their destination. They are said to be *non-rigid* if they may be unpredictably stopped by an adversary whose only limitation is the existence of $\delta > 0$, unknown to the robots, such that if the destination is at distance at most $\delta$ the robot will reach it, else it will move at least $\delta$ towards the destination.

As for their private coordinate systems, the robots are said: to have *complete agreement* if they all share the same orientation and direction of both axes; to have *one-axis agreement* if they all share the orientation and direction of just one axes (say the $Y$ axis); to have *no agreement* (or to be disoriented) if they might disagree on the orientation and direction of the axes. In all cases, there might to be agreement on the unit of distance.

As discussed in Sect. 1.2, the class of $k$-Grouping problems requires the robots to gather within finite time, at $k$ distinct points, not chosen in advance, and no longer move; this has to be done regardless of the number of robots (provided $k \leq |\mathcal{R}|$) and of their initial configuration $C(0)$ (i.e., location of the robots at time $t = 0$).

A *collective solution algorithm* for $k$-Grouping of *size* $s \geq k$ is any set $\mathcal{A} = \{A_1, ..., A_s\}$ of $s \geq k$ distinct algorithms such that: for any set of robots $\mathcal{R}$ with $|\mathcal{R}| \geq k$ and any partition $P(\mathcal{R}) = < R_1, ..., R_s >$ of $\mathcal{R}$ in $s$ non-empty subsets, if each element of $R_i$ is programmed with and executes $A_i$, then, without knowing which algorithm is being executed by the other robots and starting from any initial configuration, within finite time all robots gather at $k$ distinct points, and no longer change position.

The smallest value of $s$ among all collective solution algorithms is called the *minimum algorithm size* and denoted by $\Phi(k$-Grouping$)$.

# 3   $k$-Grouping

In this section, first, we determine $\Phi(k$-Grouping$)$ constructively by providing a set of $k$ algorithms that solve $k$-Grouping when the robots have agreement on the orientation of one axis. Then we investigate the solvability of $k$-Grouping in absence of any agreement on the coordinate system; we prove that, without any agreement on the coordinate system, no set of $k + 1$ algorithms can solve $k$-Grouping. On the other hand, the problem becomes solvable if every robot has a different algorithm.

## 3.1   Partial Agreement

First, we consider robots with agreement in the orientation of one-axis. They may not agree on the chirality. In other words, without loss of generality all robots agree on the direction of positive $y$-axis, but they may disagree on the direction of positive $x$-axis. The algorithm presented here is a general strategy that works for any $k \geq 2$. For $k = 1$, we refer the readers to the paper by Bhagat et al. [3], where they present an algorithm for 1-grouping (also known as Gathering) of robots with one-axis agreement.

### General Strategy

At a high level, the proposed collective algorithm $\mathcal{A} = \{A_1, A_2, \ldots, A_k\}$, for $k > 1$, has each robot determine in its Look phase the *smallest enclosing rectangle* (SER) of the observed configuration, where the sides of the rectangle are parallel to the axes, and use it as an input in the execution of its algorithm.

Informally, the general strategy of the collective algorithm $\mathcal{A}$ consists on having all the robots that execute the same algorithm $A_i$ to eventually move to the same line $L_i$ perpendicular to the $y$ axis, and eventually gather at a single point $M_i$ on $L_i$. The location of the destination lines is totally dynamic and may change several times during the executions of the robots' algorithms. Eventually, as we will show, the top line $L_1$ and the bottom line $L_k$ will no longer change, so neither will the height of the SER. Once this occurs (let us stress that the robots might be unaware of this fact), also the location of line $L_i$ ($1 < i < k$) is determined as being at distance $d_i = (i - 1)h/(k - 1)$ from $L_1$, where $h$ is the height of the SER, and no longer change. Then the robots executing algorithms $A_i$ eventually form a point $M_i$ on $L_i$ at distance $d_i$ from $M_1$.

The algorithm terminates once the observed configuration is composed of $k$ equidistant points on a vertical axis.

In what follows, we describe the algorithm in detail; the pseudocode is presented in Algorithm 1.

**Basic Concepts and Terminology**

Before starting the description, let us introduce some important concepts and terminology. In the following (including the algorithm description) all the geometric objects are expressed in terms of the local coordinate system of the observing robot. Let us stress that, since we are considering the weak setting where the local coordinate systems might disagree on the direction of the $x$-axis, the notion of "left" and "right" of this robot is possibly opposite that of other robots.

Consider the configuration $C(t)$ observed by a robot active at time $t$; let $Q = \{q_1, ..., q_m\}$ be the distinct positions occupied by the robots in that configuration, where $q_j = (x_j, y_j)$ ($1 \leq j \leq m$).

Given $Q$, let $l_x = \min_j(x_j)$, $r_x = \max_j(x_j)$, $b_y = \min_j(y_j)$, and $t_y = \max_j(y_j)$, where $1 \leq j \leq m \leq n$. Then the smallest enclosing rectangle $SER(Q)$ is uniquely determined by the four corners $(l_x, b_y)$, $(l_x, t_y)$, $(r_x, t_y)$, and $(r_x, b_y)$ (ref. Fig. 1).

The rectangle $SER(Q)$ is said to be *improper* if it is a horizontal line or a point (i.e., if $t_y = b_y$), *proper* otherwise. Let $h$ be the height of a proper $SER(Q)$, i.e., $h = t_y - b_y$.

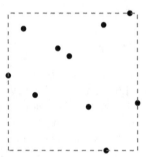

**Fig. 1.** Smallest Enclosing Rectangle of a configuration $C(t)$

In a proper $SER(Q)$, the parallel lines $L_i$ ($1 \leq i \leq k$) are defined as follows: $L_1$ is the line passing through $(l_x, t_y)$ and $(r_x, t_y)$; $L_k$ is the line passing through $(l_x, b_y)$ and $(r_x, b_y)$; $L_j$ ($1 < j < k$) is the line passing through $(l_x, l_j)$ and $(r_x, l_j)$, where $l_j = ((j - 1)b_y + (k - j)t_y)/(k - 1)$. Notice that the distance between two successive lines is $h/(k-1)$. Let $M_1$ be the midpoint of the segment of $L_1$ between $(l_x, t_y)$ and $(r_x, t_y)$.

The *final configuration* $C_f$ is the configuration where all the robots are located on a line parallel to the $y$-axis, $|Q| = k$, and the distance between consecutive points in $Q$ is $h/(k - 1)$.

---

**Algorithm 1: $k$-grouping**

---

**Input:** $k$, set of points $Q$, robot position $(p_x, p_y)$
**Output:** Destination of the robot
1  **Algorithm $A_1$:**
2    **if** $Q = C_f$ **then**
3      | destination $\leftarrow (p_x, p_y)$
4      | terminate
5    **else**
6      | compute $SER(Q)$
7      | **if** $t_y = b_y$ **then**
8        | destination $\leftarrow (p_x, p_y + 1)$
9      | **else**
10       | **if** *if there is only one robot position* $(q_x, q_y)$ *on* $L_1$ **then**
11         | destination $\leftarrow (q_x, q_y)$
12       | **else**
13         | **if** $p_y = t_y$ **then**
14           | destination $\leftarrow ((l_x + r_x)/2, p_y)$
15         | **else**
16           | destination $\leftarrow (p_x, t_y)$

17 **Algorithm $A_i$: where $1 < i \le k$**
18   **if** $Q = C_f$ **then**
19     | destination $\leftarrow (p_x, p_y)$
20     | terminate
21   **else**
22     | compute $SER(Q)$
23     | **if** $t_y = b_y$ **then**
24       | destination $\leftarrow (p_x, p_y)$
25     | **else**
26       | Let $l_i = ((i-1)b_y + (k-i)t_y)/(k-1)$
27       | **if** *if there is only one robot position* $(q_x, q_y)$ *on* $L_1$ **then**
28         | **if** $p_y = m_i$ **then**
29           | destination $\leftarrow (q_x, p_y)$
30         | **else**
31           | destination $\leftarrow (p_x, l_i)$
32       | **else**
33         | destination $\leftarrow (p_x, l_i)$

---

## Detailed Description

Let us now describe the algorithm in some details. In each round $t$, a robot activated by the scheduler, in the Look operation, observes the current configuration $C(t)$ as the set of positions $Q$. Based on this information, the robot, located at $(p_x, p_y)$, proceeds as follows.

(1) If $Q = C_f$, then it does not move, and terminates.
(2) If $Q \ne C_f$, it computes the smallest enclosing rectangle $SER(Q)$ of $Q$. What happens next depends on whether $SER(Q)$ is proper or not.

(3) Let $SER(Q)$ be proper; then both $L_1$ and $L_k$ are defined. What does the robot do depends on its algorithm.

- If the robot is executing $A_1$ and there is only one point $(q_x, q_y)$ on $L_1$, then that point becomes its destination $M_1$ in this round. If instead there is more than one point on $L_1$, the robot moves to $M_1 = ((l_x + r_x)/2, p_y)$ if it is on $L_1$, otherwise it moves parallel to $y$-axis to reach $L_1$, specifically, to the point $(p_x, t_y)$.

- If the robot is executing algorithm $A_i, i \neq 1$, first it computes the location of line $L_i$ perpendicular to the $y$-axis. If the robot is not on $L_i$, then it moves to $L_i$ by moving parallel to $y$-axis. If the robot is already located on $L_i$ and there exists exactly one point $q = (q_x, q_y)$ on $L_1$, then the robot moves to the point $(q_x, p_y)$.

(4) Finally, let $SER(Q)$ be improper; that is, $SER(Q)$ is a horizontal line or a point. In this case, if the active robot is executing $A_1$, it moves one unit distance[1] towards positive $y$-axis, otherwise it does nothing.

**Correctness**

We now show that our algorithm solves $k$-Grouping in finite time. This is done through a sequence of Lemmas; due to space limitations, these proofs are omitted. Let $C(t_0)$ be the initial configuration. We state the following lemmata with respect to $t_0$.

**Lemma 1.** *There exists $t'_0, \geq t_0$, such that smallest enclosed rectangle $SER(Q)$ of $C(t'_0)$ is proper.*

Observe that if $SER(Q)$ is proper, for that configuration all the lines $L_i$ are uniquely defined $1 \leq i \leq k$.

**Lemma 2.** *There exists $t_1 \geq t_0$ such that $L_1$ and $L_k$ remain invariant there after.*

**Lemma 3.** *There exists $t_2 \geq t_1$ such that $M_1$ remains invariant there after.*

**Lemma 4.** *There exists $t_3 \geq t_1$ such that all robots executing $A_i$, for $1 < i \leq k$ remain on $L_i$ there after.*

**Lemma 5.** *There exists $t_4 \geq \max(t_2, t_3)$ such that all robots executing $A_i$, for $1 < i \leq k$ remain at $M_i$ there after.*

**Theorem 1.** *The proposed algorithm solves $k$-Grouping in finite time with one-axis agreement under a $SSYNC$ scheduler with non-rigid movements.*

---

[1] The robots do not share a common measure of unit distance; hence, the unit distances of robots executing $A_1$ may be different from each other.

The finite amount of time elapsed from the start of the algorithm to the reaching of a final configuration can be easily determined. Under the $\mathcal{SSYNC}$ adversarial scheduler, time is measured in terms of *epochs*, where an epoch is a time interval over which all the robots are activated at least once. We have the following corollary.

**Corollary 1.** *The collective algorithm $\mathcal{A}$ solves $k$-Grouping in $O(1)$ epochs under a $\mathcal{SSYNC}$ scheduler if the robots have rigid movement and one-axis agreement, and in $O(h_{max}/\delta)$ epochs when the robots have non-rigid movement, where $h_{max} = \max(t_y - b_y, r_x - l_x)$.*

### 3.2 No Agreement

In this section, we show that robots without agreement on axes fail to achieve $k$-Grouping even with $k+1$ distinct algorithms. We also show that it is possible to achieve $k$-Grouping with an unbounded number of algorithms, specifically with each robot executing a different algorithm.

**Impossibility of $k$-Grouping**

Let $O$ be the center of the *Smallest Enclosing Circle*(SEC) of a set of point $Q$. Let $\theta$ be the smallest angle such that rotating the set of points $Q$ by $\theta$ about $O$ results in the same configuration. We define *rotational symmetry*, $\rho(Q)$ as $2\pi/\theta$.

**Theorem 2.** *With no agreement on the axes, $k$-Grouping is unsolvable under $\mathcal{FSYNC}$ with rigid movements even with $k+1$ distinct algorithms.*

*Proof.* By contradiction, let $\mathcal{A} = \{A_1, A_2, \dots, A_k, A_{k+1}\}$ be a set of collective solution algorithms; that is, starting from any initial configuration, executing $\mathcal{A}$, within finite time the robots form a configuration consisting of $k$ points, and no longer move.

Given a configuration $C(t)$, let $C_i$ be the subset of the robots executing algorithm $A_i$, and let $Q_i$ be the set of points occupied by them.

First, we define a class $\mathcal{C}_{k+1}$ of configurations. A configuration $C(t)$ belongs to $\mathcal{C}_{k+1}$, when $|C_i| = k+1$ and either (C1) or (C2) holds for each $i$.

(C1) If $|Q_i| = 1$, then all the robots executing $A_i$ are located at the center of $SEC(Q)$.
(C2) If $|Q_i| > 1$, then $\rho(Q_i) = k+1$, i.e., $Q_i$ forms a regular $k+1$-gon and the center of the polygon is the center of $SEC(Q)$.

Consider now the execution of $\mathcal{A}$ starting from a configuration $C$ in the class $\mathcal{C}_{k+1}$. Let us stress that, since the robots are oblivious, the action of an activated robot $r$ depends only on the observed set of points $Q$, its location $p = (p_x, p_y) \in Q$, and the algorithm $A_i$ it is executing. Let $u$ be the computed destination point. Observe that the same action will be performed by all robots executing $A_i$ that are in the same symmetry class of $p = (p_x, p_y)$, denoted as $\rho(p)$. We have the following cases, depending on the computed destination $u \neq p$.

(1) $u = O$: All robots in $\rho(p)$ reach the center of SEC. In fact, since $p \neq O$, by condition $(C2)$, the robots are located at the corners of the regular $k+1$-gon, they have the same view and thus the same destination $O$.

(2) $p = O$: $Q_i$ becomes a regular $k+1$-gon in $C(t+1)$. In fact, because of condition $(C1)$, all robots executing $A_i$ are located at $O$. Let the local coordinate system of each robot be rotated by $2\pi/(k+1)$, then every robot moves to a different point at distance $d$ from $O$, forming a regular $k+1$-gon in $C(t+1)$.

(3) $p \neq O$ and $u \neq O$: A new regular $k+1$-gon $Q_i$ is created. In fact, take the $k+1$-gon formed by the robots executing $A_i$. Let the local coordinate system of each of those robots be rotated by $2\pi/(k+1)$, then the destinations $u$ and $u'$ of any two robots on consecutive corners $p$ and $p'$ of the original $k+1$-gon will be also at an angle $2\pi/(k+1)$ with each other from the center, since their view were the same. Thus, the resulting $Q_i$ in $C(t+1)$ is again a regular $k+1$-gon.

**Observation 1.** *The rotational symmetry of the union of concentric regular $k+1$-gons is a multiple of $k+1$.*

Since each of the actions by any of the algorithms from the collection $\mathcal{A}$ results in another configuration with rotational symmetry $j(k+1)$ for some integer $j \geq 0$, the configurations never have a set $Q$ with $|Q| = k$, contradicting the assumed correctness of $\mathcal{A}$.

## Unbounded Possibility

Here we present a collective solution algorithm $\mathcal{A} = \{A_1, A_2, \ldots, A_n\}$, that solves $k$-Grouping for $n$ robots under $\mathcal{SSYNC}$ scheduler with non-rigid movement.

Given a configuration $C(t)$ with the corresponding set of points $Q$, $SEC(Q)$ is the *Smallest enclosing circle* of the set of points. Let $\gamma$ be the radius, and $O$ be the center of the SEC. We define the *Smallest Enclosing Annulus (SEA)* of the configuration as the area between the SEC and another circle with center $O$ and radius $\gamma/2$ (see Fig. 2).

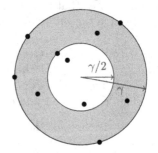

At a high level, the collective algorithm aims to form a configuration where the robots executing algorithms **Fig. 2.** Smallest Enclosing Annulus of a configuration $C(t)$

$A_1, \ldots, A_{k-1}$ move to the SEC (remain in the SEA, perhaps due to non-rigid movements), while the robots executing algorithms $A_i$, for $k \leq i \leq n$ move to the center of the SEC. This achieves $k$-Grouping by having $k-1$ robot positions on the SEA and one at the center.

The pseudocode for the algorithm is present in Algorithm 2.

---

**Algorithm 2:** $k$-Grouping with no agreement

---

**Input:** A set of points $Q$, robot position $p$, and $k$
**Output:** Destination of the robot

1 **Algorithm** $A_i$, **where** $1 \leq i < k$:
2   **if** $p \in SEA(Q)$ **then**
3     $count \leftarrow |SEA(Q)|$, i.e., the number of robot positions in $SEA(Q)$
4     **if** $count \geq k - 1$ **then**
5       $destination \leftarrow p$
6     **else**
7       $\theta \leftarrow$ smallest incident angle at $p$
8       $\theta_i \leftarrow i\theta/2k$
9       Let $u$ be the point on $SEC(Q)$ such that $\angle pOu = \theta_i$
10       $destination \leftarrow u$

11   **else**
12     $u \leftarrow$ the radial projection of $p$ on $SEC(Q)$
13     $destination \leftarrow u$

14 **Algorithm** $A_i$ **where** $k \leq i \leq n$:
15   $O \leftarrow$ center of $SEC(Q)$
16   $destination \leftarrow O$

---

# 4   Conclusion

In this paper, we investigated the problem of $k$-Grouping when robots are allowed to have multiple algorithms. We showed that under partial agreement of áxes, such as agreeing on the direction and orientation of one-axis, $k$ different algorithms are sufficient to solve $k$-Grouping, while it is impossible without agreement on the axes even when the robots have $k + 1$ algorithms. Further, we show that given $|\mathcal{R}|$ algorithms, it is possible to solve $k$-Grouping even without agreement on axes.

We leave as an conjecture that our algorithms also work under an asynchronous scheduler, possibly even when the movement of robots are non-rigid. We also conjecture that $n - 1$ algorithms are insufficient for $k$-Grouping without agreement on axes.

Observe that the problem of $k$-Grouping studied here makes no requirement on the $k$ gathering points except that they are distinct. An interesting open problem is the variation requiring the $k$ distinct points to satisfy a specific geometric pattern, such as a circle or convex hull.

# References

1. Agmon, N., Peleg, D.: Fault-tolerant gathering algorithms for autonomous mobile robots. SIAM J. Comput. **36**(1), 56–82 (2006). https://doi.org/10.1137/050645221
2. Asahiro, Y., Yamashita, M.: Minimum algorithm sizes for self-stabilizing gathering and related problems of autonomous mobile robots (extended abstract). In: 25th International Symposium on Stabilization Safety and Security of Distributed Systems, pp. 312–327 (2023)

3. Bhagat, S., Gan Chaudhuri, S., Mukhopadhyaya, K.: Fault-tolerant gathering of asynchronous oblivious mobile robots under one-axis agreement. J. Discret. Algorithms **36**, 50–62 (2016). https://doi.org/10.1016/j.jda.2015.10.005
4. Cicerone, S., Di Stefano, G., Navarra, A.: Solving the pattern formation by mobile robots with chirality. IEEE Access **9**, 88177–88204 (2021)
5. Cieliebak, M., Flocchini, P., Prencipe, G., Santoro, N.: Distributed computing by mobile robots: gathering. SIAM J. Comput. **41**(4), 829–879 (2012). https://doi.org/10.1137/100796534
6. Das, S., Flocchini, P., Santoro, N., Yamashita, M.: Forming sequences of geometric patterns with oblivious mobile robots. Distrib. Comput. **28**(2), 131–145 (2015). https://doi.org/10.1007/s00446-014-0220-9
7. Dieudonné, Y., Petit, F.: Scatter of robots. Parallel Process. Lett. **19**(1), 175–184 (2009). https://doi.org/10.1142/S0129626409000146
8. Dieudonné, Y., Petit, F.: Self-stabilizing gathering with strong multiplicity detection. Theor. Comput. Sci. **428**, 47–57 (2012). https://doi.org/10.1016/j.tcs.2011.12.010
9. Flocchini, P., Prencipe, G., Santoro, N.: Distributed Computing by Oblivious Mobile Robots. Morgan & Claypool, San Rafael (2012)
10. Flocchini, P., Prencipe, G., Santoro, N., Widmayer, P.: Arbitrary pattern formation by asynchronous oblivious robots. Theor. Comput. Sci. **407**, 412–447 (2008)
11. Flocchini, P., Prencipe, P., Santoro, N., Viglietta, G.: Distributed computing by mobile robots: uniform circle formation. Distrib. Comput. **30**, 413–457 (2017)
12. Flocchini, P.: Gathering. Chapter 4 of Distributed Computing by Mobile Entities, pp. 63–82 (2019)
13. Flocchini, P., Prencipe, G., Santoro, N. (eds.): Distributed Computing by Mobile Entities, Current Research in Moving and Computing, LNCS, vol. 11340. Springer, Cham (2019). https://doi.org/10.1007/978-3-030-11072-7
14. Izumi, T., et al.: The gathering problem for two oblivious robots with unreliable compasses. SIAM J. Comput. **41**(1), 26–46 (2012)
15. Pattanayak, D., Mondal, K., Ramesh, H., Mandal, P.: Gathering of mobile robots with weak multiplicity detection in presence of crash-faults. J. Parallel Distrib. Comput. **123**, 145–155 (2019). https://doi.org/10.1016/j.jpdc.2018.09.015
16. Suzuki, I., Yamashita, M.: Distributed anonymous mobile robots: formation of geometric patterns. SIAM J. Comput. **28**(4), 1347–1363 (1999). https://doi.org/10.1137/S009753979628292X
17. Yamashita, M., Suzuki, I.: Characterizing geometric patterns formable by oblivious anonymous mobile robots. Theor. Comput. Sci. **411**(26–28), 2433–2453 (2010). https://doi.org/10.1016/j.tcs.2010.01.037

# Broadcasting in Stars of Cliques

Akash Ambashankar$^{(\boxtimes)}$ and Hovhannes A. Harutyunyan

Concordia University, Montreal, QC H3G 1M8, Canada
akash.ambashankar@concordia.ca, haruty@cs.concordia.ca

**Abstract.** Broadcasting is an information dissemination problem in a connected network. One informed node, called the originator, must distribute a message to all other nodes by placing a series of calls along the communication lines of the network. Once a node has been informed, it contributes to the broadcasting process by distributing the message to its neighbors. Finding the broadcast time of any node in an arbitrary network is NP-complete. Polynomial time algorithms have been identified for specific topologies, while heuristics and approximation algorithms have been discovered for some others, but the problem remains open for many topologies. In this paper, we study the broadcasting problem in a topology that can be represented by the Windmill graph $Wd_{k,l}$, which contains $k$ cliques of size $l$, connected to a universal node. We also investigate the broadcast problem in *Star of Cliques*, which is an extension of the Windmill graph with arbitrary clique sizes. We present an algorithm to find the broadcast time of any node in an arbitrary star of cliques and prove the optimality of the algorithm.

**Keywords:** Broadcasting · Algorithm · Star of Cliques · Windmill graphs

## 1 Introduction

The advent of technology, particularly the Internet along with parallel and distributed computing, has revolutionized communication at an unprecedented scale, allowing instantaneous communication across vast distances. Now more than ever, innovation in communication protocols is required to enhance the efficiency, security, and scalability of networks. A lot of attention is being directed towards finding the best communication structures for parallel and distributed computing. One of the main problems investigated in this area is *broadcasting*, which is a message dissemination problem in a connected network, wherein an informed node sends the message to one of its uninformed neighbors in the network by making a call. The originator node is responsible for disseminating the message to all other nodes through a series of calls along the communication lines of the network. At each time unit, the newly informed nodes assist the originator by informing their neighbors. The process ends when all nodes are informed. Broadcasting must be completed using the least possible number of time units and is subject to the following constraints: (1) Time units are discrete.

A. A. Rescigno and U. Vaccaro (Eds.): IWOCA 2024, LNCS 14764, pp. 485–496, 2024.
https://doi.org/10.1007/978-3-031-63021-7_37

(2) Each call requires 1 time unit and involves two neighboring nodes. (3) Each node can participate in only one call per time unit. (4) Multiple calls can occur in parallel between distinct pairs of neighboring nodes.

The network can be modeled as a connected undirected graph $G = (V, E)$, with the set of vertices $V$ representing the nodes and the set of edges $E$ representing the communication lines of the network. Given a connected graph $G$ and an originator vertex $u \in V$, the minimum number of time units required to complete broadcasting in $G$ from $u$ is denoted by $b(u, G)$ and is referred to as the *broadcast time of vertex* $u$. The number of informed vertices can double at most during each time unit if each informed vertex calls a different uninformed vertex. Thus we obtain the following lower bound: $b(u, G) \geq \lceil \log n \rceil$ where $n = |V|$ is the number of nodes in the network (all logarithms presented in this paper are in base 2). By definition of the broadcasting problem, there must be at least one new informed vertex in each time unit. This leads to the following upper bound: $b(u, G) \leq n - 1$. The *broadcast time of a graph* is the maximum among the broadcast times of all vertices and is denoted by $b(G) = \max_{\forall u \in V} \{b(u, G)\}$. The set of calls used to distribute the message from originator $u$ to all other vertices is a *broadcast scheme* for vertex $u$. The broadcast scheme forms a spanning tree rooted at the originator, known as a *broadcast tree*.

Finding $b(u, G)$ and $b(G)$ for arbitrary graphs with arbitrary originators has been proven to be NP-complete [20]. The problem also remains NP-complete in more restricted families such as bounded degree graphs [4] and 3-regular planar graphs [16,18]. Research by [19] has shown that it is NP-Hard to approximate the solution of the broadcast time problem within a factor $\frac{57}{56} - \epsilon$. However, the work in [5] improves this result to within a factor of $3 - \epsilon$ and provides an approximation algorithm which produces a broadcast scheme with $O\left(\frac{\log(|V|)}{\log\log(|V|)} b(G)\right)$ rounds. This is the best approximation known for this problem. Another direction of research has been to identify exact algorithms for specific families of graphs. This was initiated by [20] with the proposal of a linear algorithm for Trees, followed by algorithms for Grids and Tori [7], Cube Connected Cycles [17] and Shuffle Exchange [15]. Eventually, exact algorithms were developed for more non-trivial topologies such as Fully Connected Trees [9], Necklace Graphs [10], $k$-cacti with constant $k$ [3], Harary-like graphs [2], and Unicyclic graphs [12,13]. For a more comprehensive introduction to broadcasting and related problems, we refer the reader to the following survey papers [8,11,14,15].

In this paper, we study the broadcasting problem in the Windmill graph $Wd_{k,l}$, a graph on $n$ vertices containing $k$ cliques of size $l$, and a universal vertex. $Wd_{k,2}$ appears as a solution to Erdős et al.'s Friendship Theorem [6] and is known as the Friendship graph. It is defined on $k$ pairs of vertices where every distinct pair has exactly one common adjacent vertex. Other popular sub-families of windmill graphs include the Dutch windmill ($k$ cliques of 2 vertices each) and the French windmill ($k$ cliques of 3 vertices each), which have been studied in the topic of graceful and harmonious graph labeling along with the radio-astronomy problem of movable antennae [1]. We also introduce *Star of Cliques*, a generalization of the Windmill graph composed of cliques with arbitrary sizes.

We present an $O(n \cdot \log n)$ algorithm to find the broadcast time of any vertex in an arbitrary star of cliques and discuss the optimality of our algorithm.

## 2  Broadcasting in Windmill Graphs

Let $Wd_{k,l}$ be a windmill graph on $n$ vertices containing $k$ cliques $C_1, C_2, \ldots, C_k$ each of size $l$, and a vertex $u$ connected to all other vertices. We refer to vertex $u$ as the universal vertex.

In any windmill graph, all vertices have equal broadcast time; at time unit 1, either vertex $u$ informs a vertex in a clique, or vice versa. In either case, by time unit 1, the universal vertex and a vertex from one of the cliques will be the only informed vertices. Since all cliques are the same size, the broadcast time is the same no matter which clique was informed first. Consequently, we opt to regard only the universal vertex as the originator for simplicity and symmetry.

**Lemma 1.** *Let $Wd_{k,l}$ be a windmill graph on $n$ vertices containing $k$ cliques $C_1, C_2, \ldots, C_k$ each of size $l$, and a universal vertex $u$. There exists an optimal broadcast scheme, wherein after informing one vertex in each of the $k-1$ cliques, the universal vertex will broadcast only to the last clique starting at time unit $k$.*

*Proof.* Let $S_{alg}$ be a broadcast scheme where during the first $k - 1$ time units, vertex $u$ informs a vertex in each clique $C_i$ for all $1 \leq i \leq k - 1$, ensuring that broadcasting in $C_i$ finishes by time $i + \lceil \log l \rceil$ without additional broadcasting from vertex $u$. Assume that starting at time unit $k$, the originator exclusively informs vertices in $C_k$ until all $l$ vertices within it are informed, effectively creating a clique $C_k \cup \{u\}$ with $l + 1$ vertices. Broadcasting in $C_k \cup \{u\}$ finishes at time $k - 1 + \lceil \log(l + 1) \rceil$, while broadcasting in $C_{k-1}$ finishes at $k - 1 + \lceil \log l \rceil$, indicating that broadcasting in $C_k$ finishes no sooner than any other clique.

By contradiction, assume there exists a optimal broadcast scheme $S_{opt}$ where the originator informs more than one clique multiple times. Since the graph has $k$ cliques and $C_k$ is the last clique that vertex $u$ informs for the first time, $C_k$ first receives the message at time unit $p$ such that $p \geq k$. Let $b(C_k)$ be the time the last vertex of $C_k$ gets informed under scheme $S_{opt}$. Then, $b_{S_{alg}}(u, Wd_{k,l}) = b(C_k)$ since $C_k$ is the last to finish broadcasting in scheme $S_{alg}$, despite being the only clique to be informed multiple times. Similarly, $b_{S_{opt}}(u, Wd_{k,l}) = b(C_k) \geq p - 1 + \lceil \log(l+1) \rceil$ since by assumption, $C_k$ is not the only clique to be informed multiple times by the originator. We can see that $b_{S_{alg}} \leq b_{S_{opt}}$ since $b_{S_{alg}} = k - 1 + \lceil \log(l + 1) \rceil$ and $k \leq p$. Hence there exists a minimum broadcast scheme where after time $k - 1$, vertex $u$ only informs vertices in clique $C_k$. Thus, $b_{S_{alg}}(u, Wd_{k,l}) = k - 1 + \lceil \log(l + 1) \rceil$.  □

According to Lemma 1, broadcasting in a Windmill graph can be described as $k - 1$ rounds of the universal vertex informing one vertex in each clique $C_1 \ldots C_{k-1}$. These informed vertices broadcast within their respective cliques while vertex $u$ acts as a part of $C_k$, informing only vertices in $C_k$ after time

---

**Algorithm 1.** Broadcast algorithm for Windmill graph from universal vertex $u$

---

**Input** Windmill graph $Wd_{k,l}$, originator vertex $u$.
**Output** A broadcast scheme for $Wd_{k,l}$.

1: **for** $i = 1 \rightarrow k$ **do**
2:     $u$ informs a vertex $v_i$ in clique $C_i$
3:     $v_i$ broadcasts in $C_i$
4: **end for**
5: **while** $C_k$ has an uninformed vertex **do**
6:     $u$ broadcasts in $C_k$
7: **end while**

---

$k - 1$. Broadcasting will finish in another $\lceil \log(l + 1) \rceil$ time units. Thus, the broadcast time of the Windmill graph is given by,

$$b(G) = k - 1 + \lceil \log(l + 1) \rceil$$

# 3   Bounds on Broadcast Time of Star of Cliques

We define a *Star of Cliques* $G_k$ to be a connected graph consisting of $k$ cliques $C_1, C_2, \ldots, C_k$, each of arbitrary sizes, and a vertex $u$ connected to all other vertices. We refer to vertex $u$ as the universal vertex. Assume $l_1 \geq l_2 \geq \cdots \geq l_k \geq 1$ where $l_i$ is the number of vertices in clique $C_i$ for all $1 \leq i \leq k$. Figure 1 depicts a star of cliques composed of 3 cliques with 10, 9, and 5 vertices respectively.

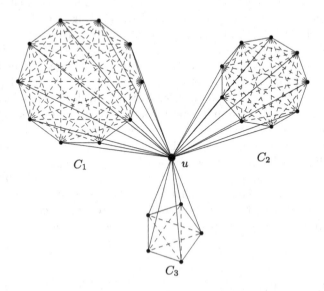

**Fig. 1.** Star of Cliques with 3 cliques of sizes 10, 9, 5

**Lemma 2.** *Let $G_k$ be a star of cliques and the universal vertex $u$ be the originator. Then, $b(u, G_k) \geq \max\limits_{1 \leq i \leq k} \{i - 1 + \lceil \log{(l_i + 1)} \rceil\}$.*

*Proof.* Consider a clique $C_p$, where $1 \leq p \leq k$. For any such $C_p$, by definition, there are at least $p$ cliques that have no fewer than $l_p$ vertices. Thus, by the pigeonhole principle, there exists a clique $C_q$ which is first informed in time unit $t_q \geq p$ and $q \leq p$. Let $b(C_q)$ be the time the last vertex of $C_q$ gets informed under any optimal broadcast scheme. Assume vertex $u$ informs other cliques up to time $t_q - 1$ and thereafter only informs vertices in $C_q$. Thus, $b(C_q) \geq t_q - 1 + \lceil \log{(l_q + 1)} \rceil \geq p - 1 + \lceil \log{(l_p + 1)} \rceil$ as $t_q \geq p$ and $l_q \geq l_p$.

We know that $b(u, G_k) = \max\limits_{1 \leq i \leq k} \{b(C_i)\} \geq \max\limits_{1 \leq i \leq k} \{i - 1 + \lceil \log{(l_i + 1)} \rceil\}$. Hence, $b(u, G_k) \geq \max\limits_{1 \leq i \leq k} \{i - 1 + \lceil \log{(l_i + 1)} \rceil\}$. □

**Lemma 3.** *Let $G_k$ be a star of cliques and the universal vertex $u$ be the originator. Then, $b(u, G_k) \leq \max\limits_{1 \leq i \leq k} \{i + \lceil \log l_i \rceil\}$.*

*Proof.* Consider the broadcast scheme in Algorithm 2 where the originator $u$ informs exactly one vertex in each clique in descending order of clique sizes; $u$ informs a vertex in clique $C_i$ at time unit $i$ for all $1 \leq i \leq k$.

---

**Algorithm 2.** Broadcast algorithm for Star of Cliques from the universal vertex $u$

---

**Input** Star of Cliques $G_k$ ($l_1 \geq l_2 \geq \cdots \geq l_k \geq 1$).
**Output** A broadcast scheme for $G_k$ with time $\max\limits_{1 \leq i \leq k} \{i + \lceil \log l_i \rceil\}$.

1: **for** $i = 1 \rightarrow k$ **do**
2:     $u$ informs a vertex $v_i$ in clique $C_i$
3:     $v_i$ broadcasts in $C_i$
4: **end for**

---

When a clique has one informed vertex, it requires another $\lceil \log l_i \rceil$ time units to inform all other vertices within the clique. Then all vertices of $C_i$ will be informed by time $i + \lceil \log l_i \rceil$. Thus, $b(u, G_k) \leq \max\limits_{1 \leq i \leq k} \{i + \lceil \log l_i \rceil\}$ □

**Lemma 4.** *Let $G_k$ be a star of cliques where the originator is the universal vertex $u$. Then, the difference between the upper and lower bounds of $b(u, G_k)$ is at most 1.*

*Proof.* Assume $C_p$ is a clique that maximizes the upper bound from Lemma 3 such that $UB = \max\limits_{1 \leq i \leq k} \{i + \lceil \log l_i \rceil\} = p + \lceil \log l_p \rceil$. Then, from the lower bound in Lemma 2 we have $LB = \max\limits_{1 \leq i \leq k} \{i - 1 + \lceil \log{(l_i + 1)} \rceil\} \geq p - 1 + \lceil \log{(l_p + 1)} \rceil \geq p + \lceil \log l_p \rceil - 1 = UB - 1$. Hence, $LB + 1 \geq UB$.

Also, assume $C_q$ is a clique that maximizes the lower bound such that $LB = \max\limits_{1 \leq i \leq k} \{i - 1 + \lceil \log{(l_i + 1)} \rceil\} = q - 1 + \lceil \log{(l_q + 1)} \rceil$. Then, $UB = \max\limits_{1 \leq i \leq k} \{i +$

$\lceil \log l_i \rceil \} \geq q + \lceil \log l_q \rceil = q - 1 + \lceil \log (2l_q) \rceil \geq q - 1 + \lceil \log (l_q + 1) \rceil = LB$. Hence, $UB \geq LB$. Putting it together, we have $LB \leq UB \leq LB + 1$. $\qquad \square$

*Remark 1.* From Lemma 4, we know $b(u, G_k)$ has two possible values. Thus, Algorithm 2 is a 1-additive approximation algorithm.

# 4    Decision Algorithm for the Broadcast Time of Star of Cliques

Let $G_k$ be a star of cliques where the originator is the universal vertex $u$. While Algorithm 2 provides a 1-additive approximation algorithm, we are left with the problem of determining the exact value of $b(u, G_k)$ for an arbitrary star of cliques. In this section, we discuss a decision algorithm that determines if broadcasting in a star of cliques from the universal vertex can be completed in $t = t' - 1$ time units where $t' = \max\limits_{1 \leq i \leq k} \{i + \lceil \log l_i \rceil\}$ is the upper bound of a given star of cliques. We also show how this decision algorithm can be used to determine the broadcast time of a given star of cliques for an arbitrary originator.

Let $CS[1 \ldots k]$ be a sorted array of binary strings, each string representing a clique in $G_k$. The base 10 value of a binary string $CS[i]$ is $l_i$ (the number of vertices in clique $C_i$), for all $1 \leq i \leq k$, such that $l_1 \geq l_2 \geq \cdots \geq l_k$. Every clique string $CS[i]$ has $t$ bits (smaller strings are left-padded with zeros). Each bit $CS[i][j]$ represents the action required by $C_i$ from vertex $u$ at time unit $j$, for $1 \leq j \leq t$. If $CS[i][j]$ is 1, vertex $u$ must inform a vertex in $C_i$ at time unit $j$; if $CS[i][j]$ is 0 no action is required.

When $u$ informs a vertex in a clique at time unit $j$ it can lead to $2^{t-j}$ vertices being informed by time unit $t$. This phenomenon is perfectly captured by binary numbers since any integer can be represented as a sum of the powers of 2. The number of vertices in $C_i$ can be represented by $l_i = 2^{p_1} + 2^{p_2} + \ldots + 2^{p_c}$ where $p_1 \ldots p_c$ can be any integers. If $CS[i][j]$ is 1, it means that $l_i \geq 2^{t-j}$, and informing $C_i$ at time unit $j$ results in $2^{t-j}$ vertices being informed by time unit $t$. Let $j_1 \ldots j_c$ be the indices of the bits within the string $CS[i]$ which are 1. Then $l_i = 2^{p_1} + 2^{p_2} + \ldots + 2^{p_c} = 2^{t-j_1} + 2^{t-j_2} + \ldots + 2^{t-j_c}$. If $u$ informs a new vertex in $C_i$ at every time unit where $CS[i][j]$ is 1, then by time unit $t$ all vertices of $C_i$ will be informed. Thus, the binary strings provide an individual broadcast scheme for each clique.

It is also interesting to note that if $u$ informs a vertex in $C_i$ when $CS[i][j]$ is 0, then $C_i$ never needs to be informed again even if any remaining bits in $CS[i]$ are 1, since $2^{t-j} > 2^{t-j+1} + 2^{t-j+2} + \ldots + 2^1 + 2^0$ (e.g.: $1000 > 0111$).

## 4.1    Broadcasting from the Universal Vertex $u$

In Algorithm 3, we present StarOfCliquesBroadcast($CS, t$), which returns *True* if broadcasting in a given star of cliques from the universal vertex $u$ can be completed in $t$ time units, and returns *False* otherwise.

**Definition 1. Unaddressed Vertices:** *When a vertex in some clique $C_p$ is informed at time $j$, this vertex becomes the root of a binomial tree $B_{t-j}$ within $C_p$. The vertices in $B_{t-j}$ are guaranteed to be informed at or before time $t$. All uninformed vertices in $C_p$ that do not belong to such a binomial tree are called unaddressed vertices.*

In the algorithm STAROFCLIQUESBROADCAST$(CS, t)$, we first check if vertex $u$ needs to inform an unaddressed vertex in $C_1$ at the current time unit, i.e. if $CS[1][1] = 1$. If so, this means that $l_1 \geq 2^{t-j}$. Hence, $u$ cannot inform any other clique in the current time unit as that would result in $C_1$ not being able to finish broadcasting, since $l_1 \geq 2^{t-j} > 2^{t-j+1} + 2^{t-j+2} + \ldots + 2^1 + 2^0$ and we cannot make up for the lost vertices. However, if $CS[2][1] = 1$ as well, then there is a conflict and the algorithm returns *False*. Note that it is sufficient to check only $CS[2]$ for conflicts since the array is in descending order and the remaining cliques $C_3 \ldots C_k$ are no bigger than $C_2$. On the other hand, if $CS[1][1] \neq 1$, this implies that no clique needs to be informed in the current time unit. But instead of keeping $u$ idle, by broadcasting to $C_1$ anyway, we ensure that $C_1$ never needs to be informed again, potentially eliminating future conflicts. Either way, we inform an unaddressed vertex in $C_1$. Then, we recursively call STAROFCLIQUESBROADCAST$(CS, t - 1)$ to begin the next iteration where $CS[i] = CS[i][2 \ldots t]$, i.e. the first bit, which represents the now elapsed time unit, is removed from every binary string. Finally, the algorithm runs until two cliques need to be informed in the same time unit, in which case it returns *False*, or we make it to the last time unit and it returns *True*.

**Complexity Analysis.** Step 7 takes $O(t)$ time to update a string if the corresponding clique is informed early. Step 13 takes $O(k \cdot t)$ time to remove the first bit of all $k$ strings. In Step 15, since only the first element of the array is out of place, the sorted array can be obtained through one iteration of insertion sort in $O(k)$ time. One iteration of the STAROFCLIQUESBROADCAST algorithm takes $O(k \cdot t)$ time. The algorithm is called $t$ times, which in total requires $O(k \cdot t^2)$ time. In addition, $O(k \cdot \log k)$ time is required to calculate the upper bound $t'$, and $O(k \cdot \log l_1)$ time is required to generate the binary strings for all cliques. The maximum possible value of $t$ is $k + \lceil \log l_1 \rceil$. We know that $k, l_1 \leq n$, since $n = l_1 + l_2 + \ldots + l_k$ is the total number of vertices in the graph. Thus, $t, k = O(n)$, which brings the overall complexity of the algorithm to $O(n^3)$.

### 4.2 Algorithm Correctness

**Lemma 5 (Greedy Choice).** *Let $G_k$ be a star of cliques where the originator is the universal vertex $u$ and $l_1 \geq l_2 \geq \cdots \geq l_k \geq 1$ where $l_i$ is the number of vertices in clique $C_i$ (excluding vertex $u$) for all $1 \leq i \leq k$. Then, there is an optimal broadcast scheme where the largest clique $C_1$ is informed in the first time unit.*

---

**Algorithm 3.** Decision algorithm for broadcasting in Star of Cliques from the universal vertex $u$ in $t$ time units. (*Note:* Arrays are indexed starting from 1)

---

**Input** $CS[1 \ldots k]$: Cliques in $G_k$ as descending sorted binary strings, $t$: target broadcast time.

**Output** $True$ if $b(u, G_k) = t$; $False$ otherwise.

```
1:  procedure STAROFCLIQUESBROADCAST(CS, t)
2:      if CS[1][1] == "1" then
3:          if CS.length > 1 and CS[2][1] == "1" then
4:              return False
5:          end if
6:      else
7:          s ← 1 + 0...0
                    ‾‾‾
                    t − 1
8:      end if
9:      if t == 1 then
10:         return True
11:     else
12:         for i ← 1 to k do
13:             CS[i] ← substring(CS[i], 2, t)
14:         end for
15:         CS.sort()
16:         return STAROFCLIQUESBROADCAST(CS, t − 1)
17:     end if
18: end procedure
```

---

*Proof.* Let $S$ be a minimum broadcast scheme such that $b(u, G_k) = t_S$. We consider the following cases based on the size of clique $C_1$:

$l_1 \geq 2^{t_S-1}$: When a vertex is informed at time $x$, it can inform a maximum of $2^{t_S-x}$ vertices within its clique. Thus, even if $u$ informs a new vertex in $C_1$ at every time unit except the first time unit, $C_1$ will have a maximum of only $2^{t_S-2} + 2^{t_S-3} + \ldots + 2^1 + 2^0 = 2^{t_S-1} - 1$ informed vertices by time $t_S$. Thus, a vertex in $C_1$ must be informed in the first time unit.

$l_1 < 2^{t_S-1}$: In some graphs, $C_1$ has to be informed in the first time unit despite $l_1 < 2^{t_S-1}$ since other cliques need to be informed in the remaining time units and we may not be able to inform even $2^{t_S-1} - 1$ vertices as seen above.

However, for the other cases where $l_1 < 2^{t_S-1}$ and it is possible to inform $C_1$ later than the first time unit and still finish in $t_S$ time units, we aim to prove that informing $C_1$ in the first time unit does not affect the broadcast time. By contradiction, let some clique $C_x$ be informed in the first time unit and $C_1$ be first informed at some time $j$ in the minimum broadcast scheme $S$. Let $Q_i$ be the set of time units where $u$ informs a vertex in $C_i$ for $1 \leq i \leq k$. We must prove that (1) $Q_1$ will be sufficient to inform all vertices in $C_x$ and (2) $Q_x$ will be sufficient to inform all vertices in $C_1$.

The first part is trivial; since $l_x \leq l_1$, $Q_1$ will be enough to inform all vertices in $C_x$. For the second part, as seen earlier, since $C_1$ doesn't have to be informed in the first time unit, $l_1 < 2^{t-1}$. Therefore time unit 1 alone is enough for all vertices in $C_1$ to be informed. Hence there exists a minimum broadcast scheme where $C_1$ is informed at time unit 1. □

**Lemma 6 (Optimal Substructure).** *Let $G_k$ be a star of cliques where the originator is the universal vertex $u$ and the set $S_j = \{C_1^j, \ldots, C_k^j\}$ be the unaddressed vertices in cliques $C_1, \ldots, C_k$ respectively at time unit $j$. Then, an ordering of $S_j$ which is an optimal broadcast scheme from the universal vertex $u$ must also include an optimal broadcast scheme for $S_{j+1}$.*

*Proof.* Let $A_j = (i_0, i_1, \ldots, i_{t-j})$ be the sequence of calls at time units $j, j + 1, \ldots, t$ where $1 \leq i_1, \ldots, i_n \leq k$. In particular, the originator informs a vertex in $C_{i_p}$ at time unit $j+p$ for all $1 \leq p \leq t-j$. Note that the sequence $A_j$ may contain multiple references to the same clique. For example, $i_0 = i_3 = i_{t-j} = 1$ means that at time units $j, j+3$ and $t$ clique $C_1$ is called. Let $S_{j+1} = \{C_1^{j+1}, \ldots, C_k^{j+1}\}$. We are left with the subproblem of finding an ordering which is an optimal broadcast scheme from $u$ in $S_{j+1}$. Let $A_{j+1} = A_j \setminus \{i_0\}$ be such an ordering, where $A_{j+1}$ is a subsequence of $A_j$. Since $A_j$ and $A_{j+1}$ start at time $j$ and $j + 1$ respectively but finish at the same time, $j + b_{A_j}(u) = (j + 1) + b_{A_{j+1}}(u)$ or $b_{A_j}(u) = b_{A_{j+1}}(u) + 1$.

We want to prove that an optimal solution to $S_j$ must also include an optimal solution to $S_{j+1}$. If a sequence $A'_{j+1}$ existed such that $b_{A'_{j+1}}(u) < b_{A_{j+1}}(u)$, then we would use $A'_{j+1}$ rather than $A_{j+1}$ in a solution to the subproblem of $S_j$, which contradicts the assumption that $A_j$ is an optimal solution, since $b_{A'_{j+1}}(u) + 1 < b_{A_{j+1}}(u) + 1 = b_{A_j}(u)$. □

Consider the star of cliques shown in Fig. 1 with three cliques of sizes 10, 9, and 5 respectively, and a universal vertex $u$. From Lemmata 2 and 3 the lower bound on the broadcast time $b(u, G_k)$ is 5 time units and the upper bound is 6. Using the procedure given in Algorithm 3, we can identify a broadcast scheme to finish broadcasting within the lower bound by informing $C_1$ earlier than required, which allows the universal vertex to broadcast to $C_2$ and $C_3$ multiple times; Vertex $u$ calls clique $C_1$ at time 1, $C_2$ at time 2 and 4, and $C_3$ at time 3 and 5, ensuring that all vertices in the graph have been informed at the lower bound.

### 4.3   Bounds on Broadcast Time from an Arbitrary Originator

Let $G_k$ be a star of cliques with universal vertex $u$. Let an arbitrary vertex $v$ be the originator where $v \neq u$ and $v \in C_i$ for $1 \leq i \leq k$. In an optimal broadcast scheme, $v$ will inform $u$ at time unit 1 since $u$ is connected to every vertex in $G_k$. If $v \in C_1$, then $b(v, G_k) = b(u, G_k)$ since in either case after time unit 1 the two informed vertices in $G_k$ will be $v$ and $u$. When $C_i \neq C_1$ but $l_i = l_1$, still $b(v, G_k) = b(u, G_k)$ since we can swap the calls of $C_i$ and $C_1$. For all other cases, we have the following bounds.

**Lemma 7.** *Let $G_k$ be a star of cliques where the originator $v \in C_i$, for $1 < i \leq k$ and $|C_i| < |C_1|$. Then, $b(v, G_k) \geq b(u, G_k \setminus C_i) + 1$.*

*Proof.* As seen earlier, $v$ informs $u$ at time 1. Since $v$ cannot contribute to broadcasting in $G_k \setminus C_i$ and $|G_k \setminus C_i| \geq |C_1| > |C_i|$, $u$ exclusively handles broadcasting in $G_k \setminus C_i$ while $v$ handles broadcasting in $C_1$. The broadcast time of $u$ in $G_k \setminus C_i$ can be determined using Algorithm 3. Moreover, since $|G_k \setminus C_i| > |C_i|$, $b(u, G_k \setminus C_i) \geq b(v, C_i)$. Hence, including time unit 1, we have $b(v, G_k) \geq b(u, G_k \setminus C_i) + 1$. □

**Lemma 8.** *Let $G_k$ be a star of cliques where the originator $v \in C_i$, for $1 < i \leq k$ and $|C_i| < |C_1|$. Then, $b(v, G_k) \leq b(u, G_k \setminus C_i) + 1$.*

*Proof.* Consider the following broadcast algorithm. First, $v$ informs $u$. Then $u$ informs vertices in $G_k \setminus C_i$ according to Algorithm 3 and $v$ informs vertices in $C_i$. As seen earlier, $b(u, G_k \setminus C_i) \geq b(v, C_i)$. Hence, including time unit 1, we have $b(v, G_k) \leq b(u, G_k \setminus C_i) + 1$. □

---

**Algorithm 4.** Broadcast algorithm for Star of Cliques from an arbitrary originator $v$

---

**Input** Star of Cliques $G_k$, originator vertex $v$
**Output** A broadcast scheme for $G_k$ from originator $v$ with time $b(v, G_k) \leq b(u, G_k \setminus C_i) + 1$ where $u$ is the universal vertex.

1: $v$ informs $u$ at time 1
2: $u$ broadcasts to $G_k \setminus C_i$ according to Algorithm 3
3: $v$ broadcasts to $C_i$

---

*Remark 2.* From Lemmata 7 and 8, we can see that when $v \neq u$ and $v \in C_i$, $b(v, G_k) = b(u, G_k \setminus C_i) + 1$ for all $1 < i \leq k$ if $l_1 > l_i$. For all other cases, $b(v, G_k) = b(u, G_k)$.

### 4.4   Improved Decision Algorithm for the Broadcast Time of Star of Cliques

Algorithm 3 can determine the $b(u, G_k)$ of an arbitrary star of cliques in $O(n^3)$ time. However, in Algorithm 5, we present an $O(n \cdot \log n)$ algorithm to compute $b(u, G_k)$ for an arbitrary star of cliques.

We use base 10 digits and store them in a max-heap called *cliqueHeap*. In each iteration, we require the two largest cliques from the top of the heap. At each time unit $j$, where $1 \leq j \leq t$, we can address $2^{t-j}$ vertices. If any clique, besides the first, needs at least $2^{t-j}$ vertices in the current time unit, then the algorithm returns *False*. Otherwise, we inform $2^{t-j}$ vertices in the first clique. We replace the first element in the heap by $l_1 - 2^{t-j}$. We repeat this process $t$ times, calling STAROFCLIQUESHEAPBROADCAST(*cliqueHeap*, $t$, $j + 1$) at the end of each iteration.

**Algorithm 5.** Heap-based decision algorithm for broadcasting in Star of Cliques from the universal vertex $u$ in $t$ time units

**Input** $cliqueHeap$: $G_k$ as a max-heap of clique sizes, $t$: target broadcast time, $j$: current time unit ($1 \leq j \leq t$).

**Output** $True$ if $b(u, G_k) = t$; $False$ otherwise.

```
 1: procedure STAROFCLIQUESHEAPBROADCAST(cliqueHeap, t, j)
 2:     l₁ ← cliqueHeap.extractMax()
 3:     l₂ ← cliqueHeap.peek()
 4:     if l₁ ≥ 2^(t-j) then
 5:         if l₂ ≥ 2^(t-j) then
 6:             return False
 7:         else
 8:             cliqueHeap.insert(l₁ − 2^(t-j))
 9:         end if
10:     else
11:         cliqueHeap.insert(0)
12:     end if
13:     if j + 1 ≤ t then:
14:         return STAROFCLIQUESHEAPBROADCAST(cliqueHeap, t, j + 1)
15:     else
16:         return True
17:     end if
18: end procedure
```

**Complexity Analysis.** Since the heap contains $k$ elements, one iteration of the algorithm requires $O(\log k)$ time due to the heap operations in Steps 2, 8, and 11. We call the algorithm $t$ times, which requires $O(t \cdot \log k)$ time for all iterations. In addition, $O(k \cdot \log k)$ time is required to sort the input and calculate the upper bound $t'$, and $O(k)$ time is required to create the heap. The maximum possible value of $t$ is $k + \lceil \log l_1 \rceil$ and $k, l_1 \leq n$, since $n = l_1 + l_2 + \ldots + l_k$ is the total number of vertices in the graph. Thus, $t, k = O(n)$, which brings the overall complexity of the algorithm to $O(n \cdot \log n)$.

## 5  Conclusion

In this paper, we study broadcasting in Windmill graphs and Star of Cliques, an extension of the Windmill graph with arbitrary clique sizes. We present an $O(n \cdot \log n)$ algorithm to find the broadcast time of any originator in an arbitrary Star of Cliques and show the optimality of the algorithm. This result can be used to design polynomial algorithms for other complex network topologies incorporating cliques such as block graphs.

## References

1. Bermond, J.C.: Graceful graphs, radio antennae and French windmills. In: Proceedings One day Combinatorics Conference, Research Notes in Mathematics, vol. 34, pp. 18–37. Pitman (1979)

2. Bhabak, P., Harutyunyan, H.A., Tanna, S.: Broadcasting in Harary-like graphs. In: 2014 IEEE 17th International Conference on Computational Science and Engineering, pp. 1269–1276. IEEE (2014)
3. Čevnik, M., Žerovnik, J.: Broadcasting on cactus graphs. J. Comb. Optim. **33**, 292–316 (2017)
4. Dinneen, M.J.: The complexity of broadcasting in bounded-degree networks. arXiv preprint math/9411222 (1994)
5. Elkin, M., Kortsarz, G.: Combinatorial logarithmic approximation algorithm for directed telephone broadcast problem. In: Proceedings of the Thirty-Fourth Annual ACM Symposium on Theory of Computing, pp. 438–447 (2002)
6. Erdős, P., Rényi, A., Sós, V.: On a problem of graph theory, studia sci. Math. Hungar. **1**, 215–235 (1966)
7. Farley, A.M., Hedetniemi, S.T.: Broadcasting in grid graphs. In: Proceedings of the 9th SE Conference on Combinatorics, Graph Theory, and Computing, Utilitas Mathematica, pp. 275–288 (1978)
8. Fraigniaud, P., Lazard, E.: Methods and problems of communication in usual networks. Discret. Appl. Math. **53**(1–3), 79–133 (1994)
9. Gholami, S., Harutyunyan, H.A., Maraachlian, E.: Optimal broadcasting in fully connected trees. J. Interconnection Netw. **23**(01), 2150037 (2023)
10. Harutyunyan, H.A., Hovhannisyan, N., Maraachlian, E.: Broadcasting in chains of rings. In: 2023 Fourteenth International Conference on Ubiquitous and Future Networks (ICUFN), pp. 506–511. IEEE (2023)
11. Harutyunyan, H.A., Liestman, A.L., Peters, J.G., Richards, D.: Broadcasting and gossiping. In: Handbook of Graph Theory, pp. 1477–1494 (2013)
12. Harutyunyan, H., Maraachlian, E.: Linear algorithm for broadcasting in unicyclic graphs. In: Lin, G. (ed.) COCOON 2007. LNCS, vol. 4598, pp. 372–382. Springer, Heidelberg (2007). https://doi.org/10.1007/978-3-540-73545-8_37
13. Harutyunyan, H.A., Maraachlian, E.: On broadcasting in unicyclic graphs. J. Comb. Optim. **16**, 307–322 (2008)
14. Hedetniemi, S.M., Hedetniemi, S.T., Liestman, A.L.: A survey of gossiping and broadcasting in communication networks. Networks **18**(4), 319–349 (1988)
15. Hromkovič, J., Jeschke, C.-D., Monien, B.: Optimal algorithms for dissemination of information in some interconnection networks. In: Rovan, B. (ed.) MFCS 1990. LNCS, vol. 452, pp. 337–346. Springer, Heidelberg (1990). https://doi.org/10.1007/BFb0029627
16. Jakoby, A., Reischuk, R., Schindelhauer, C.: The complexity of broadcasting in planar and decomposable graphs. Discret. Appl. Math. **83**(1–3), 179–206 (1998)
17. Liestman, A.L., Peters, J.G.: Broadcast networks of bounded degree. SIAM J. Discret. Math. **1**(4), 531–540 (1988)
18. Middendorf, M.: Minimum broadcast time is NP-complete for 3-regular planar graphs and deadline 2, vol. 46, pp. 281–287. Elsevier (1993)
19. Schindelhauer, C.: On the inapproximability of broadcasting time. In: Proceedings of the 3rd International Workshop on Approximation Algorithms for Combinatorial Optimization Problems, pp. 226–237 (2000)
20. Slater, P.J., Cockayne, E.J., Hedetniemi, S.T.: Information dissemination in trees. SIAM J. Comput. **10**(4), 692–701 (1981)

# Approximating Spanning Tree Congestion on Graphs with Polylog Degree

Petr Kolman[(✉)][iD]

Faculty of Mathematics and Physics, Charles University, Prague, Czech Republic
kolman@kam.mff.cuni.cz
https://kam.mff.cuni.cz/~kolman

**Abstract.** Given a graph $G$ and a spanning tree $T$ of $G$, the congestion of an edge $e \in E(T)$, with respect to $G$ and $T$, is the number of edges $uv$ in $G$ such that the unique path in $T$ connecting the vertices $u$ and $v$ traverses the edge $e$. Given a connected graph $G$, the *spanning tree congestion* problem is to construct a spanning tree $T$ that minimizes its maximum edge congestion.

It is known that the problem is NP-hard, and that *every* spanning tree is an $n/2$-approximation, but it is not even known whether an $o(n)$-approximation is possible in polynomial time; by $n$ we denote the number of vertices in the graph $G$.

We consider the problem on graphs with maximum degree bounded by $\Delta = polylog(n)$ and describe an $o(n)$-approximation algorithm; note that even on this restricted class of graphs the spanning tree congestion can be of order $n \cdot polylog(n)$.

**Keywords:** Graph sparsification · Congestion · Bisection · Spanning tree

## 1 Introduction

The construction of a spanning tree of a given graph $G$ is a cornerstone problem in graph theory and computer science, explored with various objectives over the past century [3]. In the spanning tree *congestion* problem, the objective is to construct a spanning tree $T$ of $G$ minimizing its maximum edge congestion where the congestion of an edge $e \in T$ is the number of edges $uv$ in $G$ such that the unique path between $u$ and $v$ in $T$ passes through $e$. The problem can be viewed as an extreme instance of graph sparsification (represent the connectivity of a graph by its spanning tree), or as an instance of a graph embedding problem (embed a graph into a tree $T$, with the restriction that $T$ is a spanning tree).

The problem was first considered, as far as we know, by Simonson [14] under the name Min cut (spanning) tree arrangement; the name spanning tree congestion was coined later by Ostrovskii [10]. Even though the problem has been studied for many years, its complexity is not much understood. It is known that it is NP-hard [7,12], and that *every* spanning tree is an $n/2$-approximation [11], but it is not even known whether an $o(n)$-approximation is possible in polynomial time.

A. A. Rescigno and U. Vaccaro (Eds.): IWOCA 2024, LNCS 14764, pp. 497–508, 2024.
https://doi.org/10.1007/978-3-031-63021-7_38

## 1.1    Our Results

We restrict our attention on graphs in which the degree of every vertex is at most polylogarithmic in the number of vertices. For these graphs, the spanning tree congestion ranges between 1 and $\Theta(n \cdot polylog(n))$ (cf. the lower bound (3) below). We describe a polynomial-time $o(n)$-approximation algorithm, namely an $\tilde{O}(n^{1-1/(\sqrt{\log n}+1)})$-approximation[1]. This provides a partial answer to the question (Problem 2.21) posted by Otachi [11].

The analysis of our algorithm exploits two new lower bounds on the spanning tree congestion that are of independent interest.

## 1.2    Selected Related Results

The spanning tree congestion problem is known to be NP-complete even for graphs with one vertex of unbounded degree and all other vertices of bounded degree [2]. For planar graphs, the problem is NP-hard as well [12]. Unless P=NP, no $c$-approximation with $c$ smaller than $6/5$ is possible [8]. The decision version $k$-STC of the problem, to determine whether a given graph has a spanning tree congestion at most $k$, is solvable in polynomial time for $k \leq 3$, and is NP-complete for $k \geq 5$ [8]; NP-completeness of 5-STC implies the above mentioned inapproximability bound.

The problem $k$-STC can be solved in linear time for every fixed $k$ on apex-minor-free graphs, a general class of graphs containing planar graphs, graphs of bounded treewidth, and graphs of bounded genus [2] (cf. [12]). In the same paper, the authors also show that for every fixed $d$ and $k$, there is a linear time algorithm for $k$-STC on graphs with maximum degree at most $d$; the result is based on a close connection between the spanning tree congestion and the treewidth that holds for bounded degree graphs.

Regarding algorithms for the hard instances, there is an exact exponential time algorithm for solving them [9].

For a more detailed overview of other related results, we refer to the survey paper by Otachi [11].

## 1.3    Sketch of Our Approximation

Our algorithm uses the *Divide and conquer* framework. If the graph is small enough, we use any spanning tree. Otherwise, we partition the graph by an approximation of the minimum bisection into two or more components of connectivity, each with at most $n/2$ vertices, solve the problem recursively for each of the components, and then combine the spanning trees of the components into a spanning tree of the entire graph.

The challenge is to relate the congestion of the final spanning tree to the congestion of the optimal spanning tree. To be able to do so, we prove two

---

[1] In the Big-O-Tilde notation $\tilde{O}$, we ignore polylogarithmic factors; note that for every $c > 0$, $\log^c n = o(n^{1/(\sqrt{\log n}+1)})$.

lower bounds on the spanning tree congestion. The first one provides a relation between the bisection of a graph and the spanning tree congestion (Corollary 1), the other provides a connection between the spanning tree congestion of the entire graph and the spanning tree congestion of its subgraph (Lemma 5). The analysis of the algorithm is based on the following idea: if there is a component with *large* bisection, where the exact meaning of *large* depends on the recursion level, the combination of Corollary 1 and Lemma 5 yields a *strong* lower bound on the optimal spanning tree congestion ensuring that *any* spanning tree has the desired approximation property; on the other hand, if all components that are used by the algorithm have small bisections, then it is possible to get a stronger bound on the spanning tree congestion of the spanning tree constructed by the algorithm.

## 1.4    Preliminaries

Given an undirected graph $G = (V, E)$ and a subset of vertices $S \subset V$, we denote by $E_G(S, V \setminus S)$ the set of edges between $S$ and $V \setminus S$ in $G$, and by $e_G(S, V \setminus S) = |E_G(S, V \setminus S)|$ the number of these edges; if the graph which we are referring to is clear from the context, we avoid the lower index $G$. An edge $\{u, v\} \in E$ is also denoted by $uv$ for notational simplicity; when the vertices of an edge are not important, we simply talk about an edge $e \in E$. For a subset of vertices $S \subseteq V$, $G[S]$ is the subgraph induced by $S$. By $V(G)$, we mean the vertex set of the graph $G$ and by $E(G)$ its edge set. By $d(v)$ we denote the degree of a vertex $v \in V$, and by $\Delta(i)$ the sum of the $i$ largest vertex degrees in $G$. Given a graph $G = (V, E)$ and an edge $e \in E$, $G \setminus e$ is the graph $(V, E \setminus \{e\})$.

Let $G = (V, E)$ be a connected graph and $T = (V, E_T)$ be a spanning tree of $G$. For an edge $uv \in E_T$, we denote by $S_u, S_v \subset V$ the vertex sets of the two connected components of $T \setminus uv$. We define the *congestion $c(uv)$ of the edge $uv$ with respect to $G$ and $T$* as $c(uv) = e_G(S_u, S_v)$. The *congestion $c(G, T)$ of the spanning tree $T$ of $G$* is defined as $\max_{e \in E_T} c(e)$, and the *spanning tree congestion* $\mathsf{STC}(G)$ *of $G$* is defined as the minimum value of $c(G, T)$ over all spanning trees $T$ of $G$.

A *bisection* of a graph with $n$ vertices is a partition of its vertices into two sets, $S$ and $V \setminus S$, each of size at most $\lceil n/2 \rceil$; in approximations, this requirement is sometimes relaxed to $2n/3$ (or to some other fraction). The *width of a bisection* $(S, V \setminus S)$ is $e(S, V \setminus S)$. The minimum width of a bisection of a graph $G$ is denoted $b(G)$.

There are several approximation and pseudo-approximation algorithms for the minimum bisection [1,4]. In our algorithm, the congestion of the constructed spanning tree grows exponentially with the depth of the recursion of the algorithm (cf. the proof of Lemma 7). Thus, when approximating the minimum bisection, the balance of the two parts is more important than the exact approximation factor. For this reason, we will employ the approximation algorithm by Räcke [13] and not the approximation by Arora, Rao and Vazirani [1], even though the latter has better approximation.

**Theorem 1 (Räcke [13]).** *A bisection of width within a ratio of $O(\log n)$ of the optimum can be computed in polynomial time.*

We will also need an existential result about the size of the minimum exact bisection in trees.

**Lemma 1 (Folklore, cf. [5]).** *If $T = (V, E)$ is a tree on $n$ vertices and maximum degree $\Delta$, then there exists a bisection of width at most $\Delta \cdot \log n$. Moreover, for every $k < n$, there exists a cut $S \subseteq V$ such that $|S| = k$ and $e(S, V \setminus S) \leq \Delta \cdot \log n.$[2]*

## 2   Lower Bounds

An indispensable component of the analysis of any approximation algorithm is an appropriate lower bound. In this section, we provide several known and several new lower bounds on the spanning tree congestion. Even though we do not need the old lower bounds in the analysis of our algorithm, we provide them here for comparison. The new lower bounds are not difficult to prove but we are not aware of these results; both Corollary 1 and Lemma 5 are crucial ingredients for the analysis of the algorithm in the next section.

We need two more definitions. For each pair $\{u, v\}$ of vertices of $G$ we denote by $m(u, v)$ the maximal number of edge-disjoint paths between $u$ and $v$ in $G$. The *conductance* $\phi(G)$ of a graph $G = (V, E)$ (a.k.a. the Cheeger constant of $G$) is defined as

$$\phi(G) = \min_{\substack{S \subset V \\ |S| \leq |V|/2}} \frac{e(S, V \setminus S)}{\sum_{v \in S} d(v)}.$$

**Lemma 2 (Folklore, cf. [11]).** *For every graph $G = (V, E)$ on $n$ vertices with $m$ edges,*

$$STC(G) \geq \frac{2m}{n - 1} - 1. \tag{1}$$

**Lemma 3 (Ostrovskii [10]).** *For every graph $G = (V, E)$ of maximum degree $\Delta$, minimum degree $\delta$ and diameter $diam(G)$,*

$$STC(G) \geq \max_{u, v \in V} m(u, v), \tag{2}$$

$$STC(G) \geq \frac{\delta}{\Delta} \cdot \phi(G) \cdot (n - 1), \tag{3}$$

$$STC(G) \geq \delta \cdot \phi(G) \cdot \left\lfloor \frac{diam(G)}{2} \right\rfloor. \tag{4}$$

---

[2] Both results hold even in a slightly stronger form where $\Delta \cdot \log n$ is replaced by $\Delta(\log n)$, the sum of the $\log n$ largest degrees.

**Lemma 4.** *For every graph $G = (V, E)$ on $n$ vertices with maximum degree $\Delta$, and for every $k < n$,*

$$STC(G) \geq \frac{\min\limits_{\substack{S \subseteq V \\ |S| = k}} e(S, V \setminus S)}{\Delta \cdot \log n}. \tag{5}$$

*Proof.* Let $C = \min\limits_{\substack{S \subseteq V \\ |S| = k}} e(S, V \setminus S)$ be the minimum cut-size of a $k$-size cut in $G$ and let $T$ be the spanning tree of $G$ with the minimum congestion. By Lemma 1, for the minimum $k$-size cut $S$ of $T$, $e_T(S, V \setminus S) \leq \Delta \cdot \log n$. Thus, the congestion of the tree $T$ is at least $C/(\Delta \cdot \log n)$. □

**Corollary 1.** *For every graph $G = (V, E)$ on $n$ vertices with maximum degree $\Delta$,*

$$STC(G) \geq \frac{b(G)}{\Delta \cdot \log n}. \tag{6}$$

Before stating the next lemma, we note that the spanning tree congestion of a connected subgraph of $G$ might be much larger then the spanning tree congestion of $G$; the lemma provides some bounds on the increase.

**Lemma 5.** *For every graph $G$ and every subset of vertices $S \subset V$ that induces a connected subgraph,*

$$STC(G) \geq \frac{STC(G[S])}{e(S, V \setminus S)}. \tag{7}$$

*Proof.* Consider an arbitrary subset of vertices $S \subset V$. We start by showing a slightly different bound, namely

$$STC(G) \geq \frac{STC(G[S])}{1 + \frac{e(S, V \setminus S)}{2}}. \tag{8}$$

We do so by constructing a spanning tree of $G[S]$ of congestion at most $(1 + \frac{e(S, V \setminus S)}{2}) \cdot STC(G)$. Let $T$ be the spanning tree of $G$ with congestion $STC(G)$. Let $F \subset E$ be the subset of edges $uv \in E$ with $u, v \in S$ such that the path between $u$ and $v$ in $T$ uses at least one vertex from $V \setminus S$. Note that for every edge $uv \in F$, the path between $u$ and $v$ in $T$ has to use at least two edges from $E(S, V \setminus S)$. Thus, $|F| \leq STC(G) \cdot e(S, V \setminus S)/2$.

Let $T'$ be an arbitrary extension of the forest $T[S]$ into a spanning tree of $G[S]$. It follows from the above observation that the congestion $c(G[S], T')$ of the spanning tree $T'$ of $G[S]$ is at most

$$c(G[S], T') \leq STC(G) + |F| \leq STC(G) + STC(G) \cdot \frac{e(S, V \setminus S)}{2},$$

which completes the proof of the inequality (8).

At this point, we distinguish two cases. If $e(S, V \setminus S) = 1$, then $STC(G) \geq STC(G[S]) = \frac{STC(G[S])}{e(S, V \setminus S)}$; if $e(S, V \setminus S) \geq 2$, then $1 + \frac{e(S, V \setminus S)}{2} \leq e(S, V \setminus S)$. Thus, the inequality (7) holds in both cases. □

# 3  Approximation for Graphs with Degrees Bounded by Polylog

Let $G = (V, E)$ be a connected graph on $n$ vertices with maximum degree $\Delta \leq polylog(n)$ and let $k = k(n) = \lceil \sqrt{\log n} \rceil + 1$. We construct the spanning tree of $G$ by the procedure CONSTRUCTST$(H, s, \sigma)$; the procedure is called with parameters $H = G$, $s = \frac{n}{2^{k-1}}$, $\sigma = 1$.

---

**Algorithm 1.** CONSTRUCTST$(H, s, \sigma)$

---

1: construct an approximate bisection $(S, V(H) \setminus S)$ of $H$
2: $F \leftarrow E(S, V(H) \setminus S)$; $b \leftarrow |F|$
3: **if** $b/\sigma \geq n^{1/k}$ or $|V(H)| \leq s$ **then**
4:     **return** any spanning tree of $H$
5: **for** each connected component $C$ of $H \setminus F$ **do**
6:     $T_C \leftarrow$ CONSTRUCTST$(C, s, \sigma + b)$
7: arbitrarily connect all the spanning trees $T_C$ by edges from $F$ to form
                                    a spanning tree $T$ of $H$
8: **return** $T$

---

Let $\tau$ denote the tree representing the recursive decomposition of $G$ (implicitly) constructed by the procedure CONSTRUCTST: The root $r$ of $\tau$ corresponds to the graph $G$, and the children of a non-leaf node $t \in \tau$ associated with a set $V_t$ correspond to the connected components of $G[V_t] \setminus F$ where $F$ is the set of cut edges of an approximate bisection of $G[V_t]$ computed by Räcke's algorithm (Theorem 1); we denote by $b_t = |F|$ the width of this bisection of the subgraph $G[V_t]$. The *level* $l(r)$ of the root $r$ of $\tau$ is zero, and the *level* of every child $t'$ of $t \in \tau$ is $l(t') = l(t) + 1$. We denote by $G_t = G[V_t]$ the subgraph of $G$ induced by the vertex set $V_t$, by $T_t$ the spanning tree constructed for $G_t$ by the procedure CONSTRUCTST; note that for every tree node $t \in \tau$, by construction the graph $G_t$ is connected.

A tree node $t \in \tau$ at level $l$ is *preferable* if either $l = 0$ and $b_t \geq n^{1/k}$, or $l \geq 1$ and $b_t \geq n^{1/k} \cdot \sum_{t' \in p} b_{t'}$ where $p$ is the unique path in $\tau$ between the root $r$ of $\tau$ and the immediate predecessor of the node $t$. A tree node that is not preferable is *tolerable*.

**Lemma 6.** *If there is at least one preferable node $t \in \tau$, then the spanning tree constructed by the procedure CONSTRUCTST is an $\tilde{\mathcal{O}}(n^{1-1/k(n)})$-approximation.*

*Proof.* Consider a preferable node $t \in \tau$ and let $l = l(t)$ denote its level. If $l = 0$, by the definition of the preferable node and by the $\mathcal{O}(\log n)$ bound on the approximation ratio of Räcke's algorithm for minimum bisection (Theorem 1) we know that the minimum bisection of $G$ is of size $\Omega\left(\frac{n^{1/k}}{\log n}\right)$. Thus, by Corollary 1,

$$\mathsf{STC}(G) = \Omega\left(\frac{n^{1/k}}{\Delta \cdot \log^2 n}\right).$$

If $l \geq 1$, then by the construction of the sets $V_{t'}$ by the algorithm, it follows that $e(V_t, V \setminus V_t) \leq \sum_{t' \in p} b_{t'}$ where $p$ is again the unique path between the root $r$ and the immediate predecessor of $t$. Combining this bound with Lemma 5, Corollary 1, the definition of the preferable node, and the $\mathcal{O}(\log n)$ bound on the approximation of minimum bisection, we obtain

$$\mathsf{STC}(G) \geq \frac{\mathsf{STC}(G_t)}{e(V_t, V \setminus V_t)} = \Omega\left(\frac{b(G_t)}{\Delta \cdot \log n \cdot \sum_{t' \in p} b_{t'}}\right) = \Omega\left(\frac{n^{1/k}}{\Delta \cdot \log^2 n}\right).$$

In both cases, the lower bound on $\mathsf{STC}(G)$ immediately yields the claim of the lemma as the congestion of any spanning tree $T$ of $G$ is at most $O(\Delta \cdot n)$. $\square$

**Lemma 7.** *Let $t \in \tau$ be a tolerable node at level $l = l(t)$ such that all its predecessors $t_{l-1}, \ldots, t_1, t_0 = r$ (i.e., the nodes on the path from $t$ to the root $r$ in $\tau$) are tolerable. Then (with $t_l = t$)*

$$\sum_{i=0}^{l} b_{t_i} = \mathcal{O}(b_r \cdot n^{1-1/k}). \tag{9}$$

*Proof.* By induction on the level $l$ of the node $t \in \tau$, we first prove that

$$\sum_{i=0}^{l} b_{t_i} \leq b_r \cdot (1 + n^{1/k})^l. \tag{10}$$

For $l = 0$, the relation (10) asserts simply $b_r \leq b_r$, and this holds.

For the inductive step, assume that the relation (10) holds for $l$ and we want to prove it for $l + 1$. As we are looking at a tolerable node, by the assumption of the lemma, we have $b_{t_{l+1}} < n^{1/k} \cdot \sum_{i=0}^{l} b_{t_i}$. Thus, using the inductive assumption, we derive

$$\sum_{i=0}^{l+1} b_{t_i} = \sum_{i=0}^{l} b_{t_i} + b_{t_{l+1}}$$

$$\leq \sum_{i=0}^{l} b_{t_i} \cdot (1 + n^{1/k})$$

$$\leq b_r \cdot (1 + n^{1/k})^{l+1}$$

which completes the proof of the inequality (10).

By construction, for every node $t \in \tau$, the size of $V_t$ is bounded by

$$|V_t| \leq \frac{n}{2^{l(t)}}.$$

Considering the condition in step 3 of the procedure CONSTRUCTST and our choice of the value of $s = \frac{n}{2^{k-1}}$, we conclude that $l(t) \leq k - 1$ for every node

$t \in \tau$. Combining this bound with the inequality (10), we obtain the desired estimate

$$\sum_{i=0}^{l(t)} b_{t_i} \leq b_r \cdot (1 + n^{1/k})^{k-1} = O(b_r \cdot n^{1-1/k})$$

where the last inequality uses the bound

$$\left(1 + n^{1/(\sqrt{\log n}+1)}\right)^{\sqrt{\log n}} \leq 3 \cdot n^{1-1/(\sqrt{\log n}+1)}$$

which follows from the analytical properties of the functions involved.    □

**Lemma 8.** *Given a graph $G = (V, E)$, a subset $S \subset V$, a spanning tree $T_S$ of $G[S]$, a spanning tree $T_{V \setminus S}$ of $G[V \setminus S]$, and an edge $f \in E(S, V \setminus S)$, let $T = (V, E(T_S) \cup E(T_{V \setminus S}) \cup \{f\})$. Then $T$ is a spanning tree of $G$ and*

$$c(G, T) \leq \max\{c(G[S], T_S), c(G[V \setminus S], T_{V \setminus S})\} + e(S, V \setminus S).$$

*Proof.* The fact that $T$ is a spanning tree of $G$ follows immediately from the construction of $T$. For every $e \in E(T_S)$, the congestion $c(e)$ of $e$ with respect to $G$ and $T$ is at most $c(e) \leq c(G[S], T_S) + e(S, V \setminus S)$, similarly, for every $e \in E(T_{V \setminus S})$, the congestion $c(e)$ of $e$ with respect to $G$ and $T$ is at most $c(e) \leq c(G[V \setminus S], T_{V \setminus S}) + e(S, V \setminus S)$, and the congestion of $f$ is at most $c(f) \leq e(S, V \setminus S)$.    □

**Corollary 2.** *Given a connected graph $G = (V, E)$ and a subset $F \subset E$ of edges, let $V_1, V_2, \ldots, V_k$ be the vertex sets of the connected components of $G \setminus F$ ordered in such a way that for each $l = 2, \ldots, k$, there is at least one edge $f_l$ in $G$ between $\bigcup_{i=1}^{l-1} V_i$ and $V_l$. If for each $i = 1, \ldots, k$, $T_i$ is a spanning tree of $G[V_i]$, then $T = (V, \bigcup_{i=1}^{k} E(T_i) \cup \{f_2, f_3, \ldots, f_k\})$ is a spanning tree of $G$ and*

$$c(G, T) \leq \max_{i=1,\ldots,k}\{c(G[V_i], T_i)\} + |F| .$$

*Proof.* For $i = 1, \ldots, k$, let $G_i = G[\bigcup_{j=1}^{i} V_j]$ and $T_i' = (V(G_i), \bigcup_{j=1}^{i} E(T_j) \cup \{f_2, f_3, \ldots, f_i\})$. By induction on $i$, we are going to show that $T_i'$ is a spanning tree of $G_i$ and

$$c(G_i, T_i') \leq \max_{j=1,\ldots,i}\{c(G[V_j], T_j)\} + \sum_{j=2}^{i} e(\bigcup_{l=1}^{j-1} V_l, V_j). \tag{11}$$

As $G_k = G$, $T_k' = T$ and $F \subseteq \bigcup_{j=2}^{k} E(\bigcup_{l=1}^{j-1} V_l, V_j)$, this will complete the proof.

For $i = 1$, the inequality (11) simply states that $c(G_1, T_1') \leq (G_1, T_1)$ which holds, as $T_1' = T_1$.

For the inductive step, note that by Lemma 8,

$$c(G_{i+1}, T_{i+1}') \leq \max\{c(G_i, T_i'), c(G[V_{i+1}], T_{i+1})\} + e(\bigcup_{j=1}^{i} V_j, V_{i+1}).$$

Applying the inductive assumption on $c(G_i, T_i')$ and using the fact that for any three non-negative numbers $a, b, c$, $\max\{a + b, c\} \leq \max\{a, c\} + b$, we obtain the desired bound on $c(G_{i+1}, T_{i+1}')$:

$$c(G_{i+1}, T_{i+1}') \leq \max \left\{ \max_{j=1,\ldots,i} \{c(G[V_j], T_j)\} + \sum_{j=2}^{i} e(\bigcup_{l=1}^{j-1} V_l, V_j), \; c(G[V_{i+1}], T_{i+1}) \right\}$$

$$+ e(\bigcup_{j=1}^{i} V_j, V_{i+1})$$

$$\leq \max_{j=1,\ldots,i+1} \{c(G[V_j], T_j)\} + \sum_{j=2}^{i+1} e(\bigcup_{l=1}^{j-1} V_l, V_j)$$

$\square$

**Lemma 9.** *If there is no preferable node $t \in \tau$, then the spanning tree $T$ constructed by the CONSTRUCTST procedure is an $\tilde{\mathcal{O}}(n^{1-1/k})$-approximation.*

*Proof.* For a node $t \in \tau$, let $L(t)$ denote the set of all leaves of the subtree of $\tau$ rooted in $t$, and for a leaf $s \in L(t)$, let $p(s, t)$ denote the set of all nodes on the unique path between $s$ and $t$ in the tree $\tau$. Recall that for a non-leaf node $t \in \tau$, $b_t$ is the size of the bisection of the graph $G_t$ that was used by the procedure CONSTRUCTST; for notational simplicity of the next steps, we define $b_t$ for every leaf $t \in \tau$ as well, namely as $b_t = 0$. For every node $t \in \tau$, we are going to prove the following bound:

$$c(G_t, T_t) \leq \max_{s \in L(t)} \left\{ c(G_s, T_s) + \sum_{u \in p(s,t), u \neq t} b_u \right\} + b_t \qquad (12)$$

We proceed by induction, from bottom up in the tree $\tau$. Consider first a leaf $t \in \tau$. In this case, the inequality (12) simplifies to $c(G_t, T_t) \leq c(G_t, T_t) + b_t$ which holds as $b_t = 0$. Next, consider a non-leaf node $t \in \tau$ such that the inequality (12) holds for all its children; let $t_1, \ldots, t_k$ be the children of $t$, for some $k \geq 2$. By applying Corollary 2 on the graph $G_t$, we obtain

$$c(G_t, T_t) \leq \max_{i=1,\ldots,k} \{c(G_{t_i}, T_{t_i})\} + b_t.$$

Plugging in the bound (12) for each of the subgraphs $G_{t_i}$, a simple manipulation yields the desired bound (12) for the node $t$ as well.

Note that for each leaf $s \in \tau$,

$$c(G_s, T_s) = O(\Delta \cdot n^{1-1/k}), \qquad (13)$$

and,

$$\sum_{u \in p(s,r)} b_u = O(n^{1-1/k}) \cdot b_r; \qquad (14)$$

the first bound follows from construction as the algorithms ensures $|V_s| \leq \frac{n}{2^{k-1}} \leq n^{1-1/k}$ for each leaf $s$, and the second bound is given in Lemma 7. Thus, by application of the inequality (12) for the root $r$ of the tree $\tau$, we obtain, using (13) and (14),

$$c(G,T) \leq \max_{s \in L(r)} \{c(G_s, T_s)\} + O(b_r \cdot n^{1-1/k}) = O((\Delta + b_r) \cdot n^{1-1/k}). \quad (15)$$

Exploiting once more the bound on the approximation ratio of Räcke's algorithm for the minimum bisection (Theorem 1), namely $b(G) = \Omega(\frac{b_r}{\log n})$, Corollary 1 yields

$$\mathsf{STC}(G) = \Omega\left(\frac{b_r}{\Delta \cdot \log^2 n}\right).$$

Combining this lower bound on the minimum spanning tree congestion of $G$ with the upper bound (15) on the congestion $c(G,T)$ of the spanning tree constructed by the algorithm, the proof is completed.    □

Lemmas 6 and 9 yield the main theorem.

**Theorem 2.** CONSTRUCTST *is an $\tilde{O}(n^{1-1/k})$-approximation algorithm for the minimum congestion spanning tree problem for graphs with maximum degree bounded by $\Delta = polylog(n)$.*

## 4    Conclusion

We have designed an $o(n)$-approximation algorithm for the spanning tree congestion problem for graphs with maximum degree bounded by $polylog(n)$. Apparently, there are many open problems concerning the spanning tree congestion. Here are some of them:

P1 Is the decision problem 4-STC solvable in polynomial time, or is it NP-complete [8]?

P2 Is it possible to extend the algorithm CONSTRUCTST to work for graphs of unbounded degree?

P3 Is the spanning tree congestion problem NP-hard for graphs with all degrees bounded by a constant? The known NP-hardness proofs require at least one vertex of unbounded degree.

To state the next problem, we need one more definition. Following Law and Ostrovskii [6], given a graph $G = (V, E)$ and an integer $c$, we define the value $f(G, c)$ as

$$f(G, c) = \min\{e(S, V \setminus S) \mid S \subset V, |S| = c, G[S] \text{ is connected}\}. \quad (16)$$

They observe that

$$\mathsf{STC}(G) \geq \min\left\{f(G, c) \;\middle|\; \left\lceil \frac{n-1}{\Delta} \right\rceil \leq c \leq \frac{n}{2}\right\}. \quad (17)$$

The above lower bound is based on the size of a cut $S$ satisfying three properties: i) the subgraph induced by $S$ is connected, ii) the subset $S$ has a prescribed size, iii) the number of edges $e(S, V \setminus S)$ is the smallest among all subsets satisfying the properties i) and ii); we are going to call the task of finding such a cut the *minimum connected c-cut* problem. As far as we know, if only any two of these three properties are required (i.e., minimum $c$-cut, not necessarily connected, or minimum connected cut, not necessarily of size $c$, or connected cut of size $c$, not necessarily minimum), then the problem of finding such a cut is solvable, or at least reasonably approximable, in polynomial time; however, we are not aware of any non-trivial approximation if all three requirements have to be at least approximately satisfied.

P4 Design an approximation algorithm for the minimum connected $c$-cut.

**Acknowledgments.** The author thanks Marek Chrobak for introducing him to the spanning tree congestion problem.

# References

1. Arora, S., Rao, S., Vazirani, U.V.: Expander flows, geometric embeddings and graph partitioning. J. ACM **56**(2), 5:1–5:37 (2009). Preliminary version in Proceedings of the 40th Annual ACM Symposium on Theory of Computing (STOC) (2004)
2. Bodlaender, H.L., Fomin, F.V., Golovach, P.A., Otachi, Y., van Leeuwen, E.J.: Parameterized complexity of the spanning tree congestion problem. Algorithmica **64**(1), 85–111 (2012)
3. Borůvka, O.: O jistém problému minimálním (About a certain minimal problem). Práce Moravské přírodovědecké společnosti **III**(3), 37–58 (1926)
4. Feige, U., Krauthgamer, R.: A polylogarithmic approximation of the minimum bisection. SIAM J. Comput. **31**(4), 1090–1118 (2002)
5. Fernandes, C.G., Schmidt, T.J., Taraz, A.: On minimum bisection and related cut problems in trees and tree-like graphs. J. Graph Theory **89**(2), 214–245 (2018)
6. Law, H.-F., Ostrovskii, M.: Spanning tree congestion: duality and isoperimetry; with an application to multipartite graphs. Graph Theory Notes New York **58**, 18–26 (2010)
7. Löwenstein, C.: In the complement of a dominating set. Ph.D. thesis, TU Ilmenau (2010)
8. Luu, H., Chrobak, M.: Better hardness results for the minimum spanning tree congestion problem. In: Lin, C.C., Lin, B.M.T., Liotta, G. (eds.) WALCOM 2023. LNCS, vol. 13973, pp. 167–178. Springer, Cham (2023). https://doi.org/10.1007/978-3-031-27051-2_15
9. Okamoto, Y., Otachi, Y., Uehara, R., Uno, T.: Hardness results and an exact exponential algorithm for the spanning tree congestion problem. J. Graph Algorithms Appl. **15**(6), 727–751 (2011)
10. Ostrovskii, M.: Minimal congestion trees. Discret. Math. **285**(1), 219–226 (2004)
11. Otachi, Y.: A survey on spanning tree congestion. In: Fomin, F.V., Kratsch, S., van Leeuwen, E.J. (eds.) Treewidth, Kernels, and Algorithms. LNCS, vol. 12160, pp. 165–172. Springer, Cham (2020). https://doi.org/10.1007/978-3-030-42071-0_12

# Efficient Computation of Crossing Components and Shortcut Hulls

Nikolas Alexander Schwarz[1]([✉]) and Sabine Storandt[2]

[1] Christian-Albrecht University of Kiel, Kiel, Germany
nsch@informatik.uni-kiel.de
[2] University of Konstanz, Konstanz, Germany

**Abstract.** Polygon simplification is an important building block in many geovisualization algorithms. Recently, the concept of shortcut hulls was proposed to obtain a simplified polygon that fully contains the original polygon. Given a set of potential shortcuts between the polygon vertices, the computation of the optimal shortcut hull crucially relies on identifying edge crossings among the shortcuts and computing so called crossing components. In this paper, we present novel algorithms to significantly accelerate these steps. For a simple polygon $P$ with $n$ vertices and a set of shortcuts $\mathcal{C}$, we describe an algorithm for computing all edge crossings in $\mathcal{O}(n + m + k)$, where $m := |\mathcal{C}|$ and $k$ is the number of crossings. This output-sensitive algorithm is clearly optimal and a significant improvement over general-purpose algorithms to identify edge crossings. Furthermore, we extend this algorithm to compute the crossing components in $\mathcal{O}(\min\{n + m + k, n^2\})$. As $k$ could potentially be up to $\Theta(n^4)$, this is a significant speed-up if only the crossing components are needed rather than each individual crossing. Finally, we propose a novel crossing component hierarchy data structure. It encodes the crossing components and allows to efficiently partition the polygon based thereupon. We show that our novel algorithms and data structures allow to significantly reduce the theoretical running time of shortcut hull computation.

**Keywords:** Polygon Simplification · Edge Crossings · Intersection Computation · Hierarchical Data Structure

## 1 Introduction

Identifying crossings of line segments is a classic problem in the field of computational geometry with many applications in visual computing and computer graphics, including ray shooting [14], time-series analysis and visualization [25], as well as graph clustering and rendering [8]. Shamos and Hoey [23] were the first to describe an algorithm that decides whether there is at least one crossing in an arrangement of line segments. Given $m$ line segments, they also show that the computational lower bound for this problem is $\Omega(m \log m)$. Bentley and Ottmann extended the algorithm to count and report all crossings [3]. Later,

© The Author(s), under exclusive license to Springer Nature Switzerland AG 2024
A. A. Rescigno and U. Vaccaro (Eds.): IWOCA 2024, LNCS 14764, pp. 509–522, 2024.
https://doi.org/10.1007/978-3-031-63021-7_39

their algorithm was described in more detail and with regard for degenerate cases with a running time of $\mathcal{O}(m \log m + k \log m)$ for $k$ crossings [4]. An optimal algorithm with a running time of $\mathcal{O}(m \log m + k)$ was given in [7]. However, for special cases algorithms with better performance are possible. For example, Eppstein et al. [12] showed that in a given geometric graph with $n$ nodes and $m$ edges with the number of edge crossings $k$ being sublinear in $n$ by an iterated logarithmic factor, the crossings can be computed in time $\mathcal{O}(n + k \log n)$. In [10], it was proven that if the line segments connect points that are all arranged on a circle and the cyclic sequence of the points is given, the segment crossings can be computed in linear time. They use this algorithm to efficiently compute optimal circular right angle crossing graph drawings, where all nodes are located on a circle and edge crossings have an angle of $\pi/2$.

In this paper, we focus on the following setup: Given a polygon $P$ with $n$ vertices, and a set of line segments that connect vertices of $P$, assess their crossing structure. Here, the line segment set is either contained exclusively in the interior of $P$ or exclusively in its exterior. Thus, the line segments are a subset of the interior or exterior visibility graph of the polygon vertices. This poses a generalization of the setup discussed by Dehkordi et al. [10]. The intersection structure of these visibility graphs plays an important role, e.g., in illumination computation [24] or in motion planning in polygonal environments [2]. Furthermore, a novel method for polygon simplification was recently proposed that also heavily relies on crossing computation in such line segment arrangements [5]. The resulting simplification is called a *shortcut hull*. It can be interpreted as a generalization of convex hulls, which allows to smoothly adjust the level of detail via an input parameter $\lambda$. Figure 1 depicts shortcut hulls for an example polygon with varying $\lambda$ values. The shortcut hull algorithm receives as input a polygon $P$ and a set of potential simplification edges between the polygon vertices, so-called shortcuts. The shortcut set $\mathcal{C}$ is a subset of the exterior visibility graph of the polygon. The goal is to select a sequence of crossing-free shortcuts that form a closed walk around the polygon. We will show that our newly developed algorithms significantly reduce the running time to compute shortcut hulls.

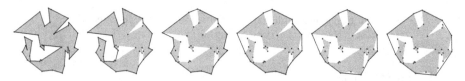

**Fig. 1.** Input polygon (green) and shortcut hulls (black outline) with varying level of detail. The right one coincides with the convex hull of the polygon. (Color figure online)

## 1.1  Related Work

The number of crossings in line segment arrangements and their intersection structure plays an important role in many application realms in computational geometry, visualization, and graph theory. A prominent example is graph drawing, where the typical objective is to find a drawing of the input graph that minimizes edge crossings [11,21,22]. Drawings with fewer intersections are easier to read and convey the structure of the input graph better to the viewer. It has been established that for large amount of edges $m$ the lower bound for the amount of edge crossings is proportional to $m^3/n^2$. This is known as the crossing lemma. The respective coefficient has been refined over time [1,18,19]. There are also many graph layouts that restrict node position. Well-studied examples are layouts where all nodes must be on the outer face, which includes the special case of circular layouts [16]. This is closely related to our setup, where nodes need to coincide with the vertices of a polygon. The class of outerplanar graphs contains the graphs that admit a planar (that is, crossing-free) drawing with all nodes being placed on the outer face [6]. For non-outerplanar graphs, crossings are enforced. Here, the goal is often to maximize the crossing angle in addition to minimizing the number of crossings to still ensure readability [9].

In a visibility graph, the nodes represent geometric objects, and edges encode visibility between the objects. The visibility graph of a simple polygon contains all edges between polygon vertices that do not intersect the exterior of the polygon. Hershberger [15] presented an algorithm to construct such a visibility graph in $\mathcal{O}(m)$ where $m \in \mathcal{O}(n^2)$ denotes the number of edges. For polygons with holes, Overmars and Welzl proposed an algorithm to compute the visibility graph in $\mathcal{O}(m \log n)$ time [17]. Pocchiola and Vegter [20] improved the algorithm to run in $\mathcal{O}(m+n \log n)$ using the notions of pseudotriangulation and visibility complexes.

## 1.2   Contribution

We provide several new results regarding crossing computation in line segment arrangements. Given a polygon $P$ with $n$ vertices, and a set of $m$ line segments that are contained in the interior (or exterior) of $P$ and whose endpoints are vertices of $P$, we design an algorithm that reports all $k$ crossings in $\mathcal{O}(n+m+k)$. This is clearly optimal and faster than the general algorithm by a factor logarithmic in $m$. Moreover, we propose a new method for crossing component computation. A naive way to obtain these components is to compute the intersection graph $G_I$ and then extract its connected components. But as the number of segments $I$ can be quadratic in $n$ and any pair of segments might cross, $G_I$ can have a size of up to $\Theta(n^4)$. Our new algorithm computes a smaller graph whose connected components are equal to those of $G_I$. We refer to this graph as the pseudo-intersection graph $G_P$. We show that the size of $G_P$ is bounded by $\mathcal{O}(n^2)$ which is directly reflected in its construction time. Thus, our new algorithm is vastly faster, especially for large inputs with many crossings.

While these results are interesting on their own, we also show their importance for efficient shortcut hull computation. The construction algorithm proposed by Bonerath et al. [5] uses the crossing components to derive a suitable partition of the polygon's interior. However, in [5], the crossing components and the respective partition were assumed to be given, and thus their efficient computation was left as an open problem. We develop a tailored data structure, called *crossing component hierarchy* that allows to process the crossing components very efficiently and present an improved theoretical running time for the complete shortcut hull algorithm based on our novel results.

## 2   Preliminaries

Throughout the paper, we always assume to be given a simple polygon $P$ as input. $P$ is defined as a closed chain of line segments that connect an ordered set of points $p_1, \ldots, p_n$. We will refer to these points also as polygon vertices and use $N := \{1, \ldots, n\}$ as their index set. Everything enclosed by the polygon is called the interior of $P$ and everything outside is called the exterior of $P$.

Two polygon vertices $p_i, p_j$ are visible from each other inside of $P$ if the line segment $\overline{p_i p_j}$ does not intersect the exterior of $P$. The visibility graph $\text{Vis}(P) = (N, E)$ of $P$ contains a node $i \in N$ for each polygon vertex $p_i$ and edges $\{i, j\} \in E$ for all mutual visible point pairs $p_i, p_j$. Two edges cross each other if a curve from one endpoint to the other cannot be drawn without a intersection (except for their end points touching each other) when respecting the constraint of not intersecting the exterior of $P$. Note that for points in general position this is equivalent to their line segment representations having a point in common apart from their end points. We will use the terms crossing and intersection interchangeably in this paper (as it is the case in most existing literature on the topic). Note, that we can also define an exterior visibility graph by demanding that line segments connecting polygon vertices do not intersect the interior of $P$. Given a set of line segments, the respective intersection graph contains a node for each segment, and edges between segment nodes if the respective pair of segments crosses. In this paper, we are interested in intersection graphs induced by a subset of the visibility graph edges.

## 3   Shortcut Hull

As a showcase application of our improved algorithms for crossing detection, we consider the computation of shortcut hulls [5]: Given a polygon $P$, a shortcut hull $Q$ is a non-crossing polygon that encloses $P$. The boundary segments of $Q$, also called shortcuts, have to be part of the exterior visibility graph of $P$. One can use the full exterior visibility graph or a subset $\mathcal{C}$ of shortcut candidates. The cost $c(Q)$ of $Q$ depends on its perimeter $\beta(Q)$, its area $\alpha(Q)$, and a user-defined parameter $\lambda \in [0, 1]$: $c(Q) = \lambda \cdot \beta(Q) + (1 - \lambda) \cdot \alpha(Q)$. The goal is to find the shortcut hull of minimum cost for given $\lambda$. Bonerath et al. present an exact algorithm for shortcut hull computation [5] with the following main steps:

1. Construct a box that fully contains $P$ (with some wiggle room) and connect its top left corner to the uppermost point of $P$ with an edge. The weakly-simple polygon that consists of the box, the inserted edge, and $P$, is called the sliced donut $\mathcal{D}$.

2. Compute an enrichment $C^+ \supset C$ with $C^+$ being a subset of the interior visibility graph of $\mathcal{D}$. For every set $C'$ of pair-wise non-crossing edges in $C$, the enrichment needs to contain a triangulation of $\mathcal{D}$ that is a superset of $C'$.
3. Use a dynamic program that iterates through triangles in $C^+$ to find the cost-optimal shortcut hull $Q \subset C$ with respect to $\lambda$.

By Bonerath et al.'s analysis [5], the above algorithm has a running time of $\mathcal{O}(n^3)$. More precisely, a running time of $\mathcal{O}(h\chi^3 + \chi n)$ is obtained where $\chi \in \mathcal{O}(n)$ denotes the spatial complexity of the regions and $h \in \mathcal{O}(n)$ the amount of regions induced by the crossing components of $C$. A region is defined as the minimal polygon that encloses a crossing component and respects the boundaries of $P$ and the spatial complexity is the maximum over the amount of vertices of each region. It is proven that for any $C$, an enrichment $C^+$ of size $\mathcal{O}(h\chi^2 + n)$ exists and that it can be computed in $\mathcal{O}(h\chi^3 + \chi n)$. This step dominates the overall running time. However, the algorithm relies on the availability of the crossing components and the regions induced by them. Computing the crossing components naively takes $\mathcal{O}(n^4)$ and thus constitutes the most expensive part of the algorithm. But as the crossing components are simply assumed to be given, the respective running time is not factored in. We present a novel method for crossing component computation in Sect. 5, and an efficient method to compute the enclosing polygons, i.e. regions in Sect. 6.

Furthermore, we propose the following modification to make the overall algorithm simpler to implement and more efficient in practice: The sliced donut, as computed in step 1, is not necessary for the algorithm to work. Instead, one can simply compute the convex hull of $P$ in $\mathcal{O}(n)$ using e. g. the algorithm by Graham [13]. Then the algorithm can be run separately for each pocket, that is, every polygon enclosed by convex hull segments and the boundary of $P$ as illustrated in Fig. 2. This also allows for solving the subproblems in parallel.

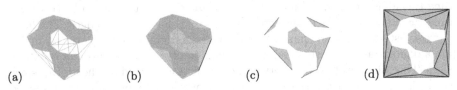

(a)                    (b)                    (c)                    (d)

**Fig. 2.** Segmentation of a polygon $P$ into smaller problems. (a) $P$ with shortcut candidates $C$. (b) Convex hull of $P$. (c) Our approach: pockets of the convex hull. (d) Sliced donut approach from [5] for comparison.

## 4    Computation of Edge Crossings

In this section, we present an optimal algorithm for computing crossings in the interior visibility graph of a polygon. Given a simple polygon with an index set $N = \{1, 2, ..., n-1, n\}$ and a subset of the visibility graph's edges $C$ with $|C| = m$, we want to compute and report all $k$ intersections among the segments in $C$. See Fig. 2a for an example of a possible input.

From here on we will consider all edges as *directed* in topological order, i.e. for any $(u,v) \in C$ it holds that $u < v$. Furthermore, we will assume that the neighborhoods of a vertex $v$, i.e. the targets of outgoing edges $N^+(v)$ and the sources of incoming edges $N^-(v)$ are represented as adjacency lists with vertices sorted in descending order. We will also use them like stacks, such that top() gets the last element in the list and pop() deletes the last element in the list. If the adjacency lists are not sorted, we can easily sort them with an algorithm similar to counting sort in $\mathcal{O}(n + m)$.

Given two edges that cross each other, one index of the second edge must lie between the indices of the first edge and the other one must lie outside. Based on that, we make the following observation:

**Observation 1.** *Given two edges $e = (a, b) \in C$ and $e' = (u, v) \in C$ with $u < a$, the following characterization holds: $a < v < b \iff e$ and $e'$ cross*

Using this observation, we design algorithm that finds all crossings. For this purpose, we take a look at the adjacency matrix, see Fig. 3. Observe that if we fix an edge $e' = (u, v)$, then all edges in the rectangle to the lower right will fulfill Observation 1. In coordinates, this is the rectangle $[u + 1, v - 1] \times [v + 1, n]$. In the example in Fig. 3 this rectangle and the corresponding edge $e' = (1, 5)$ are marked in green and blue in the matrix.

So we need an algorithm that iterates over all edges and for each edge traverses the edges in the corresponding rectangle. We do this column-wise from left to right and traverse each column from bottom to top. After we have visited an edge, we delete it. This simplifies traversing each rectangle because we will no longer have to worry about its left-hand border as all edges to its left will already have been deleted by the time we traverse it. Since constructing and traversing an adjacency matrix is computationally expensive, we simulate the method we just described with adjacency lists. Each column can be traversed using the adjacency lists for incoming edges and each rectangle can be traversed using the adjacency lists in the range $[u + 1, v - 1]$ for outgoing edges. The links between list elements are shown in red in Fig. 3, right. Furthermore, we maintain a doubly-linked list that contains all non-empty adjacency lists for outgoing edges. This allows us to skip empty adjacency lists and avoid overhead. But since we also need constant access to each element, we do not store this doubly-linked list in a conventional way. Instead, we simulate it using two arrays next and prev of size $n$. Before we run our algorithm, we fill those two arrays such that next[v] and prev[v] contain the next and previous index $u$ with $N^+(u) \neq \emptyset$ if $N^+(v) \neq \emptyset$, and $-1$ if $N^+(v) = \emptyset$.

Based on Observation 1 and the described approach to compute the intersections by a sweep over the edges, we arrive at our first main result.

**Theorem 1.** *The algorithm computes all $k$ crossings in time $\mathcal{O}(n + m + k)$.*

## 5    Computation of Crossing Components

We now present a novel algorithm for computing crossing components. Given a candidate segment set $C$ of size $m$, which as before needs to be a subset of the

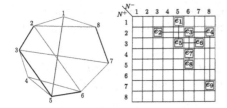

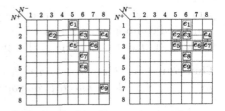

**Fig. 3.** Subset of a polygon's visibility graph with corresponding adjacency matrix

**Fig. 4.** Adjacency matrix with rectangles that allow for optimization of the algorithm

polygon's visibility graph, the naive algorithm computes the intersection graph $G_I$ of $C$ and then extracts its connected components. However, this algorithm has a running time of $\mathcal{O}(m^2)$. As $m$ might be quadratic in $n$, the running time amounts to $\mathcal{O}(n^4)$, which is impractical.

We will show that we can compute the crossing components way more efficiently by constructing a so called pseudo intersection graph $G_P$ instead of $G_I$, which is much smaller than $G_I$ but has the same connected components. To obtain $G_P$, we need the following lemma:

**Lemma 1.** *Given three edges* $\mathfrak{e} = (a, b) \in C$, $e_1 = (u_1, v_1) \in C$ *and* $e_2 = (u_2, v_2) \in C$ *with* $u_1 < a$, $u_2 < a$ *and* $a < v_1 \leq v_2$, *the following implication holds:* $e_2$ *and* $\mathfrak{e}$ *cross* $\Rightarrow e_1$ *and* $\mathfrak{e}$ *cross*

*Proof.* Let $e_2$ and $\mathfrak{e}$ intersect. $a < v_2 < b$ follows from Observation 1. According to prerequisites, $v_1 \leq v_2$ holds, therefore $v_1 < b$ follows. $a < v_1$ already holds according to the prerequisites. Thus, it follows from Observation 1 that $e_1$ and $\mathfrak{e}$ intersect.

Now, we will again take a look at the matrix representation. Observe in Fig. 4 that the yellow rectangle belonging to $e_5$ is a subset of the blue rectangle belonging to $e_1$. This is a direct consequence of Lemma 1. It also implies that all these edges are part of the same crossing component. We can exploit this as follows: When traversing a column from bottom to top, we find the first edge whose corresponding rectangle is not empty. From here on, we connect all edges in the column in the pseudo intersection graph $G_P$. When reaching the last (i.e. topmost) edge in the column, we report all edges in its corresponding rectangle as intersections. This works because all of the edges in this rectangle must belong to the same crossing component and it is a superset of all the rectangles that came before in this column.

The improvement we just described will already save us many intersection reports, but there are still plenty of edges that are visited multiple times. So we still need to improve this to construct a truly sparse graph and improve the running time asymptotically. Observe that we report many redundant edges when we traverse a row of a rectangle if that row has already been traversed as part of a previous rectangle, because these edges are already part of the same connected component in $G_P$, see Fig. 4. So instead of reporting all edges in such a row, it suffices to only report the first edge.

**Theorem 1.** *The construction yields a graph $G_P$ whose connected components are identical to the connected components of the intersection graph $G_I$.*

Once we have computed the pseudo intersection graph $G_P$, we use a graph search algorithm, such as BFS or DFS, to identify all connected components. Since $G_P$ has $m$ vertices and at most $\min\{k, n^2\}$ edges, this can be done in $\mathcal{O}(m + \min\{k, n^2\})$.

**Theorem 2.** *Crossing components can be computed in $\mathcal{O}(\min\{n + m + k, n^2\})$.*

This result is noteworthy, because it means that even for very large amounts of edge crossings $k \in \omega(n^2)$ the algorithm is still bounded by $O(n^2)$. Such large amounts of crossings are not unlikely, as the crossing lemma tells us that $k \in \Omega(m^3/n^2)$ is possible for sufficiently large $m$ [1]. So $k$ can be in the order of $\Theta(n^4)$ if $m \in \Theta(n^2)$. Nevertheless, our algorithm only has a quadratic running time in this case, and an even better one in case $k$ is small.

# 6   The Crossing Component Hierarchy

Our improved methods for crossing detection and for crossing component computation are crucial for the efficient computation of shortcut hulls. Unlike in the previous sections, we look at the exterior of the polygon. However, this is equivalent because all the relevant edges are in the interior of the pockets formed by the convex hull of the polygon.

However, the crossing components $C_1, \ldots, C_h$ themselves are not sufficient. We also need to compute for each $C_i$ the smallest subset of vertices of $P$ that avoids conflicts with the polygon's boundary and forms an enclosing polygon for $C_i$. These are referred to as *regions*. In this section, we describe how to accomplish this. We first present a basic algorithm and then introduce a data structure for acceleration, which we call the crossing component hierarchy.

## 6.1   Basic Algorithm

Our goal is now to compute the boundary of each crossing component $C_i$. As a first step, we create a list $X_i$ for each crossing component $C_i$ that contains the component vertices in ascending order. This can be achieved with a counting sort like procedure in linear time. However, the polygon described by $X_i$ does not coincide with the component boundary because the areas defined by the crossing component vertices can intersect with the input polygon's boundary, see Fig. 5a. Those areas might also intersect each other as shown in Fig. 6. As defined by Bonerath et al. [5], the boundary of a region is the shortest path (measured by vertex count) within the visibility graph around the crossing component. So if the full visibility graph is available, we can easily trace the boundary on it. But since computing the visibility graph might take time quadratic in $n$, we now describe a different way to compute the boundary.

We consider each consecutive pair $a, b$ in $X_i$ separately. (Since $X_i$ represents a polygon, the last and first entry also form a consecutive pair.) A path from $a$ to $b$ that respects the aforementioned properties must be a convex chain formed by the polygon vertices. We first define a search space by giving the two most extreme forms this path can take: The lower bound is the line segment from $a$ to $b$. If it does not intersect the polygon boundary, this is the solution. An upper bound can be constructed as follows: Consider the edges $(a, a')$ and $(b, b')$ in the crossing component with the smallest angle to the line segment $\overline{ab}$. Starting at $a$, we can follow the edge until we reach an intersection, then follow that edge and repeat this process until we reach $b$. This must be possible without encountering a polygon point because otherwise the crossing component would fall apart at such a point. The path we just traced is a conflict-free path through the plane because it follows visibility edges and as such does not intersect any polygon edges. It is also a convex chain and as such it forms an upper bound for the convex chain through polygon vertices we want to obtain.

We need to find an optimal conflict-free path within this search space. To get an intuitive idea of what this path has to look like, we first give a physical experimental method to obtain it: Span a rubber band from $a$ to $b$ tracing the upper bound described above, fixing it at the edge intersections using pins. Now remove the pins, only keeping the rubber band fixed at $a$ and $b$. The rubber band will snap into the optimal conflict-free configuration.

Now, we need to solve this problem computationally: First, we compute a polyline of points between each consecutive pair $a, b$ of all points within the area defined by the vertices of the crossing component, see Fig. 5b. To do this, we traverse all indices between each pair and check for each two indices $i \in \{a, a+1, ..., b-2, b-1\}$ and $j = i+1$ if the line segment formed by them enters or exits the area defined by the crossing component vertices. Here, we maintain which entrance occurs closest to $a$ and which exit occurs closest to $b$. The polyline is then formed by these two intersections and all vertices between them. (Note here that we only include the intersections to avoid self-intersections of the polyline. This makes the following step easier to compute. The final result can only contain polygon vertices.) For the smallest and largest index, which also form a pair $a, b$ with $a > b$, this simply wraps around, i.e., once we reach $n$, we continue with 1. Subsequently, we compute a convex chain along each polyline, see Fig. 5c. The following lemma shows the correctness of this approach.

**Lemma 2.** *Given a pair of two consecutive points of the crossing component boundary $a, b$ (i.e. $b$ follows directly after $a$ in clockwise order around the component), only points between[1] them are relevant for computing the convex chain.*

---

[1] As a reminder: As mentioned in the paragraph above Lemma 2, this refers to the indices $\{a, a+1, ..., n, 1, ...b-1, b\}$ if $a > b$.

*Proof.* We have previously defined an upper bound of the search space. All vertices not between $a$ and $b$ lie behind this upper bound, so they are not relevant for the optimal convex chain.

Doing this for all crossing components yields a running time of $\mathcal{O}(nh)$ in total. This is because in the first step we visit exactly $n$ line segments for each crossing component and perform one intersection check for each of them. For the second step, the polylines around one crossing component can at most be of length $n$ in total. Since we can compute the convex chain of each of them in linear time, this also yields $\mathcal{O}(nh)$ for all crossing components.

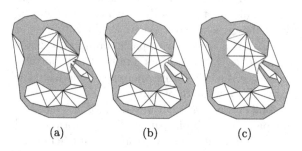
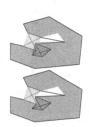

(a)        (b)        (c)

**Fig. 5.** Steps of isolating a region. (a) shows the crossing component's vertices in order with conflicts. In (b) the conflicts are resolved and in (c) each path is optimized.

**Fig. 6.** Crossing components with overlap (top) and how it is resolved (bottom).

### 6.2 Improved Algorithm

The basic algorithm described above already yields an improvement over the naive approach especially for polygons with many visibility edges but few crossing components. However, the number of crossing components $h$ could potentially be linear in $n$. In this case, the basic algorithm exhibits a quadratic running time. To get an improved running time in practice, we propose the *crossing component hierarchy* data structure.

Observe that crossing components cannot extend beyond each other. So given a pair of neighboring vertices from a crossing component, we can simply skip over other crossing components to speed up the computation. To do this efficiently, we will first need to bring them into a useful order. Note that if we represent each crossing component as an open interval $]a, b[$ of the lowest and highest vertex index it contains, then the subset relation induces a hierarchy. More precisely, this is the Hasse diagram of the partially ordered set of crossing components $(S, \subseteq)$ where $S$ contains all crossing components represented as intervals. We will now take a closer look at the properties of this Hasse diagram.

**Lemma 3.** *For two crossing components $C, D \in S$, the following property holds:* $C \nsubseteq D \wedge D \nsubseteq C \Rightarrow C \cap D = \emptyset$

*Proof.* Let $C, D \in S$ with $C \not\subseteq D \wedge D \not\subseteq C$. Assume that $C \cap D = O \neq \emptyset$. Then at least one edge of each crossing component would have to extend into $O$ creating a crossing. But then the two would not be separate crossing components in the first place. *

This means that the interval representations of two crossing components cannot overlap each other. (Note that this is only true for the interval representations. The areas defined by them can indeed overlap.) This also means that two unrelated interval representations cannot have a common subset. For the structure of the Hasse diagram, this implies that it is a forest. If we also include the interval for the convex hull pocket, we will obtain a tree. For an example, see Figs. 7a and 7b. Now we need to construct the Hasse diagram efficiently. We proceed as follows:

- Sort $S$ in descending order by the upper bounds of the intervals using counting sort.
- Sort $S$ again in ascending order by the lower bounds of the intervals.

After applying the sorting method described above, $S$ is the pre-order traversal of its Hasse diagram. This is because if we consider an interval representation of a crossing component $]a, b[$, all of its subsets come next in $S$ because for any subset $]c, d[$ it holds that $a \leq c \leq d \leq b$. These correspond exactly to its sub-tree. For any other intervals $]c, d[$, it is $c \leq a$ such that it comes before $]a, b[$ in $S$ or $c \geq b$ such that it comes after all subsets of $]a, b[$ in $S$.

From the pre-order traversal, we can easily reconstruct the Hasse diagram. We also maintain the order of siblings in this reconstruction. We can slightly extend this tree as follows: For each node, we list all indices that are part of the crossing component in order. Then, each child can be associated with a consecutive pair of indices between which its interval representation lies. This can easily be done in a merge-like manner. The resulting structure is very similar to a B-tree. We call this structure the *crossing component hierarchy*. An example is given in Fig. 7c. This entire procedure can be done in $\mathcal{O}(n)$. Once we have obtained the crossing component hierarchy, we can compute the polylines between each consecutive pair of indices as described above. But instead of iterating over all indices in between we can skip the indices that are in the intervals of the child node. For the polyline that must be computed for the pair consisting of the smallest and largest index, we can instead use its parent. Here, we need to be a bit careful, though, as parts might peak into these intervals and these conflicts also need to be resolved.

**Running Time.** A precise running time cannot be given here as it depends on the precise arrangement of candidate edges. The worst case scenario is still $\mathcal{O}(nh)$, but in cases where all multi-edge crossing components are separated from each other by at least one single-edge crossing component, this results in a running time of $O(n)$. We can also parametrize the running time as $O(nd)$, where $d$ is the maximum amount of multi-edge children of a node in the crossing

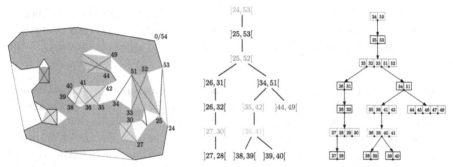

(a) Polygon with candidate set and crossing components

(b) Crossing component intervals in a Hasse diagram

(c) Crossing component hierarchy data structure

**Fig. 7.** Crossing components and their representations for the convex hull pocket on the right of the polygon

component hierarchy. This is because we have to traverse the space between them at most $d$ times. In practice, we expect $d$ to be much smaller than $h$.

### 6.3   Improved Shortcut Hull Computation

Bonerath et al. [5] describe that their algorithm traces the decomposition trees of the regions' triangulations. If we rebuild the crossing component hierarchy from the regions that we just computed, it has the same structure as these decomposition trees. This means that we can adapt Bonerath et al.'s algorithm to run directly on the crossing component hierarchy. A huge benefit of this is that it makes it unnecessary to compute a constrained triangulation before running the algorithm.

**Running Time for Shortcut Hull Computation.** For the algorithm itself, our adapted approach of running it on the crossing component hierarchy can bring a small speed-up, but the asymptotic analysis doesn't change from what was given by Bonerath et al., i.e. $\mathcal{O}(n^3)$ in the worst case, more precisely $\mathcal{O}(h\chi^3 + \chi n)$ [5]. However, contemporary approaches to do the required pre-computations have a running time in $\mathcal{O}(n^4)$ in a worst case scenario. With our novel approach, we brought this down to $\mathcal{O}(n^2)$, such that the entire computation can adhere to the running time of $\mathcal{O}(n^3)$.

## 7   Conclusions and Future Work

We have developed algorithms and methods that can take care of the pre-computations required for Bonerath et al.'s [5] algorithms for shortcut hulls. As part of this, we came up with a new output-sensitive algorithm that can find all edge crossings in linear time given a polygon and a subset of its visibility graph that also has the advantage of not being prone to numerical errors. The algorithm might also be useful for applications beyond shortcut hulls. Furthermore, we extended the algorithm to compute crossing components without finding all crossings $k$. For very large $k$, this is very useful because it limits the running time to $\mathcal{O}(n^2)$. While this is a significant improvement, it is not clear whether this is optimal. A trivial lower bound is $\Omega(m)$. In future work, this gap could be narrowed. Moreover, it would be interesting to implement the shortcut hull algorithm with our novel building blocks and to study its performance on real-world inputs.

# References

1. Ackerman, E.: On topological graphs with at most four crossings per edge. Comput. Geom. **85**, 101574 (2019). https://doi.org/10.1016/j.comgeo.2019.101574. https://www.sciencedirect.com/science/article/pii/S0925772119301154
2. Belta, C., Isler, V., Pappas, G.J.: Discrete abstractions for robot motion planning and control in polygonal environments. IEEE Trans. Rob. **21**(5), 864–874 (2005)
3. Bentley, Ottmann: Algorithms for reporting and counting geometric intersections. IEEE Trans. Comput. **C-28**(9), 643–647 (1979). https://doi.org/10.1109/TC.1979.1675432
4. de Berg, M., Cheong, O., van Kreveld, M., Overmars, M.: Computational Geometry: Algorithms and Applications, 3rd edn. Springer, Germany (2008). https://doi.org/10.1007/978-3-540-77974-2
5. Shortcut hulls: vertex-restricted outer simplifications of polygons. Comput. Geom. **112**, 101983 (2023). https://doi.org/10.1016/j.comgeo.2023.101983. https://www.sciencedirect.com/science/article/pii/S0925772123000032
6. Chartrand, G., Harary, F.: Planar permutation graphs. In: Annales de l'institut Henri Poincaré. Section B. Calcul des probabilités et statistiques, vol. 3, pp. 433–438 (1967)
7. Chazelle, B., Edelsbrunner, H.: An optimal algorithm for intersecting line segments in the plane. J. ACM **39**(1), 1–54 (1992). https://doi.org/10.1145/147508.147511
8. Cui, W., Zhou, H., Qu, H., Wong, P.C., Li, X.: Geometry-based edge clustering for graph visualization. IEEE Trans. Vis. Comput. Graph. **14**(6), 1277–1284 (2008)
9. Dehkordi, H.R., Eades, P.: Every outer-1-plane graph has a right angle crossing drawing. Int. J. Comput. Geom. Appl. **22**(06), 543–557 (2012)
10. Dehkordi, H.R., Eades, P., Hong, S.H., Nguyen, Q.: Circular right-angle crossing drawings in linear time. Theor. Comput. Sci. **639**, 26–41 (2016)
11. Eades, P., Wormald, N.C.: Edge crossings in drawings of bipartite graphs. Algorithmica **11**, 379–403 (1994)
12. Eppstein, D., Goodrich, M.T., Strash, D.: Linear-time algorithms for geometric graphs with sublinearly many edge crossings. SIAM J. Comput. **39**(8), 3814–3829 (2010)
13. Graham, R.L., Frances Yao, F.: Finding the convex hull of a simple polygon. J. Algorithms **4**(4), 324–331 (1983). https://doi.org/10.1016/0196-6774(83)90013-5. https://www.sciencedirect.com/science/article/pii/0196677483900135
14. Guibas, L., Overmars, M., Sharir, M.: Intersecting line segments, ray shooting, and other applications of geometric partitioning techniques. In: Karlsson, R., Lingas, A. (eds.) SWAT 1988. LNCS, vol. 318, pp. 64–73. Springer, Heidelberg (1988). https://doi.org/10.1007/3-540-19487-8_7

15. Hershberger, J.: An optimal visibility graph algorithm for triangulated simple polygons. Algorithmica **4**(1–4), 141–155 (1989). https://doi.org/10.1007/BF01553883
16. Klute, F., Nöllenburg, M.: Minimizing crossings in constrained two-sided circular graph layouts. J. Comput. Geom. **10**(2), 45–69 (2019)
17. Overmars, M.H., Welzl, E.: New methods for computing visibility graphs. In: Proceedings of the Fourth Annual Symposium on Computational Geometry, pp. 164–171 (1988)
18. Pach, J., Radoicic, R., Tardos, G., Toth, G.: Improving the crossing lemma by finding more crossings in sparse graphs. Discret. Comput. Geom. **36**(4), 527–552 (2006). https://doi.org/10.1007/s00454-006-1264-9
19. Pach, J., Tóth, G.: Graphs drawn with few crossings per edge. Combinatorica **17**(3), 427–439 (1997). https://doi.org/10.1007/BF01215922
20. Pocchiola, M., Vegter, G.: Topologically sweeping visibility complexes via pseudo-triangulations. Discret. Comput. Geom. **16**, 419–453 (1996)
21. Purchase, H.C.: Metrics for graph drawing aesthetics. J. Vis. Lang. Comput. **13**(5), 501–516 (2002)
22. Radermacher, M., Reichard, K., Rutter, I., Wagner, D.: Geometric heuristics for rectilinear crossing minimization. J. Exp. Algorithmics (JEA) **24**, 1–21 (2019)
23. Shamos, M.I., Hoey, D.: Geometric intersection problems. In: 17th Annual Symposium on Foundations of Computer Science (SFCS 1976), pp. 208–215 (1976). https://doi.org/10.1109/SFCS.1976.16
24. Teller, S., Hanrahan, P.: Global visibility algorithms for illumination computations. In: Proceedings of the 20th Annual Conference on Computer Graphics and Interactive Techniques, pp. 239–246 (1993)
25. Zhao, Y., Wang, Y., Zhang, J., Fu, C.W., Xu, M., Moritz, D.: KD-box: line-segment-based KD-tree for interactive exploration of large-scale time-series data. IEEE Trans. Vis. Comput. Graph. **28**(1), 890–900 (2021)

# Parameterized Complexity of Paired Domination

Nikita Andreev[1], Ivan Bliznets[2], Madhumita Kundu[4], Saket Saurabh[3,4], Vikash Tripathi[3(✉)], and Shaily Verma[5]

[1] Belgrade, Serbia
[2] University of Groningen, Groningen, The Netherlands
i.bliznets@rug.nl
[3] The Institute of Mathematical Sciences, Chennai, India
{saket,vikasht}@imsc.res.in
[4] University of Bergen, Bergen, Norway
Madhumita.Kundu@uib.no
[5] Hasso Plattner Institute, University of Potsdam, Potsdam, Germany
Shaily.Verma@hpi.de

**Abstract.** The PAIRED DOMINATION problem is one of the well-studied variants of the classical DOMINATING SET problem. In a graph $G$ on $n$ vertices, a dominating set $D$ (set of vertices such that $N[D] = V(G)$) is called a *paired dominating set* of $G$, if $G[D]$ has perfect matching. In the PAIRED DOMINATION problem, given a graph $G$ and a positive integer $k$, the task is to check whether $G$ has a paired dominating set of size at most $k$. The problem is a variant of the DOMINATING SET problem, and hence inherits most of the hardness of the DOMINATING SET problem; however, the same cannot be said about the algorithmic results. In this paper, we study the problem from the perspective of parameterized complexity, both from solution and structural parameterization, and obtain the following results.

1. We design an (non-trivial) exact exponential algorithm running in time $\mathcal{O}(1.7159^n)$.
2. It admits Strong Exponential Time Hypothesis (SETH) optimal algorithm parameterized by the treewidth (tw) of the graph $G$. The algorithm runs in time $4^{\mathsf{tw}} n^{\mathcal{O}(1)}$; and unless SETH fails, there is no algorithm running in time $(4 - \epsilon)^{\mathsf{tw}} n^{\mathcal{O}(1)}$ for any $\epsilon > 0$.
3. We design an $4^d n^{\mathcal{O}(1)}$ algorithm parameterized by the distance to cluster graphs. We complement this result by proving that the problem does not admit a polynomial kernel under this parameterization and under parameterization by vertex cover number.
4. PAIRED DOMINATION admits a polynomial kernel on graphs that exclude a biclique $K_{i,j}$.
5. We also prove that one of the counting versions of PAIRED DOMINATION parameterized by cliquewidth admits $n^{2^{\mathsf{cw}}} n^{\mathcal{O}(1)}$ time algorithm parameterized by cliquewidth (cw). However, it does not admit an FPT algorithm unless #SETH is false.

N. Andreev—Independent Researcher, Dubai, UAE.

A. A. Rescigno and U. Vaccaro (Eds.): IWOCA 2024, LNCS 14764, pp. 523–536, 2024.
https://doi.org/10.1007/978-3-031-63021-7_40

**Keywords:** Paired Domination · Parameterized Complexity · Exact Algorithms · Kernelization

# 1   Introduction

Given a graph $G$, a set $D \subseteq V(G)$ is called a *dominating* set if every vertex not in $D$ has a neighbor in $D$. Given a graph $G$ and an integer $k$, the DOMINATING SET problem asks if there exists a dominant set of size at most $k$. The combinational and algorithmic aspects of the problem are extensively studied in the literature [16,17]. To get an idea of the variants and generalizations of domination, refer to the books [14,15].

The DOMINATING SET problem generally models a guarding problem in which one needs to use the smallest number of guards to protect a certain object. This object is represented by a graph. Guards can be placed only in vertices, and if a guard occupies vertex $v$, then he is controlling vertex $v$ and all its neighbors. Under this model, it is natural to require an additional property: each guard has a partner with whom they back up each other. This model was introduced by Haynes and Slater [18]. In this case, the problem is called PAIRED DOMINATION instead of DOMINATING SET.

Given a graph $G$, a dominating set $D \subseteq V$ is called a *paired dominating set* (PD-set for short), if the induced subgraph $G[D]$ has a perfect matching. Let $M$ be a perfect matching in $G[D]$, then for an edge $uv \in M$, we say that $u$ is *paired* with $v$. Given a graph $G$ and an integer $k$, the PAIRED DOMINATION problem asks if $G$ has a PD-set of size at most $k$. It is important to note that the problem PAIRED DOMINATION does not generalize DOMINATING SET. The problem is known to be NP-complete even for restricted graph classes such as split graphs [5], bipartite graphs [5], and planar graphs [23]. For results on PAIRED DOMINATION, we refer to a survey paper [8] and the book [14]. Although there have been several findings on complexity issues, the exploration of algorithmic approaches to address NP-hardness has not been as extensive. In this paper we fill this gap by studying the problem within the framework of parameterized complexity. We note that PAIRED DOMINATION problem almost was not studied from the parameterized complexity point of view. The only result, known to us, is obtained by Hanaka et al. [13]. They studied an $r$-GROUPED DOMINATING SET and for $r = 2$ this problem coincides with the PAIRED DOMINATION problem. Hanaka et al. [13] presented an $3^{vc}n^{O(1)}$ algorithm for PAIRED DOMINATION.

As stated above, the problem is a close variant of the DOMINATING SET problem. Therefore, it is natural to find similarities and dissimilarities between DOMINATING SET and PAIRED DOMINATION. The DOMINATING SET problem is well studied in the context of parameterized complexity. The problem is known to be a canonical W[2]-hard when parameterized by the solution size [10]. The problem remains W[2]-hard even in some restricted graph classes, including bipartite and split graphs [21], where the parameter is the solution size. On the positive side, the problem is FPT on restricted graph classes such as planar graphs [1,4], nowhere dense graphs [7], $d$-degenerate graphs [2], $K_{i,j}$-free graphs [20] (a complete bipartite graph with one side of size $i$ and the other of size $j$), and graphs of bounded genus [12]. The problem is also known to be FPT with respect to structural parameters, such as treewidth [19,22], cliquewidth, and vertex cover number of a graph.

### Our Results

We design the first non-trivial exact exponential time algorithm running in time $\mathcal{O}(1.7159^n)$ to compute a minimum cardinality PD-set of a graph. The algorithm first enumerates all the minimal dominating sets of a given graph $G$ in time $\mathcal{O}(1.7159^n)$ and then for each of them uses a subroutine based on maximum matching algorithm to complete it to a PD-set of a graph. Finally, we output the one with a minimum cardinality. Designing an exact algorithm whose execution time is better than the number of minimal dominating sets of a graph $G$ ($\mathcal{O}(1.7159^n)$) is an interesting question.

Next, we consider structural parameters, treewidth (tw), vertex cover number (vc) of a graph, and distance to cluster graphs. Recall that distance to cluster graphs does not exceed vertex cover number. It is known that DOMINATING SET admits an algorithm with running time $3^{tw}n^{\mathcal{O}(1)}$; and unless SETH fails, there is no algorithm running in time $(3 - \epsilon)^{tw}n^{\mathcal{O}(1)}$ for any $\epsilon > 0$ [19]. We show a similar result for PAIRED DOMINATION. That is, we give an algorithm that runs in time $4^{tw}n^{\mathcal{O}(1)}$; and unless SETH fails, there is no algorithm running in time $(4 - \epsilon)^{tw}n^{\mathcal{O}(1)}$ for any $\epsilon > 0$. The algorithm is a classical dynamic programming algorithm over graphs of bounded treewidth with a subset convolution trick applied to join node to speed up the computation. Then we design a $4^d n^{\mathcal{O}(1)}$ algorithm where d is the distance to cluster graphs. We complement this result by proving that the problem does not admit a polynomial kernel when parameterized by the vertex cover number. As d $\leq$ vc we conclude that the PAIRED DOMINATION problem does not admit a polynomial kernel when parameterized by the distance to cluster graphs.

One of the largest family of graphs where DOMINATING SET is known to admit a FPT algorithm parameterized by the solution size $k$ is $K_{i,j}$-free graphs. We show that PAIRED DOMINATION behaves similarly to DOMINATING SET on biclique-free graphs. Towards this, we design a polynomial kernel for PAIRED DOMINATION on graphs that exclude a biclique $K_{i,j}$. The kernelization algorithm is inspired by the kernelization algorithm designed by Philip et al. [20] for the DOMINATING SET problem.

All graphs considered in this paper are simple and finite. For the basic graph theoretic notations and definitions refer to the book [9], and for the notations and definitions related to parameterized complexity refer to the book [6]. For $k \geq 1$, an integer, we use $[k]$ to denote the set of integers $\{1, 2, \ldots, k\}$.

## 2  Exact Exponential Time Algorithm

In this section, we design an exact exponential-time algorithm for PAIRED DOMINATION. Note that if a graph $G$ contains isolated vertices, then there is no PD-set in $G$; that is, $G$ is a No-instance. Therefore, we assume that the graph $G$ has no isolated vertex. We first design an algorithm that, given a graph $G$ and a dominating set $D$, constructs a PD-set $D'$ such that $D \subseteq D'$.

---

**Algorithm 1.** DOM-SET TO PD-SET

---

**Input:** A graph $G$ without an isolated vertex and a dominating set $D$ of $G$.
**Initialize:** $D' = D$;
**begin**

    Compute a maximum matching $M$ of induced subgraph $G[D']$;
    Let $A = D' \setminus V(M)$;
    **while** $(A \neq \emptyset)$ **do**
        Pick a vertex $v \in D'$;
        **if** $(N_G(v) \subseteq D')$ **then**
            $D' = D' \setminus \{v\}$;
            $A = A \setminus \{v\}$;
        **else**
            Let $u \in N_G(v) \setminus D'$;
            $D' = D' \cup \{u\}$;
            $A = A \setminus \{v\}$;
    **return** $D'$

---

**Lemma 1 (♣[1]).** *Let $G$ be a graph, $D$ be a PD-set of $G$, and $D' \subseteq D$ be a minimal dominating set of $G$. If $S$ is a PD-set of $G$ constructed from $D'$ using Algorithm 1, then $|S| \leq |D|$.*

The above lemma concludes that if $D$ is a minimum size PD-set of a graph $G$ and $D'$ is a minimal dominating set of $G$ contained in $D$, then we can obtain another PD-set $S$ of $G$ containing $D'$ such that $|S| \leq |D|$. More specifically, as $D$ is a minimum-sized PD-set of $G$, $|S| = |D|$. Thus, if for a graph $G$, $D_1, D_2, \ldots, D_r$ are all possible minimal dominating sets of $G$, and $S_1, S_2, \ldots, S_r$ are the corresponding PD-sets such that $D_i \subseteq S_i$, and $D$ is a minimum size PD-set of $G$, then $D = S_i$, where $S_i$ has minimum size among the sets $S_1, S_2, \ldots, S_r$. Fomin et al. [11] propose an algorithm that enumerates all minimal dominating sets of a graph in time $\mathcal{O}(1.7159^n)$. Formally the result is stated below.

**Theorem 1.** [11] *For a graph $G$ on $n$ vertices, all the minimal dominating sets of $G$ can be enumerated in time $\mathcal{O}(1.7159^n)$.*

Using the above results, we obtain the following theorem.

**Theorem 2.** *For a graph $G$ on $n$ vertices, PAIRED DOMINATION can be solved in time $\mathcal{O}(1.7159^n)$.*

---

[1] Proofs of the results marked with (♣) are omitted due to space constraint.

# 3 Parameterization by Treewidth and Pathwidth

In this section, we give a dynamic programming-based FPT algorithm for PAIRED DOMINATION parameterized by the treewidth of the input graph $G$ and prove matching lower bound assuming SETH. To compute a minimum cardinality PD-set of a graph $G$, we use a nice tree decomposition of $G$ with an edge introduce node. For the definition and properties of tree decomposition, see [6]. We use the following notations in our algorithm. Consider two nonempty sets, $A$ and $B$. Then $A \setminus B = \{x \in A \mid x \notin B\}$. Let $f : A \mapsto B$ be a function. Then for $a \in A$, the function $f_{a \mapsto b} : A \mapsto B$ is defined as $f_{a \mapsto b}(x) = f(x)$ for all $x \in A \setminus \{a\}$ and $f_{a \mapsto b}(a) = b$, and the function $f_{a_1 \mapsto b_1, a_2 \mapsto b_2} : A \mapsto B$ is defined as $f_{a_1 \mapsto b_1, a_2 \mapsto b_2}(x) = f(x)$ for all $x \in A \setminus \{a_1, a_2\}$ and $f_{a_1 \mapsto b_1, a_2 \mapsto b_2}(a_1) = b_1$ and $f_{a_1 \mapsto b_1, a_2 \mapsto b_2}(a_2) = b_2$. For a set $A' \subseteq A$, the *restriction* of $f$ denoted by $f_{A'}$ is defined as $f_{A'}(a) = f(a)$ for each $a \in A'$. In this case, the function $f$ is called an *extension* of $f_{A'}$.

Let $(T, \{X_t\}_{t \in V(T)})$ be a tree decomposition of the given graph $G$ with width $tw$. For a node $t \in V(T)$, let $T_t$ denotes the subtree of $T$ rooted at $t$ and $G_t = (V_t, E_t)$ be the graph associated with $T_t$, where $V_t = \bigcup_{t \in V(T_t)} X_t$ and $E_t = \{e \mid e$ is introduced in the subtree rooted at $t\})$. Note that $V_r = V(G)$ and $G_r = G$, where $r$ is the root node of $T$. We define a function $f : X_t \mapsto \{1, \hat{1}, 0, \hat{0}\}$ that labels the vertices in bag $X_t$. Each label in the set $\{1, \hat{1}, 0, \hat{0}\}$ has the following definition:

1 – vertices that are mapped to 1 are part of the partial solution $D \subseteq V_t$ and are paired with a vertex in $D$.

$\hat{1}$ – vertices that are mapped to $\hat{1}$ are part of the partial solution $D \subseteq V_t$ but are not paired in $D$.

0 – vertices that are mapped to 0 are not part of the partial solution $D \subseteq V_t$ but must be dominated by $D$.

$\hat{0}$ – the vertices mapped to $\hat{0}$ are not part of the partial solution $D \subseteq V_t$ and do not need to be dominated by $D$.

Let $t$ be a node in the tree $T$, and $X_t$ be the corresponding bag. There are $4^{|X_t|}$ possible labelling of $X_t$. A set $D \subseteq V_t$ *respects* a function $f : X_t \mapsto \{1, \hat{1}, 0, \hat{0}\}$ on $X_t$, if $D$ satisfies the following properties:

1. $D \cap X_t = f^{-1}(1) \cup f^{-1}(\hat{1})$.
2. $D$ dominates all the vertices in the set $V_t \setminus f^{-1}(\hat{0})$.
3. There is a perfect matching $M$ in $G[D \setminus f^{-1}(\hat{1})]$, that is, matching $M$ saturates all the vertices in $D \setminus f^{-1}(\hat{1})$. In this case, we say that vertices in $D \setminus f^{-1}(\hat{1})$ are paired.

For a node $t \in V(T)$ and a function $f : X_t \mapsto \{1, \hat{1}, 0, \hat{0}\}$, we define $PD[t, f]$ to be the minimum cardinality of a set $D$ such that $D$ respects $f$ on $X_t$. If for a node $t$ and labeling $f$ of $X_t$, no such minimum cardinality set $D$ exists, then we set $PD[t, f] = +\infty$. Depending on the node type, we now compute the value of $PD[t, f]$ for each node $t \in V(T)$ and $f : X_t \mapsto \{1, \hat{1}, 0, \hat{0}\}$, traversing the tree $T$ in a bottom-up manner.

1. **If $t$ is a leaf node:** By the definition of nice tree decomposition $X_t = \emptyset$. It implies that each color class is an empty set in any labeling $f$ of $X_t$. Moreover, $PD[t, f] = 0$, as $V_t = \emptyset$.

2. **If $t$ is a vertex introduce node:** Since $t$ is a vertex introduce node, $t$ has exactly one child $t'$ such that $X_t = X_{t'} \cup \{v\}$, where $v \notin X_{t'}$. We observe that $v \notin V_{t'}$ by the definition of nice tree decomposition. Since there is no edge incident on $v$ introduced at the current node $t$, $v$ cannot be dominated or paired in $G_t$. Consider a labelling $f : X_t \mapsto \{1, \hat{1}, 0, \hat{0}\}$ of $X_t$, then:

$$PD[t, f] = \begin{cases} +\infty & \text{if } f(v) = 1; \\ 1 + PD[t', f_{X_{t'}}] & \text{if } f(v) = \hat{1}; \\ +\infty & \text{if } f(v) = 0; \\ PD[t', f_{X_{t'}}] & \text{if } f(v) = \hat{0}. \end{cases}$$

3. **If $t$ is an edge introduce node:** Since $t$ is an edge introduce node, $t$ has exactly one child, say $t'$, such that $X_t = X_{t'}$ and there is an edge $uv \in E(G)$ that is introduced at node $t$. That is, $uv \in G_t$ but $uv \notin G_{t'}$. For a labeling $f : X_t \mapsto \{1, \hat{1}, 0, \hat{0}\}$, we derive the following recurrence based on the labeling of vertices $u$ and $v$.

$$PD[t, f] = \begin{cases} \min\{PD[t', f_{u \mapsto \hat{1}, v \mapsto \hat{1}}], PD[t', f]\} & \text{if } (f(u), f(v)) = (1, 1); \\ PD[t', f_{v \mapsto \hat{0}}] & \text{if } (f(u), f(v)) \in \{(1, 0), (\hat{1}, 0)\}; \\ PD[t', f_{u \mapsto \hat{0}}] & \text{if } (f(u), f(v)) \in \{(0, 1), (0, \hat{1})\}; \\ PD[t', f] & \text{otherwise.} \end{cases}$$

4. **If $t$ is a forget node:** By the definition of nice tree decomposition, $t$ has exactly one child $t'$ such that $X_t = X_{t'} \setminus \{v\}$ for some vertex $v \in X_{t'}$. Note that $v$ does not appear in any bag above $X_t$ in $T$ as $v \notin X_t$. Therefore, $v$ should be dominated (or belongs to the dominating set) in $G_{t'}$. We give the following recurrence: $PD[t, f] = \min\{PD[t', f_{v \mapsto 1}], PD[t', f_{v \mapsto 0}]\}$.

   Note that every path decomposition of a graph $G$ is also a tree decomposition with no join node. Moreover, for introduce node, introduce edge node, and forget node we considering $4^{|X_t|}$ labellings of $X_t$ and for each labelling $f$, $PD[t, f]$ can be computed in time $|V(G)|^{\mathcal{O}(1)}$. From this we can infer that PAIRED DOMINATION can be solved in time $\mathcal{O}(4^{pw}|V(G)|^{\mathcal{O}(1)})$.

5. **If $t$ is a join node:** By the definition of nice tree decomposition, $t$ has exactly two children $t_1$ and $t_2$ such that $X_t = X_{t_1} = X_{t_2}$. Let $f_1, f_2$ and $f$ be labellings of $X_{t_1}$, $X_{t_2}$, and $X_t$, respectively. We say that $f_1$ and $f_2$ are *compatible* with $f$, if the following conditions hold, for all $v \in X_t$:

   (a) $f(v) = 1$ if and only if $(f_1(v), f_2(v)) \in \{(1, \hat{1}), (\hat{1}, 1)\}$,
   (b) $f(v) = \hat{1}$ if and only if $(f_1(v), f_2(v)) = (\hat{1}, \hat{1})$,
   (c) $f(v) = 0$ if and only if $(f_1(v), f_2(v)) \in \{(0, \hat{0}), (\hat{0}, 0)\}$,
   (d) $f(v) = \hat{0}$ if and only if $(f_1(v), f_2(v)) = (\hat{0}, \hat{0})$.

Observe that there are 6 choices for a compatible triplet $(f(v), f_1(v), f_2(v))$ and so $6^{tw}$ choices for all the vertices in bag $X_t$. We use subset convolution for faster computation of the compatible functions. For this, we rewrite the above four conditions for compatible functions as follows:

(a) $f^{-1}(1) = f_1^{-1}(1) \cup f_2^{-1}(1)$,
(b) $f_1^{-1}(1) \cap f_2^{-1}(1) = \emptyset$,
(c) $f^{-1}(0) = f_1^{-1}(0) \cup f_2^{-1}(0)$,
(d) $f_1^{-1}(0) \cap f_2^{-1}(0) = \emptyset$.

We would fix the set of vertices mapped to 1 or $\hat{1}$ and apply the subset convolution to compute such functions. Let us fix $R \subseteq X_t$ and let $\mathcal{F}(R)$ be the set of all functions $f$ such that $f^{-1}(1) \cup f^{-1}(\hat{1}) = R$. Recall that we want to compute the value of $PD[t, f]$ for all $f \in \mathcal{F}(R)$. We can represent each function in $\mathcal{F}(R)$ as two sets $S \subseteq X_t \setminus R$ and $Q \subseteq R$ such that $S$ is a pre-image of 0 and $Q$ is a pre-image of 1. Thus, a function $f$ represented by $S$ and $Q$ can be defined as:

$$
\Phi_S^Q(x) = \begin{cases} 1 & \text{if } x \in Q; \\ \hat{1} & \text{if } x \in R \setminus Q; \\ 0 & \text{if } x \in S; \\ \hat{0} & \text{if } x \in X_t \setminus (R \cup S). \end{cases}
$$

Therefore, for every labeling $f \in \mathcal{F}(R)$ we have:

$$
PD[t, f] = \min_{Q \subseteq R} \left\{ \min_{\substack{A \cup B = f^{-1}(0) \\ A \cap B = \emptyset}} \left\{ PD[t_1, \Phi_A^Q] + PD[t_2, \Phi_B^Q] - |R| \right\} \right\}.
$$

We show that using subset convolution computations at join node can be done in $4^{tw} n^{\mathcal{O}(1)}$ time. The correctness proofs of each recurrence and running time analysis are omitted due to space constraints.

**Theorem 3 (♣).** *Given a graph $G$ of treewidth* tw, PAIRED DOMINATION *can be solved in* $\mathcal{O}(4^{tw} \cdot |V(G)|^{\mathcal{O}(1)})$ *time.*

Now we prove a matching lower bound.

**Theorem 4.** *Unless SETH fails, there is no algorithm for* PAIRED DOMINATION *with running time* $\mathcal{O}^*((4-\epsilon)^{pw})$ *for any $\epsilon > 0$, where pw is the pathwidth of the input graph.*

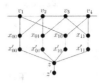

*Proof.* From SETH, it follows that there is no $\mathcal{O}^*((2 - \varepsilon)^n)$ running time algorithm for SAT, where $n$ is the number of variables in the input formula. To establish our lower bound, we transform a given CNF formula

**Fig. 1.** Gadget $G'$

$F$ with $n$ variables and $m$ clauses into a graph $G$ with a path decomposition of width at most $k = \frac{n}{2} + C$ for some constant $C$. Moreover, $F$ is satisfiable if and only if the graph $G$ has a PD-set of a special size, which we will determine later. Therefore, if we can solve PD-set in $\mathcal{O}^*(4^{\lambda k})$ time for $\lambda < 1$, then we would have a $\mathcal{O}^*(4^{\lambda(n/2+C)}) = \mathcal{O}^*((2 - \varepsilon)^n)$ running time algorithm for SAT, which leads to a contradiction. Now we proceed with the details of the transformation.

We assume that $F = C_1 \wedge C_2 \wedge \cdots \wedge C_m$. Without loss of generality, we assume that $F$ depends on variables $x_1, x_2, \ldots, x_{n-1}, x_n$, where $n = 2p$. For each $j \in [p]$ we consider a pair of variables $x_{2j-1}, x_{2j}$. For each such pair of variables, we assign a gadget $G'$ shown in Fig. 1. Consider all four possible assignments for these pairs of variables: $00, 01, 10, 11$. We associate these assignments with the following pairs of vertices $\{v_3, v_4\}, \{v_1, v_4\}, \{v_2, v_3\}, \{v_1, v_2\}$ in the corresponding gadget. We note that if an assignment $ab$ corresponds to a pair $\{v_i, v_j\}$, then the vertex $x_{ab}$ is not connected to vertices $v_i, v_j$ in the gadget. For each pair of vertices $x_{2j-1}, x_{2j}$ we create a chain of gadgets $G'$ such that the vertex $v_1$ of one gadget is connected to the vertex $v_4$ of the preceding gadget. Each such chain contains $m$ gadgets, and the $i$-th gadget corresponds to a clause $C_i$ from $F$. For each $i \in [m]$, we create a vertex named $C_i$ for the clause $C_i$. We connect the vertex $x'_{ab}$ from the $i$-th gadget of a the pair $x_{2j-1}, x_{2j}$ with the vertex $C_i$ if the assignment $x_{2j-1} = a, x_{2j} = b$ satisfies clause $C_i$. We denote obtained construction by $H$. $H$ is schematically shown in the left square of Fig. 2. We repeat this construction, $H$, exactly $\frac{3}{2}n + 1$ times, and connect these parts in a chain fashion, as shown in Fig. 2. Additionally, we add vertices $l, l', r, r'$, and edges $(l, r), (l, l'), (r, r')$. The first vertices of the leftmost $H$ (the left blue rectangle in Fig. 2) are combined into a clique and are connected respectively to $l$. Similarly, the last vertices in the right-most $H$ (the right blue rectangle in Fig. 2) generate a clique and are connected to the vertex $r$. This completes the construction. The entire construction is depicted schematically in Fig. 2. In Fig. 2, each block of four consecutive vertices corresponds to the vertices $v_1, v_2, v_3, v_4$ of the gadget $G'$. The remaining vertices from the gadget are not shown in the figure to avoid clutter.

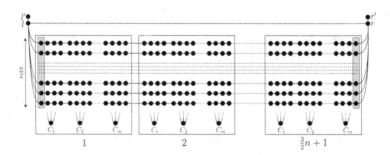

**Fig. 2.** General scheme of the constructed graph

Now, we are ready to prove the following lemma.

**Lemma 2 (♣).** *Formula $F$ has a satisfying assignment if and only if the constructed graph $G$ has a PD-set of size at most $2nm\left(\frac{3}{2}n + 1\right) + 2$.*

We have demonstrated that the constructed instance of PD-set is equivalent to the original formula $F$. Now, we need to verify that the constructed graph indeed has a pathwidth of at most $\frac{n}{2} + C$.

**Lemma 3 (♣).** *For each formula $F$, the constructed graph $G$ (see Fig. 2) has a pathwidth of at most $\frac{n}{2} + C$ for some constant $C$.*

From Lemmas 2 and 3 we deduce the statement of Theorem 4. □

## 4 Parameterization by Distance to Cluster Graphs

A graph is called a cluster graph if all its connected components are cliques (i.e., complete graphs). We say that a graph $G$ has a distance $d$ to a cluster graph if there exists a subset $X \subseteq V(G)$ such that $|X| \leq d$, and $G - X$ is a cluster graph.

It is trivial to find a PD-set in cluster graphs in polynomial time if there exists one. Hence, it is natural to consider PAIRED DOMINATION parameterized by the distance to cluster graphs.

**Theorem 5.** PAIRED DOMINATION *can be solved in $\mathcal{O}^*(4^d)$ time, where $d$ is the distance of the input graph to a cluster graph.*

*Proof.* Recall that a graph is a cluster graph if and only if it does not contain a $P_3$ (a path of length three) as an induced subgraph. Consequently, a minimum-size modulator to cluster graphs can be found in $\mathcal{O}^*(3^d)$ time, where $d$ is the distance to cluster graphs. From now on we assume that we are given a modulator to cluster graphs, denoted as $X \subset V(G)$, such that $G \setminus X$ is a cluster graph and $|X| = d$. Let us denote $G - X$ by $C$. Without loss of generality, we assume that $C = C_1 \sqcup C_2 \sqcup \cdots \sqcup C_k$ where for each $i \in [k]$ graph $C_i$ is a clique in $G - X$.

Our goal is to find a set of vertices $D$ such that $D$ is a dominating set, and there exists a perfect matching in $G[D]$. We represent $D$ as $D_1 \cup D_2$, where $D_1 = D \cap X$, $D_2 = D \cap C$. Since we do not know the actual value of $D$ or $D_1$, we consider $2^d$ potential values for $D_1 \subseteq X$. For each such set $D_1$, we find the smallest set $D_2' \subseteq C$ such that $D_1 \cup D_2'$ forms a PD-set. Subsequently, we output the smallest among these PD-sets. Note that $D_1 \cup D_2'$ is a PD-set if (i) $D_2'$ dominates the set $X \setminus N[D_1]$, (ii) $D_2'$ contains at least one vertex in a clique $C_i$ if $C_i \nsubseteq N_G(D_1)$, (iii) $G[D_1 \cup D_2']$ contains a perfect matching.

In order to find such $D_2'$ for each potential value of $D_1$, we employ a dynamic programming. We order the vertices of the cluster graph $C$ such that firstly we list all vertices of clique $C_1$, then vertices of $C_2$ and so on up to $C_k$. Let $v_1, v_2, \ldots, v_{|C|}$ be the ordering. By $C[i]$, we denote a maximal clique in $C$ that contains the vertex $v_i$, i.e., $C[i] = C_j$, if the vertex $v_i$ is contained in $C_j$. We consider a function $PDS$ with arguments $i, x, y, S', D'$, where $i \in [|C|]$, $x, y \in \{0, 1\}$, $S' \subset X \setminus N_G[D_1]$, $D' \subseteq D_1$. The function $PDS(i, x, y, S', D')$ denotes any subset $Y \subseteq \{v_1, v_2, \ldots, v_i\}$ of minimum size such that:

- Let $C[i] = C_j$ i.e., the vertex $v_i$ is contained in clique $C_j$ then for any $k < j$ we have $Y \cap C_k \neq \emptyset$ or $C_k \subseteq N_G[D_1]$;
- if $x = 1$, then $C[i] \cap Y \neq \emptyset$;
- if $y = 0$, then $G[D' \cup Y]$ has a perfect matching;
- if $y = 1$ then $G[D' \cup Y]$ has a matching that covers all vertices except one vertex from $C_j \cap Y$ (recall that $(C[i] = C_j)$);
- The set $Y$ dominates the set $S'$.

If such $Y$ does not exist, then $PDS(i, x, y, S', D') = null$. It is easy to see that the required $D_2'$ is given by $PDS(|C|, x, 0, X \setminus N_G[D_1], D_1)$, where $x = 1$ if $C[|C|] \not\subseteq N_G[D_1]$ (i.e., the last clique is not fully dominated by $D_1$), otherwise $x = 0$. Now we show how to calculate the values $PDS(i, x, y, S', D')$ using dynamic programming.

**Base Cases:** If $S' = \emptyset, x = 0, y = 0$ and there exists a perfect matching in $D'$ then $PDS(0, x, y, S', D') = \emptyset$; otherwise, $PDS(0, x, y, S', D') = null$.

To compute the value in our dynamic programming table for a cell $(i, x, y, S', D')$ we consider two cases: 1) $v_i$ is not the first vertex in the clique $C[i]$, 2) $v_i$ is the first vertex in the clique $C[i]$.

**Case 1:** Vertex $v_i$ is not the first vertex in the clique $C[i]$.

If vertex $v_i$ does not belong to $Y$, then we have $Y = PDS(i - 1, x, y, S', D')$. Otherwise, $v_i \in Y$ and we explore several potential scenarios:

1. The vertex $v_i$ has a pair $u$ among its neighbors in $D'$ (i.e., edge $v_i u$ is part of the matching in a paired dominating set). In this case,
$Y = \{v_i\} \cup PDS(i - 1, 0, y, S' \setminus N_G[v_i], D' \setminus \{u\})$.
2. The vertex $v_i$ matches with some vertex $v_j \in C[i]$, where $j < i$. If $y = 0$, then $v_j \in (Y \cap C[i]) \setminus \{v_i\}$. Hence, $Y = \{v_i\} \cup PDS(i - 1, 0, 1, S' \setminus N_G[v_i], D')$. If $y = 1$, then there is a vertex $v_k$ such that $v_k \in Y \cap C[i]$, $k \neq i, k \neq j$), and $v_k$ belongs to the paired dominating set but does not have a pair yet, i.e. its pair is a vertex $v_q$. Therefore, in $(D \cap C[i]) \setminus \{v_i\}$, there are two unpaired vertices, $v_j$ and $v_k$. In this situation, we have edges $v_j v_i$ and $v_k v_q$ as edges in the matching within the paired dominating set. However, we can replace these edges in a matching within paired dominating set with edges $v_j v_k$, $v_i v_q$. Thus, in this case, the following recurrence relation holds: $Y = \{v_i\} \cup PDS(i - 1, 0, 0, S' \setminus N_G[v_i], D')$. Hence, cases $y = 0$ and $y = 1$ can be combined by the following formula: $Y = \{v_i\} \cup PDS(i - 1, 0, 1 - y, S' \setminus N_G[v_i], D')$.
3. The vertex $v_i$ is the only unpaired vertex in the clique. In that case, $y = 1$. Similarly to the previous case, we get the following formula: $Y = \{v_i\} \cup PDS(i - 1, 0, 1 - y, S' \setminus N_G[v_i], D')$.

Combining all the subcases in this case, we obtain the following recurrence:

$$PDS(i, x, y, S', D') = \min\{PDS(i - 1, x, y, S', D'),$$
$$\{v_i\} \cup PDS(i - 1, 0, 1 - y, S' \setminus N_G[v_i], D'),$$
$$\min_{u \in N_G[v_i] \cap D'} \{\{v_i\} \cup PDS(i - 1, 0, y, S' \setminus N_G[v_i], D' \setminus u)\}.\}$$

**Case 2:** The vertex $v_i$ is the first vertex in its clique, i.e., $C[i - 1] \neq C[i]$. We introduce a new variable $x_1$. If $i > 1$ and $C[i - 1] \not\subseteq N_G[D_1]$ we set $x_1 = 1$; otherwise, we set $x_1 = 0$.

Depending on the values of $x$ and $y$, we show how to compute $PDS(i, x, y, S', D')$. In the case when $x = y = 0$, we either can take $v_i$ or not. If we do not take $v_i$ then we have $PDS(i, 0, 0, S', D') = PDS(i - 1, x_1, 0, S', D')$ since $v_i \notin Y$. If we take $v_i$, then $v_i$ must have a pair inside $D'$. Hence, in this case $PDS(i, 0, 0, S', D') = \min_{u \in N_G[v_i] \cap D'} \{v_i\} \cup PDS(i - 1, x_1, 0, S' \setminus N_G[v_i], D' \setminus u)$. Combining two cases we have $PDS(i, 0, 0, S', D') = \min\{PDS(i - 1, x_1, 0, S', D'), \min_{u \in N_G[v_i] \cap D'} \{v_i\} \cup PDS(i - 1, x_1, 0, S' \setminus N_G[v_i], D' \setminus u)\}$. If $x = 1$ and $y = 0$, then $v_i \in Y$. Hence, for similar reasons as for the case $x = y = 0$ we have $PDS(i, 1, 0, S', D') = \min_{u \in N_G[v_i] \cap D'} \{PDS(i - 1, x_1, 0, S' \setminus N_G[v_i], D' \setminus u)\}$. If $y = 1$, then $v_i \in Y$. Therefore, we have $PDS(i, 0, 1, S', D') = PDS(i, 1, 1, S', D') = \{v_i\} \cup PDS(i - 1, x_1, 0, S' \setminus N_G[v_i], D')$.

Summarizing, we have $PDS(i, x, y, S', D') =$

$$\begin{cases} \min\{PDS(i - 1, x_1, 0, S', D'), \\ \quad \min_{u \in N_G[v_i] \cap D'} \{v_i\} \cup PDS(i - 1, x_1, 0, S' \setminus N_G[v_i], D' \setminus u)\}, & \text{if } x = y = 0; \\ \min_{u \in N_G[v_i] \cap D'} \{PDS(i - 1, x_1, 0, S' \setminus N_G[v_i], D' \setminus u)\}, & \text{if } x = 1, y = 0; \\ \{v_i\} \cup PDS(i - 1, x_1, 0, S' \setminus N_G[v_i], D'), & \text{otherwise.} \end{cases}$$

In all cases we showed how to compute $PDS(i, x, y, S', D')$ based on previous values. Now we proceed with a running time analysis. Note that for a fixed $D_1 \subseteq X$, the size of the dynamic programming table is at most $|C| \cdot 2 \cdot 2 \cdot 2^{|X \setminus N_G[D_1]|} \cdot 2^{|D_1|} = \mathcal{O}^*(2^{|X|}) = \mathcal{O}^*(2^d)$. Therefore, the total running time of the algorithm is at most $\mathcal{O}^*(4^d)$.    □

To complement the previous result, we show that PAIRED DOMINATION does not admit a polynomial kernel parameterized by the distance to cluster graphs (d). Actually, we prove a stronger result using another parameter, that is, the vertex cover size of the input graph as $d \leq vc$. We derive a *polynomial parameter transformation* from the RED-BLUE DOMINATING SET problem to accomplish this task. Given a bipartite graph $G = (R, B, E)$ and an integer $k$, where $R$ and $B$ are the set of red vertices and blue vertices, respectively, the RED-BLUE DOMINATING SET problem asks to if there exists a set $R' \subseteq R$ (called a red-blue dominating set) of size at most $k$ such that $R'$ dominates all the vertices in set $B$, that is, $N(R') = B$. It is known that the RED-BLUE DOMINATING SET problem does not admit a polynomial kernel parameterized by the cardinality of $R$ unless NP $\subseteq$ co-NP/poly, see [6]. For the definitions and terminology of polynomial parameter transformation, please refer to the book [6]. Now, we are ready to prove the following result:

**Theorem 6.** *The* PAIRED DOMINATION *does not have a polynomial kernel when parameterized by vertex cover size, unless* NP $\subseteq$co-NP/poly.

*Proof.* Given an instance $(G = (R, B, E), k)$ of RED-BLUE DOMINATING SET parameterized by $|R|$, we construct an instance $(G', k', \ell)$ of PAIRED DOMINA-TION parameterized by $\ell$, where $\ell$ is the size of vertex cover as follows:

Take two copies of the set $R$, say $R_1$ and $R_2$, and two copies of $B$, say $B_1$ and $B_2$. For a vertex $r \in R$, let $r_1$ and $r_2$ be the copies of the vertex $r$ in $R_1$ and $R_2$, respectively. Similarly, for a vertex $b \in B$, let $b_1$ and $b_2$ be the copies of the vertex $b$ in $B_1$ and $B_2$, respectively. If $rb \in E(G)$ for some $r \in R$ and $b \in B$, we add edges $r_1b_1$ and $r_2b_2$ in $G'$. Furthermore, for each $r \in R$, we make $r_1$ and $r_2$ adjacent in $G'$. Next, we take a path on three vertices $a$, $b$, and $c$ and make $c$ adjacent to all vertices in $R_1 \cup R_2$. It is easy to observe that the size of a vertex cover of $G'$ is at most $2|R| + 2$ that is $\ell \leq 2|R| + 2$. Let $k' = 2k + 2$.

Next, we prove that the graph $G$ has a red-blue dominating set of size at most $k$ if and only if $G'$ has a PD-set of size at most $2k + 2$.

For the forward direction, let $S$ be a red-blue dominating set of $G'$ of size at most $k$. It is easy to see that the set $D = \{r_1, r_2 \mid r \in S\} \cup \{b, c\}$ is a PD-set of $G'$ such that $|D| \leq 2k + 2$.

For the reverse direction, consider a PD-set $D$ of $G'$ of size at most $2k + 2$. To dominate the vertex $a$, either $a$ or $b$ must be in $D$. Furthermore, a vertex in $\{a, b\}$ can only be paired with another vertex in $\{a, b, c\}$. Thus, we have $|D \cap \{a, b, c\}| \geq 2$. Therefore, if $a, b \in D$, we can safely replace the vertex $a$ with $c$ in $D$. Thus, assume that $c \in D$. Note that $|D \setminus \{a, b, c\}| \leq 2k$. Let $D_1 = D \cap (R_1 \cup B_1)$ and $D_2 = D \cap (R_2 \cup B_2)$. Then either $|D_1| \leq k$ or $|D_2| \leq k$. Without loss of generality, let us assume that $|D_1| \leq k$. Note that a vertex in $B_1$ can only be dominated by itself or a neighbor in $R_1$, and all the vertices in $R_1$ are already dominated by $c$, as $c \in D$. Thus, $D_1$ dominates the set $B_1$. Observe that if $b_1 \in B_1 \cap D_1$, then for some $r_1 \in N_{G'}(b_1)$, the updated set $D_1 = D_1 \setminus \{b_1\} \cup \{r_1\}$ also dominates all the vertices in $B_1$. Repeating this process ensures that $D_1$ dominates all vertices in $B_1$ and contains no vertex from $B_1$. Since the graph induced in $R_1 \cup B_1$ is an exact copy of $G$, the set $S = \{r \mid r_1 \in D_1\}$ is a red-blue dominating set of size at most $k$. Hence, PAIRED DOMINATION does not admit a polynomial kernel when parameterized by vertex cover.                                                                                                          □

## 5    $K_{i,j}$-Free Graphs, Counting Version Parameterized by Cliquewidth

We show that PAIRED DOMINATION behave similarly to DOMINATING SET on $K_{i,j}$-free graphs. It is known that DOMINATING SET admits a polynomial kernel on $K_{i,j}$-free graphs [20]. We adopt their technique for PAIRED DOMINATION.

**Theorem 7 (♣).** *For fixed $j \geq i \geq 1$, PAIRED DOMINATION admits a polynomial kernel on a graph that excludes $K_{i,j}$ as a subgraph.*

We also consider the counting version of PAIRED DOMINATION parameterized by cliquewidth. Specifically, we prove the following result.

**Theorem 8 (♣).** *There exists a constant $c \in \mathbf{N}$ such that the following holds: Assuming #SETH, there is no integer $k \geq 1$ such that we can count the number of minimum paired dominating sets with matchings (where the same paired dominating set is counted several times depending on how many perfect matchings it has) in time $\mathcal{O}(n^{k-c})$ on an $n$-vertex graph $G$ given together with a $k$-expression.*

Furthermore, from Theorem 8, it immediately follows that computing the number of minimum paired dominating sets with different matching inside is unlikely to be in FPT under parameterization by cliquewidth. However, it is possible to count the number of minimum dominating sets in $\mathcal{O}^*(4^{cw})$ as was shown in [3].

**Theorem 9 (♣).** *Given a graph $G$ and a $k$-expression $G$, that introduce each edge exactly once, we can compute the number of matching that generate a paired dominating sets of size $\ell$ in $\mathcal{O}(n^{2k+C})$ time for some constant $C > 0$.*

# References

1. Alber, J., Fellows, M.R., Niedermeier, R.: Polynomial-time data reduction for dominating set. J. ACM **51**(3), 363–384 (2004)
2. Alon, N., Gutner, S.: Linear time algorithms for finding a dominating set of fixed size in degenerated graphs. Algorithmica **54**(4), 544–556 (2009)
3. Bodlaender, H.L., van Leeuwen, E.J., van Rooij, J.M.M., Vatshelle, M.: Faster algorithms on branch and clique decompositions. In: Hliněný, P., Kučera, A. (eds.) MFCS 2010. LNCS, vol. 6281, pp. 174–185. Springer, Heidelberg (2010). https://doi.org/10.1007/978-3-642-15155-2_17
4. Chen, J., Fernau, H., Kanj, I.A., Xia, G.: Parametric duality and kernelization: lower bounds and upper bounds on kernel size. In: Diekert, V., Durand, B. (eds.) STACS 2005. LNCS, vol. 3404, pp. 269–280. Springer, Heidelberg (2005). https://doi.org/10.1007/978-3-540-31856-9_22
5. Chen, L., Lu, C., Zeng, Z.: Labelling algorithms for paired-domination problems in block and interval graphs. J. Comb. Optim. **19**(4), 457–470 (2010)
6. Cygan, M., et al.: Parameterized Algorithms. Springer, Cham (2015). https://doi.org/10.1007/978-3-319-21275-3
7. Dawar, A., Kreutzer, S.: Domination problems in nowhere-dense classes of graphs. In: Foundations of Software Technology and Theoretical Computer Science— FSTTCS 2009, LIPIcs. Leibniz International Proceedings in Informatics, vol. 4, pp. 157–168. Schloss Dagstuhl – Leibniz Center for Informatics, Wadern (2009)
8. Desormeaux, W.J., Henning, M.A.: Paired domination in graphs: a survey and recent results. Util. Math. **94**, 101–166 (2014)
9. Diestel, R.: Graph Theory. Graduate Texts in Mathematics, vol. 173, 4th edn. Springer, Cham (2012)
10. Downey, R.G., Fellows, M.R.: Fixed-parameter tractability and completeness I: basic results. SIAM J. Comput. **24**(4), 873–921 (1995)
11. Fomin, F.V., Grandoni, F., Pyatkin, A.V., Stepanov, A.A.: Combinatorial bounds via measure and conquer: bounding minimal dominating sets and applications. ACM Trans. Algorithms **5**(1), 9:1–9:17 (2008)
12. Fomin, F.V., Thilikos, D.M.: Fast parameterized algorithms for graphs on surfaces: linear kernel and exponential speed-up. In: Díaz, J., Karhumäki, J., Lepistö, A., Sannella, D. (eds.) ICALP 2004. LNCS, vol. 3142, pp. 581–592. Springer, Heidelberg (2004). https://doi.org/10.1007/978-3-540-27836-8_50
13. Hanaka, T., Ono, H., Otachi, Y., Uda, S.: Grouped domination parameterized by vertex cover, twin cover, and beyond. In: Mavronicolas, M. (ed.) CIAC 2023. LNCS, vol. 13898, pp. 263–277. Springer, Cham (2023). https://doi.org/10.1007/978-3-031-30448-4_19

14. Haynes, T.W., Hedetniemi, S.T., Henning, M.A. (eds.): Topics in Domination in Graphs. Developments in Mathematics, vol. 64. Springer, Cham (2020). https://doi.org/10.1007/978-3-030-51117-3

15. Haynes, T.W., Hedetniemi, S.T., Henning, M.A. (eds.): Structures of Domination in Graphs. Developments in Mathematics, vol. 66. Springer, Cham (2021). https://doi.org/10.1007/978-3-030-58892-2

16. Haynes, T.W., Hedetniemi, S.T., Slater, P.J. (eds.): Domination in Graphs: Advanced Topics, Monographs and Textbooks in Pure and Applied Mathematics, vol. 209

17. Haynes, T.W., Hedetniemi, S.T., Slater, P.J.: Fundamentals of Domination in Graphs. Monographs and Textbooks in Pure and Applied Mathematics, vol. 208. Marcel Dekker Inc., New York (1998)

18. Haynes, T.W., Slater, P.J.: Paired-domination in graphs. Netw. Int. J. $32(3)$, 199–206 (1998)

19. Lokshtanov, D., Marx, D., Saurabh, S.: Known algorithms on graphs of bounded treewidth are probably optimal. ACM Trans. Algorithms $14(2)$, Article no. 13, 30 (2018)

20. Philip, G., Raman, V., Sikdar, S.: Polynomial kernels for dominating set in graphs of bounded degeneracy and beyond. ACM Trans. Algorithms $9(1)$, 11:1–11:23 (2012)

21. Raman, V., Saurabh, S.: Short cycles make W-hard problems hard: FPT algorithms for W-hard problems in graphs with no short cycles. Algorithmica $52(2)$, 203–225 (2008)

22. van Rooij, J.M.M., Bodlaender, H.L., Rossmanith, P.: Dynamic programming on tree decompositions using generalised fast subset convolution. In: Fiat, A., Sanders, P. (eds.) ESA 2009. LNCS, vol. 5757, pp. 566–577. Springer, Heidelberg (2009). https://doi.org/10.1007/978-3-642-04128-0_51

23. Tripathi, V., Kloks, T., Pandey, A., Paul, K., Wang, H.L.: Complexity of paired domination in AT-free and planar graphs. Theor. Comput. Sci. $930$, 53–62 (2022)

# Author Index

Printed in the United States
by Baker & Taylor Publisher Services